Plant Invasions: The Role of Biotic Interactions

CABI INVASIVES SERIES

Invasive species are plants, animals or microorganisms not native to an ecosystem, whose introduction has threatened biodiversity, food security, health or economic development. Many ecosystems are affected by invasive species and they pose one of the biggest threats to biodiversity worldwide. Globalization through increased trade, transport, travel and tourism will inevitably increase the intentional or accidental introduction of organisms to new environments, and it is widely predicted that climate change will further increase the threat posed by invasive species. To help control and mitigate the effects of invasive species, scientists need access to information that not only provides an overview of and background to the field, but also keeps them up to date with the latest research findings.

This series addresses all topics relating to invasive species, including biosecurity surveillance, mapping and modelling, economics of invasive species and species interactions in plant invasions. Aimed at researchers, upper-level students and policy makers, titles in the series provide international coverage of topics related to invasive species, including both a synthesis of facts and discussions of future research perspectives and possible solutions.

Titles Available

1. *Invasive Alien Plants: An Ecological Appraisal for the Indian Subcontinent*
   Edited by J.R. Bhatt, J.S. Singh, R.S. Tripathi, S.P. Singh and R.K. Kohli
2. *Invasive Plant Ecology and Management: Linking Processes to Practice*
   Edited by Thomas A. Monaco and Roger L. Sheley
3. *Potential Invasive Pests of Agricultural Crops*
   Edited by Jorge E. Peña
4. *Invasive Species and Global Climate Change*
   Edited by Lewis H. Ziska and Jeffrey S. Dukes
5. *Bioenergy and Biological Invasions: Ecological, Agronomic and Policy Perspectives on Minimizing Risk*
   Edited by Lauren D. Quinn, David P. Matlaga and Jacob N. Barney
6. *Biosecurity Surveillance: Quantitative Approaches*
   Edited by Frith Jarrad, Sama Low-Choy and Kerrie Mengersen
7. *Pest Risk Modelling and Mapping for Invasive Alien Species*
   Edited by Robert C. Venette
8. *Invasive Alien Plants: Impacts on Development and Options for Management*
   Edited by Carol A. Ellison, K.V. Sankaran and Sean T. Murphy
9. *Invasion Biology: Hypotheses and Evidence*
   Edited by Jonathan M. Jeschke and Tina Heger
10. *Invasive Species and Human Health*
    Edited by Giuseppe Mazza and Elena Tricarico
11. *Parthenium Weed: Biology, Ecology and Management*
    Edited by Steve Adkins, Asad Shabbir and Kunjithapatham Dhileepan
12. *Biological Control in Latin America and the Caribbean: Its Rich History and Bright Future*
    Edited by Joop C. van Lenteren, Vanda H.P. Bueno, M. Gabriela Luna and Yelitza C. Colmenarez
13. *Plant Invasions: The Role of Biotic Interactions*
    Edited by Anna Traveset and David M. Richardson

Plant Invasions: The Role of Biotic Interactions

Edited by

Anna Traveset

Mediterranean Institute of Advanced Studies, CSIC-UIB, E07190 Esporles, Mallorca, Balearic Islands, Spain

and

David M. Richardson

Centre for Invasion Biology, Department of Botany and Zoology, Stellenbosch University, Stellenbosch, South Africa

CABI

CABI is a trading name of CAB International

CABI
Nosworthy Way
Wallingford
Oxfordshire OX10 8DE
UK

Tel: +44 (0)1491 832111
Fax: +44 (0)1491 833508
E-mail: info@cabi.org
Website: www.cabi.org

CABI
WeWork
One Lincoln St
24th Floor
Boston, MA 02111
USA

T: +1 (617)682-9015
E-mail: cabi-nao@cabi.org

References to Internet websites (URLs) were accurate at the time of writing.

A catalogue record for this book is available from the British Library, London, UK.

Library of Congress Cataloging-in-Publication Data

Names: Traveset, Anna, editor. | Richardson, D. M. (David M.), 1958- editor.
Title: Plant invasions : the role of biotic interactions / Anna Traveset, David M. Richardson.
Description: Boston : CAB International, 2020. | Series: Cabi invasives series ; 13 | Includes bibliographical references and index. | Summary: "Invasive species are of great environmental importance but no book before now has explored a key aspect: the nature of species interactions. This book looks at the theories, evidence and practicalities for managing invasions"-- Provided by publisher.
Identifiers: LCCN 2020018177 (print) | LCCN 2020018178 (ebook) | ISBN 9781789242171 (hardback) | ISBN 9781789242188 (ebook) | ISBN 9781789242195 (epub)
Subjects: LCSH: Plant invasions.
Classification: LCC SB613.5 .P625 2020 (print) | LCC SB613.5 (ebook) | DDC 639.9/9--dc23
LC record available at https://lccn.loc.gov/2020018177
LC ebook record available at https://lccn.loc.gov/2020018178

ISBN-13: 9781789242171 (hardback)
9781789242188 (ePDF)
9781789242195 (ePub)

Commissioning Editor: David Hemming
Editorial Assistant: Emma McCann
Production Editor: Tim Kapp

Typeset by Exeter Premedia Services Pvt Ltd, Chennai, India
Printed and bound in the UK by Severn, Gloucester

Contents

List of Contributors

Adhikari, Ashish – Department of Plant Pathology, University of Florida, Gainesville, USA. E-mail: adhikariashish@ufl.edu. ORCID: 0000-0003-1016-0617

Aizen, Marcelo A. – Instituto de Investigaciones en Biodiversidad y Medio Ambiente (INIBIOMA), Universidad Nacional del Comahue – CONICET, San Carlos de Bariloche, Rio Negro, Argentina. E-mail: maizen@comahue-conicet.gob.ar. ORCID: 0000-0001-9079-9749

Allen, Warwick J. – The Bio-Protection Research Centre, School of Biological Sciences, University of Canterbury, Christchurch, New Zealand. E-mail: warwick.j.allen@gmail.com. ORCID: 0000-0002-1859-1668

Baltzinger, Christophe – INRAE Val de Loire, Research Unit Forest Ecosystems, Nogent-sur-Vernisson, France. E-mail: christophe.baltzinger@inrae.fr. ORCID: 0000-0003-2980-6238

Caldwell, Iain R. – Australian Research Council Centre of Excellence for Coral Reef Studies, James Cook University, Townsville, Queensland, Australia. E-mail: iain.caldwell@jcu.edu.au. ORCID: 0000-0001-8148-8762

Callaway, Ragan M. – Division of Biological Sciences and the Institute on Ecosystems, University of Montana, Missoula, Montana, USA. E-mail: ray.callaway@mso.umt.edu. ORCID: 0000-0001-7342-4312

Catford, Jane A. – Department of Geography, King's College London, London, UK. E-mail: jane.catford@kcl.ac.uk. ORCID: 0000-0003-0582-5960

Cavieres, Lohengrin A. – Departamento de Botánica, Facultad de Ciencias Naturales y Oceanográficas, Universidad de Concepción, Concepción, Chile & Instituto de Ecología y Biodiversidad (IEB), Santiago, Chile. E-mail: lcaviere@udec.cl. ORCID: 0000-0001-9122-3020

Clay, Keith – Department of Ecology and Evolutionary Biology, Tulane University, New Orleans, USA. E-mail: clay@tulane.edu. ORCID: 0000-0002-3956-0887

Coetzee, Julie, A. – Centre for Biological Control, Botany Department, Rhodes University, Grahamstown, South Africa. E-mail: julie.coetzee@ru.ac.za. ORCID: 0000-0002-0364-3349

Colautti, Robert I. – Department of Biology, Queen's University, Kingston, ON K7L 3N6 Canada. E-mail: robert.colautti@queensu.ca. ORCID: 0000-0003-4213-0711

D'Antonio, Carla M. – Environmental Studies Program, University of California, Santa Barbara, USA. E-mail: dantonio@es.ucsb.edu. ORCID: 0000-0002-0395-9650

Devaney, John L. – Smithsonian Environmental Research Center, Edgewater, Maryland, USA & Department of Biology, Maynooth University, Maynooth, Co. Kildare, Ireland. E-mail: john.devaney@mu.ie. ORCID: 0000-0002-7676-0378.

Díaz Vélez, M. Celeste – Instituto Multidisciplinario de Biología Vegetal, CONICET – Universidad Nacional de Córdoba. Córdoba, Argentina. E-mail: celestemcbiol@gmail.com. ORCID: 0000-0003-3807-2969

Downs, Colleen, T. – Centre for Invasion Biology, School of Life Sciences, University of KwaZulu-Natal, Pietermaritzburg, South Africa. E-mail: downs@ukzn.ac.za. ORCID: 0000-0001-8334-1510

Emer, Carine – Department of Ecology, Institute of Biosciences, São Paulo State University, Rio Claro, São Paulo, Brazil & National Centre for Wild Birds Research and Conservation (CEMAVE - ICMBio), Floresta Nacional da Restinga de Cabedelo, Paraíba. E-mail: c.emer09@gmail.com. ORCID: 0000-0002-1258-2816

Ferreras, Ana E. – Instituto Multidisciplinario de Biología Vegetal, CONICET – Universidad Nacional de Córdoba. Córdoba, Argentina. E-mail: aferreras@imbiv.unc.edu.ar. ORCID: 0000-0003-3739-4894

Flory, S. Luke – Agronomy Department, University of Florida, Gainesville, USA. E-mail: flory@ufl.edu. ORCID: 0000-0003-3336-8613

Goss, Erica M. – Department of Plant Pathology and Emerging Pathogens Institute, University of Florida, Gainesville, USA. E-mail: emgoss@ufl.edu. ORCID: 0000-0003-3512-2107

Harmon, Philip F. – Department of Plant Pathology, University of Florida, Gainesville, USA. E-mail: pfharmon@ufl.edu. ORCID: 0000-0003-2288-5148

Heleno, Ruben, H. – Centre for Functional Ecology, Department of Life Sciences, University of Coimbra, Coimbra, Portugal. E-mail: rheleno@uc.pt. ORCID: 0000-0002-4808-4907

Hill, Martin P. – Centre for Biological Control, Department of Zoology and Entomology, Rhodes University, Grahamstown, South Africa. E-mail: m.hill@ru.ac.za. ORCID: 0000-0003-0579-5298

Holt, Robert D. – Department of Biology, University of Florida, Gainesville, USA. E-mail: rdholt@ufl.edu. ORCID: 0000-0002-6685-547X

Honor, Richard – Department of Biology, Queen's University, Kingston, ON K7L 3N6 Canada. E-mail: 14rdh@queensu.ca. ORCID: 0000-0003-3366-8851

Hui, Cang – Centre for Invasion Biology, Department of Mathematical Sciences, Stellenbosch University, Stellenbosch, South Africa & Theoretical Ecology Group, African Institute for Mathematical Sciences, Cape Town, South Africa. E-mail: chui@sun.ac.za. ORCID: 0000-0002-3660-8160

Kaiser-Bunbury, Christopher N. – Centre for Ecology and Conservation, College of Life and Environmental Sciences, University of Exeter, Cornwall Campus, Penryn, UK. E-mail: c.kaiser-bunbury@exeter.ac.uk. ORCID: 0000-0001-7254-3491

Kendig, Amy E. – Agronomy Department, University of Florida, Gainesville, Florida, USA. E-mail: aekendig@gmail.com. ORCID: 0000-0002-2774-1795

Kotanen, Peter M. – Department of Ecology and Evolutionary Biology, University of Toronto Mississauga, Mississauga, Ontario, Canada. E-mail: peter.kotanen@utoronto.ca. ORCID: 0000-0002-7842-8715

Kuebbing, Sara E. – Department of Biological Sciences, University of Pittsburgh, Pittsburgh, Pennsylvania, USA. E-mail: sara.kuebbing@pitt.edu. ORCID: 0000-0002-0834-8189

Landi, Pietro – Centre for Invasion Biology, Department of Mathematical Sciences, Stellenbosch University, Stellenbosch, South Africa. E-mail: landi@sun.ac.za. ORCID: 0000-0002-6373-2992

Lane, Brett R. – Department of Plant Pathology, University of Florida, Gainesville, USA. E-mail: b.lane@ufl.edu. ORCID: 0000-0001-5315-2225

Latombe, Guillaume – Centre for Invasion Biology, Departments of Botany & Zoology and Mathematical Sciences, Stellenbosch University, Stellenbosch, South Africa & Department of Botany and Biodiversity Research, Division of Conservation Biology, Vegetation Ecology and Landscape Ecology, University of Vienna, Vienna, Austria. E-mail: latombe.guillaume@gmail.com. ORCID: 0000-0002-8589-8387

Le Roux, Johannes J. – Department of Biological Sciences, Macquarie University, Sydney, New South Wales, Australia. E-mail: jaco.leroux@mq.edu.au. ORCID: 0000-0001-7911-9810

Lemoine, Nathan P. – Department of Biological Sciences, Marquette University, Milwaukee, Wisconsin, USA. E-mail: nathan.lemoine@marquette.edu. ORCID: 0000-0003-3436-7196

Lucero, Jacob. E. – Department of Biology, York University, Toronto, Ontario, Canada. E-mail: jelucero@yorku.ca. ORCID: 0000-0002-9535-5802

Montero-Castaño, Ana – School of Environmental Sciences, University of Guelph, Canada. E-mail: ana.montero.castano@gmail.com. ORCID: 0000-0003-2631-0085

Morales, Carolina L. – Instituto de Investigaciones en Biodiversidad y Medio Ambiente (INIBIOMA), Universidad Nacional del Comahue – CONICET, 8400 San Carlos de Bariloche, Rio Negro, Argentina. E-mail: moralesc@comahue-conicet.gob.ar. ORCID: 0000-0002-9989-5347

Msweli, Lindelwa S. – Centre for Invasion Biology, School of Life Sciences, University of KwaZulu-Natal, Pietermaritzburg, South Africa. E-mail: mswelils30@gmail.com. ORCID: 0000-0002-2879-6384

Paiaro, Valeria – Instituto Multidisciplinario de Biología Vegetal, CONICET – Universidad Nacional de Córdoba, Córdoba, Argentina. E-mail: vpaiaro@imbiv.unc.edu.ar. ORCID: 0000-0003-4448-1072

Parker, John D. – Smithsonian Environmental Research Center, Edgewater, Maryland, USA. E-mail: parkerj@si.edu. ORCID: 0000-0002-3632-7625

Pizo, Marco Aurélio – Department of Biodiversity, Universidade Estadual de São Paulo, Rio Claro, São Paulo, Brazil. E-mail: mapizof@gmail.com. ORCID: 0000-0002-3103-0371

Rehm, Evan M. – Department of Biology, Austin Peay State University, Tennessee, USA. E-mail: rehme@apsu.edu. ORCID: 0000-0002-5804-0027

Ribeiro da Silva, Fernanda – Department of Ecology, Universidade Federal de Santa Catarina, Florianópolis, Santa Catarina, Brazil. E-mail: feribs@hotmail.com. ORCID: 0000-0003-0169-0712

Richardson, David M. – Centre for Invasion Biology, Department of Botany & Zoology, Stellenbosch University, Stellenbosch, South Africa. E-mail: rich@sun.ac.za. ORCID: 0000-0001-9574-8297

Shukla, Ushma – INRAE Val de Loire, Research Unit Forest Ecosystems, Nogent-sur-Vernisson, France. E-mail: shuklaushma@gmail.com. ORCID: 0000-0001-8775-2800

Simmons, Benno I. – Centre for Ecology and Conservation, College of Life and Environmental Sciences, University of Exeter, Cornwall Campus, Penryn, UK & Department of Animal and Plant Sciences, University of Sheffield, Sheffield, UK. E-mail: benno.simmons@gmail.com. ORCID: 0000-0002-2751-9430

Smith-Ramesh, Lauren M. – National Institute for Mathematical and Biological Synthesis, Knoxville, Tennessee, USA. E-mail: laurensmithramesh@gmail.com. ORCID: 0000-0001-7520-9675

Timóteo, Sérgio – Centre for Functional Ecology, Department of Life Sciences, University of Coimbra, Calçada Martim de Freitas, 3000-456 Coimbra, Portugal. E-mail: sergio.timoteo@uc.pt. ORCID: 0000-0003-2417-3259

Traveset, Anna – Mediterranean Institute of Advanced Studies (IMEDEA-CSIC), Esporles, Mallorca, Spain. E-mail: atraveset@imedea.uib-csic.es. ORCID: 0000-0002-1816-1334

Wandrag, Elizabeth M. – School of Environmental and Rural Science, University of New England, Armidale, NSW, Australia & Institute for Applied Ecology, University of Canberra, Canberra, ACT, Australia. E-mail: elizabethwandrag@gmail.com. ORCID: 0000-0001-8140-539X

Wojan, Christopher M. – Department of Biology, Indiana University, Bloomington, USA. E-mail: cwojan@iu.edu. ORCID: 0000-0003-1783-9163

Yelenik, Stephanie G. – Pacific Island Ecosystems Research Center, U.S. Geological Survey, Hawaii, USA. E-mail: syelenik@usgs.gov. ORCID: 0000-0002-9011-0769

Acknowledgements

We thank all the chapter authors for their hard work and for their willingness to consider diverse requests from the editors.

Sincere thanks are due to the following friends and colleagues who kindly reviewed chapters and/or assisted in various ways during the preparation of the book: Marcelo A. Aizen, Bruce Anderson, Bernd Blossey, Dana M. Blumenthal, Ragan M. Callaway, Alistair Campbell, Jane A. Catford, Lohengrin A. Cavieres, Mariano Devoto, Ian A. Dickie, Jeff M. Diez, Don Drake, Carine Emer, Jennifer Firn, S. Luke Flory, Carol Frost, John F. Gaskin, Marina Golivets, Carl R. Gosper, Ruben H. Heleno, Jose L. Hierro, Patricia M. Holmes, Amy Hruska, Cang Hui, Jonathan M. Jeschke, Insu Jo, Paul Kardol, Sara E. Kuebbing, Christoph Kueffer, Martha Lopzaiza, Paula Lorenzo, Jens M. Olesen, José M. Fedriani, Marco Mello, Suzanne J. Milton, Charles E. Mitchell, Ana Montero-Castaño, Ingrid M. Parker, John D. Parker, Dean E. Pearson, Marco A. Pizo, Michelle A. Rafter, Kurt O. Reinhardt, Marc Rius, Orianne Rollin, Urs Shaffner, Andy W. Sheppard, Evan Siemann, Paula A. Tecco, Sérgio Timóteo, Jeferson Vizentin-Bugoni and Betsy von Holle.

We thank Jean Lategan for assistance in preparing the final versions of chapters.

Anna Traveset acknowledges funding from the Spanish Ministry of Science and Universities (project CGL2017-88122-P). David M. Richardson received support from the DSI-NRF Centre of Excellence for Invasion Biology, the National Research Foundation of South Africa (grant 85417) and the Oppenheimer Memorial Trust (grant 18576/03).

Glossary

Accumulation of Local Pathogens Hypothesis – see **Enemy of My Enemy Hypothesis**

Adaptation Hypothesis – The notion that **invasion** success of **non-native species** depends on adaptations to the conditions in the non-native range before and/or after **introduction**. **Non-native species** that are related to **native species** are more successful in this adaptation.

Alien species – see **non-native species**

Allelopathy – a biological phenomenon whereby an organism produces biochemicals that negatively influence the germination, growth, survival and reproduction of other organisms.

Alternate Stable Equilibrium Theory – The theory that predicts that ecosystems can exist under multiple 'states' (sets of unique biotic and abiotic conditions).

Amensalism – An interaction between two organisms of different species in which one is inhibited or destroyed while the other remains unaffected. An example of amensalism in the context of **biological invasions** is the interaction between Species A which outcompetes Species B via **allelopathy**, whereas Species B does not influence the growth of Species A.

Antagonism – An interaction between two organisms, generally of different species, that benefits one to the detriment of the other. A common example of an antagonist in the context of **biological invasions** is a **herbivore** or **pathogen** that limits the establishment of a non-native plant.

Anthropocene – The geological era during which human activity has become the dominant influence on climate and the environment.

Apparent competition – A type of **indirect interaction** that occurs through a shared predator, **herbivore** or **pathogen**. In the context of **facilitation** of non-native plants, the two apparent competitors are a native plant (Species A) and a **non-native** plant (Species B) which are not competing with each other. If a **herbivore**, pest or **pathogen** (Species C) benefits from the presence of Species B, its effect on Species A will also increase.

Apparent facilitation – see **indirect mutualism**

ASE – see **Alternate Stable Equilibrium Theory**

Biological control – In this book used to refer to the sustainable method for the management of **invasive non-native plant species** using host-specific **natural enemies** (biological control agents), mainly phytophagous insects and mites, and fungal **pathogens**, from the native range of the invasive plant to reduce population levels and/or the **impact** of the invasive plant. This is more accurately called 'classical biological control' to distinguish it from several other methods.

Biological invasions (synonyms: bioinvasions, biotic invasions, species invasions) – The phenomenon of, and suite of processes involved in determining: (i) the transport of organisms, through human activity (intentionally or accidentally, via introduction pathways) to areas outside the potential range of those organisms as defined by their natural dispersal mechanisms and biogeographical barriers; and (ii) the fate of such organisms in their new ranges, including their ability to survive, establish, reproduce, disperse, spread, proliferate, interact with **resident species** and exert influence in many ways on and in invaded ecosystems.

Biotic Acceptance Hypothesis– The view that habitats with high levels of species diversity are more vulnerable to **invasion** by **non-native plant species** (also termed the **Rich Get Richer Hypothesis)**.

Biotic Indirect Effects Hypothesis – The view that **non-native species** benefit from different indirect effects triggered by **native species**.

Biotic interactions – Those relationships that are established between at least two organisms of one or more species. The outcome of such relationships may result in the involved organisms benefitting, being harmed, or being unaffected, depending on the environmental context in which the interaction takes place.

Biotic resistance (BR) – Resistance to the establishment (or post-establishment survival, proliferation and spread) of **non-native species** due to the presence of **resident species**.

Biotic Resistance Hypothesis – A notion, derived from **Limiting Similarity Theory**, which argues that species-rich communities can withstand and even resist **biological invasions** (but see **Invasion Paradox**).

BR – see **biotic resistance**

Coevolution – Trait evolution through two (or more) species engaging in **biotic interactions** that reciprocally affect each other's fitness, posing frequency-dependent selection on each other.

Co-introduction Hypothesis – The view that the simultaneous introduction of **non-native species** and their mutualists is required to ensure the establishment or enhanced invasive performance in the new ranges.

Colonization – The immigration of a species into a new habitat and the founding of a new population.

Commensalism – A type of **direct interaction** where Species B benefits from the presence of Species A, but Species A receives no benefit from the presence of Species B. Common examples of commensalism in the context of **biological invasions** are non-native **ecosystem engineers** that alter habitats and facilitate the subsequent establishment of **non-native plants**.

Community assembly (synonym: community succession; related: community assembly rules) – The formation and succession of species assemblages in a community, driven by a combination of dispersal and **niche** processes. Processes leading to community assembly are often triggered by **disturbance,** and are increasingly influenced by **biological invasions** (cf. community disassembly).

Competition – Broadly defined as the negative influence of neighbours on an individual's performance; includes **interference competition**, **apparent competition** and **resource competition**.

Competition chain – **Competition** involving more than two species, where indirect effects may be common.

Contemporary evolution – see **rapid evolution**

Co-occurrence network – Spatial and temporal associations between sampled microbial taxa inferred from taxonomic data, typically obtained from next-generation DNA sequencing data.

Co-xenic associations – Novel interactions where both partners are non-native, but have no history of previous interactions.

Darwin's Naturalization Hypothesis – The notion that **non-native species** with close **native** relatives in their **introduced** range may have reduced chances of establishment and **invasion**; based on ideas formulated by Charles Darwin in Chapter 3 of *The Origin of Species*, borrowing ideas from Alphonse de Candolle.

Density dependence – Demographic mechanisms that regulate population dynamics, notably when the population size is at extremes, including positive density dependence that operates when the population size is low, and negative density dependence that operates when the population size is close to the carrying capacity. Density dependence often regulates population dynamics through curbed or hampered individual fitness.

Direct interaction – The effect of Species A on Species B that is not mediated through a third species (Species C). Direct interactions occur between a pair of species whether they are in isolation or within multi-species communities of varying composition (see Chapter 8, this volume).

Disease – A disorder of structure or function in an organism.

Disturbance – A temporal change, either regular or irregular (uncertain), in the environmental conditions that can trigger population fluctuations and secondary succession. Disturbance is an important driver of **biological invasions**.

Diversity–Invasibility Hypothesis – The proposition that more biologically diverse communities are less susceptible to **invasion** by **non-native species** or novel genotypes (related terms and concepts: **Biotic Resistance Hypothesis**; Diversity Resistance Hypothesis). (See also **biotic resistance, invasional meltdown.)**

Dormancy – Slowing or stopping of normal biological activity. Some dormancy mechanisms for microbes include forming spores (e.g. fungi), halting cell division (e.g. bacteria) and lysogenic reproduction (e.g. viruses).

Dynamic Equilibrium Hypothesis – The view that the establishment of a **non-native species** depends on natural fluctuations of the ecosystem, which influences the **competition** of local species.

Ecological networks – A representation of the concept of ecological communities and food webs in which species (sometimes individuals or guilds) are represented by nodes and the interactions they perform by links.

Ecosystem engineer – A species that modulates the availability of resources to other species by causing physical state changes in an ecosystem. Ecosystem engineers may be considered 'autogenic' if they change the environment through their own physical structure or 'allogenic' by transforming materials from one physical state to another. The effect of invasive ecosystem engineers on plant **invasions** is not mediated through a third species and is therefore considered a direct effect in this book.

EICA – see **Evolution of Increased Competitive Ability Hypothesis.**

EIH – see **Evolutionary Imbalance Hypothesis.**

Empty– Niche Hypothesis – The notion that **non-native species** can invade by occupying **niche** space unused by **resident species.**

Endozoochory – When seeds are ingested by animal dispersers.

Enemy Inversion Hypothesis – The view that **introduced** enemies of **non-native species** are less harmful for them in the non-native than in the native range, due to altered biotic and abiotic conditions.

Enemy of My Enemy Hypothesis (synonym: **Accumulation of Local Pathogens Hypothesis**) – The view that **introduced** enemies of a **non-native species** are less harmful to the **non-native species** than to **native species.**

Enemy Release Hypothesis (ERH) – The notion that **non-native species** have a better chance of establishing and becoming dominant when released from the negative effects of **natural enemies** that occur in their native range. The hypothesis is often accepted without recognizing that all **non-native species** will lose at least some **natural enemies** due to bottlenecks during transport.

Enemy Release Theory – A body of theory that embraces the **Enemy Release Hypothesis**, the **Shifting Defence Hypothesis** and other hypotheses that invoke the role of enemies in determining the success of a **non-native species.**

Enhanced Mutualist Hypothesis – The notion that species encounter better mutualists in the non-native range than in their native range.

Epidemic – Relatively high prevalence of a **disease** in a population at a given time.

Epizoochory – When seeds are transported in the fur or on the feathers or skin of a disperser.

ERH – see **Enemy Release Hypothesis**

Evolution of Increased Competitive Ability Hypothesis (EICA) – A concept that posits that plants **introduced** to an environment that lack their usual **herbivores** or **disease** agents will experience selection favouring individuals that allocate less energy to defence and more to growth and reproduction.

Evolutionary Imbalance Hypothesis (EIH) – The notion that successful invaders are more likely to come from biotic regions of high genetic potential (with independent lineages of large population size), experiencing a given environment for many generations and under strong **competition** from other lineages.

Evolutionary Reduced Competition Ability – The notion that a species may encounter fewer or weaker competitors following **introduction** to a new location.

Exploitative competition – A form of **competition** involving **indirect interaction** as organisms compete for common resources, such as territory, prey or food. The use of the resource by one individual will decrease the amount available for other individuals. (See also **resource competition.**)

Facilitation – Facilitative, or positive, interactions are encounters between organisms that benefit at least one of the participants and harm neither. Such interactions are considered **mutualisms** when both species derive benefit from the interaction.

Glucosinolates – Natural components of many pungent plants such as mustard, cabbage and horseradish. The pungency of those plants is due to mustard oils produced from glucosinolates when the plant material is chewed, cut or otherwise damaged.

Herbivore – An animal adapted to eat plants.

Herbivory – The consumption of plant material by animals (**herbivores**).

Host Jump Hypothesis – The notion that alien plants without co-introduced mutualists (e.g. coevolved mycorrhizal fungi) could still perform well in novel ranges by forming new associations with resident generalist or promiscuous mutualists. (See also **New Associations Hypothesis**.)

Human Release Hypothesis – The view that the abundance of **invasive non-native species** may be partly explained by the level of human activity or landscape maintenance, with intermediate levels of human activity providing optimal conditions for high abundance.

Impact – In the context of **biological invasions**, it is the description or quantification of how a **non-native species** affects the physical, chemical and biological environment.

Increased Susceptibility Hypothesis – The view that if a **non-native species** has a lower genetic diversity than the **native species**, there will be a low probability that the **non-native species** will establish due to increased pressure from enemies.

Indirect (biotic) interaction – **Impacts** of one species on the growth, fitness or population dynamics of another caused by changes in the population or behaviour of a third species. In the context of **biological invasions**, four commonly documented interactions are **apparent competition**, **indirect mutualism/commensalism**, **exploitative competition** and **trophic cascades**.

Indirect mutualism (synonym: **apparent facilitation**) – A type of **indirect interaction** that occurs through a shared mutualist or predator. In the context of **facilitation** of **non-native** plants, two **non-native** plants (Species A and B) may indirectly facilitate one another if Species A recruits a mutualist (Species C) that also facilitates Species B. Alternatively, indirect **mutualisms** could arise if the presence of a native competitor plant (Species C) reduces the competitive effect of non-native Species A on non-native Species B (also called a **competition chain**).

Infectious disease – a **disease** caused by **pathogens** that are passed between individuals and can harm the host.

Interaction motif – The simplest set of species necessary to describe a particular interaction structure and which form repeating patterns across interaction networks.

Interaction rewiring – see **interaction switching**

Interaction switching (synonym: species rewiring) – The phenomenon whereby species switch their interaction partners in an **ecological network**, either randomly or adaptively. Interaction switching can happen at different temporal scales, from rapid changes in behaviours to slow changes that coincide with phylogeographical events.

Interference competition – A form of **competition** between individuals, for example via aggression, when the individuals interfere with foraging, survival or reproduction of others, or by directly preventing their physical establishment in a portion of the habitat.

Intransitive competition – A form of **competition** in which there is no single best competitor. As in the game of rock–paper–scissors, all species can beat some competitors while all lose to some others, i.e. there are 'competitive reversals'.

Introduced – see **non-native species**.

Introduction – Movement of a species, intentionally or accidentally, due to human activity, from an area where it is **native** to a region outside that range ('**introduced**' is synonymous with **non-native**). The act of **introduction** (inoculation of propagules) may or may not lead to **invasion**.

Introduction–naturalization–invasion continuum – A conceptualization of the progression of stages and phases in the status of a **non-native** organism in a new environment which posits that the organism must negotiate a series of barriers. The extent to which a species is able to negotiate sequential barriers (mediated by **propagule pressure** and **residence time**, and frequently involving a **lag phase**) determines the organism's status as **non-native**: casual, **naturalized** or **invasive**.

Invasibility – The properties of a community, habitat or ecosystem that determine its vulnerability to **invasion**. Early studies tended to use the concept deterministically (particular systems were deemed either invasible or not), but invasibility is more appropriately considered probabilistically, and the degree of invasibility may change markedly over time due, for instance, to changes in biotic or abiotic features of the ecosystem (see also **biotic resistance**, **invasiveness**).

Invasion – The multi-stage process whereby an organism negotiates a series of potential barriers in the naturalization–invasion continuum (cf. **range expansion**).

Invasion complex – see **invasional meltdown**

Invasion ecology – The study of the causes and consequences of the **introduction** of organisms to areas outside their native range as governed by their dispersal mechanisms and biogeographical barriers. The field deals with all aspects relating to the **introduction** of organisms: their ability to establish, naturalize and invade in the target region, their interactions with **resident organisms** in their new location, and the consideration of costs and benefits of their presence and abundance with reference to human value systems. This term is often used interchangeably with 'invasion biology' in the literature (see also **invasion science**, **impact**).

Invasion paradox – The positive correlation between **native** and **non-native species** at landscape and regional scales that seemingly defies predictions from the **Biotic Resistance Hypothesis**.

Invasion science (synonym: invasion research) – The full spectrum of fields of enquiry that address issues pertaining to **non-native species** and **biological invasions**. The field embraces **invasion ecology**, but increasingly involves non-biological lines of enquiry, including economics, ethics, sociology, and inter- and transdisciplinary studies.

Invasion windows – see **opportunity windows**

Invasional meltdown (also the Invasional Meltdown Hypothesis) – A phenomenon whereby **non-native species** facilitate one another's establishment, spread and **impacts**. A related term is **invasion complex.**

Invasive species – **Non-native species** that sustain self-replacing populations over several life cycles, produce reproductive offspring, often in very large numbers at considerable distances from the parent and/or site of **introduction**, and have the potential to spread over long distances. Invasive species are a subset of **naturalized species**; not all **naturalized species** become invasive. This definition explicitly excludes any connotation of **impact**, and is based exclusively on ecological and biogeographical criteria.

Invasiveness – The features of a non-native organism, such as their life-history traits and modes of reproduction that define their capacity to invade, i.e. to overcome various barriers to **invasion**. The level of invasiveness of a species can change over time due to, for example, changes in genetic diversity through hybridization, introgression or the continued arrival of new propagules of the same species that is already established in a region, but from new and different (meta) populations, such that genetic diversity may increase. This latter concept is important in management strategies, which sometimes assume that less concern needs to be paid to the continued **introduction** of species (the continued arrival of propagules, whether accidental or intentional) that are *already* well-established in a region, overlooking the critical potential for elevated **invasiveness** over time.

Lag phase – the time between when a **non-native species** arrives in a new area and the onset of the phase of rapid, or exponential, increase. Multiple factors are frequently implicated in the persistence or dissolution of the lag phase in **invasions**, including an initial shortage of invasible sites, the absence or shortage of essential mutualists, inadequate genetic diversity and the relaxation of **competition** or **predation** (due to other alterations in the **resident biota**). Lag phases may also be the result of statistical or sampling artefacts commonly found in time series of **non-native species** records.

Limiting Similarity Theory (also Limiting Similarity Hypothesis) – A concept in theoretical ecology and community ecology that proposes the existence of a maximum level of **niche** overlap between two given species that will allow continued coexistence.

Long-distance dispersal – Dispersal of propagules over a long distance, defined either by the absolute distance travelled or by a set proportion of all propagules that disperse the farthest.

Metacommunity – In the context of **biological invasions**, multiple local communities connected at the regional level by the movement of propagules between patches and from which the regional species pool is constantly invading local communities.

Microbiome – all microorganisms found in association within a particular environment (e.g. rhizosphere), including an organism (e.g. plant) or a part of an organism (e.g. phyllosphere).

Missed Mutualisms Hypothesis – The notion that upon arrival in a new range, **non-native species** lose the beneficial mutualistic relationships that they experienced in their home range, and that this impedes **invasion**.

Mutualism – A type of **direct interaction** where Species A and Species B benefit from the presence of the other species. Examples of mutualisms implicated in plant **invasions** include plant–pollinator interactions, plant–seed disperser interactions and plant–soil organism interactions.

Mutualist Facilitation Hypothesis – The view the replacement of mutualists that are lost when plants are **introduced** to new ranges with new mutualists is key to the establishment and spread of the **non-native species**.

Native species (synonym: indigenous species) – Species that have evolved in a given area or that arrived there by natural means (via **range expansion**) without the intentional or accidental intervention of humans from an area where they are native (cf. **non-native species**).

Natural enemies – Organisms that kill, decrease the reproductive potential of or otherwise reduce the numbers of another organism. Natural enemies that limit pests are key components of integrated pest management programs, including **biological control**.

Naturalized species (synonym: established species) – Those **non-native species** that sustain self-replacing populations for several life cycles or a given period of time (10 years is advocated for plants) without direct intervention by people, or despite human intervention.

Network architecture – The structure of the adjacency matrix that form a network, depicting how the connections among nodes are organized within the network. Metrics, such as nestedness, modularity and node-degree distribution, are used to describe the matrix/network architecture.

Network modularity – The tendency of nodes to cluster in **ecological interaction networks**, i.e. networks with high modularity have high connections among nodes within specific modules but sparse connections to nodes in other modules.

Network nestedness – The organization of **ecological interaction networks** whereby specialists (e.g. pollinators that visit only few plants) interact with subsets of the species with whom generalists (e.g. pollinators that visit many different plants) interact.

New Associations Hypothesis – The view that new relationships between **non-native** and **native species** can positively or negatively influence the establishment of the **non-native species**. See **Host Jump Hypothesis**.

Niche – The way an organism fits into an ecological community or ecosystem. Through the process of natural selection, a niche is the evolutionary result of the morphological, physiological and behavioural adaptations of a species to its surroundings.

Niche partitioning – The process whereby natural selection drives competing species into different patterns of resource use or different **niches**.

Non-native species (synonyms: adventive-, alien-, exotic-, foreign-, introduced-, non-indigenous-, nonnative species) – Species whose presence in a region is attributable to human actions that enabled them to overcome fundamental biogeographical barriers (i.e. human-mediated extra-range dispersal) (cf. **native species**).

Novel Weapons Hypothesis – The idea that some **non-native plant species** may become invasive because they produce biologically active secondary metabolites that are not produced by species in invaded communities, and that such novelty provides the **non-native species** with advantages against native competitors, consumers or microbes that are not adapted to tolerate the chemical.

Operational taxonomic unit (OTU) – The definition of microbial taxa based on next-generation DNA sequencing studies, where organisms are not directly observed. An OTU is typically defined as a cluster of DNA sequence reads with 97% similarity, motivated by the expectation that these correspond approximately to species.

Opportunity windows (also Opportunity Windows Hypothesis, **invasion windows** – The idea that **invasion** success of **non-native species** increases with the availability of empty **niches** in the non-native range, and the availability of these **niches** fluctuates over space and time.

OTU – see **operational taxonomic unit**

Pathogen – An organism that causes **disease** in another organism (i.e. its host). Plant pathogens include viruses, bacteria, fungi and oomycetes.

Pathogen accumulation – An increase in **disease** prevalence (i.e. the proportion of hosts infected) or the number of **pathogen** species associated with a host over time. See **Enemy of My Enemy Hypothesis**.

Phenotypic plasticity – The ability of one genotype to produce a range of phenotypes in different environments. If the phenotypes produced in a novel environment are also associated with better performance, it is then called adaptive phenotypic plasticity.

Plant–soil feedback (PSF) – The process whereby plants alter the biotic and abiotic qualities of soil they grow in, which then alters the ability of plants to grow in that soil in the future.

Predation – Biological interaction whereby one organism, the predator, kills and eats another organism, its prey.

Propagule pressure – A concept that encompasses variation in the quantity, quality, composition and rate of supply of **non-native** organisms resulting from the transport conditions and pathways between source and recipient regions.

PSF – see **plant–soil feedback**

Range expansion – The process whereby a species spreads into new areas adjacent to their current distribution, either naturally or assisted by human-mediated changes to the environment. Differs from **invasion** in that human-mediated **colonization** across a biogeographical barrier is not implicated; the concept can be applied to both **native** and **non-native species**.

Rapid evolution (synonym: **contemporary evolution**) – **Invasive species** in their novel ranges are often under strong directional selection due to altered selection forces, combined with strong founder effects, that could lead to rapid changes in functional traits in the **invasive species** or closely interacting **native species**, often within less than a hundred generations.

Regime shift – Large, abrupt and persistent changes in the structure and function of an ecosystem, due to either a bifurcation from changing ecosystem processes or jumping basins of attraction due to a large **disturbance**.

Residence time – The time since the **introduction** of a species to a region; since the **introduction** date is usually derived from post-hoc records and is likely inaccurate, the term 'minimum residence time' is often used. The extent of **invasion** of **non-native species** generally increases with increasing residence time as species have more time to fill their potential ranges.

Resident biota/organisms – Species that are present in a community, habitat or region at the time of **introduction** of **non-native species**. The pool of **resident species** includes both **native species** and **non-native species introduced** previously. (See also **biotic resistance**.)

Resident species – In the context of **biological invasions**, **native** or **non-native species** in the **introduced** range of an invasive plant.

Resilience – Capacity of a system to maintain similar structure, functioning and feedbacks despite **disturbance** and perturbations.

Resource competition – Interaction between species or individuals vying for the same resource, i.e. water, food, shelter, territory, light or any means to maintain life and reproduce.

Resource–Enemy Release Hypothesis – The notion that fast-growing plant species adapted to high resource availability have weaker constitutive defences against enemies, and therefore incur relatively large costs when enemies are present. It is argued that these fast-growing species benefit most from **enemy release** and that the two mechanisms can act in concert to cause **invasion**; this could explain both the strong effects of resource availability on **invasion** and the extraordinary success of some **non-native species**.

Rich Get Richer Hypothesis – see **Biotic Acceptance Hypothesis**

Sampling Hypothesis – The notion that a large number of different **non-native species** is more likely to become invasive than a small number due to interspecific **competition**. The species identity of the locals is also more important than the richness in terms of the **invasion** of an area.

Seed Dispersal Effectiveness (SDE) – The contribution made by a disperser or group of dispersers to the reproductive success of a plant, both through the quantity of seeds dispersed and the quality of dispersal.

Shifting Defence Hypothesis – A variant of the **EICA** hypothesis which predicts shifts in the type of defence rather than a reduction of defences per se. (See also **EICA**.)

Social–ecological system – An ecological system intricately linked with and affected by one or more social systems.

Specialist–Generalist Hypothesis – The view that **non-native species** are more successful in a new region if the local predators are specialists and local mutualists are generalists.

Spillback – A form of **apparent competition** in which a **natural enemy** multiplies on a novel host species, but then spreads back to damage its original host; the novel host acts as the main reservoir for the enemy.

Spillover – A form of **apparent competition** in which a **natural enemy** spreads from its original host species to damage a different host; the original host acts as the main reservoir for the enemy.

Stress Gradient Hypothesis – The view that positive species interactions are more important in biologically and physically stressful habitats than in benign habitats.

Symbiosis – Refers to a mutualistic interaction (species involved benefit each other; see **mutualism**) in which there is a prolonged physical intimacy between the partner species.

Synzoochory – When seeds are transported externally but generally in the mouth, which might be followed by the hoarding of at least a fraction of those seeds.

Trait-mediated interaction – Using trait matching or trait difference to infer interaction strength.

Transmission – The acquisition of pathogenic organisms from another individual, resulting in infection.

Trophic cascade – A type of indirect effect that occurs through a consumer–resource interaction chain. For example, in the context of **non-native** plant **facilitation**, Species A (a **natural enemy** such as a predator, parasitoids or **pathogen**) reduces the abundance of its prey (Species B, a **herbivore**) which then indirectly releases a **non-native plant species** (Species C) from **herbivory**, thereby facilitating its **invasion**.

Unified framework for biological invasions – A framework that reconciles and integrates key features of the most commonly used **invasion** frameworks into a single conceptual model that is applicable to all **biological invasions**. It combines previous stage-based and barrier models and provides a terminology and categorisation for populations at different points in the **invasion** process.

Universal Tradeoff Hypothesis (UTH) – The notion that ecologically similar species of different realms have been bound to the same interspecific trade-off surface despite millions of years of independent evolution.

UTH – see **Universal Tradeoff Hypothesis**

1 Plant Invasions: The Role of Biotic Interactions – An Overview

Anna Traveset[1]* and David M. Richardson[2]

[1]Mediterranean Institute of Advanced Studies, CSIC-UIB, E07190 Esporles, Mallorca, Balearic Islands, Spain; [2]Centre for Invasion Biology, Department of Botany and Zoology, Stellenbosch University, Stellenbosch, South Africa

Abstract

Diverse biotic interactions between non-native plant species and other species from all taxonomic groups are crucial mediators of the dynamics of plant invasions. This chapter reviews the key hypotheses in invasion ecology that invoke biotic interactions to explain aspects of plant invasion dynamics. We examine the historical context of these hypotheses and assess the evidence for accepting or rejecting their predictions.

Most hypotheses invoke antagonistic interactions, mainly competition, predation, herbivory interactions and the role of pathogens. Only in the last two decades have positive (facilitative/mutualistic) interactions been explicitly included in invasion biology theory (as in ecological theory in general). Much information has accumulated in testing hypotheses relating to biotic resistance and Enemy Release Theory, although many of the emerging generalizations are still contentious. There is growing consensus that other drivers of plant invasion success, such as propagule pressure and disturbance, mediate the outcome of biotic interactions, thereby complicating our ability to make predictions, but these have rarely been assessed in both native and adventive ranges of non-native invasive species. It is also widely acknowledged that biogeographic comparisons, more than common garden experiments, are needed to shed light on many of the contradictory results. Contrasting findings have also emerged in exploring the roles of positive interactions. Despite strong evidence that such interactions are crucial in many communities, more work is needed to elucidate the factors that influence the relative importance of positive and negative interactions in different ecosystems. Different types of evidence in support of invasional meltdown have emerged for diverse habitats and across spatial scales. In light of increasing evidence that biotic indirect effects are crucial determinants of the structure, dynamics and evolution of ecological communities, both direct and indirect interactions involving native and non-native species must be considered to determine how they shape plant invasion patterns and the ecological impacts of non-native species on recipient communities.

Research that examines both biotic interactions and the factors that mediate their strength and alter interaction outcomes is needed to improve our ability to predict the effects of novel interactions between native and non-native species, and to envisage how existing invaded communities will respond to changing environmental conditions. Many opportunities exist for manipulating biotic interactions as part of integrated control strategies to reduce the extent, density and impacts of non-native plant invasions. These include the introduction of species from the native range of the non-native plant for biological control, diverse manipulations of plant—herbivore interactions and many types of interaction to enhance biotic resistance and steer vegetation recovery following non-native plant control.

*Corresponding author: atraveset@imedea.uib-csic.es

DOI: 10.1079/9781789242171.0001

1.1 The Role of Biotic Interactions in Ecology and Biogeography

1.1.1 Types of biotic interactions

Biotic interactions are those relationships that occur between at least two organisms of one or more species. The outcome of such relationships may result in the involved organisms benefiting, being harmed or being unaffected, depending on the environmental context in which the interaction takes place. Most biotic interactions involve the uptake of resources (nutrients, light, water, etc.) needed to survive, while others involve the exchange of goods and services such as protection, shelter or transport. Biotic interactions have long been categorized based on their effects on the interacting species (Fig. 1.1). Thus, *competition* exists when individuals of the interacting species are impaired due to resource or space limitation.

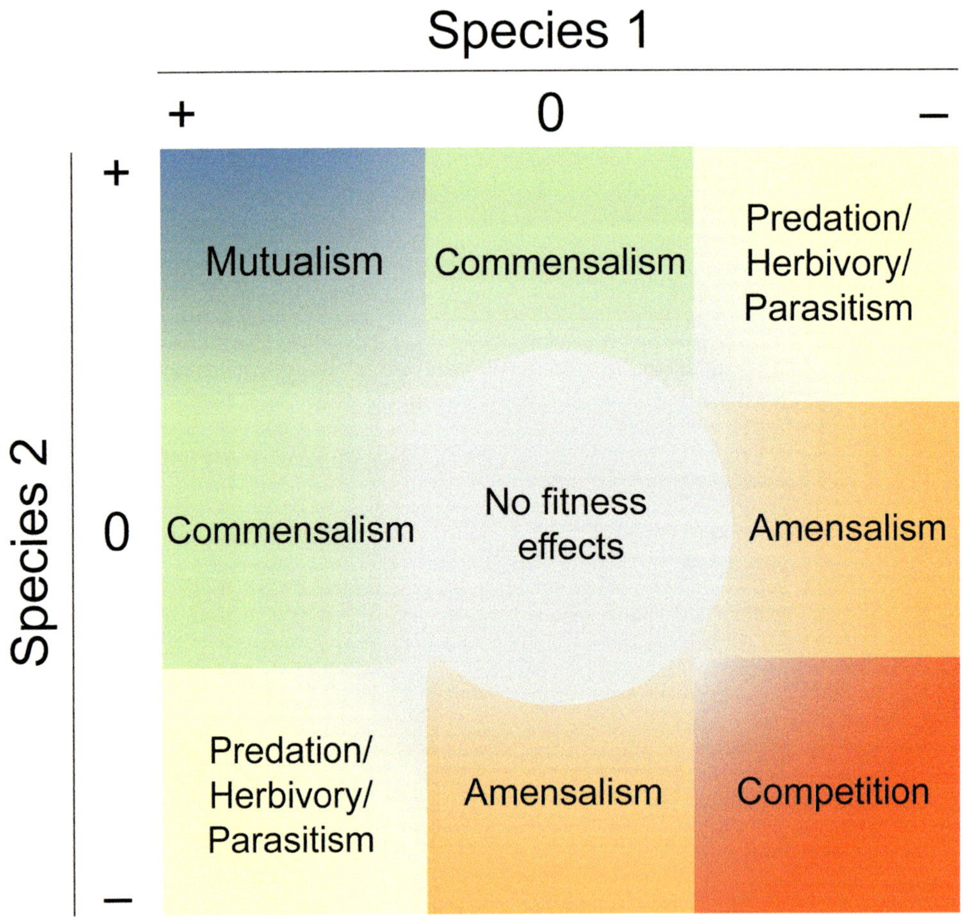

Fig. 1.1. Interaction compass displaying the range of possible interspecific biotic interactions. Signs indicate individual fitness or population growth rate: (+) positive effect, (-) negative effect and (0) neutral or no effect. The magnitude of the net interaction effect increases away from the centre. Despite such an operational categorization of interactions, a continuum of interactions exists – extending from antagonism to mutualism – depending on (a) the context, e.g. nutrient availability and density-dependent population dynamics; (b) the costs and benefits that they represent; or (c) the stage of the life history of the interacting species (adapted from Fig. 2 in Richardson *et al.*, 2000a and Fig. 1 in Pringle, 2016).

Predation is another antagonistic interaction that takes place when one individual benefits from another species by consuming it, either totally or partially, potentially leading to death of the prey. Variations of this antagonistic interaction include *parasitism* and *herbivory*, in which the consequences for the 'attacked' individual depend on the magnitude and duration of the interaction. When only one interacting species is negatively affected, the interaction is termed *amensalism*. *Mutualism*, on the other hand, is a win—win interaction involving two individuals of different species both of which benefit from the interaction. The most widespread examples of this category of interaction in plant ecology are pollination, seed dispersal and symbiosis (the latter defined as mutualism in which there is prolonged physical intimacy between partner species, e.g. mycorrhizal and endophytic fungi; Bronstein, 2015). When only one species benefits the interaction is called *commensalism*. Despite this operational categorization, there is often a continuum of interactions that extends from antagonism to mutualism depending on the context, on the costs and benefits that they represent, or on the stage of the life history of the interacting species (e.g. Pringle, 2016; Fig. 1.1). For instance, one animal species may act as a herbivore (or nectar robber) at the flowering stage, but may be a legitimate seed disperser of the same plant species later in the plant's life cycle. In the case of sexual deception, plants (mainly orchids) fool pollinators by producing structures that resemble female pollinators without providing them with any reward.

1.1.2 The influence of biotic interactions in ecological and evolutionary processes

All of the biotic interactions mentioned above potentially affect different components of individual fitness (survival, growth, reproductive success), thereby influencing demographic, ecological (e.g. population growth rate, species abundance and distribution) and evolutionary processes (e.g. natural selection, gene flow, coevolution). We cannot understand the evolution of species without considering such interactions, and biotic interactions actually promote the evolution of multiple adaptations. They are therefore among the most important drivers (if not the most important one) of the huge biodiversity on Earth. Mutualistic interactions between plants and fungal symbionts were crucial for the colonization of land and have driven the diversification of life (Kiers *et al.*, 2010). To comprehend patterns of biodiversity at local, regional or global scales, it is now widely acknowledged that we must consider not only the species that live in particular areas but also interactions among them (see below). Several studies in the last two decades have highlighted the importance of integrating biotic interactions into models to achieve more accurate predictions of the spatial and temporal variation of species distribution, and to understand community assembly patterns (e.g. Guisan and Thuiller, 2005; Boulangeat *et al.*, 2012; Godsoe *et al.*, 2017; Pearson *et al.*, 2018a). Using a novel integrated approach to interaction distribution modelling, Gravel *et al.* (2018) developed a quantitative theory to explain turnover of interactions in space and time. They propose that the ecological niche should encompass the effect of the environment on both species distribution (the Grinnellian dimension of the niche) and the ecological interactions among species (the Eltonian dimension), and call for adopting the view that community structure is best represented as a network of ecological interactions.

Besides being fundamental drivers of biodiversity and helping to explain the spatial and temporal distribution of species, biotic interactions also play fundamental roles in ecosystem functioning. They mediate most energy and nutrient assimilation and their flow through trophic chains and the decomposition of organic matter. Many plants depend, for instance, on indirect interactions with bacteria to obtain the nitrogen they need to perform photosynthesis; c. 80% of legumes (with c. 20,000 species, including many important crops) have rhizobial symbiosis. Nearly 95% of the world's plant species belong to families that are characteristically mycorrhizal (Pringle *et al.*, 2009). More than 90% of flowering plants are pollinated by animals (Ollerton *et al.*, 2011), and many plants produce seeds that rely on dispersal by animals to sustain populations and to colonize new habitats (up to 90% of plant species in the tropics require animals for dispersal; Farwig and Berens, 2012). Many plant—plant interactions are also mediated by interactions with other organisms, notably herbivores and pathogens (see below).

Interactions among species have been included in ecological theory since the early 20th century, especially after Lotka (1925) and Volterra (1926) developed their famous equations that described competition and predation between two or more species as a means to understand population growth and species coexistence in a community. Subsequent decades saw vigorous debates over Gause's (1934) 'competitive exclusion principle' which posited that two species cannot coexist on one limiting resource (Hart *et al.*, 2018). Another topic of hot debate was whether populations were regulated by density-dependent or density-independent factors. Hutchinson's (1957) formalization of the niche concept invoked only antagonistic interactions, leading to widespread acceptance that the 'realized niche' (parts of the environment that a species occupies in the presence of interactors, e.g. competitors, predators) is always smaller than the 'fundamental niche' (parts of the environment that a species can occupy in the absence of interactions with other species). Ecologists have for decades built theories of community organization that drew, explicitly or implicitly, on Hutchinson's conceptualization of the niche. MacArthur and Levins (1967) developed the concept of 'limiting similarity', the minimal niche difference between two competing species that would allow them to coexist. Community ecology progressed over the ensuing half century thanks to the development of this theoretical framework that, it was thought, provided a foundation for predicting the number and types of species in natural communities based on a functional limit to the similarity of competing species (May and McArthur, 1972; May, 1977). However, predictions of limiting similarity were found to be model-dependent, and the comparison of models including interspecific interactions gave similar results to models that excluded them (i.e. null models). Such findings challenged the overly simplistic idea that interspecific competition was the only or even the most important factor involved in structuring communities. It was not until early in the 21st century (Bruno *et al.*, 2003) that positive interactions such as mutualism or commensalism began being incorporated explicitly in theoretical models, although previous studies had already dragged facilitation into modern community ecology (e.g. Bertness and Callaway, 1994). Such work showed that the realized niche can in fact be larger than the fundamental niche, and that 'incorporating facilitative interactions into ecological theory would lead to more accurate and inclusive understanding of natural communities' (Mittelbach and McGill, 2019). Many subsequent studies have shown that positive interactions are at least as important as negative ones in mediating the structure and functioning of ecosystems (Kiers *et al.*, 2010).

Contrary to classic models that consider mutualisms to be destabilizing (May, 1981; Allesina and Tang, 2012), ecologists now recognize that these positive interactions are fundamental determinants of community-wide stability (Mougi and Kondoh, 2012; McIntire and Fajardo, 2014). The biological mechanisms that underlie the stabilizing effects of mutualisms are, however, still far from clear. Network studies have shed light on such mechanisms (Benadi *et al.*, 2012; Minoarivelo and Hui, 2016; Hale *et al.*, 2020). Benadi *et al.* (2012) were the first to consider the stability of a pollination network in a broad context, by evaluating the variation in the pollination mutualism and the degree of resource competition among plant species. They found that the effects of pollinators on the stability of the plants they visit depend on the extent to which plant resource use overlaps, and on the degree of pollinator specialization. Minoarivelo and Hui (2016) pointed out that it is necessary to distinguish ecological from evolutionary stability, and claimed that inferring network function from structure is problematic. More recently, Hale *et al.* (2020) suggested that mechanistic network theory could synthesize different types of ecological interactions, thereby elucidating how mutualism can enhance the diversity, stability and function of complex ecosystems.

The study of biotic interactions involving invasive species has particularly contributed substantially to the advancement of ecology and biogeography. In fact, non-native species provide natural 'species-addition experiments' in a wide range of habitat types which allow us to assess whether and how their effects can drive community structure not only at fine spatial scales, but also over much greater areas (from landscapes to regions and biomes). Invasion biology has provided much knowledge from different biogeographical areas and ecosystems that permit us to test the full spectrum of ecological and biogeographical hypotheses, and to formulate new ones (Sax *et al.*, 2007; also Chapter 16, this volume). Biological invasions also provide unique opportunities to study and understand processes directly, and to quantify changes in such processes over time. The geographical scale at which humans move species across the globe, enhancing biological invasions, indeed offers important insights into the complex nature of ecological systems.

The consideration of biotic interactions in invasion biology has also facilitated a better understanding of the mechanisms that allow non-native species to integrate into receptive communities, and the effects that such interactions have during the invasive process on the functioning of invaded communities (Hui and Richardson, 2017). This chapter focuses on the emerging understanding of the diverse roles of biotic interactions in invasions of non-native plants, in so doing providing the context for the other chapters in this book.

1.2 How the Study of Biotic Interactions Has Influenced Our Understanding of Plant Invasions

By definition, invasions of non-native plants have benefited from interactions with humans, at the very least through the human activities that, intentionally or accidentally, allow the species to cross major biogeographical barriers (Richardson *et al.*, 2000a). Many, probably most, invasions also benefit from diverse interactions with humans at subsequent stages of the introduction—naturalization—invasion continuum. For example, many species benefit from human-mediated dispersal opportunities within the new region (Chapters 5 and 6, this volume) and through alleviation of many potential barriers due to human disturbance of biotic and abiotic features of the environment. In what follows, we initially focus primarily on interactions among the non-native plant and other non-human organisms. There is a rich literature on the role of humans in triggering, mediating and sustaining plant invasions (Zimmermann *et al.*, 2014). We return explicitly to the role of humans as a biotic interactor affecting invasions when we address how insights on the role of interactions affect decisions on managing plant invasions (see below).

Of the many prominent hypotheses (c. 30) that have been proposed in invasion ecology (Catford *et al.*, 2009; Jeschke and Heger, 2018), a large proportion invoke species interactions. In this section, we address each of the main hypotheses, in the chronological order that they appeared in the literature (Fig. 1.2), and consider how they have influenced the understanding of the role of biotic interactions in plant invasions. Hui *et al.* (Chapter 2, this volume) also review theories and hypotheses on how biotic interactions mediate the performance and impact of non-native species on recipient communities, but their approach is different to ours. They categorize biotic interactions, direct and indirect, that emerge from both ecological fitting and coevolution during the process of invasion and community assembly. Due to the difficulty in detecting biotic interactions and measuring the dependency strength on sampling scales and population densities as well as the non-equilibrium transient dynamics of ecological networks, they call for a general framework that allows rapidly mapping the entire interaction networks in recipient ecosystems.

1.2.1 *Darwin's Naturalization Hypothesis*

Darwin (1859) was the first to suggest that non-native plant species would be more likely to establish and spread in areas that are poor in closely-related species than in those with many close relatives. The rationale behind this notion, first formalized as 'Darwin's theory' by Rejmánek (1996) and later as *Darwin's Naturalization Hypothesis* (DNH; Daehler, 2001), is that non-native plant species must compete with their close native relatives, or are more likely to be attacked by native herbivores or pathogens. This hypothesis has stimulated many studies, some of which found support for it and others that did not. For instance, considering the full list of seed plants introduced to New Zealand, Duncan and Williams (2002) reported that those with congeneric relatives are significantly more likely to naturalize, not the opposite. Their explanation was that non-native

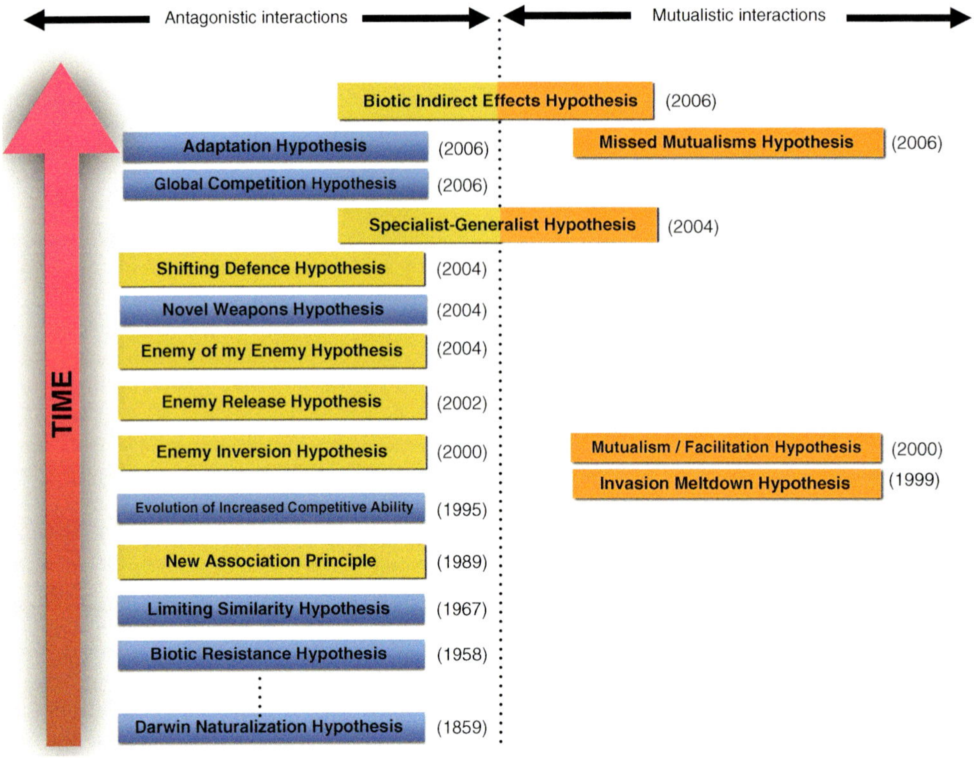

Fig. 1.2. Main hypotheses that have been advanced to explain invasion success and that have invoked biotic interactions – either antagonistic (left) or mutualistic (right). Among the antagonistic interactions, some hypotheses invoke mostly competition (blue) whereas others invoke enemies (herbivores or pathogens) (light brown). Two of the hypotheses consider both antagonistic and mutualistic interactions and are bicoloured.

species with native congeners in the receiving ecosystem are more likely to share features with them that allow them to survive there. In other words, the shared traits might pre-adapt the plants to the new environment enough to override the negative effects of competition with close relatives. This has been termed the *Adaptation Hypothesis* (AH) (Alpert, 2006; Catford *et al.*, 2009; Chapter 2, this volume), and also the *Preadaptation Hypothesis* by other authors (Cadotte *et al.*, 2018). The study by Thuiller *et al.* (2010) helped to resolve this conundrum. They reviewed and summarized studies that had tested these hypotheses and found largely inconsistent outcomes that they attributed to the use of different spatial and phylogenetic scales. To unravel the mechanisms driving biological invasions, they proposed a set of metrics derived from the niche concept to measure invader relatedness to native communities and the use of trait-based measurements of species dissimilarity. DNH, thus, is more likely to be supported at fine spatial scales. At larger spatial scales, however, AH is more likely to be supported (Strauss *et al.*, 2006; Jiang *et al.*, 2010). As different processes operate simultaneously, the relationship between invasion success and phylogenetic distance probably depends on the phylogenetic scale (Thuiller *et al.*, 2010). Cadotte *et al.* (2018) suggest that many studies that claim to have found support for DNH have used inappropriate spatial and temporal scales, have not considered different stages of invasion, and have used inappropriate metrics and analytical tests; they provide guidelines for collecting data appropriately to test contrasting hypotheses.

1.2.2 *Biotic Resistance Hypothesis* (BRH)

In his book *On the Origin of Species*, Darwin, 1859 also remarked that a 'prodigious number of plants ... never become naturalized for they cannot compete with our native plants nor resist destruction by our native animals'. A century later, Elton (1958) predicted that ecosystems with high biodiversity should be more resistant to invasion than ecosystems with low diversity. This has been termed the *Biotic Resistance Hypothesis* (also *Diversity—Invasibility Hypothesis*) and is based on the assumption that invading species are neither adapted to deal with native competitors nor defended against herbivores or pathogens in their new ranges; the recipient community thus exerts a resistance against non-native species, limiting colonization, naturalization and persistence of invaders. The role of evolutionary history in biological invasions, much emphasized in the literature on biotic resistance, was thus already acknowledged by Darwin and Elton (Chapter 9, this volume).

A milestone study that provided support for this hypothesis was the meta-analysis by Levine *et al.* (2004) which found that competition and herbivory often limit the spread of non-native species, despite rarely preventing their establishment. Shortly after this study, Parker and Hay (2005) tested BRH for 57 native and 15 non-native plants across the south-eastern USA; they found that native herbivores, in both aquatic and terrestrial systems, provide resistance to plant invasions. The effect was found to be stronger for generalist vertebrate herbivores than for invertebrate herbivores (Chapter 9, this volume). In another key review paper, Alpert (2006) predicted that altering selection pressures of native species could reduce such resistance and promote invasion.

Although much support exists for the BRH, more work is needed to clarify how biotic resistance interacts with other drivers such as propagule pressure, facilitative interactions, disturbance levels and climate change, and whether biotic resistance is a strong force in repelling plant invasions relative to such other drivers, whether the strength of such resistance varies geographically, and whether biotic resistance can be manipulated to manage plant invasions – see the review by Parker *et al.* (Chapter 9, this volume). Catford *et al.* (2020) examined the relative roles of native community diversity and climate warming on invasions, and found that the former overwhelms the latter. Zhang and van Kleunen (2019) analysed the interactions among 48 pairs of native and non-native annuals in Germany; their study revealed that interactions were generally negative (competitive) rather than facilitative, and that non-native plants outcompeted rare but not common native plants. Such findings suggest that common native competitors exert stronger biotic resistance and are thus important in determining plant invasion dynamics (Chapter 9, this volume), but more research is needed on more species and in more habitats to determine the generality of this view. Also, an important source of biotic resistance explaining invasion outcomes that is increasingly receiving attention and being incorporated into plant community theory is the effect of generalist seed predators (reviewed in Larios *et al.*, 2017).

Although it appeared much later, a key reference relating to the BRH, and predicting the opposite, is the paper by Stohlgren *et al.* (2003) which proposed '*The Rich Get Richer' Hypothesis* (later termed the *Biotic Acceptance Hypothesis* (BAH); Stohlgren *et al.*, 2006). This hypothesis posits that habitats with high levels of species diversity are more vulnerable to plant invasions. Stohlgren *et al.* (2006) found that hotspots of native plant diversity in most parts of the USA supported the largest number of non-native plant species when considered at large spatial scales. These authors, however, did not explicitly invoke species interactions but rather resource availability (in general terms) and habitat heterogeneity. Both resource availability and habitat heterogeneity increase as we move from fine scales (at which species interactions are likely to be important) to broader scales (landscape or larger) in larger areas; this provides a simple explanation for this relationship which invokes no role for biotic interactions. One of the first attempts to experimentally disentangle disturbance from diversity and productivity in invasions was the study by Pearson *et al.* (2018b).

Although different studies supported the prediction of a negative relationship between native and non-native species richness (Davies *et al.*, 2005, Davies *et al.*, 2011), there is a clear effect of scale on the direction and magnitude of diversity effects on invasion. While negative relationships were often found at fine spatial scales, relationships became positive at larger scales, resulting in what Fridley *et al.* (2007) termed *'the invasion paradox'*. This paradox still puzzles researchers (Chapter 9, this volume), but recent work has shed light on the conundrum by demonstrating that biotic resistance imposed by native species is universally common, especially when controlling for other factors such as habitat quality, distance to human activities and climate. Beaury *et al.* (2020) conclude that preserving native biodiversity should contribute to reducing the incidence and extent of invasions. Nonetheless, for biotic resistance to be useful in controlling plant invasions, it must overcome the counteracting effects of stochastic disturbance in the short term and propagule pressure in the longer term (Chapter 9, this volume). Finally, the long-standing notion that tropical areas are less invaded than temperate ones due to stronger species interactions, and thus stronger biotic resistance in the former, is tenuous once confounding factors (e.g. site area, productivity, propagule pressure) are removed (Chapter 9, this volume).

1.2.3 *Limiting Similarity Hypothesis* (LSH)

The *Limiting Similarity Hypothesis* (LSH) emerged from the 'limiting similarity' concept developed by MacArthur and Levins (1967); it predicts that successful invaders are functionally distinct from species in the recipient community, so encounter minimal competition and can fill an empty niche. LSH is supported by results from a number of studies. Here we highlight those by van Kleunen *et al.* (2010) and Ordonez *et al.* (2010) which both show that invasive species generally possess traits associated with higher performance (e.g. faster growth rates, higher seed production) compared to non-native species that are not invasive. Other important studies (e.g. Davies *et al.*, 2005, Davies *et al.*, 2011; MacDougall *et al.*, 2009; Wang *et al.*, 2013; Hulvey and Teller, 2018) have also suggested that a higher phenotypic (or phylogenetic) dissimilarity, and thus a lower niche overlap, between native and non-native species favours greater invasion success of the latter. When traits are not phylogenetically conserved, however, results can be mixed. Also, environmental—biotic filtering may select for phenotypically similar species irrespective of their origin (references in Chapter 9, this volume) and thus invasive and common native plants can often have similar trait values.

1.2.4 *Evolution of Increased Competitive Ability* (EICA) hypothesis and *Enemy Release Hypothesis* (ERH)

The *Evolution of Increased Competitive Ability* (EICA) hypothesis was first formulated by Blossey and Nötzgold (1995) (Chapter 10, this volume for details). It states that the release or reduction of negative effects induced by specialist enemies that limit populations in their home ranges enables the species to allocate freed resources (i.e. energy) to adapting and enhancing its competitive ability (by growing and reproducing at higher rates) when it arrives in an environment where specialist enemies are absent. It is closely related to the later-named *Enemy Release Hypothesis* (Keane and Crawley, 2002), and both are included in *Enemy Release Theory* (Colautti *et al.*, 2004), which also embraces the *Shifting Defence Hypothesis* (SDH) (Chapter 10, this volume) and other hypotheses that invoke enemies (see below). ERH has two main predictions: i) invasive species benefit from a loss of natural enemies, particularly those that specialize on their host (this often occurs when the invasion extends over large spatial scales); and ii) new enemies acquired in the non-native range are more likely to be generalists which also hamper native competitors. A non-native plant may thus benefit from a reduction in the number of enemies either directly – by increasing survival and reproduction – or indirectly, if generalist enemies damage the native competitors more than the invader. Many studies over the last two decades have tested such predictions, and many have found support for them. However, the ERH still remains contentious, and recent meta-analyses show contradictory results (Chapter 10, this volume). Field studies comparing native and non-native populations of the same species, i.e. biogeographical comparisons, provide more consistent support for the ERH than common garden experiments (Colautti *et al.*, 2004). Due to the lack of long-term demographical field data, it is still unclear, however, whether the biogeographic release from natural enemies results in a larger population growth or abundance in the non-native range (Chapter 10, this volume).

Besides demographic changes, enemy release involves a series of evolutionary changes for the invasive plant. In other words, the reduction of specialist enemies and a shift to a higher proportion of generalists may prompt changes in the type and amount of plant defences, implying genetic changes. This evolutionary shift seems to depend on the balance between the abundance of natural enemies in the native and introduced ranges and their effects on plant fitness (Chapter 10, this volume). This relaxed selection for costly defences in the absence of specialist enemies should lead to the evolution of genotypes that are less defended but more competitive. This idea is encapsulated in the EICA Hypothesis and its variant SDH, which predict shifts in the type of defence rather than a reduction of defences per se (Müller-Schärer *et al.*, 2004). The EICA Hypothesis has inspired much experimental research and has thus provided rich datasets amenable to meta-analysis; Honor and Colautti (Chapter 10, this volume) explore some of the reasons for the discrepancies found in different studies. Using a modelling approach grounded in eco-evolutionary theory, they also revisit some of the assumptions and predictions of enemy release theory by breaking it down into its main constituents: i) reduction in natural enemies; ii) relaxation of selection for costly defences; iii) favouring genetic variants with weaker or different defences; and d) increase in growth and reproduction. Their enemy release theory can be tested with a common-garden approach through experimental manipulation of enemy abundance. These authors call for characterizing fitness reaction norms of multiple genotypes across a range of enemy abundance to infer the historic rate of adaptive evolution or predict it into the future (Chapter 10, this volume).

1.2.5 *Invasional Meltdown Hypothesis* (IMH)

Simberloff and von Holle (1999) proposed the *Invasional Meltdown Hypothesis* (IMH) which states that direct or indirect symbiotic or facilitative relationships among non-native species cause an 'invasion domino effect' (Catford *et al.*, 2009). Invasional meltdown often occurs over several trophic levels, where one species makes a habitat or community more favourable for the other. The term and concept have been widely embraced by invasion biologists despite criticism from some quarters (Chapter 8, this volume). Against the standing paradigm on the importance of antagonistic interactions among species, including invasives, the IMH raised the idea that facilitative interactions among non-native species could reciprocally increase their invasion rate as well as the ecological impacts in the invaded community. As with the other previous hypotheses, many studies have found support for IMH but many others have not, reflecting different types of evidence for diverse habitats across a multitude of scales (reviewed in Braga *et al.*, 2018). Although it has been two decades since the IMH was proposed, and despite the frequent occurrence of multiple dominant non-native species in invaded communities, the study of interactions among non-native species and the resulting impacts is still uncommon in the invasion literature compared to the large amount of information on interactions between native and non-native species or on the impact of single invasive species. Moreover, most studies focus on direct, pairwise interactions among non-native species; only a few have been carried out at the community or ecosystem levels. We therefore have limited empirical data on how indirect interactions affect patterns of invasion in communities comprised of multiple native and non-native species (Chapter 8, this volume). Consequently, we still do not know whether facilitative interactions are more prevalent and important than those that hinder invasions (Simberloff, 2006). Although the term 'invasional meltdown' is often used loosely to refer to any form of facilitation among non-native species, Simberloff and von Holle (1999) defined the term with reference to community-level phenomena that may arise from facilitative population-level processes (Chapter 8, this volume). Moreover, to adequately assess the prevalence of the phenomenon of invasional meltdown, we need to consider direct and indirect interactions across non-native and native species in many communities, and determine how they shape invasion patterns and the ecological impacts of non-native species.

1.2.6 *Enemy Inversion Hypothesis* (EIH) and *Enemy of My Enemy Hypothesis* (EEH)

In 2004, a new variant of the ERH appeared – the *Enemy Inversion Hypothesis* (Colautti *et al.*, 2004). Its premise was that when natural enemies of non-native species are also introduced into new range they can be less effective, or may even have an opposite effect, due to changed abiotic conditions and/or other novel biotic interactions (Pearson and Callaway, 2003). These authors found that the negative effects of the non-native invasive *C. maculosa* on native plants became even stronger after gallflies (whose larvae inhabit *Centaurea* flowerheads) were introduced to control the invasive plant and became a favoured meal for deer mice. This interaction arose because the gallflies provided food subsidies that doubled or tripled deer mouse populations (Pearson and Fletcher, 2008), which dramatically increased the impacts of seed predation by deer mouse on native plant recruitment (Pearson and Callaway, 2008) and elevated the prevalence of hantavirus (a zoonotic disease lethal to humans) in deer mouse populations (Pearson and Callaway, 2006). How prevalent such restructuring of multispecies interactions is within communities remains unexplored, probably because of the challenges inherent in studying multiple interactions in communities (Colautti *et al.*, 2004). Yet another Hypothesis associated with the ERH – and considered a special case of the EIH – is the *Enemy of My Enemy Hypothesis* (Colautti *et al.*, 2004; Eppinga *et al.*, 2006). Based on plant—soil feedback models, it argues that non-native species accumulate local pathogens that limit the invader's abundance, but are more detrimental to native species, resulting in apparent competition. Most reported cases supporting the EEH deal with invasive animal species, and empirical evidence for the relative importance of the EEH is still very poor (Colautti *et al.*, 2004), though there is accumulating evidence in plant—herbivore—plant systems (Chapter 17, this volume). The EIH and EEH are associated with the New Associations Hypothesis (NAH), previously termed the 'new associations principle' by Hokkanen and Pimentel (1989). These hypotheses predict that non-native species represent naïve hosts (with few or no defences) to native enemies, and that they establish new relationships with species in recipient ecosystems that may enhance or impede invasion success. Another hypothesis related to enemies in the recipient community is the *Specialist—Generalist Hypothesis* (S-GH) formulated by Callaway and Ridenour (2004) and later formalized by Sax *et al.* (2007). This hypothesis predicts that invasion success is greatest when enemies in recipient community are specialists (unable to prey on non-native species) and when native mutualists (or facilitators of invasions) are generalists. Subsequent studies, however, have found that the frequency of specialists vs generalists of both antagonists (predators/pathogens) and mutualists vary notably across different types of communities. This hypothesis remains contentious and has found only partial support (Parker *et al.*, 2006).

1.2.7 *Mutualist / Facilitation Hypothesis*

In 2000, the first clear conceptualization of the role of mutualisms as mediators of plant invasions was published (Richardson *et al.*, 2000b). This study supported the view that many non-native plants rely on mutualisms (pollination, seed dispersal, plant—fungal and bacterial interactions) in their new environments to overcome barriers to establishment and become naturalized and potentially invasive. The paper reviewed many studies that revealed that the spread of many non-native plants, particularly woody species, are well served by generalist pollinators and that pollinator limitation is not a significant barrier to establishment or spread. It also showed that the seeds of the most notorious invasive plants are largely dispersed by animals, especially birds and mammals (Chapters 5 and 6, this volume). Microbe—plant interactions, despite prevalent low specificity, can be important in mediating invasions of some plant species (Chapters 3 and 19, this volume). The *Mutualist Facilitation Hypothesis* offers an explanation for invasional meltdown through linked feedbacks (non-natives alter habitats and non-natives benefit mutualists; Fig. 1 in Richardson *et al.*, 2000b). This framing stimulated many studies to determine the roles of different types of mutualistic interactions on non-native plants in a wide range of ecosystems (Traveset and Richardson, 2014).

Another hypothesis proposed within the larger framework established by Richardson *et al.* (2000b) was termed the *Missed Mutualisms Hypothesis* (MMH) by Alpert (2006) and also by Mitchell *et al.* (2006); it posited that upon arrival in a new range, non-native species lose the beneficial mutualistic relationships that they experienced in their home range, and that this impedes invasion. There are indeed reported cases of invasive plants that have not been able to establish and become invasive until their mutualistic partner has also been introduced; such cases involve mostly plant—soil microbe mutualisms (e.g. *Casuarina* sp. or *Morella faya* which need *Frankia* bacteria strains). Pines (genus *Pinus*) are an example of an invasive group of trees that need ectomycorrhizas to establish. Dickie *et al.* (2010) found that most mycorrhizal species associated with the invasive *Pinus contorta* in New Zealand were also non-native, thus providing support for co-invasion with mutualists rather than for novel associations with native mycorrhizas. A recent study by Moyano *et al.* (2020), also focusing on the genus *Pinus*, found that those species that depend more on mutualistic ectomycorrhizal fungi are more invasive (more likely to establish), a finding that defies predictions of the MMH.

The number of studies examining how mutualistic interactions contribute to plant invasions has grown exponentially in recent years. Five chapters in this volume – those by Montero-Castaño and Traveset (Chapter 4), Díaz *et al.* (Chapter 5), Baltzinger *et al.* (Chapter 6), Cavieres (Chapter 7) and Callaway and Lucero (Chapter 3) – review this literature. They all emphasize the importance of considering positive interactions when seeking a mechanistic understanding of plant invasions. The only study that we know of that assesses the role of biotic interactions in structuring the invasive biota of a whole region is that by Le Roux *et al.* (2020) for South Africa. Such insights are important for elucidating how ecological factors via biotic interactions scale up to shape biogeographical patterns (Richardson and Pyšek, 2012).

The 'other side of the coin' – how non-native species influence the mutualistic interactions prevailing in the receptive communities – was thoroughly reviewed by Traveset and Richardson (2006) (and updated in Traveset and Richardson, 2011, 2014), who described the different mechanisms whereby biotic interactions established with the non-native species in the new environment can disrupt native mutualistic interactions with potential consequences for community structure and function. The pattern emerging so far is that high interaction generalization is the norm in most ecosystems, and that this allows non-native species to infiltrate the new environments with few barriers. On the other hand, mutualistic disruptions due to invasions are known to be pervasive, and subsequent cascading effects are increasingly being reported (Traveset and Richardson, 2014). The impacts of non-native plants on plant—pollinator interactions are dealt with in this volume in the chapters by Aizen and Morales (Chapter 13). Impacts of plant invasions on plant—seed dispersal interactions are reviewed by Heleno (Chapter 14), while impacts on plant—microbe interactions are covered by Callaway and Lucero (Chapter 3).

1.2.8 *Novel Weapons Hypothesis* (NWH)

The *Novel Weapons Hypothesis* (NWH), first proposed by Callaway and Aschehoug (2000), and later named by Callaway and Ridenour (2004), was an important step towards elucidating invasive success and the evolution of increased competitive ability. It posits that non-native species release allelopathic chemicals that inhibit and repress potential competitors in their new ranges. Native species are not adapted to the novel biochemical weapons, and this enhances the competitive ability and success of the non-native species. These authors found that root exudates – that are ineffective against resident neighbours because of adaptation – might have an inhibitory effect on native plants in the recipient communities. They also predicted that the selective advantage of possessing a novel weapon could result in rapid evolution of that weapon. This hypothesis provided an alternative explanation to the 'growth *vs* defence' trade-off assumed in the evolution of competitive ability in invasive plants; it predicts that selection acts directly on those traits, providing a competitive advantage in their new habitats. While the role of allelopathy via direct phytotoxicity has been hotly debated in recent decades, many studies suggest that allelopathy mediated by indirect biotic interactions, particularly via fungal mutualists, is an important driver of some plant invasions (Chapter 15, this volume). This has been clearly demonstrated for *Centaurea maculosa* in North America. The novel weapons influence not only native plants but also other organisms (Cappuccino and Arnason, 2006); they can reduce herbivory or suppress mycorrhizal fungi that are beneficial to native competitors but not necessarily to the invasive plant (Callaway *et al.*, 2008). Native species can adapt to the novel compounds with time (Lankau *et al.*, 2009). The possibility that chemical compounds produced by native plants may contribute to resistance against the invasion of non-native plants has received very little attention (van Kleunen *et al.*, 2018). Smith-Ramesh (Chapter 15, this volume) reviews the growing body of evidence and discusses future avenues of research on allelopathy and plant invasions, emphasizing the mechanisms by which allelopathic invaders disrupt biotic interactions and how this can alter the structure of interaction networks in the invaded habitat.

1.2.9 *Global Competition Hypothesis (GCH)*

Alpert (2006) introduced the *Global Competition Hypothesis* (GCH) which suggests that the likelihood that the non-native species pool will include a species with traits enabling it to outcompete native species increases with the number of species introduced. This notion, analogous to the sampling effect that is sometimes used to explain the positive relationship between biodiversity and ecosystem functioning, assumes that particular life-history strategies and traits of the invasive species, combined with the ability to avoid enemies, make a species a better competitor for a shared resource (Blumenthal, 2006). The GCH has been invoked to explain the high degree of invasion on oceanic islands by dominant species from the mainland which would have had the opportunity to sampling more variation than the island biota (Pyšek and Richardson, 2006). Wandrag and Catford (Chapter 16, this volume) revise this and other hypotheses that invoke competitive interactions in plant invasions, providing the mechanisms as well as the predicted outcome for each one. These authors call for a predictive framework for identifying and managing plant invasions that includes a better understanding of the influence that individuals have on each other's performance.

1.2.10 *Biotic Indirect Effects Hypothesis (BIEH)*

Also in 2006, White and colleagues introduced a new hypothesis – the *Biotic Indirect Effects Hypothesis* (BIEH) – which invoked facilitation of invasion mediated by indirect community interactions, i.e. how 'a' alters the effect that 'b' has on 'c'. Such complex interactions have long been overlooked in ecology although there has been increased interest in recent years, partly because of the increased computational power available for network modelling, in determining their role in the structure, dynamics and evolution of ecological communities (Chapter 17, this volume). The four most commonly documented indirect effects are: i) apparent competition; ii) indirect mutualism/commensalism; iii) exploitative competition for a biotic resource; and iv) trophic cascades (White *et al.*, 2006). Indirect biotic interactions, which can be mediated by both biotic and abiotic factors, are defined as impacts of one species on the growth, fitness or population dynamics of another species through changes in the population or behaviour of a third species. Chapter 17, this volume, focuses on the biotic factors that mediate impacts of invasive plants via indirect biotic interactions, and proposes several potential future research directions and questions that remain to be answered to understand the relevance of indirect interactions in mediating success and impacts of invasive species. The indirect effects that herbivores have on plant invasions, together with the direct ones, are dealt with by Kotanen (Chapter 12, this volume). His review concludes that herbivore damage to plant invaders is pervasive, though not universal or uniform. The type and significance of herbivore damage, and the implications for invasion dynamics, depend on many factors; these include the residence time of the non-native plant species, its evolutionary history, and its receptive community. It remains unclear how often herbivory in the invaded range can strongly affect populations of both non-native plants and their native competitors (see also the review by Maron and Crone, 2006). Further research on the influence of herbivores on the demography of non-native plants is needed to assess when and how often damage has consequences at the population level.

1.3 The Effect of Multiple Stressors on Biotic Interactions Influencing Plant Invasions

The introduction and invasion success of non-native plant species is typically associated with environmental change and human activities (Pyšek *et al.*, 2010). These factors can influence the biotic interactions among species in the recipient community, for instance by affecting the strength and direction of competition (Chapter 16, this volume), by mediating the direct and indirect effect of enemies on non-native and native plants (Chapters 11 and 17, this volume), and through the many facilitative/mutualistic interactions that take place in the community. There is evidence that changes in resource supply and disturbance, both of which are associated with human activities, affect competitive outcomes between native and non-native plants. Non-natives tend to outcompete natives when resource supply is high (Seabloom *et al.*, 2015). Disturbance (including complete biomass removal through soil tilling, herbicide use, fire or partial biomass removal through grazing) is known to provide windows of opportunity for the invasion of non-native plant species that can rapidly exploit available resources, such as light or water (Seabloom *et al.*, 2006). In turn, some non-native plants have caused changes in vegetation structure to such an extent that fire regimes are altered, leading to massive ecosystem-level impacts (Brooks *et al.*, 2004), including regime shifts (Gaertner *et al.*, 2014). Impacts include disruptions of many biotic interactions, making restoration to preinvasion conditions much more difficult, if not impossible.

The type and strength of biotic interactions that influence plant invasions depends, to a large extent, on human activities. Humans influence the locations to which species are introduced, including through climate matching of non-native plant species with target sites of introduction. Moreover, humans tend to introduce propagules of non-native species repeatedly, and often in very large numbers. The type and frequency of human-mediated disturbance can also determine the benefits that non-native plants derive from the loss of native ones. The likelihood of non-native plants outcompeting native plants increases if disturbance: i) directly removes native competitors or hinders their vigour; or ii) increases resource availability due to decreased resource uptake by the native species. Often, these two disturbance effects are difficult to tease apart (Chapter 16, this volume).

Global change drivers such as climate change and agricultural intensification are continuously creating new species combinations, simultaneously modifying the conditions under which such species interact, and the outcomes of such interactions (Ricciardi *et al.*, 2017). Understanding the processes that influence where non-native plants establish and how they interact with other species in recipient communities is crucial for effective management of invasions and their impacts. There are cases where non-native species themselves act as drivers of change; there are many examples of non-native plant species that drive regime shifts in invaded ecosystems (Gaertner *et al.*, 2014; Shackleton *et al.*, 2018). Human-mediated land use, together with species-intrinsic traits (e.g. their competitive ability, high fruit production), might interactively influence the dominance of non-native species in some situations, thereby exacerbating the effects of global change. Invasive plants can alter animal communities, trophic interactions, primary productivity, nutrient cycling and disturbance regimes (e.g. Mack *et al.*, 2000), and such effects may in turn be mediated by the other species that are interacting with invasive plants (Chapter 11, this volume). For instance, pathogens interacting with invasive plants can: i) influence soil formation and water regulation; ii) lead to tree mortality which can increase the intensity of forest fires; or iii) alter nutrient cycling through food webs. Moreover, invasive species and pathogens can both affect the composition of soil microbial communities, but their combined effects are unclear (Chapter 11, this volume).

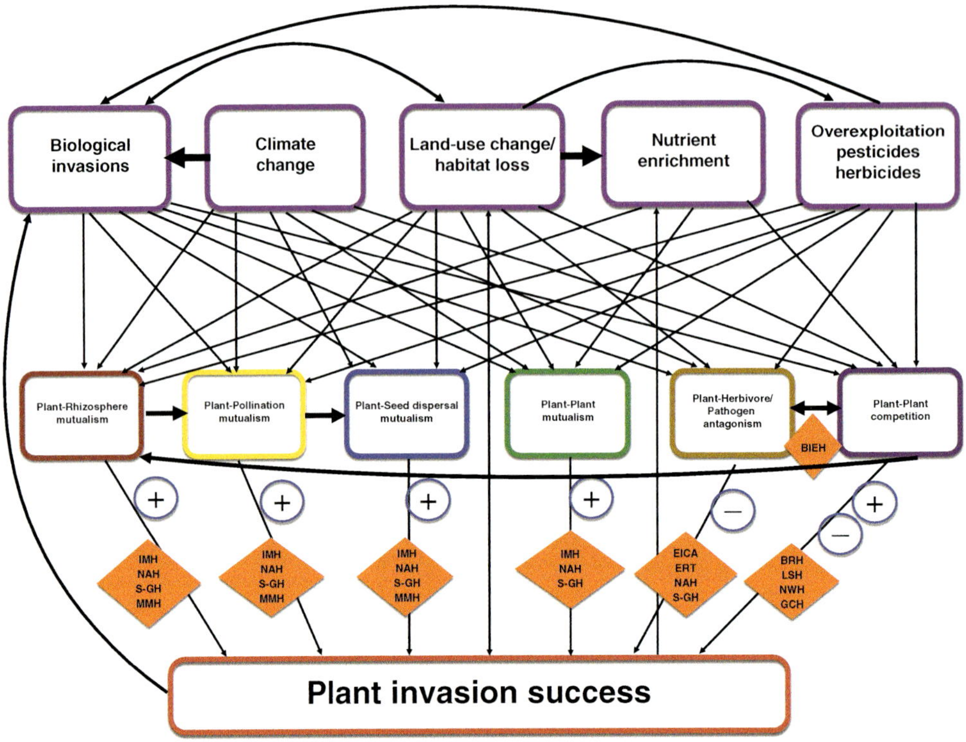

Fig. 1.3. Conceptual model showing the different drivers of global change that influence plant invasion success through their direct or indirect effects on biotic interactions. The drivers are not completely independent; some can have an effect on others (see text for examples), and plant invasion success can in turn further enhance biological invasions, land use change and nutrient enrichment. Likewise, some biotic interactions are linked (directly or indirectly) to others. The sign of the effects of biotic interaction type on plant invasion success indicates whether the interaction is positive (mutualism/facilitation) or negative (antagonistic interaction). The outcome of plant–plant competition may be positive or negative depending on whether the non-native or the native plant is outcompeted. The different hypotheses associated with each prediction that invoke biotic interactions in explaining plant invasion success are shown in the diamonds. Hypothesis acronyms: BIEH: Biotic Indirect Effects Hypothesis; BRH: Biotic Resistance Hypothesis; EICA: Evolution of Increased Competitive Ability Hypothesis; ERT: Enemy Release Theory, which embraces other hypotheses regarding enemy release (see text); GCH: Global Competition Hypothesis; IMH: Invasional Meltdown Hypothesis; LSH: Limiting Similarity Hypothesis; MMH: Missed Mutualisms Hypothesis; NAH: New Associations Hypothesis; and NWH: Novel Weapons Hypothesis.

Understanding when and how species respond to changing abiotic conditions and biotic interactions is essential to allow us to predict the outcomes of novel interactions between natives and non-natives, and to envisage how existing invaded communities will respond to changing environmental conditions. Research, like that of Catford *et al.* (2020) that simultaneously examines how shifts in both biotic (i.e. community composition) and abiotic conditions alter interaction outcomes is much needed, although challenging. Fig. 1.3 provides a conceptual model showing the different drivers of global change that can influence plant invasion success mediated through biotic interactions. Some of these biotic interactions may in turn influence each other (e.g. plant—plant competition may be mediated by herbivores), and a plant invasion may feed back to enhance or disrupt the effect of some global change drivers.

1.4 Experimental Approaches in the Study of Biotic Interactions and Plant Invasions

Many approaches have been used in the study of plant invasions (van Kleunen *et al.*, 2018). Although most studies focus on individual invasive species, comparative approaches are essential to uncover and understand the factors that determine invasion success. Developing experimental and modelling approaches to examine competition between natives and non-natives, for instance, has received considerable effort and has proved useful in advancing the understanding the role of competition in shaping species distributions (Chapter 16, this volume). On the other hand, understanding how biotic interactions have affected and will continue to affect plant invasions requires us to consider both the evolutionary history and current eco-evolutionary dynamics of such interactions, as they may change over the course of an invasion as a result of ecological and evolutionary processes. The type, strength and outcome of biotic interactions depend on the traits and evolutionary history of both the non-native plant and its potential interactors in the introduced range. The context-dependency of the outcomes of biotic interactions presents a challenge in the quest to predict the invasion success and impact of particular plant invaders on a given recipient community. Teasing apart the direct and indirect drivers of plant invasions and their impacts is particularly challenging. Structural equation modelling may help in isolating different drivers of plant invasions and testing their relative effects. A good example of the potential of this method was provided by Linders *et al.* (2019). Establishing research networks that replicate experiments globally (Borer *et al.*, 2014) also has the potential to contribute significantly to unveiling patterns and mechanisms driving plant invasions.

Heger and Jeschke (2018) proposed a new hierarchy-of-hypotheses (HoH) approach for synthesizing the large amount of published data and for testing specific hypotheses to increase our ability to explain and predict biological invasions. They suggest adding this HoH approach to the ecological toolbox as an additional method to approach complexity and context dependence. However, complex phenomena like biological invasions are unlikely to be strictly hierarchical because invasion success is determined by multiple factors that operate idiosyncratically in time and space. For this reason, we suggest that major advances in the field are more likely to come from the application of networks of models rather than a hierarchy of hypotheses (Scheiner and Fox, 2018).

Frost *et al.* (2019) proposed a holistic approach, using network theory, that simultaneously considers all relevant 'effects and effectors' to understand aspects of invasion from propagule pressure through to invader establishment, spread and eventual ecological impacts. They argued that network theory can inform us which species are most likely to become invasive, which ecosystems are most invasible and what impacts an invader will have on the recipient community. Some species-level network characteristics can tell us, for instance, how vulnerable a species is to enemy-release or how likely it is that a new species will become invasive.

Assessing how network characteristics of invaders vary through time (as they are becoming more abundant in the community) can also give insights on the mechanisms that allow them to invade. Likewise, understanding how network structure relates to the niche space available for invasion allows us to assess how resistant or vulnerable communities are to invasion (Minoarivelo and Hui, 2016; Chapter 2, this volume). For instance, highly nested antagonistic networks appear to be more susceptible to invasion due to high interaction asymmetry. However, the opposite pattern occurs in mutualistic networks. Modularity, another important network metric, can also be informative regarding the susceptibility of a community to invasion, although findings are still contentious (Frost *et al.*, 2019). Nestedness and modularity seem to have opposing effects on the stability of antagonistic and mutualistic networks; thus, higher modularity stabilizes antagonistic networks but destabilizes mutualistic ones, whereas the opposite occurs for nestedness (Thébault and Fontaine, 2010). On the other hand, the impact of invasive plants on network structure has also yielded different results; some studies have found that they either increase or decrease network structural properties whereas most studies reported no significant effect (Chapter 18, this volume). Due to the high degree of context-dependency apparent among these studies, Hui and Richardson (2019) proposed a theoretical framework based on community stability theory to shift from static and dynamic networks to complex adaptive networks when investigating invasibility and biotic resistance of recipient communities. This theoretical framework provides specific hypotheses to test regarding invasions in ecological networks. However, empirical tests of emerging theory are still very rare, especially in terms of manipulative experiments (Frost *et al.*, 2019). Likewise, an approach of gradients of invasion by plant species can help to clarify the inconsistent results that have emerged to date for the effect of plant invasions on networks, the existence of thresholds from which network effects are detectable, and how species-level metrics and species roles change as networks become more invaded (Chapter 18, this volume).

Research on the functional role of invasive species in ecological networks or on their impact on ecosystem functioning is still in its infancy, but a network framework looks promising for advancing our understanding of plant invasions (Chapter 18, this volume). For instance, a network approach can be useful to predict which functional changes will follow an invasion based on how invader effects on other species' abundances will spread throughout the network and – by means of coextinction modelling – what the consequences of the removal of a keystone species will be for network function or stability (e.g. Traveset *et al.*, 2015). The findings so far indicate that although invasive species do not necessarily impact network structure or stability in the recipient community, they can take up central positions (acting as a hub or as an important connector) and thus play a relevant role in the network (e.g. Traveset *et al.*, 2015). Moreover, invaders are also involved in indirect effects (e.g. apparent competition) that can be unveiled by means of a network approach (Frost *et al.*, 2019). Lastly, networks are also useful for studying the effectiveness of biological control methods as well as other forms of invasive species management (Chapters 21 and 22, this volume). The increased use of multilayer networks – which include more than a single ecological process at a time) – is also promising, although there are challenges because of the difficulty of accounting simultaneously for different types of interactions. Such an approach has the potential to provide insights into community assembly and disassembly, and that may also allow us to design effective restoration strategies in invaded communities (Chapters 18 and 23, this volume).

Rapid advances are being made in molecular approaches in the study of biological invasions, and exciting opportunities are emerging for addressing important questions. In Chapter 19 (this volume), Le Roux summarizes the use of such approaches to improve our understanding of plant—microbial interactions, the structure and dynamics of the networks they underlie, and how these respond to perturbation by non-native species. Specifically, new developments in next-generation sequencing (NGS) techniques are overcoming the limitations relating to the identification and quantification of microbes. They also provide access to cost-effective sequencing platforms and high-quality big data. It is now possible to better understand microbial diversity and function and to build more precise ecological networks (Toju, 2015). Using molecular approaches, numerous interactions can be easily sampled over short periods and large spatial scales compared to traditional observational studies. Nevertheless, the frequent non-linear relationship between biomass and barcode abundance, together with contamination and sequencing errors, remain big challenges that need to be overcome (Chapter 19, this volume).

1.5 Considering Biotic Interactions to Improve Our Ability to Manage Plant Invasions

The preceding sections show that non-native plants readily forge many types of associations and engage in diverse interactions with resident species, both native and non-native. Such associations and interactions mediate the invasiveness of non-native species, invasibility of ecosystems and the impact that non-natives have in ecosystems that have been invaded. How can we use the rapidly improving knowledge of the diverse roles of biotic interactions in plant invasions to manage plant invasions more effectively?

As discussed earlier in the chapter, by far the most important biotic actor influencing plant invasions is *Homo sapiens*. All invasions are caused by humans and perceived by humans, almost all are influenced in diverse ways at different stages by different human-mediated actions, and humans decide whether to manage invasions and, if so, how. There is a vast literature on 'invasion science' – 'the full spectrum of fields of enquiry that address issues pertaining to alien species and biological invasions ... [This discipline] embraces invasion ecology but increasingly involves non-biological lines of enquiry, including economics, ethics, sociology and interdisciplinary studies' (Hui and Richardson, 2017). It is beyond the scope of this chapter and book to review all aspects of the human dimensions of plant invasions. Rather, we focus on non-human actors implicated in influential biotic interactions affecting plant invasions, and the direct role that humans can play in changing the trajectory of plant invasions by influencing such interactions.

First, emerging insights on the role of interactions have some, albeit limited, value for improving our ability to screen species for their capacity to establish and spread in new environments. Elucidation of the roles of mutualisms conceptualized by Richardson *et al.* (2000b) – animal-mediated pollination and seed dispersal, and symbioses between plant roots and microbiota – has greatly improved our understanding of plant invasion dynamics. However, research over the past two decades has confirmed the key findings and predictions of that study – that specialized associations very rarely thwart invasions and that the potential for the absence of specialized symbionts to hinder establishment or spread of non-native species is decreasing rapidly as levels of homogenization of regional biotas increases. For example, whereas the absence of mycorrhizal symbionts was a potent barrier to the establishment of non-native pines in the southern hemisphere until the 17th century, widespread plantings and intentional or accidental inoculation of soils with microbiota has all but eliminated this barrier to establishment and spread. Also, many studies have shown the capacity for species pairs that evolved in isolation to form novel associations that assist non-natives to negotiate various barriers to becoming naturalized or invasive (e.g. cockatoos dispersing pines in Australia; Richardson *et al.*, 2000b). The main message to emerge from many studies is that the absence of specialist symbionts very seldom limits the capacity of non-native plants to naturalize or become invasive. One finding that is useful for screening new introductions is that plants with fleshy fruits containing small seeds must always be considered high-risk introductions because they are readily assimilated into seed-dispersal networks.

Various options exist for manipulating biotic interactions between non-native plants and the resident biota in invaded ecosystems, some of which have potential for reducing the abundance of the non-natives. The most obvious intervention is to introduce natural enemies of the non-native species from their native range, the aim being to eliminate or reduce the benefits that the non-natives have enjoyed by being freed from the negative effects of co-evolved antagonists. This is the domain of classical biological control which involves locating host-specific natural enemies in the native range, evaluating their safety through host-specificity testing, releasing them in the invaded range and monitoring the outcome (Pitcairn, 2011). Biological control agents are themselves non-native species and are subject to the same types of barriers as any other alien species. Many species that are introduced as biological control agents fail to establish, and many of those that do establish have very limited, if any, effect on the abundance, distribution or impact of the target invasive plant species, and some become invasive species themselves (Chapter 20, this volume). Some biological control agents have, however, had profound effects on the target invasive species, causing dramatic declines in abundance, density and impacts (Van Driesche *et al.*, 2010). The textbook example of such effects was the biological control of *Opuntia* species in Australia using several introduced insect species (Dodd, 1940). Biological control will never eradicate a non-native plant, but aims to reduce its extent, density and abundance to below a 'nuisance level'. In many cases, careful integration of biological control with other types of management has yielded major social and economic benefits (van Wilgen *et al.*, 2020). Nonetheless, biological control of invasive non-native plants remains contentious (e.g. Havens *et al.*, 2019).

Most control operations that target non-native plant species involve a combination of measures, including mechanical, chemical and biological methods, and what is often termed 'cultural control' (Clout and Williams, 2009). In the first three, the aim is to introduce levels of human-mediated 'predation' to reduce the capacity of the non-native plants to establish, reproduce, spread and persist. Such interventions focus on various phases of the life cycle of the non-native plant, depending on the biology of the species and limitations imposed by the availability of resources, features of the invaded ecosystems, and other factors. Cultural control involves a wide range of interventions that modify biotic or abiotic features of the environment with the aim of killing, harming or limiting the opportunities of non-native species to establish, reproduce, spread or to have negative impact. For non-native plant invasions, the most commonly applied interventions involve manipulating patterns, levels and regimes of grazing or browsing by native animals or (usually non-native) livestock, modifying fire, nutrient or irrigation regimes, and various approaches aimed at maintaining, improving or restoring the vigour of native (or other, less problematic, non-native) species, thereby enhancing biotic resistance (Chapter 9, this volume).

Given that non-native plants are readily assimilated into pollination and seed-dispersal networks, is it realistic to think that human actions can exert sufficient influence on these networks over large enough spatial scales to eliminate or disadvantage non-native plant species enough to have a meaningful effect on their capacity to progress along the introduction—naturalization—invasion continuum? Although it is theoretically possible to manage the dispersal vectors directly, for example by modifying the behaviour of frugivores by manipulating the landscape, canopy and availability of fruits (Gosper *et al.*, 2005; Buckley *et al.*, 2006), we could find no evidence of such interventions being pursued effectively over large scales. A key reason for this is the conflict of interest that arises when interventions aimed at managing the non-native plants have a high risk of negatively affecting features of the invaded ecosystems. For example, where invasive non-native plants form part of the diet of native frugivores, interventions directed to interfere with this interaction are very likely to impact populations of the native frugivores, with potentially detrimental impacts to aspects of ecosystem functioning (Díaz Vélez *et al.*, 2018). Such conflicts are frequently amplified when other aspects of global change, such as habitat modification, have reduced native frugivore populations (Elise Müller de Lima *et al.*, 2015). The feasibility of proposed interventions, such as providing alternative sources of food for frugivores by boosting the availability of native plants that fruit at the same time as the non-natives (Wotton and McAlpine, 2015) remains to be tested. Such measures are unlikely to contribute substantially to control efforts, except possibly over very small spatial scales. Knowledge of the dynamics of interactions between frugivores and fleshy-fruited species in an invaded ecosystem can, however, be useful for predicting likely trajectories of spread for non-native fleshy-fruited species that are assimilated in frugivore—plant networks (Chapter 22, this volume). This may be useful for directing management interventions, for example to identify sites for surveillance or priority action in special habitats. Although Dickie *et al.* (2017) state that 'understanding plant—fungal interactions can improve all stages of management of linked plant—fungal invasions, ranging from risk assessment to ecosystem restoration' we could find no examples in the literature to show the application of such knowledge to prevent or reduce invasions.

Climate change has the potential to radically alter the dynamics of biotic interactions influencing invasions of non-native plants (Traveset and Richardson, 2014). Complex synergistic interactions between climate and many other factors have been shown to affect plant invasion dynamics in many ways (e.g. Richardson and Bond, 1991). Such complexities must be considered when forecasting invasion trajectories and when developing management strategies. Reeves (2017) provides a useful summary of the challenges in this regard when considering potential effects of climate change on biological control of invasive plants.

1.6 Profitable Avenues for Future Research

As in other facets of invasion science, the study of interactions has contributed to the emergence of a plethora of hypotheses and theories. We have reviewed the key ones in this chapter. Much attention is being given to the critical evaluation of these hypotheses with a view to arriving at a more manageable set to guide progress in the field (Catford *et al.*, 2009; Jeschke and Heger, 2018). Much work remains to be done to fine-tune our understanding of the many types of biotic interactions that are pivotal components of many of the most prominent theories and hypotheses. In this regard, it would be useful to convene 'global networks' (sensu Packer *et al.*, 2017) to focus on pivotal areas pertaining to the biotic interactions reviewed in this chapter and elsewhere in the book. Several taxa and systems have emerged as 'model systems' (sensu Kueffer *et al.*, 2013) for the study of key aspects of biotic interactions that mediate plant invasions. Prominent examples include *Lythrum salicaria* for issues pertaining to enemy release (Chapter 10, this volume), *Centaurea* species for elucidation of the Novel Weapons Hypothesis (Chapter 3, this volume), and *Alliaria petiolata* for understanding how allelopathy mediates invasion (Chapter 15, this volume). There is much scope for identifying additional model taxa, as well as 'invasion syndromes' (sensu Novoa *et al.*, 2020) to serve as foci for research to shed more light on key unknowns, such as the relative importance of certain interactions compared to disturbance, propagule pressure and other factors.

As discussed in this chapter and elsewhere in the book (notably in Chapters 2, 18 and 22), network theory has much potential to improve our understanding of the role of biotic interactions in plant invasions. There is, however, an urgent need for more experimental studies that incorporate both positive and negative biotic interactions simultaneously for a range of non-native plants across all major categories of ecosystems.

Given the increasing extent and impacts of plant invasions, there is an urgent need to find more effective and sustainable management approaches. Giving more consideration to biotic interactions is crucial for improving management. Biological control is well established as a cost-effective element of integrated management for many invasive plant species in many types of ecosystems (Chapter 20, this volume). More attention must be directed at rolling out such programmes, especially to parts of the world with limited resources to allocate to more expensive types of interventions. There are biological control success stories for many of the world's most widespread and damaging plant invasions, and there are many opportunities to replicate such successes in new areas. Hill and Coetzee (Chapter 20, this volume) stress the need for '... better understanding and integration of the top-down and bottom-up biotic interactions in the biological control context ... if the final goal of habitat restoration is to be realized'.

Many opportunities exist to increase the resistance of ecosystems to invasion by maintaining or restoring key processes and by limiting disturbance. Biotic interactions are crucial mediators of ecosystem recovery following management directed at removal invasive non-native plants (Chapters 21 and 23, this volume). Restoration of ecosystems after control of invasive species can be fast-tracked by manipulating biotic interactions, for example through facilitation of recovery of native species and the alleviation of problems associated with legacy effects and secondary invasions. Developing innovative and effective interventions requires consideration of multiple trophic levels. Preventing invasional meltdowns and regime shifts should be a key management priority, especially in the face of continuing high rates of non-native species introductions (Seebens *et al.*, 2017).

This volume presents 22 chapters (besides this one) that explore the current understanding of the roles of biotic interactions in invasions of non-native plants. We have structured the book in seven parts: background; positive and negative interactions in the soil; mutualistic interactions that promote plant invasions; antagonistic interactions that hinder plant invasions; consequences of plant invasions for biotic interactions among native species; novel techniques and experimental approaches in the study of plant invasions; and biotic interactions and the management of ecosystems invaded by non-native plants. Fifty-two authors from 14 countries (as determined by their current institutional affiliations) contributed to the chapters. Although the geographical spread of authors mirrors the well-documented geographical bias in the study of plant invasions (Pyšek *et al.*, 2008), many authors have experience in numerous regions of the world. Every effort was made to provide a comprehensive global coverage of research that has addressed biotic interactions affecting plant invasion. This introductory chapter has identified some key information gaps, but each of the other chapters provides its own list of priorities for further research. We hope that this book will stimulate many more studies on biotic interactions that will improve our understanding of the processes that mediate plant invasions and will help us to develop more effective measures to manage these invasions.

Acknowledgements

We acknowledge funding from the Spanish Ministry of Science and Universities (project CGL2017-88122-P), the DSI-NRF Centre of Excellence for Invasion Biology and the National Research Foundation of South Africa (grant 85417) and the Oppenheimer Memorial Trust (grant 18576/03) to DMR. Jane Catford, Cang Hui, Dean Pearson, Warwick Allen, Ray Callaway and Jacob Lucero provided useful comments on an early draft of the MS.

References

Allesina, S. and Tang, S. (2012) Stability criteria for complex ecosystems. *Nature* 483(7388), 205–208. DOI: 10.1038/nature10832.

Alpert, P. (2006) The advantages and disadvantages of being introduced. *Biological Invasions* 8(7), 1523–1534. DOI: 10.1007/s10530-005-5844-z.

Beaury, E.M., Finn, J.T., Corbin, J.D., Barr, V. and Bradley, B.A. (2020) Biotic resistance to invasion is ubiquitous across ecosystems of the United States. *Ecology Letters* 23(3), 476–482. DOI: 10.1111/ele.13446.

Benadi, G., Blüthgen, N., Hovestadt, T. and Poethke, H.-J. (2012) Population dynamics of plant and pollinator communities: stability reconsidered. *The American Naturalist* 179(2), 157–168. DOI: 10.1086/663685.

Bertness, M.D. and Callaway, R.M. (1994) Positive interactions in communities. *Trends in Ecology & Evolution* 9, 191–193.

Blossey, B. and Nötzgold, R. (1995) Evolution of increased competitive ability in invasive non-Indigenous plants: a Hypothesis. *Journal of Ecology* 83, 887–889.

Blumenthal, D.M. (2006) Interactions between resource availability and enemy release in plant invasion. *Ecology Letters* 9, 887–895.

Borer, E.T., Harpole, W.S., Adler, P.B., Lind, E.M., Orrock, J.L. *et al*. (2014) Finding generality in ecology: a model for globally distributed experiments. *Methods in Ecology and Evolution* 5, 65–73.

Boulangeat, I*et al*. (2012) Accounting for dispersal and biotic interactions to disentangle the drivers of species distributions and their abundances. *Ecology Letters* 15, 584–593.

Braga, R.R., Gómez-Aparicio, L., Heger, T., Vitule, J.R.S. and Jeschke, J.M. (2018) Structuring evidence for invasional meltdown: broad support with biases and gaps. *Biological Invasions* 20, 923–936.

Bronstein, J.L. (2015) *Mutualisms*. Oxford University Press, Oxford.

Brooks, M.L., D'Antonio, C.M., Richardson, D.M., Grace, J.B., Keeley, J.E. *et al*. (2004) Effects of invasive alien plants on fire regimes. *BioScience* 54, 677–688.

Bruno, J.F., Stachowicz, J.J. and Bertness, M.D. (2003) Inclusion of facilitation into ecological theory. *Trends in Ecology & Evolution* 18, 119–125.

Buckley, Y.M., Anderson, S., Catterall, C.P., Corlett, R.T., Engel, T. *et al*. (2006) Management of plant invasions mediated by frugivore interactions. *Journal of Applied Ecology* 43, 848–857.

Cadotte, M.W., Campbell, S.E., Li, S., Sodhi, D.S. and Mandrak, N.E. (2018) – Preadaptation and naturalization of nonnative species: Darwin's two fundamental insights into species invasion. *Annual Review of Plant Biology* 69, 661–684.

Callaway, R.M. and Aschehoug, E.T. (2000) Invasive plants versus their new and old neighbors: a mechanism for exotic invasion. *Science* 290(5491), 521–523. DOI: 10.1126/science.290.5491.521.

Callaway, R.M., Cipollini, D., Barto, K., Thelen, G.C., Hallett, S.G. *et al.* (2008) Novel weapons: invasive plant suppresses fungal mutualists in America but not in its native Europe. *Ecology* 89(4), 1043–1055. DOI: 10.1890/07-0370.1.
Callaway, R.M. and Ridenour, W.M. (2004) Novel weapons: invasive success and the evolution of increased competitive ability. *Frontiers in Ecology and the Environment* 2(8), 436–443. DOI: 10.1890/1540-9295(2004)002[0436:NWISAT]2.0.CO;2.
Cappuccino, N. and Arnason, J.T. (2006) Novel chemistry of invasive exotic plants. *Biology Letters* 2, 189–193.
Catford, J.A., Dwyer, J.M., Palma, E., Cowles, J.M. and Tilman, D. (2020) Community diversity outweighs effect of warming on plant colonization. *Global Change Biology* 26(5), 3079–3090. DOI: 10.1111/gcb.15017.
Catford, J.A., Jansson, R. and Nilsson, C. (2009) Reducing redundancy in invasion ecology by integrating hypotheses into a single theoretical framework. *Diversity and Distributions* 15(1), 22–40. DOI: 10.1111/j.1472-4642.2008.00521.x.
Clout, M.N. and Williams, P.A. (eds) (2009) *Invasive Species Management. A Handbook of Principles and Techniques*. Oxford University Press, Oxford.
Colautti, R.I., Ricciardi, A., Grigorovich, I.A. and MacIsaac, H.J. (2004) Is invasion success explained by the enemy release Hypothesis? *Ecology Letters* 7, 721–733.
Daehler, C.C. (2001) Darwin's Naturalization Hypothesis revisited. *American Naturalist* 158, 324–330.
Darwin, C. (1859) *On the Origin of Species by Means of Natural Selection, or the Preservation of Favoured Races in the Struggle for Life*. Murray, London.
Davies, K.F., Cavender-Bares, J. and Deacon, N. (2011) Native communities determine the identity of exotic invaders even at scales at which communities are unsaturated. *Diversity and Distributions* 17(1), 35–42. DOI: 10.1111/j.1472-4642.2010.00721.x.
Davies, K.F., Chesson, P., Harrison, S., Inouye, B.D., Melbourne, B.A. *et al.* (2005) Spatial heterogeneity explains the scale dependence of the native–exotic diversity relationship. *Ecology* 86(6), 1602–1610. DOI: 10.1890/04-1196.
Dickie, I.A., Bolstridge, N., Cooper, J.A. and Peltzer, D.A. (2010) Co-invasion by *Pinus* and its mycorrhizal fungi. *New Phytologist* 187, 475–484.
Dickie, I.A., Bufford, J.L. and Cobb, R.C. *et al.* (2017) The emerging science of linked plant-fungal invasions. *New Phytologist* 215, 1314–1332.
Dodd, A.P. (1940) *The Biological Campaign Against Prickly Pear*. Commonwealth Prickly Pear Board, Brisbane.
Duncan, R.P. and Williams, P.A. (2002) Ecology: Darwin's naturalization Hypothesis challenged. *Nature* 417, 608–609.
Díaz Vélez, M.C., Sérsic, A.N., Traveset, A. and Paiaro, V. (2018) The role of frugivorous birds in fruit removal and seed germination of the invasive alien *Cotoneaster franchetii* in central Argentina. *Austral Ecology* 43(5), 558–566. DOI: 10.1111/aec.12592.
Elise Müller de Lima, R., de Sá Dechoum, M. and Castellani, T.T. (2015) Native seed dispersers may promote the spread of the invasive Japanese raisin tree (*Hovenia dulcis* Thunb.) in seasonal deciduous forest in southern Brazil. *Tropical Conservation Science* 8(3), 846–862. DOI: 10.1177/194008291500800318.
Elton, C.S. (1958) *The Ecology of Invasions by Animals and Plants*. Methuen, London.
Eppinga, M.B., Rietkerk, M., Dekker, S.C. and De Ruiter, P.C. (2006) Accumulation of local pathogens: a new Hypothesis to explain exotic plant invasions. *Oikos* 114, 168–176.
Farwig, N. and Berens, D.G. (2012) Imagine a world without seed dispersers: a review of threats, consequences and future directions. *Basic and Applied Ecology* 13, 109–115.
Fridley, J.D., Stachowicz, J.J., Naeem, S., Sax, D.F., Seabloom, E.W. *et al.* (2007) The invasion paradox: reconciling pattern and process in species invasions. *Ecology* 88, 3–17.
Frost, C.M., Allen, W.J., Courchamp, F., Jeschke, J.M., Saul, W.C. *et al.* (2019) Using network theory to understand and predict biological invasions. *Trends in Ecology & Evolution* 34, 831–843.
Gaertner, M., Biggs, R., Te Beest, M., Hui, C., Molofsky, J. *et al.* (2014) Invasive plants as drivers of regime shifts: identifying high priority invaders that alter feedback relationships. *Diversity and Distributions* 20, 733–744.
Gause, G.F. (1934) *The Struggle for Existence*. Williams and Wilkins, Baltimore, MD, USA.
Godsoe, W., Jankowski, J., Holt, R.D. and Gravel, D. (2017) Integrating biogeography with contemporary niche theory. *Trends in Ecology & Evolution* 32, 488–499.

Gosper, C.R., Stansbury, C.D. and Vivian-Smith, G. (2005) Seed dispersal of fleshy-fruited invasive plants by birds: contributing factors and management options. *Diversity and Distributions* 11, 549–558.
Gravel, D., Baiser, B. and Dunne, J. *et al.* (2018) Bringing Elton and Grinnell together: a quantitative framework to represent the biogeography of ecological interaction networks. *Ecography* 41, 1–15.
Guisan, A. and Thuiller, W. (2005) Predicting species distribution: offering more than simple habitat models. *Ecology Letters* 8, 993–1009.
Hale, K.R.S., Valdovinos, F.S. and Martínez, N.D. (2020) Pollinators in food webs: mutualistic interactions increase diversity, stability, and function in multiplex networks. bioRxiv preprint first posted online Oct. 3, 2019.
Hart, S.P., Freckleton, R.P. and Levine, J.M. (2018) How to quantify competitive ability. *Journal of Ecology* 106, 1902–1909.
Havens, K., Jolls, C.L., Knight, T.M. and Vitt, P. (2019) Risks and rewards: assessing the effectiveness and safety of classical invasive plant biocontrol by arthropods. *BioScience* 69, 247–258.
Heger, T. and Jeschke, J.M. (2018) The hierarchy-of-Hypotheses approach. In: Jeschke, J.M. and Heger, T. (eds) *Invasion Biology: Hypotheses and Evidence, CABI*. Wallingford, pp. 14–18.
Hokkanen, H.M.T. and Pimentel, D. (1989) New associations in biological control: theory and practice. *The Canadian Entomologist* 121(10), 829–840. DOI: 10.4039/Ent121829-10.
Hui, C. and Richardson, D.M. (2017) *Invasion Dynamics*. Oxford University Press, Oxford.
Hui, C. and Richardson, D.M. (2019) How to invade an ecological network. *Trends in Ecology & Evolution* 34(2), 121–131. DOI: 10.1016/j.tree.2018.11.003.
Hulvey, K.B. and Teller, B.J. (2018) Site conditions determine a key native plant's contribution to invasion resistance in grasslands. *Ecology* 99, 1257–1264.
Hutchinson, G.E. (1957) Concluding remarks. *Cold Spring Harbor Symposium on Quantitative Biology* 22, 415–427.
Jiang, L., Tan, J. and Pu, Z. (2010) An experimental test of Darwin's naturalization Hypothesis. *American Naturalist* 175, 415–423.
Jeschke, J.M. and Heger, T. (eds) (2018) *Invasion Biology: Hypotheses and Evidence*. CABI, Wallingford.
Keane, R.M. and Crawley, M.J. (2002) Exotic plant invasions and the enemy release Hypothesis. *Trends in Ecology & Evolution* 17, 164–170.
Kiers, E.T., Palmer, T.M., Ives, A.R., Bruno, J.F. and Bronstein, J.L. (2010) Mutualisms in a changing world: an evolutionary perspective. *Ecology Letters* 13, 1459–1474.
Kueffer, C., Pyšek, P. and Richardson, D.M. (2013) Integrative invasion science: model systems, multi-site studies, focused meta-analysis, and invasion syndromes. *New Phytologist* 200, 615–633.
Lankau, R.A., Nuzzo, V., Spyreas, G. and Davis, A.S. (2009) Evolutionary limits ameliorate the negative impact of an invasive plant. *Proceedings of the National Academy of Sciences USA* 106, 15362–15367.
Larios, L., Pearson, D.E. and Maron, J.L. (2017) Incorporating the effects of generalist seed predators into plant community theory. *Functional Ecology* 31, 1856–1867.
Le Roux, J.J., Clusella-Trullas., Mairal, M., Mokotjomela, T., Richardson, D.M. *et al.* (2020) Biotic interactions as mediators of biological invasions: insights from South Africa. In: Van Wilgen, B.W., Measey, J., Richardson, D.M., Wilson, J.R.U., Zengeya, T, G.J. (eds) *Biological Invasions in South Africa*. Springer, Berlin, pp. 387–427.
Levine, J.M., Adler, P.B. and Yelenik, S.G. (2004) A meta-analysis of biotic resistance to exotic plant invasions. *Ecology Letters* 7, 975–989.
Linders, T.E.W., Schaffner, U., Eschen, R., Abebe, A., Choge, S.K. *et al.* (2019) Direct and indirect effects of invasive species: biodiversity loss is a major mechanism by which an invasive tree affects ecosystem functioning. *Journal of Ecology* 107, 2660–2672.
Lotka, A.J. (1925) *Elements of Physical Biology. Williams and.* Wilkins, Baltimore.
MacArthur, R.H. and Levins, R. (1967) The limiting similarity, convergence, and divergence of coexisting species. *American Naturalist* 101, 377–385.
MacDougall, A.S., Gilbert, B. and Levine, J.M. (2009) Plant invasions and the niche. *Journal of Ecology* 97, 609–615.
Mack, R.N., Simberloff, D., Mark Lonsdale, W., Evans, H., Clout, M. *et al.* (2000) Biotic invasions: causes, epidemiology, global consequences, and control. *Ecological Applications* 10, 689–710.
Maron, J.L. and Crone, E. (2006) Herbivory: effects on plant abundance, distribution and population growth. *Proceedings of the Royal Society B* 273, 2575–2584.
May, R.M. (1977) Thresholds and breakpoints in ecosystems with a multiplicity of stable states. *Nature* 269, 471–477.

May, R.M. (1981) Patterns in multi-species communities. In: May, R.M. (ed.) *Theoretical Ecology: Principles and Applications. Blackwell, Oxford*, pp. 197–227.

May, R.M. and McArthur, R.H. (1972) Niche overlap as a function of environmental variability. *Proceedings of the National Academy of Sciences USA* 69, 1109–1113.

McIntire, E.J.B. and Fajardo, A. (2014) Facilitation as a ubiquitous driver of biodiversity. *New Phytologist* 201, 403–416.

Minoarivelo, H.O. and Hui, C. (2016) Invading a mutualistic network: to be or not to be similar. *Ecology and Evolution* 6, 4981–4996.

Mitchell, C.E., Agrawal, A.A., Bever, J.D., Gilbert, G.S., Hufbauer, R.A. *et al.* (2006) Biotic interactions and plant invasions. *Ecology Letters* 9, 726–740.

Mittelbach, G.G. and McGill, B.J. (2019) *Community Ecology*, 2nd Edition. University Press, Oxford, Oxford.

Mougi, A. and Kondoh, M. (2012) Diversity of interaction types and ecological community stability. *Science* 337, 349–351.

Moyano, J., Rodriguez-Cabal, M.A. and Nuñez, M.A. (2020) Highly invasive tree species are more dependent on mutualisms. *Ecology* 101(5), e02997. DOI: 10.1002/ecy.2997.

Müller-Schärer, H., Schaffner, U. and Steinger, T. (2004) Evolution in invasive plants: implications for biological control. *Trends in Ecology & Evolution* 19, 417–422.

Novoa, A., Richardson, D.M., Pyšek, P., Meyerson, L.A., Bacher, S. and Canavan, S. *et al.* (2020) Invasion syndromes: a systematic approach for predicting biological invasions and facilitating effective management. *Biological Invasions* 22(5), 1801–1820. DOI: 10.1007/s10530-020-02220-w.

Ollerton, J., Winfree, R. and Tarrant, S. (2011) How many flowering plants are pollinated by animals? *Oikos* 120, 321–326.

Ordonez, A., Wright, I.J. and Olff, H. (2010) Functional differences between native and alien species: a global-scale comparison. *Functional Ecology* 24, 1353–1361.

Packer, J.G., Meyerson, L.A., Richardson, D.M., Brundu, G., Allen, W. *et al.* (2017) Global network for invasion science: benefits, challenges and guidelines. *Biological Invasions* 19, 1081–1096.

Parker, J.D., Burkepile, D.E. and Hay, M.E. (2006) Opposing effects of native and exotic herbivores on plant invasions. *Science* 311, 1459–1461.

Parker, J.D. and Hay, M.E. (2005) Biotic resistance to plant invasions? native herbivores prefer non-native plants. *Ecology Letters* 8, 959–967.

Pearson, D.E. and Callaway, R.M. (2003) Indirect effects of host-specific biological control agents. *Trends in Ecology & Evolution* 18(9), 456–461. DOI: 10.1016/S0169-5347(03)00188-5.

Pearson, D.E. and Callaway, R.M. (2006) Biological control agents elevate hanta virus by subsidizing mice. *Ecology Letters* 9, 443–450.

Pearson, D.E. and Callaway, R.M. (2008) Weed-biocontrol insects reduce native-plant recruitment through second-order apparent competition. *Ecological Applications* 18(6), 1489–1500. DOI: 10.1890/07-1789.1.

Pearson, D.E. and Fletcher, R.J. Jr. (2008) Mitigating exotic impacts: restoring native deer mouse populations elevated by an exotic food subsidy. *Ecological Applications* 18, 321–334.

Pearson, D.E., Ortega, Y.K., Eren, Özkan. and Hierro, J.L. (2018a) Community assembly theory as a framework for biological invasions. *Trends in Ecology & Evolution* 33(5), 313–325. DOI: 10.1016/j.tree.2018.03.002.

Pearson, D.E., Ortega, Y.K., Villarreal, D., Lekberg, Y., Cock, M.C. *et al.* (2018b) The fluctuating resource hypothesis explains invasibility, but not exotic advantage following disturbance. *Ecology* 99(6), 1296–1305. DOI: 10.1002/ecy.2235.

Pitcairn, M.J. (2011) Biological control, of plants. In: Simberloff, D. and Rejmánek, M. (eds) *Encyclopedia of Biological Invasions, University of California Press, Berkeley*, pp. 63–67.

Pringle, A., Bever, J.D., Gardes, M., Parrent, J.L., Rillig, M.C. *et al.* (2009) Mycorrhizal symbioses and plant invasions. *Annual Review of Ecology Evolution & Systematics* 40, 699–715.

Pringle, E.G. (2016) Orienting the interaction compass: resource availability as a major driver of context dependence. *PLoS Biology* 14(10), e2000891. DOI: 10.1371/journal.pbio.2000891.

Pyšek, P., Jarošík, V., Hulme, P.E., Kühn, I., Wild, J. *et al.* (2010) Disentangling the role of environmental and human pressures on biological invasions across Europe. *Proceedings of the National Academy of Sciences USA* 107(27), 12157–12162. DOI: 10.1073/pnas.1002314107.

Pyšek, P. and Richardson, D.M. (2006) The biogeography of naturalization in alien plants. *Journal of Biogeography* 33(12), 2040–2050. DOI: 10.1111/j.1365-2699.2006.01578.x.

Pyšek, P., Richardson, D.M., Pergl, J., Jarošík, V., Sixtová, Z. *et al.* (2008) Geographical and taxonomical biases in invasion ecology. *Trends in Ecology & Evolution* 23(5), 237–244. DOI: 10.1016/j.tree.2008.02.002.

Reeves, J.L. (2017) *Climate change effects on biological control of invasive plants by insects*. CAB Reviews 12.

Rejmánek, M. (1996) A theory of seed plant invasiveness: the first sketch. *Biological Conservation* 78, 171–181.

Ricciardi, A., Blackburn, T.M., Carlton, J.T., Dick, J.T.A., Hulme, P.E. *et al.* (2017) Invasion science: a horizon scan of emerging challenges and opportunities. *Trends in Ecology & Evolution* 32, 464–474.

Richardson, D.M. and Bond, W.J. (1991) Determinants of plant distribution: evidence from pine invasions. *The American Naturalist* 137(5), 639–668. DOI: 10.1086/285186.

Richardson, D.M., Allsopp, N., D'Antonio, C.M., Milton, S.J. and Rejmánek, M. (2000a) Plant invasions: the role of mutualisms. *Biological Reviews* 75(1), 65–93. DOI: 10.1017/S0006323199005435.

Richardson, D.M., Pyšek, P., Rejmánek, M., Barbour, M.G., Panetta, D.F. *et al.* (2000b). *Naturalization and invasion of alien plants - concepts and definitions Diversity and Distributions* 6, 93–107.

Richardson, D.M. and Pyšek, P. (2012) Naturalization of introduced plants: ecological drivers of biogeographic patterns. *New Phytologist* 196, 383–396.

Sax, D.F., Stachowicz, J.J., Brown, J.H., Bruno, J.F., Dawson, M.N. *et al.* (2007) Ecological and evolutionary insights from species invasions. *Trends in Ecology & Evolution* 22, 465–471.

Scheiner, S.M. and Fox, G.A. (2018) A hierarchy of hypotheses or a network of models. In: Jeschke, J.M. and Heger, T. (eds) *Invasion Biology: Hypotheses and Evidence*. CABI, Wallingford, UK, pp. 30–37.

Seabloom, E.W., Borer, E.T., Buckley, Y.M., Cleland, E.E., Davies, K.F. *et al.* (2015) Plant species' origin predicts dominance and response to nutrient enrichment and herbivores in global grasslands. *Nature Communications* 6, 7710. DOI: 10.1038/ncomms8710.

Seabloom, E.W., Williams, J.W., Slayback, D., Stoms, D.M., Viers, J.H. *et al.* (2006) Human impacts, plant invasion, and imperiled plant species in California. *Ecological Applications* 16(4), 1338–1350. DOI: 10.1890/1051-0761(2006)016[1338:HIPIAI]2.0.CO;2.

Seebens, H., Blackburn, T.M., Dyer, E.E., Genovesi, P., Hulme, P.E. *et al.* (2017) No saturation in the accumulation of alien species worldwide. *Nature Communications* 8(1), 14435. DOI: 10.1038/ncomms14435.

Shackleton, R.T., Biggs, R., Richardson, D.M. and Larson, B.M.H. (2018) Social-ecological drivers and impacts of invasion-related regime shifts: consequences for ecosystem services and human wellbeing. *Environmental Science and Policy* 89, 300–314.

Simberloff, D. (2006) Invasional meltdown 6 years later: important phenomenon, unfortunate metaphor, or both? *Ecology Letters* 9, 912–919.

Simberloff, D. and von Holle, B. (1999) Positive interactions of non-Indigenous species: Invasional meltdown? *Biological Invasions* 1, 21–32.

Stohlgren, T.J., Barnett, D.T. and Kartesz, J.T. (2003) The rich get richer: patterns of plant invasions in the United States. *Frontiers in Ecology and the Environment* 1, 11–14.

Stohlgren, T., Jarnevich, C., Chong, G.W. and Evangelista, P.H. (2006) Scale and plant invasions: a theory of biotic acceptance. *Preslia* 78, 405–426.

Strauss, S.Y., Lau, J.A. and Carroll, S.P. (2006) Evolutionary responses of natives to introduced species: what do introductions tell us about natural communities? *Ecology Letters* 9, 357–374.

Thuiller, W., Gallien, L., Boulangeat, I., De Bello, F., Münkemüller, T. *et al.* (2010) Resolving Darwin's naturalization conundrum: A quest for evidence. *Diversity and Distributions* 16, 461–475.

Thébault, E. and Fontaine, C. (2010) Stability of ecological communities and the architecture of mutualistic and trophic networks. *Science* 329, 853–856.

Toju, H. (2015) High-throughput DNA barcoding for ecological network studies. *Population Ecology* 57, 37–51.

Traveset, A., Olesen, J.M., Nogales, M., Vargas, P., Jaramillo, P. *et al.* (2015) Bird–flower visitation networks in the Galápagos unveil a widespread interaction release. *Nature Communications* 6, 6376.

Traveset, A. and Richardson, D.M. (2006) Biological invasions as disruptors of plant reproductive mutualisms. *Trends in Ecology & Evolution* 21(4), 208–216. DOI: 10.1016/j.tree.2006.01.006.

Traveset, A. and Richardson, D.M. (2011) Mutualisms: key drivers of invasions ... key casualties of invasions. In: Richardson, D.M. (ed.) *Fifty Years of Invasion Ecology. The Legacy of Charles Elton*. Wiley-Blackwell, Oxford, pp. 143–160.

Traveset, A. and Richardson, D.M. (2014) Mutualistic interactions and biological invasions. *Annual Review of Ecology, Evolution, and Systematics* 45(1), 89–113. DOI: 10.1146/annurev-ecolsys-120213-091857.

Van Driesche, R.G., Carruthers, R.I., Center, T., Hoddle, M.S., Hough-Goldstein, J. *et al.* (2010) Classical biological control for the protection of natural ecosystems. *BiolControl* 54, S2–S33. DOI: 10.1016/j.biocontrol.2010.03.003.

van Kleunen, M., Bossdorf, O. and Dawson, W. (2018) The ecology and evolution of alien plants. *Annual Review of Ecology, Evolution, and Systematics* 49(1), 25–47. DOI: 10.1146/annurev-ecolsys-110617-062654.

van Kleunen, M., Weber, E. and Fischer, M. (2010) A meta-analysis of trait differences between invasive and non-invasive plant species. *Ecology Letters* 13(2), 235–245. DOI: 10.1111/j.1461-0248.2009.01418.x.

van Wilgen, B.W., Raghu, S., Sheppard, A.W. and Schaffner, U. (2020) Quantifying the social and economic benefits of the biological control of invasive alien plants in natural ecosystems. *Current Opinion in Insect Science* 38, 1–5. DOI: 10.1016/j.cois.2019.12.004.

Volterra, V. (1926) *Variazione e Fluttuazioni del Numero d'Individui in Specie Animali Conviventi*. Memoria della Regia Accademia Nazionale dei Liincei, ser 6(2), 31-113. Reprinted (1931). In: Chapman, R.N. (ed.) *Animal Ecology*. McGraw-Hill, New York.

Wang, J., Ge, Y., Zhang, C.B., Bai, Y. and Du, Z.K. (2013) Dominant functional group effects on the invasion resistance at different resource levels. *PLoS ONE* 8(10), e77220. DOI: 10.1371/journal.pone.0077220.

White, E.M., Wilson, J.C. and Clarke, A.R. (2006) Biotic indirect effects: a neglected concept in invasion biology. *Diversity and Distributions* 12, 443–455.

Wotton, D.M. and McAlpine, K.G. (2015) Seed dispersal of fleshy-fruited environmental weeds in New Zealand. *New Zealand Journal of Ecology* 39, 135–169.

Zhang, Z. and van Kleunen, M. (2019) Common alien plants are more competitive than rare natives but not than common natives. *Ecology Letters* 22(9), 1378–1386. DOI: 10.1111/ele.13320.

Zimmermann, H., Brandt, P., Fischer, J., Welk, E. and von Wehrden, H. (2014) The human release Hypothesis for biological invasions: human activity as a determinant of the abundance of invasive plant species. *F1000Research* 3(109), 109. DOI: 10.12688/f1000research.3740.2.

2 The Role of Biotic Interactions in Invasion Ecology: Theories and Hypotheses

Cang Hui[1,2]*, Pietro Landi[1] and Guillaume Latombe[1,3,4]

[1]Centre for Invasion Biology, Department of Mathematical Sciences, Stellenbosch University, Matieland 7602, South Africa; [2]Theoretical Ecology Group, African Institute for Mathematical Sciences, Cape Town 7945, South Africa; [3]Centre for Invasion Biology, Department of Botany and Zoology, Stellenbosch University, Matieland 7602, South Africa; [4]Department of Botany and Biodiversity Research, Division of Conservation Biology, Vegetation Ecology and Landscape Ecology, University of Vienna, 1030 Vienna, Austria

Abstract

Changes in biotic interactions in the native and invaded range can enable a non-native species to establish and spread in novel environments. Invasive non-native species can in turn generate impacts in recipient systems partly through the changes they impose on biotic interactions; these interactions can lead to altered ecosystem processes in the recipient systems. This chapter reviews models, theories and hypotheses on how invasion performance and impact of introduced species in recipient ecosystems can be conjectured according to biotic interactions between native and non-native species. It starts by exploring the nature of biotic interactions as ensembles of ecological and evolutionary games between individuals of both the same and different groups. This allows us to categorize biotic interactions as direct and indirect (i.e. those involving more than two species) that emerge from both coevolution and ecological fitting during community assembly and invasion. We then introduce conceptual models that can reveal the ecological and evolutionary dynamics between interacting non-native and resident species in ecological networks and communities. Moving from such theoretical grounding, we review 20 hypotheses that have been proposed in invasion ecology to explain the invasion performance of a single non-native species, and seven hypotheses relating to the creation and function of assemblages of non-native species within recipient ecosystems. We argue that, although biotic interactions are ubiquitous and quintessential to the assessment of invasion performance, they are nonetheless difficult to detect and measure due to strength dependency on sampling scales and population densities, as well as the non-equilibrium transient dynamics of ecological communities and networks. We therefore call for coordinated efforts in invasion science and beyond, to devise and review approaches that can rapidly map out the entire web of dynamic interactions in a recipient ecosystem.

*Corresponding author: chui@sun.ac.za

DOI: 10.1079/9781789242171.0002

2.1 The Nature of Biotic Interactions

Biotic interactions are ubiquitous and quintessential for the survival of any species, and this is certainly true for non-native plant species in a novel environment (Richardson *et al.*, 2000; Hui and Richardson, 2017). Besides habitat suitability, invasion performance depends on biotic interactions in a profound way. Plant invaders interact not only with other plant species within the same guild via resource competition (Chapter 16, this volume), but also with species from other guilds such as herbivores and pollinators (Chapter 13, this volume). We will first unpack the nature of biotic interactions before explaining related invasion hypotheses and theories, spanning from direct pairwise interactions to ecological networks and non-native biomes.

To survive and complete its life cycle, an individual organism must constantly 'play games' (i.e. interact) with co-occurring individuals from the same and other species, for resources and opportunities. This individual, carrying a specific set of ecological and evolutionary strategies, will likely receive a payoff (positive or negative) from these games that depends not only on its own strategies, but also those of its opponents, as well as the environmental context over a relevant spatial and temporal scale. The payoff can normally be calculated as the benefit minus the cost of each game, which could eventually affect its fitness (a performance measure of survival and reproduction). If the benefit exceeds the cost (i.e. a positive payoff), the game is considered beneficial to the focal player; otherwise, it is considered detrimental. Players in a game can choose strategies that improve, or at least conserve, their expected payoffs but without the group-level average payoff in mind – a source for individual-group conflicts.

Once individuals have been sorted into groups, we could calculate the average or overall payoff for a group when its members are playing games with individuals of another group (inter-specific) or of the same group (intra-specific). *Biotic interactions* can thus be defined as the ensembles of such one-on-one games after grouping individuals according to specific labels. Typical biotic interactions include those that are mutually beneficial (e.g. pollination) or detrimental (e.g. resource competition), as well as those that are beneficial to one group but detrimental to another (e.g. herbivory). Any such group-level classification of interactions will inevitably overlook the complexity of individual strategies within and between groups (Wolf and Weissing, 2012). As each individual has a unique set of strategies/traits, the performance of its embedded group depends not only on the performance of the average strategy but also the plasticity, diversity and distribution of individual strategies within the group (Dall *et al.*, 2004). For instance, there are plenty of free-riders and antagonistic bacteria in root nodules that are part of the overall symbiotic interaction between invasive legumes and rhizobia (Keet *et al.*, 2017).

The type of biotic interaction can change with the quality of environment (e.g. a harsh environment favours facilitative interaction while a conducive environment promotes antagonism) and density (e.g. herbivores switching preferred food from optimal foraging; van Baalen *et al.*, 2001). Some strategies can be modified instantaneously by the individual (e.g. a predator can change its foraging behaviour through learning; Zhang and Hui, 2014), while most are encoded in its genome and can only be naturally selected via relatively slow evolutionary processes over generations. Consequently, annual plant invaders can adapt to novel environments faster than perennial plant invaders, resulting in different life-history strategies being targeted by selection (Gioria *et al.*, 2018). As the group-level payoffs are dependent on the spatiotemporal scales of the ensembles, the strength of biotic interactions has also to be quantified for two specific groups over confined space and time for involved individuals.

In an ecological network, numerous pairwise and higher-order interactions between resident and introduced species are in play, simultaneously and instantaneously, potentially forming complex reinforcing and balancing feedbacks that drive the adaptive cycles (panarchy) of complex system evolution (Gunderson and Holling, 2003). Biotic interactions that foster reinforcing feedbacks, through for instance ecosystem engineering and intransitive interactions (Gaertner *et al.*, 2014; Gallien *et al.*, 2018), can thus greatly affect the instability and invasibility of recipient ecosystems. Consequently, the recipient ecosystem is often susceptible to the invasion of non-native species with specific functional and demographic traits (Hui *et al.*, 2016). Importantly, such a system is not isolated, but is embedded in a much larger system (e.g. a meta-community) with a constant influx of non-native species from the regional or external pool and outflux of extirpated species, driven by many intertwined factors (e.g. global environmental change, disturbance and biological invasions). This highlights that an invaded community functions as an open dynamical system (Hui and Richardson, 2019a).

The nature of biotic interactions, depicted here as ensembles of individual interactions (being scale dependent, intertwined, embedded, transient and dynamic) has immediate implications for understanding the invasion performance and invasiveness of non-native species in an introduced ecosystem – and this has given birth to a plethora of invasion theories and hypotheses that invoke biotic interactions. What we intend to present here is a list of models and related invasion hypotheses. These models are generic, to afford explicit investigation of their structures and dynamics using tractable mathematics. Nevertheless, they can also be considered building blocks for designing more complex and realistic models. Multiple hypotheses (Catford *et al.*, 2009; Enders *et al.*, 2018) have been proposed to explain the success (or failure) of non-native species in overcoming the different barriers along the introduction—naturalization—invasion continuum (Richardson and Pyšek, 2006; Blackburn *et al.*, 2011). Most of these hypotheses invoke the role of biotic interactions. The undertone of many hypotheses is why an introduced species can perform better than a resident native species that has very similar traits.

2.2 Direct Interactions Between Natives and Non-natives

The direct impact of one species on another when not mediated or transmitted through a third species (pairwise interactions), is defined as *direct interaction*. Non-native plants can engage with resident species in a diverse array of direct interactions (Richardson *et al.*, 2000). There are five types of direct interactions, classified according to the combinations of average payoffs for each species: interference competition, antagonism, mutualism, commensalism and amensalism. Introducing a species into an ecological community/network has repercussions that may ripple across the ecosystem.

As a baseline, the strength of biotic interaction of species B on A can be defined as the change in population growth rate or demographic fitness of A, in response to the increase of population size or density of B. For example, we can depict the population dynamics of a native species (N) and a non-native invader (A) using a simplified Lotka—Volterra model in the following vector format:

$$\begin{bmatrix} \dot{N} \\ \dot{A} \end{bmatrix} = \begin{bmatrix} r_N - r_N\alpha_{NN}N & -r_N\alpha_{NA}N \\ -r_A\alpha_{AN}A & r_A - r_A\alpha_{AA}A \end{bmatrix} \begin{bmatrix} N \\ A \end{bmatrix} \quad (2.1)$$

where r_N and r_A represent the intrinsic population growth rate for the native and the non-native species respectively; α_{NN} and α_{AA} the coefficients of negative density dependence for self-regulation; α_{NA} expresses the per-capita interaction strength of the non-native species on the native species, and α_{AN} is the per-capita interaction strength of the native species on the non-native species.

To explore the conditions for stable coexistence of the native and the non-native species, we could examine the *invasibility criterion* that requires either species to be able to increase from low density while the other one stays at its equilibrium, which gives us two conditions for coexistence: $\alpha_{NN} > \alpha_{AN}$ and $\alpha_{AA} > \alpha_{NA}$. That is, the necessary condition for coexistence is that intraspecific interaction strength needs to be greater than interspecific interaction strength. For the case of biological invasion, we only need to satisfy the condition for the non-native species to invade while the native species stands still: $\alpha_{NN} > \alpha_{AN}$, meaning that the self-regulation of the native species needs to be stronger than its resistance against the non-native species, to allow the non-native species to invade successfully. In particular, when the native is sitting at its equilibrium and the non-native species at an initial low density ($A = \varepsilon$), the native population size declines at a rate of $r_N\varepsilon\alpha_{NA}/\alpha_{NN}$ and the non-native species grows at a rate of $r_A\varepsilon(1-\alpha_{NA}\varepsilon-\alpha_{AN}/\alpha_{NN})$. Evidently, once the invasibility criterion for the

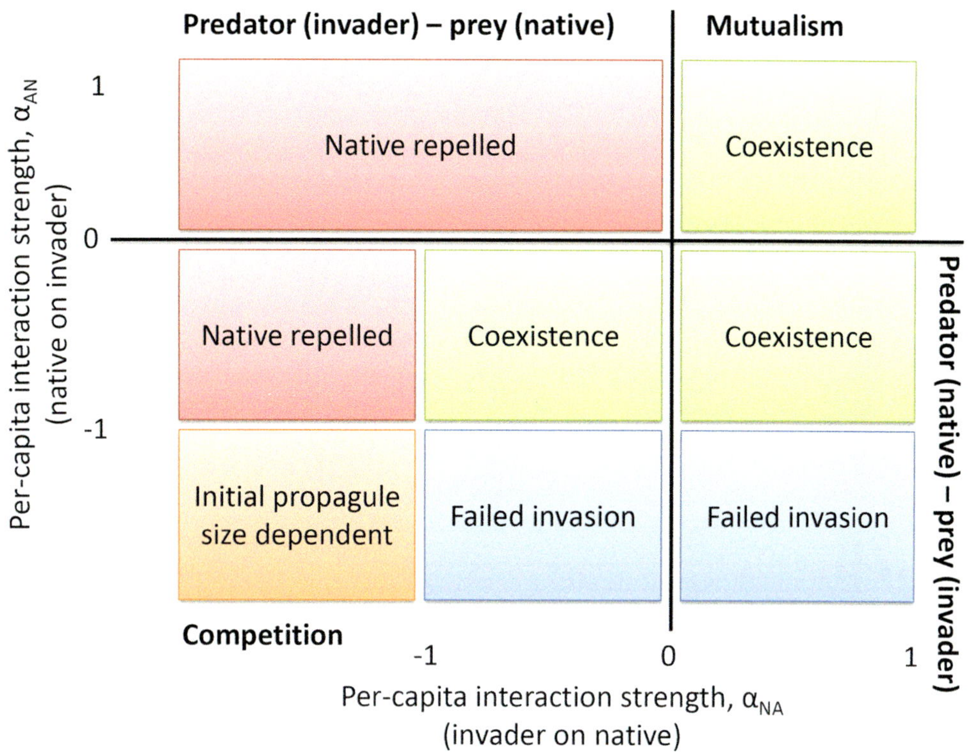

Fig. 2.1. The division of biotic interactions based on per-capita interaction strengths. The qualitative results from the biotic interaction (Equation 2.2, with intraspecific interaction strengths assumed to be 1) are given in the blocks; the zero lines separating the four biotic interactions indicate commensalism and amensalism. Redrawn from Fig. 6.1 in Hui and Richardson, 2017.

non-native is satisfied ($\alpha_{NN} > \alpha_{AN}$), the non-native species with a higher intrinsic growth rate (large r_A), a larger initial propagule size (large ε), less self-regulation (small α_{AA}) and receiving less resistance from the native species (small α_{AN}) will expand its population faster. On the other hand, the native species with a higher intrinsic growth rate (large r_N), less self-regulation (smaller α_{NN}) and experiencing a stronger negative impact from the non-native species (larger α_{NA}) will suffer a more rapid population decline. After setting the intraspecific effects to 1, we could summarize the dynamical behaviour of this simple model in Fig. 2.1.

As direct interactions impose clear payoffs on each species, a great number of invasion hypotheses have been developed for a plethora of mostly ecological scenarios (and a few evolutionary scenarios) where introduced species in their invaded ranges have typically left behind those directly interacting partners, mutualists or antagonists, in their original communities, but face a new set of species that could initiate direct interactions. Enders *et al.* (2018) listed 33 invasion hypotheses, of which 19 relate explicitly to biotic interactions (Table 2.1). For instance, the Enemy of My Enemy (Eppinga *et al.*, 2006), the Enemy Reduction (Colautti *et al.*, 2004) and the Enemy Release (Keane and Crawley, 2002) hypotheses are based on the assumption that a cross-guild antagonistic interaction with an enemy is lower in the introduced than in the historical range. This could be the case either because the enemy has a higher negative effect on native species (in which case the absolute effect of the enemy on the non-native species is not necessarily lower in the introduced compared to the historical range, but is lower relative to other species of the same guild), or because the enemy is less abundant or absent. This should result in a non-native species

Table 2.1. Single-species hypotheses involving biotic interactions. Definitions taken from Catford *et al.* (2009) and Enders *et al.* (2018).

Processes	Hypothesis	Definition
Based on cross-guild biotic interactions only	Enemy of My Enemy aka Accumulation of Local Pathogens Hypothesis (Eppinga *et al.*, 2006)	Introduced enemies of a non-native species are less harmful to the non-native than to native species.
	Enemy Reduction Hypothesis (Colautti *et al.*, 2004)	The partial release of enemies in the non-native range contributes to invasion success.
	Enemy Release Hypothesis (Keane and Crawley, 2002)	The absence of enemies in the exotic range is a cause of invasion success.
	Increased Susceptibility Hypothesis (Colautti *et al.*, 2004)	If a non-native species has a lower genetic diversity than the native species, there will be a low probability that the non-native species will establish itself due to increased pressure from enemies.
	Missed Mutualisms Hypothesis (Colautti *et al.*, 2004; Mitchell *et al.*, 2006)	Non-native species suffer from missing mutualists in their adventive ranges.
	Specialist–Generalist Hypothesis (Callaway *et al.*, 2004)	Non-native species are more successful in a new region if the local predators are specialists and local mutualists are generalists.
	New Associations Hypothesis (Colautti *et al.*, 2004)	New relationships between non-native and native species can positively or negatively influence the establishment of the non-native species.
	Invasional Meltdown Hypothesis (Simberloff and Von Holle, 1999; Sax *et al.*, 2007)	The presence of non-native species in an ecosystem facilitates invasion by additional species, increasing their likelihood of survival or ecological impact.
Based on within-guild biotic interactions	Novel Weapons Hypothesis (Callaway and Ridenour, 2004)	In the non-range, non-native species can have a competitive advantage against native species because they possess a novel weapon, i.e. a trait that is new to the resident community of native species and therefore affects them negatively.
	Biotic Indirect Effects Hypothesis (Callaway *et al.*, 2004)	Non-native species benefit from different indirect effects triggered by native species.

Continued

Table 2.1. Continued

Processes	Hypothesis	Definition
Based on biotic interactions and abiotic selection	Empty Niche Hypothesis (MacArthur, 1970)	The invasion success of non-native species increases with the availability of empty niches in the non-native range.
	Enemy Inversion Hypothesis (Colautti *et al.*, 2004)	Introduced enemies of non-native species are less harmful for them in the non-native than in the native range, due to altered biotic and abiotic conditions.
	Resource—Enemy Release Hypothesis (Blumenthal, 2006)	The non-native species is released from its natural enemies and can allocate more energy to reproduction, and invasion success increases with the availability of resources.
	Darwin's Naturalization Hypothesis (Darwin, 1859)	The invasion success of non-native species is higher in areas that are poor in closely related species than in areas that are rich in closely related species.
	Limiting Similarity Hypothesis (MacArthur and Levins, 1967)	The invasion success of non-native species is high if they highly differ from native species, and it is low if they are similar to native species.
	Sampling Hypothesis (Crawley *et al.*, 1999)	A large number of different non-native species is more likely to become invasive than a small number due to interspecific competition. The species identity of the locals is also more important than the richness in terms of the invasion of an area.
	Adaptation Hypothesis (Duncan and Williams, 2002)	The invasion success of non-native species depends on the adaptation to the conditions in the non-native range before and/or after the introduction. Non-native species that are related to native species are more successful in this adaptation.
Based on biotic interactions, abiotic selection and drift	Opportunity Windows Hypothesis (Johnstone, 1986)	The invasion success of non-native species increases with the availability of empty niches in the non-native range, and the availability of these niches fluctuates over space and time.
	Dynamic Equilibrium Hypothesis (Huston, 1979)	The establishment of a non-native species depends on natural fluctuations of the ecosystem, which influences the competition of local species.
Based on biotic interactions and evolution	Evolution of Increased Competitive Ability Hypothesis (Blossey and Nötzold, 1995)	After having been released from natural enemies, non-native species will allocate more energy in growth and/or reproduction (this reallocation is due to genetic changes), which makes them more competitive.

experiencing a decline in damage and having an enhanced performance in the introduced range. On the other hand, the Novel Weapons Hypothesis posits that non-native species can have traits that negatively affect the native species of the resident community (Callaway and Ridenour, 2004; see also Chapter 3, this volume). Some non-native plants release phytotoxic biochemical compounds that directly harm native plant species (Callaway and Ridenour, 2004), or adopt indirect interactions to harm fungi involved in mutualistic interactions with the native plant species (Callaway *et al.*, 2008).

We could also write the *relative growth ratio* of the non-native to the native species when the population size of the non-native is trivial, while the native species sits at its equilibrium:

$$\frac{1}{A/N}\frac{d(A/N)}{dt} = \frac{\dot{A}}{A} - \frac{\dot{N}}{N} = (r_A - r_N) + r_A\left(\frac{r_N}{r_A} - \frac{\alpha_{AN}}{\alpha_{NN}}\right) \quad (2.2)$$

For successful invasion, we need to ensure this ratio is positive. Note, the right-hand side of the equation includes two parts (*sensu* Chesson, 2000): fitness difference ($r_A{-}r_N$) and stabilising force ($\alpha_{AN} > \alpha_{NN}$, with $\alpha_{AN} = \alpha_{NN}$ suggesting the case of complete niche overlap, and $\alpha_{AN} = 0$ the case of complete niche segregation). When the two species have the same per-capita growth rate, there is no fitness difference (i.e. a neutral process) and the invader relies on the stabilizing force to invade successfully. We can therefore expand the scenario of invasion and coexistence (Fig. 2.2). The simple conceptualization of interactions between native and non-native species provides a starting point for discussing the types of biotic interaction that are relevant when considering invasion dynamics.

Many invasion hypotheses assume that invasion success depends on a combination of biotic (within- or cross-guild) interactions and abiotic selection, i.e. taking a broad perspective on niche theory (Chapter 1, this volume). However, these combinations are often quite loosely defined. For example, the Resource–Enemy Release Hypothesis (Blumenthal, 2006) is similar to the Enemy Release Hypothesis, except that it assumes that invasion success is maximised when resources are high, and therefore assumes that both cross-guild biotic interactions and potentially abiotic selection (if the resource is inorganic) influence invasion success. It therefore combines, in an additive way,

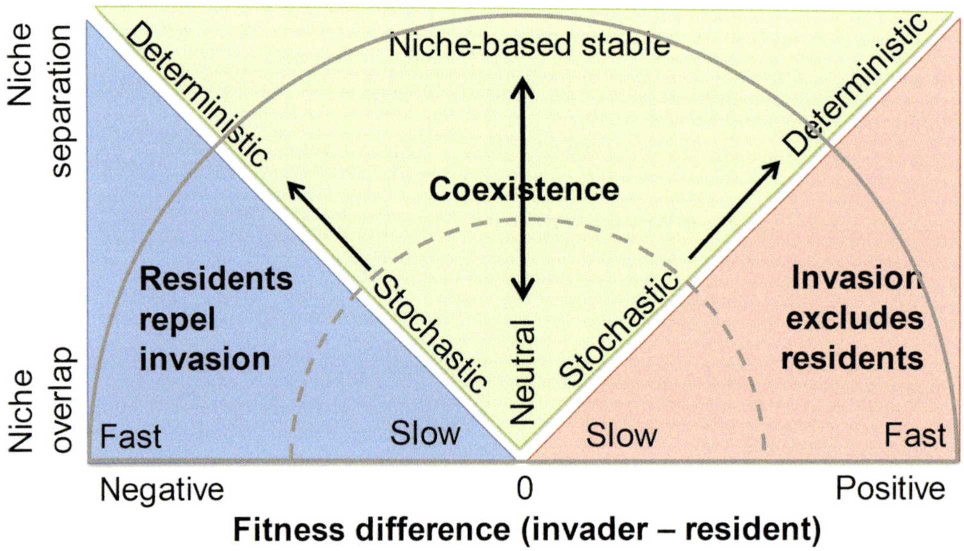

Fig. 2.2. Multiple coexistence mechanisms for explaining the invasion performance in a community, from the results of equation (2.3) and its extended version. (Lower centre): closeness to the origin means that the fate of invasion is governed by stochastic processes; (lower left or right): with overlapped niches and big fitness differences, successful invaders of this kind normally have superior competitive abilities and can rapidly exclude resident species, causing big impacts; (upper centre): with similar fitness but separated niches, the invasiveness of the invader is low; (upper left or right): with different fitness and separated niches, the establishment is assured for invaders having higher fitness. Redrawn from Fig. 8.2 in Hui and Richardson, 2017, inspired by MacDougall *et al.* (2009).

two existing and independent invasion hypotheses: the Enemy Release and Increased Resource Availability (Sher and Hyatt, 1999). In contrast, the Enemy Inversion Hypothesis (Colautti *et al.*, 2004) assumes that biotic and abiotic processes may influence the cross-guild biotic interactions between a non-native species and its introduced enemy (often used as a biocontrol agent), e.g. by providing alternative resources to the latter (see Pearson *et al.*, 2000).

2.3 Trait-mediated Interactions and Coevolution

The interactions described in the previous section are not necessarily fixed, but can vary in time due to various factors, with potentially important implications for biological invasions. As biotic interactions are ensembles of individual games, the forces that can change the value and distribution of individual payoffs fundamentally drive the changes in biotic interactions. The default position is therefore to consider changes in biotic interactions in the light of coevolution (Ehrlich and Raven, 1964). This has led many to consider the role of coevolution and phylogenetic constraints in explaining the topology and function of ecological networks, and we could formulate evolutionary dynamics through considering trait-mediated interactions. Such coevolutionary forces can drive the change in biotic interactions largely through interference with trait-mediated interaction strength. Within the ecological timescales that are relevant to invasion management, coevolution can have a great effect on the ecological experience of a species by constraining its fundamental niche via trait-mediated biotic interactions, and therefore drastically changing its realized niche.

If demographic rates and interaction strengths are functions of some phenotypic traits (e.g. body size), we can rewrite the model above to depict trait-mediated interactions between the native and the non-native species:

$$\begin{bmatrix} \dot{N} \\ \dot{A} \end{bmatrix} = \begin{bmatrix} r(x) - r(x)a(x,x)N & -r(x)a(x,y)N \\ -r(y)a(y,x)A & r(y) - r(y)a(y,y)A \end{bmatrix} \begin{bmatrix} N \\ A \end{bmatrix} \tag{2.3}$$

where x is the trait of the native and y the trait of the invader. We could then describe the trait evolution according to the framework of *adaptive dynamics* (Hui *et al.*, 2018). Here we introduce a more flexible framework. Native trait dynamics triggered by invasion can be described by $\dot{x} = \mu_N s_x$, where μ_N is a factor scaling the speed of trait relative to ecological dynamics, and s_x is the selection gradient on the native trait x. This selection gradient is the slope of the fitness f_N, i.e. $s_x = \frac{\partial f_N}{\partial x}$, with $f_N = r(x)(1-a(x,x)N-a(x,y)A)$ the per-capita growth rate of the natives.

In the initial phase of invasion, the native species sits at its equilibrium, while the non-native species starts from a small population size ($A = \varepsilon$). We thus have $f_N = -r(x)a(x,y)\varepsilon$ and consequently for the trait evolutionary dynamics of the native species we have $\dot{x}=0$ if the propagule pressure is small ($\varepsilon = 0$). As for the trait dynamics of the invader, we have $\dot{y} = \mu_A s_y$, where similarly $s_y = \frac{\partial f_A}{\partial y}$ and $f_A = r(y)(1 - a(y,y)A - a(y,x)N)$. In the initial phase of invasion with small propagule pressure, we have the trait evolutionary dynamics of the non-native species,

$$\dot{y} = \mu_A \left(\frac{\partial r(y)}{\partial y}\right)\left(1 - a(y,x)/a(x,x)\right) - r(y)/a(x,x)\frac{\partial a(y,x)}{\partial y} \tag{2.4}$$

This means that the native species does not initially come under directional selection from the invasion. In contrast, the trait of the non-native species can experience a certain level of directional selection. With complete niche segregation, the interaction imposed from the native species on the non-native species is negligible ($a(y,x) = 0$), and the trait dynamics then become $\dot{y} = \mu_A \frac{\partial r(y)}{\partial y}$, suggesting trait evolution towards the direction that increases the invader's growth rate. With complete niche overlap, the interaction imposed from the native species to the non-native species is equal to its intraspecific interaction strength ($a(y,x) = a(x,x)$), and trait evolution halts $\dot{y} = 0$. With partial niche overlap/segregation ($a(y,x) < a(x,x)$), trait evolution of the non-native species is under the trade-off of increasing its growth rate and reducing the interspecific pressure from the native species.

This then allows us to explore trait evolution and directional selection driven by trait-mediated interactions, immediately after the invasion. The Evolution of Increased Competitive Ability (EICA; Blossey and Nötzold, 1995) Hypothesis assumes that release from native enemies and, therefore, changes in biotic interactions, can enable a non-native plant species to allocate more energy in growth or reproduction, thereby conferring it a competitive advantage over native species. It has been observed that post-introduction evolution is indeed common for non-native plants, but not necessarily for the purpose hypothesized by EICA (Felker-Quinn *et al.*, 2013). More investigations on the combination of biotic interactions and evolution are therefore needed. In contrast, the Increased Susceptibility Hypothesis suggests the opposite of enemy reduction, i.e. an increase in susceptibility of non-native species to enemies in the introduced range can occur if the non-native species has a lower genetic diversity than the native species, therefore decreasing the chance to have individuals with specific defence mechanisms (Colautti *et al.*, 2004).

Darwin's Naturalization Hypothesis (Darwin, 1859), the Adaption Hypothesis (Duncan and Williams, 2002) and Limiting Similarity (MacArthur and Levins, 1967) are all based on the assumption that similarity between functional traits or phylogeny (acting as a proxy for functional traits) of native and non-native species influence the outcome of invasion events. Darwin's Naturalization Hypothesis and the Adaption Hypothesis both assume that a non-native species with similar traits to native species will more easily invade a novel environment, which can be explained by the two kinds of species relying on the same biotic but also abiotic niche processes. In contrast, Limiting Similarity assumes the opposite, that trait difference will enable a non-native species to fill available niches, and is therefore similar to the Empty Niche Hypothesis. Limiting Similarity has mostly been tested for plants and is overall well supported in the literature (Jeschke and Erhard, 2018). These hypotheses can nonetheless encompass such a breadth of processes that they are difficult to precisely test empirically. Especially, the choice of the appropriate spatial scale and the invasion stage can be crucial to support one or another hypothesis (de Bello *et al.*, 2013).

2.4 Ecological Fitting via Interaction Rewiring

Although coevolution can explain the change, and perhaps more importantly the emergence, of biotic interactions, it cannot explain the rapid fluctuation and turnover of biotic interactions observed in many natural communities (Petanidou *et al.*, 2008). Janzen (1985) proposed the concept of *ecological fitting* to explain the establishment of novel interactions between species which had shared little evolutionary history. According to ecological fitting, no drastic evolutionary change is needed to allow invading species to fit into local assemblages in their invaded ranges; instead, readjustment from resident species through ecological fitting could provide sufficient opportunity for the establishment of novel biotic interactions. Many successful invasive non-native plant species have promiscuous interactions and experience host jumps in their novel range (Richardson *et al.*, 2000), thus providing excellent evidence to support the establishment of novel interactions through ecological fitting.

Ecological fitting requires species to switch their partners according to changes of context, such as the temporal presence/absence of resources and new mutualists from disturbance and biological invasion. This opportunity-seeking nature of selecting and switching interaction partners is essential for the survival of consumers that are competing for available resources in their individual games. Consumers tend to prefer to select highly profitable and abundant resources, rather than all available resources, so as to optimize the energy intake rate (Stephens and Krebs, 1986) and minimize the risk of resource scarcity (Zhang and Hui, 2014). This leads to two phenomena: the 'zero-one' rule that a predator will only consume highly profitable prey, but ignore less profitable ones even when encountered, and the 'prey-switching' phenomenon, where the consumption of a resource is disproportionately large when the resource is abundant relative to other resources, and disproportionately small when it is relatively rare (Murdoch, 1969). This means that introduced plants are afforded a window of time when resident herbivores would simply ignore the presence of such rare resources, especially if the herbivores are adaptive foragers (Jaworski *et al.*, 2013).

Adaptive interaction switching can occur at three different levels (Hui and Richardson, 2017). At the individual level, it is important to choose with what to interact (e.g. habitat or diet selection) or to avoid (e.g. applying anti-predation strategies). Such preferential interactions could simply arise from optimal or adaptive foraging where individuals aim to maximize their energy intake rate while minimising the risk of starvation and predation (Zhang and Hui, 2014). Consequently, interaction switching can occur rapidly for individuals at a pace even faster than the typical ecological timescale (e.g. host switching in parasites; van Baalen *et al.*, 2001). At the population level, successful interactions often require involved species to possess a certain level of matching or complementary traits, such as a matching phenology for flowering time in plants, and the foraging activity of insect pollinators (Waser, 2015). Through adaptively changing the level of trait complementarity and plasticity, populations can adjust interaction strength to their own advantage (Santamaría and Rodríguez-Gironés, 2007). At the species level, species can completely lose or regain the possibility of interactions by developing incompatible traits or forbidden links, and setting up barriers to exploitation (Jordano *et al.*, 2003).

Without coevolved mutualists, interaction promiscuity is essential for invasion success, especially for compulsory mutualists (Aizen *et al.*, 2012; Heleno *et al.*, 2013). Species often switch their interaction partners in mutualistic networks (Olesen *et al.*, 2008). Non-native plants may benefit from forming novel mutualisms with native soil microbes capable of forming associations with a wide range of hosts, thereby facilitating establishment and spread (Reinhart and Callaway, 2006), as demonstrated by the similar nitrogen-fixing bacterial communities in invasive legume nodules across invaded and native ranges (Birnbaum *et al.*, 2016). In a pollination network, pollinators continually switch the plant species with which they interact in response to environmental disturbances and the availability of resources (Whittall and Hodges, 2007), whereas plants can also adjust phenology (e.g. flowering time) and morphology (e.g. flower heterostyly) to affect their pollinators (Kaiser-Bunbury *et al.*, 2010). Interaction promiscuity and adaptive rewiring can have a profound influence on the stability of ecological networks, and models implementing adaptive rewiring can explain large variations in many non-random patterns (e.g. modularity and nestedness) in both mutualistic and antagonistic networks (Zhang *et al.*, 2011; Nuwagaba *et al.*, 2015; Nnakenyi *et al.*, 2019).

Several hypotheses on the effect of biotic interactions on invasion performance (e.g. Host Jump, New Associations, Enemy Release and EICA; see later sections) have been developed based on the behaviour of adaptive interaction-switching. This means that successful invaders, especially those requiring high levels of symbiotic specificity, often rely on the co-introduction of coevolved mutualists to perform well in their invaded range (Traveset and Richardson, 2014; Le Roux *et al.*, 2017; Mohammed *et al.*, 2018). For instance, the Missed Mutualism Hypothesis (Colautti *et al.*, 2004; Mitchell *et al.*, 2006) predicts the complement of the Enemy Release Hypothesis; it predicts that a species benefiting from mutualistic interactions in its native range must establish a similar type of positive interaction in the introduced range in order to invade. In that case, specialized interactions would prevent or limit the ability of a non-native species to invade. Missed mutualism is one of the main reasons for invasion failure by plants (Traveset and Richardson, 2014). However, widespread host generalism appears to make this mechanism relatively rare in nature, be it for plants or other taxa (Mitchell *et al.*, 2006).

2.5 Ecological Networks and Opportunity Niches

In any ecological community/network, there is a multitude of direct interactions; this explains why little empirical support has emerged for many hypotheses that seek to explain invasion performance in a highly simplified scenario. The stability of complex natural communities with a multitude of direct interactions can be greatly affected by the type of biotic interactions within it (Tylianakis and Morris, 2017). For instance, mutualistic interactions between species of two guilds in a bipartite network can form a different community topology—stability relationship, distinct from the one which emerged from antagonistic cross-guild interactions (Thébault and Fontaine, 2010). Since most ecological communities are multilayer networks (Pilosof *et al.*, 2017), the proportion of different types of direct interactions becomes an important indicator of network stability (Mougi and Kondoh, 2012), while the ratio of the total number of mutualistic to antagonistic partners to a species can greatly affect its persistence (Melian *et al.*, 2009). Together, composition of interaction types, network size and connectance, and distribution of interaction strength, can all greatly affect community stability, known as the *stability criterion* (Allesina and Tang, 2012; Landi *et al.*, 2018).

When an ecological network is invaded by a number of non-native species, we can formulate this invaded network using a joint dynamical system extending the previous two-species system of equations (Hui and Richardson, 2019b):

$$\begin{bmatrix} \dot{N} \\ \dot{A} \end{bmatrix} = \begin{bmatrix} M & P \\ R & L \end{bmatrix} \begin{bmatrix} N \\ A \end{bmatrix} \qquad (2.5)$$

Note here, $N = [N_1, N_2, \ldots, N_m]^T$ is the vertical vector of population size (abundance) for m number of native (resident) species; $A = [A_1, A_2, \ldots, A_n]^T$ is the vertical vector of population size for n number of non-native species. The submatrix $M_{m\times m}$ represents the native-to-native interaction block; the submatrix $P_{m\times n}$ is the non-native-to-native impact block; the submatrix $R_{n\times m}$ represents the native-to-non-native resistance block; the submatrix $L_{n\times n}$ represents the non-native-to-non-native (invasional meltdown) block. Biological invasions can push ecological communities towards marginal stability and eventual instability, and any factors that violate the stability criterion thus enhance the community's susceptibility to invasion. Consequently, each species' response to the invasion is revealed by the major axis (the eigenvector of the lead eigenvalue) of the joint matrix $J = \{\{M, P\}, \{R, L\}\}$, which can be estimated as the first principal component (PC1) from a principal component analysis of this joint matrix. If the coefficient of PC1 for species i is positive, the species will increase; otherwise, it will decline. The major axis can thus assess which non-native species can invade or be repelled, and which resident species can be favoured or suppressed as a result of such biological invasions.

Invasion performance and invasibility are related to the loss of network stability or, more precisely, network instability. Several invasion hypotheses regarding the invasibility of recipient ecosystems have been proposed to identify these aggravators of ecosystem instability (Hui *et al.*, 2016). With trait-mediated interactions in an ecological network, we could explore those specific non-native traits that could experience greater population growth (invasion performance) and cause greater impacts in the recipient communities (Fig. 2.3). The Specialist—Generalist Hypothesis (Callaway *et al.*, 2004) relies on a combination of the two previous mechanisms to explain biological invasions: the enemies present in the introduced range must be specialists, and therefore less likely to affect novel species with which they have not coevolved; in contrast, the mutualists should be generalists, to benefit the non-native species (Sax *et al.*, 2007). Finally, the Invasional Meltdown Hypothesis, which posits that the presence of some non-native species can favour invasion by other species (Simberloff and Von Holle, 1999), is well supported in the literature, especially for plants (Braga *et al.*, 2018; see Chapter 8, this volume). Invasional meltdown can be caused by various processes, including direct biotic interactions (e.g. pollination, host—parasite interactions, herbivory, etc.), but also changes in the abiotic environment through increased disturbance or altered fire regimes.

Besides direct interactions, an ecological network also harbours a great diversity of *indirect interactions* (White *et al.*, 2006; Chapter 17, this volume). The first type of indirect interaction is known as an *interaction chain*, in which a donor species affects the abundance of a transmitter and has an effect on a recipient. These are thus combinations of at least two pairwise direct interactions. Negative indirect interactions include exploitative competition and apparent competition. Positive indirect interactions include keystone predation, trophic cascade, competitive mutualism and indirect mutualism. For instance, trophic cascade (or food chain mutualism) takes place when a plant species is positively affected by the decrease in herbivory caused by a predator reducing herbivore numbers (Bosc *et al.*, 2018). Indirect mutualism (or facilitation) comprises a more complex four-species

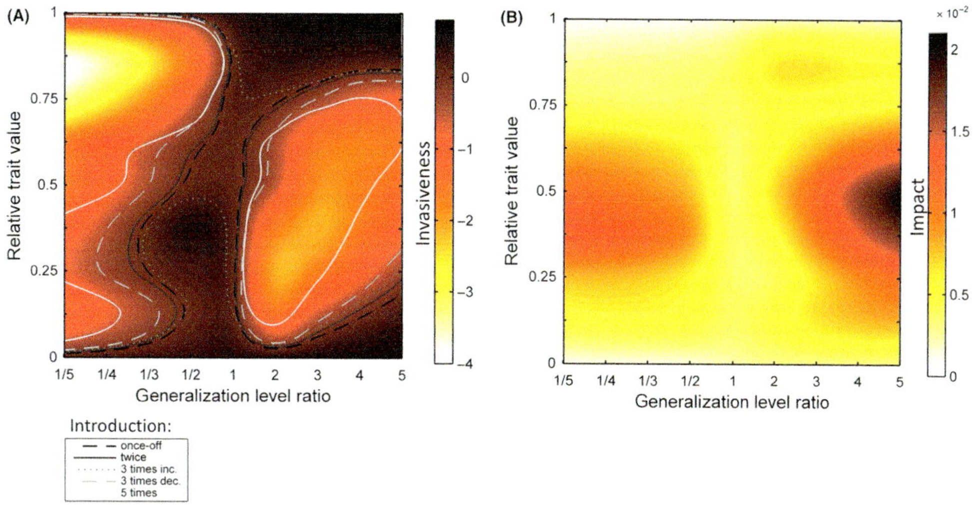

Fig. 2.3. (A) Invasiveness and (B) impact of the invader as a function of its relative trait value and generalization level ratio, relative to those of the native community. Invasiveness and impact values represent the average over 100 medium-sized networks. Lines represent the zero level of invasiveness under different introduction modes. From Minoarivelo and Hui (2016), published under CC BY 4.0.

system with two competitors which do not interact directly, but consume different prey species competing with each other; if one consumer increases its consumption, its prey decreases, which in turn reduces competition with the second prey, providing more food for the second consumer species (Wootton, 1994). A peculiar type of indirect interaction is intransitive competition, where species are arranged in a loop rather than forming a hierarchy of competitive interactions (Levine *et al.*, 2017). Intransitive competition can emerge from community assembly through trait evolution (Gallien *et al.*, 2018), posed as either a stabilizing or destabilizing force. Compared to direct interactions, interaction chains involve more species and consider more specific context. However, interaction chains have not been widely invoked in invasion hypotheses but should be when invasion hypotheses of direct interactions are violated.

The second type of indirect interaction, *interaction modification*, occurs when the donor alters some other attribute of the transmitter, such as behaviour or trait plasticity (Werner and Peacor, 2003). These are higher-order interactions (HOIs), where more than two individuals have to be present for the interaction to occur (Wootton, 1994; Terry *et al.*, 2017). HOIs are only starting to receive attention in theoretical ecology, such as HOIs in pollination—herbivory meta-communities (Mayfield and Stouffer, 2017) and the formation of hypernetworks that contain hyperlinks connecting more than two nodes (Golubski *et al.*, 2016). The inclusion of HOIs has been shown to improve our predictive power in explaining species' performance in communities (Letten and Stouffer, 2019), and can profoundly affect, largely facilitating, community stability and species coexistence (Bairey *et al.*, 2016; Grilli *et al.*, 2017). Almost all species are involved simultaneously in both mutualistic and antagonistic interactions (Strauss *et al.*, 2006). For instance, herbivory can affect pollination in diverse ways (Lucas-Barbosa *et al.*, 2016): it can reduce pollination through altering floral display, activating chemical defences, and decreasing availability and quality of rewards for pollinators (Adler *et al.*, 2006). In contrast, root herbivory has been shown to increase pollinator visits to flowers of the mustard *Sinapis arvensis* (Poveda *et al.*, 2005). In addition, secondary metabolites of plants induced by herbivores can serve as antibiotics for pollinators, thereby enhancing pollination and plant reproduction (Richardson *et al.*, 2016). Reciprocally, pollination can also affect herbivory. Several floral traits for pollination (e.g. corolla size and colour, nectar and floral scent) have been shown to attract herbivores (Adler *et al.*, 2006). HOIs could result in most invasion hypotheses that rely on direct interactions failing to capture the interaction complexity, especially those subject to the processes of coevolution in invaded communities. We expect that further development in invasion models to include HOIs could improve our ability to predict invasion performance.

Some invasion models have highlighted factors that can impose evolutionary barriers and create empty niche opportunities – largely through biotic interactions that can drastically change the invasion fitness landscape and give rise to unfilled pockets of positive fitness in trait space. For instance, an introduced species with a similar niche to a resident species might only suffer from weak biotic resistance – because the presence of the similar niche would ensure the introduced species would easily flourish. In contrast, an introduced species with its niche/trait sitting between the niches/traits of two resident species faces an uncertain outcome: either there is an empty niche to allow invasion, or no niche available for invasion (Hui *et al.*, 2016). The imposed evolutionary barriers prohibit trait radiation in the resident species of a community; consequently, this would constrain the distribution of functional traits in the community and potentially create empty niches (note, disturbance can also create temporally available empty niches). Such empty niches set up by evolutionary barriers also highlight the priority effect: the traits and sequence/history of invasion can greatly affect how the invasibility of a community unfolds. Non-native species could encounter empty realized niches with positive invasion fitness that have not been exploited (from disturbance) or are incapable of being exploited (from evolutionary barriers) by resident species; this could greatly increase the non-natives' chances of invasion success (Elton, 1958; MacArthur, 1970). The Sampling Hypothesis (Crawley *et al.*, 1999) is the counterpart of the empty niche: it assumes that a wide species pool will lead to more invasion events, because there will be a higher chance that a species will have the characteristics to fill an empty niche.

All previous hypotheses assume static native environments, i.e. that they are typified by a set of constant characteristics, which determine the biotic processes that will enable a non-native species to invade. However, ecosystems are dynamic and their composition and environmental conditions fluctuate in time. It is therefore logical to conclude that biotic interactions between non-natives or non-native and native species, and the resulting invasion success, may vary accordingly. The Opportunity Windows Hypothesis (Johnstone, 1986) is an extension of the Empty Niche Hypothesis which assumes that niche availability fluctuates both in space and time, therefore offering different invasion opportunities at different locations and different points in time. Similarly, the Dynamic Equilibrium Hypothesis (Huston, 1979) assumes that abiotic and biotic conditions fluctuate, influencing the competition of local species, therefore potentially offering invasion opportunities to non-native species. The importance of the dynamic aspect of ecosystems for biological invasions is apparent, for example, for herbivorous insects relying on plants with annual cycles, which must rely on specific periods during which host plants are suitable, such as budburst (Ward and Masters, 2007).

2.6. Non-native Biomes and Compositional Turnover

Most invasion hypotheses aim to explain the invasion success of single species in a novel environment. However, several recent studies have highlighted the necessity to consider non-native species (of different invasion stages) as a cluster, providing important insights on macroecology and island biogeography, as well as community-level assembly with regards to the native community (Rouget *et al.*, 2015; Roura-Pascual *et al.*, 2016). Such studies are based on the comparison of three main patterns: non-native and native species richness; non-native and native species turnover (i.e. the comparison of variations in species composition); and trait and functional composition. The presence of strong niche processes between non-native and native species during the invasion process should generate some clear large-scale spatial patterns, whereas the absence of any correspondence between non-native and native community patterns would suggest that other processes are at play,

Table 2.2. Non-native community hypotheses (definitions follow Rouget *et al.*, 2015).

Hypotheses	Definition
The Weed Shaped Hole (Mack *et al.*, 2000; Buckley *et al.*, 2007; Rouget *et al.*, 2015)	Invasions are facilitated by certain levels of disturbance; these may be anthropogenic or natural (and so an inherent function of the invasibility of an ecosystem).
The Biome Decides (Alpert *et al.*, 2000; Rouget *et al.*, 2015)	Native communities differ inherently in invasibility, such that there is selectivity regarding which invasive species can and cannot invade.
Goldilocks (Rouget *et al.*, 2015)	Native communities and invasive species share broadly similar abiotic requirements, such that they occupy the same niches.
A New World Order (regime shift) (Simberloff and Von Holle, 1999; Sax *et al.*, 2007; Rouget *et al.*, 2015)	A new set of boundaries is formed by suites of interacting native and non-native species. These new associations lead to novel ecosystems that need not be similar in nature to those in native communities.
Something In The Way You Move (Wilson *et al.*, 2009; Donaldson *et al.*, 2014; Rouget *et al.*, 2015)	Distribution patterns in invasions are dominated by drivers associated with introduction histories of non-native species.
Random Tessellation (Daleo *et al.*, 2009; Rouget *et al.*, 2015)	The distributions of non-native species is inherently idiosyncratic. When many species are considered it appears that the relative clustering of distributions is simply constrained by geometry.
Decreasing Stochasticity (Hui *et al.*, 2013)	Stochasticity is especially important in early stages of invasions, and older assemblages (comprising species that arrived in the environment a long time ago) should display more deterministic composition than new ones.

such as introduction history (Something In The Way You Move, Table 2.2), or a mix of many different processes specific to each particular species (Random Tessellation, Table 2.2). Inferring processes from patterns is nonetheless a complicated task, as different processes can generate the same patterns, and clear signals in the three patterns may be generated by either biotic (The Biome Decides, Table 2.2) or abiotic selection processes (Goldilocks, Table 2.2) (Rouget *et al.*, 2015). It is therefore important to account for the effect of abiotic processes to infer biotic ones.

It has been shown that the turnover of non-native plant communities was correlated with that of native plant communities in a network of nature reserves, although native and non-native species turnover was driven by different abiotic variables. This suggests that biotic processes, although unaccounted for in surveys, could potentially be important mediators of community composition (Latombe *et al.*, 2018). In contrast, non-native and native species richness was mostly explained by abiotic variables, therefore suggesting the presence of abiotic factors. Abiotic and biotic processes may therefore play complementary roles in determining the number of non-native species and their identity, respectively. In a similar study, the richness and turnover of non-native and native ant communities were correlated in the Pacific islands, but were also driven by the same abiotic variables, therefore suggesting that abiotic, rather than biotic processes, were at play in structuring non-native community assembly (Latombe *et al.*, 2019).

The importance of biotic processes can be related to the rarity or commonness of non-native species. Using zeta diversity, a compositional measure of species turnover that can differentiate the role of common vs rare species in community assemblage patterns (Hui and McGeoch, 2014; McGeoch *et al.*, 2019), Latombe *et al.* (2018) showed that the relationship between non-native and native plant species turnover in the network of nature reserves was only observed for rare species, and not for common ones. Moreover, the importance of biotic processes to the structure of non-native communities is likely to increase with time, as the structure of young communities is likely to be driven by introduction histories, whereas deterministic abiotic and biotic processes will enable the species that compose them to progress along the introduction—naturalization—invasion continuum (Richardson and Pyšek, 2006; Hui *et al.*, 2013).

To our knowledge, the explicit disentanglement of abiotic and biotic processes has mostly been done for compositional diversity (richness and turnover) patterns, but such an approach for functional diversity is still lacking (e.g. Ordonez *et al.*, 2010; Tecco *et al.*, 2010). As a starting point, using functional similarity in trait space as the proxy of interaction strength, the realized interaction strength and the consequence of such interactions differ between plants at different invasion stages: naturalized non-invasive plants are shown to be functionally similar to co-occurring native plant species, while invasive plants, in contrast, occupy the periphery of the plant functional trait space (Divíšek *et al.*, 2018). This highlights the great challenge to invasion ecologists – to predict the impacts of invasion-induced biotic interactions we need to measure the interaction strength between pairs of species as a function of their encounter rate (the product of their local densities) and the level of their trait disparity/matching. The critical knowledge needed for us to unpack the invasibility of the recipient ecosystem and forecast invasion dynamics, both for resident and introduced species, is thus the community matrix of both density-dependent and trait-mediated interaction strength (Hui and Richardson, 2019b). Other chapters in this book discuss more empirical attempts in this direction.

Acknowledgements

We are grateful to Jonathan Jeschke, Sérgio Timóteo, David M. Richardson, Anna Traveset and one anonymous reviewer for constructive comments. This work is supported by the South African Research Chair Initiative and the National Research Foundation (No. 89967) and the DSI-NRF Centre of Excellence for Invasion Biology.

References

Adler, L.S., Wink, M., Distl, M. and Lentz, A.J. (2006) Leaf herbivory and nutrients increase nectar alkaloids. *Ecology Letters* 9(8), 960–967. DOI: 10.1111/j.1461-0248.2006.00944.x.

Aizen, M.A., Sabatino, M. and Tylianakis, J.M. (2012) Specialization and rarity predict nonrandom loss of interactions from mutualist networks. *Science* 335(6075), 1486–1489. DOI: 10.1126/science.1215320.

Allesina, S. and Tang, S. (2012) Stability criteria for complex ecosystems. *Nature* 483(7388), 205–208. DOI: 10.1038/nature10832.

Alpert, P., Bone, E. and Holzapfel, C. (2000) Invasiveness, invasibility and the role of environmental stress in the spread of non-native plants. *Perspective in Plant Ecology, Evolution and Systematics* 3(1), 52–66. DOI: 10.1078/1433-8319-00004.

Bairey, E., Kelsic, E.D. and Kishony, R. (2016) High-Order species interactions shape ecosystem diversity. *Nature Communications* 7, 12285. DOI: 10.1038/ncomms12285.

Birnbaum, C., Bissett, A., Thrall, P.H. and Leishman, M.R. (2016) Nitrogen-Fixing bacterial communities in invasive legume nodules and associated soils are similar across introduced and native range populations in Australia. *Journal of Biogeography* 43(8), 1631–1644. DOI: 10.1111/jbi.12752.

Blackburn, T.M., Pyšek, P., Bacher, S., Carlton, J.T., Duncan, R.P. *et al.* (2011) A proposed unified framework for biological invasions. *Trends in Ecology & Evolution* 26(7), 333–339. DOI: 10.1016/j.tree.2011.03.023.

Blossey, B. and Nötzold, R. (1995) Evolution of increased competitive ability in invasive nonindigenous plants: a hypothesis. *Journal of Ecology* 83(5), 887–889. DOI: 10.2307/2261425.

Blumenthal, D.M. (2006) Interactions between resource availability and enemy release in plant invasion. *Ecology Letters* 9(7), 887–895. DOI: 10.1111/j.1461-0248.2006.00934.x.
Bosc, C., Roets, F., Hui, C. and Pauw, A. (2018) Interactions among predators and plant specificity protect herbivores from top predators. *Ecology* 99(7), 1602–1609. DOI: 10.1002/ecy.2377.
Braga, R.R., Gómez Aparicio, L., Heger, T., Vitule, J.R.S. and Jeschke, J.M. (2018) Invasional meltdown hypothesis. In: Jeschke, J.M. and Heger, T. (eds) *Invasion Biology: Hypotheses and Evidence. CAB International*. Wallingford, UK, pp. 79–91.
Buckley, Y.M., Bolker, B.M. and Rees, M. (2007) Disturbance, invasion and re-invasion: managing the weed-shaped hole in disturbed ecosystems. *Ecology Letters* 10, 809–817.
Callaway, R.M., Cipollini, D., Barto, K., Thelen, G.C., Hallett, S.G. *et al*. (2008) Novel weapons: invasive plant suppresses fungal mutualists in America but not in its native Europe. *Ecology* 89(4), 1043–1055. DOI: 10.1890/07-0370.1.
Callaway, R.M. and Ridenour, W.M. (2004) Novel weapons: invasive success and the evolution of increased competitive ability. *Frontiers in Ecology and the Environment* 2(8), 436–443. DOI: 10.1890/1540-9295(2004)002[0436:NWISAT]2.0.CO;2.
Callaway, R.M., Thelen, G.C., Rodriguez, A. and Holben, W.E. (2004) Soil biota and exotic plant invasion. *Nature* 427(6976), 731–733. DOI: 10.1038/nature02322.
Catford, J.A., Jansson, R. and Nilsson, C. (2009) Reducing redundancy in invasion ecology by integrating hypotheses into a single theoretical framework. *Diversity and Distributions* 15(1), 22–40. DOI: 10.1111/j.1472-4642.2008.00521.x.
Chesson, P. (2000) Mechanisms of maintenance of species diversity. *Annual Review of Ecology and Systematics* 31(1), 343–366. DOI: 10.1146/annurev.ecolsys.31.1.343.
Colautti, R.I., Ricciardi, A., Grigorovich, I.A. and MacIsaac, H.J. (2004) Is invasion success explained by the enemy release hypothesis? *Ecology Letters* 7(8), 721–733. DOI: 10.1111/j.1461-0248.2004.00616.x.
Crawley, M.J., Brown, S.L., Heard, M.S. and Edwards, G.R. (1999) Invasion-resistance in experimental grassland communities: species richness or species identity? *Ecology Letters* 2, 140–148.
Daleo, P., Alberti, J. and Iribarne, O. (2009) Biological invasions and the neutral theory. *Diversity and Distributions* 15(4), 547–553. DOI: 10.1111/j.1472-4642.2009.00576.x.
Dall, S.R.X., Houston, A.I. and McNamara, J.M. (2004) The behavioural ecology of personality: consistent individual differences from an adaptive perspective. *Ecology Letters* 7(8), 734–739. DOI: 10.1111/j.1461-0248.2004.00618.x.
Darwin, C. (1859) *On the origin of species by means of natural selection*. John Murray, London.
de Bello, F., Vandewalle, M., Reitalu, T., Lepš, J., Prentice, H.C. *et al*. (2013) Evidence for scale- and disturbance-dependent trait assembly patterns in dry semi-natural grasslands. *Journal of Ecology* 101(5), 1237–1244. DOI: 10.1111/1365-2745.12139.
Divíšek, J., Chytrý, M., Beckage, B., Gotelli, N.J., Lososová, Z. *et al*. (2018) Similarity of introduced plant species to native ones facilitates naturalization, but differences enhance invasion success. *Nature Communications* 9(1), 4631. DOI: 10.1038/s41467-018-06995-4.
Donaldson, J.E., Hui, C., Richardson, D.M., Robertson, M.P., Webber, B.L. *et al*. (2014) Invasion trajectory of alien trees: the role of introduction pathway and planting history. *Global Change Biology* 20, 1527–1537.
Duncan, R.P. and Williams, P.A. (2002) Darwin's naturalization hypothesis challenged. *Nature* 417(6889), 608–609. DOI: 10.1038/417608a.
Ehrlich, P.R. and Raven, P.H. (1964) Butterflies and plants: a study in coevolution. *Evolution* 18(4), 586–608. DOI: 10.1111/j.1558-5646.1964.tb01674.x.
Elton, C.S. (1958) *The Ecology of Invasions by Animals and Plants*. Methuen, London.
Enders, M., Hütt, M. and Jeschke, J.M. (2018) Drawing a map of invasion biology based on a network of hypotheses. *Ecosphere* 9, e02146.
Eppinga, M.B., Rietkerk, M., Dekker, S.C., De Ruiter, P.C. and Van der Putten, W.H. (2006) Accumulation of local pathogens: a new hypothesis to explain exotic plant invasions. *Oikos* 114, 168–176.
Felker-Quinn, E., Schweitzer, J.A. and Bailey, J.K. (2013) Meta-analysis reveals evolution in invasive plant species but little support for evolution of increased competitive ability (EICA). *Ecology and Evolution* 3, 739–751.
Gaertner, M., Biggs, R., Te Beest, M., Hui, C., Molofsky, J. *et al*. (2014) Invasive plants as drivers of regime shifts: identifying high priority invaders that alter feedback relationships. *Diversity and Distributions* 20, 733–744.

Gallien, L., Landi, P., Hui, C. and Richardson, D.M. (2018) Emergence of weak-intransitive competition through adaptive diversification and eco-evolutionary feedbacks. *Journal of Ecology* 106(3), 877–889. DOI: 10.1111/1365-2745.12961.
Gioria, M., Pyšek, P. and Osborne, B.A. (2018) Timing is everything: does early and late germination favor invasions by herbaceous alien plants? *Journal of Plant Ecology* 11, 4–16.
Golubski, A.J., Westlund, E.E., Vandermeer, J. and Pascual, M. (2016) Ecological networks over the edge: Hypergraph trait-mediated indirect interaction (TMII) structure. *Trends in Ecology & Evolution* 31, 344–354.
Grilli, J., Barabas, G., Michalska-Smith, M.J. and Allesina, S. (2017) Higher-Order interactions stabilize dynamics in competitive network models. *Nature* 548, 210–213.
Gunderson, L.H. and Holling, C.S. (2003) *Panarchy: Understanding transformation in human and natural systems*. Island Press, Washington, D.C., USA.
Heleno, R.H., Olesen, J.M. and Nogales, M. *et al.* (2013) Seed dispersal networks in the Galápagos and the consequences of alien plant invasions. *Proceedings of the Royal Society B* 280, 20122112.
Hui, C., Richardson, D.M., Landi, P., Minoarivelo, H.O., Garnas, J. *et al.* (2016) Defining invasiveness and invisibility in ecological networks. *Biological Invasions* 18, 971–983.
Hui, C., Richardson, D.M., Pyšek, P., Le Roux, J.J., Kučera, T. *et al.* (2013) Increasing functional modularity with residence time in the co-distribution of native and introduced vascular plants. *Nature Communications* 4, 2454. DOI: 10.1038/ncomms3454.
Hui, C. and McGeoch, M.A. (2014) Zeta diversity as a concept and metric that unifies incidence-based biodiversity patterns. *The American Naturalist* 184(5), 684–694. DOI: 10.1086/678125.
Hui, C., Minoarivelo, H.O. and Landi, P. (2018) Modelling coevolution in ecological networks with adaptive dynamics. *Mathematical Methods in the Applied Sciences* 41, 8407–8422.
Hui, C. and Richardson, D.M. (2019a) Network invasion as an open dynamical system: response to Rossberg and Barabás. *Trends in Ecology & Evolution* 34, 386–387.
Hui, C. and Richardson, D.M. (2019b) How to invade an ecological network. *Trends in Ecology & Evolution* 34, 121–131.
Hui, C. and Richardson, D.M. (2017) *Invasion Dynamics*. Oxford University Press, Oxford.
Huston, M. (1979) A general hypothesis of species diversity. *American Naturalist* 113, 81–101.
Janzen, D.H. (1985) On ecological fitting. *Oikos* 45, 308–310.
Jaworski, C.C., Bompard, A. and Genies, L. *et al.* (2013) Preference and prey switching in a generalist predator attacking local and invasive alien pests. *PLoS ONE* 8, e82231.
Jeschke, J.M. and Erhard, F. (2018) Darwin's Naturalization and Limiting Similarity Hypotheses. In: Jeschke, J.M. and Heger, T. (eds) *Invasion Biology: Hypotheses and Evidence. CAB International*. Wallingford, UK, pp. 140–146.
Johnstone, I.M. (1986) Plant invasion windows: a time-based classification of invasion potential. *Biological Reviews* 61, 369–394.
Jordano, P., Bascompte, J. and Olesen, J.M. (2003) Invariant properties in coevolutionary networks of plant-animal interactions. *Ecology Letters* 6, 69–81.
Kaiser-Bunbury, C.N., Muff, S. and Memmott, J. *et al.* (2010) The robustness of pollination networks to the loss of species and interactions: a quantitative approach incorporating pollinator behaviour. *Ecology Letters* 13, 442–452.
Keane, R.M. and Crawley, M.J. (2002) Exotic plant invasions and the enemy release hypothesis. *Trends in Ecology & Evolution* 17, 164–170.
Keet, J.H., Ellis, A.G., Hui, C. and Le Roux, J.J. (2017) Legume-rhizobium symbiotic promiscuity does not determine plant invasiveness. *Annals of Botany* 119, 1319–1331.
Landi, P., Minoarivelo, H.O., Brannstrom, A., Hui, C. and Dieckmann, U. (2018) Complexity and stability of ecological networks: a review of the theory. *Population Ecology* 60, 319–345.
Latombe, G., Richardson, D.M., Pyšek, P., Kučera, T. and Hui, C. (2018) Drivers of species turnover vary with species commonness for native and alien plants with different residence times. *Ecology* 99, 2763–2775.
Latombe, G., Roura-Pascual, N. and Hui, C. (2019) Similar compositional turnover but distinct insular environmental and geographical drivers of native and exotic ants in two oceans. *Journal of Biogeography* 46, 2299–2310.
Le Roux, J.J., Hui, C., Keet, J.H. and Ellis, A. (2017) Co-Introduction versus ecological fitting pathways to the establishment of effective mutualisms during biological invasions. *New Phytologist* 215, 1354–1360.

Letten, A.D. and Stouffer, D.B. (2019) The mechanistic basis for higher-order interactions and non-additivity in competitive communities. *Ecology Letters* 22, 423–436.
Levine, J.M., Bascompte, J., Adler, P.B. and Allesina, S. (2017) Beyond pairwise mechanisms of species coexistence in complex communities. *Nature* 546, 56–64.
Lucas-Barbosa, D., Sun, P., Hakman, A., van Beek, T.A., van Loon, J.J.A. *et al.* (2016) Visual and odour cues: plant responses to pollination and herbivory affect the behaviour of flower visitors. *Functional Ecology* 30, 431–441.
MacArthur, R. (1970) Species packing and competitive equilibrium for many species. *Theoretical Population Biology* 1, 1–11.
MacArthur, R. and Levins, R. (1967) The limiting similarity, convergence, and divergence of coexisting species. *American Naturalist* 101, 377–385.
MacDougall, A.S., Gilbert, B. and Levine, J.M. (2009) Plant invasions and the niche. *Journal of Ecology* 97, 609–615.
Mack, R.N., Simberloff, D., Mark Lonsdale, W., Evans, H., Clout, M. *et al.* (2000) Biotic invasions: causes, epidemiology, global consequences, and control. *Ecological Applications* 10, 689–710.
Mayfield, M.M. and Stouffer, D.B. (2017) Higher-order interactions capture unexplained complexity in diverse communities. *Nature Ecology and Evolution* 1, 0062.
McGeoch, M.A., Latombe, G., Andrew, N.R., Nakagawa, S., Nipperess, D.A. *et al.* (2019) Measuring continuous compositional change using decline and decay in zeta diversity. *Ecology* 100, e02830.
Melian, C.J., Bascompte, J., Jordano, P. and Krivan, V. (2009) Diversity in a complex ecological network with two interaction types. *Oikos* 118, 122–130.
Minoarivelo, H.O. and Hui, C. (2016) Invading a mutualistic network: to be or not to be similar. *Ecology and Evolution* 6, 4981–4996.
Mitchell, C.E., Agrawal, A.A., Bever, J.D., Gilbert, G.S., Hufbauer, R.A. *et al.* (2006) Biotic interactions and plant invasions. *Ecology Letters* 9, 726–740.
Mohammed, M.M.A., Landi, P., Minoarivelo, H.O. and Hui, C. (2018) Frugivory and seed dispersal: extended bi-stable persistence and reduced clustering of plants. *Ecological Modelling* 380, 31–39.
Mougi, A. and Kondoh, M. (2012) Diversity of interaction types and ecological community stability. *Science* 337, 349–351.
Murdoch, W.W. (1969) Switching in general predators: experiments on predator specificity and stability of prey populations. *Ecological Monographs* 39, 335–354.
Nnakenyi, C.A., Traveset, A., Heleno, R., Minoarivelo, H.O. and Hui, C. (2019) Fine-tuning the nested structure of pollination networks by adaptive interaction switching biogeography and sampling effect in the Galápagos islands. *Oikos* 128, 1413–1423.
Nuwagaba, S., Zhang, F. and Hui, C. (2015) A hybrid behavioural rule of adaptation and drift explains the emergent architecture of antagonistic networks. *Proceedings of the Royal Society B: Biological Sciences* 282, 20150320.
Olesen, J.M., Bascompte, J. and Elberling, H. *et al.* (2008) Temporal dynamics in a pollination network. *Ecology* 89, 1573–1582.
Ordonez, A., Wright, I.J. and Olff, H. (2010) Functional differences between native and alien species: a global-scale comparison. *Functional Ecology* 24, 1353–1361.
Pearson, D.E., McKelvey, K.S. and Ruggiero, L.F. (2000) Non-target effects of an introduced biological control agent on deer mouse ecology. *Oecologia* 122, 121–128.
Petanidou, T., Kallimanis, A.S. and Tzanopoulos, J. *et al.* (2008) Long-term observation of a pollination network: fluctuation in species and interactions, relative invariance of network structure and implications for estimates of specialization. *Ecology Letters* 11, 564–575.
Pilosof, S., Porter, M.A., Pascual, M. and Kefi, S. (2017) The multilayer nature of ecological networks. *Nature Ecology and Evolution* 1, 0101.
Poveda, K., Steffan-Dewenter, I., Scheu, S. and Tscharntke, T. (2005) Effects of decomposers and herbivores on plant performance and aboveground plant-insect interactions. *Oikos* 108, 503–510.
Reinhart, K.O. and Callaway, R.M. (2006) Soil biota and invasive plants. *New Phytologist* 170, 445–457.
Richardson, D.M., Allsopp, N., D'Antonio, C.M., Milton, S.J. and Rejmánek, M. (2000) Plant invasions–the role of mutualisms. *Biological Reviews* 75, 65–93.
Richardson, L.L., Bowers, M.D. and Irwin, R.E. (2016) Nectar chemistry mediates the behavior of parasitized bees: consequences for plant fitness. *Ecology* 97, 325–337.
Richardson, D.M. and Pyšek, P. (2006) Plant invasions: merging the concepts of species invasiveness and community invasibility. *Progress in Physical Geography* 30, 409–431.

Rouget, M., Hui, C., Renteria, J., Richardson, D.M. and Wilson, J.R.U. (2015) Plant invasions as a biogeographical assay: vegetation biomes constrain the distribution of invasive alien species assemblages. *South African Journal of Botany* 101, 24–31.

Roura-Pascual, N., Sanders, N. and Hui, C. (2016) The distribution and diversity of insular ants: do exotic species play by different rules? *Global Ecology and Biogeography* 25, 642–654.

Santamaría, L. and Rodríguez-Gironés, M.A. (2007) Linkage rules for plant-pollinator networks: trait complementarity or exploitation barriers? *PLoS Biology* 5, 354–362.

Sax, D.F., Stachowicz, J.J., Brown, J.H., Bruno, J.F., Dawson, M.N. *et al.* (2007) Ecological and evolutionary insights from species invasions. *Trends in Ecology & Evolution* 22, 465–471.

Sher, A.A. and Hyatt, L.A. (1999) The disturbed resource-flux invasion matrix: a new framework for patterns of plant invasion. *Biological Invasions* 1, 107–114.

Simberloff, D. and Von Holle, B. (1999) Positive interactions of nonindigenous species: invasional meltdown? *Biological Invasions* 1, 21–32.

Stephens, D.W. and Krebs, J.R. (1986) *Foraging Theory*. Princeton University Press, Princeton, NJ.

Strauss, S.Y., Lau, J.A. and Carroll, S.P. (2006) Evolutionary responses of natives to introduced species: what do introductions tell us about natural communities? *Ecology Letters* 9, 357–374.

Tecco, P.A., Díaz, S., Cabido, M. and Urcelay, C. (2010) Functional traits of alien plants across contrasting climatic and land-use regimes: do aliens join the locals or try harder than them? *Journal of Ecology* 98, 17–27.

Terry, J.C.D., Morris, R.J. and Bonsall, M.B. (2017) Trophic interaction modifications: an empirical and theoretical framework. *Ecology Letters* 20, 1219–1230.

Thébault, E. and Fontaine, C. (2010) Stability of ecological communities and the architecture of mutualistic and trophic networks. *Science* 329, 853–856.

Traveset, A. and Richardson, D.M. (2014) Mutualistic interactions and biological invasions. *Annual Review of Ecology, Evolution, and Systematics* 45, 89–113.

Tylianakis, J.M. and Morris, R.J. (2017) Ecological networks across environmental gradients. *Annual Review of Ecology, Evolution, and Systematics* 48, 25–48.

van Baalen, M., van Rijn, P.C.J., Krivan, V. and Sabelis, M.W. (2001) Alternative food, switching predators, and the persistence of predator–prey systems. *American Naturalist* 157, 512–524.

Ward, N.L. and Masters, G.J. (2007) Linking climate change and species invasion: an illustration using insect herbivores. *Global Change Biology* 13(8), 1605–1615. DOI: 10.1111/j.1365-2486.2007.01399.x.

Waser, N.M. (2015) Competition for pollination and the evolution of flowering time. *American Naturalist* 185(6), 3–5. DOI: 10.1086/681255.

Werner, E.E. and Peacor, S.D. (2003) A review of trait-mediated indirect interactions in ecological communities. *Ecology* 84(5), 1083–1100. DOI: 10.1890/0012-9658(2003)084[1083:AROTII]2.0.CO;2.

White, E.M., Wilson, J.C. and Clarke, A.R. (2006) Biotic indirect effects: a neglected concept in invasion biology. *Diversity and Distributions* 12(4), 443–455. DOI: 10.1111/j.1366-9516.2006.00265.x.

Whittall, J.B. and Hodges, S.A. (2007) Pollinator shifts drive increasingly long nectar spurs in columbine flowers. *Nature* 447(7145), 706–709. DOI: 10.1038/nature05857.

Wilson, J.R.U., Dormontt, E.E., Prentis, P.J., Lowe, A.J. and Richardson, D.M. (2009) Something in the way you move: dispersal pathways affect invasion success. *Trends in Ecology & Evolution* 24(3), 136–144. DOI: 10.1016/j.tree.2008.10.007.

Wolf, M. and Weissing, F.J. (2012) Animal personalities: consequences for ecology and evolution. *Trends in Ecology & Evolution* 27(8), 452–461. DOI: 10.1016/j.tree.2012.05.001.

Wootton, J.T. (1994) The nature and consequences of indirect effects in ecological communities. *Annual Review of Ecology and Systematics* 25(1), 443–466. DOI: 10.1146/annurev.es.25.110194.002303.

Zhang, F. and Hui, C. (2014) Recent experience-driven behaviour optimizes foraging. *Animal Behaviour* 88, 13–19. DOI: 10.1016/j.anbehav.2013.11.002.

Zhang, F., Hui, C. and Terblanche, J.S. (2011) An interaction switch predicts the nested architecture of mutualistic networks. *Ecology Letters* 14(8), 797–803. DOI: 10.1111/j.1461-0248.2011.01647.x.

3 Soil Biota and Non-native Plant Invasions

Ragan M. Callaway[1]* and Jacob E. Lucero[2]

[1]Division of Biological Sciences and the Institute on Ecosystems, University of Montana, Missoula, MT, USA; [2]Department of Biology, York University, Toronto, Ontario, Canada

Abstract

The trajectory of plant invasions – for better or for worse – can be tied to interactions between plants and the soil community. Here, we highlight five broad ways in which belowground interactions can influence the trajectory of biological invasions by non-native plant species. First, many non-native plant species in their non-native ranges can interact very differently with the resident soil community than do native species. Second, non-native plant species often interact very differently with the soil community in their non-native ranges than in their native ranges, which can result in enemy release from antagonistic interactions. Third, non-native plant species can cultivate a soil community that disproportionately harms native competitors in invaded communities. Fourth, antagonistic soil biota in invaded communities can reduce the performance of non-native plant species, resulting in meaningful biotic resistance against invasion. Fifth, besides or in addition to antagonistic interactions with soil biota, soil mutualisms can promote the success of invasive plant species i) when mutualists co-invade with non-native plant species that require obligate specialist mutualists, ii) when mutualists enhance the performance of non-native plant species in their non-native ranges, and iii) when biotic interactions in the invaded community suppress the soil mutualists of native plant species. We conclude that management practices aimed at manipulating plant—soil interactions have considerable potential to help control plant invasions, but further work is needed to understand the spatial, temporal, taxonomic and biogeographic drivers of context dependence in interactions among plants and soil biota.

3.1 Introduction

Biological invasions by non-native plant species degrade biodiversity and ecosystem function worldwide (Simberloff *et al.*, 2013) at great economic cost (Pimentel *et al.*, 2005). Many factors contribute to plant invasions, including biotic interactions that may or may not occur in the non-native range of the invaders (Mitchell *et al.*, 2006; Richardson *et al.*, 2014). The presence or absence of these interactions can influence invasion trajectories by facilitating the performance, abundance and population growth of non-native species (Griffith, 2010; Richardson *et al.*, 2014); by disproportionately harming native competitors (Lucero and Callaway, 2018); or by providing biotic resistance against invasion (Byun *et al.*, 2015; St. Clair *et al.*, 2016). Notably, soil biota can mediate each of these outcomes (Callaway *et al.*, 2004a, b; Inderjit and van der Putten, 2010; Dawson and Schrama, 2016).

Soil biota comprise a broad and complex group of consumers and mutualists that have strong negative and beneficial effects on plants. Broadly, soil biota affect plants as pathogens, through root-fungus and other mutualisms, and by driving the nutrient cycles on which plants depend (Packer and Clay, 2000; Van der Putten *et al.*, 2001). These effects and the reciprocal effects of plants on soil biota contribute to a wide range of complex and dynamic feedback interactions between plants and the microbial

*Corresponding author: ray.callaway@mso.umt.edu

 Plant Invasions: The Role of Biotic Interactions (eds Anna Traveset and David M. Richardson)
DOI: 10.1079/9781789242171.0003

communities that develop around their roots (i.e. plant—soil feedbacks – PSFs; Van der Putten *et al.*, 1993; Bever, 1994). Positive feedbacks are thought to occur when plant species disproportionately accumulate biota around and in their roots that have beneficial effects on the plants that cultivate them, such as mycorrhizal fungi and nitrogen fixers. Positive feedbacks, though logically not long sustainable, are thought to decrease plant community diversity because of the advantages gained by a few select species (Bever *et al.*, 1997; Bever, 2002). Negative feedbacks are thought to occur when plant species disproportionately accumulate pathogenic biota in their rhizospheres, which can create conditions that are increasingly hostile to the plants that cultivate the pathogens (Van der Putten *et al.*, 1993; Bever, 1994; Klironomos, 2002). Negative feedbacks are thought to enhance community diversity by increasing species turnover rates and reducing mono-dominance (Reinhart, 2012; Maron *et al.*, 2016). In this chapter we first examine soil biota and non-native plant invasions from this 'black box' soil community framework, in the context of plant—soil feedbacks, and then break down some components of the black box that appear to be connected to the invasion. But by no means do we exhaust all taxa that might affect invasion, or the literature on the taxa we target.

It helps to consider soil biota from the perspective of the leading explanation for successful biological invasions by non-native plant species – the Enemy Release Hypothesis (Keane and Crawley, 2002). This is the idea that translocation to a distant range allows some non-native plant species to escape top-down controls imposed by consumers in the native range. Generally, escape is explored in the context of specialist consumers, but recently, studies have shown that non-native invaders can also escape from the less discriminatory effects of generalist consumers (e.g. Vermeij *et al.*, 2009; Lucero *et al.*, 2019; but see Schultheis *et al.*, 2015). Others have found that native generalists often attack or prefer native plant species over non-native invasive plant species (e.g. Knight *et al.*, 2009; Vermeij *et al.*, 2009; Tallamy *et al.*, 2010; Schaffner *et al.*, 2011; Lucero and Callaway, 2018; but see Parker *et al.*, 2006). Interestingly, escaping generalists is explicitly predicted by the Enemy Release Hypothesis (Keane and Crawley, 2002; Joshi and Vrieling, 2005).

In this framework of generalist effects on invasive plant species, a substantial number of studies have shown that the biogeographic origin of soil biota and plant species affects the strength and direction of PSFs. Non-native invasive plant species commonly experience weaker PSFs in soil from their non-native ranges than native species (e.g. Kulmatiski *et al.*, 2008; Lekberg *et al.*, 2018). In this context it is important to note that not all non-native plant species are invasive – here we reserve the term 'invaders' for non-native plant species that exhibit greater density, abundance or community dominance in their non-native ranges than in their native ranges (see Hierro *et al.*, 2005). For the interpretation of empirical results, non-native invaders can be further classified with respect to the *intensity* of their success and impacts, with some being 'weak invaders' and others 'strong invaders' (Simberloff, 1981; Ortega and Pearson, 2005). This distinction between types of invaders is important in the context of the mechanisms we explore here.

Evidence for biogeographic patterns in the ecology of soil biota has emerged from comparisons of the performance of non-native invasive species in soil from native and non-native ranges and from comparisons of the effects of soil biota on non-native invasive species vs native species in recipient communities. These biogeographic patterns, and by extension non-native plant invasions themselves, are the primary focus of this chapter. Our goals are to explore soil biota as contributors to the trajectory of non-native plant invasions and call attention to the role of evolutionary history on community structure through the lens of geographic isolation (see Darwin, 1859; Wallace, 1876; Ma *et al.*, 2016). Many of the subjects tackled in this chapter have been well reviewed elsewhere, and we refer to many of these reviews; however, the chapter does not come close to a complete review of the relevant literature. Instead, we have tried to review and summarize the most salient and timely issues that seem most important to the role of soil biota in non-native plant invasion. To do this, we first examine comparisons of non-native and native plant species within just one range; second we look at biogeographic comparisons that compare belowground processes in both native and non-native ranges; third we focus more closely on soil antagonists and biotic resistance; and finally we discuss the effects of soil mutualists.

3.2 Non-native vs Native Plant Comparisons

Comparisons of the direction and intensity of the effects of soil biota on native and non-native species in the new range of non-natives provide the earliest and most abundant evidence for enemy release from soil biota. As an aside, the potential for enemy release from soil biota is also interesting because it questions the influential 'everything is everywhere' paradigm for microbes (Baas Becking, 1934; Rout and Callaway, 2012). To our knowledge, the first experimental comparison of PSFs among native and non-native plants was with old field species in southern Ontario (Klironomos, 2002). Klironomos compared the growth of five rare native species to that of five non-native invasive species when grown in soil 'cultivated', or trained, by conspecifics. All five native species generated strong *negative* PSFs, indicating the presence of inhibitory soil biota that targeted those particular native species. In striking contrast, four of the five non-native invasive species generated strong *positive* feedbacks and none generated negative feedbacks, indicating the absence of inhibitory soil biota that could target those species. Klironomos also conducted experiments with isolated mycorrhizae, typically plant mutualists, and found that four of five native plant species exhibited positive feedbacks vs two of five for the non-native invasive plant species. Sterilization of trained soil eliminated all feedbacks for all species, implicating biota rather than nutrients. Agrawal *et al.* (2005) experimented with ten congeneric pairs of native and non-native plant species and found that soil microbes reduced the growth of native species twice as much as non-native congeners. Interestingly, differences in the response of native vs non-native species to soil microbes was much greater than differences in response to above-ground herbivores or pathogens, 'indicating that interactions with underground microbes provided the most consistent evidence for enemy release ... ' relative to evidence for the effects of above-ground herbivores and pathogens. This difference in growth reduction occurred for five of the ten pairs, and importantly, many of the non-native species used were not highly invasive, yet still showed an effect of biogeographical origin. Using species and soils from Kalahari in Botswana, van der Putten *et al.* (2007) compared PSFs for the non-native invasive grass, *Cenchrus biflorus* (native to the more northern Sahel savannas of Africa and India) with those of two dominant native grasses, *Eragrostis lehmanniana* and *Aristida meridionalis*. The non-native grass had neutral to positive PSFs, whereas both native species showed neutral to negative PSFs.

Over the last 15 years, the pace of publications on comparisons between natives and non-natives using soil in the non-native ranges of the non-natives have accelerated and been thoroughly reviewed (e.g. Kulmatiski *et al.*, 2008; Suding *et al.*, 2013, Inderjit and Cahill, 2015; Dawson and Schrama, 2016; Lekberg *et al.*, 2018). To our knowledge, the first of these was by Kulmatiski *et al.* (2008), which included a meta-analysis on 329 experiments from 45 independent studies, or subsets of these studies, and incorporated a comparison of non-native invasive and native species. Some studies had a biogeographic range component as well. They found no mean positive effects for any group of species, but for two of three analytical models, non-native invasive species (34 experiments) generated much weaker negative feedbacks than native species (148 experiments). Interestingly, there were no differences between the effect sizes on non-native species that were non-invasive and native species. Kulmatiski *et al.* (2008) interpreted their results as 'evidence that PSFs may help explain why some non-native species become invasive while others do not ... suggesting that the success of these [invasive] plants can be explained by a release from soil-based enemies'. In contrast, their meta-analysis indicated that non-native species that are not invasive experienced more biotic resistance (*sensu* Elton 1958) than the non-native invasive species.

Because non-native invasive species often have unusually strong competitive effects on natives (Vilà *et al.*, 2011; Bernard-Verdier and Hulme, 2019; Pearse *et al.*, 2019), studying plant—plant competition in the context of PSFs might be of value. Lekberg *et al.* (2018)

conducted a review and meta-analysis of studies that integrated competitive and PSF effects. There was a significant negative relationship between the strength of PSFs for target species and the intensity of their competitive effects on neighbours. In other words, strong competitors tended to experience stronger negative PSFs. Across taxa, PSFs were predominantly negative, similar in their intensity to competition, and functioned in an additive or synergistic manner with competition. Only four non-native species were included in the data set (used in 11 experiments) of Lekberg *et al.* (2018), but even so, all non-native species in all experiments generated PSFs on conspecifics that were either neutral or positive. Lekberg found very few species, native or not, in the positive PSF plus strongest competitor quadrant of the graph (potential 'Hutchinsonian demons' free from trade-offs), but the highly invasive *Acacia dealbata* (Rodríguez-Echeverría *et al.*, 2013; native to Oceania but widely invasive elsewhere) demonstrated the most positive combination of PSF and competitive effect of any species.

One set of studies not identified in the Lekberg *et al.* (2018) literature search examined PSFs and competitive interactions between the non-native invasive *Centaurea stoebe* and two congeners, the North American native *Festuca idahoensis* and the European *Festuca ovina* (Callaway *et al.*, 2004a; Reinhart and Callaway, 2006; see also 'Biogeographic comparisons ' below). When competing with *F. ovina* in European (native) soils, *C. stoebe* grew much better in soils that had been trained by *F. ovina* than soils that had been trained by other *C. stoebe* plants. In contrast, when *C. stoebe* competed with *F. idahoensis* in North American soils, it grew better in soils trained by conspecifics than soils trained by *F. idahoensis*. Sterilization of soil eliminated feedbacks in both cases. Manipulation of soil biota in the field with fungicide also indicates that biota may substantially affect the way natives and invasive *C. stoebe* interact (Callaway *et al.*, 2004b; see also Marler *et al.*, 1999). In contrast, Crawford and Knight (2017) found that the invader *Lespedeza cuneata* generated strong soil biota driven positive feedbacks in its non-native range, but that competition from native species eliminated this feedback.

As noted by others (see Kulmatiski, 2017), inference for invasions derived from PSFs is probably most limited by the difficulty of conducting true feedback studies in the field. In part, this is because coordinating all components of a solid PSF experiment cannot be done effectively in the field. One can train soil in common gardens, but cannot then effectively sterilize soil as controls in the field because of rapid microbial recolonization from surrounding soil. Nevertheless, there have been impressive efforts to understand PSFs and non-native invaders in the field. Dostál *et al.* (2013) integrated experiments with PSFs and competition from natives, as described above, to understand a temporal trend of decline in the abundance and impact of *Heracleum mantegazzianum*, an invader in Europe, over time. In a common garden experiments they found that PSFs became less negative with time since invasion, and this tilted competitive interactions towards native species. Kulmatiski (2017) trained field soil for four years in common garden experiments with a dominant invader, *Centaurea diffusa*, and a dominant native, *Pseudoroegneria spicata*, and then planted 'one native and one non-native plant community' into these plots. After three more years, he modelled the results and found that both species cultivated soils that favoured themselves, e.g. positive feedbacks. In contrast to many much shorter-term greenhouse experiments, his model predicted that non-native plants would dominate even without PSFs, due to competitive outcomes.

In summary, there is substantial and growing evidence that non-native invasive plant species in their non-native ranges can interact very differently with the resident soil community than do native species and non-native, non-invasive species. These differences may help explain why some non-native species become explosive invaders while others do not. In addition, disproportionately escaping the effects of negative belowground interactions relative to native plant species may help explain the strong community-level impacts of non-native invasive plant species.

3.3 Biogeographic Comparisons

The most powerful evidence for enemy release, regardless of the type of enemy, comes from biogeographic experiments – those conducted in parallel in both the native and invasive range of non-native invasive plant species (Maron and Vilá, 2001). This is in part because of enemy release, described above (Keane and Crawley, 2002). Thus, without comparative biogeographic experiments, high levels of attack or suppression in the non-native range might be interpreted as enemy escape, but similar attack in the native range would raise questions about escape or release. Biogeographic studies of generalist herbivore pressure on non-native, invasive marine algae in Hawaii provide an excellent example of this. Vermeij *et al.* (2009) studied two species, *Acanthophora spicifera* and *Hypnea musciformis*, that are native to the Caribbean and were introduced to Hawaii. In experiments in Hawaii, they found that native herbivores there preferred these non-native species over native marine algae, suggesting no role for enemy escape in the invasion. But, when the intensity of herbivory was compared in both ranges, herbivore attack was much lower for both invasive algae in Hawaii than in the Caribbean, which clearly supports enemy release. In the context of antagonistic soil biota, the Enemy Release Hypothesis predicts stronger negative effects of soil antagonists and/or more negative PSFs in the native range than in the non-native range. However, replicating simultaneous, coordinated experiments in the native and non-native ranges of non-native invaders presents logistical challenges. Not surprisingly, relatively few studies have experimentally examined the effects of natural enemies in this way (but see DeWalt *et al.*, 2004; Williams *et al.*, 2008; Vermeij *et al.*, 2009; Lucero *et al.*, 2019).

None the less, a handful of biogeographically explicit studies have demonstrated enemy release from the effects of antagonistic soil biota. For instance, Reinhart *et al.* (2003) reported that the invasion of *Prunus serotina*, a native of North America, into north-western Europe was strongly promoted by soil biota, as European soil biota facilitated the growth of *P. serotina*. In contrast, soil biota that accumulated around *P. serotina* in the native range was highly inhibitory. These results correspond with Janzen-Connell-like (Janzen, 1970) suppression of *P. serotina* seedlings in proximity to conspecific adults in the native range (Packer and Clay, 2000, 2003) and large distances between saplings and adult trees (Reinhart *et al.*, 2003). In contrast, in the non-native range of the Netherlands where soil biota had positive effects on *P. serotina*, saplings and trees were much closer together (Reinhart *et al.*, 2003). Reinhart *et al.* (2010) found that pathogenic *Pythium* spp. from the native range caused far more root rot and mortality of *P. serotina* seedlings from both North American and European populations than *Pythium* from the non-native range. Considered together, this is very good evidence for enemy release. In experiments with *C. stoebe*, a perennial forb native to Eurasia but invasive in North America, Callaway *et al.* (2004a) found that sterilizing soil from many sites in part of the native range (France) increased total biomass by almost a factor of three. Sterilization of soil from part of the non-native range (Montana) increased total biomass by less than 20%. In native French soil, *C. stoebe* cultivated increasingly negative PSFs, whereas in Montana soil, *C. stoebe* cultivated increasingly positive PSFs. This apparent escape corresponds with a far greater abundance of *C. stoebe* in its non-native range than in its native range (Pearson *et al.*, 2018). Gundale *et al.* (2014) tested the effects of soil biota on growth of *Pinus contorta*, using soils from the native range of Canada and the non-native range of Sweden. They found that *P. contorta* grew larger in Swedish soil than Canadian soil due to differences in soil biota, further supporting the idea that the effects of soil biota can switch from negative to positive following introduction to a new biogeographic range. Interestingly, this switch for *P. contorta* results in more highly invasive behaviour in the southern hemisphere than in the northern hemisphere (Davis *et al.*, 2019; see below), suggesting that patterns of enemy release may be region-specific.

Some tests of the Enemy Release Hypothesis with respect to soil biota have yielded more nuanced results. Reinhart and Callaway (2004) tested the hypothesis in the context of mutualisms (see below) by examining density patterns and how soil sterilization affected the growth of *Acer negundo* (native to North America but invasive in Europe) and *Acer platanoides* (native to Europe but invasive in North America) seedlings in their respective native and non-native ranges. Distances from focal trees to the nearest *Acer* conspecifics were 56–77% less in their non-native ranges than in their native ranges. In addition, the effects of soil sterilization differed between the native and non-native ranges of *Acer* taxa. But there was no clear evidence for a feedback-related escape from soil biota. Instead, soil associated with conspecifics in the non-native ranges *decreased* the performance of *Acer* seedlings, the opposite of what might be predicted by the Enemy Release Hypothesis. But in contrast to this, soil associated with other non-*Acer* species in the non-native ranges increased the performance of *Acer* seedlings, and there was a 'net or regional negative effect of resident soil biota on growth' in the native ranges of *Acers*. In other words, they did not find a switch from strong negative PSFs in native ranges to weak negative, neutral or positive PSFs in non-native ranges, but instead they found a somewhat regional, i.e. not host-specific, stronger negative effect of soil biota on *Acer* species in their native ranges than in their non-native ranges. This corresponded with soil biota associated with dominant native *Acer* species enhancing invasion by non-native *Acer* species. In sum, these findings suggest apparent escape from harmful biota occurring at regional scales in the non-native range. Roughly similar biogeographic differences in both positive and negative effects of soil biota were found by Yang *et al.* (2013) for *Triadica sebifera* (previously *Sapium sebiferum*), a tree native to China but invasive in the south-eastern USA. The net effect of soil biota from the native range on *T. sebifera* was neutral to negative, consistent with enemy release. In contrast, soil biota from the non-native range had positive effects and, when grown in soil from the USA, mycorrhizal colonization on the roots of *T. sebifera* was higher than in soils from China – results more consistent with the Enhanced Mutualist Hypothesis (Reinhart and Callaway, 2006; see below).

The only other evidence that we know of for these very broad scale biogeographic effects of soil biota comes from studies of native range vs non-native range PSFs for six European invaders in North America. Maron *et al.* (2014) compared PSFs with soil biota collected from 11 grasslands in Europe and 10 in North America and found that four of the six species (*C. stoebe*, *Euphorbia esula*, *Linaria vulgaris* and *Potentilla recta*) tested exhibited substantially reduced shoot biomass in 'live' vs sterile soil from Europe. In contrast, North American soils produced no PSFs for any of the species. Differences in PSFs corresponded with far more pathogenic root lesions on plants grown in European soil vs plants grown in North American soil. For two of the three species that were tested for root biomass (*E. esula* and *Leucanthemum vulgare*), PSFs from native range soil were much more inhibitory than soil from the non-native range. Taken together, these findings are consistent with the Enemy Release Hypothesis. Importantly, negative feedbacks generated by European soil varied tremendously among sites (e.g. from +0.03 to −0.96 for *P. recta*), and this variation did not correspond with the presence of the target species at the field sites. Such variation suggests that highly replicated samples might be necessary to detect significant biogeographical effects (this was not the case for *C. stoebe*), and strong negative PSFs from sites without the target species suggest the possibility of escape from soil biota that are widely distributed in native ranges.

Most studies have not explored how the duration of training events might affect the direction and intensity of PSFs, but de la Peña *et al.* (2010) compared short- and long-term training of soils with *Carpobrotus edulis* (iceplant), a native to South Africa, but invasive in many coastal environments around the world. They used soil from the native range and from Portugal and Spain in the non-native range and found that when training was short-term, PSFs were negative in the two non-native ranges and positive in the native range, the opposite of many other biogeographic studies of PSFs. In other words, there was an initial 'natural resistance to invasion by the soil biota'. But long-term training reversed this pattern: such training in non-native soil developed soil biota that improved the growth of *C. edulis* and inhibited the growth of natives.

Not all explicitly biogeographic and widely sampled studies have found differences in the effects of native and non-native range soils. For example, Beckstead and Parker (2003) asked whether the strong inhibitory effects of soil biota from Europe on *Ammophila arenaria* were relaxed in soil from the coastal dunes of California where the species is highly invasive. They found no evidence of release from natural enemies in North American soil, as its effects on the growth *A. arenaria* were strongly negative and nearly equal to that in the native range. Unusually broad biogeographical studies of PSFs on *Centaurea solstitialis* by Andonian *et al.* (2011) found no evidence for enemy release in comparisons of soils from native Eurasian sites and non-native sites spanning North and South America and found negative PSFs in all regions. Importantly, *C. solstitialis* generated stronger negative feedbacks in ranges where it is the most invasive, vs neutral PSFs where it is not invasive. Using the same species and regions, but in experiments that replaced PSFs with soil sterilization treatments, Andonian *et al.* (2012) reported that biogeographic differences in negative PSFs between native and non-native ranges were consistent with enemy release. However, there were very strong negative PSFs in Argentina, where *C. solstitialis* is highly invasive, indicating exceptions to the rule. Hierro *et al.* (2006) found that sterilization of native range Eurasian soils caused a 75% increase in the total biomass of *C. solstitialis*, but sterilization of soils from the non-native ranges of California and Argentina has no effect on the invader. Birnbaum and Leishman (2013) found no differences in the effects of soil from the native western Australia distributions of four *Acacia* species and the invasive *Paraserianthes lophantha*, a relative of *Acacia*, to the effects of soil from eastern Australia where these species do not occur. They suggested that this might be due to 'the widespread occurrence of acacias and their associated soil microbial communities throughout the Australian continent'. However, considering the widespread distribution of soil microbes on continents (see discussion in Rout and Callaway, 2012) and the regional range effects reported by Reinhart and Callaway (2004) and Maron *et al.* (2014), relevant biota may have occurred in both western and eastern Australia (but see van Grunsven *et al.*, 2007). This appeared to be the case for N-fixing symbiotic biota on these leguminous trees as nodule formation was similar for soils from both ranges (also see Callaway *et al.*, 2011).

In sum, biogeographic comparisons of the effects of natural enemies in the native and non-native ranges non-native invasive plant species can provide crucial insight into mechanisms of invasion and are essential for testing the Enemy Release Hypothesis. With respect to soil biota, such studies are relatively rare, and existing studies have yielded mixed results. Most have found strong evidence for enemy release from antagonistic soil biota, but a few have found more nuanced or negative results. In addition, the relative importance of enemy release compared to other non-mutually exclusive mechanisms of invasion (reviewed by Catford *et al.*, 2009) is unclear.

3.4 Soil Antagonists and Biotic Resistance

Despite frequent reports of relatively weak PSFs in the non-native ranges of non-native invaders relative to native ranges (see above), there is evidence that specific components or net effects of soil biota in the non-native ranges of non-native species can mount strong biotic resistance to invasion. For example, Flory *et al.* (2011) reported that infections by pathogens commonly found on native plant species in the south-eastern USA inhibited the performance of the non-native invader *Microstegium vimineum*. Nijjer *et al.* (2007) conducted experiments using soil from the non-native range, *T. sebifera* (a tree native to China but highly invasive across south-eastern North America), and several native tree species. They grew trees in soils collected near conspecifics or near other species. *T. sebifera* seedlings survived less and were smaller in conspecific soils than other soils. This was not the case for native species. They also applied fungicide or sterilized soils which eliminated the negative effects of *T. sebifera* soil on *T. sebifera* seedlings. Nijjer *et al.*'s results suggest that antagonistic soil biota in conspecific soils provided biotic resistance against *T. sebifera* invasion by reducing invader but not native performance. However, Yang *et al.* (2013) reported the opposite pattern in the same system, suggesting that biotic resistance via negative PSFs can be highly conditional (see Maron *et al.*, 2015).

Biotic resistance via negative PSFs is also apparent in the case of the non-native invader *P. recta*. Callaway *et al.* (2013) found that *P. recta* easily invaded common garden monocultures of other genera. Similarly, monocultures of the native congener *Potentilla arguta* in the garden were also highly invaded by *P. recta*. Interestingly, the biomass of *P. recta* was 50% greater when grown in soils from monocultures of *F. idahoensis*, a native grass, than in soils from monocultures of either *P. arguta* or *P. recta*. Sterilizing soil from

F. idahoensis rhizospheres did not affect *P. recta*, but sterilizing soil from *P. arguta* and *P. recta* increased the biomass of *P. recta* by 108 and 90%, respectively. These findings suggest that antagonistic biota cultured in congeneric but not *F. idahoensis* soils provided biotic resistance against *P. recta* invasion. In a PSF experiment, soil trained by *F. idahoensis* resulted in a positive feedback for *P. recta*. In contrast, soils trained independently by each of two native *Potentilla* species (*P. arguta* and *Potentilla gracilis*), *P. recta* or *Dasiphora fruticosa* (formerly *Potentilla fruticosa*) decreased the biomass of *P. recta*, indicating biotic resistance via negative PSFs in non-native range soil. These results show the typical negative feedbacks for an invader in the non-native range, but also highlight the potential for native congeners (both native and non-native) to resist invasion by accumulating soil biota harmful to invaders.

In addition, non-congeneric native species can generate biotic resistance by cultivating soil biota that disproportionately antagonize non-native invasive species. For example, in the Netherlands, Vermeij *et al.* (2009) found that native *L. vulgare* harboured high densities of root-feeding nematodes that spilled over and strongly reduced the abundance and species richness of a number of invasive species in experimental gardens. This example is one of very few documented cases of biotic resistance via apparent competition. Interestingly, biotic resistance via antagonistic soil biota can depend upon environmental context. In China, Chen *et al.* (2012) reported that soil-mediated biotic resistance against *Coreopsis grandiflora* (a forb native to North America) invasion depended on light levels corresponding to different irradiances in forest understories. At high light intensities, suppression of soil biota via fungicide improved the performance of *C. grandiflora*, but at low light, suppression of soil biota had no effect. Thus, soil biota generated biotic resistance against *C. grandiflora* invasion in high- but not low-light conditions.

In sum, soil antagonists can limit the performance of non-native invasive plant species, resulting in meaningful biotic resistance against invasion. Such biotic resistance may be amplified in diverse assemblages as a part of classic diversity-function relationships, and many studies have found that diverse native communities can resist non-native invasion better than depauperate ones (Ruijven *et al.*, 2003; Levine *et al.*, 2004; Byun *et al.*, 2013). Propagule pressure (Hollebone and Hay, 2007), abiotic disturbance (Stewart *et al.*, 2009) and loss of consumer pressure (Pearson *et al.*, 2011) can overwhelm biotic resistance and accelerate invasions, but biotic resistance via antagonistic interactions with soil biota, when it occurs, should be considered an important ecosystem service. In particular invaded systems, little is known about which native species or species combinations tend to cultivate soil communities that mount the strongest biotic resistance, but this would be an important topic for applied work.

3.5 Soil Antagonists Facilitating Invasion

In some cases, non-native invaders gain advantages over native competitors by stimulating soil biota that harm the latter. Eppinga *et al.* (2006) modelled interactions between the California dune-invader *A. arenaria* and soil biota, and found that the accumulation of local pathogens by *A. arenaria* could inhibit native plant species. They proposed the Accumulation of Local Pathogens Hypothesis as a process by which invasions might succeed. *Chromolaena odorata*, one of the world's most invasive tropical plant species, appears to do just this. Mangla and Callaway (2008) found that invasive *C. odorata* in India accumulated high levels of generalist pathogens from the genus *Fusarium* in its rhizosphere, apparently with minor consequences for itself but with inhibitory consequences for native species. *C. odorata* root leachate experimentally added to uninvaded soils increased *Fusarium* spore density and increased the inhibitory effects of the soils. This *C. odorata*-pathogen scenario is an example of spillover (Flory and Clay, 2013), which has also been observed in other systems. For example, *Avena fatua*, an non-native invasive grass in California, can harbour barley yellow mosaic virus that spills over to and decreases the abundance of the native competitors *Lolium multiflorum* and *Digitaria sanguinalis*. Similarly, *Bromus tectorum*, an annual grass species native to Eurasia but highly invasive across western North America, can dramatically increase the abundance of the seed pathogen

Pyrenophora semeniperda in invaded soils, which can spill over to and severely deplete the densities of native seed banks (Beckstead *et al*., 2010). Interestingly, such spillover is unlikely in the native range of *B. tectorum* because *P. semeniperda* is extremely rare in Eurasia (Stewart *et al*., 2009). These examples of pathogen spillover can also be characterized as cases of apparent competition (Power and Mitchell, 2004).

Taken together, these studies demonstrate the capacity for non-native invasive plant species to cultivate soil biota that disproportionately harm native competitors (e.g., Li *et al*., 2014). Such accumulation of local pathogens is one mechanism whereby non-native invaders can displace natives and impact non-native communities. However, this body of work is relatively small, and many knowledge gaps exist. For instance, it remains unclear how the effects of pathogen spillover (e.g. Beckstead *et al*., 2010) change over time, and it is also unclear which environmental conditions exacerbate or inhibit such spillover. Furthermore, whether biogeographic context (i.e. whether interactions occur in the native or non-native range of a focal invader) influences the direction and intensity of pathogenic spillover has attracted little attention.

3.6 Soil Mutualists Facilitating Invasion

Just as biogeographic escape from antagonistic soil biota in the native range can promote the success of non-native invasive plant species, so too can novel or altered associations with beneficial soil biota in the non-native range. There are several non-mutually exclusive ways this can occur. First, the Enhanced Mutualism Hypothesis attributes the success of some non-native species, with at least a facultative dependence on soil mutualisms, to either biogeographically novel (Marler *et al*., 1999) or enhanced (Reinhart and Callaway, 2006) soil mutualisms formed in the non-native range relative to the native range. Additionally, invaders may alter or eliminate soil mutualisms in non-native communities that harm native plant species (Grove *et al*., 2017). For both scenarios, an important conceptual consideration is the necessity for non-native species requiring obligate specialist mutualists to bring them along if they are to establish and invade non-native communities (see Traveset and Richardson, 2014). Importantly, none of these possibilities is as clear cut conceptually or empirically as the Enemy Release Hypothesis. This is because soil 'mutualists' can have parasitic effects and, depending on abiotic conditions (Johnson *et al*., 1997), can have strikingly different and unpredictable effects on invasion (see above). It is also because empirical outcomes are highly inconsistent (Traveset and Richardson, 2014). The objectives of this section are to discuss i) why understanding the roles of soil mutualisms in non-native plant invasions is not as tidy as for pathogens and PSFs in general, ii) co-invasions involving non-native plant species and their obligate specialist mutualists, iii) the potential for soil mutualists to enhance the performance of non-native species in their non-native range, and iv) the potential for non-native plant species to suppress the soil mutualists of native plant species.

3.6.1 Mutualists and the black box of soil

With the very different potential effects of soil mutualisms in mind, PSFs are generally measured and considered as net effects produced by both parasitic/pathogenic and mutualistic soil biota. Such net effects are the crucial determinants of the overall ecological roles of PSFs in plant communities. Yet, these components are difficult to explore separately and mechanistically. For example, Reinhart and Callaway (2006) suggested that biogeographic variation in the strength and/or direction of PSFs might be due to differences in the effects of either pathogens or mutualists. Breaking down PSFs this way would assert the Enemy Release Hypothesis when there are greater negative effects of pathogens in native ranges than in non-native ranges and similar effects of mutualists in both ranges. In contrast, greater positive effects of mutualist biota in non-native ranges than in native ranges and similar impacts of pathogens between ranges would be consistent with the Enhanced Mutualisms Hypothesis.

Most PSF experiments, de facto, take the black box approach by manipulating whatever creatures (pathogenic bacteria and fungi, neutral bacteria and fungi, decomposers, protists, viruses and macro-organisms) are in the soil at the time it is collected. The 'curse of the black box' (Cortois and De Deyn, 2012) emerges from the difficult choice between holistic or reductionist approaches. Holistic approaches, such as PSF studies, provide insight into the general function of natural systems, but without a great degree of mechanistic understanding. Reductionist approaches can provide mechanistic or management insights, but assume that components of the soil biota have the same effects with and without other components of the soil biota, which can be a very problematic assumption (see discussion in Cortois and De Deyn, 2012).

Thus, black box approaches might be the best for explaining plant invasions. For example, in a biogeographically explicit experiment, Callaway *et al.* (2011) attempted to peer into the black box of soils from the native, 'expanded' (within the same continent), and non-native ranges of *Robinia pseudoacacia*. This species is native to the Appalachian and Ozark Mountains of North America but has been introduced throughout the rest of North America and much of Europe, where in can be invasive in places. Testing one group of soil mutualists, N-fixing rhizobia, Callaway *et al.* (2011) found that *R. pseudoacacia* formed nodules with a broad phylogenetic range of N-fixing bacteria in all soils, and leaf N did not differ among the different sources of soil. This suggests that the global expansion of *R. pseudoacacia* has probably not been limited by a lack of appropriate mutualistic N-fixers (also see Birnbaum and Leishman, 2013). In contrast, Callaway *et al.* (2011) found biogeographic patterns for another group of mutualists – arbuscular mycorrhizal fungi (AMF) from the native range stimulated stronger positive feedbacks than AMF from the expanded or invasive ranges. With respect to pathogens, *Pythium* taxa in soil in the native range were not more pathogenic than those from non-native ranges. But, when mutualist and antagonistic soil community members were considered together, feedbacks produced by the total soil biota were more negative from soils from the native range than from the other ranges. These results suggest that biogeographic differences in pathogen effects may be more important for invasiveness than those for mutualists, and that escape from the *net* negative effects of the whole soil community may be key to understanding some invasions and might not easily be distinguished in studies of isolated components of the soil biota.

3.6.2 Co-invasion with mutualists

Despite the importance of black box studies of whole soil communities in invasion, explicit focus on mutualists alone has provided important insight into non-native invasions. First, if a non-native species leaves obligate mutualists behind it will not naturalize, much less become invasive. Leaving behind obligate mutualists is a distinct possibility because non-native plant species are often translocated as seeds separate from their native soils (Kowarik and der Lippe, 2008). For soil mutualists, the best examples of co-invasion come from species in the family Pinaceae, which was formerly biogeographically restricted roughly to the northern hemisphere (Laurasia) before the Anthropocene, but now have been introduced throughout the southern hemisphere (Gondwana). As discussed in Dickie *et al.* (2010), early attempts to plant *Pinus* in the southern hemisphere were unsuccessful without their obligate ectomycorrhizal mutualists (reviewed in Pringle *et al.*, 2009). Now, plantations of Pinaceae species in the southern hemisphere are dominated by non-native ectomycorrhizal fungi (see references in Dickie *et al.*, 2010), and *Pinus* species commonly co-invade with two host-specific fungal partners: *Rhizopogon* and *Suillus* (Ashkannejhad and Horton, 2006; Collier and Bidartondo, 2009). Dickie *et al.* (2010) also tested whether co-invasion or acquisition of new ectomycorrhizal mutualists were most likely in the invasion of *P. contorta* in New Zealand. They compared ectomycorrhizal fungi associated with *P. contorta* with those fungi associated with the co-occurring native *Nothofagus solandri*. They found that *P. contorta* was colonized overwhelmingly by non-native fungi and globally distributed native fungi, whereas *N. solandri* was colonized much more by native fungi. They concluded that co-invasion of *P. contorta* and its fungal mutualists was the most likely way mutualisms were established in the non-native range (see Gundale *et al.*, 2014; Davis *et al.*, 2019 above). In Argentina, Nuñez *et al.* (2009) found lower mycorrhizal colonization on the roots of *Pseudotsuga menziesii*, *P. contorta* and *Pinus ponderosa* – North American Pinaceae introduced to Argentina – and fewer relevant fungal mutualists in the soil far from plantations of the Pinaceae than in soil near the plantations. The low abundance of these mutualists away from plantations appeared to limit invasion.

Importantly, such co-invasion with mycorrhizal fungi is not the rule. Moora *et al.* (2011) identified AMF on the palm *Trachycarpus fortunei* in native and non-native ranges and found that globally common fungal taxa were most over-represented in the non-native range. These geographically widespread and non-host-specific fungal species appeared to be newly acquired in the non-native range (also see Parker *et al.*, 2007; Swift *et al.* (2019)), but the mechanisms underlying these novel associations are unclear.

Co-invasion is suspected for other invasive plant—soil biota mutualisms. The degree of specialization among nitrogen-fixing rhizobia and leguminous plants in natural systems is still being disentangled, but there is evidence for both the promiscuous acquisition of any rhizobia by some legume species, but more tightly co-evolved legume—rhizobia symbioses for other species (see discussion in Rodríguez-Echeverría *et al.*, 2012). A good example of the latter is mono-dominance of non-native *Bradyrhizobia* in nodules of invasive populations of Australian *Acacia* species in Portugal (Rodríguez-Echeverría *et al.*, 2011; also see Rodríguez-Echeverría *et al.*, 2007; Wei *et al.*, 2009).

The Eurasian legume, *Medicago polymorpha*, and the rhizobial bacteria *Ensifer medicae*, co-invaded the New World roughly 250 years ago (Rome *et al.*, 1996; see discussion in Porter *et al.*, 2011). *Medicago polymorpha* is highly dependent on *E. medicae* for forming functional root nodules in both the native and invaded range (Silva *et al.*, 2007), and *E. medicae* is found primarily on *Medicago* species. The genus *Medicago* is not native to the New World, thus *E. medicae* is suspected to have co-invaded with *M. polymorpha*. Porter *et al.* (2011) found evidence that this rather specific mutualistic relationship showed signs of local adaption to different soil types in the non-native range. They reported that *M. polymorpha* collected from serpentine soils allocated more mass to symbiotic root nodules when grown in in serpentine soil than did *M. polymorpha* collected from non-serpentine soils, and vice versa. They argued that these co-invading mutualists had 'undergone [the] rapid evolution of divergence for soil-specific investment in the mutualism'. Such evolutionary divergence of tight invasive mutualists remains a very intriguing line of research, but regardless, a limited literature supports the notion that plant species with specialized mutualists require co-invasion to succeed.

There is even evidence that AMF, despite their mutualistic tendency, can produce indirect resistance to invasion. Waller *et al.* (2016) explored whether populations of *Centaurea solstitialis*, an annual species native to Eurasia but invasive in North America, from plant populations from both the native and non-native ranges, differed in the way they interacted with AMF in non-native-range soil and the North American soils cultivated by the native bunchgrass, *Stipa pulchra*. When alone, *C. solstitialis* from both ranges derived only weak benefits from AM fungi. But, AMF suppressed *C. solstitialis* when in competition with *S. pulchra*, and this effect was greater for genotypes from the native range of the invader. Thus AMF provided biotic resistance, but through its indirect effect on competition. Some invasive species may respond by exhibiting reduced dependency on mutualists in their non-native ranges (Seifert *et al.*, 2009; Bunn *et al.*, 2014).

3.6.3 Soil mutualists and enhanced performance of non-natives

For a number of reasons, soil 'mutualists' can either disproportionately improve or inhibit the performance of non-native invasive plant species (Levine *et al.*, 2004; Wolfe and Klironomos, 2005; Reinhart and Callaway, 2006; Pringle *et al.*, 2009); even within the same system over time (Birnbaum and Leishman, 2013). Variable effects of soil mutualists can be apparent through succession (Lau and Suwa, 2016) or potentially along abiotic gradients (e.g. Johnson *et al.*, 1997; Klironomos, 2003). In this context, it is important to note that mycorrhizae, generally referred to as mutualists here and elsewhere, can be highly parasitic depending on species combinations and abiotic factors (Johnson *et al.*, 1997; Klironomos, 2003). In a large comparison of fungal sources and plant species, Klironomos, 2003 found that a number of different mycorrhizal species stimulated large numbers of positive, neutral and negative growth responses in 64 different plant species, clearly showing that a simple classification of mycorrhizal fungi as mutualists is too simplistic (also see Johnson *et al.*, 1997).

Many studies show that soil microbes can improve the performance of non-native invasive plant species, often more than that of co-occurring natives (see 'Non-native vs native plant comparisons', above), but in this section we focus on studies that more explicitly target soil biota that generally function as mutualists. In his classic comparison of rare native species to non-native invasive species, Klironomos (2002) (noted above) found that isolated mycorrhizae generated positive feedbacks for four of five native species but only two of five for the invasive species. However, a recent review of the role of AMF in plant invasion (Aslani *et al.*, 2019) generally concluded that non-native and native plant species do not show striking differences in responses to or colonization by AMF (see also Lekberg *et al.*, 2013; Bunn *et al.*, 2015; Reinhart *et al.*, 2017). On the other hand, Pringle *et al.* (2009) reanalyzed data from Wilson and Hartnett (1998) for plant species in the Kansas prairie and found that non-native (but not necessarily invasive) species were much less responsive to, and perhaps more reliant on, mycorrhizae than native species. Regardless, many invasive species appear to benefit from mutualisms with AMF (see review by Shah *et al.*, 2009) as they do with ectomycorrhizal fungi (see 'Co-invasion with mutualists', above). Menzel *et al.* (2017) reported that facultatively mycorrhizal non-native plant species (but not necessarily invasive species) occupied larger distributional ranges in Germany that than non-mycorrhizal non-natives and obligate mycorrhizal non-native species. Callaway *et al.* (2011), see 'Mutualists and the black box of soil', above) found that AMF from the native range stimulated greater growth responses by *R. pseudoacacia* than AMF from non-native ranges around the world, but this cannot explain the tree's invasiveness. The team of de Souza *et al.* (2018) tested how AMF collected from the root zones of invasive and native plant species affected the development of invasive plant species. They found that native and invasive species cultivated different AMF taxa, that invasive species suppressed AMF diversity and that all AMF sources promoted the growth and phosphorus uptake of invasive species. However, AMF from the rhizospheres of the non-native species increased their growth more than AMF from the rhizospheres of native species. Zhang *et al.* (2010) reported that *Solidago canadensis*, which is native to North America but invasive in China, also promoted the abundance of AMF species that in turn promoted the growth of *S. canadensis* more than the AMF community cultivated by native species.

AMF may also have indirect benefits for invaders. Chen *et al.* (2019) studied the effects of two arbuscular mycorrhizal fungal species on *Wedelia trilobata*, a South American invader in China. They found that both fungal symbionts increased *W. trilobata* growth, a direct effect, but also reduced root infections by the pathogenic fungus *Rhizoctonia solani*, an indirect effect.

Other kinds of symbiotic biota in plants, apparently at least partially recruited from the soil, can also strongly promote the growth and the impact of non-native invaders. *Sorghum halepense* harbours endophytic bacteria in its rhizomes that solubilize phosphate, produce iron siderophores and produce the growth hormone indole acetic acid in vitro (Rout and Chrzanowski, 2009; Rout *et al.*, 2013). They found that inhibiting these bacteria with an antibiotic strongly reduced plant growth rate, biomass and rhizome production. In the field, *S. halepense* has substantial positive effects on biogeochemical cycles and accumulation of nitrogen, phosphate and iron. Although the microbes were not identified, Parepa *et al.* (2013) grew native plant species in different fractions of the biota in European soil and then invaded these mesocosms with *Fallopia x bohemica*, an invasive hybrid. They found that *Fallopia* benefited from the soil biota more than any of the native species, and this shifted competition in favour of *Fallopia*. Experimental additions of activated carbon eliminated these growth and competitive advantages, indicating that the mechanisms for the positive soil biota effects were chemically mediated. In China, Sun and He (2010) found that soil microbes from the rhizospheres of invasive *S. canadensis* and native *Stipa bungaena* both elicited facilitative effects on the emergence of *S. canadensis* but not *S. bungaena*. Similarly, these soil biota increased the competitive advantage of *S. canadensis* over *S. bungaena*.

Importantly, the positive effects of soil mutualists on non-native invasive plants can change over time. For example, Lau and Suwa (2016) found that soils from areas invaded by *Vicia villosa* increased growth of the leguminous invader more than biota from uninvaded sites. These effects were also stronger for soil biota from recently invaded sites than soil from sites with older invasions.

3.6.4 Suppression of mutualists of native species

In a thorough review of invasion-mediated disruptions of native plant-AMF mutualisms, Grove *et al.* (2017) reported that 'invaders can disrupt native mutualisms if they are non-mycorrhizal, associate with a different type of mycorrhiza, or associate with different taxa of the same type of mycorrhiza as the native plants'. They also noted that invaders that increase soil nutrients (see Liao *et al.*, 2008; McLeod *et al.*, 2016) can disrupt native plant-AMF mutualisms. In an example of the effects of non-mycorrhizal plants, the European invader of North American forests, *Alliaria petiolata*, suppresses the growth of North American species by disrupting their relationships with AMF (Stinson *et al.*, 2006; Cantor *et al.*, 2011). However, as noted above, if these effects are to explain the invasion of *A. petiolata*, then such effects must be stronger in North America than in Europe. Subsequent experiments found far stronger inhibitory effects of *A. petiolata* on mycorrhizas in North American soils than on mycorrhizas in European soils, a biogeographical effect that appeared to be due to specific flavonoid fractions in *A. petiolata* extracts (Callaway *et al.*, 2008). The suppression of North American mycorrhizal fungi corresponded with strong inhibition of North American plant species that rely on these fungi, whereas congeneric European plants were weakly affected. These results indicate that phytochemicals that did not disturb mycorrhizal symbionts in the native range acted as 'novel weapons' (Callaway and Aschehoug, 2000; Barto *et al.*, 2010; Uesugi and Kessler, 2013) that were harmful to naïve native mutualists in the non-native range. Barto *et al.* (2011) explored the potential of a novel weapons scenario as a mechanism driving plant invasion and found reduced AMF colonization in patches of forest invaded by *A. petiolata*, but no effects on AMF richness. Lankau *et al.* (2009) and Lankau (2011a,b) explored these effects in evolutionary detail and found evidence that that the impact of *A. petiolata* on AMF varied among individuals depending on allelochemical concentrations. They then put this in an ecological/evolutionary context and showed a decline in the phytotoxin production of *A. petiolata* and its indirect, soil biota-based, impact on native species aver a 50-year chronosequence, an apparent response to natural selection for decreased phytotoxin production. This evolutionary context corresponded with results showing that *A. petiolata* from sites with high densities of native competitors produced more of the toxic compounds than *A. petiolata* from sites with low densities of natives (Lankau, 2012). Furthermore, when he collected a native competitor from sites with more toxic ecotypes of *A. petiolata*, they were more tolerant to the invader, 'suggesting coevolutionary dynamics between the species', much of which may have been determined by plant-AMF relationships and disruptions of those relationships. We know of very few such studies that link non-native invasions to mutualisms, and then to rapid evolution of these relationships, but invaded systems may provide other opportunities to do so.

In a clear example of the effects of invaders *with* AMF mutualisms on native species that also form AMF relationships, Vogelsang and Bever (2009) measured AMF-mediated competition between the native *Gnaphalium californicum* and the invasive *Carduus pycnocephalus* in southern California. They found that *C. pycnocephalus* outcompeted *G. californicum* by disrupting the latter's interactions with AMF. Similarly, in Europe, Zubek *et al.* (2016) found that three invasive species, one of which was non-mycorrhizal, disrupted mutualisms between AMF and native plant species, and that nutrient uptake by natives was reduced when grown in soil cultivated by the invaders. And in China, Zhang *et al.*, 2010Yuan *et al.* (2014) reported evidence that secondary metabolites released by *S. canadensis*, native to North America, increase the competitive effect of *S. canadensis* on the native *Kummerowia striata* by enhancing AMF mutualisms favourable to the invader.

In summary, there is substantial, albeit inconsistent, evidence that soil mutualisms can influence the success of non-native invasive plant species. Specifically, mutualisms have important implications for plant invasions i) when mutualists co-invade with non-native plant species that require obligate specialist mutualists, ii) when mutualists enhance the performance of non-native species in their non-native range, and iii) when biotic interactions in invaded communities suppress the soil mutualists of native plant species. Importantly, examining the role of soil mutualists in the context of non-native plant invasion is relatively difficult because the effects of soil 'mutualists' are exceptionally variable with respect to time, space, species (both plant and mutualist), environmental context and even experimental design. Understanding the drivers of such context dependence is an important objective for future research in both basic and applied frameworks.

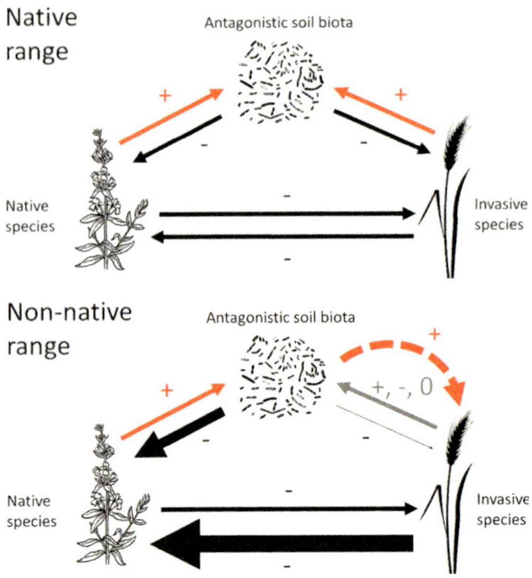

Fig. 3.1. Plant–soil interactions can influence the trajectory of biological invasions by non-native plant species. For example, invasive species can experience less inhibition from antagonistic soil biota in the non-native range relative to the native range (i.e. enemy release). In addition, native species in the non-native range of invaders can experience greater inhibition from antagonistic soil biota than invasive competitors. Both scenarios promote the competitive dominance of invaders in the non-native range, and the latter results in a positive indirect effect of antagonistic soil biota on invasive species. Importantly, the effects of invasive species on soil biota in non-native ranges are idiosyncratic and can range from positive to negative. Arrows and +/– signs indicate the direction and sign of interactions; solid arrows indicate direct interactions; dashed arrows indicate indirect interactions; arrow width indicates interaction strength. Positive interactions are coloured red, variable interactions are coloured grey and negative interactions are coloured black, for emphasis.

3.7 Conclusions

Though not comprehensive, we have reviewed studies from invaded ecosystems globally that have considered interactions between non-native invasive plant species and soil biota, with an emphasis on soil microbes. This rich literature strongly indicates that the trajectory of plant invasions – for better or for worse – can be tied to interactions between plants and the soil community (Fig. 3.1). Specifically, we have highlighted five ways in which belowground interactions can influence plant invasions. First, non-native invasive plant species in their non-native ranges can interact very differently with the resident soil community than do native species and non-native non-invasive species. Disproportionately escaping the effects of negative belowground interactions relative to native plant species and non-invasive non-natives may help explain the strong community-level impacts of non-native invasive plant species and may help explain why some non-native species become explosive invaders while others do not. Second, non-native invasive plant species can interact very differently with the soil community in their native and non-native ranges. Biogeographic comparisons of the effects of belowground natural enemies in the native and non-native ranges of non-native invaders have found strong, albeit inconsistent, evidence for enemy release from antagonistic soil biota, which lends critical mechanistic insight into the invasion process. Third, non-native invasive plant species can cultivate a soil community that disproportionately harms native competitors in invaded communities. This accumulation of local pathogens is a mechanism whereby non-native invaders can displace native competitors in recipient communities. Fourth, antagonistic soil biota can reduce the performance of non-native plant species, resulting in meaningful biotic resistance against invasion. Such biotic resistance is a mechanism of invasion failure/attenuation and should thus be treated as an important ecosystem service worthy of conservation. Fifth, besides or in addition to antagonistic interactions with soil biota, soil mutualisms can promote the success of invasive plant species i) when mutualists co-invade with non-native plant species that require obligate specialist mutualists, ii) when mutualists strongly enhance the performance of non-native plant species in their non-native range, and iii) when biotic interactions in the invaded community suppress the soil mutualists of native plant species.

In light of this large and growing literature, we conclude that management practices aimed at manipulating plant—soil biota interactions have considerable potential to help control plant invasions. However, the successful manipulation of soil communities to manage plant invasions has been hampered by the relative difficulty of experimentally studying the effects of below-ground biota from both holistic (i.e. 'black box') and reductionist perspectives. A great deal of further work is needed to understand the spatial, temporal, taxonomic, biogeographic and experimental drivers of context dependence in plant—soil biota interactions.

References

Agrawal, A.A., Kotanen, P.M., Mitchell, C.E., Power, A.G., Godsoe, W. *et al.* (2005) Enemy release? an experiment with congeneric plant pairs and diverse above- and belowground enemies. *Ecology* 86(11), 2979–2989. DOI: 10.1890/05-0219.

Andonian, K., Hierro, J.L., Khetsuriani, L., Becerra, P., Janoyan, G. *et al.* (2011) Range-expanding populations of a globally introduced weed experience negative plant-soil feedbacks. *PLoS ONE* 6(5), e20117. DOI: 10.1371/journal.pone.0020117.

Andonian, K., Hierro, J.L., Khetsuriani, L., Becerra, P.I., Janoyan, G. *et al.* (2012) Geographic mosaics of plant-soil microbe interactions in a global plant invasion. *Journal of Biogeography* 39(3), 600–608. DOI: 10.1111/j.1365-2699.2011.02629.x.

Ashkannejhad, S. and Horton, T.R. (2006) Ectomycorrhizal ecology of Pinus contorta on the Oregon dunes. *The New Phytologist* 169, 331–339.

Aslani, F., Juraimi, A.S., Ahmad-Hamdani, M.S., Alam, M.A., Hasan, M.M. *et al.* (2019) The role of arbuscular mycorrhizal fungi in plant invasion trajectory. *Plant and Soil* 441(1–2), 1–14. DOI: 10.1007/s11104-019-04127-5.

Baas Becking, L.G.M. (1934) *Geobiologie of inleiding tot de milieukunde Diligentia Wetensch.* van Stockum's Gravenhange.

Barto, E.K., Antunes, P.M., Stinson, K., Koch, A.M., Klironomos, J.N. *et al.* (2011) Differences in arbuscular mycorrhizal fungal communities associated with sugar maple seedlings in and outside of invaded garlic mustard forest patches. *Biological Invasions* 13(12), 2755–2762. DOI: 10.1007/s10530-011-9945-6.

Barto, E.K., Powell, J.R. and Cipollini, D. (2010) How novel are the chemical weapons of garlic mustard in North American forest understories? *Biological Invasions* 12(10), 3465–3471. DOI: 10.1007/s10530-010-9744-5.

Beckstead, J., Meyer, S.E., Connolly, B.M., Huck, M.B. and Street, L.E. (2010) Cheatgrass facilitates spillover of a seed bank pathogen onto native grass species. *Journal of Ecology* 98(1), 168–177. DOI: 10.1111/j.1365-2745.2009.01599.x.

Beckstead, J. and Parker, I.M. (2003) Invasiveness of *AMMOPHILA arenaria*: release from soil-borne pathogens? *Ecology* 84(11), 2824–2831. DOI: 10.1890/02-0517.

Bernard-Verdier, M. and Hulme, P.E. (2019) Alien plants can be associated with a decrease in local and regional native richness even when at low abundance. *Journal of Ecology* 107(3), 1343–1354. DOI: 10.1111/1365-2745.13124.

Bever, J.D. (1994) Feedback between plants and their soil communities in an old field community. *Ecology* 75(7), 1965–1977. DOI: 10.2307/1941601.

Bever, J.D. (2002) Negative feedback within a mutualism: host-specific growth of mycorrhizal fungi reduces plant benefit. *Proceedings of the Royal Society of London. Series B: Biological Sciences* 269(1509), 2595–2601. DOI: 10.1098/rspb.2002.2162.

Bever, J.D., Westover, K.M. and Antonovics, J. (1997) Incorporating the soil community into plant population dynamics: the utility of the feedback approach. *The Journal of Ecology* 85(5), 561–573. DOI: 10.2307/2960528.

Birnbaum, C. and Leishman, M.R. (2013) Plant-soil feedbacks do not explain invasion success of *Acacia* species in introduced range populations in Australia. *Biological Invasions* 15(12), 2609–2625. DOI: 10.1007/s10530-013-0478-z.

Bunn, R.A., Lekberg, Y., Gallagher, C., Rosendahl, S. and Ramsey, P.W. (2014) Grassland invaders and their mycorrhizal symbionts: a study across climate and invasion gradients. *Ecology and Evolution* 4(6), 794–805. DOI: 10.1002/ece3.917.

Bunn, R.A., Ramsey, P.W. and Lekberg, Y. (2015) Do native and invasive plants differ in their interactions with arbuscular mycorrhizal fungi? A meta-analysis. *Journal of Ecology* 103(6), 1547–1556.

Byun, C., de Blois, S. and Brisson, J. (2013) Plant functional group identity and diversity determine biotic resistance to invasion by an exotic grass. *Journal of Ecology* 101(1), 128–139. DOI: 10.1111/1365-2745.12016.

Byun, C., de Blois, S. and Brisson, J. (2015) Interactions between abiotic constraint, propagule pressure, and biotic resistance regulate plant invasion. *Oecologia* 178(1), 285–296. DOI: 10.1007/s00442-014-3188-z.

Callaway, R.M. and Aschehoug, E.T. (2000) Invasive plants versus their new and old neighbors: a mechanism for exotic invasion. *Science* 290(5491), 521–523. DOI: 10.1126/science.290.5491.521.

Callaway, R.M., Bedmar, E.J., Reinhart, K.O., Silvan, C.G. and Klironomos, J. (2011) Effects of soil biota from different ranges on *Robinia* invasion: acquiring mutualists and escaping pathogens. *Ecology* 92(5), 1027–1035. DOI: 10.1890/10-0089.1.

Callaway, R.M., Cipollini, D., Barto, K., Thelen, G.C., Hallett, S.G. *et al.* (2008) Novel weapons: invasive plant suppresses fungal mutualists in America but not in its native Europe. *Ecology* 89(4), 1043–1055. DOI: 10.1890/07-0370.1.

Callaway, R.M., Montesinos, D., Williams, K. and Maron, J.L. (2013) Native congeners provide biotic resistance to invasive *Potentilla* through soil biota. *Ecology* 94(6), 1223–1229. DOI: 10.1890/12-1875.1.

Callaway, R.M., Thelen, G.C., Rodrigues, A. and Holben, W.E. (2004a) Release from inhibitory soil biota in Europe may promote exotic plant invasion in North America. *Nature* 427, 731–733.

Callaway, R.M., Thelen, G.C., Barth, S., Ramsey, P.W. and Gannon, J.E. (2004b) Soil fungi alter interactions between the invader *Centaurea maculosa* and North American natives. *Ecology* 85(4), 1062–1071. DOI: 10.1890/02-0775.

Cantor, A., Hale, A., Aaron, J., Traw, M.B. and Kalisz, S. (2011) Low allelochemical concentrations detected in garlic mustard-invaded forest soils inhibit fungal growth and AMF spore germination. *Biological Invasions* 13(12), 3015–3025. DOI: 10.1007/s10530-011-9986-x.

Catford, J.A., Jansson, R. and Nilsson, C. (2009) Reducing redundancy in invasion ecology by integrating hypotheses into a single theoretical framework. *Diversity and Distributions* 15(1), 22–40. DOI: 10.1111/j.1472-4642.2008.00521.x.

Chen, H., Chang, R.Y., Wang, R.Q. and Liu, J. (2012) The biotic resistance of soil biota on the invasive weed *Coreopsis grandiflora* varies with irradiance levels. *Advanced Materials Research* 573-574, 1044–1051. DOI: 10.4028/www.scientific.net/AMR.573-574.1044.

Chen, Q., Wu, W.-W., Qi, S.-S., Cheng, H., Li, Q. *et al.* (2019) Arbuscular mycorrhizal fungi improve the growth and disease resistance of the invasive plant *Wedelia trilobata*. *Journal of Applied Microbiology* 28. DOI: 10.1111/jam.14415.

Collier, F.A. and Bidartondo, M.I. (2009) Waiting for fungi: the ectomycorrhizal invasion of lowland heathlands. *Journal of Ecology* 97(5), 950–963. DOI: 10.1111/j.1365-2745.2009.01544.x.

Cortois, R. and De Deyn, G.B. (2012) The curse of the black box. *Plant Soil* 350(1-2), 27–33. DOI: 10.1007/s11104-011-0963-z.

Crawford, K.M. and Knight, T.M. (2017) Competition overwhelms the positive plant–soil feedback generated by an invasive plant. *Oecologia* 183(1), 211–220. DOI: 10.1007/s00442-016-3759-2.

Darwin, C. (1859) *On the origin of species by means of natural selection, or preservation of favoured races in the struggle for life*. John Murray, London.

Davis, K.T., Callaway, R.M., Fajardo, A., Pauchard, A., Nuñez, M.A. *et al*. (2019) Severity of impacts of an introduced species corresponds with regional eco-evolutionary experience. *Ecography* 42(1), 12–22. DOI: 10.1111/ecog.04014.

Dawson, W. and Schrama, M. (2016) Identifying the role of soil microbes in plant invasions. *Journal of Ecology* 104(5), 1211–1218. DOI: 10.1111/1365-2745.12619.

de la Peña, E., de Clercq, N., Bonte, D., Roiloa, S., Rodríguez-Echeverría, S. *et al*. (2010) Plant-soil feedback as a mechanism of invasion by *Carpobrotus edulis*. *Biological Invasions* 12(10), 3637–3648. DOI: 10.1007/s10530-010-9756-1.

de Souza, T.A.F., de Andrade, L.A., Freitas, H. and da Silva Sandim, A. (2018) Biological invasion influences the outcome of plant-soil feedback in the invasive plant species from the Brazilian semi-arid. *Microbial Ecology* 76(1), 102–112. DOI: 10.1007/s00248-017-0999-6.

DeWalt, S.J., Denslow, J.S. and Ickes, K. (2004) Natural-enemy release facilitates habitat expansion of the invasive tropical shrub *Clidemia hirta*. *Ecology* 85(2), 471–483. DOI: 10.1890/02-0728.

Dickie, I.A., Bolstridge, N., Cooper, J.A. and Peltzer, D.A. (2010) Co-invasion by *Pinus* and its mycorrhizal fungi. *New Phytologist* 187(2), 475–484. DOI: 10.1111/j.1469-8137.2010.03277.x.

Dostál, P., Müllerová, J., Pyšek, P., Pergl, J. and Klinerová, T. (2013) The impact of an invasive plant changes over time. *Ecology Letters* 16(10), 1277–1284. DOI: 10.1111/ele.12166.

Eppinga, M.B., Rietkerk, M., Dekker, S.C., De Ruiter, P.C., Van der Putten, W.H. *et al*. (2006) Accumulation of local pathogens: a new hypothesis to explain exotic plant invasions. *Oikos* 114(1), 168–176. DOI: 10.1111/j.2006.0030-1299.14625.x.

Flory, S.L. and Clay, K. (2013) Pathogen accumulation and long-term dynamics of plant invasions. *Journal of Ecology* 101(3), 607–613. DOI: 10.1111/1365-2745.12078.

Flory, S.L., Kleczewski, N. and Clay, K. (2011) Ecological consequences of pathogen accumulation on an invasive grass. *Ecosphere* 2(10), art120. DOI: 10.1890/ES11-00191.1.

Griffith, A.B. (2010) Positive effects of native shrubs on *Bromus tectorum* demography. *Ecology* 91(1), 141–154. DOI: 10.1890/08-1446.1.

Grove, S., Haubensak, K.A., Gehring, C. and Parker, I.M. (2017) Mycorrhizae, invasions, and the temporal dynamics of mutualism disruption. *Journal of Ecology* 105(6), 1496–1508. DOI: 10.1111/1365-2745.12853.

Gundale, M.J., Kardol, P., Nilsson, M.-C., Nilsson, U., Lucas, R.W. *et al*. (2014) Interactions with soil biota shift from negative to positive when a tree species is moved outside its native range. *New Phytologist* 202(2), 415–421. DOI: 10.1111/nph.12699.

Hierro, J.L., Maron, J.L. and Callaway, R.M. (2005) A biogeographical approach to plant invasions: the importance of studying exotics in their introduced and native range. *Journal of Ecology* 93(1), 5–15. DOI: 10.1111/j.0022-0477.2004.00953.x.

Hierro, J.L., Villarreal, D., Eren, O., Graham, J.M. and Callaway, R.M. (2006) Disturbance facilitates invasion: the effects are stronger abroad than at home. *The American Naturalist* 168(2), 144–156. DOI: 10.1086/505767.

Hollebone, A.L. and Hay, M.E. (2007) Propagule pressure of an invasive crab overwhelms native biotic resistance. *Marine Ecology Progress Series* 342, 191–196. DOI: 10.3354/meps342191.

Inderjit and Cahill, J.F. (2015) Linkages of plant-soil feedbacks and underlying invasion mechanisms. *AoB PLANTS* 7, plv022. DOI: 10.1093/aobpla/plv022.

Inderjit and van der Putten, W.H. (2010) Impacts of soil microbial communities on exotic plant invasions. *Trends in Ecology & Evolution* 25(9), 512–519. DOI: 10.1016/j.tree.2010.06.006.

Janzen, D.H. (1970) Herbivores and the number of tree species in tropical forests. *The American Naturalist* 104(940), 501–528. DOI: 10.1086/282687.

Johnson, N.C., Graham, J.-H. and Smith, F.A. (1997) Functioning of mycorrhizal associations along the mutualism-parasitism continuum. *New Phytologist* 135(4), 575–585. DOI: 10.1046/j.1469-8137.1997.00729.x.

Joshi, J. and Vrieling, K. (2005) The enemy release and EICA hypothesis revisited: incorporating the fundamental difference between specialist and generalist herbivores. *Ecology Letters* 8(7), 704–714. DOI: 10.1111/j.1461-0248.2005.00769.x.

Keane, R.M. and Crawley, M.J. (2002) Exotic plant invasions and the ERH. *Trends in Ecology & Evolution* 17, 164–170.

Klironomos, J.N. (2002) Feedback with soil biota contributes to plant rarity and invasiveness in communities. *Nature* 417(6884), 67–70. DOI: 10.1038/417067a.

Klironomos, J.N. (2003) Variation in plant response to native and exotic arbuscular mycorrhizal fungi. *Ecology* 84(9), 2292–2301. DOI: 10.1890/02-0413.

Knight, T.M., Dunn, J.L., Smith, L.A., Davis, J. and Kalisz, S. (2009) Deer facilitate invasive plant success in a Pennsylvania forest understory. *Natural Areas Journal* 29(2), 110–116. DOI: 10.3375/043.029.0202.

Kowarik, I. and der Lippe, M. (2008) Pathways in plant invasions. In: Nentwig, W. (ed.) *Biological Invasions. Ecological Studies (Analysis and Synthesis)*. 193. Springer, Berlin, Heidelberg.

Kulmatiski, A. (2017) Community-level plant–soil feedbacks explain landscape distribution of native and non-native plants. *Ecology and Evolution*.

Kulmatiski, A., Beard, K.H., Stevens, J.R. and Cobbold, S.M. (2008) Plant-soil feedbacks: a meta-analytical review. *Ecology Letters* 11(9), 980–992. DOI: 10.1111/j.1461-0248.2008.01209.x.

Lankau, R.A. (2011a) Resistance and recovery of soil microbial communities in the face of *Alliaria petiolata* invasions. *New Phytologist* 189(2), 536–548. DOI: 10.1111/j.1469-8137.2010.03481.x.

Lankau, R.A. (2011b) Intraspecific variation in allelochemistry determines an invasive species' impact on soil microbial communities. *Oecologia* 165(2), 453–463. DOI: 10.1007/s00442-010-1736-8.

Lankau, R.A. (2012) Coevolution between invasive and native plants driven by chemical competition and soil biota. *Proceedings of the National Academy of Sciences* 109(28), 11240–11245. DOI: 10.1073/pnas.1201343109.

Lankau, R.A., Nuzzo, V., Spyreas, G. and Davis, A.S. (2009) Evolutionary limits ameliorate the negative impact of an invasive plant. *Proceedings of the National Academy of Sciences* 106(36), 15362–15367. DOI: 10.1073/pnas.0905446106.

Lau, J.A. and Suwa, T. (2016) The changing nature of plant–microbe interactions during a biological invasion. *Biological Invasions* 18(12), 3527–3534. DOI: 10.1007/s10530-016-1245-8.

Lekberg, Y., Bever, J.D., Bunn, R.A., Callaway, R.M., Hart, M.M. *et al.* (2018) Relative importance of competition and plant-soil feedback, their synergy, context dependency and implications for coexistence. *Ecology Letters* 21(8), 1268–1281. DOI: 10.1111/ele.13093.

Lekberg, Y., Gibbons, S.M., Rosendahl, S. and Ramsey, P.W. (2013) Severe plant invasions can increase mycorrhizal fungal abundance and diversity. *The ISME Journal* 7(7), 1424–1433. DOI: 10.1038/ismej.2013.41.

Levine, J.M., Adler, P.B. and Yelenik, S.G. (2004) A meta-analysis of biotic resistance to exotic plant invasions. *Ecology Letters* 7(10), 975–989. DOI: 10.1111/j.1461-0248.2004.00657.x.

Li, H.-N., Xiao, B., Liu, W.-X. and Wan, F.-H. (2014) Changes in soil biota resulting from growth of the invasive weed, *Ambrosia artemisiifolia* L. (Compositae), enhance its success and reduce growth of co-occurring plants. *Journal of Integrative Agriculture* 13(9), 1962–1971. DOI: 10.1016/S2095-3119(13)60569-9.

Liao, C., Peng, R., Luo, Y., Zhou, X., Wu, X. *et al.* (2008) Altered ecosystem carbon and nitrogen cycles by plant invasion: a meta-analysis. *New Phytologist* 177(3), 706–714. DOI: 10.1111/j.1469-8137.2007.02290.x.

Lucero, J.E. and Callaway, R.M. (2018) Native granivores reduce the establishment of native grasses but not invasive *Bromus tectorum*. *Biological Invasions* 20(12), 3491–3497. DOI: 10.1007/s10530-018-1789-x.

Lucero, J.E., Schaffner, U., Asadi, G., Bagheri, A., Rajabov, T. *et al.* (2019) Enemy release from the effects of generalist granivores can facilitate *Bromus tectorum* invasion in the Great Basin Desert. *Ecology and Evolution* 9(15), 8490–8499. DOI: 10.1002/ece3.5314.

Ma, C., Li, S.-peng., Pu, Z., Tan, J., Liu, M. *et al.* (2016) Different effects of invader–native phylogenetic relatedness on invasion success and impact: a meta-analysis of Darwin's naturalization hypothesis. *Proceedings of the Royal Society B: Biological Sciences* 283(1838), 20160663. DOI: 10.1098/rspb.2016.0663.

Mangla, S.I. and Callaway, R.M. (2008) Exotic invasive plant accumulates native soil pathogens which inhibit native plants. *The Journal of Ecology* 96, 58–67.

Marler, M.J., Zabinski, C.A. and Callaway, R.M. (1999) Mycorrhizae indirectly enhance competitive effects of an invasive forb on a native bunchgrass. *Ecology* 80(4), 1180–1186. DOI: 10.1890/0012-9658(1999)080[1180:MIECEO]2.0.CO;2.

Maron, J.L., Klironomos, J., Waller, L. and Callaway, R.M. (2014) Regional scale sampling reveals that invasive plants escape from suppressive soil biota. *The Journal of Ecology* 102, 19–27.

Maron, J.L., Luo, W., Callaway, R.M. and Pal, R.W. (2015) Do exotic plants lose resistance to pathogenic soil biota from their native range? A test with *Solidago gigantea*. *Oecologia* 179(2), 447–454. DOI: 10.1007/s00442-015-3350-2.

Maron, J.L., Laney Smith, A., Ortega, Y.K., Pearson, D.E. and Callaway, R.M. (2016) Negative plant-soil feedbacks increase with plant abundance, and are unchanged by competition. *Ecology* 97(8), 2055–2063. DOI: 10.1002/ecy.1431.

Maron, J.L. and Vilá, M. (2001) When do herbivores affect plant invasion? Evidence for the natural enemies and biotic resistance hypotheses. *Oikos* 95(3), 361–373. DOI: 10.1034/j.1600-0706.2001.950301.x.

McLeod, M.L., Cleveland, C.C., Lekberg, Y., Maron, J.L., Philippot, L. *et al.* (2016) Exotic invasive plants increase productivity, abundance of ammonia-oxidizing bacteria and nitrogen availability in intermountain grasslands. *Journal of Ecology* 104(4), 994–1002. DOI: 10.1111/1365-2745.12584.

Menzel, A., Hempel, S., Klotz, S., Moora, M., Pyšek, P. *et al.* (2017) Mycorrhizal status helps explain invasion success of alien plant species. *Ecology* 98(1), 92–102. DOI: 10.1002/ecy.1621.

Mitchell, C.E., Agrawal, A.A., Bever, J.D., Gilbert, G.S., Hufbauer, R.A. *et al.* (2006) Biotic interactions and plant invasions. *Ecology Letters* 9(6), 726–740. DOI: 10.1111/j.1461-0248.2006.00908.x.

Moora, M., Berger, S., Davison, J., Öpik, M., Bommarco, R. *et al.* (2011) Alien plants associate with widespread generalist arbuscular mycorrhizal fungal taxa: evidence from a continental-scale study using massively parallel 454 sequencing. *Journal of Biogeography* 38(7), 1305–1317. DOI: 10.1111/j.1365-2699.2011.02478.x.

Nijjer, S., Rogers, W.E. and Siemann, E. (2007) Negative plant–soil feedbacks may limit persistence of an invasive tree due to rapid accumulation of soil pathogens. *Proceedings Royal Society B* 274, 2621–2627.

Nuñez, M.A., Horton, T.R. and Simberloff, D. (2009) Lack of belowground mutualisms hinders Pinaceae invasions. *Ecology* 90(9), 2352–2359. DOI: 10.1890/08-2139.1.

Ortega, Y.K. and Pearson, D.E. (2005) Weak vs. strong invaders of natural plant communities: assessing invasibility and impact. *Ecological Applications* 15(2), 651–661. DOI: 10.1890/04-0119.

Packer, A. and Clay, K. (2000) Soil pathogens and spatial patterns of seedling mortality in a temperate tree. *Nature* 404(6775), 278–281. DOI: 10.1038/35005072.

Packer, A. and Clay, K. (2003) Soil pathogens and *Prunus* SEROTINA seedling and SAPLING growth near conspecific trees. *Ecology* 84(1), 108–119. DOI: 10.1890/0012-9658(2003)084[0108:SPAPSS]2.0.CO;2.

Parepa, M., Schaffner, U. and Bossdorf, O. (2013) Help from under ground: soil biota facilitate knotweed invasion. *Ecosphere* 4(2), art31. DOI: 10.1890/ES13-00011.1.

Parker, J.D., Burkepile, D.E. and Hay, M.E. (2006) Opposing effects of native and exotic herbivores on plant invasions. *Science* 311(5766), 1459–1461. DOI: 10.1126/science.1121407.

Parker, M.A., Wurtz, A.K. and Paynter, Q. (2007) Nodule symbiosis of invasive *Mimosa pigra* in Australia and in ancestral habitats: a comparative analysis. *Biological Invasions* 9(2), 127–138. DOI: 10.1007/s10530-006-0009-2.

Pearse, I.S., Sofaer, H.R., Zaya, D.N. and Spyreas, G. (2019) Non-Native plants have greater impacts because of differing per-capita effects and nonlinear abundance–impact curves. *Ecology Letters* 22, 1214–1220.

Pearson, D.E., Callaway, R.M. and Maron, J.L. (2011) Biotic resistance via granivory: establishment by invasive, naturalized, and native asters reflects generalist preference. *Ecology* 92(9), 1748–1757. DOI: 10.1890/11-0164.1.

Pearson, D.E., Eren, Özkan., Ortega, Y.K., Villarreal, D., Şentürk, M. *et al.* (2017) Are exotic plants more abundant in the introduced versus native range? *Journal of Ecology* 106(2), 727–736. DOI: 10.1111/1365-2745.12881.

Pimentel, D., Zuniga, R. and Morrison, D. (2005) Update on the environmental and economic costs associated with alien-invasive species in the United States. *Ecological Economics* 52(3), 273–288. DOI: 10.1016/j.ecolecon.2004.10.002.

Porter, S.S., Stanton, M.L. and Rice, K.J. (2011) Mutualism and adaptive divergence: co-invasion of a heterogeneous grassland by an exotic legume-rhizobium symbiosis. *PLoS ONE* 6(12), e27935. DOI: 10.1371/journal.pone.0027935.

Power, A. G. and Mitchell, C. E. (2004) Pathogen spillover in disease epidemics. *The American Naturalist* 164(S5), S79–S89. DOI: 10.1086/424610.

Pringle, A., Bever, J.D., Gardes, M., Parrent, J.L., Rillig, M.C. *et al.* (2009) Mycorrhizal symbioses and plant invasions. *Annual Review of Ecology, Evolution, and Systematics* 40(1), 699–715. DOI: 10.1146/annurev.ecolsys.39.110707.173454.
Reinhart, K.O. (2012) The organization of plant communities: negative plant–soil feedbacks and semiarid grasslands. *Ecology* 93(11), 2377–2385. DOI: 10.1890/12-0486.1.
Reinhart, K.O. and Callaway, R.M. (2004) Soil biota facilitate exotic *Acer* invasions in Europe and North America. *Ecological Applications* 14(6), 1737–1745. DOI: 10.1890/03-5204.
Reinhart, K.O. and Callaway, R.M. (2006) Soil biota and invasive plants. *New Phytologist* 170(3), 445–457. DOI: 10.1111/j.1469-8137.2006.01715.x.
Reinhart, K.O., Lekberg, Y., Klironomos, J. and Maherali, H. (2017) Does responsiveness to arbuscular mycorrhizal fungi depend on plant invasive status? *Ecology and Evolution* 7(16), 6482–6492. DOI: 10.1002/ece3.3226.
Reinhart, K.O., Packer, A., Van der Putten, W.H. and Clay, K. (2003) Plant-soil biota interactions and spatial distribution of black cherry in its native and invasive ranges. *Ecology Letters* 6(12), 1046–1050. DOI: 10.1046/j.1461-0248.2003.00539.x.
Reinhart, K.O., Tytgat, T., Van der Putten, W.H. and Clay, K. (2010) Virulence of soil-borne pathogens and invasion by *Prunus serotina*. *New Phytologist* 186(2), 484–495. DOI: 10.1111/j.1469-8137.2009.03159.x.
Richardson, D.M., Hui, C., Nuñez, M.A. and Pauchard, A. (2014) Tree invasions: patterns, processes, challenges and opportunities. *Biological Invasions* 16(3), 473–481. DOI: 10.1007/s10530-013-0606-9.
Rodríguez-Echeverría, S., Afonso, C., Correia, M., Lorenzo, P. and Roiloa, S.R. (2013) The effect of soil legacy on competition and invasion by *Acacia dealbata* Link. *Plant Ecology* 214(9), 1139–1146. DOI: 10.1007/s11258-013-0238-2.
Rodríguez-Echeverría, S., Crisóstomo, J.A. and Freitas, H. (2007) Genetic diversity of rhizobia associated with *Acacia longifolia* in two stages of invasion of coastal sand dunes. *Applied and Environmental Microbiology* 73(15), 5066–5070. DOI: 10.1128/AEM.00613-07.
Rodríguez-Echeverría, S., Fajardo, S., Ruiz-Díez, B. and Fernández-Pascual, M. (2012) Differential effectiveness of novel and old legume-rhizobia mutualisms: implications for invasion by exotic legumes. *Oecologia* 170(1), 253–261. DOI: 10.1007/s00442-012-2299-7.
Rodríguez-Echeverría, S., Le Roux, J.J., Crisóstomo, J.A. and Ndlovu, J. (2011) Jack-of-all-trades and master of many? How does associated rhizobial diversity influence the colonization success of Australian *Acacia* species? *Diversity and Distributions* 17(5), 946–957. DOI: 10.1111/j.1472-4642.2011.00787.x.
Rome, S., Fernandez, M.P., Brunel, B., Normand, P. and Cleyet-Marel, J.C. (1996) *Sinorhizobium medicae* sp. nov., isolated from annual *Medicago* spp. *International Journal of Systematic Bacteriology* 46(4), 972–980. DOI: 10.1099/00207713-46-4-972.
Rout, M.E. and Callaway, R.M. (2012) Interactions between exotic invasive plants and soil microbes in the rhizosphere suggest that 'everything is not everywhere'. *Annals of Botany* 110(2), 213–222. DOI: 10.1093/aob/mcs061.
Rout, M.E. and Chrzanowski, T.H. (2009) The invasive *Sorghum halepense* harbors endophytic N_2-fixing bacteria and alters soil biogeochemistry. *Plant and Soil* 315(1-2), 163–172. DOI: 10.1007/s11104-008-9740-z.
Rout, M.E., Chrzanowski, T.H., Westlie, T.K., DeLuca, T.H., Callaway, R.M. *et al.* (2013) Bacterial endophytes enhance invasive plant competition. *American Journal of Botany* 100, 1726–1737.
Ruijven, J., De Deyn, G.B. and Berendse, F. (2003) Diversity reduces invasibility in experimental plant communities: the role of plant species. *Ecology Letters* 6(10), 910–918. DOI: 10.1046/j.1461-0248.2003.00516.x.
Schaffner, U., Ridenour, W.M., Wolf, V.C., Bassett, T., Müller, C. *et al.* (2011) Plant invasions, generalist herbivores, and novel defense weapons. *Ecology* 92(4), 829–835. DOI: 10.1890/10-1230.1.
Schultheis, E.H., Berardi, A.E. and Lau, J.A. (2015) No release for the wicked: enemy release is dynamic and not associated with invasiveness. *Ecology* 96(9), 2446–2457. DOI: 10.1890/14-2158.1.
Seifert, E.K., Bever, J.D. and Maron, J.L. (2009) Evidence for the evolution of reduced mycorrhizal dependence during plant invasion. *Ecology* 90(4), 1055–1062. DOI: 10.1890/08-0419.1.
Shah, M.A., Reshi, Z.A. and Khasa, D.P. (2009) Arbuscular mycorrhizas: drivers or passengers of alien plant invasion. *The Botanical Review* 75(4), 397–417. DOI: 10.1007/s12229-009-9039-7.
Silva, C., Kan, F.L. and Martínez-Romero, E. (2007) Population genetic structure of *Sinorhizobium meliloti* and *S. medicae* isolated from nodules of *Medicago* spp. in Mexico. *FEMS Microbiology Ecology* 60(3), 477–489. DOI: 10.1111/j.1574-6941.2007.00301.x.

Simberloff, D. (1981) Community effects of introduced species. In: Nitecki, T.H. (ed.) *Biotic Crises in Ecological and Evolutionary Yime. Academic Press*. New York, pp. 53–81.

Simberloff, D., Martin, J.-L., Genovesi, P., Maris, V., Wardle, D.A. *et al.* (2013) Impacts of biological invasions: what's what and the way forward. *Trends in Ecology & Evolution* 28(1), 58–66. DOI: 10.1016/j.tree.2012.07.013.

St. Clair, S.B., O'Connor, R., Gill, R. and McMillan, B. (2016) Biotic resistance and disturbance: rodent consumers regulate post-fire plant invasions and increase plant community diversity. *Ecology* 97(7), 1700–1711. DOI: 10.1002/ecy.1391.

Stewart, T.E., Allen, P.S. and Meyer, S.E. (2009) First report of *Pyrenophora* seminiperda in turkey and Greece. *Plant Disease* 93(12), 1351. DOI: 10.1094/PDIS-93-12-1351B.

Stinson, K.A., Campbell, S.A., Powell, J.R., Wolfe, B.E., Callaway, R.M. *et al.* (2006) Invasive plant suppresses the growth of native tree seedlings by disrupting belowground mutualisms. *PLoS Biology* 4(5), e140. DOI: 10.1371/journal.pbio.0040140.

Suding, K.N., Stanley Harpole, W., Fukami, T., Kulmatiski, A., MacDougall, A.S. *et al.* (2013) Consequences of plant-soil feedbacks in invasion. *Journal of Ecology* 101(2), 298–308. DOI: 10.1111/1365-2745.12057.

Sun, Z.-K. and He, W.-M. (2010) Evidence for enhanced mutualism hypothesis: *Solidago canadensis* plants from regular soils perform better. *PLoS ONE* 5(11), e15418. DOI: 10.1371/journal.pone.0015418.

Swift, S., Munroe, S., Im, C., Tipton, L. and Hynson, N.A. (2019) Remote tropical island colonization does not preclude symbiotic specialists: new evidence of mycorrhizal specificity across the geographic distribution of the Hawaiian endemic orchid Anoectochilus sandvicensis. *Annals of Botany* 123(4), 657–666. DOI: 10.1093/aob/mcy198.

Tallamy, D.W., Ballard, M. and D'Amico, V. (2010) Can alien plants support generalist insect herbivores? *Biological Invasions* 12(7), 2285–2292. DOI: 10.1007/s10530-009-9639-5.

Traveset, A. and Richardson, D.M. (2014) Mutualistic interactions and biological invasions. *Annual Review of Ecology, Evolution, and Systematics* 45(1), 89–113. DOI: 10.1146/annurev-ecolsys-120213-091857.

Uesugi, A. and Kessler, A. (2013) Herbivore exclusion drives the evolution of plant competitiveness via increased allelopathy. *New Phytologist* 198(3), 916–924. DOI: 10.1111/nph.12172.

Van der Putten, W.H., Van Dijk, C. and Peters, B.A.M. (1993) Plant-Specific soil-borne diseases contribute to succession in foredune vegetation. *Nature* 362(6415), 53–56. DOI: 10.1038/362053a0.

Van der Putten, W.H., Vet, L.E.M., Harvey, J.A. and Wäckers, F.L. (2001) Linking above- and belowground multitrophic interactions of plants, herbivores, pathogens, and their antagonists. *Trends in Ecology & Evolution* 16(10), 547–554. DOI: 10.1016/S0169-5347(01)02265-0.

van der Putten, W.H., Kowalchuk, G.A., Brinkman, E.P., Doodeman, G.T.A., van der Kaaij, R.M. *et al.* (2007) Soil feedback of exotic savanna grass relates to pathogen absence and mycorrhizal selectivity. *Ecology* 88(4), 978–988. DOI: 10.1890/06-1051.

van Grunsven, R.O.Y.H.A., Van der Putten, W.I.M.H., Bezemer, T.M., Tamis, W.I.L.L.M., Berendse, F. *et al.* (2007) Reduced plant—soil feedback of plant species expanding their range as compared to natives. *Journal of Ecology* 95(5), 1050–1057. DOI: 10.1111/j.1365-2745.2007.01282.x.

Vermeij, M.J.A., Smith, T.B., Dailer, M.L. and Smith, C.M. (2009) Release from native herbivores facilitates the persistence of invasive marine algae: a biogeographical comparison of the relative contribution of nutrients and herbivory to invasion success. *Biological Invasions* 11(6), 1463–1474. DOI: 10.1007/s10530-008-9354-7.

Vilà, M., Espinar, J.L., Hejda, M., Hulme, P.E., Jarošík, V. *et al.* (2011) Ecological impacts of invasive alien plants: a meta-analysis of their effects on species, communities and ecosystems. *Ecology Letters* 14(7), 702–708. DOI: 10.1111/j.1461-0248.2011.01628.x.

Vogelsang, K.M. and Bever, J.D. (2009) Mycorrhizal densities decline in association with nonnative plants and contribute to plant invasion. *Ecology* 90(2), 399–407. DOI: 10.1890/07-2144.1.

Wallace, A.R. (1876) *The Geographical Distribution Of Animals; With A Study Of The Relations Of Living And Extinct Faunas As Elucidating The Past Changes Of The Earth's Surface*. Harper & Brothers, New York.

Waller, L.P., Callaway, R.M., Klironomos, J.N., Ortega, Y.K. and Maron, J.L. (2016) Reduced mycorrhizal responsiveness leads to increased competitive tolerance in an invasive exotic plant. *Journal of Ecology* 104(6), 1599–1607. DOI: 10.1111/1365-2745.12641.

Wei, G., Chen, W., Zhu, W., Chen, C., Young, J.P.W. *et al.* (2009) Invasive *Robinia pseudoacacia* in China is nodulated by *Mesorhizobium* and *Sinorhizobium* species that share similar nodulation genes with native American symbionts. *FEMS Microbiology Ecology* 68(3), 320–328. DOI: 10.1111/j.1574-6941.2009.00673.x.

Williams, J.L., Auge, H. and Maron, J.L. (2008) Different gardens, different results: native and introduced populations exhibit contrasting phenotypes across common gardens. *Oecologia* 157(2), 239–248. DOI: 10.1007/s00442-008-1075-1.

Wilson, G.W. and Hartnett, D.C. (1998) Interspecific variation in plant responses to mycorrhizal colonization in tallgrass prairie. *American Journal of Botany* 85(12), 1732–1738. DOI: 10.2307/2446507.

Wolfe, B.E. and Klironomos, J.N. (2005) Breaking new ground: soil communities and exotic plant invasion. *BioScience* 55(6), 477–487. DOI: 10.1641/0006-3568(2005)055[0477:BNGSCA]2.0.CO;2.

Yang, Q., Carrillo, J., Jin, H., Shang, L., Hovick, S.M. *et al.* (2013) Plant–soil biota interactions of an invasive species in its native and introduced ranges: implications for invasion success. *Soil Biology and Biochemistry* 65, 78–85. DOI: 10.1016/j.soilbio.2013.05.004.

Yuan, Y., Tang, J., Leng, D., Hu, S., Yong, J.W.H. *et al.* (2014) An invasive plant promotes its arbuscular mycorrhizal symbioses and competitiveness through its secondary metabolites: indirect evidence from activated carbon. *PLoS ONE* 9(5), e97163. DOI: 10.1371/journal.pone.0097163.

Zhang, Q., Yang, R., Tang, J., Yang, H., Hu, S. *et al.* (2010) Positive feedback between mycorrhizal fungi and plants influences plant invasion success and resistance to invasion. *PLoS ONE* 5(8), e12380. DOI: 10.1371/journal.pone.0012380.

Zubek, S., Majewska, M.L., Błaszkowski, J., Stefanowicz, A.M., Nobis, M. *et al.* (2016) Invasive plants affect arbuscular mycorrhizal fungi abundance and species richness as well as the performance of native plants grown in invaded soils. *Biology and Fertility of Soils* 52(6), 879–893. DOI: 10.1007/s00374-016-1127-3.

4 Pollination Interactions Promoting Plant Invasions

Ana Montero-Castaño[1]* and Anna Traveset[2]

[1]School of Environmental Sciences, University of Guelph, Canada; [2]Mediterranean Institute of Advanced Studies (IMEDEA-CSIC), Esporles, Mallorca, Spain

Abstract

Most plant species rely on, or benefit from, animal pollination. Therefore, pollination interactions are expected to play a key role in the reproduction and invasion success of non-native plants in their new areas. Understanding this role will allow us to better predict certain plant invasions. Also, it will allow us to explore the potential of invasion management measures based on disrupting or avoiding pollination interactions.

In this chapter we review the available information on reproductive systems and their degree of dependence on animal-mediated pollination of non-native plant species. We review the characteristics of resident pollinators feeding on non-native plants and the different environmental setups that allow or impede non-native plants to reproduce in their new areas. Finally, we explore the scarce literature on invasion management measures based on disrupting pollination interactions and discuss their potentiality.

Evidence so far shows that animal pollination does not usually act as an effective barrier to invasion. Most introduced plants are able to receive suitable pollination service from resident pollinators, while others are able to minimize their reliance on pollinators through different mechanisms (e.g. selfing or asexual reproduction). The environmental settings where the introduction occurs (for instance, the presence of neighbours with similar or dissimilar flower morphologies), can play an important role on the success or failure of non-native plants overcoming reproductive barriers.

Although it seems that most introduced plants do not face pollination barriers, we consider that, for certain species, the disruption or avoidance of pollination interactions as control or prevention measures deserve further exploration.

4.1 Introduction

Mutualistic interactions can be important at all stages of the introduction—naturalisation—invasion continuum and at all phases of a plant's life cycle (Traveset and Richardson, 2014). In the case of non-native plant species that, at least to certain degree, rely on sexual reproduction and on outcrossing (i.e. on receiving pollen from different individuals for ovule fertilization), the pollination interactions they establish in their new areas determine their reproductive output and eventually their invasion success (Parker, 1997; Gross *et al.*, 2010; Rodger *et al.*, 2010). The fruit set of ca. 90% of flowering plants and 75% of the world's main food crops depends on, or benefits from, animal pollination (Klein *et al.*, 2007; Ollerton *et al.*, 2011). Therefore, pollination interactions are expected to play a key role in the naturalization—invasion processes of most plants introduced to new areas. Such interactions are determined both by the traits of the species involved in them and by environmental factors (Morales and Aizen, 2006), influencing both the quantitative and the qualitative components of the pollination process (Aizen and Harder, 2007). Understanding the role of pollination interactions in the reproductive success

*Corresponding author: ana.montero.castano@gmail.com

(eds Anna Traveset and David M. Richardson)
DOI: 10.1079/9781789242171.0004

of non-native species in their new areas can provide new insights for invasion management; specifically, about the potential of avoiding or disrupting pollination interactions as prevention, control or eradication measures.

In this chapter, we review the increasing knowledge on the relevance of pollination interactions promoting or constraining plant invasions in different ecosystems across the globe, with the goal of assessing the following:

1. To what extent do invasive plants depend on pollinators for their reproduction and when does such dependence influence invasion?
2. Which are the species traits and the main functional groups of the resident pollinators visiting invasive plants?
3. How do environmental factors affect pollination interactions in plant invasions?
4. What are the implications of plant—pollinator interactions for the management of plant invasions?

4.2 To What Extent Do Invasive Plants Depend on Pollinators for Their Reproduction and When Does Such Dependence Influence Invasion?

The extent to which pollination interactions are relevant in plant invasions will depend on different traits of the introduced plant: i) its reliance on sexual reproduction; ii) its degree of self-compatibility; iii) its dependence on pollinators for pollen movement from anthers to stigmas; and iv) its plasticity for the above-mentioned traits. Then, if introduced plants rely on sexual reproduction and require pollinators for pollen movement, invasion success may be contingent on e) the plant's degree of ecological specialization to pollinator species or functional groups and f) its plasticity for this trait.

Despite the importance of these life history traits of introduced plants for our understanding, prediction and management of invasions, in most cases we still lack quantitative information on the type of reproduction, breeding system and pollination requirements, even for the most invasive species (Barrett, 2011). Therefore, the incidence of these traits in the invasive flora will also remain unknown until more experimental studies are conducted. In Table 4.1 we show the information available about these traits for the world's worst plant invasive species, according to Lowe *et al.* (2000), in their native and invaded ranges. Overall, the table shows important knowledge gaps about the breeding systems of these species, mostly in their native ranges. It also shows that most of the species we have information for use animal pollen vectors. Whether the role of these pollen vectors is necessary or facultative in the reproduction and invasion success of these species remains unknown in most cases.

4.3 Sexual Reproduction, Self-pollination and Vector-mediated Fertilization in Invasive Floras

Nearly 90% of flowering plants depend on, or benefit from, animal pollination (Ollerton *et al.*, 2011) and thus show a certain degree of sexual reproduction (see Table 4.1). However, some plants are able to reproduce vegetatively, reducing or eliminating their reliance on pollinators. Vegetative reproduction relates to invasion success in different habitats and biogeographical areas (Pyšek *et al.*, 1995; Godfrey *et al.*, 2004; Lloret *et al.*, 2005). For example, the geophyte *Oxalis pes-caprae* has invaded the Mediterranean basin by reproducing vegetatively through bulbs (Vilà *et al.*, 2006). Until recently, this tristylous (i.e. with three morphs that differ in the lengths of the pistil and stamens), and intramorph incompatible species native from South Africa, was not able to reproduce sexually in the invaded area because only the short-styled morph was introduced (Costa *et al.*, 2017).

On the other hand, in the case of introduced plant species that have sexual reproduction, self-pollination (or selfing, i.e. autonomous ovule fertilization in self-compatible species) is likely to promote invasiveness by assuring reproduction when mates and pollinators are limiting (Baker's Law: Baker, 1955). Indeed, selfing ability and invasiveness have been found to positively correlate in different studies (van Kleunen *et al.*, 2008; Pyšek *et al.*, 2011; Razanajatovo *et al.*, 2016). Nevertheless, obligate outcrossing does not necessarily imply invasion failure. In fact, there are plant invaders that are obligate outcrossers

Table 4.1 Traits for the world's worst plant invasive species.

Species	Range	If sexual reproduction: Reproduction type	Compatibility	Pollen–vector dependence	Vector type	If animal vector: Principal pollinator species or functional group	Generalization level	Pollen limitation	References
Spartina anglica Common cord-grass	Native								Nehring and Adsersen, 2006; Li *et al.*, 2008
	Invasive	Sexual and asexual	Self-incompatible	Pollen–vector dependent	Wind			Yes	
Eichhornia crassipes Water hyacinth	Native	Sexual and asexual	Self-compatible	Pollen–vector dependent	Animal	*Ancyloscelis gigas*	Specialist		el Seed and Obeid, 1975; Liu *et al.*, 2013; Barrett, 1980
	Invasive	Sexual and asexual	Self-compatible	Pollen–vector dependent	Animal	*Apis mellifera*	Generalist	Yes	
Spathodea campanulata African tulip tree	Native	Sexual	Self-incompatible	Pollen–vector dependent	Animal	*Acridotheres tristis*, *A. fuscus*, *Corvus splendens*, *C. macrorhynchos*, *Dicrurus adsimilis* and *Psittacula krameri*			Bittencourt *et al.*, 2003
	Invasive	Sexual	Self-incompatible	Pollen–vector dependent	Animal	Birds			
Acacia mearnsii Black wattle	Native	Sexual	Self-incompatible	Pollen–vector dependent	Animal	Insects		Yes	Alves and Marins-Corder, 2009; Beck-Pay, 2012
	Invasive	Sexual	Self-incompatible	Pollen–vector dependent	Animal	*Macrodactylus suturalis*			
Schinus terebinthifolius Brazilian pepper tree	Native	Sexual	Self-incompatible	Pollen–vector dependent	Animal	*Apis mellifera*, Halictidae and *Palpada vinetorum*		Yes	Farizel and Cristina, 2008; Cesario and Gaglianone, 2013
	Invasive	Sexual	Self-incompatible	Pollen–vector dependent	Animal				
Imperata cylindrica Cogon grass	Native	Sexual and asexual	Self-incompatible	Pollen–vector dependent					MacDonald, 2007
	Invasive	Sexual and asexual	Self-incompatib e	Pollen–vector dependent					

Continued

Table 4.1 Continued

Species	Range	Reproduction type	If sexual reproduction: Compatibility	Pollen–vector dependence	Vector type	If animal vector: Principal pollinator species or functional group	Generalization level	Pollen limitation	References
Pinus pinaster Cluster pine	Native	sexual	Self-incompatible	Pollen–vector dependent	Wind				González-Martínez *et al.*, 2007; Fernandes *et al.*, 2008
	Invasive	Sexual	Self-incompatible	Pollen–vector dependent	Wind				
Opuntia stricta Erect pricklypear	Native	Sexual and asexual	Self-compatible	Pollen–vector dependent	Animal	*Agapostemon splendens*, *Megachile brevis* and *Apis mellifera*	Generalist	No	Bartomeus and Vilà, 2009; Spears, 2009
	Invasive	Sexual and asexual	Self-compatible	Pollen–vector dependent	Animal	*Apis mellifera*, *Xylocopa violacea* and other Hymenoptera	Generalist	No	
Morella faya Fire tree	Native	Sexual	Self-incompatible	Pollen–vector dependent	Wind				ISSG, 2020 Species profile: *Morella faya*; Global Invasive Species Database, 2020c
	Invasive	Sexual	Self-incompatible	Pollen–vector dependent	Animal and wind	*Apis mellifera*	Generalist		
Arundo donax Giant reed	Native	Asexual							ISSG, 2020 Species profile: *Arundo donax*; Global Invasive Species Database, 2020a; Balogh *et al.*, 2012
	Invasive	Asexual							
Ulex europaeus Gorse	Native	Sexual	Self-compatible	Pollen–vector dependent	Animal	*Apis mellifera* and *Bombus* sp.	Generalist		Bowman *et al.*, 2008; Atlan *et al.*, 2015
	Invasive	Sexual	Self-compatible	Pollen–vector dependent	Animal	*Apis mellifera* and *Bombus* sp.	Generalist		
Hiptage benghalensis Hiptage	Native	Sexual and asexual	Self-compatible	Pollen–vector dependent	Animal	*Apis dorsata*	Generalist	Yes	Ren *et al.*, 2013; ISSG, 2020 Species profile: *Hiptage benghalensis;* Global Invasive Species Database, 2020b
	Invasive	Sexual and asexual							
Fallopia japonica Japanese knotweed	Native	Sexual and asexual	Self-incompatible	Pollen–vector dependent	Animal and wind	Insects			Forman and Kesseli, 2003; Zhou *et al.*, 2003
	Invasive	Sexual and asexual	Self-incompatible						

Continued

Table 4.1 Continued

Species	Range	If sexual reproduction: Reproduction type	Compatibility	Pollen–vector dependence	If animal vector: Vector type	Principal pollinator species or functional group	Generalization level	Pollen limitation	References
Hedychium gardnerianum Kahili ginger	Native								Cordeiro and Silva, 2003
	Invasive	Sexual and asexual							
Clidemia hirta Koster's curse	Native	Sexual	Self-compatible	Pollen–vector dependent	Animal	Apidae and Halictidae bees			Ferreira *et al*., 1994
	Invasive								
Pueraria montana var. *lobata* Kudzy	Native	Sexual and asexual			Animal				Thornton, 2004; Geerts *et al*., 2016
	Invasive	Sexual and asexual	Self-compatible	Pollen–vector dependent	Animal	*Apis mellifera*, *Megachile sculpturalis* and *Xylocopa virginica*	Generalist		
Lantana camara Lantana	Native								Vivian-Smith and Panetta, 2009
	Invasive	Sexual and asexual	Self-compatible		Animal	Butterflies and thrips			
Euphorbia esula Leafy spurge	Native								Morrow, 1979
	Invasive	Sexual and asexual	Self-compatible	Pollen–vector dependent	Animal	Insects			
Laucaena leucocephala Leucaena	Native								Hughes, 1998
	Invasive	Sexual	Self-compatible		Animal	Insects	Generalist		
Melaleuca quinquenervia Melaleuca	Native								Van *et al*., 2005
	Invasive	Sexual		Pollen–vector dependent	Animal	Birds, insects and mammals			
Prosopis glandulosa Mesquite	Native								Solbrig and Cantino, 1975; Pasiecznik and Felker, 1992
	Invasive	Sexual	Self-incompatible		Animal		Generalist		
Miconia calvescens Miconia	Native								Meyer, 2010
	Invasive	Asexual	Self-compatible	Selfer	Wind			No	

Continued

Table 4.1 Continued

Species	Range	If sexual reproduction: Reproduction type	Compatibility	Pollen–vector dependence	Vector type	If animal vector: Principal pollinator species or functional group	Generalization level	Pollen limitation	References
Mikania micrantha Mile-a-minute weed	Native								Macanawai *et al.*, 2015
	Invasive	Sexual and asexual			Animal, water and wind	*Apis mellifera* and other Hymenoptera, flies and butterflies	Generalist		
Mimosa pigra Mimosa	Native								Lukitsch and Elliott, 2012
	Invasive	Sexual	Self-compatible		Wind				
Ligustrum robustum Privet	Native	Sexual and asexual	Self-incompatible		Animal				Lavergne *et al.*, 1999; Swarbrick *et al.*, 1999
	Invasive	Sexual and asexual			Animal		Generalist		
Ceropia peltada Pumpwood	Native								
	Invasive								
Lythrum salicaria Purple loosestrife	Native	Sexual	Self-incompatible	Pollen–vector dependent	Animal			Yes	Thompson *et al.*, 1987; Burns *et al.*, 2019
	Invasive	Sexual	Self-incompatible		Animal	Insects		No	
Cinchona pubescens Quinine tree	Native								Starr *et al.*, 2003
	Invasive	Sexual	Self-incompatible		Animal	Bees, butterflies and flies			
Ardisia elliptica Shoebutton ardisia	Native								Pascarella, 1997
	Invasive	Sexual	Self-compatible		Animal	Small bees			
Chromolaena odorata Siam weed	Native								Gautier, 1993; Witkowski, 2002; Lakshmi and Raju, 2011
	Invasive	Sexual and asexual			Animal	Insects	Generalist		
Psidium cattleianum Strawberry guava	Native	Sexual	Self-compatible		Animal		Generalist		Huenneke and Vitousek, 1990; Ferreira da Costa *et al.*, 2015
	Invasive	Sexual	Self-compatible		Animal		Generalist		

Continued

Table 4.1 Continued

Species	Range	If sexual reproduction				If animal vector			References
		Reproduction type	Compatibility	Pollen–vector dependence	Vector type	Principal pollinator species or functional group	Generalization level	Pollen limitation	
Tamarix ramosissima Tamarisk	Native	Sexual and asexual	Self-compatible		Animal	Bees		Yes	Bowser, 1957; Gaskin and Schaal, 2002; Gaskin and Schaal, 2003; Chen and Zhao, 2015
	Invasive	Sexual and asexual	Self-compatible		Animal			Yes	
Sphaegneticola trilobata Wedelia	Native								
	Invasive								
Rubus ellipticus Yellow Himalayan raspberry	Native								
	Invasive								

Reproductive characteristics of the world's worst plant invasive species, according to Lowe *et al.* (2000), in their native and invaded ranges. Characteristics that influence the role of pollinator interactions in the reproductive success of these invasive species. This table informs us about whether pollinator interactions may or may not play a role in the reproductive success of these invasive species. However, the magnitude of that role, i.e. the contribution of pollinator interactions to the overall reproductive output of the species, is beyond the scope of this table.Information for the table was gathered in January 2020 from CABI Invasive Species Compendium (CABI, 2020), the European Network of Invasive Alien Species (NOBANIS) database (www.nobanis.org), the Global Invasive Species Database (GISD) (www.iucngisd.org/gisd) and a literature search in the Web of Science with the following keyword combinations: (Latin species name AND reproduct*), (Latin species name AND pollinat*) and (Latin species name AND breed*).For each species and distribution range (native and invaded), the table pools the information from different studies, which does not imply that species show the same reproductive strategy (or pollinator species or functional group) along their distribution ranges. For example, *Fallopia japonica* seems to reproduce exclusively asexually in its invaded range except in one site in the USA where it also shows sexual reproduction. For representation purposes, compatibility and pollen–vector dependence are treated in the table as categorical. However, these variables are continuous rather than categorical. Therefore, species reported as self-incompatible in the table might show certain degree of compatibility, and vice versa. Also, species reported as pollen–vector dependent might show certain selfing rates (if self-compatible) and vice versa. Grey cells in the table represent gaps of information. Finally, notice that the extrapolation of the incidence of these traits to other invasive flora must be done cautiously, as the 34 species explored here are only a small representation of the global invasive flora.

owing to self-incompatibility or dioecy, or that are self-compatible but not capable of autonomous self-pollination. Notable cases include the self-incompatible and tristylous *Lythrum salicaria* invading North America (Mal *et al.*, 1992), the self-incompatible *Paederia foetida* invading southern USA and Hawaii (Liu *et al.*, 2006), the dioecious and mainly insect—vector dependent *Salix cinerea* invading Australia (Hopley and Young, 2015), the poorly self-compatible *Senecio inaequidens* invading Europe (López-García and Maillet, 2005), and the self-compatible but pollinator dependent *Cytisus scoparius* invading most temperate regions in the world (Simpson *et al.*, 2005), among others (see Table 4.1). The invasion success of these plant species that require cross-pollination implies that they have been able to find suitable pollen vectors in their new areas. Abiotic factors such as water (hydrophily) and wind (anemophily) can move the pollen of some plant species, reducing or eliminating their reliance on pollinators. However, and though we lack information about the incidence of each type of pollination vectors in the invasive flora, the majority of the world's most invasive woody plants are visited by animals (see Table S.1. in Traveset and Richardson, 2014 and Table 4.1) and, thus, likely to be biotically pollinated (zoophily) in their new areas.

Moreover, breeding systems can show certain degree of phenotypic plasticity and even contemporary evolution. Irrespective of the underlying mechanism, invasions frequently cause transitions to uni-parental reproduction (e.g. clonality and selfing) (reviewed in Barrett, 2011). Selfing requires self-compatibility, which is a quantitative rather than a qualitative trait. Variation along the self-compatibility to self-incompatibility continuum is observed among plant populations (Atlan *et al.*, 2015) and even within populations (Daehler, 1998). This variation has the potential to influence evolutionary changes if there is a selection pressure (Barrett *et al.*, 2008). Pollen limitation, owing to the lack of pollinators or mates, is indeed considered a selective factor favouring, among other adaptive solutions, the evolution of self-compatibility (Harder and Aizen, 2010).

Finally, not only the assurance of reproduction (owing to selfing or to the presence of suitable pollen vectors in the case of outcrossers), but also its quality must be considered when assessing the demographical consequences of the different reproductive strategies of invaders. For instance, inbreeding depression, i.e. reduced population fitness compared to outbred progeny, is frequently associated with selfing (Husband and Schemske, 1996). However, outcrossers can also suffer genetic depletion; for example, the probability of genes with homozygous recessive maladaptive alleles is higher than in selfing populations where such alleles are often purged (Dart and Eckert, 2013). Despite the importance of the qualitative aspects of reproduction, only few studies have investigated them in invasion contexts (e.g. Rodger, 2011; Mullarkey *et al.*, 2013).

Rodger (2011) assessed the potential contribution of selfing to the invasiveness of the Australian tree *Acacia dealbata* in South Africa. Although trees are usually outcrossers (Barrett *et al.*, 1996), *A. dealbata* showed high selfing rates in its invaded range. It also showed high inbreeding depression over seed development and progeny growth and survival. Inbreeding depression could thus detract from the reproductive assurance benefit of selfing. However, the author argues that in invasion contexts, where introduced individuals might be isolated from conspecifics and where outcrossing might not be possible, selfing might still promote invasion as far as there are chances for some selfed progeny to survive. Hence, whether non-native plants benefit from self-pollination will depend on the environmental context in which the introduction occurs. The interaction between the environmental context and selfing has in fact been suggested as one of the mechanisms explaining the 'genetic paradox' of some invasions (Schrieber and Lachmuth, 2017). Such a paradox refers to the success of some invasive species despite experiencing inbreeding depression, genetic depletion or drift loads. The influence of environmental conditions in the output of pollination interactions in invasion processes is discussed later in this chapter.

4.4 Specialist Pollination in Invasive Floras

Plants are more likely to attract pollinators in their new range if they have generalist pollination systems. This enables them to quickly integrate into resident plant—pollinator networks, either serviced by native or non-native pollinators (e.g. Memmott and Waser, 2002; Morales and Aizen, 2006; Traveset and Richardson, 2014). However, pollen limitation can still occur due to low quality pollination service. In the case of non-natives visited by generalist pollinators, one of the factors that may compromise the quality of pollen is the deposition of pollen from other species, that is, heterospecific pollen (Mitchell *et al.*, 2009). Heterospecific pollen can reduce reproductive success by chemically or physically (e.g. stigma clogging) interfering with conspecific pollen germination, pollen tube growth and/or ovule fertilization (Morales and Traveset, 2008). Few studies have assessed the role of heterospecific pollen deposition on the reproduction of non-native species. Little available evidence shows that non-native plants receive higher heterospecific pollen loads than natives, and despite this, their reproductive success is not affected (Suárez-Mariño *et al.*, 2019). Whether this is a general pattern and which are the underlying mechanisms explaining the lower vulnerability of non-native plants deserves further exploration.

Besides, other factors can reduce heterospecific pollen deposition even in generalist pollination systems. For instance, some insect pollinators, including the ubiquitous honeybee (*Apis mellifera*), show high flower constancy within single foraging bouts (Grüter and Ratnieks, 2011). That could explain, for example, the high percentage of stigmas (94.4%) in invasive populations of *Hedysarum coronarium* (syn. *Sulla coronaria*) in Menorca (Balearic Islands) that only receive conspecific pollen despite being mainly visited by the super-generalist honeybee (Montero-Castaño *et al.*, 2014). Moreover, some pollinators show phenotypic specialisations for transporting different species of pollen in different body parts that in subsequent conspecific visits will touch the stigmas, thus reducing heterospecific pollen deposition (Armbruster, 2017).

Not only plants with generalist pollination systems infiltrate resident plant—pollinator communities: specialist species can also be integrated in different ways. Some may find their specific pollinators or pollinator functional groups if these have wide distribution ranges (Rodger *et al.*, 2010) or if they have also been introduced in the same area (leading to a potential 'invasion complexes' see Olesen *et al.*, 2002). For example, the main pollinator of *Lilium formosanum* is the moth *Agrius convolvuli*, whose wide distribution area covers both the native (Taiwan) and the invaded (South Africa) distribution areas of this lily (Rodger *et al.*, 2010).

In addition, specialist plants often link to the most generalist pollinators (Bascompte *et al.*, 2006). For instance, invasive populations of *Opuntia dillenii* in Tenerife (Canary Islands) are pollinated by few insect species, the most important one being the highly generalist endemic *Bombus canariensis* (Padrón *et al.*, 2009). This asymmetry of mutual dependences between interacting species contributes to assure a pollination service to specialist plants introduced outside their native areas.

Other specialist non-native plants may generalize or acquire completely new types of pollinator species or functional groups. Interestingly, though most plants are adapted to a particular pollinator functional group, they often have a secondary pollinator functional group that, though less efficiently, can likely contribute to their reproductive success when the primary group is absent (Rosas-Guerrero *et al.*, 2014). For example, bird-pollinated flowers are often efficiently pollinated by bees, as observed for *Fuchsia magellanica*, which is pollinated by hummingbirds in its native range in South America and by insects in invaded areas in Europe (Valentine, 1978). *Nicotiana glauca* is also pollinated by hummingbirds in its native range in South America, whereas in invaded areas of South Africa is pollinated by sunbirds (Issaly *et al.*, 2020). In the Galápagos, where insects are scarce, land-bird species often effectively visit the flowers of non-native plants, such as guava (Hervías-Parejo and Traveset, 2018).

Moreover, species visited by a small number of resident pollinators can still receive adequate pollination service and have high reproductive success, as Thompson and Knight (2018) found for ten non-native species in North America.

Finally, even self-pollinated plant species can be included in resident plant—pollinator communities, promoting their invasion through increased outcrossing and seed set if autonomous selfing does not result in the fertilization of all ovules (Aizen and Harder, 2007).

4.5 Species Traits and Main Functional Groups of the Resident Pollinators Visiting Invasive Plants

Pollination services are provided mainly by insects, but different species of birds, lizards, bats and small mammals can also act as effective pollinators (Olesen and Valido, 2003; Quesada *et al.*, 2003; Hervías-Parejo and Traveset, 2018). Whether one group of pollinators will be more likely to integrate a particular plant invader than another will depend on different pollinator and plant traits. For example, some pollinator traits are body size, proboscis length (in the case of insects) and gape size (in the case of birds or lizards); while in the case of plants, flower shape, size, floral display or type of floral resource are among the most influential. An example of how flower traits may sort the functional groups of pollinators visiting them is the European legume *C. scoparius*, which requires insect visitors large enough to push the keel down and split the petals to access the pollen rewards and allow fertilization (Parker, 1997).

In most documented cases, pollination in the non-native range is carried out by resident pollinators belonging to the same or similar functional groups as those pollinating plants in their native ranges. Examples include *H. coronarium*, pollinated by some of the same hymenopteran and coleopteran species in native and introduced areas in Spain (Montero-Castaño *et al.*, 2014), and *Rhododendron ponticum*, pollinated by bumblebees both in its native (Spain) and introduced range (Ireland) (Stout *et al.*, 2006). See Table 4.1 for more examples. Non-native plants pollinated by birds in their native ranges are also usually well served by birds in their invading areas (Geerts and Pauw, 2009, Issaly *et al.*, 2020). However, for most invasive plants worldwide, we have no information on what species visit them in their native or introduced range (Table 4.1), or how efficient they are as pollinators.

Information about the generalization level of the resident pollinators integrating non-native plants in their diets requires community-level approaches. Although such studies are still scarce, available evidence shows that non-native plants are usually visited by generalist resident pollinators (Traveset and Richardson, 2014). The prevalence of generalist pollinators in most communities may partly explain why pollen limitation is not important for most plant invaders (Pyšek *et al.*, 2011). In addition, super-generalist pollinators such as the honeybee and bumblebees (*Bombus* spp.), which have been extensively introduced worldwide, often play a key role in such integration (Stout *et al.*, 2002; Goulson, 2003; Simpson *et al.*, 2005; Gross *et al.*, 2010; Geslin *et al.*, 2017; Table 4.1), creating invasion complexes (Chapter 13, this volume). This is the case for *Impatiens glandulifera* in the UK, *Opuntia* spp. in Tenerife and Menorca and *H. coronarium* also in Menorca that were mainly visited by honeybees (Lopezaraiza-Mikel *et al.*, 2007; Padrón *et al.*, 2009; Montero-Castaño *et al.*, 2014).

However, the fact that generalist pollinators more frequently introduce non-native plants into pollination networks might simply be a result of their abundance and wide diets and not of a preferential linking. In fact, Stouffer *et al.* (2014) analysed the 25 invaded pollination networks available at the time and found that specialist pollinators are far more likely to interact with non-native plants than would be expected by chance. Therefore, from an ecological and evolutionary point of view, such specialist pollinators integrating non-native plants in their diets deserve further attention.

4.6 Environmental Conditions Affecting the Output of Pollination Interactions in Invasion Processes

The sole presence of suitable pollinators in the new range is insufficient to explain the pollination service received by a non-native plant. Plant—pollinator interactions occur under specific environmental conditions that modulate their frequency and quality. Besides, such conditions can change along the different stages of the invasion process; i.e. arrival, establishment, persistence and expansion – see Richardson *et al.* (2000a) and Pyšek *et al.* (2004).

4.7 Common Environmental Settings at the Initial Stages of Invasion Processes

The potential environmental settings are as diverse as introduction events. However, due to the main introduction pathways (e.g. horticulture), some environmental conditions are relatively common at the initial stages of invasion processes. Many plant introductions begin with low numbers of highly related individuals isolated from other conspecific populations (Liebhold and

Bascompte, 2003) and occur in human-altered habitats (Grass *et al.*, 2013). Such conditions have direct implications on the pollination services that non-native plants receive in their new areas.

Small plant stands as such might not be attractive for resident pollinators, though in the case of many introduced ornamentals, that can be to some extent counteracted by their showy and rewarding floral displays (Lloret *et al.*, 2005). Nevertheless, pollination attraction in multispecies assemblages depends on the relationship among co-flowering species, such as their relative abundances or relatedness (Carvalheiro *et al.*, 2014). Small, non-native plant stands can benefit from neighbours if they share pollinators that the neighbours attract – what was first named by Thomson (1978) as the 'magnet effect'. Thus, small stands of non-natives may benefit from the effect of native neighbours acting as magnet species (Fig. 4.1a, b), as was for example reported for the non-native *Echium vulgare* in Chile (Carvallo *et al.*, 2013). The opposite effect, i.e. native neighbours monopolising pollinator visits, could also be expected. However, the scarcity of studies exploring the effect of native plants on their non-native neighbours' pollination prevents us discussing if there is a general trend. In invasion contexts, there is more information available on the effect of non-native plants on the pollination of their native neighbours. This information

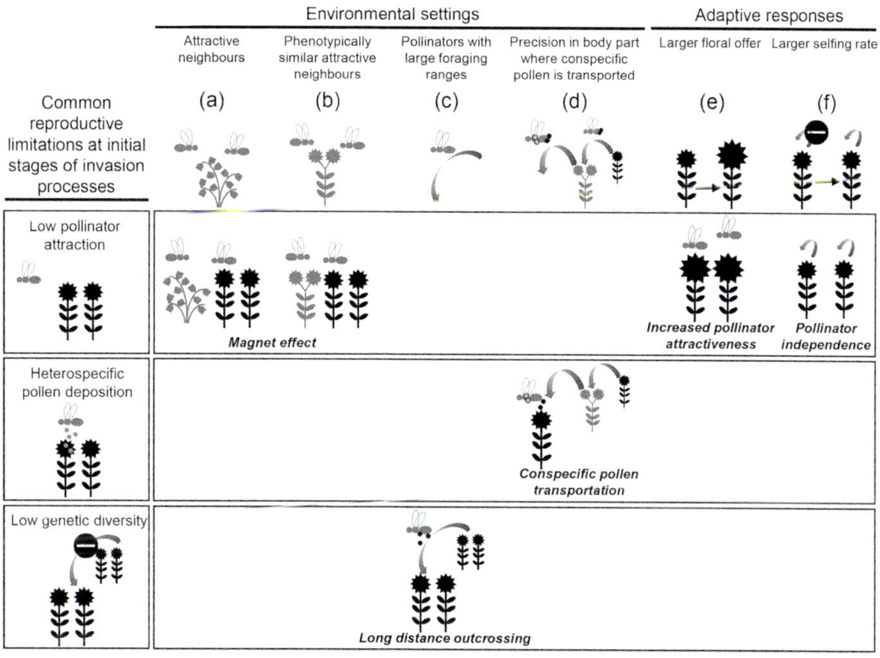

Fig. 4.1. Limitations to overcome the reproductive barriers that obligate outcrossing introduced plant species (represented in black) are often faced at the initial stages of the invasion processes due to their low numbers, high genetic relatedness and isolation. Some variants of these initial environmental settings, as well as adaptive responses of the introduced species, can provide ways to overcome reproductive barriers through different underlying mechanisms (in bold italics). Highly attractive (a) and phenotypically similar neighbours (b) can attract pollinators to small stands of introduced patches through magnet effects. The presence of pollinators with large flying and foraging distances (c) can enable cross-pollination with less genetically similar conspecifics, while the presence of pollinators with precise and specific body parts where they transport conspecific pollen can reduce the risk of heterospecific pollen deposition (d). In addition, introduced species can show phenotypic and genetic adaptive responses to the lack of pollinators and mates by showing larger floral displays to attract pollinators (e) or by expressing higher rates of selfing (f).

was reviewed and synthesized by Morales and Traveset (2009) who found that pollinator visits to natives decreased with the relative abundance of their non-native neighbours and with their phenotypic similarity (flower symmetry and colour) with them. Therefore, we can indirectly infer that non-natives usually usurp pollinator visits from natives, mostly from those phenotypically similar (Fig. 4.1b). However, this conclusion has to be taken cautiously for non-natives at initial stages of invasion because study cases might be biased towards invasive species that have successfully established and are already at advanced stages of their invasion.

Nevertheless, if the attraction of pollinators increases with the phenotypic similarity between non-native and native neighbours, we would expect attracted pollinators to more frequently conduct interspecific foraging bouts (increasing the risk of heterospecific pollen deposition) because their ability to differentiate floral resources can be reduced. In fact, in experimental mixed plots in Canada with the invasive *Lythrum salicaria* and its relative native *L. alatum*, 33–65% of pollinator movements generated heterospecific pollen deposition (Brown *et al.*, 2002). In these cases, the precision in the body part where pollen is transported (Armbruster, 2017), rather than flower constancy, could reduce the risk of heterospecific pollen deposition (Fig. 4.1d).

Even if pollen on non-native stigmas is conspecific, its genetic quality can also be compromised by the lack of suitable mates, reducing the long-term viability of introduced populations through inbreeding and genetic drift, as in other founding events. Introduced populations of self-incompatible, dioecious or heterostylous species are especially vulnerable to these founder effects (Dlugosch and Parker, 2008a). *Lythrum salicaria* is also an example of tristylous species that possesses an incompatibility system that reduces intra-morph fertilization. Monomorphic and isolated populations in invaded areas in Canada are pollen limited. However, this species presents partial self-compatibility that enables sexual reproduction in these populations until migrants of the other morphs arrive (Fig. 4.1f). Besides, pollinators with large foraging ranges, like bumblebees, may allow long distance inter-morph pollen flow among isolated populations (Fig. 4.1c) (Balogh and Barrett, 2018).

Other introduced species cope with founder effects through phenotypic adaptations and even rapid evolutionary changes. In a recent global meta-analysis, Burns *et al.* (2019) found that some invaders might be able to compensate the reduced reproductive success per flower due to pollen limitation by producing many flowers. Whether these changes had a genetic base or not was not explored. In the case of isolated populations of the invasive *Hypericum canariense* in California and Hawaii, despite their diminished genetic diversity compared to their native range in the Canary Islands, researchers found rapid and apparently adaptive genetic changes that contribute to their outcrossing rates and fitness in the invaded range (Dlugosch and Parker, 2008b). Invasive populations have increased growth rates and larger floral displays (Fig. 4.1d) compared to the native populations, and they have adapted their flowering phenologies to the latitude of the invaded areas as well.

Despite the above-mentioned mechanisms to overcome the lack of suitable mates and diminished genetic diversity in founding events, successful introductions are more consistently observed following multiple introductions. This is because, apart from reducing the chances of establishment failure owing to stochastic effects, multiple introductions can influx genetic variation and novel genetic combinations (Tobin *et al.*, 2011).

Finally, the fact that many invasion processes start in human-altered habitats such as urban and agricultural areas could also be associated with a risk of pollen limitation due to depauperate pollinator communities (Hernandez *et al.*, 2016). However, the abundance of ornamental plants, the presence of gardens and highly fertilized areas and the high spatial heterogeneity of some urban areas, can counteract the species loss associated with urbanization (McKinney, 2008). On the other hand, the super-generalist honeybee has been introduced in agricultural and in human-altered areas all around the world (Moritz *et al.*, 2005; MacIvor *et al.*, 2017) and may reduce or negate pollen limitation. For example, *H. coronarium*, that escaped from crops in Menorca where it is highly pollen limited, received 80–93% of its visits from the honeybee (Montero-Castaño *et al.*, 2014).

4.8 Changing Environmental Conditions

The above-mentioned environmental conditions commonly found at the initial stages of invasions change over time and along the invasion process. In fact, successful non-native species that become invasive are highly responsible for those changes. They can alter their recipient environments in different ways, such as modifying species composition, abundance and the diversity of invaded plant communities (Pyšek *et al.*, 2012). This can also cause changes in their own pollination interactions (Chapter 13, this volume), for instance, through density-dependent effects. Large monospecific stands of non-native species increase the floral offer in the invaded area, potentially leading to a shortage of pollinators per floral unit, as was observed in invasive populations of *C. scoparius* in Argentina (Cavallero *et al.*, 2018). Another common output in advanced stages of invasion processes is the reduction of species diversity (Vilà *et al.*, 2011). As plant diversity positively correlates with pollinator diversity (Hegland and Boeke, 2006), invasive populations could experience a decrease of pollinator visits, though that could be rapidly compensated by a positive response by pollinators to the increasing abundance of the invasive (at least until certain floral offer at which there is a shortage of pollinators and the floral offer is so high that the abundance of pollinators per flower decreases, see example above of *C. scoparius*). At larger spatial scales, reduced plant diversity could be detrimental as the pollination services provided by diverse communities are more likely to remain stable under changing conditions (Ghazoul, 2006). In addition, in the long term, a largely monotypic diet comprising non-native floral resources may also be detrimental for the non-native plant through negative consequences to the health of the pollinator community (Vanbergen *et al.*, 2018).

4.9 Combined Effect of Other Global Change Drivers on Plant Invasions

Plant invasions often occur simultaneously with other global change drivers such as landscape alteration, climate change, pathogen spread and use of pesticides (Goulson *et al.*, 2015). Drivers affecting pollination interactions (including those involving non-native species) can nullify each other or have additive and even synergistic effects (González-Varo *et al.*, 2013). However, these interacting effects have been poorly explored using experiments.

Many invasion processes are associated with different aspects of landscape alteration (e.g. habitat fragmentation, habitat loss, habitat isolation, land use intensification, etc.). One of the few empirical studies exploring the combined effect of these two drivers was conducted by Grass *et al.* (2013) in South Africa. Along two gradients comprising non-native relative abundance and land use intensity, the authors found a turnover of pollinator species towards more generalist ones. Contrary to what they expected, they did not find an interactive effect between the two drivers but an additive one, probably due to the low integration of non-native plants in the recipient plant—pollinator communities in the study area.

Any driver affecting the abundance, richness or behaviour of pollinators in invaded areas or the performance of the non-native plants can potentially interact with the invasion process of insect-pollinated plants. Landscape alteration can affect pollinators by changing the availability of and access to food and nesting resources. A meta-analysis showed that pollinator abundance and visitation rates were generally negatively affected by landscape alteration, though effects differed among pollinator taxa and habitat types (Montero-Castaño and Vilà, 2012).

Climate change can alter flowering periods, flight times and distribution areas; moreover, plants and insects may not respond equally, potentially decoupling their interaction. In the Mediterranean basin, insect phenology showed a steeper advance than plant phenology (Gordo and Sanz, 2005) and plant—butterfly interactions showed phenological asynchronies during a period of 17 years associated with aridity (Donoso *et al.*, 2016). Climate change can also alter pollinator—plant interactions through effects on pollinator body size, energetic demands, or on the quantity and quality of food resources in the invaded area (Schweiger *et al.*, 2010), or originating morphological and physiological mismatches between non-native plants and resident pollinators (Tylianakis *et al.*, 2008). The spread of pathogens and the exposition to pesticides reduce pollinator populations and can have sub-lethal effects such as alterations in the foraging behaviour of pollinators (Feltham *et al.*, 2014; Koch *et al.*, 2017).

Finally, invasion processes can influence (and usually facilitate) each other, establishing invasion complexes or meltdowns (Olesen *et al.*, 2002). In some invaded areas, non-native pollinators show preferences for non-native plants, associations that may enhance the successful establishment of both non-native species (Morales and Aizen, 2002). Invasional meltdowns are discussed with more detail in Chapter 8, this volume.

4.10 Implications of Plant–Pollinator Interactions for Management of Plant Invasions

Understanding the role of reproductive barriers in invasion processes is essential for effective management decision making and risk assessment of non-native species (Bufford and Daehler, 2014). Thus, it is essential to consider the role of pollination interactions in overcoming these reproductive barriers. From a theoretical viewpoint, if pollination interactions can favour certain plant invasion processes, effective eradication or prevention measures could be based on disrupting or avoiding such pollination interactions.

4.10.1 Eradication

For eradication measures based on disrupting pollination interactions to be fully efficient, target invasive plant species should exclusively depend on pollinators for their reproduction and preventing the visits by suitable pollinators needs to be feasible. However, even given the first condition, the second one is difficult to meet when invasive plants are generalists, able to switch to new pollinators, long-lived species, or when invaded areas and seed banks are already large. These limitations might explain why we lack examples of eradication measures of invasive plants consisting of disrupting pre-ovule-fertilization; by contrast, disrupting mating processes is an eradication method widely used with other invasive taxa such as insects (e.g. Brockerhoff *et al.*, 2010). Under these scenarios, disrupting other life-stages such as seed dispersal, germination or factors such as resource availability, may have larger detrimental effects on the demography of invasive plants (Lloret *et al.*, 2005; Swope and Parker, 2012).

Even under scenarios in which eradication is not feasible, if an invasive plant lacks the capacity for vegetative reproduction and it is an obligate or facultative outcrosser, its seed production will be susceptible to pollination limitation. If there is a demographic link between seed production and population growth, disrupting pollination interactions could be achieved as a control method (Parker and Haubensak, 2002). If managed pollinators are the main pollinators, invasions may be controlled via their regulation. Disruption of pollination interactions has been suggested for several species such as *Lupinus arboreus* (Stout *et al.*, 2002), *Phyla canescens* (Gross *et al.*, 2010), *Lantana camara* and *C. scoparius* in mainland Australia and Tasmania; *C. scoparius* in NW USA, New Zealand and Argentina and *Solanum torvum* in the USA (Fig. 4.2).

Lantana camara in invaded areas in eastern Australia shows a strong positive correlation between its seed set and honeybee abundance, which is the only pollinator recorded in some of the invaded sites. Therefore, disrupting honeybee visits could have detrimental effects on the species' invasiveness, which could be achieved by removing honeybee hives from the vicinity of invaded areas (Goulson and Derwent, 2004).

Cytisus scoparius, native to Europe, has become invasive in many temperate regions worldwide. Seedling mortality is negligible and high rates of seed mortality (60 ± 99%) are needed to decrease population growth (Sheppard *et al.*, 2002). Therefore, control measures focused on post-fertilization mortality seem unfeasible, while measures to reduce seed production could be more effective (Potter *et al.*, 2009). In invaded areas in southern Argentina, the decrease in the abundance of the invasive *Bombus terrestris* following a volcanic eruption in 2011 resulted in a reduction in flower visitation to *C. scoparius* (Morales *et al.*, 2014), supporting the idea that disrupting pollination interactions could play a key role controlling this invader.

Similarly, in areas of Australia and New Zealand invaded by *C. scoparius*, the only pollinator able to trip its papilionate flowers is the honeybee. Simpson *et al.* (2005) estimated that destroying feral honeybee hives in invaded areas of Australia could reduce *C. scoparius* fruit production by

	Lantana camara	*Cytisus scoparius*			*Solanum torvum*

Breeding system	• Mostly sexual • Mostly self-incompatible, though 2.2% flowers capable of selfing • Dependent on pollinators for ovule fertilisation	• Sexual • Self-compatible • Dependent on pollinators for ovule fertilisation			• Sexual (capable of vegetative propagation in mist chambers) • Self-compatible • Dependent on pollinators for ovule fertilisation
Invaded area	Queensland, Australia	Southern Argentina	Australia	New Zealand	Florida, USA
Main pollinator	*Apis mellifera*	*Bombus terrestris*	*Apis mellifera*	*Apis mellifera*	*Euglossa viridissima*
Proposed control measure	Removal of honeybee hives in invaded areas	Control of *B. terrestris* populations and new introductions	Removal of feral honeybee hives	Removal (or avoidance of the introduction) of honeybee hives plus biological control with the seed-feeder *Bruchidius villosus*	Exclusion of *E. viridissima* visits
Expected results and limitations	99% fruit set reduction in populations with honeybee as the only pollinator. Lower reductions in areas with other pollinators such as butterflies	Not quantified. Poorly feasible due to *B. terrestris* rapid spread in the area	0.7 to 27% fruit set reduction	Extinction of some invading populations	75% fruit set reduction. Poorly feasible due to the naturalisation of *E. viridissima* in the invaded area
References	Goulson and Derwent, 2004	Morales and Aizen, 2002; Morales *et al.*, 2013; Morales *et al.*, 2014	Simpson *et al.*, 2005	Paynter *et al.*, 2010	Liu and Pemberton, 2009

Fig. 4.2. Examples of three invasive plant species for which control methods based on disrupting their pollination interactions have been proposed. Information about their breeding systems and pollination in the invaded areas is given. The proposed control method in those specific areas, together with the expected results and limitations, are also provided. Photos: *Lantana camara* and *Cytisus scoparius*, Ana Montero-Castaño; modified '*Solanum torvum* 5161065' by Forest and Kim Starr, licensed under CC BY 3.0.

between 0.7 and 27%. In New Zealand, the seed-feeder biological control agent *Bruchidius villosus* can reduce *C. scoparius* seed set by an average of 73%. Both strategies combined could reduce seed rain (seeds that reach the ground) to levels where models suggest that most populations will become extinct (Paynter *et al.*, 2010).

Solanum torvum, in invaded areas in Florida, receives visits from a few native bees and one introduced orchid bee, *Euglossa viridissima*. In open areas, where only native bees visit the plant, it suffers pollen limitation. By contrast, in forested areas, where the introduced *E. viridissima* frequently visits the plant, *S. torvum* is not pollen limited. In an experiment conducted in these forested areas, *S. torvum* fruit set was reduced by 75% when *E. viridissima* visits were excluded (Liu and Pemberton, 2009), suggesting that managing the introduction of the bee could contribute to the control of *S. torvum* invasions.

By disrupting pollinator interactions, and thus gene flow among invasive conspecifics, not only the quantity but the quality of visits is reduced. Inbreeding can limit the number of seeds that mature or can reduce the adaptive potential of new individuals. However, the latter is unlikely to prevent the expansion or evolution of novel invasive behaviour (Dlugosch and Parker, 2008a).

4.10.2 Prevention

Some authors have suggested avoiding the introduction of managed and generalist pollinators (mainly honeybees and bumblebees) as a simple and effective preventive measure against the invasion of already introduced pollinator-dependent plants (e.g. Simpson *et al.*, 2005; Dafni *et al.*, 2010). This measure will be more efficient in those areas where managed pollinators are still absent and/or where no native pollinators occupy similar functional niches. One of those areas is mainland Australia, where there are no native bumblebees, but where bumblebee introductions could facilitate invasiveness for several introduced plant species that currently lack efficient pollinators (Dafni *et al.*, 2010). Depauperate native bee faunas also make other areas such as the archipelagos in the South West Pacific region, South America or the Canary Islands, especially vulnerable to the introduction of commercial bumblebees.

Avoiding the introduction of commercial managed pollinators, or removing their feral colonies, is a conservation measure that has also been suggested for preventing negative effects i) on native pollinator faunas, due to competition for floral and nest resources and spread of pathogens (Aizen *et al.*, 2019); ii) on native flora reproduction (Valido *et al.*, 2019); and iii) on crop pollination (Sáez *et al.*, 2014).

Similarly, avoiding the introduction of suitable pollinators can be an effective measure for preventing the establishment of introduced species with specialized pollination systems (Richardson *et al.*, 2000b), although cases of specialized plants becoming more generalist in their introduced areas have been described (e.g. *F. magellanica*; see above). The most studied example of an introduced plant species that only became invasive after its specialized pollinator was also introduced is the fig, *Ficus microcarpa*. This species is native to Asia and Australasia and has invaded tropical and subtropical areas where it was sterile until the arrival of its specialized pollinator, the agaonid wasp *Parapristina verticillatawas* (Nadel *et al.*, 1992). In Hawaii island, the introduced milkweed *Calotropis gigantea* does not produce fruits naturally. The carpenter bee species that pollinate *C. gigantea* in its native range are absent in Hawaii, as well as other functionally equivalent pollinators, preventing its expansion (Bufford and Daehler, 2014). In other areas, its range has expanded in association with pollinators functionally equivalent to the large carpenter bees (Keigh and Longman, 2004; Cavalcante and Major, 2006).

4.11 Conclusions

Evidence so far shows that biotic pollination does not usually act as an effective barrier to invasion. A good fraction of introduced plants are able to overcome the requirement for biotic pollination in their non-native ranges by being able to self, and another substantial fraction of species appear to have been able to link in with either native or non-native pollinators, despite being unable to self-pollinate.

At the initial stages of the invasion process, many non-native plants can face the lack of pollinators and mates due to founder effects. However, certain environmental settings such as the presence of abundant and phenologically similar neighbours in the invaded areas, as well as the potential of some species for phenotypic or genetic adaptive responses, allow introduced species to overcome pollen limitation.

Although most introduced plants do not face pollination barriers, this chapter has provided examples showing that disrupting or avoiding pollination interactions could be achieved as control or prevention measures for certain select species. For these cases, risk assessment protocols considering plant breeding systems would further improve invasion predictions. To advance our understanding of the role of pollination interactions in the plant invasion process, and in invasion management, we need more studies comparing the breeding system and pollination ecology of the plant invader in both the native and the new range.

References

Aizen, M.A. and Harder, L.D. (2007) Expanding the limits of the pollen-limitation concept: effects of pollen quantity and quality. *Ecology* 88(2), 271–281. DOI: 10.1890/06-1017.

Aizen, M.A., Smith-Ramírez, C., Morales, C.L., Vieli, L., Sáez, A. *et al.* (2019) Coordinated species importation policies are needed to reduce serious invasions globally: the case of alien bumblebees in South America. *Journal of Applied Ecology* 56(1), 100–106. DOI: 10.1111/1365-2664.13121.

Alves, E.M.S. and Marins-Corder, M.P. (2009) Reproductive biology of *Acacia Mearnsii* De Wild. (Fabaceae) IV: Flower visitors. *Revista Arvore* 33, 443–450.

Armbruster, W.S. (2017) The specialization continuum in pollination systems: diversity of concepts and implications for ecology, evolution and conservation. *Functional Ecology* 31(1), 88–100. DOI: 10.1111/1365-2435.12783.

Atlan, A., Schermann-Legionnet, A., Udo, N. and Tarayre, M. (2015) Self-incompatibility in *Ulex europaeus*: Variations in native and invaded regions. *International Journal of Plant Sciences* 176(6), 515–524. DOI: 10.1086/681669.

Baker, H.G. (1955) Self-compatibility and establishment after long-distance dispersal. *Evolution* 9, 347–349.

Balogh, C.M. and Barrett, S.C.H. (2018) The influence of floral morph ratios and low plant density on mating and fertility in a Tristylous colonizing species. *Botany* 96(8), 533–545. DOI: 10.1139/cjb-2018-0032.

Balogh, E., Herr, J.M., Czakó, M. and Márton, L. (2012) Defective development of male and female gametophytes in *Arundo donax* L. (POACEAE). *Biomass and Bioenergy* 45, 265–269. DOI: 10.1016/j.biombioe.2012.06.010.

Barrett, S.C.H. (1980) Sexual reproduction in Eichhornia crassipes (water hyacinth). I. fertility of clones from diverse regions. *The Journal of Applied Ecology* 17(1), 101–112.

Barrett, S.C.H. (2011) Why reproductive systems matter for the invasion biology of plants. In: Richardson, D.M. (ed.) *Fifty years of invasion ecology: The legacy of Charles Elton. Wiley-Blackwell*. Oxford, UK, pp. 195–210.

Barrett, S.C.H., Colautti, R.I. and Eckert, C.G. (2008) Plant reproductive systems and evolution during biological invasion. *Molecular Ecology* 17(1), 373–383. DOI: 10.1111/j.1365-294X.2007.03503.x.

Barrett, S.C.H., Harder, L.D. and Worley, A.C. (1996) The comparative biology of pollination and mating in flowering plants. *Philosophical Transactions of the Royal Society B* 351, 1271–1280.

Bartomeus, I. and Vilà, M. (2009) Breeding system and pollen limitation in two supergeneralist alien plants invading Mediterranean shrublands. *Australian Journal of Botany* 57(2), 109–115. DOI: 10.1071/BT08169.

Bascompte, J., Jordano, P. and Olesen, J.M. (2006) Asymmetric coevolutionary networks facilitate biodiversity maintenance. *Science* 312(5772), 431–433. DOI: 10.1126/science.1123412.

Beck-Pay, S.L. (2012) Identification of pre-zygotic reproductive and morphological barriers limiting controlled crossed seed production of triploid *Acacia mearnsii*. *South African Journal of Botany* 79, 51–61. DOI: 10.1016/j.sajb.2011.11.002.

Bittencourt, N.S., Gibbs, P.E. and Semir, J. (2003) Histological study of post-pollination events in *Spathodea campanulata* beauv. (Bignoniaceae), a species with late-acting self-incompatibility. *Annals of Botany* 91, 827–834.

Bowman, G., Tarayre, M. and Atlan, A. (2008) How is the invasive gorse *Ulex europaeus* pollinated during winter? A lesson from its native range. *Plant Ecology* 197, 197–206.

Bowser, C.W. (1957) Introduction and spread of the undesirable tamarisks in the Pacific Southwest section of the United States and comments concerning the plants' influence upon the Indigenous vegetation. *Transactions of the American Geophysical Union* 38, 415–416.

Brockerhoff, E.G., Liebhold, A.M., Richardson, B. and Suckling, D.M. (2010) Eradication of invasive forest insects: concepts, methods, costs and benefits. *New Zealand Journal of Forestry Science* 40, 117–135.

Brown, B.J., Mitchell, R.J. and Graham, S.A. (2002) Competition for pollination between an invasive species (purple loosestrife) and a native congener. *Ecology* 83, 2328–2336.

Bufford, J.L. and Daehler, C.C. (2014) Sterility and lack of pollinator services explain reproductive failure in non-invasive ornamental plants. *Diversity and Distributions* 20, 975–985.

Burns, J.H., Bennett, J.M., Li, J., Xia, J., Arceo-Gómez, G. *et al.* (2019) Plant traits moderate pollen limitation of introduced and native plants: a phylogenetic meta-analysis of global scale. *New Phytologist* 223, 2063–2075.

CABI (2020) Invasive Species Compendium. Available at: www.cabi.org/isc (accessed 15 January 2020).

Carvalheiro, L.G., Biesmeijer, J.C., Benadi, G., Fründ, J., Stang, M. *et al.* (2014) The potential for indirect effects between co-flowering plants via shared pollinators depends on resource abundance, accessibility and relatedness. *Ecology Letters* 17, 1389–1399.

Carvallo, G.O., Medel, R. and Navarro, L. (2013) Assessing the effects of native plants on the pollination of an exotic herb, the blueweed *Echium vulgare* (Boraginaceae). *Arthropod-Plant Interactions* 7, 475–484.

Cavalcante, A. and Major, I. (2006) Invasion of alien plants in the Caatinga biome. *Ambio* 35, 141–143.

Cavallero, L., Morales, C.L., Montero-Castaño, A., Gowda, J.H. and Aizen, M.A. (2018) Scale-dependent effects of conspecific flower availability on pollination quantity and quality in an invasive shrub. *Oecologia* 188, 501–513.

Cesario, L.F. and Gaglianone, M.C. (2013) Pollinators of *Schinus terebinthifolius* Raddi (Anacardiaceae) in vegetational formations of Restinga in northern Rio de Janeiro State. *Bioscience Journal*29, 458–467.

Chen, M. and Zhao, X.-Y. (2015) Pollen limitation and breeding system of Tamarix ramosissima (Ledeb) in patched habitats. *South African Journal of Botany* 99(July 2015), 36–43.

Cordeiro, N. and Silva, L. (2003) Seed production and vegetative growth of *Hedychium gardnerianum* Ker-Gawler (Zingiberaceae) in São Miguel Island (Azores). *Life and Marine Sciences* 20A, 31–36.

Costa, J., Ferrero, V., Castro, M., Loureiro, J., Navarro, L. *et al.* (2017) Variation in the incompatibility reactions in tristylous *Oxalis pes-caprae*: large-scale screening in South African native and Mediterranean basin invasive populations. *Perspectives in Plant Ecology Evolution and Systematics* 24, 25–36.

Daehler, C.C. (1998) Variation in self-fertility and the reproductive advantage of self-fertility for an invading plant (*Spartina alterniflora*). *Evolutionary Ecology* 12, 553–568.

Dafni, A., Kevan, P., Gross, C.L. and Goka, K. (2010) *Bombus terrestris*, pollinator, invasive and pest: An assessment of problems associated with its widespread introductions for commercial purposes. *Applied Entomology and Zoology* 45, 101–113.

Dart, S. and Eckert, C.G. (2013) Experimental and genetic analyses reveal that inbreeding depression declines with increased self-fertilization among populations of a coastal dune plant. *Journal of Evolutionary Biology* 26, 587–599.

Dlugosch, K.M. and Parker, I.M. (2008a) Founding events in species invasions: genetic variation, adaptive evolution, and the role of multiple introductions. *Molecular Ecology* 17, 431–449.

Dlugosch, K.M. and Parker, I.M. (2008b) Invading populations of an ornamental shrub show rapid life history evolution despite genetic bottlenecks. *Ecology Letters* 11, 701–709.

Donoso, I., Stefanescu, C., Martínez-Abraín, A. and Traveset, A. (2016) Phenological asynchrony in plant–butterfly interactions associated with climate: a community-wide perspective. *Oikos* 125, 1434–1444.

el Seed, M.T. and Obeid, M. (1975). *Sexual reproduction of Eichhornia crassipes (Mart.) Solms in the Nile. Weed Research* 15(1), 7–12.

Farizel, L. and Cristina, M. (2008) O biologia floral e fenologia reprodutiva de *Schinus terebinthifolius* Raddi. *Acta Botanica Brasilica* 22, 828–833.

Feltham, H., Park, K. and Goulson, D. (2014) Field realistic doses of pesticide imidacloprid reduce bumblebee pollen foraging efficiency. *Ecotoxicology* 23, 317–323.

Fernandes, L., Rocheta, M., Cordeiro, J., Pereira, S., Gerber, S. *et al.* (2008) Genetic variation, mating patterns and gene flow in a *Pinus pinaster* Aiton clonal seed orchard. *Annals of Forest Science* 65, 706–706.

Ferreira, S.A.N., Antonio, I.C. and Jansen, M.R.A. (1994) Biología reproductiva de *Clidemia hirta* (L.) D. Don (Melastomataceae). *Acta Amazonica* 24(3-4), 183–188. DOI: 10.1590/1809-43921994243188.

Ferreira da Costa, C., Krupek, R. and Krawczyk, A. (2015) Diversidade de visitantes florais e biologia reprodutiva do Araçá (*Psidium cattleianum* Sabine) em fragmento de mata e área urbana. *Bioikos, Campinas* 29, 11–18.

Forman, J. and Kesseli, R.V. (2003) Sexual reproduction in the invasive species *Fallopia japonica* (Polygonaceae). *American Journal of Botany* 90, 586–592.

Gaskin, J.F. and Schaal, B.A. (2002) Hybrid Tamarix widespread in U.S. invasion and undetected in native Asian range. *Proceedings of the National Academy of Sciences* 99(17), 26, 11256–11259.

Gaskin, J.F. and Schaal, B.A. (2003) Molecular phylogenetic investigation of US. invasive Tamarix. *Systematic Botany* 28(1), 86–95.

Gautier, L. (1993) Reproduction of a pantropical weed: *Chromolaena odorata* (L.) King and Robinson (Asteraceae). *Candollea* 48, 179–193.

Geerts, S., Mashele, B.V., Visser, V. and Wilson, J.R.U. (2016) Lack of human-assisted dispersal means *Pueraria montana* var. *lobata* (kudzu vine) could still be eradicated from South Africa. *Biological Invasions* 18, 3119–3126.

Geerts, S. and Pauw, A. (2009) African sunbirds hover to pollinate an invasive hummingbird-pollinated plant. *Oikos* 118, 573–579.

Geslin, B., Gauzens, B., Baude, M., Dajoz, I., Fontaine, C. *et al.* (2017) Massively introduced managed species and their consequences for plant-pollinator interactions. *Advances in Ecological Research* 57, 147–199.

Ghazoul, J. (2006) Floral diversity and the facilitation of pollination. *Journal of Ecology* 94, 295–304.
Global Invasive Species Database (2020a) Species Profile: *Arundo Donax*. Available at: www.iucngisd.org/gisd/species.php?sc=112 (accessed 15 January 2020).
Global Invasive Species Database (2020b) Species Profile: *Hiptage Benghalensis*. Available at: www.iucngisd.org/gisd/species.php?sc=87 (accessed 15 January 2020).
Global Invasive Species Database (2020c) Species Profile: *Morella Faya*. Available at: www.iucngisd.org/gisd/species.php?sc=100 (accessed 15 January 2020).
Godfrey, R., Lepschi, B. and Mallinson, D. (2004) Ecological filtering of exotic plants in an Australian sub-alpine environment. *Journal of Vegetation Science* 15, 227–236.
González-Martínez, S.C., Gerber, S., Cervera, M.T., Martínez-Zapater, J.M., Alía, R. *et al.* (2007) Selfing and sibship structure in a two-cohort stand of maritime pine (*Pinus pinaster* Ait.) using nuclear SSR markers. *Annals of Forest Science* 64, 219–228.
González-Varo, J.P., Biesmeijer, J.C., Bommarco, R., Potts, S.G. and Schweiger, O. (2013) Combined effects of global change pressures on animal-mediated pollination. *Trends in Ecology & Evolution* 28, 524–530.
Gordo, O. and Sanz, J.J. (2005) Phenology and climate change: a long-term study in a Mediterranean locality. *Oecologia* 146, 484–495.
Goulson, D. (2003) Effects of introduced bees on native ecosystems. *Annual Review of Ecology, Evolution and Systematics* 34, 1–26.
Goulson, D. and Derwent, L.C. (2004) Synergistic interactions between an exotic honeybee and an exotic weed: pollination of *Lantana camara* in Australia. *Weed Research* 44(3), 195–202. DOI: 10.1111/j.1365-3180.2004.00391.x.
Goulson, D., Nicholls, E., Botías, C. and Rotheray, E.L. (2015) Bee declines driven by combined stress from parasites, pesticides, and lack of flowers. *Science* 347(6229), 1255957. DOI: 10.1126/science.1255957.
Grass, I., Berens, D.G. and Farwig, N. (2013) Additive effects of exotic plant abundance and land-use intensity on plant–pollinator interactions. *Oecologia* 173, 913–923.
Gross, C.L., Gorrell, L., Macdonald, M.J. and Fatemi, M. (2010) Honeybees facilitate the invasion of *Phyla canescens* (Verbenaceae) in Australia – no bees. *no seed! Weed Research* 50, 364–372.
Grüter, C. and Ratnieks, F.L.W. (2011) Flower constancy in insect pollinators: adaptive foraging behavior or cognitive limitation? *Communicative and Integrative Biology* 4, 633–636.
Harder, L.D. and Aizen, M.A. (2010) Floral adaptation and diversification under pollen limitation. *Philosophical Transactions of the Royal Society B: Biological Sciences* 365(529–543).
Hegland, S.J. and Boeke, L. (2006) Relationships between the density and diversity of floral resources and flower visitor activity in a temperate grassland community. *Ecological Entomology* 31, 532–538.
Hernandez, J.L., Frankie, G.W. and Thorp, R.W. (2016) Ecology of urban bees: a review of current knowledge and directions for future study. *Cities and the Environment* 2, 1–15.
Hervías-Parejo, S. and Traveset, A. (2018) Pollination effectiveness of opportunistic Galápagos birds compared to that of insects: from fruit set to seedling emergence. *American Journal of Botany* 105, 1142–1153.
Hopley, T. and Young, A.G. (2015) Knowledge of the reproductive ecology of the invasive *Salix cinerea,* in its invaded range, assists in more targeted management strategies. *Australian Journal of Botany* 63, 477–483.
Huenneke, L.F. and Vitousek, P.M. (1990) Seedling and clonal recruitment of the invasive tree *Psidium cattleianum*: Implications for management of native Hawaiian forests. *Biological Conservation* 53, 199–211.
Hughes, C.E. (1998) *Leucaena*: A genetic resources handbook. Tropical Forestry Papers 37. *Oxford Forestry Institute, University of Oxford, Oxford, UK.*
Husband, C.B. and Schemske, D.W. (1996) Evolution of the magnitude and timing of inbreeding. *Evolution* 50, 54–70.
ISSG (2020) Global Invasive Species Database. Available at: www.iucngisd.org/gisd (accessed 15 January 2020).
Issaly, E.A., Sérsic, A.N., Pauw, A., Cocucci, A.A., Traveset, A. *et al.* (2020) Reproductive ecology of the bird-pollinated *Nicotiana glauca* across native and introduced ranges with contrasting pollination environments. *Biological Invasions* 22, 485–498.
Keigh, G. and Longman, V. (2004) The naturalized vascular plants of Western Australia 1: checklist, environmental weeds and distribution in IBRA regions. *Plant Protection Quarterly* 19, 12–32.

Klein, A., Vaissière, B.E., Cane, J.H., Steffan-Dewenter, I., Cunningham, S.A. *et al.* (2007) Importance of pollinators in changing landscapes for world crops. *Proceedings of the Royal Society B* 274, 303–313.
Koch, H., Brown, M.J. and Stevenson, P.C. (2017) The role of disease in bee foraging ecology. *Current Opinion in Insect Science* 21, 60–67.
Lakshmi, P. and Raju, A. (2011) *Chromolaena odorata* (L.) King and Robinson (Asteraceae), an important nectar source for adult butterflies. *Journal of Threatened Taxa* 3, 1548–1549.
Lavergne, C., Rameau, J.C. and Figier, J. (1999) The invasive woody weed *Ligustrum robustum* subsp. *walkeri* threatens native forests on La Réunion. *Biological Invasions* 1, 377–392.
Li, H., An, S., Zhi, Y., Yan, C., Zhao, L. *et al.* (2008) Protogynous, pollen limitation and low seed production reasoned for the dieback of *Spartina anglica* in coastal China. *Plant Science* 174, 299–309.
Liebhold, A. and Bascompte, J. (2003) The Allee effect, stochastic dynamics and the eradication of alien species. *Ecology Letters* 6, 133–140.
Liu, H. and Pemberton, R.W. (2009) Solitary invasive orchid bee outperforms co-occurring native bees to promote fruit set of an invasive *Solanum*. *Oecologia* 159(3), 515–525. DOI: 10.1007/s00442-008-1232-6.
Liu, H., Pemberton, R.W. and Stiling, P. (2006) Native and introduced pollinators promote a self-incompatible invasive woody vine (*Paederia foetida* L) in Florida. *The Journal of the Torrey Botanical Society* 133(2), 304–311. DOI: 10.3159/1095-5674(2006)133[304:NAIPPA]2.0.CO;2.
Liu, W., Wang, Y., Chen, Q. and Yu, S. (2013) Pollination of invasive Eichhornia crassipes (Pontederiaceae) by the introduced honeybee (Apis mellifera L.) in South China. *Plant Systematics and Evolution* 299(5), 817–825.
Lloret, F., Médail, F., Brundu, G., Camarda, I., Moragues, E. *et al.* (2005) Species attributes and invasion success by alien plants on Mediterranean islands. *Journal of Ecology* 93, 512–520.
Lukitsch, B. and Elliott, L. (2012) Developing solutions to evolving weed problems. In: Eldershaw, V. (ed.) *Proceedings of the18th Australasian Weeds Conference*. Melbourne, Victoria, Australia, pp. 364–366.
López-García, M.C. and Maillet, J. (2005) Biological characteristics of an invasive South African species. *Biological Invasions* 7, 181–194.
Lopezaraiza-Mikel, M.E., Hayes, R.B., Whalley, M.R. Memmott, J, Martin, R. (2007) The impact of an alien plant on a native plant-pollinator network: an experimental approach. *Ecology Letters* 10(7), 539–550. DOI: 10.1111/j.1461-0248.2007.01055.x.
Lowe, S., Browne, M. and Boudjelas, S. (2000) 100 of the world's worst invasive alien species. A selection from the Global Invasive Species Database. Published by the Invasive Species Specialist Group (ISSG) a specialist group of the Species Survival Commission (SSC) of the World Conservation Union (IUCN), 12.
Macanawai, A.R., Day, M.D. and Adkins, S.W. (2015) Effects of age, length, and pattern of burial on survival of *Mikania micrantha* stem sections. *Pacific Science* 69(1), 95–102. DOI: 10.2984/69.1.7.
MacDonald, G.E. (2007) Cogongrass (*Imperata cylindrica*): Biology, distribution, and impacts in the southeastern U.S. In: Loewenstein, N.J. and Miller, J.H. (eds) *Proceedings of the Regional Cogongrass Conference: A Cogongrass Management Guide. Confronting the Cogongrass Crisis Across the South*. Alabama, U.S, pp. 10–23.
MacIvor, J.S., Roberto, A.N., Sodhi, D.S., Onuferko, T.M. and Cadotte, M.W. (2017) Honey bees are the dominant diurnal pollinator of native milkweed in a large urban Park. *Ecology and Evolution* 7, 8456–8462.
Mal, T.K., Lovett-Doust, J., L, L.-D. and Mulligan, G.A. (1992) The biology of Canadian weeds. 100. *Lythrum salicaria. Canadian Journal of Plant Science* 72, 1305–1330.
McKinney, M.L. (2008) Effects of urbanization on species richness: a review of plants and animals. *Urban Ecosystems* 11, 161–176.
Memmott, J. and Waser, N.M. (2002) Integration of alien plants into a native flower–pollinator visitation web. *Proceedings of the Royal Society of London B* 269, 2395–2399.
Meyer, J.Y. (2010) The *Miconia Saga*, 20 years of study and control in French Polynesia (1988-2008). In: Loope, L.L., Meyer, J.Y., Hardesty, B.Y. and Smith, C.W. (eds) *Proceedings of the International Miconia Conference*. Hawaii, USA, 4-7 May 2009, Keanae, Maui.
Mitchell, R.J., Flanagan, R.J., Brown, B.J., Waser, N.M. and Karron, J.D. (2009) New frontiers in competition for pollination. *Annals of Botany* 103, 1403–1413.
Montero-Castaño, A. and Vilà, M. (2012) Impact of landscape alteration and invasions on pollinators: a meta-analysis. *Journal of Ecology* 100(4), 1–10. DOI: 10.1111/j.1365-2745.2012.01968.x.

Montero-Castaño, A., Vilà, M. and Ortiz-Sánchez, F.J. (2014) Pollination ecology of a plant in its native and introduced areas. *Acta Oecologica* 56, 1–9. DOI: 10.1016/j.actao.2014.01.001.

Morales, C.L. and Aizen, M.A. (2002) Does invasion of exotic plants promote invasion of exotic flower visitors? A case study from the temperate forests of the southern Andes. *Biological Invasions* 4(1/2), 87–100. DOI: 10.1023/A:1020513012689.

Morales, C.L. and Aizen, M.A. (2006) Invasive mutualisms and the structure of plant-pollinator interactions in the temperate forests of north-west Patagonia, Argentina. *Journal of Ecology* 94(1), 171–180. DOI: 10.1111/j.1365-2745.2005.01069.x.

Morales, C.L. and Traveset, A. (2008) Interspecific pollen transfer: magnitude, prevalence and consequences for plant fitness. *Critical Reviews in Plant Sciences* 27(4), 221–238. DOI: 10.1080/07352680802205631.

Morales, C.L. and Traveset, A. (2009) A meta-analysis of impacts of alien vs. native plants on pollinator visitation and reproductive success of co-flowering native plants. *Ecology Letters* 12(7), 716–728. DOI: 10.1111/j.1461-0248.2009.01319.x.

Morales, C.L., Sáez, A., Arbetman, M.P., Cavallero, L. and Aizen, M.A. (2014) Detrimental effects of volcanic ash deposition on bee fauna and plant-pollinator interactions. *Ecología Austral* 24(1), 042–050. DOI: 10.25260/EA.14.24.1.0.36.

Moritz, R.F.A., Hartel, S. and Neumann, P. (2005) Global invasions of the western honeybee (*Apis mellifera*) and the consequences for biodiversity. *Ecoscience* 12, 289–301.

Morrow, L. (1979) Studies on the reproductive biology of leafy spurge (*Euphoriba esula*). *Weed Science* 27, 106–109.

Mullarkey, A.A., Byers, D.L. and Anderson, R.C. (2013) Inbreeding depression and partitioning of genetic load in the invasive biennial *Alliaria petiolata* (Brassicaceae). *American Journal of Botany* 100, 509–518.

Nadel, H., Frank, J.H. and Knight, R.J. (1992) Escapees and accomplices: the naturalization of exotic *Ficus* and their associated faunas in Florida. *The Florida Entomologist* 75(1), 29–38. DOI: 10.2307/3495478.

Nehring, S. and Adsersen, H. (2006) NOBANIS – Invasive Alien Species Fact Sheet – *Spartina Anglica*. Available at: www.nobanis.org (accessed 15 January 2020).

Olesen, J.M., Eskildsen, L.I. and Venkatasamy, S. (2002) Invasion of pollination networks on oceanic islands: importance of invader complexes and endemic super generalists. *Diversity Distributions* 8(3), 181–192. DOI: 10.1046/j.1472-4642.2002.00148.x.

Olesen, J.M. and Valido, A. (2003) Lizards as pollinators and seed dispersers: an island phenomenon. *Trends in Ecology & Evolution* 18(4), 177–181. DOI: 10.1016/S0169-5347(03)00004-1.

Ollerton, J., Winfree, R. and Tarrant, S. (2011) How many flowering plants are pollinated by animals? *Oikos* 120(3), 321–326. DOI: 10.1111/j.1600-0706.2010.18644.x.

Padrón, B., Traveset, A., Biedenweg, T., Díaz, D., Nogales, M. *et al.* (2009) Impact of alien plant invaders on pollination networks in two archipelagos. *PLoS ONE* 4(7), e6275. DOI: 10.1371/journal.pone.0006275.

Parker, I.M. (1997) Pollinator limitation of *Cytisus scoparius* (Scotch Broom), an invasive exotic shrub. *Ecology* 78(5), 1457–1470. DOI: 10.1890/0012-9658(1997)078[1457:PLOCSS]2.0.CO;2.

Parker, I.M. and Haubensak, K.A. (2002) Comparative pollinator limitation of two non-native shrubs: do mutualisms influence invasions? *Oecologia* 130(2), 250–258. DOI: 10.1007/s004420100799.

Pascarella, J.B. (1997) Breeding systems of *Ardisia* Sw. (Myrsinaceae). *Brittonia* 49(1), 45–53. DOI: 10.2307/2807693.

Pasiecznik, N.M. and Felker, P. (1992) Mechanical cleaning of *Prosopis* seed. *Nitrogen Fixing Tree Research Reports* 10, 186–188.

Paynter, Q., Main, A., Hugh Gourlay, A., Peterson, P.G., Fowler, S.V. *et al.* (2010) Disruption of an exotic mutualism can improve management of an invasive plant: varroa mite, honeybees and biological control of Scotch broom *Cytisus scoparius* in New Zealand. *Journal of Applied Ecology* 47(2), 309–317. DOI: 10.1111/j.1365-2664.2010.01784.x.

Potter, K.J.B., Kriticos, D.J., Watt, M.S. and Leriche, A. (2009) The current and future potential distribution of *Cytisus scoparius* : a weed of pastoral systems, natural ecosystems and plantation forestry. *Weed Research* 49(3), 271–282. DOI: 10.1111/j.1365-3180.2009.00697.x.

Pyšek, P., Jarošík, V., Chytrý, M., Danihelca, J., Kühn, I. *et al.* (2011) Successful invaders co-opt pollinators of native flora and accumulate insect pollinators with increasing residence time. *Ecological Monographs* 81(2), 277–293. DOI: 10.1890/10-0630.1.

Pyšek, P., Jarošík, V., Hulme, P.E., Pergl, J., Hejda, M. *et al.* (2012) A global assessment of invasive plant impacts on resident species, communities and ecosystems: the interaction of impact measures, invading species' traits and environment. *Global Change Biology* 18(5), 1725–1737. DOI: 10.1111/j.1365-2486.2011.02636.x.

Pyšek, P., Prach, K. and milauer, P. (1995) Relating invasion success to plant traits: an analysis of the Czech alien flora. In: Pyšek, P., Prach, K., Rejmánek, M. and Wade, M. (eds) *Plant invasions: general aspects and special problems. SPB Academic.* The Netherlands, Amsterdam, pp. 39–60.

Pyšek, P., Richardson, D.M., Rejmánek, M., Webster, G.L., Williamson, M. *et al.* (2004) Alien plants in checklists and floras: towards better communication between taxonomists and ecologists. *TAXON* 53(1), 131–143. DOI: 10.2307/4135498.

Quesada, M., Stoner, K.E., Rosas-Guerrero, V., Palacios-Guevara, C. and Lobo, J.A. (2003) Effects of habitat disruption on the activity of nectarivorous bats (Chiroptera: Phyllostomidae) in a dry tropical forest: implications for the reproductive success of the neotropical tree *Ceiba grandiflora*. *Oecologia* 135(3), 400–406. DOI: 10.1007/s00442-003-1234-3.

Razanajatovo, M., Maurel, N., Dawson, W., Essl, F., Kreft, H. *et al.* (2016) Plants capable of selfing are more likely to become naturalized. *Nature Communications* 7, 13313.

Ren, M.-X., Zhong, Y.-F. and Song, X.-Q. (2013) Mirror-image flowers without buzz pollination in the Asian endemic *Hiptage benghalensis* (Malpighiaceae). *Botanical Journal of the Linnean Society* 173(4), 764–774. DOI: 10.1111/boj.12101.

Richardson, D.M., Allsopp, N., D'Antonio, C.M., Milton, S.J. and Rejmánek, M. (2000a) Plant invasions--the role of mutualisms. *Biological Reviews of the Cambridge Philosophical Society* 75(1), 65–93. DOI: 10.1017/S0006323199005435.

Richardson, D.M., Pyšek, P., Rejmánek, M., Barbour, M.G., Panetta, F.D. *et al.* (2000b) Naturalization and invasion of alien plants: concepts and definitions. *Diversity and Distributions* 6(2), 93–107. DOI: 10.1046/j.1472-4642.2000.00083.x.

Rodger, J.G. (2011) Consequences of self-fertilization for fecundity and progeny performance in invasive plants. PHD thesis. University of KwaZulu-Natal, Pietermaritzburg, South Africa.

Rodger, J. G., van Kleunen, M. and Johnson, S. D. (2010) Does specialized pollination impede plant invasions? *International Journal of Plant Sciences* 171(4), 382–391. DOI: 10.1086/651226.

Rosas-Guerrero, V., Aguilar, R., Martén-Rodríguez, S., Ashworth, L., Lopezaraiza-Mikel, M. *et al.* (2014) A quantitative review of pollination syndromes: do floral traits predict effective pollinators? *Ecology Letters* 17(3), 388–400. DOI: 10.1111/ele.12224.

Sáez, A., Morales, C.L., Ramos, L.Y. and Aizen, M.A. (2014) Extremely frequent bee visits increase pollen deposition but reduce drupelet set in raspberry. *Journal of Applied Ecology* 51(6), 1603–1612. DOI: 10.1111/1365-2664.12325.

Schrieber, K. and Lachmuth, S. (2017) The genetic paradox of invasions revisited: the potential role of inbreeding × environment interactions in invasion success. *Biological Reviews* 92(2), 939–952. DOI: 10.1111/brv.12263.

Schweiger, O., Biesmeijer, J.C., Bommarco, R., Hickler, T., Hulme, P.E. *et al.* (2010) Multiple stressors on biotic interactions: how climate change and alien species interact to affect pollination. *Biological Reviews* 85(4), no–795. DOI: 10.1111/j.1469-185X.2010.00125.x.

Sheppard, A.W., Hodge, P., Paynter, Q. and Rees, M. (2002) Factors affecting invasion and persistence of broom *Cytisus scoparius* in Australia. *Journal of Applied Ecology* 39(5), 721–734. DOI: 10.1046/j.1365-2664.2002.00750.x.

Simpson, S.R., Gross, C.L. and Silberbauer, L.X. (2005) Broom and honeybees in Australia: an alien liaison. *Plant Biology* 7(5), 541–548. DOI: 10.1055/s-2005-865855.

Solbrig, O. and Cantino, P. (1975) Reproductive adaptations in *Prosopis* (Leguminosae-Mimosoideae). *Journal of the Arnold Arboretum* 56, 185–210.

Spears, E.E. (2009) Island and mainland pollination ecology of *Centrosema virginianum* and *Opuntia stricta*. *Journal of Ecology* 75, 351–362.

Starr, F., Starr, K. and Loope, L. (2003) *Cinchona pubescens: Quinine tree, Rubiaceae*. Plants of Hawaii Report.

Stouffer, D.B., Cirtwill, A.R., Bascompte, J. and Bartomeus, I. (2014) How exotic plants integrate into pollination networks. *Journal of Ecology* 102(6), 1442–1450. DOI: 10.1111/1365-2745.12310.

Stout, J.C., Kells, A.R. and Goulson, D. (2002) Pollination of the invasive exotic shrub *Lupinus arboreus* (Fabaceae) by introduced bees in Tasmania. *Biological Conservation* 106(3), 425–434. DOI: 10.1016/S0006-3207(02)00046-0.

Stout, J.C., Parnell, J.A.N., Arroyo, J. and Crowe, T.P. (2006) Pollination ecology and seed production of *Rhododendron ponticum* in native and exotic habitats. *Biodiversity and Conservation* 15(2), 755–777. DOI: 10.1007/s10531-004-1065-5.

Suárez-Mariño, A., Arceo-Gómez, G., Sosenski, P. and Parra-Tabla, V. (2019) Patterns and effects of heterospecific pollen transfer between an invasive and two native plant species: the importance of pollen arrival time to the stigma. *American Journal of Botany* 106(10), 1308–1315. DOI: 10.1002/ajb2.1361.

Swarbrick, J.T., Timmins, S.M. and Bullen, K.M. (1999) The biology of Australian weeds. 36. Ligustrum lucidum Aiton and Ligustrum sinense Lour. *Plant Protection Quarterly* 14(4), 122–130.

Swope, S.M. and Parker, I.M. (2012) Complex interactions among biocontrol agents, pollinators, and an invasive weed: a structural equation modeling approach. *Ecological Applications* 22(8), 2122–2134. DOI: 10.1890/12-0131.1.

Thompson, D.Q., Stuckey, R.L. and Thompson, E.B. (1987) Spread, impact, and control of purple loosestrife (*Lythrum salicaria*) in North American wetlands. United States Fish and Wildlife Service, Fish and Wildlife Research No.2. *Washington DC, USA*.

Thompson, A.H. and Knight, T.M. (2018) Exotic plant species receive adequate pollinator service despite variable integration into plant-pollinator networks. *Oecologia* 187(1), 135–142. DOI: 10.1007/s00442-018-4096-4.

Thomson, J.D. (1978) Effects of stand composition on insect visitation in two-species mixtures of *Hieracium*. *American Midland Naturalist* 100(2), 431–440. DOI: 10.2307/2424843.

Thornton, M.R. (2004) Arthopod fauna associated with kudzu (*Pueraria montana* var. *lobata* Willd) in North Carolina. *MsC Thesis. North Carolina State University*.

Tobin, P.C., Berec, L. and Liebhold, A.M. (2011) Exploiting Allee effects for managing biological invasions. *Ecology Letters* 14(6), 615–624. DOI: 10.1111/j.1461-0248.2011.01614.x.

Traveset, A. and Richardson, D.M. (2014) Mutualistic interactions and biological invasions. *Annual Review of Ecology, Evolution, and Systematics* 45(1), 89–113. DOI: 10.1146/annurev-ecolsys-120213-091857.

Tylianakis, J.M., Didham, R.K., Bascompte, J. and Wardle, D.A. (2008) Global change and species interactions in terrestrial ecosystems. *Ecology Letters* 11(12), 1351–1363. DOI: 10.1111/j.1461-0248.2008.01250.x.

Valentine, D.H. (1978) The pollination of introduced species, with special reference to the British Isles and the genus *Impatiens*. In: Richards, A.J. (ed.) *The pollination of flowers by insects. Academic Press for the Linnean Society of London*. London, UK, pp. 117–123.

Valido, A., Rodríguez-Rodríguez, M.C. and Jordano, P. (2019) Honeybees disrupt the structure and functionality of plant-pollinator networks. *Scientific Reports* 9(1), 1–11. DOI: 10.1038/s41598-019-41271-5.

Van, T.K., Rayamajhi, M.B. and Center, T.D. (2005) Seed longevity of *Melaleuca quinquenervia*: a burial experiment in South Florida. *Journal of Aquatic Plant Management* 43, 39–42.

van Kleunen, M., Manning, J.C., Pasqualetto, V. and Johnson, S.D. (2008) Phylogenetically independent associations between autonomous self-fertilization and plant invasiveness. *The American Naturalist* 171(2), 195–201. DOI: 10.1086/525057.

Vanbergen, A.J., Espíndola, A. and Aizen, M.A. (2018) Risks to pollinators and pollination from invasive alien species. *Nature Ecology & Evolution* 2(1), 16–25. DOI: 10.1038/s41559-017-0412-3.

Vilà, M., Bartomeus, I., Gimeno, I., Traveset, A. and Moragues, E. (2006) Demography of the invasive geophyte *Oxalis pes-caprae* across a Mediterranean island. *Annals of Botany* 97(6), 1055–1062. DOI: 10.1093/aob/mcl052.

Vilà, M., Espinar, J.L., Hejda, M., Hulme, P.E., Jarošík, V. *et al*. (2011) Ecological impacts of invasive alien plants: a meta-analysis of their effects on species, communities and ecosystems. *Ecology Letters* 14(7), 702–708.

Vivian-Smith, G. and Panetta, F.D. (2009) Lantana (*Lantana camara*) Seed Bank Dynamics: Seedling Emergence and Seed Survival. *Invasive Plant Science and Management* 2(2), 141–150. DOI: 10.1614/IPSM-08-130.1.

Witkowski (2002) Invasion intensity and regeneration potential of the non-native invasive plant *Chromolaena odorata* at St Lucia, South Africa. In: Zachariades, C., Muniappan, R. and Strathie, L.W. (eds) *Proceedings of the Fifth International Workshop on Biological Control and Management of Chromolaena odorata*. Durban, South Africa, pp. pp. 106–117.

Zhou, Z., Miwa, M., Nara, K., Wu, B., Nakaya, H. *et al*. (2003) Patch establishment and development of a clonal plant, *Polygonum cuspidatum*, on Mount Fuji. *Molecular Ecology* 12(6), 1361–1373. DOI: 10.1046/j.1365-294X.2003.01816.x.

5 Seed Dispersal Interactions Promoting Plant Invasions

M. Celeste Díaz Vélez*, Ana E. Ferreras and Valeria Paiaro

Instituto Multidisciplinario de Biología Vegetal, CONICET-Universidad Nacional de Córdoba. Córdoba, Argentina

Abstract

Animal dispersers are essential for many non-native plants since they facilitate seed movement and might promote seed germination and seedling establishment, thereby increasing their chances of invasion. This chapter reviews the published literature on seed dispersal of non-native plant species by native and/or non-native animals. The following questions are addressed: i) Are interactions between non-native plants and their animal dispersers evenly studied worldwide? ii) Which are the distinctive traits (i.e. geographical origin, life form, dispersal strategy and propagule traits) of non-native plants that are dispersed by animals? iii) Which are the most studied groups of dispersers of non-native plants around the world? iv) Does the literature provide evidence for the Invasional Meltdown Hypothesis (non-native plant—non-native disperser facilitation)? v) What is the role of animal dispersers at different stages of the non-native plant regeneration process? Our dataset of 204 articles indicates that geographical distribution of the studies was highly heterogeneous among continents, with the highest number coming from North America and the lowest from Asia and Central America. Most of the non-native plants involved in dispersal studies were woody species from Asia with fleshy fruits dispersed by endozoochory. More than the half of the animal dispersal agents noted were birds, followed by mammals, ants and reptiles. The dominance of bird-dispersal interactions over other animal groups was consistent across geographical regions. Although most of the studies involved only native dispersers, interactions among non-native species were detected, providing support for the existence of invasional meltdown processes. Of the total number of reviewed articles reporting seed removal, 74% evaluated seed dispersal, but only a few studies included seed germination (35.3%), seedling establishment (5.4 %) or seed predation (23.5 %). Finally, we discuss some research biases and directions for future studies in the area.

5.1 Introduction

Mutualistic interactions are essential for the maintenance of biodiversity and ecosystem services worldwide (Bronstein, 2015). However, mutualisms can also favour the success of non-native species, which in turn can alter the structure and functioning of ecosystems (Traveset and Richardson, 2014). Although humans are the most important dispersal vectors of these species (Mack *et al.*, 2000), finding other animal dispersers in new areas is essential for many plants to ensure their establishment and to increase their chances of population growth (Farwig and Berens, 2012). The disperser might move non-native seeds to sites that provide the necessary conditions for seed germination and seedling establishment (Farwig and Berens, 2012). Accordingly, a recent study comparing a total of 45 traits between invasive and non-invasive non-native woody species in the USA found that four out of the six distinctive traits were related to seed dispersal (Nuñez-Mir *et al.*, 2019). Thus, mutualistic interactions between non-native plants and animal dispersers, non-native or native, might be important drivers of

*Corresponding author: celestemcbiol@gmail.com

 Plant Invasions: The Role of Biotic Interactions (eds Anna Traveset and David M. Richardson)
DOI: 10.1079/9781789242171.0005

the invasion success of many non-native plant species worldwide.

In this chapter we review the published literature on seed dispersal of non-native plant species by animals. After compiling data from the literature, we assess the geographical distribution of the reviewed studies. Then, we analyse general information regarding features of the non-native plants involved in dispersal interactions, including geographical origins, life forms, main dispersal strategies (i.e. endozoochory, synzoochory and epizoochory) and propagule traits. We follow with data about the different groups of animals involved in non-native seed dispersal (i.e. birds, mammals, ants and reptiles), providing information on their relative importance and dispersal strategies around the world, and their role in plant invasion ecology. We describe how the literature on animal seed dispersal provides evidence for the Invasional Meltdown Hypothesis (non-native plant—non-native disperser facilitation). Then, we evaluate measures of seed dispersal effectiveness and the role of animal dispersers at the different stages of the non-native plant regeneration process, specifically seed germination, seedling establishment and seed predation. Finally, we propose some directions for future studies on the dispersal of non-native plant species.

5.2 Methods: Data set

We searched for peer-reviewed publications of seed dispersal of non-native plant species by animals in the Scopus scientific literature database (March 2019). The search comprised the following keyword combinations within article title, abstract or keywords: (bird OR mammal OR bat OR ant OR fish OR amphibian OR reptile) AND ('non-native' OR invasi* OR alien OR exotic) AND (diaspore OR fruit OR seed). Only for mammals, we also included a particular reference to bats due to the great importance of this group as dispersers and the few studies that were obtained by searching only with 'mammal' as keyword. The search yielded a total of 1100 articles, of which we selected those that include a non-native plant (invasive or non-invasive) in which at least seed removal was measured and a minimum of one removal agent (native or non-native) was identified. Our final database covered 204 articles including a total of 844 non-native plant species and 672 animal species. For each selected article we then compiled information on the invaded area (country, continent), non-native plants (i.e. species, life form and geographical origin), animal dispersers (i.e. group, species and status – native or non-native), and dispersal type (i.e. endozoochory, synzoochory or epizoochory). The results are presented as percentages which were calculated considering the total amount of records found in our database for each category (i.e. the number of records could be more than one per article; for example when two groups of dispersers were found within the same article). We also assessed if the studies recorded seed dispersal (providing data on seed deposition in the soil, faeces, mist nets or seed rain), seed predation, seed germination and seedling establishment. In the case of germination, we evaluated the use of control treatments and the assessment of mechanisms underlying germination responses of dispersed seeds (de-inhibition or scarification). Percentages of studies assessing each of these aspects were calculated based on the total number of articles. Although our review differentiated studies that involved only fruit or seed removal from those that assessed actual dispersal, for simplicity, here we use the term *dispersal* when we refer to the whole database in the text.

5.3 Results and Discussion

5.3.1 Are interactions between non-native plants and their animal dispersers evenly studied worldwide?

There is an important heterogeneity in the global distribution of studies assessing mutualisms between non-native plants and their dispersers (Fig. 5.1; see also Chapter 6, this volume). Our search showed that North America accounted for the highest number of studies, most of which (near 80%) were from the USA, followed by Europe, South America, Oceania and Africa (with similar numbers of studies). The position of Africa was largely determined by South Africa, which alone accounted for near 70% of research efforts in this continent. Asia accounted

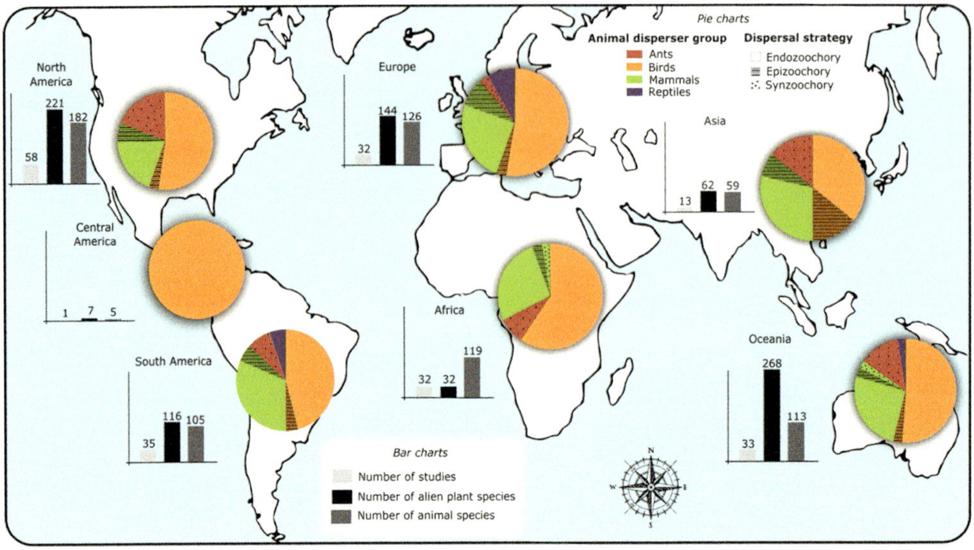

Fig. 5.1. Geographical structure of data on dispersal of non-native plants obtained from the Scopus scientific literature database. Number of studies, plant species and animal species in each continent (or continental region in the case of the Americas) are presented in bar charts. Each species was counted only once per region, regardless of the number of studies or interactions it was involved in. The proportional contribution of different groups of animal dispersers and their dispersal type (i.e. endozoochory, synzoochory and epizoochory) across geographical regions are shown in pie charts.

for a lower number of studies (less than half than other continents), whereas only one study was from Central America. According to global databases about non-native plants distribution worldwide (Weber, 2003; Richardson and Rejmánek, 2011; Rejmánek and Richardson, 2013) and distributions of plants that depend on animals as dispersal vectors (Farwig and Berens, 2012), Oceania is the continent with the highest number of non-native plant species and a tropical area with high plant dependence on animal dispersers. Thus, the poor number of studies in our database suggests low research intensity in this continent. However, in accordance with global databases (Weber, 2003; Richardson and Rejmánek, 2011), Oceania in our database showed the highest number of non-native species studied, even more than North America with almost double number of studies (Fig. 5.1). The discrepancy between the number of studies and the number of plant species studied could be explained by differences in the number of species per study. In fact, Oceania has the higher number of studies with more than ten non-native plant species, which account for 220 of the 268 plants studied in this continent (Table 5.1). North America, on the other hand, exhibits high research effort, considering that it has the highest number of dispersal studies, even though in global databases it has a similar or even lower number of non-native plant species in comparison with other regions (Weber, 2003; Richardson and Rejmánek, 2011) and that plants in this region are less dependent on animal dispersal than plants in tropical regions (Farwig and Berens, 2012). All in all, we observed that the number of non-native plant species studied in each region does not match the number of studies (Fig. 5.1), which could be an indirect proxy of the level of detail at which different non-native plant species were studied in different continents.

The number of animal species studied in each region was lower than the number of non-native plant species, except in Africa, where animal species were three times the number of studied non-native plants (Fig. 5.1). In this case, only three studies accounted for 77 of the 119

Table 5.1. Number of studies with more than ten species of plants or animals in each continent (or continental region in the case of the Americas) and number of species involved in such studies (in parentheses).

Continent/continental region	N° studies with more than 10 species of plants (n° spp.)	N° studies with more than 10 species of animals (n° spp.)
Africa	0	3 (77)
Asia	2 (43)	2 (31)
Europe	5 (93)	5 (74)
North America	6 (159)	6 (148)
Oceania	7 (220)	3 (59)
South America	2 (53)	4 (65)
Central America	0	0

animal species studied (Table 5.1). From these three studies, one studied many non-native plants interacting with their dispersers, but the other two studied *Acacia cyclops* and *A. saligna* which alone interact with 53 species of dispersers. This showed, on the one hand, the higher number of interactions that non-native species could exhibit in their recipient communities and, on the other hand, the importance of increasing studies in Africa.

Finally, the studies included in our database that were performed on islands accounted for 27%, which could be explained by the fact that islands are areas with a high abundance of non-native plants (Richardson and Rejmánek, 2011) and with plant species highly dependent on particular animal species (Schleuning *et al.*, 2014).

Overall, the discrepancies in research efforts across continents highlight the need for further studies, particularly in tropical regions, where animals are known to disperse up to 90% of the seeds (Farwig and Berens, 2012), and some understudied temperate regions, for example northern areas within Asia.

5.3.2 Which Are the Distinctive Traits of Non-native Plants Dispersed by Animals?

Geographical origins

According to our database (from the 125 articles that mentioned the origin of the non-native plants), most non-native plant species involved in dispersal studies were from Asia (36.3%). Europe (15.6%) and North America (15%) were the second most frequent donors of non-native plants, followed by South and Central America (10 and 9.4%, respectively). Finally, only 7.5 and 6.3% of the plant species included in our database were from Oceania and Africa, respectively. The predominance of Asian and European species is typical of non-native floras worldwide (e.g. Weber, 2003; Stohlgren *et al.*, 2011). The positions of donor regions are clearly influenced by historical/cultural factors over the past few centuries (Mack *et al.*, 2003). For example, European colonial history has played an important role in the dissemination of plants around the world (Crosby, 1986). More recently, intentional exchange for specific purposes, such as forestry, agriculture and ornamental use, has dominated the introduction of non-native plants (Hulme *et al.*, 2008; Taylor, 2009). This modern plant commerce might explain the high frequency of movement of species from the floristically rich Asian region (particularly China) to the rest of the world (Richardson and Rejmánek, 2011).

Life forms

Most of the studies in our database recorded dispersal interactions of woody non-native plants (72%), including trees, shrubs and vines. In many parts of the world, woody species dominate the lists of non-native plants and are among the most conspicuous and damaging invasive species (Pyšek *et al.*, 2008; Richardson and Rejmánek, 2011). In addition, non-native trees and shrubs are mostly dispersed by animals, mainly birds (Richardson and Rejmánek, 2011). Trees were the most studied non-native lifeform (41% of studies), followed by shrubs (27%). Herbs were the next most common lifeform in dispersal studies (21%), while non-native grasses and vines were much less represented in our database (7 and 4%, respectively).

Dispersal strategies

Animal seed dispersal can occur through different mechanisms that can be summarized into three major types: i) endozoochory (i.e. when seeds are ingested by the disperser); ii) synzoochory (i.e. when seeds are transported externally, but generally in the mouth, which might be followed by the hoarding of at least a fraction of those seeds; Gómez *et al.*, 2019); and iii) epizoochory, also named ectozoochory (i.e. when seeds are transported in dispersers' fur, feathers, and/or skin). Epizoochory, which is a more specific type of interaction than endozoochory because it is associated with a narrower range of dispersal seed morphologies (Couvreur *et al.*, 2005), is characteristic of many non-native plants found in disturbed areas (Castillo-Flores and Calvo-Irabién, 2003; Dovrat *et al.*, 2012). For example, Chuong *et al.* (2016) found that most non-native seeds dispersed by cattle in California rangelands have epizoochorous adaptations. In our database, epizoochory was the least frequent type of interaction, accounting for 9.3% of the dispersal interactions, coincident with the global percentage of angiosperms that showed this dispersal mode (10%; Kulbaba *et al.*, 2009). The number of studies dealing with synzoochory was 13.7%, while most studies in our database recorded endozoochory interactions (77%), confirming the overwhelming relevance of this type of dispersal for plant invasion (Daniel Kissling *et al.*, 2009; Richardson and Rejmánek, 2011; see also Chapter 6, this volume).

Fruit and seed traits

The success of non-native plant species is partly attributed to fruit or seed traits that favour effective seed dispersal (Buckley *et al.*, 2006). Indeed, at least two-thirds of the most widespread invasive woody plants have fleshy fruits or produce other structures to attract animal dispersers (Traveset and Richardson, 2014). One valuable approach used to assess the role of propagule traits in promoting invasion is to compare co-occurring native and non-native plants in terms of fruit traits and dispersal performance (Daehler, 2003). In this regard, a study that characterized up to 38 traits of fruit morphology, pulp nutrients and phenology of 34 vertebrate-dispersed non-native species in Australia found that non-native fruits had significantly smaller seeds, and pulp contained more sugar and higher variability in nitrogen content than indigenous species (Gosper and Vivian-Smith, 2010). Another study found that some invasive species produced fruits with lower water content, resulting in a higher relative yield and energy content (Kueffer *et al.*, 2009). However, there is a recent study in subtropical Andean Forest that found that fruits of native and non-native species are very similar in their nutrient content (Rojas *et al.*, 2019). In several studies, non-native plants were found to produce more fruits than their native congeners, possibly contributing to their greater dispersal (Traveset and Richardson, 2006; Ferreras and Galetto, 2010; *but see* Drummond, 2005). In addition, non-native species tended to have longer fruiting seasons than natives (Gosper and Vivian-Smith, 2010) or to fruit when native fruit production is limited, favouring more readily consumption of non-native fruits, contributing to their more rapid spread (Gurvich *et al.*, 2005; Chimera and Drake, 2010; Tecco *et al.*, 2013; but see Vergara-Tabares *et al.*, 2016).

Regarding propagule traits that guarantee their transport by the different groups of animal dispersers, a survey based on traits of c. 200 non-native plant species (Cronk and Fuller, 1995) showed that 25% were adapted to dispersal by birds, 14% by mammals and 1% by ants, whereas 45% had no obvious adaptations for animal dispersal. Non-native plants dispersed by frugivorous birds usually have showy (black, blue or red) small-sized fleshy fruits (Richardson *et al.*, 2000; Gosper and Vivian-Smith, 2009). In epizoochorous non-native seeds dispersed by mammals, diaspore size, shape and presence of adhesive structures have been shown to influence dispersal distance (Castillo-Flores and Calvo-Irabién, 2003). Non-native plant species dispersed by ants usually present elaiosomes or caruncles in their seeds (Jongejans *et al.*, 2015).

5.3.3 Which Are the Most Studied Groups of Dispersers of Non-native Plants Around the World?

From the total of articles dealing with seed removal (except those dealing only with predation), over a half included birds (55.3 %), followed by mammals (30.5%), ants (11.5%) and reptiles (2.7%). Importantly, we did not find studies evaluating non-native plant dispersal by amphibians or fishes. The dominance of bird-dispersal interactions over other animal groups remained consistent (i.e. >50%) across geographical regions (Fig. 5.1), which was expected considering that they are the most diverse and well-distributed group of seed dispersers (Daniel Kissling *et al.*, 2009). In each region, avian dispersal was followed by mammals which accounted for between 25 and 36% of the total dispersal interactions. Only in North America the number of studies on mammals was similar to the number of studies on dispersal by ants, which in other regions represented less than 14% of the total number of interactions, being especially low in Europe (Fig. 5.1). Dispersal by reptiles was studied in only three of the six regions considered, representing less than 5% of the total dispersal interactions in Oceania and South America; whereas in Europe, dispersal by reptiles reached 8% of the total dispersal interactions. The 18.6% of studies involves non-native plant species dispersed by animals from two or three different groups. For example, in the western Mediterranean, Padrón *et al.* (2011) found that three species of the non-native *Opuntia* were dispersed by different species of mammals (e.g. stone martens, gulls, wild boards, rabbits), birds and tortoises.

Regarding dispersal strategies of non-native plants adopted by the different disperser groups, endozoochory was the main dispersal mechanism registered for birds and mammals in all the geographical regions, and the only dispersal mode registered for reptiles (Fig. 5.1). Epizoochory by mammals was studied in all regions where interactions with this animal group were registered, whereas epizoochory by birds was not registered in studies from Africa. Asia showed the highest proportion of studies dealing with epizoochory, and particularly by birds. Finally, while synzoochory by mammals was registered in a very low proportion in Africa and Oceania, this dispersal strategy involved the total of ant dispersal interactions.

Birds

The finding that birds are the main group of dispersers of non-native plants worldwide (Fig. 5.1) is in agreement with earlier analyses (Richardson and Rejmánek, 2011) and is not surprising because birds are among the most efficient long-distance vectors of dispersal (Vittoz and Engler, 2007). Our database showed that bird dispersal of non-native seeds was mostly through endozoochory (93%). Frugivorous birds vary in body sizes, feeding behaviour and seed retention times. Therefore, depending on the bird species, they can play an important role in medium- and long-distance seed dispersal (Mokotjomela *et al.*, 2013a; García-Álvarez *et al.*, 2015). For example, in South Africa Mokotjomela *et al.* (2013a) found that potential dispersal distances of non-native plant species were 9.4 km if seeds were consumed by the small *Zosterops capensis*, 17.8 km by the medium-sized *Pycnonotus capensis* and 21.2 km by the large *Colius striatus*. Moreover, García-Álvarez *et al.* (2015) found that migratory water birds such as ducks and geese disperse non-native seeds with maximal dispersal distances of well over 1000 km. Seabirds and migratory birds are also important epizoochorous dispersers of non-native species among oceanic islands, although only a few studies have tested this (e.g. Choi *et al.*, 2010). For example, Aoyama *et al.* (2012) found viable seeds of nine non-native and native plant species adhering to feathers of seabirds, and reported a positive relationship between the distributions of plant species and seabirds in the Japanese Ogasawara Islands. Our database showed that the birds that disperse non-native plant species belong to several taxonomic families. Most of them are generalist frugivores or omnivores of small and medium body sizes. Most bird dispersers are Passeriformes in the families Turdidae, Sturnidae, Zosteropidae, Mimidae, Pycnonotidae, Sylviidae and Corvidae. In addition, there are medium-sized bird dispersers of the families Coliidae, Columbidae and Musophagidae. Finally, we found a single study evaluating seed dispersal by a large, flightless bird, i.e. *Casuarius casuarius* (Westcott *et al.*, 2008).

Mammals

Mammals comprise a diverse group of seed dispersers and the second most important group of dispersers in our database (Fig. 5.1). Dispersal of non-native seeds by mammals was mainly through endozoochory (80.3%, see also Chapter 6, this volume). In most records of interactions between mammals and non-native plants, consumption of non-native fruits was described as a complement of a more diverse diet (Dovrat *et al.*, 2012; Barrios-Garcia and Simberloff, 2013). However, in some cases, fruits of non-native species represent a high proportion of the ingested fruits (e.g. Silva *et al.*, 2005; Williams *et al.*, 2008). Indeed, bats are a cosmopolitan order with many frugivorous species, which can sometimes disperse comparatively more seeds than birds (Jordaan *et al.*, 2012). In addition, mammals are important epizoochorous dispersers of non-native plants (15.5% of the registered interactions in this group of dispersers, Fig. 5.1). In this sense, Nelis *et al.* (2010) found that rabbits in Robison Crusoe Island, Chile, carried nearly seven times as many seeds of non-native species than of native species. Mammals vary greatly in body size and gut retention times, and can travel from a few metres to kilometres (Stiles, 2000; see also Chapter 6, this volume). Some mammals can travel long distances in a day, enabling non-native species expansion to new areas. For example, Williams *et al.* (2008) in Connecticut, USA, found that white tailed deer dispersed over 350 viable seeds of non-native species per day, with a mean daily travel distance of 568 m and a maximum of 5932 m. In our database, multiple mammalian families were found to disperse non-native plants. The most common were from the order Chiroptera (e.g. bats) with 13.8% of registered interactions, followed by members of the families Canidae (e.g. foxes, coyotes) with 10.1%, Leporidae (e.g. rabbits, cottontails) 10.1%, Sciuridae (e.g. squirrels) 8.3%, Suidae (e.g. boars, pigs) 7.3%, Didelphidae (e.g. possums) 6.4% and Cervidae (e.g. deer) 6.4%. In addition, 14.7% of the studies recorded different species of Muridae (e.g. mice), with both disperser and predator roles. This group was responsible for synzoochory dispersal recorded in mammals (i.e. 4.2%). Our review may underestimate the importance of mammals as dispersers of non-native plants, considering that some authors could use the family, order or species in the title, abstract or keywords of their articles. For example, dispersal by cattle was seemingly underestimated in our search, considering that they are one of the most common mammals around the world and that there are studies signalling them as responsible for the dispersal of many non-native species (Chuong *et al.*, 2016; see Chapter 6, this volume).

Ants

Ants are important dispersers of non-native plants, particularly for those seeds that present elaiosomes or caruncula (Jongejans *et al.*, 2015; Gammans *et al.*, 2018). However, this group also disperses seeds that lack a clear dispersal syndrome. They are generally involved in secondary seed dispersal of non-native plants, removing seeds that have either fallen from the plant (Bebawi and Campbell, 2004) or that were previously dispersed by other animals (de Lima *et al.*, 2015). Compared to other dispersers, ants carry seeds only for short distances (maximum of about 5 m, French and Major, 2001). Thus, they are unlikely to be responsible for non-native seed introduction into new areas, but might contribute to the persistence and local spread of non-native species. For instance, *Acacia saligna*, invasive in South Africa, interacts with 13 ant species (French and Major, 2001). Only a few studies of interactions between ants and non-native plants measured both dispersal and some proxy of germination/recruitment (Prior *et al.*, 2015).

Reptiles

The roles of tortoises, turtles, lizards and iguanas as seed dispersers have rarely been considered, especially for non-native plants (Valido and Olesen, 2007; Traveset *et al.*, 2016). In all the studies of our database reporting reptiles as potential seed dispersers of non-native plants, they appear on islands together with other functional groups of dispersers, especially birds and mammals (Heleno *et al.*, 2011, 2013). For example, two species of native lizards (*Microlophus* spp.) and the giant tortoise *Chelonoidis nigra*, together with two bird species, were important dispersers of highly invasive plant species in the Galápagos (Heleno *et al.*, 2013). The fact that this group of animals is mostly studied on islands is probably because these habitats are generally characterized by depauperate bird and mammal faunas and high densities of reptiles, especially lizards (Traveset and Sáez, 1997; Rodda and Dean-Bradley, 2002).

Invasional meltdown (facilitation among non-native species) in seed dispersal interactions

The Invasional Meltdown Hypothesis postulates that two or more non-native species facilitate each other's establishment or exacerbate each other's impact on native species, accelerating invasion rates and increasing the number of non-native species (Simberloff and Holle, 1999; Von Holle, 2011; see Chapter 8, this volume). Of the 172 reviewed studies that specified the origin of the animal dispersers (excluding those focusing on seed predation only), 66% involved only native dispersers, 20% included both native and non-native dispersers, whereas only 14% involved only non-native species of dispersers. Notwithstanding, only a few studies mentioned or tested specifically the invasional meltdown effect (e.g. López-Darias and Nogales, 2008; Prior *et al.*, 2015). For example, Bourgeois *et al.* (2005) found that non-native mammals (rats and rabbits) enhanced seed germination of the non-native plant *Carpobrotus* sp., which, in return, provided a rich food source to these animals during the dry summer season on French Mediterranean islands. Additionally, many non-native bird species (e.g. *Pycnonotus cafer, Pycnonotus jocosus, Psittacula krameri, Sturnus vulgaris* and *Zosterops japonicus*) have been reported to disperse non-native plants (e.g. Mandon-Dalger *et al.*, 2004; Thabethe *et al.*, 2015). For instance, it was recently observed in Oahu (Hawaii) that non-native bird species are the only dispersers of native plants, but they disperse a much higher proportion of non-native plant species (Vizentin-Bugoni *et al.*, 2019). The interaction of a non-native plant with a non-native disperser seems to be very common in ants (30.8% from the total of studies that have ants as a disperser group). Many non-native ant species have invaded different habitats around the world and there is evidence that non-native ants seemingly favour non-native plant dispersal (Rowles and O'Dowd, 2009) at the expense of seed dispersal of native species (Ness and Bronstein, 2004). For example, Prior *et al.* (2015) found that a non-native ant enhanced seed removal of both native and non-native plants, but the relative benefit was higher for the non-native plant, which failed to recruit under adult plants.

5.3.4 What Is the Role of Animal Dispersers in the Different Stages of the Non-native Plant Regeneration Process?

Seed dispersal effectiveness (SDE) is the contribution made by a disperser or a group of dispersers to a plant's reproductive success, both through the quantity of seeds dispersed and the quality of dispersal (Schupp *et al.*, 2010). In our database, the quantity of dispersed seeds – i.e. number of disperser visits and/or number of seed removed per visit – was measured mainly as the number of seeds in pellets, faeces, regurgitates and gut contents, and, less frequently, on seed rain (e.g. Chimera and Drake, 2010; Parejo *et al.*, 2013; Silvestre *et al.*, 2016). In turn, the quality of seed dispersal by endozoochory was measured in the reviewed studies mainly as the germination percentage of gut-passed seeds (e.g. Mokotjomela *et al.*, 2016; Díaz Vélez *et al.*, 2018) – i.e. an indicator of the process acting on seeds during their passage through dispersers' mouth, beak or digestive tract. This SDE component was less frequently determined as the quality of seed deposition, including both the deposition pattern (e.g. defecation or regurgitation) and the type of deposition microsites (e.g. Silva *et al.*, 2005). Quantity and quality of SDE were rarely evaluated together in seed dispersal studies of non-native species (but see Silva *et al.*, 2005; Mokotjomela *et al.*, 2016).

Studying the influence of seed dispersers on postdispersal stages of the regeneration process of non-native plants (i.e. seed predation, germination, seedling establishment) might contribute to a deeper understanding of the SDE provided by each disperser. In our database, of the 204 articles reporting seed removal, a high proportion (74%) evaluated seed dispersal (Fig. 5.2) – i.e. studies providing some evidence of seed arrival at a new place (e.g. seed presence in faeces). However, fewer studies evaluated germination (35.3%) and/

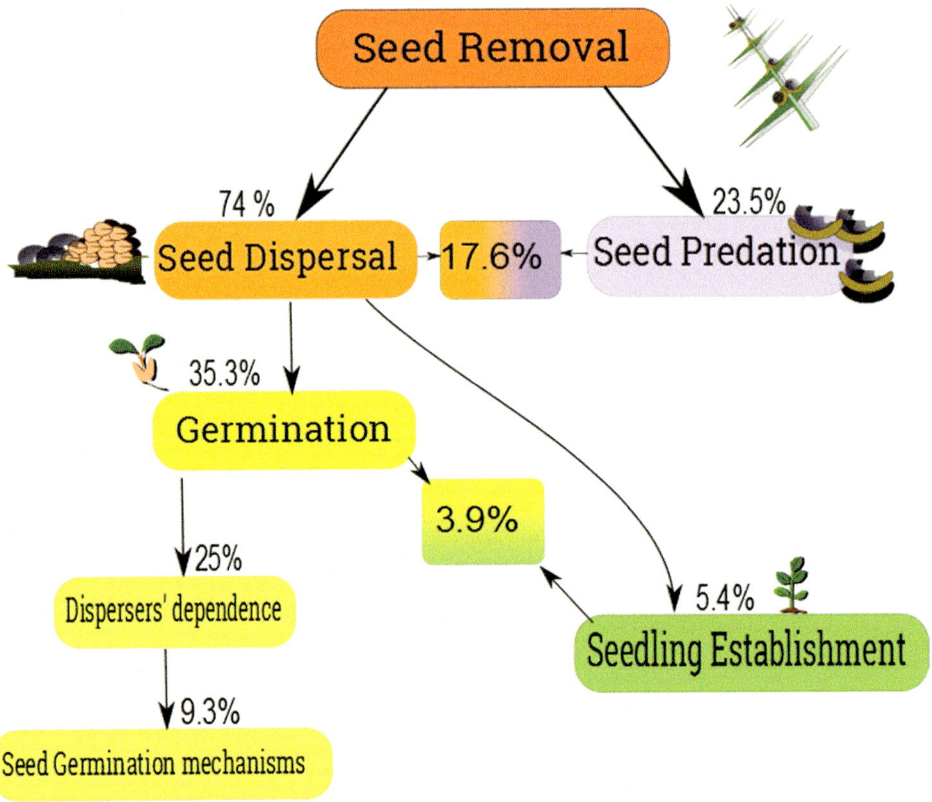

Fig. 5.2. Percentage of studies focusing on each of the different stages of the regeneration process. Each percentage was calculated from the total number of studies. 'Dispersers' dependence' refers to seed germination studies that consider a control treatment, and 'Seed germination mechanisms' stands for those that distinguished between scarification and/or de-inhibition mechanisms. Intermediate boxes with a mix of two colours indicate the percentage of studies that evaluate both stages of the regeneration process.

or seedling establishment (5.4%) of those dispersed seeds, to have a better understanding of the quality of the dispersal service. Finally, only a small proportion of the studies analysed seed predation (23.5%).

Seed germination

From the whole pool of studies, only 35.3% include experiments of seed germination (Fig. 5.2). Within the endozoochorous studies evaluating germination of non-native seeds, half found that the passage of seeds through the digestive tract of the dispersers increased seed germination (e.g. Jordaan *et al.*, 2011). This influence at the germination stage certainly increases the invasion success of those plants. However, only a few studies (9.3%, Fig. 5.2) tested seed germination comparing three treatments (i.e. gut-passed seeds, manually extracted seeds and intact fruits) to determine the different mechanisms acting on gut-passed seeds (de-inhibition or scarification, e.g. Bebawi and Campbell, 2004; Díaz Vélez *et al.*, 2018). Seed de-inhibition, which implies that animal fruit ingestion increases germination via the separation of seeds from the fruit pulp, was the most frequent mechanism in non-native species (e.g. Jordaan *et al.*, 2012; Thabethe *et al.*, 2015). For example, in the mountainous region of central Argentina, *Turdus chiguanco*, the main bird disperser of the non-native shrub *Cotoneaster franchetii*, has a positive effect on seed germination through a de-inhibition mechanism (Díaz Vélez *et al.*, 2018). Further studies are needed to test whether the prevalence of de-inhibition over scarification (which involves chemical or mechanical modifications of the seed coat) is a general trend among non-native plants dispersed by endozoochory. In addition, experimental comparisons disentangling such effects may improve the knowledge of the dependence of non-native species on animals as seed dispersers.

Seedling establishment

The quality of seed deposition, i.e. the deposition of seeds on substrates suitable for seedling growth, is important to determine SDE. However, this factor is rarely considered in seed dispersal studies, and even less with non-native species (Fig. 5.2). For example, in Chile, Silva *et al.* (2005) recorded that 59% of the native fox faeces with the non-native *Schinus molle* were deposited in unfavourable sites (i.e. open sites, dry microsites, etc.). The particular behaviour of some dispersers could either enhance or limit the chances of establishment of some non-native species. An example of a benefit has been described for the dispersal of the non-native *Passiflora tripartita* by pigs, which generally excrete in rooting sites where competition from native plants is low (Beavon and Kelly, 2014). In turn, dispersers might limit the non-native plant's chances of establishment. For example, *Opuntia stricta* has higher chances of seed predation by a gerbil when seeds are within baboon faeces rather than in the soil (Dudenhoeffer and Hodge, 2018).

Seed predation

Since dispersers could be key for non-native species invasion, seed predation could be important for limiting their establishment and expansion (Maron and Vila, 2001; Larios *et al.*, 2017). In fact, a study of coniferous invasion in Argentina observed that seed predators can be an important component of biological resistance to non-native plant invasions (Nuñez *et al.*, 2008). Besides the resistance of native biota, a non-native seed predator can also limit a non-native plant invasion. For example, in South Africa, ingestion by the invasive parakeet species caused a greater reduction in seed germination than the native turaco species (Thabethe *et al.*, 2015). Importantly, the number of animals that could act both as seed dispersers and predators is not small. Clear examples are seed-caching animals, parrots (Psittacidae), pigeons and doves (Gosper and Gosper, 2008; Thabethe *et al.*, 2015; Gómez *et al.*, 2019). In this sense, some studies suggested that frugivores represent a continuum in terms of their contribution to seed dispersal (Heleno *et al.*, 2011; Gómez *et al.*, 2019). Additionally, many studies addressing seed dispersal might be underestimating seed predation, since many dispersers prey upon some of the seeds they disseminate (Bobadilla *et al.*, 2016). In our database, a low percentage of the articles evaluate this interaction (23.5%, Fig. 5.2). All in all, the net effect of seed predators on population expansion of non-native species is unclear and might depend on the degree of dependence of the species on their seeds (Eriksson and Ehrlén, 1992). Mathematical models predict that an impact of at least 90% seed loss is needed to reduce population growth of non-native species (Parker, 2000). However, some studies suggest that seed predators might affect invasion rate (Maron and Vila, 2001; Nuñez *et al.*, 2008).

5.4 Conclusions and Future Directions

Studies on seed dispersal of non-native species are still scarce. Although humans are the main drivers of non-native species movement around the world, most non-native plants need to find a dispersal agent in invaded areas to establish self-sustaining populations. The probability of becoming invasive and the rates of spread will partly depend on the effectiveness of the dispersal agent that the plants encounter. In this sense, it is remarkable that most studies reviewed here did not consider seed germination or seedling establishment. Including these stages in seed dispersal studies will be important to fully understand SDE and provide useful management strategies to prevent new invasion foci and limit spread of non-native plant species. Moreover, a research bias, with some understudied biogeographic regions or dispersal interactions, precludes a balanced synthesis. In this sense, more research on some disperser groups such as large flightless birds (e.g. emu) is needed. Also, there is a need to conduct studies in Central America, and to have more studies in Oceania, Asia and Africa (excluding South Africa). In fact, many of the patterns observed could be a product of a high research effort in the temperate northern hemisphere (North America and Europe). Importantly, interactions among non-native species were detected, suggesting the occurrence of invasional meltdown mechanisms. Therefore, interaction dispersal networks distinguishing non-native plant species will be very useful to increase the effectiveness of management strategies (e.g. Heleno *et al.*, 2013). Finally, although several studies in our search compared the dispersal process of non-native and native species, very few have compared invasive vs non-invasive non-natives, or invasive species in their native vs non-native distribution range (but see Lundgren *et al.*, 2013; Mokotjomela *et al.*, 2013b; Montesinos *et al.*, 2018). Such comparisons would be useful to more fully understand the role of dispersal by animals in the success of plant invasions.

Acknowledgements

We acknowledge the Consejo Nacional de Investigaciones Científicas y Técnicas (CONICET) and the Universidad Nacional de Córdoba (UNC). We thank Paula A. Tecco and Andres Issaly for useful advice during the writing of this review and Jorgelina Brasca for the English revision. Financial support was provided by SECyT (33620180100201CB) and BID2015 PICT 3089.

References

Aoyama, Y., Kawakami, K. and Chiba, S. (2012) Seabirds as adhesive seed dispersers of alien and native plants in the oceanic Ogasawara Islands, Japan. *Biodiversity and Conservation* 21(11), 2787–2801. DOI: 10.1007/s10531-012-0336-9.

Barrios-Garcia, M.N. and Simberloff, D. (2013) Linking the pattern to the mechanism: how an introduced mammal facilitates plant invasions. *Austral Ecology* 38(8), 884–890. DOI: 10.1111/aec.12027.

Beavon, M.A. and Kelly, D. (2014) Dispersal of banana passionfruit (*Passiflora tripartita var. mollissima*) by exotic mammals in New Zealand facilitates plant invasiveness. *New Zealand Journal of Ecology* 39, 43–49.

Bebawi, F.F. and Campbell, S.D. (2004) Interactions between meat ants (*Iridomyrmex spadius*) and bellyache bush (*Jatropha gossypiifolia*). *Agriculture* 44, 1157–1164.

Bobadilla, S.Y., Benitez, V.V. and Guichón, M.L. (2016) Asiatic *Callosciurus* squirrels as seed dispersers of exotic plants in the Pampas. *Current Zoology* 62(3), 215–219. DOI: 10.1093/cz/zow009.

Bourgeois, K., Suehs, C.M., Vidal, E. and Médail, F. (2005) Invasional meltdown potential: facilitation between introduced plants and mammals on French Mediterranean islands. *EcoScience* 12(2), 248–256. DOI: 10.2980/i1195-6860-12-2-248.1.

Bronstein, J.L. (2015) *Mutualisms*. Oxford University Press, Oxford.

Buckley, Y.M., Anderson, S., Catterall, C.P., Corlett, R.T., Engel, T. *et al.* (2006) Management of plant invasions mediated by frugivore interactions. *Journal of Applied Ecology* 43(5), 848–857. DOI: 10.1111/j.1365-2664.2006.01210.x.

Castillo-Flores, A.A. and Calvo-Irabién, L.M. (2003) Animal dispersal of two secondary-vegetation herbs into the evergreen rain forest of south-eastern Mexico. *Journal of Tropical Ecology* 19(3), 271–278. DOI: 10.1017/S0266467403003304.

Chimera, C.G. and Drake, D.R. (2010) Patterns of seed dispersal and dispersal failure in a Hawaiian dry forest having only introduced birds. *Biotropica* 42(4), 493–502. DOI: 10.1111/j.1744-7429.2009.00610.x.

Choi, C.Y., Nam, H.Y. and Chae, H.Y. (2010) Exotic seeds on the feathers of migratory birds on a stopover island in Korea. *Journal of Ecology and Field Biology* 33, 19–22.

Chuong, J., Huxley, J., Spotswood, E.N., Nichols, L., Mariotte, P. *et al*. (2015) Cattle as dispersal vectors of invasive and introduced plants in a California annual grassland. *Rangeland Ecology & Management* 69(1), 52–58. DOI: 10.1016/j.rama.2015.10.009.

Couvreur, M., Cosyns, E., Hermy, M. and Hoffmann, M. (2005) Complementarity of epi- and endozoochory of plant seeds by free ranging donkeys. *Ecography* 28(1), 37–48. DOI: 10.1111/j.0906-7590.2005.04159.x.

Cronk, Q.C.B. and Fuller, J.L. (1995) *Invasive Plants: The Threat to Natural Ecosystems Worldwide*. Chapman and Hall, London.

Crosby, A.W. (1986) *Ecological imperialism: The Biological Expansion of Europe, 900–1900*. Cambridge University Press, Cambridge.

Daehler, C.C. (2003) Performance comparisons of co-occurring native and alien invasive plants: implications for conservation and restoration. *Annual Review of Ecology, Evolution, and Systematics* 34(1), 183–211. DOI: 10.1146/annurev.ecolsys.34.011802.132403.

Daniel Kissling, W., Böhning-Gaese, K. and Jetz, W. (2009) The global distribution of frugivory in birds. *Global Ecology and Biogeography* 18(2), 150–162. DOI: 10.1111/j.1466-8238.2008.00431.x.

de Lima, R.E.M., de Sá Dechoum, M. and Castellani, T.T. (2015) Native seed dispersers may promote the spread of the invasive Japanese raisin tree (Hovenia dulcis Thunb.) in seasonal deciduous forest in southern Brazil. *Tropical Conservation Science* 88, 846–862.

Dovrat, G., Perevolotsky, A. and Ne'eman, G. (2012) Wild boars as seed dispersal agents of exotic plants from agricultural lands to conservation areas. *Journal of Arid Environments* 78, 49–54. DOI: 10.1016/j.jaridenv.2011.11.011.

Drummond, B.A. (2005) The selection of native and invasive plants by frugivorous birds in Maine. *Northeastern Naturalist* 12(1), 33–44. DOI: 10.1656/1092-6194(2005)012[0033:TSONAI]2.0.CO;2.

Dudenhoeffer, M. and Hodge, A.-M.C. (2018) Opposing forces of seed dispersal and seed predation by mammals for an invasive cactus in central Kenya. *African Journal of Ecology* 56(2), 179–184. DOI: 10.1111/aje.12504.

Díaz Vélez, M.C., Sérsic, A.N., Traveset, A. and Paiaro, V. (2018) The role of frugivorous birds in fruit removal and seed germination of the invasive alien *Cotoneaster franchetii* in central Argentina. *Austral Ecology* 43(5), 558–566. DOI: 10.1111/aec.12592.

Eriksson, O. and Ehrlén, J. (1992) Seed and microsite limitation of recruitment in plant populations. *Oecologia* 91(3), 360–364. DOI: 10.1007/BF00317624.

Farwig, N. and Berens, D.G. (2012) Imagine a world without seed dispersers: a review of threats, consequences and future directions. *Basic and Applied Ecology* 13(2), 109–115. DOI: 10.1016/j.baae.2012.02.006.

Ferreras, A.E. and Galetto, L. (2010) From seed production to seedling establishment: important steps in an invasive process. *Acta Oecologica* 36(2), 211–218. DOI: 10.1016/j.actao.2009.12.005.

French, K. and Major, R.E. (2001) Effect of an exotic *Acacia* (Fabaceae) on ant assemblages in South African fynbos. *Austral Ecology* 26(4), 303–310. DOI: 10.1046/j.1442-9993.2001.01115.x.

Gammans, N., Drummond, F. and Groden, E. (2018) Impacts of the invasive European red ant (*Myrmica rubra* (L.): Hymenoptera; Formicidae) on a myrmecochorous system in the northeastern United States. *Environmental Entomology* 47(4), 908–991. DOI: 10.1093/ee/nvy069.

García-Álvarez, A., van Leeuwen, C.H.A., Luque, C.J., Hussner, A., Vélez-Martín, A. *et al*. (2015) Internal transport of alien and native plants by geese and ducks: an experimental study. *Freshwater Biology* 60(7), 1316–1329. DOI: 10.1111/fwb.12567.

Gómez, J.M., Schupp, E.W. and Jordano, P. (2019) Synzoochory: the ecological and evolutionary relevance of a dual interaction. *Biological Reviews* 94(3), 874–902. DOI: 10.1111/brv.12481.

Gosper, C.R. and Gosper, D.G. (2008) Foods of pigeons and doves in fragmented landscapes of subtropical eastern Australia. *Australian Field Ornithology* 25, 76–86.

Gosper, C.R. and Vivian-Smith, G. (2009) The role of fruit traits of bird-dispersed plants in invasiveness and weed risk assessment. *Diversity and Distributions* 15(6), 1037–1046. DOI: 10.1111/j.1472-4642.2009.00599.x.

Gosper, C.R. and Vivian-Smith, G. (2010) Fruit traits of vertebrate-dispersed alien plants: smaller seeds and more pulp sugar than Indigenous species. *Biological Invasions* 12(7), 2153–2163. DOI: 10.1007/s10530-009-9617-y.

Gurvich, D.E., Tecco, P.A. and Díaz, S. (2005) Plant invasions in undisturbed ecosystems: the triggering attribute approach. *Journal of Vegetation Science* 16(6), 723–728. DOI: 10.1111/j.1654-1103.2005.tb02415.x.

Heleno, R., Blake, S., Jaramillo, P., Traveset, A., Vargas, P. *et al.* (2011) Frugivory and seed dispersal in the Galápagos: what is the state of the art? *Integrative Zoology* 6(2), 110–129. DOI: 10.1111/j.1749-4877.2011.00236.x.

Heleno, R.H., Olesen, J.M., Nogales, M., Vargas, P. and Traveset, A. (2013) Seed dispersal networks in the Galápagos and the consequences of alien plant invasions. *Proceedings of the Royal Society B* 280(1750), 2012–2112. DOI: 10.1098/rspb.2012.2112.

Hulme, P.E., Bacher, S., Kenis, M., Klotz, S., Kühn, I. *et al.* (2008) Grasping at the routes of biological invasions: a framework for integrating pathways into policy. *Journal of Applied Ecology* 45, 403–414.

Jongejans, E., Silverman, E.J., Skarpaas, O. and Shea, K. (2015) Post-dispersal seed removal of *Carduus nutans* and *C. acanthoides* by insects and small mammals. *Ecological Research* 30, 173–180.

Jordaan, L.A., Johnson, S.D. and Downs, C.T. (2011) The role of avian frugivores in germination of seeds of fleshy-fruited invasive alien plants. *Biological Invasions* 13(8), 1917–1930. DOI: 10.1007/s10530-011-0013-z.

Jordaan, L.A., Johnson, S.D. and Downs, C.T. (2012) Wahlberg's epauletted fruit bat (*Epomophorus wahlbergi*) as a potential dispersal agent for fleshy-fruited invasive alien plants: effects of handling behaviour on seed germination. *Biological Invasions* 14(5), 1–10. DOI: 10.1007/s10530-011-0131-7.

Kueffer, C., Kronauer, L. and Edwards, P.J. (2009) Wider spectrum of fruit traits in invasive than native floras may increase the vulnerability of oceanic islands to plant invasions. *Oikos* 118(9), 1327–1334. DOI: 10.1111/j.1600-0706.2009.17185.x.

Kulbaba, M.W., Tardif, J.C. and Staniforth, R.J. (2009) Morphological and ecological relationships between burrs and furs. *The American Midland Naturalist* 161(2), 380–39. DOI: 10.1674/0003-0031-161.2.380.

Larios, L., Pearson, D.E. and Maron, J.L. (2017) Incorporating the effects of generalist seed predators into plant community theory. *Functional Ecology* 31(10), 1856–1867. DOI: 10.1111/1365-2435.12905.

López-Darias, M. and Nogales, M. (2008) Effects of the invasive Barbary ground squirrel (*Atlantoxerus getulus*) on seed dispersal systems of insular xeric environments. *Journal of Arid Environments* 72(6), 926–939. DOI: 10.1016/j.jaridenv.2007.12.006.

Lundgren, J.G., Saska, P. and Honěk, A. (2013) Molecular approach to describing a seed-based food web: the post-dispersal granivore community of an invasive plant. *Ecology and Evolution* 3(6), 1642–1652. DOI: 10.1002/ece3.580.

Mack, R.N., Simberloff, D., Mark Lonsdale, W., Evans, H., Clout, M. *et al.* (2000) Biotic invasions: causes, epidemiology, global consequences, and control. *Ecological Applications* 10(3), 689–710. DOI: 10.1890/1051-0761(2000)010[0689:BICEGC]2.0.CO;2.

Mack, R.N., Ruiz, G.M. and Carlton, J.T. (2003) Global plant dispersal, naturalization, and invasion: pathways, modes, and circumstances. In: Ruis, G.M. and Carlton, J.T. (eds) *Invasive species: Vectors and Management Strategies*. Island Press, Washington D.C, pp. pp. 3–30.

Mandon-Dalger, I., Clergeau, P., Tassin, J., Rivière, J.-N. and Gatti, S. (2004) Relationships between alien plants and an alien bird species on reunion island. *Journal of Tropical Ecology* 20(6), 635–642. DOI: 10.1017/S0266467404001774.

Maron, J.L. and Vila, M. (2001) When do herbivores affect plant invasion? Evidence for the natural enemies and biotic resistance hypotheses. *Oikos* 95(3), 361–373. DOI: 10.1034/j.1600-0706.2001.950301.x.

Mokotjomela, T.M., Musil, C.F. and Esler, K.J. (2013a) Potential seed dispersal distances of native and non-native fleshy fruiting shrubs in the South African Mediterranean climate region. *Plant Ecology* 214(9), 1127–1137. DOI: 10.1007/s11258-013-0237-3.

Mokotjomela, T.M., Musil, C.F. and Esler, K.J. (2013b) Frugivorous birds visit fruits of emerging alien shrub species more frequently than those of native shrub species in the South African Mediterranean climate region. *South African Journal of Botany* 86, 73–78. DOI: 10.1016/j.sajb.2013.02.004.

Mokotjomela, T.M., Downs, C.T., Esler, K. and Knight, J. (2016) Seed dispersal effectiveness: a comparison of four bird species feeding on seeds of invasive Acacia cyclops in South Africa. *South African Journal of Botany* 105, 259–263. DOI: 10.1016/j.sajb.2016.04.005.

Montesinos, D., Castro, S. and Rodríguez-Echeverría, S. (2018) Invasive acacias experience higher ant seed removal rates at the invasion edges. *Web Ecology* 12(1), 33–37. DOI: 10.5194/we-12-33-2012.

Nelis, L.C. Wootton, J.T, Timothy Wootton, J. (2010) Treatment-based Markov chain models clarify mechanisms of invasion in an invaded grassland community. *Proceedings of the Royal Society B* 277(1681), 539–547. DOI: 10.1098/rspb.2009.1564.

Ness, J.H. and Bronstein, J.L. (2004) The effects of invasive ants on prospective ant mutualists. *Biological Invasions* 6, 445–461.

Nuñez, M.A., Simberloff, D. and Relva, M.A. (2008) Seed predation as a barrier to alien conifer invasions. *Biological Invasions* 10, 1389–1398.

Nuñez-Mir, G.C., Guo, Q., Rejmánek, M., Iannone III, B.V. and Fei, S. (2019) Predicting invasiveness of exotic woody species using a traits-based framework. *Ecology*, e02797.

Padrón, B., Nogales, M., Traveset, A., Vil, M., Martínez-Abraín, A. *et al.* (2011) Integration of invasive *Opuntia* spp. by native and alien seed dispersers in the Mediterranean area and the Canary Islands. *Biological Invasions* 13, 831–844.

Parejo, S.H., Ceia, R.S., Ramos, J.A., Sampaio, H.L. and Heleno, R.H. (2013) Tiptoeing between restoration and invasion: seed rain into natural gaps within a highly invaded relic forest in the Azores. *European Journal of Forest Research* 133, 383–390.

Parker, I.M. (2000) Invasion dynamics of *Cytisus scoparius*: A matrix model approach. *Ecological Applications* 10, 726–743.

Prior, K.M., Robinson, J.M., Meadley Dunphy, S.A. and Frederickson, M.E. (2015) Mutualism between co-introduced species facilitates invasion and alters plant community structure. *Proceedings of the Royal Society B* 282.

Pyšek, P., Richardson, D.M., Pergl, J., Jarošík, V., Sixtova, Z. *et al.* (2008) Geographical and taxonomic biases in invasion ecology. *Trends in Ecology & Evolution* 23, 237–244.

Rejmánek, M. and Richardson, D.M. (2013) Trees and shrubs as invasive alien species – 2013 update of the global database. *Diversity and Distributions* 19, 1093–1094.

Richardson, D.M., Allsopp, N., D'Antonio, C.M., Milton, S.J. and Rejmánek, M. (2000) Plant invasions–the role of mutualisms. *Biological Reviews* 75, 65–93.

Richardson, D.M. and Rejmánek, M. (2011) Trees and shrubs as invasive alien species–a global review. *Diversity and Distributions* 17, 788–809.

Rodda, G.H. and Dean-Bradley, K. (2002) Excess density compensation of island herpetofaunal assemblages. *Journal of Biogeography* 29, 623–632.

Rojas, T.N., Gallo, M.C.F., Vergara-Tabares, D.L., Nazaro, M.G., Zampini, I.C. *et al.* (2019) Being popular or freak: how alien plants integrate into native plant-frugivore networks. *Biological Invasions* 21, 2589–2598.

Rowles, A.D. and O'Dowd, D.J. (2009) New mutualism for old: indirect disruption and direct facilitation of seed dispersal following Argentine ant invasion. *Oecologia* 158, 709–716.

Schleuning, M., Böhning-Gaese, K., Dehling, D.M. and Burns, K.C. (2014) At a loss for birds: Insularity increases asymmetry in seed-dispersal networks. *Global Ecology and Biogeography* 23, 385–394.

Schupp, E.W., Jordano, P. and Gómez, J.M. (2010) Seed dispersal effectiveness revisited: a conceptual review. *New Phytologist* 188, 333–353.

Silva, S.I., Bozinovic, F. and Jaksic, F.M. (2005) Frugivory and seed dispersal by foxes in relation to mammalian prey abundance in a semiarid thornscrub. *Austral Ecology* 30, 739–746.

Silvestre, S.M., da Rocha, P.A., da Cunha, M.A., Santana, J.P. and Ferrari, S.F. (2016) Diet and seed dispersal potential of the white-lined bat, *Platyrrhinus lineatus* (Geoffroy, E., 1810), at a site in northeastern Brazil. *Studies on Neotropical Fauna and Environment* 51, 37–44.

Simberloff, D. and Holle, V, V. (1999) Positive interactions of nonindigenous species: Invasional meltdown? *Biological Invasions* 1, 21–32.

Stiles, E.W. (2000) Animals as seed dispersers. In: Fenner, M. (ed.) *Seeds: The Ecology of Regeneration in Communities, CABI.* Wallingford, UK, pp. 111–123.

Stohlgren, T.J., Pyšek, P., Kartesz, J., Nishino, M., Pauchard, A. *et al.* (2011) Widespread plant species: natives versus aliens in our changing world. *Biological Invasions* 13, 1931–1944.

Taylor, J.M. (2009) *The Global Migration of Ornamental Plants: How the World Got into your Garden.* Missouri Botanical Garden Press, St. *Louis*.

Tecco, P.A., Urcelay, C., Diaz, S., Cabido, M. and Pérez-Harguindeguy, N. (2013) Contrasting functional trait syndromes underlay woody alien success in the same ecosystem. *Austral Ecology* 38, 443–451.
Thabethe, V., Wilson, A.L., Hart, L.A. and Downs, C.T. (2015) Ingestion by an invasive parakeet species reduces germination success of invasive alien plants relative to ingestion by indigenous turaco species in South Africa. *Biological Invasions* 17, 3029–3039.
Traveset, A. and Richardson, D.M. (2006) Biological invasions as disruptors of plant reproductive mutualisms. *Trends in Ecology & Evolution* 21, 208–216.
Traveset, A. and Richardson, D.M. (2014) Mutualistic interactions and biological invasions. *Annual Review of Ecology, Evolution, and Systematics* 45, 89–113.
Traveset, A. and Sáez, E. (1997) Pollination of Euphorbia dendroides by lizards and insects: spatio-temporal variation in patterns of flower visitation. *Oecologia* 111(2), 241–248. DOI: 10.1007/PL00008816.
Traveset, A., Nogales, M., Vargas, P., Rumeu, B., Olesen, J.M. *et al.* (2016) Galápagos land iguana (*Conolophus subcristatus*) as a seed disperser. *Integrative Zoology* 11, 207–213.
Valido, A. and Olesen, J.M. (2007) The importance of lizards as frugivores and seed dispersers. In: Dennis, A.J., Schupp, E.W., Green, R.J. and Westcott, D.A. (eds) *Seed dispersal: Theory and its application in a changing world*. CAB International, UK, Wallingford, pp. 124–147.
Vergara-Tabares, D.L., Badini, J. and Peluc, S.I. (2016) Fruiting phenology as a "triggering attribute" of invasion process: Do invasive species take advantage of seed dispersal service provided by native birds? *Biological Invasions* 18, 677–687.
Vittoz, P. and Engler, R. (2007) Seed dispersal distances: a typology based on dispersal modes and plant traits. *Botanica Helvetica* 117, 109–124.
Vizentin-Bugoni, J., Tarwater, C.E., Foster, J.T., Drake, D.R., Gleditsch, J.M. *et al.* (2019) Structure, spatial dynamics, and stability of novel seed dispersal mutualistic networks in Hawaii. *Science* 364, 78–82.
Von Holle, B.V. (2011) Invasional meltdown. In: Simberloff, D. and Rejmánek, M. (eds) *Encyclopedia of Biological Invasions*. University of California Press, Berkeley, pp. 360–364.
Weber, E. (2003) *Invasive Plant Species of the World: A Reference Guide to Environmental Weeds*. CABI Publishing, Wallingford, UK.
Westcott, D.A., Setter, M., Bradford, M.G., McKeown, A. and Setter, S. (2008) Cassowary dispersal of the invasive pond apple in a tropical rainforest: the contribution of subordinate dispersal modes in invasion. *Diversity and Distributions* 14, 432–439.
Williams, S.C., Ward, J.S. and Ramakrishnan, U. (2008) Endozoochory by white-tailed deer (*Odocoileus virginianus*) across a suburban/woodland interface. *Forest Ecology and Management* 255, 940–947.

6 Ungulates as Dispersal Vectors of Non-native Plants

Christophe Baltzinger[1,2*], Ushma Shukla[1], Lindelwa S. Msweli[2] and Colleen T. Downs[2]

[1]INRAE Val de Loire, Research Unit Forest Ecosystems, Nogent-sur-Vernisson, France; [2]Centre for Invasion Biology and Centre for Functional Biodiversity, School of Life Sciences, University of KwaZulu-Natal, Pietermaritzburg, South Africa

Abstract

Ungulates are present worldwide with 257 recorded species, including livestock. They cover different functional gradients, be it feeding regime, digestive strategy, body size, body mass, fur characteristics or sociality. All these specificities may intervene at different stages of animal-mediated plant dispersal. Ungulates move diaspores from both native and non-native plants, through endo- and epizoochory. Initially introduced by humans, non-native plants bearing specific traits can be carried over long distances and to new environments by ungulates. These vectors can further free local resources necessary for the germination and the subsequent growth of the released diaspores. We first looked at trait-based plant community changes at different timescales in the presence of different native ungulates. We then reviewed the literature on endozoochory, regurgitation and fur-epizoochory assisted by ungulates, focusing on the dispersal of non-native plants. We made an overall assessment of ungulate-mediated non-native plant dispersal by biogeographical zone and dispersal mode, and then provided additional information on plant growth form and taxonomy, vectors and associated modes of dispersal. Results are presented for four main ungulate families: Cervidae, Bovidae, Suidae and Equidae. For each family, we highlight our findings either by ungulate if sufficiently represented (e.g. *Odocoileus virginianus, Bison bison, Bos taurus*) or by group of species. According to their feeding regime, grazers dispersed solely forbs and graminoids whereas omnivores also dispersed plants from other growth forms (i.e. cactus, vine, shrub and tree). Numerous non-native plants are dispersed by ungulates around the world, but this is probably the visible part of the iceberg, as only 32 ungulates (i.e. 12%) have been studied as vectors so far, suggesting their overall contribution is certainly underrated.

6.1 Introduction of Non-native Plants

'It is both interesting and troubling to note that not only do feral pigs ... create disturbed sites where non indigenous plants can establish, but they both further enhance the naturalization process by dispersing seeds to the site they disturb' (Schiffman, 1997). Non-native invasive plants displace and replace local, native plants, being one of the main causes of biodiversity erosion. Plant invasions generally occur in two phases. The first includes the introduction of the non-native species, through a long-distance dispersal event, often facilitated by or associated with anthropic activities. The second phase involves its naturalization and expansion. Non-native plants can expand their ranges in recipient ecosystems, resulting in different patterns of spread (Hui and Richardson, 2017). One of the reasons why non-native plants are generally successful in invading recipient ecosystems concerns the Enemy Release Hypothesis (Keane and Crawley, 2002). Ecosystems are enemy-free for non-native

*Corresponding author: christophe.baltzinger@inrae.fr

(eds Anna Traveset and David M. Richardson)
DOI: 10.1079/9781789242171.0006

newcomers, providing them with a competitive advantage over local species. Numerous vectors can drive the introduction, naturalization and expansion of non-natives.

Plants can benefit from ungulates for dispersal distance, seed germination (deinhibition, scarification and fertilization effect through endozoochory) and establishment (Chapter 5, this volume). For instance, by its toxicity at certain development stages (seeds and first leaves), common cocklebur (*Xanthium strumarium*; Botha *et al.*, 2014) escapes ungulate herbivory and gains competitive advantage over more palatable plants, consumed by large herbivores.

Some native ungulates, contrary to other large mammals, are relatively abundant (Russell *et al.*, 2001). They generally have highly diverse plant-based diets, meaning that they regularly interact with plants during feeding bouts, and can thus consume diaspores that they release remotely. They generally cover long distances, from hectometres to kilometres, on a regular basis within their home ranges (Pellerin *et al.*, 2016). Historically, domestic ungulates also dispersed plants over exceptional distances via ancient transhumance routes (Manzano and Malo, 2006).

Interactions between ungulates and plants fit in the general framework of ecosystem engineering effects (Wilby *et al.*, 2001). They include transport (seed dispersal), physical and chemical engineering processes affecting soil properties and germination of seeds present on the ground or in the soil seed bank. This framework also includes trophic interactions that modulate the spatial distribution of plants, plant community composition and other taxonomic groups (insects, birds) by cascading effects. Ungulates are dominant interacting agents and their management is of great interest for plant community dynamics and ecosystem functioning.

Ungulates are involved in diverse internal and external plant dispersal mechanisms that include primary mechanisms: endozoochory, regurgitation and fur-epizoochory; and several secondary dispersal mechanisms: hoof-epizoochory, transfer-epizoochory (Liehrmann *et al.*, 2018) and seed dispersal by dung beetles. Each of these mechanisms or pathways (Liddle and Elgar, 1984) differentially affects the fate of the conveyed diaspores (Baltzinger *et al.*, 2019).

In a recent review, Albert *et al.* (2015a) showed that ungulate-mediated seed dispersal was a generalist ecological process, affecting an important part of the accessible flora (~44%) dispersed by temperate ungulates. Different ungulates dispersed the same plants by similar or diverse dispersal mechanisms. However, they also showed that different mechanisms of ungulate-mediated dispersal relied on the interplay between plants/diaspores and dispersal vector traits (diaspore releasing height and ungulate body size for fur-epizoochory, or diaspore morphology and ungulate feeding guild for endozoochory; Albert *et al.*, 2015b).

The high taxonomic (257 species, Baltzinger *et al.*, 2019) and functional diversity of native, introduced and domestic ungulates along with the diversity of plant dispersal mechanisms determine the characteristics, diversity and abundance of plants dispersed (Fig. 6.1). Fur characteristics, body size and use of open vs. forested areas are the main predictors in fur-epizoochory. For transfer-epizoochory, the more frequent the degree of social interactions, the higher the probability of transfer from one individual to another. Endozoochory is more a matter of feeding regime, digestive strategy and animal body mass. Seed regurgitation by ruminants concerns mostly fleshy-fruited plants, and seeds too large to pass through the orifice between the rumen and the omasum. Ungulates, on a world scale, contribute to the dispersal, range expansion and success of non-native plants.

6.2 Spatio-temporal Changes in Plant Communities and Characteristics of Winner Species

Once introduced, plants can spread at different spatial scales depending on the timescale considered. This expansion is linked to dispersal events. The use of retrospective approaches (Wang and Smith, 2002) helps infer the underlying ecological processes that explain current plant distribution patterns compared with their historical distributions. Longitudinal approaches are based on long-term vegetation resurveys of permanent plots (Boulanger *et al.*, 2011; Box 6.1). They allow the assessment of changes in plant abundance, occurrence and spatial expansion. Coupled with censuses of ungulate abundance (hunting bags, browsing indices) or with erection

Fig. 6.1. Ungulate seed dispersers, e.g. Cervidae (1) white-tailed deer, *Odocoileus virginianus* (Anticosti Island, Canada) (2) roe deer, *Capreolus capreolus* (Sologne, France); Bovidae (3) impala, *Aepyceros melampus* (Hluhluwe-iMfolozi Park, South Africa) (4) European bison, *Bison bonasus* (Prioksko-Terrasny Nature Reserve, Russia); Suidae (5) warthog, *Phacochoerus africanus* (Hluhluwe-iMfolozi Park, South Africa) (6) wild pig, *Sus scrofa* (Sologne, France); Equidae (7) plains zebra, *Equus quagga* (Kruger National Park, South Africa) (8) donkey, *Equus asinus* (Mkhuze, South Africa). Pictures by Christophe Baltzinger.

of fences excluding ungulates, they can provide indirect evidence of the role ungulates play in the observed plant community changes. To date, there are relatively few studies linking plant distribution patterns to ungulate-mediated long distance dispersal events (Vickery *et al.*, 1986).

The spatial scale of these community changes is partly dependent on initial interplot distances and is not necessarily optimal to highlight animal-mediated dispersal events. Longitudinal approaches have been used to detect winners and losers (McKinney and Lockwood, 1999), identify their associated attributes and the potential causes of the observed changes. Winners are species that increase in abundance, and expand their spatial distribution. They generally share specific traits, including small size, high fecundity, rapid dispersal and being generalist species, although this is not always the case (Box 6.1, hound's tongue *Cynoglossum germanicum*, a rare and specialist plant species). A contrary set of traits characterizes loser species. In agreement with McKinney and Lockwood (1999), we thus expect specialist plants with a restricted ecological niche to be more sensitive to habitat changes than more widespread ubiquitous plants that should cope better with various environmental conditions.

Temporal windows chosen to track modifications in plant communities allow us to assess the speed of changes and generally cover decades (10-year and 30-year in Boulanger *et al.*, 2011, 2018; 50-year e.g. in Vild *et al.*, 2017). These authors undertook studies to infer the potential role of varying deer population abundance in shaping plant communities.

Following a ten-year period, using a network of 82 pairs of fenced and unfenced plots covering all major forest types at the French national scale, Boulanger *et al.* (2018) demonstrated an increase in ruderal and epizoochorous plants in the presence of red deer (*Cervus elaphus*), roe deer (*Capreolus capreolus*) and wild pig (*Sus scrofa*), at intermediate densities. However, no non-native plants were detected as winners.

Following a 30-year period, Boulanger *et al.* (2011) showed the expansion of *C. germanicum* (Box 6.1), a rare plant species, epizoochorous, toxic and dependent on soil disturbance for germination and establishment. Other plants also increased their frequency over this time period, lots of them being epizoochorous (graminoids *Brachypodium sylvaticum*, *Bromus benekenii*, *Carex sylvatica*, *Hordelymus europaeus* and forbs *Galium odoratum*, *Geum urbanum*, *Stachys sylvatica*). Of 169 native plants censused, we recorded 57 winners, 27 losers, and the remnants showed no change. Moreover, the temporal changes observed in plant communities depended on a reduction in deer herbivory (Boulanger, 2010).

Two studies, in the Czech Republic (Vild *et al.*, 2017) and in the USA (Wiegmann and Waller, 2006) looked at changes in vegetation

Box 6.1. Hound's tongue (*Cynoglossum sp.*, Boraginaceae), plants advantaged in the presence of ungulates.

Cynoglossum sp. (Fig. 6.2) are epizoochorous plants, locally both rare (*C. germanicum* in France) and invasive (*C. officinale* in British Columbia; Clerck-Floate, 1997). For instance, *C. virginianum*, is a native plant in Virginia (USA) known to be dispersed by white-tailed deer (Shiponeni and Milton, 2006). Couvreur *et al.* (2004) showed that donkeys dispersed *C. officinale* within a network of nature reserves connected by grazing domestic ungulates. *Cynoglossum sp.* all share common attributes favouring their expansion in the presence of biotic dispersal vectors, and they bear glochids on their seeds that cling to animal fur. Seeds from *C. creticum* can remain a long time in animal fur (60% of remaining seeds on sheep and 10% on goat after 48 hours; Shmida and Ellner, 1983). However, Whigham *et al.*, 1993 mention that most of the seeds remain attached to the plant and disperse at a maximum distance matching the plant height. These plants are also unpalatable to deer, and are known to be highly toxic to cattle and horses because of their high content of pyrrolizidine alkaloids, which are hepatotoxins (De Jong *et al.*, 1990). Moreover, seedling germination and seedling establishment of *C. officinale* is favoured on disturbed sites (De Jong *et al.*, 1990); specifically *C. germanicum* frequently occurs on deer bedsites below oak trees. These plant traits (epizoochorous, toxic and ruderal) led to the amazing range expansion of *C. germanicum* in the 88 km^2 Arc-en-Barrois forest in France, within a 30-year time period (Boulanger *et al.*, 2011; Fig. 6.3).

Fig. 6.2. *Cynoglossum officinale*, Domaine National de Chambord, France, June 2017. Picture by Christophe Baltzinger.

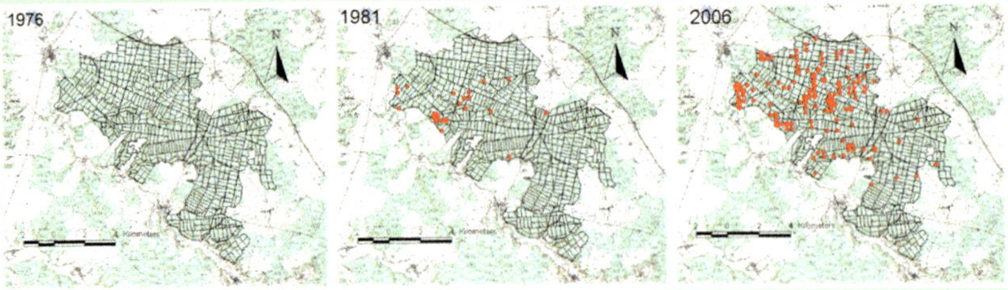

Fig. 6.3. Cynoglossum germanicum spatial distribution patterns in 1976, 1981 and 2006, based on the long-term monitoring of permanent plots, one plot every 10 ha on a square grid. A red circle indicates the presence of the species; its size varies with its local abundance. The plant was absent from the plot network, when the monitoring began in 1976 (Boulanger *et al.*, 2011).

C. germanicum was probably introduced with exogenous soil during forest plantation some time between 1976 and 1981. The presence of *C. germanicum* was further significantly associated with a deer frequentation index. These plant characteristics are typical of what would define an invasive species (Vavra *et al.*, 2007) in the presence of deer. A rough assessment of the yearly dispersal distance fitting the distribution pattern of *C. germanicum* in 2006 averages 360 $m.yr^{-1}$. Highlighting such changes in plant spatio-temporal distribution patterns rely on the long-term monitoring of permanent plots (see section 6.2, 'Spatio-temporal Changes in Plant Communities and Characteristics of Winner Species', p. 106).

community over a 50-year period. Vild *et al.* (2017) showed an increase in ruderal, nitrophilous and light-demanding species. Among the winners, they highlighted four non-native plants, three hemerochorous forbs (*Amaranthus retroflexus, Conyza canadensis, Solanum nigrum*) dispersed by ungulates. Nevertheless, contrary to short-term studies, there was a reduction in epizoochorous plants while ruderal and early successional plants benefited. Wiegmann and Waller (2006) identified five non-native plants: three forbs (*Galeopsis tetrahit, Hieracium aurianticum, Veronica officinalis*) and two graminoids (*Poa nemoralis, Poa pratensis*) among the 21 winners (~25% non-native species). These non-native plants, present at the start of the monitoring, expanded over the 50 years. The authors suggested that white-tailed deer (*O. virginianus*) was probably a major driver of the changes observed over the 50 years as most winners were either browsing resistant or tolerant in comparison with the more sensitive to deer loser species. Shiponeni and Milton (2006) also concluded that 'chronic high deer density facilitated increased abundances of several non-native invasive plants' that were browse tolerant. Similarly, Knight *et al.* (2009) demonstrated that two highly invasive plants (*Alliaria petiolata* and *Microstegium vimineum*) were favoured in the presence of white-tailed deer (see also examples in Chapter 5, this volume). However, although often winners, non-native plants do not explain the main changes observed in plant communities.

6.3 Fluctuating Resources in Plant Communities and Ungulate-mediated Non-native Plant Dispersal

The way ungulates locally modulate the abundance and spatial heterogeneity of resources explains how they can shape plant assemblages. The regional plant species pool is sieved into local plant communities (Lortie *et al.*, 2004) via four major ecological filters namely: i) the ability of the plant to reach a new environment (dispersal phase); ii) the environmental filter (tolerance to local abiotic conditions); iii) plant-plant interactions and iv) other biotic interactions. Davis *et al.* (2000) proposed a theory of invasibility based on the fluctuations of resources in plant communities, which offers a general framework to integrate different ungulate-mediated ecological processes within ecosystems. This theory also provides a convincing framework of how ungulates might favour non-native plant invasion.

Invasion of plant communities by newcomers depends on three main elements: the diaspore arrival, the characteristics of the novel individual and the susceptibility of the local environment to be colonized by this novel individual. This last element relies on the transient release or addition of local resources for the establishment and growth of the novel individual. Olff and Ritchie (1998) proposed such an outline to describe the effects of herbivores in grasslands. Additionally, this invasibility scheme is generic and applies to any arriving individual, whether a native colonizer or a non-native invader.

For Davis *et al.* (2000), the novel environment is never at equilibrium, subject to regular fluctuations in resource availability, and thus hosts transient communities. A plant community is therefore prone to invasion any time resources are released or added (light, nutrients, water or even physical space). The susceptibility of local environments to be colonized fluctuates with time, matching windows of opportunity (Myster, 1993), and should be highest immediately after resource release, especially because resident vegetation requires time to use resources efficiently. Resources become available following various events, whether the temporary reduction in resource uptake by the resident vegetation, the addition of external resources or a combination of both (see Fig. 1 in Davis *et al.*, 2000). Introducing grazers is recognized as one favourable situation increasing opportunities for newcomers to colonize (Rico *et al.*, 2014). In a meta-analysis on seed addition experiments, Myers and Harms (2009) showed that the addition of similar number of different diaspores of various sizes increased local species richness and that this effect was even stronger in previously disturbed sites.

6.3.1 Diaspore Arrival

The capacity for diaspores to be dispersed is a critical step in biological invasions (Pyšek and Richardson, 2010). Wild ungulates can convey diaspores over long distances through different dispersal mechanisms (see examples in Chapter 5, this volume). Through endozoochory, diaspores can be clumped into faeces, with diverse and numerous diaspores, whereas they may be more isolated, assumed less diverse and without faecal matrix through regurgitation or fur-epizoochory (Baltzinger *et al.*, 2019). Diaspore arrival mediated by ungulates is considered directed dispersal (Wenny, 2001). Ungulates are selective herbivores, searching for preferred food items. Picard *et al.* (2016) found that plants dispersed by endozoochory came from open habitats where dispersal vectors used to feed. Ungulates generally display high within-home range spatial fidelity (red and roe deer; Richard *et al.*, 2014), regularly using the same paths or core areas. Regurgitation often occurs at ruminating sites (cleaned bed sites) where diaspores are released (Delibes *et al.*, 2019). Hoof-epizoochory can be an advantage to colonize novel environments as diaspores are released in specific microhabitats (scraped ground, or within hoof prints).

6.3.2 Characteristics of Dispersed Diaspores

Characteristics of diaspores conveyed by ungulates also depend upon the dispersal mechanisms (Albert *et al.*, 2015a; see Chapter 5, this volume). In their review, Albert's team compared ten characteristics between dispersed and non-dispersed plants. They first showed that five characteristics were common to endozoochory, fur-epizoochory and hoof-epizoochory: ungulates preferentially dispersed nitrophilous plants, with persistent seed bank, from open habitats, bearing elongated diaspores or dry fruits. In contrast with regurgitation, which concerned plants bearing fleshy fruits with large seeds (Delibes *et al.*, 2019).

In comparison with endozoochory, epizoochory (fur- and hoof-) was more likely for diaspores released relatively high in the vegetation. More specifically, fur-epizoochory advantaged diaspores with a hooked or an elongated appendage, whereas hoof-epizoochory targeted relatively light diaspores without hooked appendages. Zoochory is an interactive process, which depends on specific interactions between the ungulate vector and the dispersed plant and diaspore traits (Albert *et al.*, 2015b). Plant trait selectivity through ungulate-mediated dispersal was stronger for epizoochory (fur- then hoof-) than for endozoochory (Albert *et al.*, 2015a).

6.3.3 Local Transient Availability of Resources in the Novel Environment

Ungulates can lower or constrain the use of local resources by the resident vegetation. They feed on above-ground vegetation (leaves, buds, bark pieces) or below-ground parts (roots, bulbs, rhizomes). They can also damage saplings through fraying or breaking woody stems. They thus limit resource uptake but also indirectly provide access for light to the understorey lower stratum by lowering the physical obstruction of the upper layers.

They also reduce competition for resources, when they scrape the ground for food (mushrooms, roots), thermal comfort (bed sites) or territorial marking at the expense of the resident vegetation, which is damaged or even killed. Other disturbances (e.g. wild/controlled fires) can have similar effects and plant regrowth will be attractive to herbivores. Freed space acts as physical trap (e.g. hoofprints) for arriving diaspores offering specific microclimatic and competition-free conditions. Deer tracks are considered preferential colonization paths (Lefcort and Pettoello, 2012).

Ungulates also release nutrients through defecation and miction, and this can be reinforced if animals feed in fertilized croplands (Seagle, 2003). Milotić and Hoffmann (2016a, b) showed that the addition of faeces lowered the germination rate of grassland species but later favoured their growth and flowering. Abundant domestic cattle defecation can also kill resident vegetation in constrained areas and release more nutrients in the longer term.

6.4 Different Pathways for Introduction and Dispersal of Non-native Plants

Non-native diaspores are introduced to novel environments by numerous human-induced and long-distance pathways (Schiffman, 1997). Some introductions are accidental by contamination of seed lots, fodder or potted seedlings (Box 6.1) or by adherence to a mobile vector (livestock, vehicles, travellers). Other introductions are intentional by the use of non-native ornamental or cultivated plants, for erosion control, for medicinal use or timber production.

Domestic herbivores are considered mobile links connecting different reserves via external seed dispersal (Couvreur *et al.*, 2004). Rotational shepherding explained patch connectivity for 27 plants, zoochorous or not (Rico *et al.*, 2014). Domestic ungulates could act as potential rewilding tools and non-native vectors in grazed habitats. Wild boars conveyed non-native plants from agricultural lands to conservation areas, but few established (Dovrat *et al.*, 2012); in Iran, only wild boar dispersed a non-native plant (Karimi *et al.*, 2018). Non-native introduction is a main concern of horse riding activities along trails within protected areas; however Gower (2008) showed no non-native plant carried by horses established in the field. Whereas for American bison in Canada, Sigaud (2018) located the more likely introduction points at the interface between croplands and Prince Albert National Park. Ungulate paths are considered roads for plant colonization, for better or worse (Lefcort and Pettoello, 2012).

Plants with long-distance dispersal mechanisms (endozoochory, epizoochory, anemochory) were more likely to be naturalized in new areas, suggested Malo and Suárez (1997). Furthermore, endozoochory showed up as the most effective dispersal mechanism (see Chapter 5). However, plants are more likely to be introduced across continents by fur-epizoochory, due to the mismatch between duration of gut passage time (even though few seeds can be released seven days later, Doucette *et al.*, 2001) and length of travel. Once introduced, the authors further showed that zoochory was probably the most efficient mechanism for the spread of plants, already established or present as seeds within fodder for livestock (e.g. sheep and goats considered responsible for the spread of non-native forbs and grasses to remote oceanic islands).

6.5 Which, Where, How and by Whom?

We reviewed the existing literature on ungulate-mediated non-native plant dispersal to identify *which* plant taxa were dispersed (family, genus and species) with their growth forms; *where* they occurred (by ecozone and country); *how* they were dispersed (endozoochory, fur-epizoochory or both); and *by whom* (which ungulate family, species, feeding regime and fur type).

Used datasets include non-native plants dispersed by native, domestic and introduced ungulates. We did not consider studies without non-native plants or focused on a single non-native. For instance, we excluded European studies used in the review by Ansong and Pickering (2013) on weeds dispersed by horses at the world scale, because the invasive weeds cited were not non-native species. A data set corresponds to the pool of non-native plants dispersed by a specific ungulate in a given study area. We recorded 57 data sets, concerning 26 out of 32 ungulate dispersal vectors, Cervidae, Bovidae, Suidae and Equidae (Table 6.1, Fig. 6.4), six ecozones and ten countries (Table 6.2). Endozoochory largely outweighed fur-epizoochory (n = 45 vs n = 12, Table 6.2), already noted by Baltzinger *et al.* (2019) and Díaz-Vélez *et al.* (see Chapter 5, this volume).

Some studies were dedicated to specific interactions, white-tailed deer and honeysuckle (Vellend, 2002), common cocklebur and domestic ungulates (Box 6.2; Liddle and Elgar, 1984),

Table 6.1. List of the dispersal vectors by family of ungulates, species name and common name, associated with their feeding regime (CS, concentrate selector; IMF, intermediate mixed feeder; GRE, grass and roughage eater; OM, omnivore) and the dispersal mechanism they are involved in: endozoochory (endo), regurgitation (regu), fur-epizoochory (fur-epi).

Species name by family	Common name	Feeding regime**	Dispersal mechanism
Cervidae			
Axis porcinus	hog deer	IMF	endo
Cervus elaphus	red deer / American elk	IMF	endo
Cervus nippon	Sika deer	IMF	endo
Dama dama	fallow deer	IMF	endo
Odocoileus virginianus	white-tailed deer	CS	endo
Odocoileus hemionus	mule deer	CS	endo
Rusa marianna	Philippine sambar	IMF	endo
Rusa unicolor	sambar	IMF	endo
Bovidae			
Aepyceros melampus	impala	IMF	endo
*Alcephalus buselaphus**	red hartebeest	GRE	endo
*Antidorcas marsupialis**	springbok	CS	endo
Antilope cervicapra	blackbuck	GRE	endo
Bison bison	American bison	GRE	endo, fur-epi
Bos taurus	domestic cattle	GRE	endo, fur-epi
Capra hircus	goat	IMF	endo
Cephalophus natalensis	red duiker	CS	endo
Connochaetes gnou, C. taurinus	wildebeest	GRE	endo
*Damaliscus dorcas**	bontebok	GRE	endo
*Oryx gazella**	gemsbok	GRE	endo
Ovis aries	sheep	GRE	endo, fur-epi
*Pelea capreolus**	grey rhebok	CS	endo
Philantomba monticola	blue duiker	CS	endo, regu, fur-epi
Sylvicapra grimmia	grey duiker	CS	endo, regu, fur-epi
Taurotragus oryx	eland	IMF	endo, regu, fur-epi
Tragelaphus angasii	nyala	CS	endo, regu, fur-epi
Tragelaphus scriptus	bushbuck	CS	endo, regu, fur-epi
Tragelaphus strepticeros	greater kudu	CS	endo, fur-epi
Suidae			
Sus scrofa, S. s. lybica	wild pig	OM	endo, fur-epi
Potamochoerus larvatus	bushpig	OM	endo
Equidae			
Equus asinus	donkey	GRE	endo
Equus caballus, E. c. ferus	horse, feral horse	GRE	endo, fur-epi
Equus burchelli, Equus zebra	zebra	GRE	endo

*species included in the category other bovids; Shiponeni and Milton, 2006; **Hofmann, 1989; Milton and Dean, 2001; Hempson *et al.*, 2015; Ahrestani *et al.*, 2016.

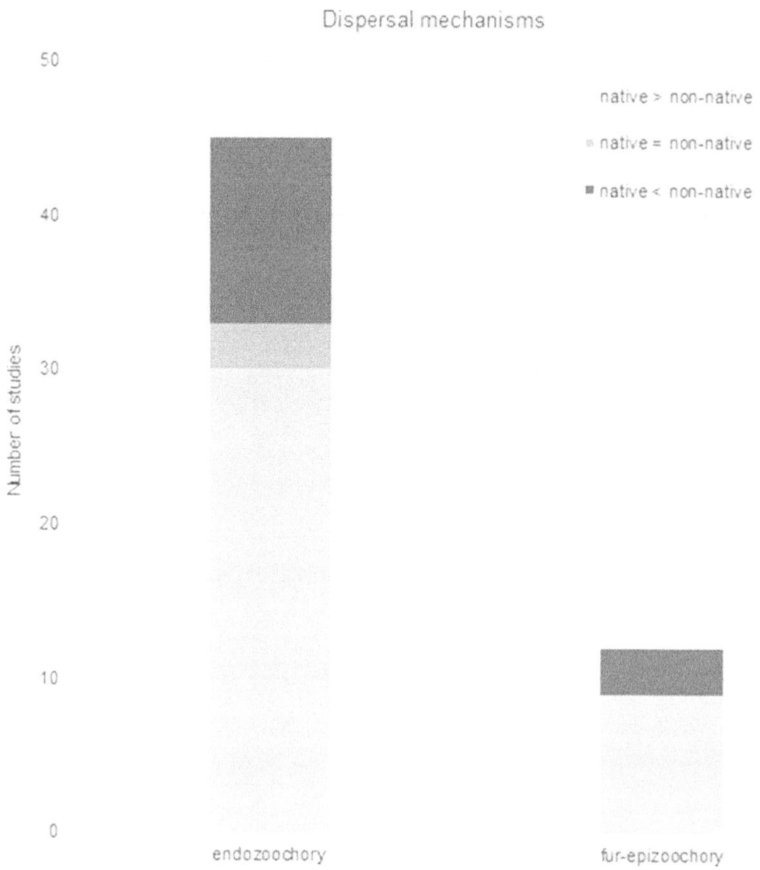

Fig. 6.4. Number of studies by dispersal modes (endozoochory and fur-epizoochory) divided into three categories (native > non-native, native = non-native, native < non-native).

Table 6.2. Report of the datasets collected, addressing non-native plant dispersal by ungulates, according to their biogeographical zone (ecozone, country) and the dispersal mode involved.

Ecozone	Country	Endozoochory	Fur-epizoochory	
Australasia	Australia	2	-	**2**
Afrotropic	South Africa	13	5	**18**
Nearctic	Canada	1	-	**1**
	USA	20	6	**26**
Neotropic	Argentina	1	-	**1**
Oceania	Guam (USA)	2	-	**2**
Palearctic	Czech Republic	2	-	**2**
	Finland	1	-	**1**
	Sweden	1	-	**1**
	Iran	1	-	**1**
	Israel	1	1	**2**
		45	**12**	

Box 6.2. Pathways in diaspore dispersal for the epizoochorous common cocklebur (*Xanthium strumarium*, Asteraceae).

Common cocklebur *Xanthium strumarium* is a non-native invasive species in Europe, largely distributed along the banks of rivers like the Loire (France, Fig. 6.5). It produces large amounts of large hooked diaspores (see Fig. 1 in Liddle and Elgar, 1984), with adaptations for epizoochorous transport. They usually bear more than 50 hooks and release their diaspores high above ground (~0.75 m). Liehrmann *et al*. (2018) assessed the external retention time of these diaspores in relation to the fur characteristics of different social ungulate vectors (red deer, dwarf goat and donkey), their intraspecific interactions and individual grooming behaviour.

Cocklebur diaspores can be dispersed through multiple pathways (see Fig. 6 in Liddle and Elgar, 1984): barochory, hydrochory (e.g. flooding events) and fur-epizoochory (in the mane and tail of horses, or any body part of sheep). They can also be released as a catapult from the long plant stalk (diaspore releasing height). Water can also move diaspores secondarily when burrs lay on the ground.

Hydrochory can move diaspores downstream for long distances, whereas ungulate-mediated dispersal can potentially transport diaspores in any direction (including upstream). Liehrmann *et al*. (2018) experimentally showed that a red deer hind released a burr nearly 3 km from its attachment site. Liddle and Elgar (1984) showed that burrs remained in the fur of different domestic ungulates (cattle, sheep and horses), and diaspore retention functions fitted generated fat-tailed distributions, and potential long-distance dispersal (Liehrmann *et al*., 2018).

This plant is also toxic at certain stages of its development. Only, the two encapsuled seeds in the burrs and the cotyledonary leaves contain a toxic compound, the carboxyatractyloside. The seeds and the burrs are thus rarely consumed by the ungulates. Botha *et al*. (2014) concluded that it 'rarely causes poisoning in cattle', which confers a competitive advantage for this plant over palatable neighbouring plants. Various studies report *Xanthium sp*. being dispersed by different ungulates (American bison, cattle, various southern African bovids and horse, see Tables 6.5, 6.6, 6.7 and 6.9). Liehrmann *et al*. (2018) further proved for the first time that intraspecific interactions in the three ungulates they studied allowed diaspore transfer from one individual to another one, diversifying its dispersal pathway and potential distance covered in relation with the group size. Although, Liddle and Elgar (1984) did not specifically test if trampled burrs would further germinate, they experimentally showed that horses contributed to the burial of 18% of the burrs left on animal tracks, suggesting that ungulates may secondarily favour their germination and seedling recruitment.

Fig. 6.5. *Xanthium strumarium*, banks of the Loire river, Gien, France, 8 September 2019. Picture by Christophe Baltzinger.

hound's tongue and cattle (Clerck-Floate, 1997) or sambar (*Rusa unicolor*) and Himalayan honeysuckle (*Leycesteria formosa*; Eyles, 2002). Whereas in others, some dispersal vectors were pooled: red deer and sika deer (*Cervus nippon*; Lepková *et al.*, 2018); white-tailed and mule deer (*Odocoileus hemionus*; Bartuszevige and Endress, 2008); different bovids (Shiponeni and Milton, 2006; Haarmeyer *et al.*, 2010), whatever their feeding regimes.

We listed non-native plants dispersed by Artiodactyla and Perissodactyla from four families (Cervidae, Bovidae, Suidae and Equidae). Each subsection generally accounted for the most-studied dispersal vectors, or for vectors with endozoochory and fur-epizoochory data. Within Cervidae, the second subsection focuses on cervids introduced to Australia along with a discussion on other cervids from elsewhere in the world. Three subsections present the Bovidae, one for American bison, one for domestic cattle (*B. taurus*) and one for the communities of South African bovids (including blackbuck *Antilope cervicapra* in India). Wild pig, both in its native range and as an introduced species, with few mentions for bush pig (*Potamochoerus larvatus*) in South Africa, represent Suidae. Domestic horses, with few mentions of feral horses, donkeys and zebras (*Equus burchelli*, *E. zebra*) describe Equidae.

We checked plant names using the Taxonomic Name Resolution Service v4.0 (Boyle *et al.*, 2013), providing updated information on family, genus and species. Each subsection and associated table listed the plants dispersed by growth form: cactus, forbs, graminoids (Cyperaceae, Juncaceae and Poaceae), shrubs, trees and vines; by dispersal vectors and mechanisms (endozoochory, regurgitation and fur-epizoochory) and associated references.

The USA (Nearctic) and South Africa (Afrotropic) were the most studied geographic areas for non-native plant dispersal for both endozoochory and fur-epizoochory. The USA was also registered as the most studied geographic area in a previous chapter analysing a diverse set of seed dispersers (see Chapter 5, this volume). Indomalaya and Antarctic were not represented. The prevalence of the USA was linked to the number of referenced studies, whereas South Africa benefited from the diversity of ungulate communities and number of study sites (Milton *et al.*, 1990). First, significantly more studies had more native plant than non-native plant taxa dispersed, both for endozoochory ($\chi^2_{(2)} = 25.2$, $p = 3.410^{-6}$) and fur-epizoochory ($\chi^2_{(2)} = 10.5$, $p = 5.310^{-3}$, Fig. 6.4). The proportion of non-native plants varied from 2 out of 102 (~2%) to 4 out of 4 (100%) for endozoochory and from 1 out of 17 (~6%) to 7 out of 7 (100%) for fur-epizoochory.

Secondly, in terms of fur-epizoochory and hair characteristics, non-native plants predominated native plants in only 3 out of 12 cases. These three cases included American bison, with wavy hair and domestic cattle with straight hair, whereas sheep, with the most epizoochory efficient curled hair, systematically dispersed more native plants than non-native ones (Fig. 6.6).

Thirdly, concerning endozoochory and the ungulates feeding guilds, we observed systematically more studies with more native than non-native plants dispersed (Fig. 6.7). Most datasets ($n = 19$) concerned grass and roughage eaters (cattle, bison, wildebeest: *Connochaetes gnou/taurinus*, sheep, horse, zebra); secondly ($n = 10$) concentrate selectors (white-tailed deer, greater kudu *Tragelaphus strepticeros* and bushbuck *Tragelaphus scriptus*) and intermediate mixed feeders ($n = 9$), with nearly as many cases as different dispersal vectors, and finally omnivores ($n = 5$).

6.6 Non-Native Plant Dispersal by Cervidae

6.6.1 White-tailed and mule deer in the USA

Overabundant deer populations in North America have driven lots of research (Russell *et al.*, 2001) and white-tailed deer, a medium-sized browser cervid, was the most studied dispersal vector within its native range. We recorded nine studies concerning endozoochory (Table 6.3), but none

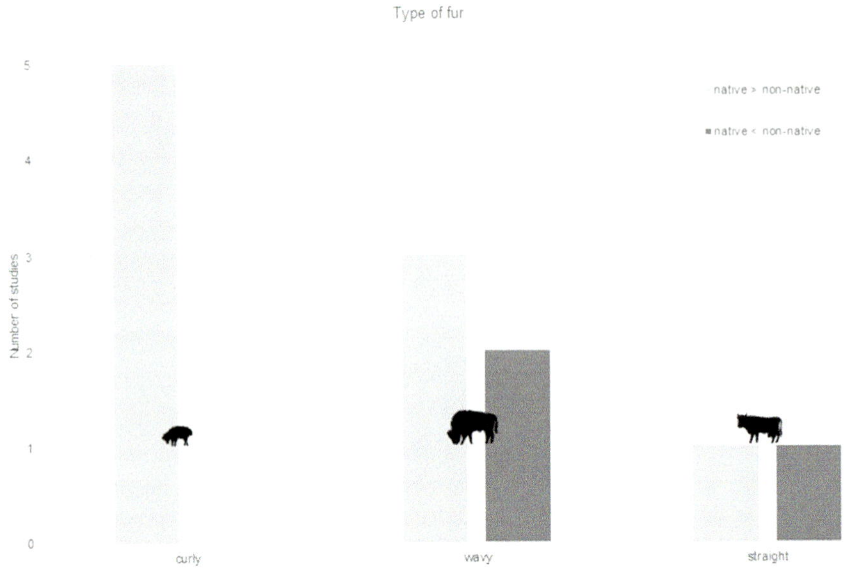

Fig. 6.6. Number of fur-epizoochory studies divided into two categories (native > non-native, native < non-native) as a function of fur type (curly – sheep, wavy – bison, straight – cattle) and associated dispersal vectors.

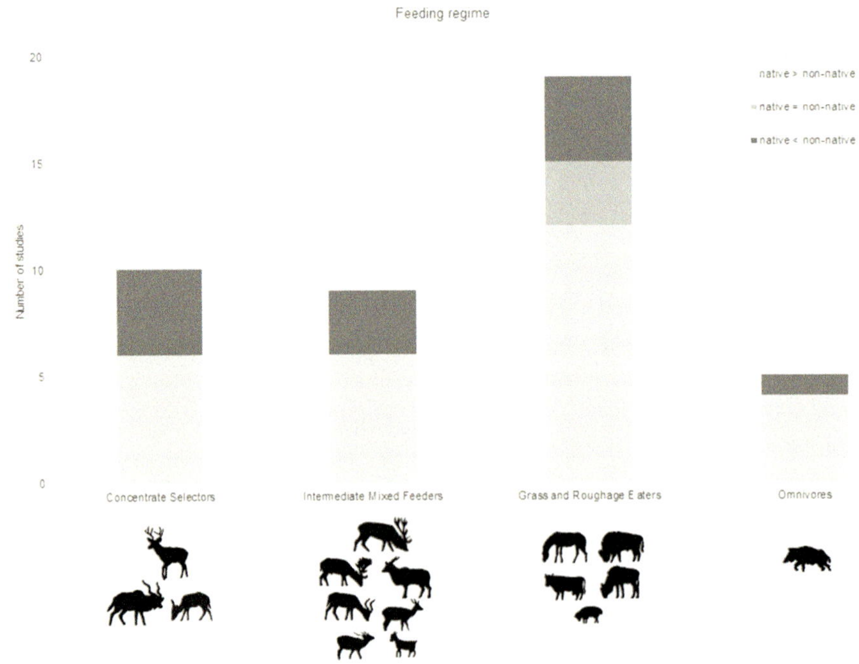

Fig. 6.7. Number of endozoochory studies divided into three categories (native>non-native, native=non-native, native<non-native) as a function of the dispersal vector feeding regime. Concentrate selectors are represented by white-tailed deer, greater kudu and bushbuck ; intermediate mixed feeders by red deer, sika deer, fallow deer, hog deer, Philippine deer, goat, eland and impala ; grass and roughage eaters by bison, cattle, sheep, wildebeest, horse, donkey and zebra; omnivores include wild pig and bushpig. Latin names are indicated in Table 6.1.

Table 6.3. List of non-native plants referenced as being dispersed by white-tailed deer (*Odocoileus virginianus*) and mule deer (*Odocoileus hemionus* [c]) by endozoochory, ordered by growth form (forb, graminoid, shrub, tree, vine) and family.

Growth form	Family	Plant species	References
Forb	Amaranthaceae	*Amaranthus hybridus*	b
		Amaranthus palmeri	h
		Amaranthus retroflexus	b, e
		Amaranthus viridis	g*
		Chenopodium album	b, d, e, h
		Chenopodium glaucum	b
		Chenopodium pumilio	d
	Asteraceae	*Artemisia vulgaris*	d
		Cirsium arvense	e
		Galinsoga quadriradiata	h
		Gnaphalium uliginosum	b
		Leucanthemum vulgare	b
		Matricaria matricarioides	b
		Sonchus asper	b, d
		Taraxacum officinale	b, c, e
	Brassicaceae	*Capsella bursa-pastoris*	b, d
		Thlaspi arvense	c
	Caryophyllaceae	*Cerastium fontanum*	d
		Cerastium glomeratum	c
		Silene latifolia subsp. alba	d
		Spergularia media	b
		Stellaria media	b, d
	Commelinaceae	*Murdannia nudiflora*	g
	Convolvulaceae	*Jacquemontia tamnifolia*	g*
	Fabaceae	*Coronilla varia*	b
		Kummerowia striata	g
		Lotus corniculatus	b, e
		Medicago lupulina	b
		Medicago minima	g
		Melilotus officinalis subsp. alba	b
		Melilotus sp.	e
		Securigera varia	d
		Trifolium dubium	g
		Trifolium pratense	b, d
		Trifolium repens	d, i
		Trifolium sp.	d
		Vicia tetrasperma	b
		Vicia sp.	d

Continued

Table 6.3. Continued

Growth form	Family	Plant species	References
	Lamiaceae	*Nepeta cataria*	d
	Lythraceae	*Lythrum salicaria*	b
	Plantaginaceae	*Linaria vulgaris*	b
		Plantago major	b
		Veronica officinalis	b, d, i
		Veronica persica	d
		Veronica serpyllifolia	b
	Polygonaceae	*Persicaria maculosa*	h
		Polygonum caespitosum	d
		Polygonum lapathifolium	b
		Polygonum persicaria	b, d
		Rumex acetosella	i
		Rumex obtusifolius	e
	Portulacaceae	*Portulaca amilis*	g
		Portulaca grandiflora	b
		Portulaca oleracea	b, d, h
	Rosaceae	*Potentilla norvegica*	b
		Potentilla recta	b, c, h
	Rubiaceae	*Galium mollugo*	b
		Oldenlandia corymbosa	g*
	Solanaceae	*Capsicum sp.*	d
		Lycopersicon esculentum	b
		Petunia sp.	d
		Solanum dulcamara	b, d
		Solanum nigrum	d
		Solanum physalifolium	d
	Urticaceae	*Urtica dioica*	h
Graminoid	Cyperaceae	*Cyperus compressus*	g*
		Kyllinga brevifolia	g*
	Poaceae	*Agrostis capillaris*	d
		Agrostis gigantea	b, e
		Agrostis stolonifera	b
		Anthoxanthum odoratum	i
		Dactylis glomerata	b
		Digitaria ischaemum	b, e
		Digitaria sanguinalis	b, e, g
		Echinochloa crus-galli	e
		Elymus repens	d
		Holcus lanatus	i
		Microstegium vimineum	d

Continued

Table 6.3. Continued

Growth form	Family	Plant species	References
		Panicum miliaceum	d
		Pennisetum glaucum	d
		Phleum pratense	b
		Poa annua	b
		Poa compressa	b, d
		Poa pratensis	b, c, h
		Puccinellia distans	b
		Setaria viridis	e
		Sorghum bicolor	b, e
		Sorghum halepense	d
		Triticum aestivum	b
		Zea mays	d
Shrub	Caprifoliaceae	*Lonicera aff. ×bella*	a, b
		Lonicera japonica	d
		Lonicera maackii	f
		Lonicera morrowii	a, e
		Lonicera tatarica	a, e
	Elaeagnaceae	*Elaeagnus umbellata*	d, i
	Rhamnaceae	*Rhamnus cathartica*	b
	Rosaceae	*Rosa multiflora*	b, d, i
		Rubus idaeus	b
		Rubus phoenicolasius	d
Tree	Araceae	*Phellodendron japonicum*	d
	Rosaceae	*Malus sp.*	d
		Pyrus sp.	b
Vine	Vitaceae	*Ampelopsis brevipedunculata*	d

References: Vellend, 2002 [a]; Myers *et al*., 2004 [b]; Bartuszevige and Endress, 2008 [c]; Williams *et al*., 2008 [d]; Blyth *et al*., 2013 [e]; Guiden *et al*., 2015 [f]; Pile *et al*., 2015 [g] (* cryptogenic origin); Guiden, 2017 [h]; Flaherty *et al*., 2018 [i].

on fur-epizoochory. Bartuszevige and Endress (2008) did not distinguish white-tailed from mule deer, and two studies focused on invading honeysuckles (*Lonicera* sp.; Vellend, 2002; Guiden *et al*., 2015). We recorded 104 non-native plants, dispersed across different states of the USA. These taxa were divided into five growth forms: 65 forbs, 25 graminoids, ten shrubs, three trees and one vine (Table 6.3). Forbs dispersed belonged to 16 different families, predominantly Fabaceae, Asteraceae and Amaranthaceae. Twenty-three of the graminoids dispersed were Poaceae. Shrubs belonged to four families, with five different *Lonicera* species (Caprifoliaceae). *Chenopodium album* was the unique plant dispersed in four studies, whereas seven plants appeared three times: *Taraxacum officinale*, *Veronica officinalis*, *Portulaca oleracea*, *Potentilla recta*, *Digitaria sanguinalis*, *Poa pratensis* and *Rosa multiflora*.

Table 6.4. List of non-native plants referenced as being dispersed by hog deer (*Axis porcinus*), red deer, American elk (*Cervus elaphus*), fallow deer (*Dama dama*), sika deer (*Cervus nippon*), sambar (*Rusa unicolor*), Philippine sambar (*Rusa marianna*) by endozoochory, ordered by growth form (forb, graminoid, shrub, tree, vine) and family.

Growth form	Family	Plant species	Dispersal vector	References
Forb	Amaranthaceae	*Chenopodium album*	fallow deer, red deer, sika deer	e, g
	Asteraceae	*Erigeron annuus*	red deer, sika deer	g
		Hypochaeris glabra	hog deer	c
		Lapsana communis	red deer, sika deer	g
		Oncosiphon piluliferum	hog deer	c
		Tanacetum vulgare	red deer, sika deer	g
	Brassicaceae	*Thlaspi arvense*	American elk	b
	Caryophyllaceae	*Cerastium glomeratum*	American elk, hog deer	b, c
		Spergularia rubra	hog deer	c
		Stellaria media	American elk, hog deer	b, c
	Fabaceae	*Medicago minima*	hog deer	c
		Medicago polymorpha	hog deer	c
		Trifolium hybridum	American elk	b
		Trifolium repens	American elk	b
		Trifolium sp.	fallow deer	e
	Gentianaceae	*Centaurium erythraea*	hog deer	c
	Geraniaceae	*Geranium pusillum*	red deer, sika deer	g
	Malvaceae	*Modiola caroliniana*	fallow deer	e
	Papaveraceae	*Papaver somniferum*	fallow deer	e
	Plantaginaceae	*Plantago coronopus*	hog deer	c
		Plantago lanceolata	hog deer	c
		Veronica arvensis	American elk	b
	Polygonaceae	*Polygonum aviculare*	hog deer	c
		Rumex acetosella	American elk, hog deer	b, c
		Rumex conglomeratus	hog deer	c
	Primulaceae	*Anagallis arvensis*	hog deer	c
	Ranunculaceae	*Ranunculus muricatus*	hog deer	c
	Scrophulariaceae	*Verbascum thapsus*	fallow deer	e
	Verbenaceae	*Verbena bonariensis*	fallow deer	e
		Verbena sp.	fallow deer	e
Graminoid	Cyperaceae	*Cyperus eragrostis*	fallow deer	e
	Juncaceae	*Juncus bufonius*	hog deer	c
		Juncus tenuis	red deer, sika deer	g
	Poaceae	*Apera interrupta*	American elk	b
		Briza minor	hog deer	c
		Bromus hordeaceus	hog deer	c

Continued

Table 6.4. Continued

Growth form	Family	Plant species	Dispersal vector	References
		Bromus inermus	American elk	b
		Eragrostis cilianensis	fallow deer	e
		Festuca ovina	American elk	b
		Poa annua	hog deer	c
		Poa compressa	American elk	b
		Poa pratensis	American elk	b
		Vulpia bromoides	hog deer	c
Shrub	Caprifoliaceae	*Leycesteria formosa*	sambar	a
	Rosaceae	*Rubus fruticosus*	sambar	d
Tree	Caricaceae	*Carica papaya*	Philippine sambar	f
	Lamiaceae	*Vitex parviflora*	Philippine sambar	f
Vine	Asteraceae	*Mikania micrantha*	Philippine sambar	f
	Passifloraceae	*Passiflora suberosa*	Philippine sambar	f

References: Eyles, 2002 (Australia) [a]; Bartuszevige and Endress, 2008 (USA) [b]; Davis *et al.*, 2010 (Australia) [c]; Forsyth and Davis, 2011 (Australia) [d]; Claridge *et al.*, 2016 (Australia) [e]; Gawel *et al.*, 2018 (USA Guam) [f]; Lepková *et al.*, 2018 (Czech Republic) [g].

6.6.2 Cervidae Introduced to Australia and Other Deer Species

Australia hosts no native deer species. Fallow deer *Dama dama*, hog deer *Axis porcinus*, red deer, chital *Axis axis*, Philippine sambar *Rusa marianna* and sambar have been introduced to different parts, with great invasion potential (Davis *et al.*, 2016). Non-native plants dispersed by non-native deer in Australia appeared in four different studies, focusing on three of the six introduced cervids (fallow deer, hog deer and sambar), all being intermediate mixed feeders (Table 6.4). Thirty-one non-native plants, mainly forbs (n = 22), were dispersed by endozoochory, each of the 13 forb families represented by maximum three non-native plants. Graminoids comprised Cyperaceae, Juncaceae and five Poaceae. Only sambar dispersed non-native shrubs. Again, the current knowledge is sparse by the lack of information on the other three non-native ungulates. More information has been collected on horses (see Equidae subsection) and other herbivores (Calviño-Cancela, 2011) dispersing non-native plants in Australia.

The other deer species are also intermediate mixed feeders, favouring endozoochorous seed dispersal (Table 6.4). However, few data were collected on non-native plants they dispersed in different areas. Plant dispersal by red deer has been well studied in Europe (Baltzinger *et al.*, 2019) but with rare mentions of non-native plants. We found 22 non-native plants, including 12 forbs, six graminoids, two trees and two vines dispersed by red deer, American elk, Sika deer and Philippine sambar. Some non-native plants listed are dispersed by red deer in their native range (e.g. *Stellaria media*, *Rumex acetosella*, *Poa pratensis*). Again, the potential for non-native plants dispersal by cervids is underrated by a global lack of studies, even more for fur-epizoochory. Cervidae like white-tailed deer (Table 6.3) or red deer generally have abundant populations with potential major interactions with plant populations both antagonistic (herbivory) and mutualistic (zoochory).

Table 6.5. List of non-native plants referenced as being dispersed by American bison (*Bison bison*) by endozoochory (endo) and fur-epizoochory (fur-epi), ordered by growth form (forb, graminoid) and family.

Growth form	Family	Plant species	Dispersal mode	References
Forb	Amaranthaceae	*Amaranthus albus*	endo	d
		Amaranthus graecizans	endo, fur-epi	c, d
		Amaranthus retroflexus	endo	d
		Axyris amaranthoides	endo	d
		Chenopodium album	endo, fur-epi	c
	Apiaceae	*Daucus carota*	endo, fur-epi	c
		Pastinaca sativa	endo, fur-epi	c
		Torilis arvensis	endo, fur-epi	b
	Asteraceae	*Arctium minus*	fur-epi	c
		Cirsium arvense	endo	d
		Sonchus arvensis	endo	d
		Taraxacum officinale	endo, fur-epi	c, d
		Xanthium spinosum	fur-epi	a
		Xanthium strumarium	fur-epi	a, b, c
	Brassicaceae	*Capsella bursa-pastoris*	endo, fur-epi	b, c
		Thlaspi arvense	endo	d
	Caryophyllaceae	*Silene noctiflora*	endo	d
	Fabaceae	*Medicago lupulina*	endo	b, d
		Medicago polymorpha	fur-epi	a
		Medicago sativa	endo, fur-epi	a, d
		Medicago sp.	fur-epi	a
		Melilotus officinalis	endo, fur-epi	b, d
		Melilotus sp.	endo	c
		Trifolium pratense	endo, fur-epi	c, d
		Trifolium repens	endo	c, d
	Geraniaceae	*Erodium botrys*	fur-epi	a
		Erodium cicutarium	fur-epi	a
		Erodium sp.	fur-epi	a
	Lamiaceae	*Galeopsis tetrahit*	endo	d
		Marrubium vulgare	fur-epi	a
	Plantaginaceae	*Plantago major*	endo	c, d
	Polygonaceae	*Polygonum convolvulus*	endo	d
		Polygonum lapathifolium	endo	d
		Rumex crispus	endo, fur-epi	c
	Rosaceae	*Potentilla norvegica*	endo	d
Graminoid	Poaceae	*Avena sp.*	fur-epi	a
		Brachypodium distachyon	fur-epi	a
		Bromus inermis	endo, fur-epi	c, d
		Bromus japonicus	endo, fur-epi	c

Continued

Table 6.5. Continued

Growth form	Family	Plant species	Dispersal mode	References
		Bromus sp1.	fur-epi	a
		Bromus sp2.	endo, fur-epi	b
		Cynodon dactylon	endo, fur-epi	b
		Dactylis glomerata	fur-epi	c
		Digitaria sanguinalis	endo, fur-epi	c
		Echinochloa crus-galli	endo, fur-epi	c
		Festuca arundinacea	endo, fur-epi	c
		Hordeum sp.	fur-epi	a
		Phalaris arundinacea	endo, fur-epi	c
		Phleum pratense	endo	d
		Poa pratensis	endo, fur-epi	c
		Setaria faberi	endo, fur-epi	c
		Setaria glauca	endo	d

References: Constible *et al.*, 2005 [a]; Rosas *et al.*, 2008 [b]; Eyheralde, 2015 [c]; Sigaud, 2018 [d].

6.7 Non-native Plant Dispersal by Bovidae

6.7.1 American bison in North America

American bison, a grass and roughage eater, with relatively wavy fur, was identified as dispersal vector four times, balanced between endozoochory and fur-epizoochory. All studies included non-native plants, and occurred from 2005 to 2018, in the USA and Canada, in its native range as well as introduced on Santa Catalina Island (Constible *et al.*, 2005). The European bison (*Bison bonasus*), intensively studied in Poland (Białowieża; Jaroszewicz *et al.*, 2009), is a model for rewilding areas, only dispersing (n=178) native plants. American bison dispersed 52 non-native plants, 35 forbs (mainly Fabaceae, Asteraceae and Amaranthaceae) and 17 Poaceae, as expected from its feeding regime. Seventeen species were solely dispersed via endozoochory (~33%), 14 solely via fur-epizoochory (~27%), but 21 by both mechanisms (~40%) with strong implications for dispersal effectiveness. Common cocklebur, which bears epizoochorous diaspores (Table 6.5, Box 6.2) was identified three times, whereas 10 additional plants were recovered twice, seven of them dispersed by both dispersal mechanisms, and three by endozoochory only.

6.7.2 Domestic Cattle

Domestic cattle, a grass and roughage eater, with straight hair (for the races studied) was studied as a non-native plant dispersal vector five times; twice in the USA, once in Argentina, Sweden and Canada (focused on hound's tongue epizoochory, Box 6.1). Cattle dispersed 61 non-woody non-native plants including 36 forbs and 25 Poaceae, similar to American bison with its feeding regime. Most plants were dispersed solely by endozoochory (45 plants,~74%), 10 (~16%) by fur-epizoochory, and 6 (~10%) by both dispersal mechanisms, which is relatively lower than for American bison. Fabaceae occurred frequently, with nine non-native plants (5 *Trifolium sp.*) dispersed mostly by endozoochory. *Stellaria media* was dispersed at three sites, *Trifolium repens* and *Cynodon dactylon* were dispersed at two study sites. With the present worldwide use of cattle, either for milk or meat production, the potential for native (e.g. Auffret and Cousins, 2013) and non-native plant dispersal is certainly underestimated (Table 6.6).

Table 6.6. List of non-native plants referenced as being dispersed by domestic cattle (*Bos taurus*) by endozoochory (endo) and fur-epizoochory (fur-epi), ordered by growth form (forb, graminoid) and family.

Growth form	Family	Plant species	Dispersal mode	References
Forb	Apiaceae	*Torilis arvensis*	fur-epi	[e]
	Asteraceae	*Matricaria discoidea*	endo	[e]
		Picris echioides	endo, fur-epi	[e]
		Silybum marianum	fur-epi	[e]
		Xanthium spinosum	fur-epi	[e]
	Boraginaceae	*Cynoglossum officinale*	fur-epi	[a]
		Myosotis stricta	endo	[b]
	Brassicaceae	*Alyssum alyssoides*	endo	[b]
		Capsella bursa-pastoris	endo	[e]
	Caryophyllaceae	*Cerastium glomeratum*	endo	[b]
		Stellaria media	endo	[b,d,e]
	Fabaceae	*Lotus tenuis*	endo	[c]
		Medicago lupulina	endo	[b]
		Medicago polymorpha	endo	[e]
		Trifolium hirtum	endo, fur-epi	[e]
		Trifolium hybridum	endo	[b]
		Trifolium repens	endo	[b, c]
		Trifolium subterraneum	endo	[e]
		Trifolium sp.	endo	[b]
		Vicia villosa	endo	[e]
	Geraniaceae	*Erodium* sp.	fur-epi	[e]
		Geranium molle	endo	[e]
	Lamiaceae	*Mentha pulegium*	endo	[c]
	Lythraceae	*Lythrum hyssopifolia*	endo	[c]
	Plantaginaceae	*Plantago lanceolata*	endo	[e]
		Veronica arvensis	endo	[b]
		Veronica chamaedrys	endo	[b]
		Veronica persica	endo	[e]
	Polygonaceae	*Polygonum aviculare*	endo	[c]
		Rumex acetosella	endo	[b]
		Rumex crispus	endo	[c]
	Portulacaceae	*Portulaca oleracea*	endo	[c]
	Primulaceae	*Anagallis arvensis*	endo	[c]
	Rubiaceae	*Galium parisiense*	endo	[e]
		Sherardia arvensis	endo	[e]
	Urticaceae	*Urtica dioica*	endo	[d]
Graminoid	Poaceae	*Aegilops truncialis*	fur-epi	[e]
		Agrostis avenacea	endo	[c]
		Agrostis gigantea	endo	[b]

Continued

Table 6.6. Continued

Growth form	Family	Plant species	Dispersal mode	References
		Apera interrupta	endo	b
		Avena sp.	endo, fur-epi	e
		Brachypodium distachyon	endo, fur-epi	e
		Bromus hordeaceus	fur-epi	e
		Bromus inermus	endo	b
		Cynodon dactylon	endo	c, e
		Dactylus glomerata	endo	b
		Echinochloa crus-galli	endo	c
		Elymus caput-medusae	fur-epi	e
		Festuca ovina	endo	b
		Festuca perennis	endo, fur-epi	e
		Festuca temulenta	endo, fur-epi	e
		Hordeum murinum	fur-epi	e
		Lolium multiflorum	endo	c
		Lolium perenne	endo	c
		Parapholis incurva	endo	c
		Phalaris aquatica	fur-epi	e
		Poa annua	endo	c
		Poa compressa	endo	b
		Poa pratensis	endo	b
		Polypogon monspeliensis	endo	e
		Setaria pumila	endo	e

References: Clerck-Floate, 1997 [a]; Bartuszevige and Endress, 2008 [b]; Vignolio and Fernandez, 2010 [c]; Auffret and Cousins, 2013 [d]; Chuong *et al.*, 2016 [e].

6.7.3 Native and Domestic Bovids in Southern Africa

Southern Africa hosts very diverse ungulate communities, but information on ungulate-mediated seed dispersal is relatively scarce (Table 6.7). However, most studies published so far include non-native plants and involve different native and/or domestic bovids. Bovid communities in southern Africa are also functionally highly diverse in terms of feeding guilds (Hempson *et al.*, 2015), body size, body mass, fur type, habitat use and sociality. Consequently, they are likely to disperse a wide diversity of plants bearing different traits. Altogether, southern African bovids dispersed 46 non-native taxa including one cactus, 27 forbs, seven Poaceae, five shrubs and six trees (two Fabaceae). Non-native plants were dispersed by two domestic bovids (goat and sheep), more than ten different native bovids of various feeding guilds (concentrate selectors, intermediate mixed feeders and grass and roughage eaters), by endozoochory, regurgitation and fur-epizoochory. Amaranthaceae with at least four different *Atriplex sp.* and Asteraceae with three *Bidens sp.* dispersed by fur-epizoochory and endozoochory (*Bidens bipinnata* by impala *Aepyceros melampus*) were the most represented. Plants from all growth forms (except vines) were dispersed. Birds also dispersed *Solanum mauritianum* and *Lantana camara*, two invasive non-native fleshy-fruited plants. In South Africa, there is a great opportunity for

Table 6.7. List of non-native plants referenced as being dispersed by native and domestic Bovidae in South Africa by endozoochory (endo), regurgitation (regu) and fur-epizoochory (fur-epi), ordered by growth form (cactus, forb, graminoid, shrub, tree) and family. Except *Prosopis juliflora* dispersed by blackbuck by endozoochory in India.

Growth form	Family	Plant species	Dispersal mode	Dispersal vector*	References
Cactus	Cactaceae	*Opuntia ficus-indica*	endo	bushbuck	b
Forb	Amaranthaceae	*Atriplex lindleyi*	endo, fur-epi	goat, kudu, sheep	a, c, f
		Atriplex muelleri	fur-epi	sheep	a
		Atriplex semibaccata	endo	goat, kudu, sheep	c, f
		Atriplex sp.	endo	goat, sheep	f
		Chenopodium album	endo	goat, sheep	f
		Chenopodium sp.	endo	goat, sheep	f
		Gomphrena celosioides	endo	impala	d
		Salsola sp.	endo	goat, sheep	f
	Asteraceae	*Bidens bipinnata*	fur-epi, endo	bushbuck, eland, grey duiker, impala	d, i
		Bidens biternata	fur-epi	bushbuck	i
		Bidens pilosa	fur-epi	blue duiker, grey duiker	i
		Conyza sp.	endo	wildebeest, other bovids	e
		Xanthium strumarium	fur-epi	bushbuck, grey duiker, kudu, nyala	i
	Caryophyllaceae	*Drymaria cordata*	fur-epi	blue duiker	i
		Spergularia media	endo	goat, sheep	f
		Spergularia sp.	endo	wildebeest, other bovids	e
		Stellaria media	endo	eland, other bovids	e
		Stellaria sp.	endo	wildebeest, other bovids	e
	Fabaceae	*Medicago polymorpha*	fur-epi	sheep	a
		Medicago sativa	endo	goat	c
		Medicago sp.	endo	other bovids	e
		Trifolium sp.	endo	wildebeest, other bovids	e
	Geraniaceae	*Erodium moschatum*	endo	wildebeest, other bovids	e
	Oxalidaceae	*Oxalis corniculata*	endo	eland, impala	d
	Papaveraceae	*Argemone mexicana*	fur-epi	sheep	a
		Argemone ochroleuca	endo	kudu	c
	Primulaceae	*Anagallis arvensis*	endo	other bovids	e
Graminoid	Poaceae	*Briza maxima*	endo	eland, wildebeest, other bovids	e
		Bromus diandrus	endo	wildebeest	e

Continued

Table 6.7. Continued

Growth form	Family	Plant species	Dispersal mode	Dispersal vector*	References
		Bromus murinum	fur-epi	sheep	a
		Bromus pectinatus	endo	wildebeest	e
		Lolium sp.	endo	eland, wildebeest, other bovids	e
		Poa annua	endo	eland, wildebeest, other bovids	e
		Vulpia myuros	endo	eland, wildebeest, other bovids	e
Shrub	Verbenaceae	*Lantana camara*	endo	bushbuck, kudu, nyala	i
	Caricaceae	*Papaya carica*	endo	bushbuck	i
	Myrtaceae	*Psidium cattelinaum*	endo	bushbuck, grey duiker	i
		Psidium guajava	endo	bushbuck, grey duiker	i
	Solanaceae	*Solanum mauritianum*	endo	bushbuck, grey duiker	h, i
Tree	Anacardiaceae	*Mangifera indica*	regu	bushbuck, nyala	i
	Fabaceae	*Acacia melanoxylon*	endo	bushbuck	i
		Gleditsia triacanthos	regu	bushbuck, grey duiker, eland	i
		Prosopis juliflora	endo	blackbuck	g
	Meliaceae	*Melia azedarach*	endo	bushbuck, blue duiker, red duiker	i
	Moraceae	*Morus alba*	endo	blue duiker, red duiker	i
	Rosaceae	*Eriobotrya japonica*	regu	bushbuck, blue duiker	i

References: Milton *et al.*, 1990 [a]; Castley *et al.*, 2001 [b]; Milton and Dean, 2001 [c]; Slater and Du Toit, 2002 [d]; Shiponeni and Milton, 2006 [e]; Haarmeyer *et al.*, 2010 [f]; Jadeja *et al.*, 2013 [g]; Msweli, 2018 [h]; Nichols, 2018 [i].
*Other bovids in Shiponeni and Milton (2006) include springbok (*Antidorcas marsupialis*), bontebok (*Damaliscus dorcas*), red hartebeest (*Alcephalus buselaphus*), grey rhebok (*Pelea capreolus*) and gemsbok (*Oryx gazella*)

ungulate-mediated dispersal studies, based on the functional diversity of both plant and potential dispersal vectors. In India, blackbuck, a grass and roughage eater, is also known to disperse mesquite (*Prosopis juliflora*, Fabaceae), a highly invasive non-native (Jadeja *et al.*, 2013). In South Africa, other livestock prevent pre-dispersal destruction of *Prosopis* sp. seeds by bruchids and favour their integration in the soil seed bank.

6.8 Non-native Plant Dispersal by Suidae

Suidae (wild boar and bush pig) were grouped as frugivore opportunistic, omnivorous dispersal vectors. Only Dovrat *et al.* (2012) combined endozoochory and fur-epizoochory, highlighting three forbs dispersed by both mechanisms (*Amaranthus blitoides*, *Amaranthus blitum* and *Conyza sp.*). Eight studies occurred in the native

Table 6.8. List of non-native plants referenced as being dispersed by Suidae, wild pig (*Sus scrofa, S. s. lybica*) and bushpig (*Potamochoerus larvatus*) by endozoochory (endo) and fur-epizoochory (fur-epi) ordered by growth form (cactus, forb, graminoid, shrub, tree, vine) and family.

Growth form	Family	Plant species	Dispersal mode	Dispersal vector	References
Cactus	Cactaceae	*Opuntia ficus-indica*	endo	bushpig	a
		Opuntia maxima	endo	wild pig	b
		Opuntia sp.	endo	wild pig	c
Forb	Amaranthaceae	*Amaranthus blitoides*	endo, fur-epi	wild pig	c
		Amaranthus blitum	endo, fur-epi	wild pig	c
		Amaranthus cruentus	endo	wild pig	c
		Amaranthus spinosus	endo	wild pig	c
	Apiaceae	*Daucus carota*	endo	wild pig	f
	Asteraceae	*Bidens tripartita*	fur-epi	wild pig	c
		Conyza canadensis	endo	wild pig	e
		Conyza sp.	endo, fur-epi	wild pig	c
		Lapsana communis	endo	wild pig	f
		Tanacetum vulgare	endo	wild pig	f
	Euphorbiaceae	*Euphorbia hirta*	endo	wild pig	c
	Solanaceae	*Solanum lycopersicum*	endo	wild pig	c
Graminoid	Juncaceae	*Juncus tenuis*	endo	wild pig	f
Shrub	Asteraceae	*Chromolaena odorata*	endo	wild pig	d
	Myrtaceae	*Psidium cattelinaum*	endo	bushpig	g
		Psidium guajava	endo	bushpig	g
	Solanaceae	*Solanum mauritianum*	endo	bushpig	g
Tree	Caricaceae	*Carica papaya*	endo	wild pig	d
	Fabaceae	*Acacia cyclops*	endo	bushpig	a
		Acacia saligna	endo	wild pig	c
		Leucaena leucocephala	endo	wild pig	d
	Moraceae	*Ficus benghalensis*	endo	wild pig	c
		Ficus religiosa	endo	wild pig	c
		Morus sp.	endo	wild pig	c
	Rutaceae	*Citrus sinensis*	endo	wild pig	h
Vine	Cucurbitaceae	*Coccinia grandis*	endo	wild pig	d
	Passifloraceae	*Passiflora suberosa*	endo	wild pig	d

References: Castley *et al.*, 2001 (South Africa) [a]; Padrón *et al.*, 2011 (Spain) [b]; Dovrat *et al.*, 2012 (Israël) [c] (USA Guam) [d]; Karimi *et al.*, 2018 (Iran) [e]; Lepková *et al.*, 2018 (Czech Republic) [f]; Nichols, 2018 (South Africa) [g]; Peris *et al.*, 2019 (Brazil) [h]

Table 6.9. List of non-native plants dispersed by Equidae including horse (*Equus caballus)*, feral horse (*E. c. ferus*), donkey (*E. asinus)* and zebra *(E. burchelli, E. zebra*) by endozoochory (except *Xanthium strumarium* by fur-epizoochory) ordered by growth form (forb, graminoid, tree) and family.

Growth form	Family	Plant species	Dispersal vector	References
Forb	Amaranthaceae	*Amaranthus retroflexus*	horse	g
		Amaranthus spinosus	horse	d
		Chenopodium album	horse	i
		Chenopodium ambrosioides	horse	d
	Apiaceae	*Daucus carota*	horse	d
	Araliaceae	*Hydrocotyle sp.*	horse	c
	Asteraceae	*Achillea millefolium*	horse	i
		Arthemis cotula	horse	h
		Hypochaeris glabra	horse	h
		Lactuca serriola	zebra	f
		Matricaria chamomilla	horse	d
		Pseudognaphalium luteoalbum	horse	h
		Soliva sessilis	horse	h
		Taraxacum officinale	horse	d
		Tripleurospermum inodorum	horse	i
		Xanthium strumarium	horse (fur-epi)	a
	Brassicaceae	*Alyssum desertorum*	feral horse	j
		Capsella bursa-pastoris	horse	g
		Cardamine hirsuta	horse	d
		Hirschfeldia incana	horse	h
		Sisymbrium sp.	feral horse	j
	Caryophyllaceae	*Cerastium glomeratum*	horse	c, d
		Spergularia rubra	horse	g
		Spergularia sp.	zebra	f
		Stellaria media	horse	c, h, i
	Fabaceae	*Kummerowia striata*	horse	d
		Medicago minima	horse	c
		Medicago polymorpha	horse	c, h
		Medicago sativa	donkey	e
		Medicago truncatula	horse	c
		Medicago sp.	zebra	f
		Melilotus indicus	horse	c
		Melilotus sp.	horse	d
		Trifolium arvense	horse	c
		Trifolium glomeratum	horse	c
		Trifolium hirtum	horse	h
		Trifolium pratense	horse	d
		Trifolium repens	horse	b, c, d, i
		Trifolium subterraneum	horse	c

Continued

Table 6.9. Continued

Growth form	Family	Plant species	Dispersal vector	References
		Trifolium sp.	zebra	f
	Geraniaceae	*Erodium moschatum*	zebra	f
	Lamiaceae	*Mentha pulegium*	horse	h
		Prunella vulgaris	horse	d
	Linaceae	*Linum bienne*	horse	h
		Linum usitatissimum	horse	h
	Lythraceae	*Lythrum hyssopifolium*	horse	h
	Malvaceae	*Malva nicaeensis*	horse	h
	Molluginaceae	*Mollugo verticillata*	horse	d
	Oxalidaceae	*Oxalis sp.*	horse	c
	Plantaginaceae	*Plantago lanceolata*	horse, zebra	b, c, d, f
		Plantago major	horse	d
		Veronica perigrina	horse	d
		Veronica serpyllifolia	horse	i
	Polygonaceae	*Polygonum arenastrum*	horse	g
		Polygonum aviculare	horse	d, h
		Polygonum cespitosum	horse	d
		Polygonum convolvus	horse	d
		Rumex acetosella	horse	b, h
		Rumex crispus	horse	d
		Rumex obtusifolius	horse	d
		Rumex sp.	horse	c
	Portulacaceae	*Portulaca oleracea*	horse	d, h
	Rubiaceae	*Galium aparine*	horse	c
		Galium murale	horse	h
	Scrophulariaceae	*Verbascum thapsus*	horse	d
Graminoid	Juncaceae	*Juncus bufonius*	horse	c
	Poaceae	*Agrostis capillaris*	horse	c
		Aira caryophyllea	horse	b, h
		Avena barbata	horse	c
		Avena fatua	horse	c
		Avena sativa	horse	g
		Avena sp.	horse	d
		Briza maxima	zebra	f
		Briza minor	horse	c
		Bromus diandrus	horse, zebra	b, c, f
		Bromus inermis	horse	g
		Bromus hordeaceus	horse	b
		Bromus pectinatus	zebra	f
		Bromus rubens	horse	c
		Bromus tectorum	horse, feral horse	g, j

Continued

Table 6.9. Continued

Growth form	Family	Plant species	Dispersal vector	References
		Cenchrus sp.	horse	c
		Dactylis glomerata	horse	b, g
		Digitaria ischaemum	horse	d
		Digitaria sanguinalis	horse	d
		Eleusine indica	horse	d
		Festuca arundinacea	horse	d
		Holcus lanatus	horse	b, c
		Hordeum marinum	horse	h
		Hordeum sp.	horse	c
		Lolium multiflorum	horse	h
		Lolium perenne	horse	b, c
		Lolium rigidum	horse	c
		Lolium sp1.	zebra	f
		Lolium sp2.	zebra	f
		Phalaris paradoxa	horse	c
		Poa annua	horse, zebra	c, d, f, h
		Poa pratensis	horse	g, h
		Setaria viridis	horse	g
		Vulpia bromoides	horse	b
		Vulpia myuros	horse, zebra	f, h
Tree	Fabaceae	*Prosopis sp.*	donkey	e

References: Liddle and Elgar, 1984 [a]; Whinam *et al.*, 1994 [b]; Weaver and Adams, 1996 [c]; Campbell and Gibson, 2001 [d]; Milton and Dean, 2001 [e]; Shiponeni and Milton, 2006 [f]; Wells and Lauenroth, 2007 [g]; Quinn *et al.*, 2008 [h]; Törn *et al.*, 2010 [i]; King *et al.*, 2019 [j]

range of the Suidae concerned (Iran, Israel, Czech Republic and Spain for wild boar; South Africa for bush pig) as well as those introduced on islands (Guam, USA). Altogether Suidae dispersed 30 non-native taxa, representing all growth forms (three cactus; 12 forbs, one graminoid, four shrubs, eight trees and two vines) as expected from their generalist feeding regime. Suidae feed on various above- and below-ground parts, being able to transport diaspores and vegetative parts like rhizomes. Non-native plants producing fleshy fruits were dispersed (*Opuntia sp., Psidium sp., Solanum mauritianum, Citrus sinensis*, Table 6.8). European studies highlighted wild boar both as a seed predator and as a seed disperser, but rarely for non-native plants. Feral pig is well distributed worldwide, which requires more attention. Wild pig is also more effective through fur-epizoochory (Picard and Baltzinger, 2012) than endozoochory (Picard *et al.*, 2016).

6.9 Non-native Plant Dispersal by Equidae

Domestic and feral horses, zebra and donkey represented Equidae. They are all grass and roughage eaters with relatively straight fur. Horses have been quite intensively studied regarding non-native plant dispersal, in relation to horse-riding practices, mostly in the USA and Australia, but see Törn *et al.* (2010) in Finland. Table 6.9 listed plants considered non-native in the country where they were dispersed. Weeds mentioned by Ansong and Pickering (2013) in their review and retrieved from European studies are not non-natives. Most non-native plants were dispersed by endozoochory and by domestic horses (King *et al.*, 2019 for feral horses).

We gathered 101 non-native plants dispersed by Equidae (Table 6.9), nearly twice more forbs ($n = 65$) than graminoids ($n = 35$, 34 Poaceae and 1 Juncaceae). Among 18 families of non-native forbs dispersed by Equidae, Fabaceae ($n = 15$, 7 *Trifolium* and five *Medicago sp.*), Asteraceae ($n = 10$) and Polygonaceae ($n = 8$) were the most represented. *T. repens*, *Plantago lanceolata* and *Poa annua* were identified four times vs. three times for *Stellaria media* and *Bromus diandrus*, and 84 species appeared only once. Common cocklebur was the single non-native plant dispersed by fur-epizoochory (Table 6.9, Box 6.2). Similar to domestic cattle, the role of Equidae (mainly domestic horses) in non-native plant dispersal is probably underrated. However, Gower (2008) showed that not a single non-native plant present in horse dung germinated, nor established in the field, questioning the real contribution of horse riding in the invasion of local plant communities by non-native plants.

6.10 Conclusions

Shiponeni and Milton (2006) stated that 'antelopes could play a role in restoration via seed dispersal from natural vegetation to abandoned fields' in southern Africa, and 'also influence the rate of vegetation recovery in disturbed ecosystems by supplying seeds of both indigenous and non-indigenous species'. The debate remains open whether ungulate-mediated plant dispersal is more an opportunity than a threat.

Although numerous non-native plants are dispersed by native, introduced and domestic ungulates, non-native plant dispersal and invasion mediated by large ungulates are certainly underrated. Potential ungulate vectors are distributed worldwide as livestock, abundant deer populations in their native range or introduced in areas devoid of native ungulates like New Zealand or Australia (Davis *et al.*, 2016). Wilson and Reeder (2005) censused 257 native ungulates, in comparison with the 32 (~12%) reviewed species known to be involved in non-native plant dispersal. Some families are not even mentioned (Rhinocerotidae and Giraffidae). Their effectiveness in plant dispersal is also linked to their own population dynamics and range expansion, with both expanding and declining ungulates.

This review also highlighted the low number of studies dedicated to either epizoochory, or both mechanisms, acting as complementary pathways for plant dispersal. Albert *et al.* (2015a) demonstrated that epizoochory applied the most selective filter on the regional flora requiring more attention in future studies.

Some non-native plants are known to be dispersed in their native range by ungulates, suggesting that plant and diaspore traits identified by Albert *et al.* (2015a) and Picard *et al.* (2016) might be relatively similar between non-native and native plants. This remains to be tested.

Identifying the effects of such vectors in non-native plants' long-distance dispersal and invasion is not an easy task and requires combining long-term vegetation surveys with spatially explicit modelling. Moreover, most plants are dispersed by multiple vectors, and assessing their relative effectiveness is challenging. Finally, ungulates act not only as dispersal vectors but also as selective herbivores, physical and chemical engineers (Wilby *et al.*, 2001), and non-native plant invasion can result from the combination of these actions (Box 6.1).

Acknowledgements

CB's work on this chapter was supported by funding as a Research Fellow from the DSI-NRF Centre of Excellence for Invasion Biology at Stellenbosch University, the French Embassy in Pretoria and the MiDi Network. US is a PhD student funded by INRAE and Centre-Val de Loire Region. CTD and LSM were funded by the National Research Foundation, the University of KwaZulu-Natal and the DSI-NRF Centre of Excellence for Invasion Biology. We thank colleagues from the workshop organized during summer 2018 in Pietermaritzburg, and Anna Traveset, Jose M. Fedriani and Suzanne J. Milton for their comments. We are grateful to Geoff Nichols for his botanical help and Marie Baltzinger for her drawings.

References

Ahrestani, F.S., Heitkönig, I.M.A., Matsubayashi, H. and Prins, H.H.T. (2016) Grazing and browsing by large herbivores in South and Southeast Asia. In: Ahrestani, F.S. and Sankaran, M.E. (eds) *The Ecology of Large Herbivores in South and Southeast Asia*. Springer, Dordrecht, The Netherlands, pp. 99–120. DOI: 10.1007/978-94-017-7570-0_4

Albert, A., Auffret, A.G., Cosyns, E., Cousins, S.A.O., D'hondt, B. *et al.* (2015a) Seed dispersal by ungulates as an ecological filter: a trait-based meta-analysis. *Oikos* 124(9), 1109–1120. DOI: 10.1111/oik.02512.

Albert, A., Mårell, A., Picard, M. and Baltzinger, C. (2015b) Using basic plant traits to predict ungulate seed dispersal potential. *Ecography* 38(5), 440–449. DOI: 10.1111/ecog.00709.

Ansong, M. and Pickering, C. (2013) A global review of weeds that can germinate from horse dung. *Ecological Management & Restoration* 14(3), 216–223. DOI: 10.1111/emr.12057.

Auffret, A.G. and Cousins, S.A.O. (2013) Grassland connectivity by motor vehicles and grazing livestock. *Ecography* 36(10), 1150–1157. DOI: 10.1111/j.1600-0587.2013.00185.x.

Baltzinger, C., Karimi, S. and Shukla, U. (2019) Plants on the move: Hitch-hiking with ungulates distributes diaspores across landscapes. *Frontiers in Ecology and Evolution* 7, 38. DOI: 10.3389/fevo.2019.00038.

Bartuszevige, A.M. and Endress, B.A. (2008) Do ungulates facilitate native and exotic plant spread? seed dispersal by cattle, elk and deer in northeastern Oregon. *Journal of Arid Environments* 72(6), 904–913. DOI: 10.1016/j.jaridenv.2007.11.007

Blyth, L.H., Ouborg, L.J., Johnson, D.M. and Anderson, L.J. (2013) The short-term germination and establishment success of deer-dispersed seeds in mesic temperate forests. *The Journal of the Torrey Botanical Society* 140(3), 334–348. DOI: 10.3159/TORREY-D-13-00006.1

Botha, C.J., Lessing, D., Rösemann, M., van Wilpe, E. and Williams, J.H. (2014) Analytical confirmation of Xanthium strumarium poisoning in cattle. *Journal of Veterinary Diagnostic Investigation* 26(5), 640–645. DOI: 10.1177/1040638714542867.

Boulanger, V. (2010) Pression d'Herbivorie et Dynamique des Communautés Végétales: Influence à Court et Moyen Termes des Populations de Cervidés sur la Diversité des Communautés Végétales en Forêt. Doctoral dissertation. Université Henri Poincaré, Nancy, France.

Boulanger, V., Baltzinger, C., Saïd, S., Ballon, P., Ningre, F. *et al.* (2011) Deer-mediated expansion of a rare plant species. *Plant Ecology* 212(2), 307–314. DOI: 10.1007/s11258-010-9823-9.

Boulanger, V., Dupouey, J.-L., Archaux, F., Badeau, V., Baltzinger, C. *et al.* (2018) Ungulates increase forest plant species richness to the benefit of non-forest specialists. *Global Change Biology* 24(2), e485–e495. DOI: 10.1111/gcb.13899.

Boyle, B., Hopkins, N., Lu, Z., Raygoza Garay, J.A., Mozzherin, D. *et al.* (2013) The taxonomic name resolution service: an online tool for automated standardization of plant names. *BMC Bioinformatics* 14, 16. DOI: 10.1186/1471-2105-14-16.

Calviño-Cancela, M. (2011) Seed dispersal of alien and native plants by vertebrate herbivores. *Biological Invasions* 13(4), 895–904. DOI: 10.1007/s10530-010-9877-6.

Campbell, J.E. and Gibson, D.J. (2001) The effect of seeds of exotic species transported via horse dung on vegetation along trail corridors. *Plant Ecology* 157(1), 23–35. DOI: 10.1023/A:1013751615636

Castley, J.G., Bruton, J.-S., Kerley, G.I.H. and McLachlan, A. (2001) The importance of seed dispersal in the Alexandria coastal Dunefield, South Africa. *Journal of Coastal Conservation* 7(1), 57–70. DOI: 10.1007/bf02742468

Claridge, A.W., Hunt, R., Thrall, P.H. and Mills, D.J. (2016) Germination of native and introduced plants from scats of Fallow Deer (*Dama dama*) and Eastern Grey Kangaroo (*Macropus giganteus*) in a south-eastern Australian woodland landscape. *Ecological Management & Restoration* 17(1), 56–62. DOI: 10.1111/emr.12193

Clerck-Floate, R.D. (1997) Cattle as dispersers of Hound's-Tongue on rangeland in southeastern British Columbia. *Journal of Range Management* 50(3), 239–243. DOI: 10.2307/4003722.

Constible, J.M., Sweitzer, R.A., Vuren, D.H.V., Schuyler, P.T. and Knapp, D.A. (2005) Dispersal of non-native plants by introduced bison in an island ecosystem. *Biological Invasions* 7(4), 699–709. DOI: 10.1007/s10530-004-5859-x.

Couvreur, M., Christiaen, B., Verheyen, K. and Hermy, M. (2004) Large herbivores as mobile links between isolated nature reserves through adhesive seed dispersal. *Applied Vegetation Science* 7(2), 229–236. DOI: 10.1111/j.1654-109X.2004.tb00614.x.

Davis, M.A., Grime, J.P. and Thompson, K. (2000) Fluctuating resources in plant communities: a general theory of invasibility. *Journal of Ecology* 88(3), 528–534. DOI: 10.1046/j.1365-2745.2000.00473.x.

Davis, N.E., Forsyth, D.M. and Coulson, G. (2010) Facilitative interactions between an exotic mammal and native and exotic plants: hog deer (*Axis porcinus*) as seed dispersers in south-eastern Australia. *Biological Invasions* 12(5), 1079–1092. DOI: 10.1007/s10530-009-9525-1.

Davis, N.E., Bennett, A., Forsyth, D.M., Bowman, D.M.J.S., Lefroy, E.C. *et al.* (2016) A systematic review of the impacts and management of introduced deer (family Cervidae) in Australia. *Wildlife Research* 43(6), 515–532. DOI: 10.1071/WR16148.

De Jong, T.J., Klinkhamer, P.G.L. and Boorman, L.A. (1990) Cynoglossum officinale L. *The Journal of Ecology* 78(4), 1123–1144. DOI: 10.2307/2260956.

Delibes, M., Castañeda, I. and Fedriani, J.M. (2019) Spitting seeds from the cud: a review of an endozoochory exclusive to ruminants. *Frontiers in Ecology and Evolution* 7, 265. DOI: 10.3389/fevo.2019.00265.

Doucette, K.M., Wittenberg, K.M. and McCaughey, W.P. (2001) Seed recovery and germination of reseeded species fed to cattle. *Journal of Range Management* 54(5), 575–581. DOI: 10.2307/4003587.

Dovrat, G., Perevolotsky, A. and Ne'eman, G. (2012) Wild boars as seed dispersal agents of exotic plants from agricultural lands to conservation areas. *Journal of Arid Environments* 78, 49–54. DOI: 10.1016/j.jaridenv.2011.11.011.

Eyheralde, P.G. (2015) Bison-mediated seed dispersal in a Tallgrass prairie reconstruction. Doctoral dissertation. Iowa State University, Ames, Iowa/USA.

Eyles, D. (2002) *Sambar Deer (Cervus unicolor) as a Potential Seed Vector for the Spread of the Environmental Weed Himalayan Honeysuckle (Leycesteria formosa) at Mount Buffalo National Park. Honours dissertation.* University of Melbourne, Parkville, Australia.

Flaherty, K.L., Rentch, J.S. and Anderson, J.T. (2018) Wetland seed dispersal by white-tailed deer in a large freshwater wetland complex. *AoB PLANTS* 10(1), plx074. DOI: 10.1093/aobpla/plx074.

Forsyth, D.M. and Davis, N.E. (2011) Diets of non-native deer in Australia estimated by macroscopic versus microhistological rumen analysis. *The Journal of Wildlife Management* 75(6), 1488–1497. DOI: 10.1002/jwmg.179.

Gawel, A.M., Rogers, H.S., Miller, R.H. and Kerr, A.M. (2018) Contrasting ecological roles of non-native ungulates in a novel ecosystem. *Royal Society Open Science* 5(4), 170151. DOI: 10.1098/rsos.170151.

Gower, S.T. (2008) Are horses responsible for introducing non-native plants along forest trails in the eastern United States? *Forest Ecology and Management* 256(5), 997–1003. DOI: 10.1016/j.foreco.2008.06.012.

Guiden, P.W. (2017) Spatial heterogeneity in white-tailed deer activity increases seed dispersal of shade-intolerant plants near forest edges in fragmented forests. *The Journal of the Torrey Botanical Society* 144(4), 371–378. DOI: 10.3159/TORREY-D-16-00045.1

Guiden, P., Gorchov, D.L., Nielsen, C. and Schauber, E. (2015) Seed dispersal of an invasive shrub, Amur honeysuckle (*Lonicera maackii*), by white-tailed deer in a fragmented agricultural-forest matrix. *Plant Ecology* 216(7), 939–950. DOI: 10.1007/s11258-015-0480-x.

Haarmeyer, D.H., Bösing, B.M., Schmiedel, U. and Dengler, J. (2010) The role of domestic herbivores in endozoochorous plant dispersal in the arid Knersvlakte, South Africa. *South African Journal of Botany* 76(2), 359–364. DOI: 10.1016/j.sajb.2009.12.001.

Hempson, G.P., Archibald, S. and Bond, W.J. (2015) A continent-wide assessment of the form and intensity of large mammal herbivory in Africa. *Science* 350(6264), 1056–1061. DOI: 10.1126/science.aac7978.

Hofmann, R.R. (1989) Evolutionary steps of ecophysiological adaptation and diversification of ruminants: a comparative view of their digestive system. *Oecologia* 78(4), 443–457. DOI: 10.1007/BF00378733.

Hui, C. and Richardson, D.M. (2017) *Invasion Dynamics*. Oxford University Press, Oxford. DOI: 10.1093/acprof:oso/9780198745334.001.0001.

Jadeja, S., Prasad, S., Quader, S. and Isvaran, K. (2013) Antelope mating strategies facilitate invasion of grasslands by a woody weed. *Oikos* 122(10), 1441–1452. DOI: 10.1111/j.1600-0706.2013.00320.x.

Jaroszewicz, B., Pirożnikow, E. and Sagehorn, R. (2009) Endozoochory by European bison (Bison bonasus) in Białowieża primeval forest across a management gradient. *Forest Ecology and Management* 258(1), 11–17. DOI: 10.1016/j.foreco.2009.03.040.

Karimi, S., Hemami, M.-R., Tarkesh Esfahani, M., Akhani, H. and Baltzinger, C. (2018) Complementary endozoochorous seed dispersal by large mammals in the Golestan National Park, Iran. *Seed Science Research* 28(4), 294–302. DOI: 10.1017/S0960258518000351.

Keane, R. and Crawley, M.J. (2002) Exotic plant invasions and the enemy release hypothesis. *Trends in Ecology & Evolution* 17(4), 164–170. DOI: 10.1016/S0169-5347(02)02499-0.

King, S.R.B., Schoenecker, K.A. and Manier, D.J. (2019) Potential spread of cheatgrass (*Bromus tectorum*) and other invasive species by feral horses (*Equus ferus caballus*) in Western Colorado. *Rangeland Ecology & Management* 72(4), 706–710. DOI: 10.1016/j.rama.2019.02.006.

Knight, T.M., Dunn, J.L., Smith, L.A., Davis, J. and Kalisz, S. (2009) Deer facilitate invasive plant success in a Pennsylvania forest understory. *Natural Areas Journal* 29(2), 110–116. DOI: 10.3375/043.029.0202.

Lefcort, H. and Pettoello, C.L. (2012) White-Tailed deer trails are associated with the spread of exotic forbs. *Natural Areas Journal* 32(2), 159–165. DOI: 10.3375/043.032.0204.

Lepková, B., Horčičková, E. and Vojta, J. (2018) Endozoochorous seed dispersal by free-ranging herbivores in an abandoned landscape. *Plant Ecology* 219(9), 1127–1138. DOI: 10.1007/s11258-018-0864-9.

Liddle, M.J. and Elgar, M.A. (1984) Multiple pathways in diaspore dispersal, exemplified by studies of Noogoora Burr (*Xanthium occidentale* Bertol., compositae). *Botanical Journal of the Linnean Society* 88(4), 303–315. DOI: 10.1111/j.1095-8339.1984.tb01578.x.

Liehrmann, O., Jégoux, F., Guilbert, M.-A., Isselin-Nondedeu, F., Saïd, S. *et al.* (2018) Epizoochorous dispersal by ungulates depends on fur, grooming and social interactions. *Ecology and Evolution* 8(3), 1582–1594. DOI: 10.1002/ece3.3768.

Lortie, C.J., Brooker, R.W., Choler, P., Kikvidze, Z., Michalet, R. *et al.* (2004) Rethinking plant community theory. *Oikos* 107(2), 433–438. DOI: 10.1111/j.0030-1299.2004.13250.x.

Malo, J.E. and Suárez, F. (1997) Dispersal mechanism and transcontinental naturalization proneness among Mediterranean herbaceous species. *Journal of Biogeography* 24(3), 391–394. DOI: 10.1046/j.1365-2699.1997.00107.x.

Manzano, P. and Malo, J.E. (2006) Extreme long-distance seed dispersal via sheep. *Frontiers in Ecology and the Environment* 4(5), 244–248. DOI: 10.1890/1540-9295(2006)004[0244:ELSDVS]2.0.CO;2.

McKinney, M.L. and Lockwood, J.L. (1999) Biotic homogenization: a few winners replacing many losers in the next mass extinction. *Trends in Ecology & Evolution* 14(11), 450–453. DOI: 10.1016/S0169-5347(99)01679-1.

Milotić, T. and Hoffmann, M. (2016a) Cost or benefit for growth and flowering of seedlings and juvenile grassland plants in a dung environment. *Plant Ecology* 217(8), 1025–1042. DOI: 10.1007/s11258-016-0629-2.

Milotić, T. and Hoffmann, M. (2016b) Reduced germination success of temperate grassland seeds sown in dung: consequences for post-dispersal seed fate. *Plant Biology* 18(6), 1038–1047. DOI: 10.1111/plb.12506.

Milton, S.J. and Dean, W.R.J. (2001) Seeds dispersed in dung of insectivores and herbivores in semi-arid southern Africa. *Journal of Arid Environments* 47(4), 465–483. DOI: 10.1006/jare.2000.0727.

Milton, S.J., Siegfried, W.R. and Dean, W.R.J. (1990) The distribution of epizoochoric plant species: a clue to the prehistoric use of arid Karoo rangelands by large herbivores. *Journal of Biogeography* 17(1), 25–34. DOI: 10.2307/2845185.

Msweli, L.S. (2018) *Effects of indigenous southern African ungulates on seed germination of the Alien Invasive Lantana (Lantana camara) and Bugweed (Solanum mauritianum). Honours dissertation*. University of KwaZulu-Natal, Pietermaritzburg, South Africa.

Myers, J.A. and Harms, K.E. (2009) Seed arrival, ecological filters, and plant species richness: a meta-analysis. *Ecology Letters* 12(11), 1250–1260. DOI: 10.1111/j.1461-0248.2009.01373.x.

Myers, J.A., Vellend, M., Gardescu, S. and Marks, P.L. (2004) Seed dispersal by white-tailed deer: implications for long-distance dispersal, invasion, and migration of plants in eastern North America. *Oecologia* 139(1), 35–44. DOI: 10.1007/s00442-003-1474-2.

Myster, R.W. (1993) Tree invasion and establishment in old fields at Hutcheson Memorial forest. *The Botanical Review* 59(4), 251–272. DOI: 10.1007/BF02857418.

Nichols, G. (2018) List of plant species dispersed by native ungulates in South Africa. *Workshop on Seed Dispersal by Native Ungulates from Southern Africa*, School of Life Sciences, Pietermaritzburg, South Africa, July 20-21, 2018.

Olff, H. and Ritchie, M.E. (1998) Effects of herbivores on grassland plant diversity. *Trends in Ecology & Evolution* 13(7), 261–265. DOI: 10.1016/S0169-5347(98)01364-0.
Padrón, B., Nogales, M., Traveset, A., Vilà, M., Martínez-Abraín, A. *et al.* (2011) Integration of invasive Opuntia spp. by native and alien seed dispersers in the Mediterranean area and the Canary Islands. *Biological Invasions* 13(4), 831–844. DOI: 10.1007/s10530-010-9872-y.
Pellerin, M., Picard, M., Saïd, S., Baubet, E. and Baltzinger, C. (2016) Complementary endozoochorous long-distance seed dispersal by three native herbivorous ungulates in Europe. *Basic and Applied Ecology* 17(4), 321–332. DOI: 10.1016/j.baae.2016.01.005.
Peris, J.E., Malara, T.M., Borges, R., Falconi, J.R., Peña, L. *et al.* (2019) Reunion overseas: introduced wild boars and cultivated orange trees interact in the Brazilian Atlantic forest. *Revista de Biología Tropical* 67(4), 901–912. DOI: 10.15517/rbt.v67i4.35862.
Picard, M. and Baltzinger, C. (2012) Hitch-hiking in the wild: should seeds rely on ungulates? *Plant Ecology and Evolution* 145(1), 24–30. DOI: 10.5091/plecevo.2012.689.
Picard, M., Chevalier, R., Barrier, R., Boscardin, Y. and Baltzinger, C. (2016) Functional traits of seeds dispersed through endozoochory by native forest ungulates. *Journal of Vegetation Science* 27(5), 987–998. DOI: 10.1111/jvs.12418.
Pile, L.S., Wang, G.G., Polomski, R., Yarrow, G. and Stuyck, C.M. (2015) Potential for nonnative endozoochorous seed dispersal by white-tailed deer in a southeastern maritime forest. *Invasive Plant Science and Management* 8(1), 32–43. DOI: 10.1614/IPSM-D-14-00027.1
Pyšek, P. and Richardson, D.M. (2010) Invasive species, environmental change and management, and health. *Annual Review of Environment and Resources* 35(1), 25–55. DOI: 10.1146/annurev-environ-033009-095548.
Quinn, L.D., Kolipinski, M., Coelho, V.R., Davis, B., Vianney, J.-M. *et al.* (2008) Germination of invasive plant seeds after digestion by horses in California. *Natural Areas Journal* 28(4), 356–362. DOI: 10.3375/0885-8608(2008)28[356:GOIPSA]2.0.CO;2.
Richard, E., Saïd, S., Hamann, J.-L. and Gaillard, J.-M. (2014) Daily, seasonal, and annual variations in individual home-range overlap of two sympatric species of deer. *Canadian Journal of Zoology* 92(10), 853–859. DOI: 10.1139/cjz-2014-0045.
Rico, Y., Boehmer, H.J. and Wagner, H.H. (2014) Effect of rotational shepherding on demographic and genetic connectivity of calcareous grassland plants. *Conservation Biology* 28(2), 467–477. DOI: 10.1111/cobi.12186.
Rosas, C.A., Engle, D.M., Shaw, J.H. and Palmer, M.W. (2008) Seed dispersal by *Bison bison* in a tallgrass prairie. *Journal of Vegetation Science* 19(6), 769–778. DOI: 10.3170/2008-8-18447.
Russell, F.L., Zippin, D.B. and Fowler, N.L. (2001) Effects of white-tailed deer (Odocoileus virginianus) on plants, plant populations and communities: a review. *The American Midland Naturalist* 146(1), 1–26. DOI: 10.1674/0003-0031(2001)146[0001:EOWTDO]2.0.CO;2.
Schiffman, P.M. (1997) Animal-mediated dispersal and disturbance: driving forces behind alien plant naturalization. In: Luken, J.O. and Thieret, J.W. (eds) *Assessment and Management of Plant Invasions*. Springer, New York, NY, pp. 87–94. DOI: 10.1007/978-1-4612-1926-2_8.
Seagle, S.W. (2003) Can ungulates foraging in a multiple-use landscape alter forest nitrogen budgets? *Oikos* 103(1), 230–234. DOI: 10.1034/j.1600-0706.2003.12287.x.
Shiponeni, N.N. and Milton, S.J. (2006) Seed dispersal in the dung of large herbivores: implications for restoration of Renosterveld shrubland old fields. *Biodiversity and Conservation* 15(10), 3161–3175. DOI: 10.1007/s10531-005-6317-5.
Shmida, A. and Ellner, S. (1983) Seed dispersal on pastoral grazers in open mediterranean chaparral, Israel. *Israel Journal of Plant Sciences* 32(3), 147–159. DOI: 10.1080/0021213X.1983.10676970.
Sigaud, M. (2018) Le bison des prairies entre aire protégée et terres agricoles: causes, conséquences et perspectives de gestion. Doctoral dissertation. Université Laval, Québec, Canada.
Slater, K. and Du Toit, J.T. (2002) Seed dispersal by chacma baboons and syntopic ungulates in southern African savannas. *South African Journal of Wildlife Research* 32(1), 75–79.
Törn, A., Siikamäki, P. and Tolvanen, A. (2010) Can horse riding induce the introduction and establishment of alien plant species through endozoochory and gap creation? *Plant Ecology* 208(2), 235–244. DOI: 10.1007/s11258-009-9701-5.
Vavra, M., Parks, C.G. and Wisdom, M.J. (2007) Biodiversity, exotic plant species, and herbivory: the good, the bad, and the ungulate. *Forest Ecology and Management* 246(1), 66–72. DOI: 10.1016/j.foreco.2007.03.051.

Vellend, M. (2002) A pest and an invader: white-tailed deer (*Odocoileus virginianus* Zimm.) as a seed dispersal agent for honeysuckle shrubs (*Lonicera* L.). *Natural Areas Journal* 22(3), 230–234.
Vickery, R.K., Phillips, D.R. and Wonsavage, P.R. (1986) Seed dispersal in Mimulus guttatus by wind and deer. *American Midland Naturalist* 116(1), 206–208. DOI: 10.2307/2425954.
Vild, O., Hédl, R., Kopecký, M., Szabó, P., Suchánková, S. *et al*. (2017) The paradox of long-term ungulate impact: increase of plant species richness in a temperate forest. *Applied Vegetation Science* 20(2), 282–292. DOI: 10.1111/avsc.12289.
Wang, B.C. and Smith, T.B. (2002) Closing the seed dispersal loop. *Trends in Ecology & Evolution* 17(8), 379–386. DOI: 10.1016/S0169-5347(02)02541-7.
Weaver, V. and Adams, R. (1996) Horses as vectors in the dispersal of weeds into native vegetation. *11th Australian Weeds Conference*, Weed Science Society of Victoria Inc., Melbourne, Australia, 30 September - 3 October 1996, pp. 383–387.
Wells, F.H. and Lauenroth, W.K. (2007) The potential for horses to disperse alien plants along recreational trails. *Rangeland Ecology & Management* 60(6), 574–577. DOI: 10.2111/06-102r1.1
Wenny, D.G. (2001) Advantages of seed dispersal: a re-evaluation of directed dispersal. *Evolutionary Ecology Research* 3(1), 51–74.
Whigham, D.F., O'Neill, J. and Cipollini, M. (1993) The role of tree gaps in maintaining the population structure of a woodland herb: *Cynoglossum virginianum* L. *Plant Species Biology* 8(2-3), 107–115. DOI: 10.1111/j.1442-1984.1993.tb00062.x.
Whinam, J., Cannell, E.J., Kirkpatrick, J.B. and Comfort, M. (1994) Studies on the potential impact of recreational horseriding on some alpine environments of the Central Plateau, Tasmania. *Journal of Environmental Management* 40(2), 103–117. DOI: 10.1006/jema.1994.1007.
Wiegmann, S.M. and Waller, D.M. (2006) Fifty years of change in northern upland forest understories: Identity and traits of "winner" and "loser" plant species. *Biological Conservation* 129(1), 109–123. DOI: 10.1016/j.biocon.2005.10.027.
Wilby, A., Shachak, M. and Boeken, B. (2001) Integration of ecosystem engineering and trophic effects of herbivores. *Oikos* 92(3), 436–444. DOI: 10.1034/j.1600-0706.2001.920305.x.
Williams, S.C., Ward, J.S. and Ramakrishnan, U. (2008) Endozoochory by white-tailed deer (*Odocoileus virginianus*) across a suburban/woodland interface. *Forest Ecology and Management* 255(3-4), 940–947. DOI: 10.1016/j.foreco.2007.10.003.
Wilson, D.E. and Reeder, D.M. (2005) *Mammal Species of the World - A Taxonomic and Geographic Reference*, 3rd edn. Johns Hopkins University Press, Baltimore, Maryland.

7 The Role of Plant–Plant Facilitation in Non-native Plant Invasions

Lohengrin A. Cavieres*

Departamento de Botánica, Facultad de Ciencias Naturales y Oceanográficas, Universidad de Concepción, Concepción, Chile; Instituto de Ecología y Biodiversidad (IEB), Santiago, Chile

Abstract

Biological invasions are one the most important drivers of the current environmental changes generating important biodiversity losses. Although several hypotheses have been proposed to understand the mechanisms underpinning biological invasions, most of them relate to negative interactions among native and invasive species, where the capacity for many invasive species to reduce diversity is often attributed to a greater competitiveness. However, neighbouring species can also show facilitative interactions, where the presence of one species can facilitate another directly by improving environmental conditions or indirectly through negative effects on a third party species. This chapter reviews the scientific literature on plant invasion, seeking examples of where facilitative interactions either among native and non-native plant species or among non-native species were demonstrated. There are several examples of native species that directly facilitate a non-native species, while examples of native species having a negative effect either on a native or a non-native species that compete with a target non-native, generating a net indirect facilitative effect of the native on the target non-native, are less numerous. Direct facilitation among non-native species has been reported as part of the 'invasional meltdown' phenomenon (Chapter 8, this volume). There are cases where non-native species can have a negative effect on a native species that competes with a target non-native, generating a net indirect facilitative effect among the non-natives. Finally, a non-native species can have a direct facilitative effect on native species, which might have important implications in restoration.

7.1 Introduction

Biological invasions have dramatically altered the natural world, threatening native species and communities, affecting human health and costing enormous amounts of money to control (Simberloff *et al.*, 2013; Blackburn *et al.*, 2019). Three major processes characterize plant invasions: rapid local population increase; the establishment of local dominance (which could include the formation of monocultures); and/or rapid range expansion (Gurevitch *et al.*, 2011). These three processes differ in both causes and impacts, but when populations of a non-native species demonstrate one or more of these defining characteristics (i.e. rapid local population increase, establishment of monocultures or range expansion) scientists, managers and the general public recognize that an invasion has occurred or is ongoing.

More than two dozen hypotheses for biological invasions have been proposed, based on specific individual drivers of invasions (reviewed by Catford *et al.*, 2009; Chapters 1 and 2, this volume). It is now clear that no one individual explanation is sufficient to account for all biological invasions and that the individual hypotheses partially overlap in mechanisms and

*lcaviere@udec.cl

 Plant Invasions: The Role of Biotic Interactions (eds Anna Traveset and David M. Richardson)
DOI: 10.1079/9781789242171.0007

predictions. Different mechanisms may contribute to the invasion of different species, or to the same species in different places or at different times, as well as the possibility that they may also act simultaneously.

Much research has been done with the purpose of unravelling the mechanisms responsible for successful invasion, with many of the suggested mechanisms relating to negative interactions among native and non-native species, such as competition (Vilà and Weiner, 2004; Gioria and Osborne, 2014) and allelopathy (Hierro and Callaway, 2003; Chapter 15, this volume). In particular, the capacity for many invasive species to reduce diversity and to form nearly monospecific stands in their non-native ranges has often been attributed to a superior capacity of invasive species to compete for limited resources (Vilà and Weiner, 2004; Gioria and Osborne, 2014).

In contrast to competition, facilitation is a biotic interaction in which at least one of the species involved benefits from the presence of the other(s) (Callaway, 2007). The presence of one species can facilitate another directly by improving environmental conditions. Commonly reported improvements of the environmental conditions underlying the facilitation of one species on another include: cooler temperatures and lower evaporative demand, increased soil moisture and greater availability of nutrients (for a comprehensive review of facilitation mechanisms see Callaway, 2007). In addition, facilitation may occur indirectly by decreasing herbivory and/or competition pressure (e.g. Callaway *et al.*, 2005). Facilitative interactions among plants play an important role in determining the richness and structure of natural communities (e.g. Brooker *et al.*, 2008; Cavieres *et al.*, 2014).

In recent years, however, the potentially important role of positive species interactions for invasion processes has received attention (Simberloff and Von Holle, 1999; Richardson *et al.*, 2000; Bruno *et al.*, 2005; Traveset and Richardson, 2014). The most commonly reported mechanisms whereby facilitative interactions determine invasion processes are non-native animals pollinating and dispersing invasive plants, and non-native animals and plants modifying the environment in ways that favour the spread of other non-native species (Simberloff and Von Holle, 1999; Richardson *et al.*, 2000; Traveset and Richardson, 2014). For instance, introduced animals disperse seeds of invasive non-native plant species, increasing the magnitude of the invasion process and generating an 'invasional meltdown' (Simberloff and Von Holle, 1999; Simberloff, 2006). However, despite recognition of the role of positive interactions in invasions, examples of direct positive interactions among plants (i.e. facilitation) during invasions, either between non-native species or between native and non-native species, are less numerous (but see examples below). Native plant species facilitating non-native species or non-natives facilitating other non-natives add substantially to our view of the mechanisms underlying the invasibility of plant communities and the tools that managers might use in the management of the invasive species. Hence, there is value in analysing the different ways in which native and non-native species can establish facilitative interactions among them and the underlying mechanisms.

For this chapter, I reviewed the scientific literature on plant invasion-seeking examples of where facilitative interactions either among native and non-native plant species or among non-native species were demonstrated. The examples were selected having in mind the different types of interactions that native and non-native plants can establish either directly or indirectly. As indicated in Fig. 7.1, native species can directly facilitate a non-native species, and several examples demonstrate this type of interaction. In other cases, a native species can have a negative effect either on a native or a non-native species that competes with a target non-native, generating a net indirect facilitative effect of the native on the target non-native. On the other hand, a non-native species can directly facilitate other non-native species. Alternatively, a non-native species can have a negative effect on a native species that competes with a target non-native, generating a net indirect facilitative effect among the non-natives. Finally, a non-native species can have a direct facilitative effect on native species, which, as noted above, might have important implications in restoration.

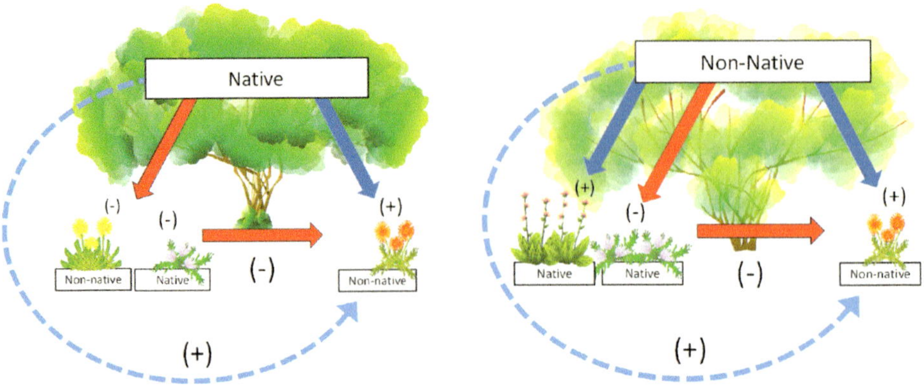

Fig. 7.1. Diagram illustrating the different types of facilitative interactions between native and non-native plant species. On the left, a native species can directly facilitate a non-native species. In other cases, a native species can have a negative effect either on a native or a non-native species that compete with the target non-native, generating a net indirect facilitative effect of the native on the non-native. On the right, a non-native species can directly facilitate other non-native species. In other cases, a non-native species can have a negative effect on a native species that competes with the target non-native, generating a net indirect facilitative effect of the among the non-natives. Finally, a non-native species can have a direct facilitative effect on native species.

7.2 Native Species Facilitating Non-native Species

In certain habitats, native species generate habitat patches with different environmental conditions than the surrounding unmodified habitat, and the presence of such habitat patches can affect the distribution and performance of non-native species (see examples below) as well as native species (e.g. Callaway, 2007). Thus, if non-native plants are sensitive to the stress amelioration provided by native species, facilitative interactions between these species (i.e. native nurses and non-native beneficiaries) could increase the susceptibility of plant communities to biological invasions.

In a pioneer study, Maron and Connors (1996) analysed the interaction between the native nitrogen-fixing shrub, *Lupinus arboreus* and the non-native grass *Bromus diandrus* in the invasion of Californian coastal prairies. The native shrub grows rapidly, fertilizing sandy soils with nitrogen-rich litter. The dense lupine canopy blocks light, restricting vegetative growth under bushes, but insect herbivory kills lupines, opening exposed nitrogen-rich sites within the plant community. However, in experimental patches where *L. arboreus* was removed, pool sizes of exchangeable ammonium and nitrate nitrogen were higher than that in adjacent sites free of *L. arboreus*. Seedlings of the non-native grass accumulated greater root and shoot biomass when grown in a greenhouse in soil collected under removed lupines compared to seedlings grown in soil collected at least 1 m away from the native shrub. In the field, above-ground plant biomass was more than twice as great in removed lupine patches than in adjacent lupine-free grasslands, although removed lupine patches contained fewer plant species. Thus, this native lupine species facilitated the growth of a non-native invasive species. However, it is worth noting that in this example there was no clear direct interaction between the native and the non-native species, but instead an exploitation by the non-native species of the environmental modifications generated by the native after its death.

In shrublands dominated by Chenopodiaceae species in southern Australia, Lenz and Facelli (2003) studied the interaction between the non-native succulent *Orbea variegata* (Asclepiadaceae) and native chenopod shrubs. To assess the different mechanisms by which native shrubs facilitated the non-native species, these authors conducted a series of field experiments to determine the role of shrub canopies and microclimatic modifications to soils on the survival of transplanted *O. variegata* individuals. Their results indicated that the reduction of light and temperature under shrubs, but not increased nutrient level nor water availability, were the main facilitative mechanisms. Although the non-native *O. variegata* can germinate and establish away from shrubs, native shrub canopies improved the growth and survival of *O. variegata*, promoting the invasive process of this non-native species. Rodríguez-Buriticá and Miriti (2009) showed that in the Joshua tree NP in California, the non-native grass species *Schismus barbatus* grows better when associated with the native shrub *Ambrosia dumosa*. However, seedlings of the native shrub were outcompeted at high densities of the invasive grass, indicating a negative feedback in the relationship between the native nurse and the facilitated non-native. Griffith (2010) showed that the understorey of the native shrubs *Artemisia tridentata* and *Purshia tridentata* has higher soil fertility and less extreme microclimate that facilitate the reproduction and population growth of the invasive grass *Bromus tectorum*.

In the Atacama Desert, Northern Chile, Madrigal-González *et al.* (2013) showed that the non-native annual *Mesembryanthemum crystallinum* had a strong positive spatial association (facilitation) with the native cactus *Eulychnia acida* (Fig. 7.2). In addition, the above-ground biomass of *M. crystallinum* individuals was fourfold higher under the influence of *E. acida*. Soil collected under the native cactus contained more available nutrients (N, P and K), water, and soluble salts than soils collected away from it. Thus, the environmental modifications generated by the native cactus seem to facilitate the establishment and growth of the non-native herb. Lucero *et al.* (2019), in a four-year observational study, examined the effects of native nurse shrubs on the relative abundance of the annual plant community that included native and non-native species in a desert ecosystem at Carrizo Plain National Monument in California. Although the effect of shrubs varied by year and by the identity of annual species, shrubs consistently enhanced the abundance of the annual plant community and facilitated both native and non-native species. However, non-native annuals were facilitated 2.75 times more intensely than native annuals. There were no experimental evaluations of the mechanisms by which the native shrubs better facilitated the non-native annuals, which is an important knowledge gap in general. If native and non-native species capitalize on different mechanisms of facilitation, the potential management of the invasive non-natives could be easier than in situations where both groups of species benefited from the same mechanisms.

In alpine plant communities of the central Chilean Andes the perennial non-native herb *Taraxacum officinale* is a conspicuous species invading native communities (Fig. 7.2). Cavieres *et al.* (2005) showed that *T. officinale* is positively associated with the cushion plant *Azorella madreporica* at 3200 m in elevation. In general, due to its low stature and compact form, alpine cushion plants create microhabitat conditions that differ from their surroundings, providing more stable substrates with higher nutrient and water content, milder temperatures, etc. (Cavieres *et al.*, 2006, Cavieres *et al.*, 2014). Survival of seedlings, as well as the net-photosynthetic rates of *T. officinale* individuals, were higher within cushions than in open areas away from them, suggesting that the microclimatic modifications generated by this native cushion facilitates the establishment and performance of this non-native invasive species. In another study, Cavieres *et al.* (2008) showed that the facilitative interaction between *T. officinale* and native nurse cushion species was species-specific. In sites where the native nurses *Azorella madreporica* and *A. ruizii* (syn. *Laretia acaulis*) coexist, the non-native herb was more associated with *A. madreporica*, where it also had a higher photochemical efficiency and reproductive output. Soil temperature and moisture, as well as soil nitrogen and phosphorous content, were similar within the two cushion species; however, potassium was two times higher in the soil beneath *A. ruizii* than *A. madreporica*. Thus, although the native cushion species facilitated the establishment and performance of a non-native invasive species, this interaction is highly species-specific due to differences in soil nutrient content beneath the different benefactor species.

Acer negundo is non-native species in the Middle Rhone floodplains forest of France. Saccone *et al.* (2010) transplanted juveniles of

Fig. 7.2. (A) *Ephedra* sp. with the non-native *Bromus rubens* underneath it, and surrounded outside by natives (photograph by Jacob Lucero); (B) *Baccharis magellanica* facilitating the non-native tree *Pinus contorta* in Patagonia; (C) *Copiapoa* sp. surrounded by the invasive *Mesembryanthemum crystallinum* (photograph by Rafael Garcia); (D) Cushion plant *Azorella madreporica* facilitating the non-native *Taraxacum officinale* and *Cerastium arvense* in the Andes.

this non-native species within three communities (*Salix*, *Acer* and *Fraxinus* stands) as well as in experimental gaps, with and without the herb layer of the forests. These authors reported that *A. negundo* seedlings were directly facilitated for survival in the *Salix* and *Acer* communities where the reduction of flooding disturbance by the web of interlocking stems was the driving mechanism of facilitation. Although it is more frequent to find examples of direct facilitative interactions of native species on non-natives in harsh unproductive environments, there are examples from more productive environments, indicating that this phenomenon is not privative of harsh environments.

7.3 Indirect Facilitation by Natives

Positive direct effects of native species on non-native species are well documented, but benefits that native species may provide to non-natives through indirect channels are less recorded or appreciated (Chapter 17, this volume). From work outside the literature on invasion ecology it is clear that indirect interactions, where the effect of one species on another is mediated through a third intermediate species, are both exceedingly common and ecologically influential (Levine, 1999; Michalet *et al.*, 2015).

As in the case of direct facilitative interaction between native and non-native species, indirect beneficial effects to non-natives should promote successful invasion, an outcome that is less prevalent in the invasion literature than the reverse (i.e. effects of invaders on native communities). Some positive indirect effects of native species on invasive species have been fairly well documented, including chains of mutualists: dispersal or pollination mutualisms or predators that consume the native competitor of a non-native species (White *et al.*, 2006). Recently, Northfield *et al.* (2018) analysed the scientific literature on biological invasions, and identified a total of eight possible pathways that may lead native species to indirectly benefit non-native species. In this chapter, I focus on cases where one native species competes strongly with another native species that would otherwise competitively resist the non-native species (Fig. 7.1). As far as I am aware, there are no examples where a native species indirectly facilitates a non-native by suppressing a native competitor, a phenomenon that certainly deserves more attention.

Metlen *et al.* (2013) and Metlen and Callaway (2015) nicely show that the non-native species *Centaurea stoebe* and *B. tectorum* have a different microhabitat distribution due to the presence of the native species *Pinus ponderosa*. While *C. stoebe* was much less abundant under *P. ponderosa* canopy than in surrounding open grasslands, *B. tectorum* was more abundant under the native species. Although the more fertile soils found beneath the native species facilitated both non-native species and did not alter competitive outcomes among them, *Pinus* litter shifted competitive outcomes in favour of *B. tectorum* by more strongly inhibiting *C. stoebe* chemically. Thus, a native species (*P. ponderosa*) modified interactions between two other non-native species by negatively affecting one species more than the other, hence indirectly facilitating invasion of the non-native. Recently, Fill *et al.* (2019) showed that in greenhouse experiments the indirect interactions between a native shrub and an invasive legume species alter the community composition of native forb species. As mentioned, more research is needed to unveil the importance and the mechanisms by which native species may be indirectly facilitating non-native species.

7.4 Facilitation by Natives and the Invasibility of Communities

The characteristics of both resident communities and the invading species are important for determining the outcomes of biological invasions (Catford *et al.*, 2009). From the resident community perspective, species diversity can resist invasions, often attributed to more complete niche occupation and greater competition for resources (Fridley *et al.*, 2007; Theoharides and Dukes, 2007). In other words, highly diverse communities are thought to efficiently use all the available resource niches, leaving less for invaders. Thus, they should be less prone to invasions (Tilman, 2004; Chapter 16, this volume). In contrast, facilitation of invaders by natives can cause invasibility of plant communities to deviate from the near universal prediction of a decline of invasibility with increasing native richness; whereas facilitation of natives by natives may increase native community resistance. Bulleri *et al.* (2008) reviewed the consequences of facilitation of non-native species by native nurses in the invasibility of communities based on the availability of resources and how efficient species are in the use of those resources. They reported that facilitation always increases invasibility when resources are scarce and the diversity of native species relatively high. This can be explained because at higher native species richness, there are higher chances of having an effective facilitator species when resources are limiting.

In the context of facilitation, there is a scale-based paradox in the relationship between species richness and biological invasions (Sax and Brown, 2000; Stohlgren *et al.*, 2003). Generally, the effect of native diversity on invasions is negative at local scales but breaks down or even reverses at larger scales (Sax and Brown, 2000). Fridley *et al.* (2007) identified eight processes as important factors underlying this paradox, and native species that directly facilitate non-native species was listed as one of them (also see Von Holle, 2005). Interestingly, Von Holle found that while non-native species were facilitated by native trees in floodplain forests of south-western Virginia, USA, in non-riparian sites there was a negative relationship between native richness and invasion success. Recently, Northfield *et al.* (2018), using a modelling approach, and by applying concepts and approaches taken from the coexistence literature to invasion ecology, hypothesized that positive indirect effects of competitors on potential invaders were more common in communities characterized by higher species richness and resource availability. In addition, the effects of indirect interactions on species invasions at larger scales (i.e. greater numbers of resources and species) were stronger between species that are more specialized in their resource use patterns. There are few studies that address the role of facilitative interactions in explaining the invasion paradox, and it is thus a poorly occupied niche (see Altieri *et al.*, 2010 for an example in intertidal communities).

7.5 Non-natives Facilitating Other Non-natives

Non-native plants also have the potential to modify both the abiotic and biotic environments in a way that facilitates the establishment or growth of their own offspring or other non-natives (e.g. Reinhart *et al.*, 2006; Proença *et al.*, 2019).

Reinhart *et al.* (2006) showed that in montane forests in Montana (USA), seedlings and saplings of the non-native tree *Acer platanoides* had higher densities, recruitment rates, survival rates and experienced less photoinhibition and water stress beneath conspecifics than beneath adjacent *Pseudotsuga menziesii* native trees. In this case, there was more soil moisture in *A. platanoides* patches than in *P. menziesii*-dominated forests. In mixed-grass prairies of the Dakotas and eastern Montana, Jordan *et al.* (2008) showed that the invasive grasses *Bromus inermis* and *Agropyron cristatus* modified soil conditions in a way that promoted their own growth. Both *A. cristatus* and *B. inermis* changed soil microbes over several cycles of growth, and the resultant microbiota increased seedling growth of both non-natives compared to the responses to soil microbiota associated with native species. Interestingly, soil conditioning by *A. cristatus* facilitated *B. inermis* seedling growth, while *B. inermis* had a facilitative effect on other non-native species (*Euphorbia esula*). Thus, there was evidence of inter-specific facilitation among the non-natives. In addition, Jordan *et al.* (2008) showed that each invasive species had strongly antagonistic effects on native forbs. For example, soil conditioning by *A. cristatus* reduced the growth of the native herbs *Aster ericoides* and *Ratibida columnifera* while *B. inermis* had antagonistic effects on *Linum lewisii* and *R. columnifera*. This study is a clear example of the importance of the soil biota in the mechanisms by which facilitative and antagonistic effects operated in explaining invasion success, and supports the hypothesis that plant invasion can be promoted by positive feedback mediated by living components of soil (see also Kending *et al.*, Callaway and Lucero, this volume).

Vitousek and Walker (1989) provided one of the first examples of facilitation among non-native species. These authors showed that in poor volcanic soils of Hawaii, the non-native nitrogen-fixing shrub *Morella faya* increased nitrogen content in the soil where the seedlings of the non-native shrub *Psidium cattleianum* accumulated more biomass than when grown in soil from under the native *Metrosideros polymorpha*. The grass *Poa pratensis* and the N-fixing herb *Melilotus officinalis* are non-natives to the Great Plains of the USA. Dornbusch *et al.* (2018) investigated how N enrichment from *M. officinalis* affected the above-ground biomass production of *P. pratensis* and of the native grass *Pascopyrum smithii*. In a controlled greenhouse environment, soils from *M. officinalis* contained ≈ 340% more plant-available N than untreated soils, and when seedlings of the native and non-native grass species were grown in intraspecific pairs, the increase in the non-native biomass relative to untreated pots (520%) was double the increase in the native biomass (260%). This indicated that the non-native grass could use increased soil N to produce proportionally more above-ground biomass than the native grass competitor. Tecco *et al.* (2006, 2007) studied the influence of the invasive shrub *Pyracantha angustifolia* on the recruitment of other non-native woody species in a secondary shrubland in Córdoba mountains, Argentina. They found that both native and non-native species richness were higher under *P. angustifolia* than under other microhabitats. In particular, the non-native shrub *Ligustrum lucidum* was four times as dense under the canopy of *P. angustifolia*, where the seedling survival was higher than in open areas or under native nurse shrubs. This facilitative interaction seemed to be related to *P. angustifolia*'s denser shading, and the mechanical protection of its canopy against herbivory.

Kuebbing and Nuñez (2015) conducted a bibliographic revision of the interactions among non-native species, where they quantified the types of interactions and the factors that affected interaction outcomes. They reviewed 65 studies comprising 201 observations. A a meta-analysis showed that negative and neutral interactions among non-natives prevailed. Although facilitative interactions were less frequently reported and positive mean effect sizes were rare, plant characteristics such as nitrogen fixation, life cycle (annual or perennial) and functional group did drive positive interactions. Positive interactions were three times more frequent when a neighbouring non-native was a nitrogen fixer and 3.5 times lower when a neighbouring non-native was an annual. Woody plants were two or four times more likely to have positive interactions relative to grasses or herbs, respectively. Thus, although less reported in the literature so far, in some cases and sites facilitation may play an important role in the different stages of the invasion process (see below).

7.6 Indirect Facilitative Effects Among Non-natives

As in the case of indirect facilitative interactions between native and non-native species, indirect effects among non-natives also occur and are important for invasion success, but are less frequently reported. Saccone *et al.* (2010), in the Middle Rhone floodplain forest of France, found that in stands of the non-native *A. negundo*, tree canopies strongly decreased herbaceous competitive species, hence, by reducing the belowground competitive effect of the herb layer, adult *A. negundo* indirectly facilitated the growth of its own seedlings. Similarly, Siemann and Rogers (2003) showed that seedlings of the invasive *Sapium sebiferum* were indirectly facilitated by the canopy of adult *S. sebiferum* by decreasing root competition from prairie grasses in the shade.

Flory and Bauer (2014) experimentally evaluated the indirect facilitation between the non-native invasive grass *Microstegium vimineum* on the success of the non-native *Alliaria petiolata* by introducing seeds of this last non-native species into replicated plots previously invaded experimentally by *M. vimineum*. Although *A. petiolata* more readily colonized plots without *M. vimineum*, it produced more biomass (327% greater) and more siliques per plant (234% more) in *M. vimineum*-invaded plots. The facilitation of *A. petiolata* by *M. vimineum* was associated with a reduction in the biomass of resident species and significantly greater light availability, indicating that both the biotic and abiotic modification generated by the invasion of *M. vimineum* facilitated the secondary plant invasion of *A. petiolata*.

7.7 Facilitation Among Non-natives and the Invasibility of Communities

From the examples mentioned above, facilitation among non-native species may have important consequences for the invasibility of plant communities.

O'Loughlin and Green (2017) defined 'secondary invasion' as when the invasion success of one non-native species (the secondary invader) is completely contingent on the presence, influence and impact of one or more other non-native species (primary invaders). Primary invaders are non-native species that can successfully invade without the alteration of ecosystem properties by other non-native species, while a secondary invader is defined as any non-native species that is unable to successfully invade due to some inhibiting property of the recipient ecosystem. A secondary invader only becomes so after primary invaders, through their presence and influence, alter that inhibiting property and invasion can proceed. In accordance with the evidence commented on above, facilitative interactions among non-natives play a crucial role in the invasibility of communities by generating opportunities for secondary invasion to occur. This increased invasibility may occur either by direct or indirect facilitative interactions among the non-native species, increasing the chances that non-native species can outcompete and locally lead to the extinction of native species.

Simberloff and Von Holle (1999) introduced the term 'invasional meltdown' to describe the process by which non-native species facilitate one another's invasion in various ways, increasing the likelihood of survival and/or magnitude of impact and potentially leading to an accelerating increase in the number of non-native species and cumulative impact. These authors highlighted the importance that positive interactions can have on the invasion process, despite the primacy of evaluations of the importance of negative interactions in the invasion literature. Originally, the most commonly reported mechanisms for invasional meltdown were non-native animals pollinating and dispersing invasive plants, and non-native animals and plants modifying the environment in ways that favour the recruitment of other non-native species (Simberloff and Von Holle, 1999; Richardson *et al.*, 2000; Simberloff, 2006). Braga *et al.* (2018) and Braga *et al.* (2019) conducted a revision of the literature that cited the study Simberloff and Von Holle (1999), and found that out of the 208 relevant publications for the analyses (half of them on plants and algae), 65 (31%) indicated that at least one non-native species had a positive effect on another non-native species with no reciprocal positive effect on the former, and another 47 studies (23%) found that interactions among non-natives was mutually positive – strong evidence for invasional meltdown.

7.8 Non-native Facilitating Native Species

An interesting but largely unappreciated type of facilitative interaction is that where non-native plants have the potential to modify both the abiotic and biotic environments in ways that facilitate the establishment and/or growth of native species (Fig. 7.1).

For example, Svriz *et al.* (2013) studied the effects of the non-native species *Rosa rubiginosa* in the recruitment of native and non-native species in disturbed shrublands of Argentinean Patagonia. Although *R. rubiginosa* harboured other non-native species beneath its canopy, all native species of the community were found only beneath the canopy of this non-native species. This seemed to be related to higher soil moisture and nitrogen content, as well as the herbivory protection provided beneath *R. rubiginosa*. In the Argentinean pampas, the vegetative growth and aboveground biomass of the native perennial grass *Paspalum dilatatum* increased in the presence of the non-native herb *Lotus tenuis* (Quinos *et al.*, 1998), while in Nairobi National Park the invasive *Opuntia ficus-indica* facilitates native species by protecting them from the native herbivores (Oduor *et al.*, 2018). These examples indicate that the facilitative effects of non-native species on native ones can alter the natural structure of plant communities (see also Ignace and Chesson, 2014).

Rodriguez (2006) conducted a literature review and found that facilitative interactions between non-native and native species occur in a wide range of habitats, having cascading effects across trophic levels, and restructuring communities in ways that can lead to evolutionary changes. Habitat modification by the non-natives was the most frequently documented mechanism, reflecting the drastic changes that non-native species generated in the invasion process. Nonetheless, focusing attention on facilitative effects of non-native species on native ones not only develops a broader understanding of ecological dynamics, but also draws our attention to some cases where these interactions may become valuable tools for management and conservation. For instance, in tropical forests of Indonesia, natural regeneration of native tree species in the understorey of fast-growing plantation tree species was higher than in adjacent natural riverine forest and non-cultivated grasslands, providing potential for a rapid restoration of that forest ecosystem (Otsamo, 2000). By means of seedling transplant experiments, Becerra and Montenegro (2013) showed that in the Mediterranean-type climate zone of central Chile, the non-native species *Pinus radiata* has the potential to facilitate native tree and shrub species beneath its canopy. Although this facilitative effect was weaker than that exerted by native nurses, the presence of *P. radiata* individuals on degraded areas brings opportunities for more effective restoration programmes (see also D'Antonio and Meyerson, 2002, and Yelenik *et al.*, this volume).

7.9 Changes in Facilitative Effects

Harsh environments are those where the occurrence of abiotic conditions that create rapid plant mortality (e.g. frost, extreme heat, and drought) are common. Evaluations of factors influencing ecosystem susceptibility to plant invasion support the idea that abiotic conditions affect invasions of non-native species – more environmentally harsh sites have lower numbers of invaders (Stohlgren *et al.*, 2003; Pauchard *et al.*, 2009).

Along environmental severity gradients, facilitative interactions are likely to be more frequent as the environmental conditions become harsher, while negative competitive interactions should be more frequent on milder zones along the severity gradients (Bertness and Callaway, 1994; Brooker and Callaghan, 1998). The former is known as the Stress Gradient Hypothesis (SGH) and a number of studies have demonstrated it along environmental severity gradients in natural communities (e.g. Callaway *et al.*, 2002; He *et al.*, 2013; Cavieres *et al.*, 2014). The reasoning was that harsh environments may restrict plants from acquiring resources, and any amelioration of these conditions will favour growth to the extent that it outweighs the negative, competitive impact of growing in close associations. Thus, it seems likely that facilitative interactions among native and non-native species, or among non-native plant species can also shift along environmental severity gradients, having important implications on the invasibility of communities as indicated earlier.

In riparian habitats of flood plain forests of south-western Virginia, USA, Von Holle (2005) showed that, contrary to expectations, non-native species were facilitated by native trees and did not interact with the understorey resident native species. Interestingly, however, in upland zones of these forests, native species competed with non-native species, indicating that interactions among them change with the environmental context.

The nitrogen fixing invasive species *M. officinalis* acts as a nurse species for both native and non-native species in South Dakota, USA (Van Riper and Larson, 2009). In Badlands sparse vegetation, these authors reported that there was a consistent, strong and positive relationship between *M. officinalis* cover and native and non-native species cover. In contrast, in wheatgrass prairies, *M. officinalis* showed no facilitative effects, but rather had a weak competitive effect on other species. These results suggest both facilitative and competitive roles for *M. officinalis* depending on the vegetation type it invades.

Kleinhesselink *et al.* (2014) assessed the ability of the native shrub (*Ericameria ericoides*) to act as a nurse species across a stress gradient in a coastal dune in northern California, USA. This native shrub increased organic matter and decreased wind speeds within its canopy and the growth of the dominant invasive grass *B. diandrus* was facilitated both by shrub's canopy and by soil from under shrubs. Experimental removal of native shrubs had 40% less non-native biomass than controls. However, counter to the SGH, the effects of shrubs on the physical environment and individual plant growth did not increase across the established stress gradient at this site. Reisner *et al.* (2015) showed that along an environmental stress gradient in the Northern Great Basin floristic province of Oregon, the native shrub *A. tridentata* negatively interacted with native species in the low stress end of the gradient, but showed facilitative interactions in the high stress end. However, the interaction with the non-natives *B. tectorum* and *Lepidium perforatum* was negative along the entire gradient.

The cushion plant *Azorella monantha* positively interacts with two non-native plant species (*Cerastium arvense* (Caryophyllaceae) and *T. officinale*) along an elevation gradient in the Andes of central Chile (Badano *et al.*, 2007). Nevertheless, the cushion plant had greater facilitative effects on both non-natives at higher elevations. Indeed, at the higher elevation site, *C. arvense* was only detected within cushions, suggesting that cushions are expanding the altitudinal distribution range of this non-native species. These results suggest that ecosystem engineering by native species could promote biological invasions in harsh environments, leading to higher abundances of invaders than those expected in the absence of engineers. The amelioration of extreme abiotic conditions by cushions can indeed make stressful habitats invasible by non-native plant species, with larger positive effects on their performance as environmental harshness increases.

7.10 Concluding Remarks

Native and non-native species have direct and indirect facilitative interactions, with important consequences for the invasibility of communities. However, our knowledge of the mechanisms underlying these interactions is still meagre, and our ability to predict when and how facilitative interactions promote non-native species is poor. It is important to gather more information about how often facilitation occurs and the consequences for recipient communities. Thus, it seems imperative that invasion biologists incorporate facilitative interactions between native and non-native as well as among non-native species into future ecological research if we wish to determine their relative strength in comparison to competitive effects, and the response of native populations and communities. Predicting positive non-native interactions is an important tool for determining habitat susceptibility to a particular invasion and for prioritizing management of non-natives with a higher likelihood of facilitative interactions.

The literature indicates that facilitation of non-native species is likely to be more important in stressful or environmentally severe habitats, weakening the natural barrier to invasion related with the environmental severity. With climate change, some currently environmental harsh habitats may be no longer harsh, while other may be even harsher. Thus, knowledge of all interactions where non-native species are involved, including facilitation, will be important as we strive to ameliorate the consequences of climate change for future invasions and the management or eradication of non-native invaders already present.

References

Altieri, A.H., van Wesenbeeck, B.K., Bertness, M.D. and Silliman, B.R. (2010) Facilitation cascade drives positive relationship between native biodiversity and invasion success. *Ecology* 91(5), 1269–1275. DOI: 10.1890/09-1301.1.

Badano, E. I., Villarroel, E., Bustamante, R.O., Marquet, P.A. and Cavieres, L.A. (2007) Ecosystem engineering facilitates invasions by exotic plants in high-Andean ecosystems. *Journal of Ecology* 95(4), 682–688. DOI: 10.1111/j.1365-2745.2007.01262.x.

Becerra, P.I. and Montenegro, G. (2013) The widely invasive tree *Pinus radiata* facilitates regeneration of native woody species in a semi-arid ecosystem. *Applied Vegetation Science* 16(2), 173–183. DOI: 10.1111/j.1654-109X.2012.01221.x.

Bertness, M.D. and Callaway, R. (1994) Positive interactions in communities. *Trends in Ecology & Evolution* 9(5), 191–193. DOI: 10.1016/0169-5347(94)90088-4.
Blackburn, T.M., Bellard, C. and Ricciardi, A. (2019) Alien versus native species as drivers of recent extinctions. *Frontiers in Ecology and the Environment* 17(4), 203–207. DOI: 10.1002/fee.2020.
Braga, R.R., Gómez-Aparicio, L., Heger, T., Vitule, J.R.S. and Jeschke, J.M. (2018) Structuring evidence for invasional meltdown: broad support but with biases and gaps. *Biological Invasions* 20(4), 923–936. DOI: 10.1007/s10530-017-1582-2.
Braga, R.R., Gómez-Aparicio, L., Heger, T., Vitule, J.R.S. and Jeschke, J.M. (2019) Invasional meltdown hypothesis. In: Jeschke, J.M. and T, Heger. (eds) *Invasion Biology: Hypotheses and Evidence. CABI*. Wallingford, pp. 79–91.
Brooker, R.W. and Callaghan, T.V. (1998) The balance between positive and negative plant interactions and its relationship to environmental gradients: a model. *Oikos* 81(1), 196–207. DOI: 10.2307/3546481.
Brooker, R.W., Maestre, F.T., Callaway, R.M., Lortie, C.L., Cavieres, L.A. *et al*. (2008) Facilitation in plant communities: the past, the present, and the future. *The Journal of Ecology* 96, 18–34.
Bruno, J.F., Fridley, J.D., Bromberg, K. and Bertness, M.D. (2005) Insights into biotic interactions from studies of species invasions. In: Sax, D.F., Stachowicz, J.J. and Gaines, S.D. (eds) *Species Invasions: Insights into Ecology, Evolution, and Biogeography*. MA, Sinauer Sunderland, pp. 13–40.
Bulleri, F., Bruno, J.F. and Benedetti-Cecchi, L. (2008) Beyond competition: incorporating positive interactions between species to predict ecosystem invasibility. *PLoS Biology* 6(6), e162. DOI: 10.1371/journal.pbio.0060162.
Callaway, R.M. (2007) *Positive Interactions and Interdependence in Plant Communities*. Springer, Berlin.
Callaway, R.M., Brooker, R.W., Choler, P., Kikvidze, Z., Lortie, C.J. *et al*. (2002) Positive interactions among alpine plants increase with stress. *Nature* 417(6891), 844–848. DOI: 10.1038/nature00812.
Callaway, R.M., Kikodze, D., Chiboshvili, M. and Khetsuriani, L. (2005) Unpalatable plants protect neighbors from grazing and increase plant community diversity. *Ecology* 86(7), 1856–1862. DOI: 10.1890/04-0784.
Catford, J.A., Jansson, R. and Nilsson, C. (2009) Reducing redundancy in invasion ecology by integrating hypotheses into a single theoretical framework. *Diversity and Distributions* 15(1), 22–40. DOI: 10.1111/j.1472-4642.2008.00521.x.
Cavieres, L.A., Quiroz, C.L., Molina-Montenegro, M.A., Muñoz, A.A. and Pauchard, A. (2005) Nurse effect of the native cushion plant *Azorella monantha* on the invasive non-native *Taraxacum officinale* in the high-Andes of central Chile. *Perspectives in Plant Ecology, Evolution and Systematics* 7(3), 217–226. DOI: 10.1016/j.ppees.2005.09.002.
Cavieres, L.A., Badano, E.I., Sierra-Almeida, A., Gómez-González, S. and Molina-Montenegro, M.A. (2006) Positive interactions between alpine plant species and the nurse cushion plant *Laretia acaulis* do not increase with elevation in the Andes of central Chile. *New Phytologist* 169(1), 59–69. DOI: 10.1111/j.1469-8137.2005.01573.x.
Cavieres, L.A., Brooker, R.W., Butterfield, B.J., Cook, B.J., Kikvidze, Z. *et al*. (2014) Facilitative plant interactions and climate simultaneously drive alpine plant diversity. *Ecology Letters* 17(2), 193–202. DOI: 10.1111/ele.12217.
Cavieres, L.A., Quiroz, C.L. and Molina-Montenegro, M.A. (2008) Facilitation of the non-native *Taraxacum officinale* by native nurse cushion species in the high Andes of central Chile: are there differences between nurses? *Functional Ecology* 22, 148–156.
Dornbusch, M.J., Limb, R.F. and Gasch, C.K. (2018) Facilitation of an exotic grass through nitrogen enrichment by an exotic legume. *Rangeland Ecology & Management* 71(6), 691–694. DOI: 10.1016/j.rama.2018.06.009.
D'Antonio, C. and Meyerson, L.A. (2002) Exotic plant species as problems and solutions in ecological restoration: a synthesis. *Restoration Ecology* 10(4), 703–713. DOI: 10.1046/j.1526-100X.2002.01051.x.
Fill, J.M., Pearson, E., Knight, T.M. and Crandall, R.M. (2019) An invasive legume increases perennial grass biomass: an indirect pathway for plant community change. *PLoS One* 14(1), e0211295. DOI: 10.1371/journal.pone.0211295.
Flory, S.L. and Bauer, J.T. (2014) Experimental evidence for indirect facilitation among invasive plants. *Journal of Ecology* 102(1), 12–18. DOI: 10.1111/1365-2745.12186.
Fridley, J.D., Stachowicz, J.J., Naeem, S., Sax, D.F., Seabloom, E.W. *et al*. (2007) The invasion paradox: reconciling pattern and process in species invasions. *Ecology* 88(1), 3–17. DOI: 10.1890/0012-9658(2007)88[3:TIPRPA]2.0.CO;2.

Gioria, M. and Osborne, B.A. (2014) Resource competition in plant invasions: emerging patterns and research needs. *Frontiers in Plant Science* 5, 501. DOI: 10.3389/fpls.2014.00501.

Griffith, A.B. (2010) Positive effects of native shrubs on *Bromus tectorum* demography. *Ecology* 91(1), 141–154. DOI: 10.1890/08-1446.1.

Gurevitch, J., Fox, G.A., Wardle, G.M., Inderjit Taub, D. (2011) Emergent insights from the synthesis of conceptual frameworks for biological invasions. *Ecology Letters* 14(4), 407–418. DOI: 10.1111/j.1461-0248.2011.01594.x.

He, Q., Bertness, M.D. and Altieri, A.H. (2013) Global shifts towards positive species interactions with increasing environmental stress. *Ecology Letters* 16(5), 695–706. DOI: 10.1111/ele.12080.

Hierro, J.L. and Callaway, R.M. (2003) Allelopathy and exotic plant invasion. *Plant and Soil* 256(1), 29–39. DOI: 10.1023/A:1026208327014.

Ignace, D.D. and Chesson, P. (2014) Removing an invader: evidence for forces reassembling a Chihuahuan desert ecosystem. *Ecology* 95(11), 3203–3212. DOI: 10.1890/14-0456.1.

Jordan, N.R., Larson, D.L. and Huerd, S.C. (2008) Soil modification by invasive plants: effects on native and invasive species of mixed-grass prairies. *Biological Invasions* 10(2), 177–190. DOI: 10.1007/s10530-007-9121-1.

Kleinhesselink, A.R., Magnoli, S.M. and Cushman, J.H. (2014) Shrubs as ecosystem engineers across an environmental gradient: effects on species richness and exotic plant invasion. *Oecologia* 175(4), 1277–1290. DOI: 10.1007/s00442-014-2972-0.

Kuebbing, S.E. and Nuñez, M.A. (2015) Negative, neutral, and positive interactions among nonnative plants: patterns, processes, and management implications. *Global Change Biology* 21(2), 926–934. DOI: 10.1111/gcb.12711.

Lenz, T.I. and Facelli, J.M. (2003) Shade facilitates an invasive stem succulent in a chenopod shrubland in South Australia. *Austral Ecology* 28(5), 480–490. DOI: 10.1046/j.1442-9993.2003.01304.x.

Levine, J.M. (1999) Indirect facilitation: evidence and predictions from a riparian community. *Ecology* 80(5), 1762–1769. DOI: 10.1890/0012-9658(1999)080[1762:IFEAPF]2.0.CO;2.

Lucero, J.E., Noble, T., Haas, S., Westphal, M., Butterfield, H. S. *et al.* (2019) The dark side of facilitation: native shrubs facilitate exotic annuals more strongly than native annuals. *NeoBiota* 44, 75–93. DOI: 10.3897/neobiota.44.33771.

Madrigal-González, J., Cea, A.P., Sánchez-Fernández, L.A., Martínez-Tillería, K.P., Calderón, J.E. *et al.* (2013) Facilitation of the non-native annual plant *Mesembryanthemum crystallinum* (Aizoaceae) by the endemic cactus *Eulychnia acida* (Cactaceae) in the Atacama Desert. *Biological Invasions* 15(7), 1439–1447. DOI: 10.1007/s10530-012-0382-y.

Maron, J.L. and Connors, P.G. (1996) A native nitrogen-fixing shrub facilitates weed invasion. *Oecologia* 105(3), 302–312. DOI: 10.1007/BF00328732.

Metlen, K.L., Aschehoug, E.T. and Callaway, R.M. (2013) Competitive outcomes between two exotic invaders are modified by direct and indirect effects of a native conifer. *Oikos* 122(4), 632–640. DOI: 10.1111/j.1600-0706.2012.20792.x.

Metlen, K.L. and Callaway, R.M. (2015) Native North American pine attenuates the competitive effects of a European invader on native grasses. *Biological Invasions* 17(4), 1227–1237. DOI: 10.1007/s10530-014-0790-2.

Michalet, R., Chen, S.., An, L.., Wang, X.., Wang, Y. *et al.* (2015) Communities: are they groups of hidden interactions? *Journal of Vegetation Science* 26(2), 207–218. DOI: 10.1111/jvs.12226.

Northfield, T.D., Laurance, S.G.W., Mayfield, M.M., Paini, D.R., Snyder, W.E. *et al.* (2018) Native turncoats and indirect facilitation of species invasions. *Proceedings of the Royal Society B* 285(1871), 20171936. DOI: 10.1098/rspb.2017.1936.

Oduor, A.M.O., Long, H., Fandohan, A.B., Liu, J. and Yu, X. (2018) An invasive plant provides refuge to native plant species in an intensely grazed ecosystem. *Biological Invasions* 20(10), 2745–2751. DOI: 10.1007/s10530-018-1757-5.

O'Loughlin, L.S. and Green, P.T. (2017) Secondary invasion: when invasion success is contingent on other invaders altering the properties of recipient ecosystems. *Ecology and Evolution* 7(19), 7628–7637. DOI: 10.1002/ece3.3315.

Otsamo, R. (2000) Secondary forest regeneration under fast-growing forest plantations on degraded *Imperata cylindrica* grasslands. *New Forests* 19(1), 69–93. DOI: 10.1023/A:1006688022020.

Pauchard, A., Kueffer, C., Dietz, H., Daehler, C.C., Alexander, J. *et al.* (2009) Ain't no mountain high enough: plant invasions reaching new elevations. *Frontiers in Ecology and the Environment* 7(9), 479–486. DOI: 10.1890/080072.

Proença, B., Nez, T., Poli, A., Ciutat, A., Devaux, L. *et al.* (2019) Intraspecific facilitation explains the spread of the invasive engineer *Spartina anglica* in Atlantic salt marshes. *Journal of Vegetation Science* 30(2), 212–223. DOI: 10.1111/jvs.12720.

Quinos, P.M., Insausti, P. and Soriano, A. (1998) Facilitative effect of Lotustenuis on Paspalumdilatatum in a lowland grassland of Argentina. *Oecologia* 114(3), 427–431. DOI: 10.1007/PL00008819.

Reinhart, K.O., Maestre, F.T. and Callaway, R.M. (2006) Facilitation and inhibition of seedlings of an invasive tree (*Acer platanoides*) by different tree species in a mountain ecosystem. *Biological Invasions* 8(2), 231–240. DOI: 10.1007/s10530-004-5163-9.

Reisner, M.D., Doescher, P.S. and Pyke, D.A. (2015) Stress-gradient hypothesis explains susceptibility to *Bromus tectorum* invasion and community stability in North America's semi-arid *Artemisia tridentata wyomingensis* ecosystems. *Journal of Vegetation Science* 26(6), 1212–1224. DOI: 10.1111/jvs.12327.

Richardson, D.M., Allsopp, N., D'Antonio, C.M., Milton, S.J. and Rejmánek, M. (2000) Plant invasions--the role of mutualisms. *Biological Reviews of the Cambridge Philosophical Society* 75(1), 65–93. DOI: 10.1017/S0006323199005435.

Rodriguez, L.F. (2006) Can invasive species facilitate native species? evidence of how, when, and why these impacts occur. *Biological Invasions* 8(4), 927–939. DOI: 10.1007/s10530-005-5103-3.

Rodríguez-Buriticá, S. and Miriti, M.N. (2009) Biting the hand that feeds: the invasive grass *Schismus barbatus* (Poaceae) is facilitated by, but reduces establishment of, the native shrub *Ambrosia dumosa* (Asteraceae). *Journal of Vegetation Science* 20(2), 241–250. DOI: 10.1111/j.1654-1103.2009.05588.x.

Saccone, P., Pagès, J.-P., Girel, J., Brun, J.-J. and Michalet, R. (2010) Acer negundo invasion along a successional gradient: early direct facilitation by native pioneers and late indirect facilitation by conspecifics. *New Phytologist* 187(3), 831–842. DOI: 10.1111/j.1469-8137.2010.03289.x.

Sax, D.F. and Brown, J.H. (2000) The paradox of invasion. *Global Ecology and Biogeography* 9(5), 363–371. DOI: 10.1046/j.1365-2699.2000.00217.x.

Siemann, E. and Rogers, W.E. (2003) Changes in light and nitrogen availability under pioneer trees may indirectly facilitate tree invasions of grasslands. *Journal of Ecology* 91(6), 923–931. DOI: 10.1046/j.1365-2745.2003.00822.x.

Simberloff, D. (2006) Invasional meltdown 6 years later: important phenomenon, unfortunate metaphor, or both? *Ecology Letters* 9(8), 912–919. DOI: 10.1111/j.1461-0248.2006.00939.x.

Simberloff, D., Martin, J.-L., Genovesi, P., Maris, V., Wardle, D.A. *et al.* (2013) Impacts of biological invasions: what's what and the way forward. *Trends in Ecology & Evolution* 28(1), 58–66. DOI: 10.1016/j.tree.2012.07.013.

Simberloff, D. and Von Holle, B. (1999) Positive interactions of nonindigenous species: invasional meltdown? *Biological Invasions* 1(1), 21–32. DOI: 10.1023/A:1010086329619.

Stohlgren, T.J., Barnett, D.T. and Kartesz, J.T. (2003) The rich get richer: patterns of plant invasions in the United States. *Frontiers in Ecology and the Environment* 1(1), 11–14. DOI: 10.1890/1540-9295(2003)001[0011:TRGRPO]2.0.CO;2.

Svriz, M., Damascos, M.A., Zimmermann, H. and Hensen, I. (2013) The exotic shrub *Rosa rubiginosa* as a nurse plant. Implications for the restoration of disturbed temperate forests in Patagonia, Argentina. *Forest Ecology and Management* 289, 234–242. DOI: 10.1016/j.foreco.2012.09.037.

Tecco, P.A., Díaz, S., Gurvich, D.E., Perez-Harguindeguy, N., Cabido, M. *et al.* (2007) Facilitation and interference underlying the association between the woody invaders *Pyracantha angustifolia* and *Ligustrum lucidum*. *Applied Vegetation Science* 10(2), 211–218. DOI: 10.1111/j.1654-109X.2007.tb00519.x.

Tecco, P.A., Gurvich, D.E., Díaz, S., Pérez-Harguindeguy, N. and Cabido, M. (2006) Positive interaction between invasive plants: the influence of *Pyracantha angustifolia* on the recruitment of native and exotic woody species. *Austral Ecology* 31(3), 293–300. DOI: 10.1111/j.1442-9993.2006.01557.x.

Theoharides, K.A. and Dukes, J.S. (2007) Plant invasion across space and time: factors affecting nonindigenous species success during four stages of invasion. *New Phytologist* 176(2), 256–273. DOI: 10.1111/j.1469-8137.2007.02207.x.

Tilman, D. (2004) Niche tradeoffs, neutrality, and community structure: a stochastic theory of resource competition, invasion, and community assembly. *Proceedings of the National Academy of Sciences* 101(30), 10854–10861. DOI: 10.1073/pnas.0403458101.

Traveset, A. and Richardson, D.M. (2014) Mutualistic interactions and biological invasions. *Annual Review of Ecology, Evolution, and Systematics* 45(1), 89–113. DOI: 10.1146/annurev-ecolsys-120213-091857.

Van Riper, L.C. and Larson, D.L. (2009) Role of invasive *Melilotus officinalis* in two native plant communities. *Plant Ecology* 200(1), 129–139. DOI: 10.1007/s11258-008-9438-6.
Vilà, M. and Weiner, J. (2004) Are invasive plant species better competitors than native plant species? - evidence from pair-wise experiments. *Oikos* 105(2), 229–238. DOI: 10.1111/j.0030-1299.2004.12682.x.
Vitousek, P.M. and Walker, L.R. (1989) Biological invasion by *Myrica faya* in Hawai'i: plant demography, nitrogen fixation, ecosystem effects. *Ecological Monographs* 59(3), 247–265. DOI: 10.2307/1942601.
Von Holle, B. (2005) Biotic resistance to invader establishment of a southern Appalachian plant community is determined by environmental conditions. *Journal of Ecology* 93(1), 16–26. DOI: 10.1111/j.0022-0477.2004.00946.x.
White, E.M., Wilson, J.C. and Clarke, A.R. (2006) Biotic indirect effects: a neglected concept in invasion biology. *Diversity and Distributions* 12(4), 443–455. DOI: 10.1111/j.1366-9516.2006.00265.x.

8 How Direct and Indirect Non-native Interactions Can Promote Plant Invasions, Lead to Invasional Meltdown and Inform Management Decisions

Sara E. Kuebbing*

Department of Biological Sciences, University of Pittsburgh, Pittsburgh, Pennsylvania, USA

Abstract

In 1999, Daniel Simberloff and Betsy Von Holle introduced the term 'invasional meltdown'. The term and the concept have been embraced and critiqued but have taken a firm hold within the invasion biology canon. The original formulation of the concept argued two key points: first, biologists rarely study how non-natives interact with one another. Second, nearly all the conceptual models about the success and impact of invasive species are predicated on the importance of competitive interactions and an implicit assumption that non-natives should interfere with establishment, spread and impact of other non-natives. In response, Simberloff and Von Holle called for more research on invader interactions and proposed an alternative consequence of non-native species interactions – invasional meltdown – where facilitative interactions among non-natives could increase the invasion rate or ecological impacts in invaded systems.

This chapter outlines the primary pathways in which direct and indirect interactions among non-natives could lead to invasional meltdown. It provides examples of how different types of interactions among non-natives could lead to net positive effects on the invasion success of non-native plants or the impact of non-native plants on invaded ecosystems. Direct effects are by far the most commonly explored form of non-native-non-native interaction, primarily focusing on plant mutualisms with pollinators, seed dispersers or soil microbial mutualists. There are, however, also examples of non-native plants that benefit from commensal and even herbivorous interactions with other non-natives. Indirect interactions among non-natives are very infrequently studied. Although examples are scarce, non-natives may indirectly benefit other non-native plants through trophic cascades, apparent competition and indirect mutualisms. It remains unclear whether indirect effects are important pathways to invasional meltdown. More work is needed on studying ecosystems that are invaded by multiple non-native species and we need to consider the full range of interactions among non-natives that could either stymie or promote their spread, population growth and impact. Only then can we address how common facilitative interactions are relative to competitive interactions among non-natives or provide robust suggestions on how to manage ecosystems.

8.1 Introduction

There are few hypotheses in invasion biology that consider interactions among non-native species. The most prominent is the Invasional Meltdown Hypothesis (Catford *et al.*, 2009; see also Chapter 7, this volume; Fig. 8.1). During an invasional meltdown, non-natives facilitate the probability of establishment, spread or increase in abundance of other non-natives, potentially

*sara.kuebbing@pitt.edu

(eds Anna Traveset and David M. Richardson)
DOI: 10.1079/9781789242171.0008

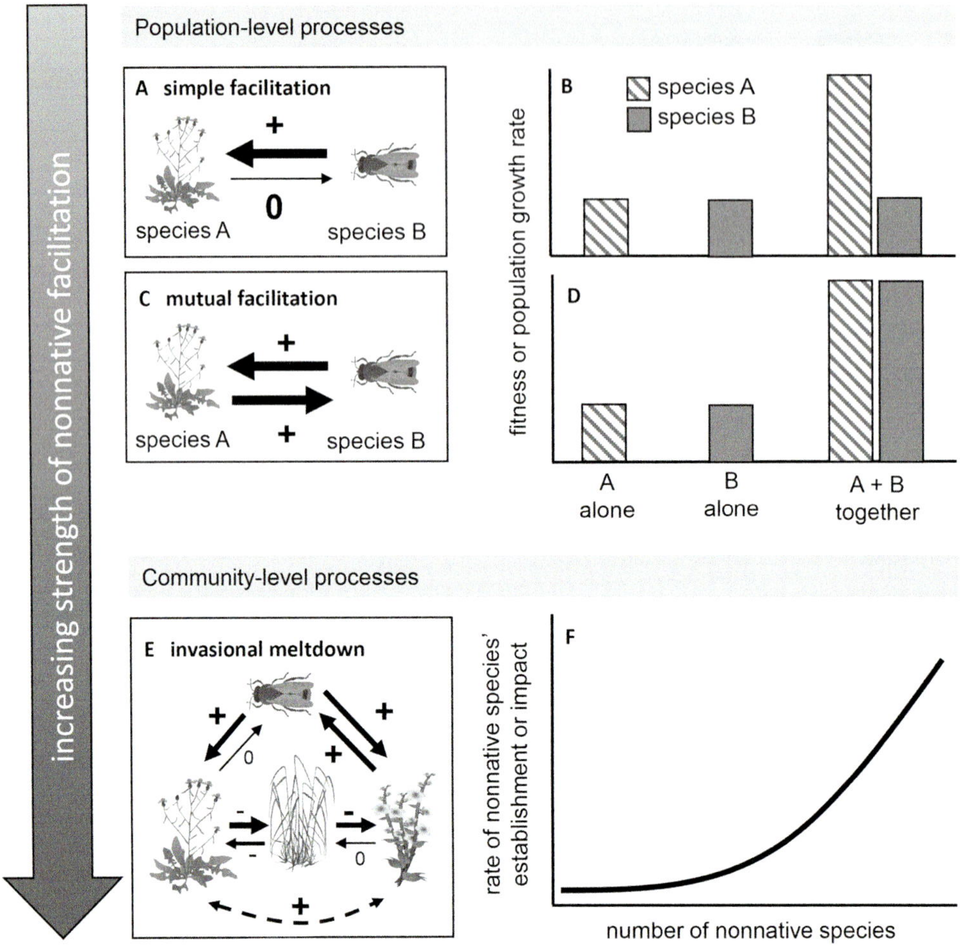

Fig. 8.1. The term 'invasional meltdown' is sometimes broadly defined as any form of facilitation among non-native species (A, C, E). The original definition of the term by Simberloff and Von Holle (1999), however, narrowly defined invasional meltdown as a community-level phenomenon (E) that may arise from facilitative population-level processes (A, C). Different types of ecological data on population size or fitness (B, D) or community invasion rates or impacts (F) are required for distinguishing between population- and community-level processes. Symbols for diagrams courtesy of the Integration and Application Network (ian.umces.edu/symbols).

causing synergistic increases in the impact of non-natives (Simberloff and Von Holle, 1999; Simberloff, 2006). In the original framing of the concept, a range of faciliatory interactions were considered potential pathways to invasional meltdown (Simberloff and Von Holle, 1999; Simberloff, 2006). The weakest interaction to promote a meltdown is 'simple facilitation' or a one-way positive interaction among non-natives that only benefits one of the two interacting non-natives (Fig. 8.1a). A stronger interaction that could promote a meltdown is 'mutual facilitation' where both non-natives benefit from the other (Fig. 8.1c). These interactions are population-level processes that are supported by an increase in a non-native's population growth rate or fitness in the presence of another non-native relative to when the non-native is found alone (Fig. 8.1b,d). By themselves, these types of positive interactions among non-natives do not constitute a meltdown, although they could aid in the process.

An invasional meltdown scenario constitutes a community-level process where facilitation among non-natives could lead to a runaway 'auto-catalytic process that would accelerate the replacement of native communities, perhaps to the point of no return' (Simberloff, 2006; Fig. 8.1e). In a full meltdown, the net effect of all interactions among non-natives leads to an increasing rate of establishment, spread or impact of more non-native species (Fig. 8.1f). In other words, as the number of non-natives increases in a community, the probability that more non-natives will establish or have greater impacts will increase exponentially. The vast majority of invasional meltdown studies focus on how non-native interactions facilitate the establishment, spread or population growth of non-natives while only a handful of studies have attempted to understand whether co-occurring non-natives generate additive or non-additive impacts (Kuebbing *et al.*, 2014, Kuebbing *et al.*, 2016; Pearson *et al.*, 2016a; Tekiela and Barney, 2017).

Since the introduction of the concept, some biologists advocate that discussions of meltdowns focus only on instances that meet the 'full' invasional meltdown criteria (Gurevitch, 2006) because these are the most consequential (and potentially interesting) outcomes of invasion in ecological communities. For this chapter, I discuss the full range of interactions – from weak to strong – that could lead to invasional meltdowns (Fig. 8.1). I do so in part to identify the multiple pathways that could lead to meltdowns and highlight the range of interactions that may occur among two or more non-native species. I also do so because in many cases gathering the necessary community-level data to support invasional meltdowns may be unethical if it requires experimentally introducing new non-native species (Simberloff, 2006; Fig. 8.1c) or data may be impossible to collect. For example, in Mauritius, the non-native common bulbul bird (*Pycnonotus jocosus*) preferentially feeds on fleshy fruits of non-native shrubs. Although spread of many non-native plants coincided with the bird's introduction, this occurred over a century ago and no one collected bird or plant population data so we cannot confirm a meltdown occurred (Linnebjerg *et al.*, 2010). Perhaps in this instance, hindsight of these 'historic' meltdowns provides a strong caution to carefully monitor new introductions of 'meltdown-prone' systems. The bulbul was recently introduced to La Réunion Island where it is spreading the same fleshy-fruited non-native shrubs into primary successional forests. For these early-successional ecosystems, only time and continued monitoring will tell whether these bird-plant facilitative interactions lead to invasional meltdowns (Potgieter *et al.*, 2014). While few of the examples included in this chapter are evidence of 'full' invasional meltdown (Fig. 8.1c), they are valuable to consider for laying a foundation for the many types of interactions that could be involved in an invasional meltdown.

Recent studies have quantitatively reviewed empirical support for the Invasional Meltdown Hypothesis (Braga *et al.*, 2018a,b) and the alternative hypothesis to invasional meltdown – the Biotic Resistance Hypothesis – is reviewed in detail in Chapter 9 (this volume). Thus, I intentionally focus on providing examples of how different types of interactions among non-natives could lead to net positive effects on the invasion success of non-native plants or the impact of non-native plants on invaded ecosystems. I do not attempt to quantify or assess whether these potential pathways to invasional meltdown are more frequent or more important than competitive or negative interactions for a few reasons. First, although it has been two decades since the introduction of the invasional meltdown concept, the study of interactions among non-native species and the resulting impacts of co-occurring invaders is still infrequent and subordinate in the literature relative to the study of interactions between native and non-native species or the impact of single invasive species (Kuebbing *et al.*, 2013; Jackson, 2015; Kuebbing and Nuñez, 2016). Second, studies on non-native interactions typically focus on direct, pairwise interactions among non-native species (White *et al.*, 2006; Chapter 17, this volume) and we have limited empirical data on how indirect interactions affect patterns of invasion in communities comprised of multiple native and non-native species. Because of these two limitations, it is still premature to answer one of the key questions posed by Simberloff and Von Holle (1999): 'Are faciliatory interactions more frequent or more important than hindering ones?' (Simberloff, 2006).

8.2 Direct Interactions Among Non-native Species

One of the easiest invasional meltdown mechanisms to detect occurs when a non-native plant is directly facilitated by another non-native species. Here, I define direct interaction as an interaction between two species that is not mediated through a third species (Strauss, 1991). Non-native species can facilitate non-native plants through mutualistic, commensalistic or herbivorous interactions (Fig. 8.2). Non-native plants engage in mutualisms – where both non-natives facilitate the other – with their pollinators, seed dispersers and soil microbial symbionts (Richardson *et al.*, 2000; Traveset and Richardson, 2014; Fig. 8.2a). In commensalisms, a plant's fitness is enhanced while the other non-native is neither benefited nor harmed. Nearly all the examples of commensalisms among non-natives involve non-native ecosystem engineers that modify the environment to the benefit of a non-native plant (Fig. 8.2b). Interactions between non-native plants and herbivores are rarely considered in discussions of invasional meltdown because herbivory is not typically thought to be beneficial to a plant. However, for some non-natives, herbivory can facilitate the spread or growth of the non-native plant (Fig. 8.2c).

8.3 Mutualisms and Invasional Meltdown

One of the most obvious patterns in plant invasional meltdown studies is that mutualistic interactions are more likely to occur between plants and species in other trophic levels (Simberloff and Von Holle, 1999; Richardson *et al.*, 2000; Simberloff, 2006; Chapter 7, this volume). Early examples of facilitation focused on interactions occurring aboveground, between non-native plants, pollinators and seed dispersers (Fig. 8.2a). Today, we are accumulating examples of belowground interactions that facilitate invasions, and more examples are likely to be unearthed in the coming years (Nuñez and Dickie, 2014; Traveset and Richardson, 2014).

8.4 Non-native Plants and Pollinators

There is ample evidence that non-native plants routinely benefit from non-native pollinators (Richardson *et al.*, 2000; Goulson, 2003; Traveset and Richardson, 2014) but limited evidence that non-native pollinators are receiving large population-level benefits. Early writing foreshadowed that widely introduced super-generalist pollinators, like the European honeybee (*Apis mellifera*) and European bumblebee (*Bombus terrestris*), may facilitate the establishment of many non-native plants (Richardson *et al.*, 2000). Because these non-native pollinators are domestic species introduced for crop pollination (Goulson, 2003) they likely receive less benefits than the non-native plants they pollinate, forming simple facilitative interactions (Fig. 8.1a). Many non-native plants receive a large proportion of their pollination services – sometimes their only pollination services – from one or both of these two pollinators. Super-generalist non-native pollinators are frequently integral components in non-native plant—pollinator community networks and facilitate non-native plants relative to native plants in these complex interaction webs (Richardson *et al.*, 2000; Morales and Aizen, 2006; Aizen *et al.*, 2008; Bartomeus *et al.*, 2008; Padrón *et al.*, 2009; Abe *et al.*, 2011; Morales *et al.*, 2017). Interestingly, these *generalist* pollinators can also facilitate the spread of non-native plants with *specialist* pollination syndromes. Hummingbirds pollinate the banana passionfruit vine (*Passiflora tripartita* var. *mollissima*) in its native range in the South American Andes. Surprisingly, it is one of the most invasive vines in New Zealand (Beavon and Kelly, 2012, 2015) even though hummingbirds do not occur on the island. Instead, non-native bees act as surrogate pollinators for the vine; preventing bee access to flowers reduced the vine's fruit set by nearly 70% (Beavon and Kelly, 2012).

This is likely an incomplete list of examples of non-native plant–pollinator pairs because the pollination biology of non-native plants and bees is infrequently studied (Goulson, 2003; Russo, 2016) and bees are not the only possible pollinators of non-native plants. The global establishment rate of non-native invertebrates has

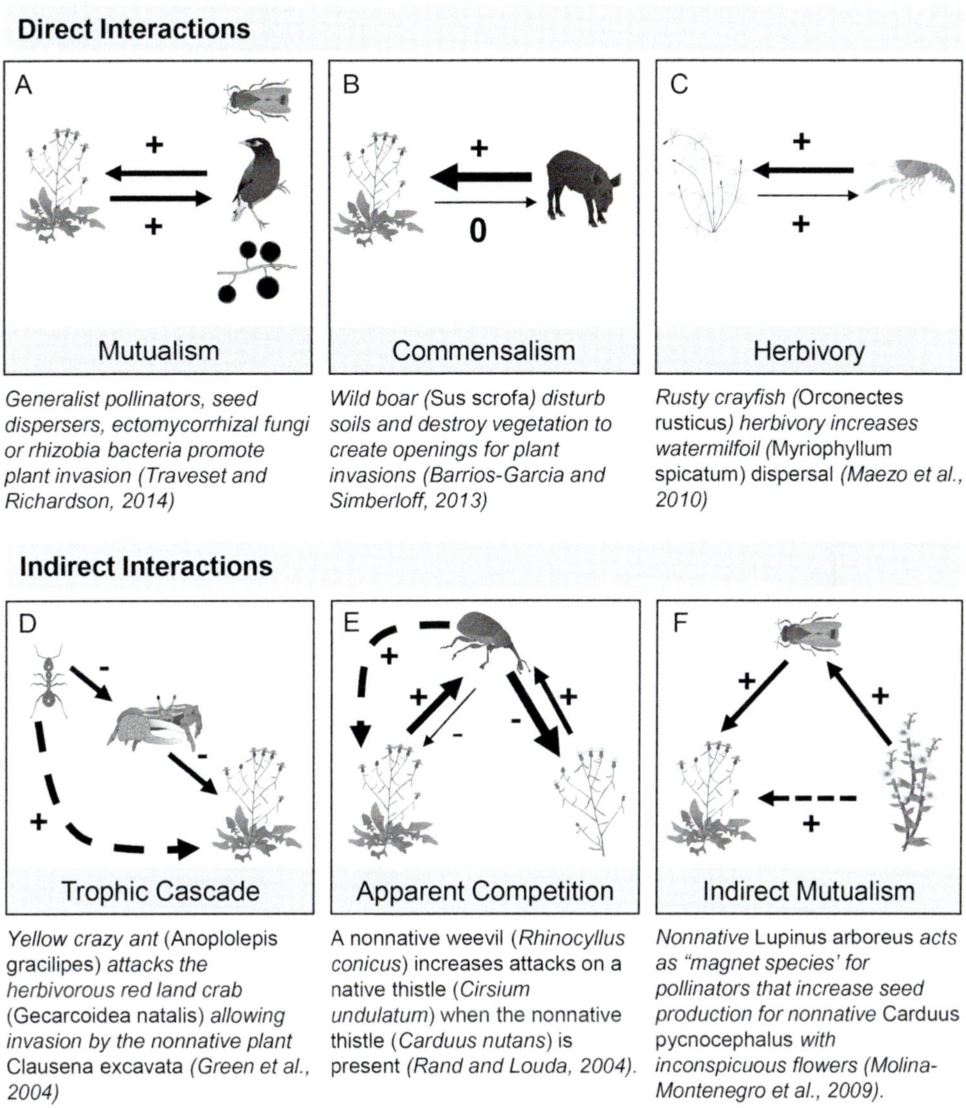

Fig. 8.2. An invasional meltdown is a community-level phenomenon where the *net effect* of all interactions is positive and leads to an increasing rate of establishment or accelerating impact of non-native species (Simberloff and Von Holle, 1999; Simberloff, 2006). There are many possible types of direct (bold lines) and indirect interactions (dashed lines) between non-native species that could lead to an invasional meltdown. Line width indicates the relative strength of the interaction. Symbols for diagrams courtesy of the Integration and Application Network (ian.umces.edu/symbols) and PhyloPic (www.phylopic.org).

more than doubled in the past quarter-century (Hulme, 2009). Some proportion of these invertebrates will be generalist pollinators, which creates a formidable palate of potentially facilitative non-native plant—pollinator interactions for ecologists to explore.

There are limited examples of specialized non-native plant–pollinator mutualisms, which are more likely to lead to 'full' invasional meltdowns (Fig. 8.1e). To my knowledge, the only example of a 'full' plant–pollinator meltdown is the tightly coevolved plant—insect pairing of *Ficus* trees and their host-specific wasps in Florida, USA (Ramírez and Montero, 1988; McKey and Kaufmann, 1991). This may be because plants with specialized pollination syndromes should be far less likely to become invasive relative to plants with generalist pollination syndromes (Baker, 1965, 1974; Richardson *et al.*, 2000). This does not mean 'full' plant—pollinator meltdowns do not occur, it just means they may be infrequent.

An innovative approach to identify potential plant—pollinator invasional meltdowns is studying non-natives *early* in the invasion process. Many non-native plants are cultivated (van Kleunen *et al.*, 2018) but have not invaded outside of cultivation, potentially owing to a lack of pollinators (Bufford and Daehler, 2014; Moodley *et al.*, 2016). This was the case in Hawaii, USA, where cultivated milkweed (*Calotropis gigantea*) only fruited when it was hand pollinated, suggesting a missing mutualist is limiting its spread (Bufford and Daehler, 2014). Although tracking failed invasions is difficult (Zenni and Nuñez, 2013), it may provide a promising avenue for detecting future meltdowns (Bufford and Daehler, 2014).

8.5 Non-native Plants and Seed Dispersers

There are many documented examples of non-native animals dispersing non-native plant seeds (Richardson *et al.*, 2000; Traveset and Richardson, 2014; Martin-Albarracin *et al.*, 2015). As with non-native pollinators, non-native animals with wide diet breadths are more likely to successfully invade a new range (Martin-Albarracin *et al.*, 2015) and most non-native seed dispersers include generalist birds (Bartuszevige and Gorchov, 2006; Foster and Robinson, 2007; Kawakami *et al.*, 2009; Spotswood *et al.*, 2012; Martin-Albarracin *et al.*, 2015) and mammals (Davis *et al.*, 2010; Calvino-Cancela, 2011; Beavon and Kelly, 2015; Bobadilla *et al.*, 2016; Calvino-Cancela, 2011). Of those examples of non-native animals that disperse seeds of non-native plants, there is limited evidence that the plants are dispersal-limited or that the animals are food-limited. While a few studies show that seed germination rates increased after passage through non-native dispersers' guts (Bartuszevige and Gorchov, 2006; Beavon and Kelly, 2015; Bobadilla *et al.*, 2016), it was not tested whether native seed dispersers provided the same service or if increased germination rates led to increased population growth rates of non-native plants. This suggests that while non-native plants and animals may frequently facilitate one another, many of these interactions do not constitute 'full' invasional meltdowns.

There are a few circumstances, however, where generalist seed dispersers could promote an invasional meltdown. Plant–disperser meltdowns may be more likely to occur when the fruiting phenology of non-native plants differs from native plants. In this scenario, non-native plants may boost the population size and spread the rate of a non-native seed disperser. For example, in the Mascarene Islands in the Indian Ocean, the non-native red-whiskered bulbul (*P. jocosus*) invaded the island much more rapidly than during prior introduction events in other regions (Clergeau and Mandon-Dalger, 2001). One hypothesis for the bird's rapid spread is that the island's large population of non-native plants that produce year-round fruit increased the bird's population growth and spread (Mandon-Dalger *et al.*, 2004). Indeed, the bulbul's population size is larger in sites that contain fleshy-fruited non-natives with year-round fruit relative to sites dominated by native plants with seasonal fruit (Mandon-Dalger *et al.*, 2004). Seeds of the most invasive non-native plants are prevalent in bulbul faecal samples and the seed germination rates of some non-natives increase significantly after passage through the bulbul gut (Mandon-Dalger *et al.*, 2004; Linnebjerg *et al.*, 2010). Similarly, in Mediterranean dunes the non-native succulent hottentot fig (*Carpobrotus edulis*) produces a water and energy-rich food source during the dry season when natives are not producing fruit. The fig is eaten by many non-native mammals including the European rabbit (*Oryctolagus cuniculus*), the Norway rat (*Rattus rattus*) and the wood mouse (*Apodemus sylvaticus*), which disperse seeds and enhance seed germination relative to uneaten fruit (Bourgeois *et al.*, 2004, Bourgeois *et al.*, 2005).

Another form of plant–disperser meltdown may occur through positive feedback loops that arise when a non-native seed disperser has strong preferences for non-native fruit relative to native fruit (Martin-Albarracin *et al.*, 2015; Traveset and Richardson, 2014). If a seed disperser prefers non-native fruit, they will spread more non-native seed and increase the non-native plant's abundance. As the plant becomes more abundant, it becomes the dominant food source and perpetuates the feedback. A bevy of studies demonstrate non-native dispersers prefer non-native plant fruit (Foster and Robinson, 2007; Lafleur *et al.*, 2007; Rowles *et al.*, 2009; Chimera and Drake, 2010; Spotswood *et al.*, 2012; Prior *et al.*, 2015; MacFarlane *et al.*, 2016) and that non-native plants comprise a large proportion of all seeds found in faecal samples of non-native dispersers (Kawakami *et al.*, 2009; Chimera and Drake, 2010; Linnebjerg *et al.*, 2010; Spotswood *et al.*, 2012). These positive feedback loops could be when non-native animals act as both seed predators and seed dispersers. For example, in the Hawaiian Islands, non-native rats (*Rattus* spp.) may act as seeds dispersers for the most widespread non-native plants, but seed predators for many native plants (Shiels and Drake, 2011; Shiels, 2011).

Seed-dispersing non-native ants demonstrate this same pattern of disperser-preference-driven feedback loops. The invasive Argentine ant (*Linepithema humile*) is one of the most aggressive and globally widespread invasive insects and is a poor seed disperser for native plants (Rodriguez-Cabal *et al.*, 2009). In coastal scrublands in south-eastern Australia, the Argentine ant accounts for 92% of all ant—seed interactions and disperses significant fewer seeds of the native tree *Acacia retinodes* and significantly more seeds of the invasive South African shrub, *Polygala myrtifolia*, relative to native ant dispersal patterns at uninvaded sites (Rowles *et al.*, 2009). Similarly, in north-eastern US deciduous forests, the non-native ant *Myrmica rubra* preferred seeds of the non-native herb, *Chelidonium majus*, relative to three other common native plants (Prior *et al.*, 2015). When the non-native ant was present, the non-native plant had eight times more seedlings and produced almost two times more flowers than when a native seed-dispersing ant was present (Prior *et al.*, 2015).

The majority of examples of non-native plants benefiting from seed dispersal services focus on consumptive dispersal of non-native plant seed (endozoochory). However, animals can also disperse seeds though non-consumptive means (exozoochory), although this dispersal mechanism is infrequently studied (van Leeuwen, 2018). It has been suggested that non-native livestock or other wide-ranging mammals may promote long-distance spread of non-natives (Simberloff and Von Holle, 1999), but the only demonstrated example of this is from Santa Catalina Island, California, USA, where non-native bison (*Bison bison*) carry viable seeds of the non-native forbs *Marrubium vulgare* and *Xanthium strumarium* (Constible *et al.*, 2005).

8.6 Non-native Plants and Soil Microbial Mutualists

Some of the most exciting, and potentially devastating, interactions among non-natives may be occurring belowground. For some non-native plant taxa – primarily those that affiliate with nitrogen-fixing bacteria or ectomycorrhizal fungi – there is now clear evidence that the absence or presence of soil mutualists can prevent or promote plant invasions (Richardson *et al.*, 2000; Nuñez *et al.*, 2009) or facilitate the invasion of plants and microbes (Nuñez *et al.*, 2009; Dickie *et al.*, 2010; Rodríguez-Echeverría, 2010; Hayward *et al.*, 2015a). Many barriers to detecting microbial symbionts have disappeared with the advent of new molecular tools that can identify soil microbial taxa (Widder *et al.*, 2016). There are now clear patterns in non-native plant–soil mutualisms. First, humans are frequently and successfully co-introducing non-native plants with their soil microbial mutualists (Richardson *et al.*, 2000; Nuñez and Dickie, 2014); second, plant–microbe co-introduction events are facilitating the spread of both plants and microbes; and third, there are likely synergistic impacts on native soil microbes and native plant—microbe interactions that could fundamentally alter the composition of soil microbial communities, plant communities and perhaps even above-ground plant–pollinator or plant–disperser mutualisms (Rodríguez-Echeverría and Traveset, 2015).

It is likely that we will see an increase in examples of invasional meltdown – both historic and contemporary – as we continue exploring interactions between non-native soil microbes and plants. Co-introduction of non-native plants and their soil mutualist partners may be much more common than co-introductions of plants and pollinators or seed dispersers because many non-native plants with obligate belowground mutualisms were intentionally introduced as economic commodities (Richardson *et al.*, 2000; Richardson and Rejmánek, 2011; Ambrosini *et al.*, 2016). For example, non-native *Pinus* are over-represented relative to their global taxonomic diversity (Richardson and Rejmánek, 2011). This is likely because humans intentionally co-introduce *Pinus* and their ectomycorrhizal fungi mutualists in forestry plantations, which has enabled *Pinus* species to overcome barriers to invasion (Richardson and Higgins, 1998). Humans have transported at least 200 ectomycorrhizal fungal species around the globe and the majority of these recorded introductions were intentional with forestry plantations of *Pinus* or *Eucalyptus* species (Vellinga *et al.*, 2009). Additionally, even if soil microbe mutualists are not intentionally introduced, they are incredibly robust to a range of environmental conditions and survive transport through a variety of pathways including free-living in soils, on the roots of live plants or even on the exterior of dried plant seeds (Perez-Ramirez *et al.*, 1998; Stępkowski *et al.*, 2005; Pringle *et al.*, 2009; Litchman, 2010).

While co-introduction of plants and their mutualists does not necessarily lead to co-invasion and a full invasional meltdown, it sets the stage for one. Many intentionally introduced ectomycorrhizal fungi fail to establish in soils outside of the original forestry plantation (Nuñez *et al.*, 2009). In some instances, however, a single 'pioneer' ectomycorrhizal fungus can initiate a plant invasion (Hayward *et al.*, 2015b). For example, the fungal genus *Suillus* associates with tree roots on the leading edge of *Pinus* invasions in Patagonia (Hayward *et al.*, 2015b; Ureclay *et al.*, 2017), New Zealand (Dickie *et al.*, 2010) and Hawaii (Hynson *et al.*, 2013). *Suillus* spp. are good 'pioneer' fungi because they produce many fruiting bodies that are ingested and dispersed by large mammals and their spores are hardy and persist in the soil until a suitable host plant establishes (Hayward *et al.*, 2015b). Wind-dispersed *Pinus* seeds that land on *Suillus*-inhabited soils can establish and then provide a suitable soil habitat for later-successional non-native ectomycorrhizal fungi (Hayward *et al.*, 2015a). In some cases, the spread of non-native ectomycorrhizal fungi from plantations can facilitate plant invasions to environments in which they are not typically found, such as high-elevation mountains that typically experience lower levels of invasion (Ureclay *et al.*, 2017). Lower fungal diversity in the soils around invading *Pinus* species relative to native ectomycorrhizal trees (Dickie *et al.*, 2010) or *Pinus* in their native range (Hynson *et al.*, 2013) supports the idea that this 'leapfrog' invasion pattern may be common in *Pinus* invasions.

The potential for invasional meltdown does not end with tree–fungal mutualisms. Non-native mammals can facilitate invasion by dispersing fungal spores to new locations. Non-native wild boar (*Sus scrofa*) and fallow deer (*Dama dama*) in Patagonia, non-native brushtail possum (*Trichosurus vulpecula*) in New Zealand and red deer (*Cervus elaphus*) in both regions, disperse a wide range of non-native, but not native, ectomycorrhizal fungal spores (Nuñez *et al.*, 2013; Wood *et al.*, 2015). In both regions, these non-native mammals are functionally unique because there are no native large mammals to disperse fungal spores (Nuñez *et al.*, 2013; Wood *et al.*, 2015).

Plant—microbe invasional meltdowns also seem common among leguminous plants that form mutualistic relationships with symbiotic nitrogen-fixing bacteria, or rhizobia. One of the first recorded plant—rhizobia co-invasions was in the Hawaiian Islands, USA, where the non-native shrub *Morella faya* and *Frankia* rhizobia invaded early successional and nutrient-poor soils (Turner and Vitousek, 1987). There are now many more examples of plant—rhizobia co-invasion. South and Central American mimosa trees (*Mimosa* spp.) predominately affiliate with American lineages of *Cupriavidus* and *Burkholderia* rhizobia in their non-native range in the Philippines (Andrus *et al.*, 2012), Taiwan (Chen *et al.*, 2005) and Yunnan Province in southern China (Liu *et al.*, 2012). Partnerships between European legumes and *Bradyrhizobium* bacteria have been found in Scotch broom (*Cytisus scoparius*) populations in north-western USA (Horn *et al.*, 2014), seven non-native clover species (*Trifolium* spp.) in New Zealand (McGinn *et al.*, 2016), and invasive lupines (*Lupinus albus, L. angustifolius, L. cosentinii* and *L. luteus*) and serradella (*Ornithopus* spp.) in Australia and South Africa (Stępkowski *et al.*, 2005). Finally, nitrogen-fixing *Acacia* species that evolved in Australia are some of the most widely introduced woody plants (Richardson and Rejmánek, 2011) and are nearly always found with Australian *Bradyrhizobium* that were likely co-introduced with the *Acacia* (Rodríguez-Echeverría *et al.*, 2011). Thus far, non-native *Bradyrhizobium* have been detected in non-native *Acacia* tree roots and soil in western Australia (Birnbaum *et al.*, 2012, Birnbaum *et al.*, 2014, 2016), Portugal (Crisóstomo *et al.*, 2013; Rodríguez-Echeverría, 2010; Rodríguez-Echeverría *et al.*, 2012), and South Africa (Ndlovu *et al.*, 2013).

Many of these non-native legumes do not affiliate with native rhizobia (Weir *et al.*, 2004; La Pierre *et al.*, 2017), suggesting that without the co-introduction of the non-native rhizobia, plant invasion would not occur. For example, there was nearly no overlap in *Bradyrhizobium* genotypes between three invasive European legumes (*Genista monspessulana, Spartium junceum* and *Ulex europaeus*) and six native legumes (*Acmispon glaber, A. heermannii, A. micranthus, A. strigosus, Lupinus arboreus* and *L. bicolor*) in California, USA (La Pierre *et al.*, 2017). This was also the case for a suite of non-native legumes in New Zealand (Weir *et al.*, 2004).

Co-invasion of non-native legumes and rhizobial mutualists may also have large population impacts on co-occurring native plants. Native legumes do interact with non-native rhizobia, which may potentially disrupt co-evolved native plant—rhizobial mutualisms (Rodríguez-Echeverría, 2010). More concerning, new evidence suggests that native and non-native rhizobia can readily hybridize. Rhizobia can evolve rapidly through lateral gene transfer that may foster invasional meltdowns. In some regions invaded by non-native legumes, rhizobial strains in the soils combine genetic elements of native and non-native bacterial genomes. The 'core' native bacterial genome contains basic housekeeping genes for the species with evolved tolerance to unique soil conditions, while the 'accessory' non-native bacterial genomes include the specific signalling genes that stimulate a plant to produce nodules for the bacteria (Remigi *et al.*, 2016; Andrews *et al.*, 2018). These novel, hybrid rhizobia can invade soils that a non-native rhizobial lineage could not (Wei *et al.*, 2009; Liu *et al.*, 2012), increase soil rhizobial diversity for plants in their non-native relative to their native range (Ndlovu *et al.*, 2013; Horn *et al.*, 2014) and potentially disrupt native legume–rhizobia symbioses (Rodríguez-Echeverría, 2010; Le Roux *et al.*, 2017).

8.6.1 Commensalisms and invasional meltdowns

Non-native animals and plants can directly facilitate other non-native plants by modifying the environment (Simberloff and Von Holle, 1999; Fig. 8.2b). Non-native species can impact nutrient and carbon cycling, fire and hydrological regimes or the physical structure of an ecosystem (Liao *et al.*, 2008; Ehrenfeld, 2010; Simberloff, 2011) earning some non-natives the label 'ecosystem engineer' (Jones *et al.*, 1994). Ecosystem engineers are broadly defined as species that modulate the availability of resources to other species by modification of the biotic or abiotic environment. Non-native species can facilitate plant invasions as allogenic engineers – by transforming materials from one physical state to another – or as autogenic engineers – by changing the environment through their own physical structure (Jones *et al.*, 1994). While ecologists disagree whether interactions mediated through environmental modification constitute direct or indirect interactions (Strauss, 1991; Wootton, 1994), for this chapter, I consider these commensalisms to be direct interactions because they are not mediated by the presence of a third species. Regardless of whether these habitat-modifying non-natives are directly or indirectly affecting non-native plants, they represent a viable pathway to invasional meltdown. Nearly all of these examples constitute 'weak' facilitative interactions among non-natives (Fig. 8.1a) because there is typically no evidence that the non-native plant facilitates the ecosystem-modifying invader.

8.7 Allogenic Ecosystem Engineers and Invasional Meltdown

Some of the first examples of non-natives modifying ecosystems were of nitrogen-fixing non-native plants that invaded nutrient-poor soils, increased plant-available soil nitrogen and facilitated the establishment of other non-native plants (Simberloff and Von Holle, 1999). This same scenario has now been documented a few more times. The annual herbaceous legume, partridge pea (*Chamaecrista nictitans*), facilitates invasion of non-native fountain grass (*Pennisetum setaceum*) into native dry Hawaiian grasslands (Carino and Daehler, 2002). Nutrient-poor oak (*Quercus* spp.) forests that are invaded by black locust (*Robinia pseudoacacia*) in north-eastern USA (Vilà and Weiner, 2004) or *Acacia dealbata* in north-western Spain (González-Muñoz *et al.*, 2012) had higher plant-available soil nitrogen levels and percentage of non-native plants relative to oak-dominated forest soils.

Invasive invertebrate pests and microbial pathogens cause dramatic alterations to ecosystems by eliminating entire species from forests (Lovett *et al.*, 2016; Kenis *et al.*, 2017). The loss of foundational canopy trees increases light availability to the forest floor, alters carbon and nutrient cycles and affects the abundance of other non-plant species (Ellison *et al.*, 2005). Approximately 63% of the USA's forests (334 million hectares) are at risk of substantial tree mortality by non-native forest pests (Lovett *et al.*, 2016). There is a growing list of non-native species that invade and persist in closed-canopy forests (Martin *et al.*, 2008) and may be primed to spread more rapidly with large-scale tree canopy losses. For example, the invasion of the non-native insect Emerald ash borer (*Agrilus planipennis*) in midwestern USA forests leads to complete ash tree (*Fraxinus* spp.) mortality in less than five years. Loss of the ash, a once dominant canopy tree, creates canopy gaps that are quickly colonized by a variety of non-native shrubs that increase in abundance much more rapidly than native shrubs (Klooster *et al.*, 2018).

Non-native animals can also modify ecosystems thorough trampling or destroying vegetation or digging and disrupting soils. The wild boar (*S. scrofa*) is globally one of the most widespread invasive mammals. Boar are omnivorous with feeding habits that include destructive soil rooting practices that eliminate vegetation and disturb soils (Barrios-Garcia and Ballari, 2012). In Patagonia, Argentina, non-native plants had twice the rate of establishment and biomass in boar-rooting patches relative to undisturbed sites and boar-dispersed non-native plants into their rooting patches through their droppings (Barrios-Garcia and Simberloff, 2013). Similarly, in Texas, USA, the non-native tallow tree (*Sapium sebiferum*) was more than twice as abundant in areas with boar than areas without boar (Siemann *et al.*, 2009).

Similar to wild boar disruption of soils, the introduction of non-native earthworms in northern North America has led to rapid and dramatic impacts on soil structure, litter decomposition rates and nutrient cycling, and plant community composition (Hendrix and Bohlen, 2002; Cameron *et al.*, 2016; Craven *et al.*, 2017). One of the most conspicuous patterns is that non-native earthworm abundance tends to be positively related to non-native plant abundance and negatively related to native plant abundance (Kourtev *et al.*, 1999; Heneghan *et al.*, 2007; Nuzzo *et al.*, 2009; Clause *et al.*, 2015; Dávalos *et al.*, 2015; Craven *et al.*, 2017; Mueller *et al.*, 2018; Lloyd *et al.*, 2019). Proposed mechanisms for these patterns include earthworm disruption of important native plant–mycorrhizal mutualisms (Lawrence *et al.*, 2003; Paudel *et al.*, 2016), increase of soil nutrient cycling rates or changes in soil pH that preferentially promote non-native plant growth (Belote and Jones, 2009; Lloyd *et al.*, 2019) and the ingestion and deposition of non-native seeds into nutrient-rich earthworm burrows that enhances non-native germination and establishment (Quackenbush *et al.*, 2012; Clause *et al.*, 2015 but see Roth *et al.*, 2015). The majority of studies on non-native plant—earthworm interactions are observational, probably because removing earthworms is challenging and introducing earthworms is unethical. However, non-native plant removal experiments provide evidence that non-native plants facilitate non-native earthworms. Removal of the non-native shrub Chinese privet (*Ligustrum sinense*) in south-eastern US forests led to decreased soil pH and a significance decrease in non-native and a fourfold increase in native earthworm abundance (Lobe *et al.*,

2014). Similarly, the removal of the non-native woody shrubs buckthorn (*Rhamnus cathartica*) and bush honeysuckle (*Lonicera* x *bella*) in northern USA forests reduced nutrient-rich leaf litter inputs and non-native earthworm biomass by 50% (Madritch and Lindroth, 2009).

While it is premature to conclude that non-native earthworms and plants are benefiting from each other and not a symptom of other ecosystem disturbances (Dávalos *et al.*, 2015), these are notable patterns worth exploring. Promising avenues of study include experimental manipulations of non-native plant abundance or establishing observational plots along the front lines of earthworm invasions to capture ecosystem and plant community properties before and after invasion. Additionally, it is likely that the impact of non-native earthworms is not restricted to North America. Humans have introduced invasive earthworms to many other regions of the globe that have yet to be studied (James and Hendrix, 2004).

8.8 Autogenic Ecosystem Engineers and Invasional Meltdown

Non-native plants generally have higher net primary productivity and produce more shoot biomass that native species (Liao *et al.*, 2008; Vilà *et al.*, 2011), which increases the likelihood that they can alter an ecosystem's structure through increasing total plant biomass (Simberloff, 2011). Non-native plants can protect other non-natives from predation, ameliorate harsh environmental conditions or provide physical structures that enhance the growth of other non-native plants. For example, emergent non-native cattails (*Typha* spp.) in Michigan, USA, wetlands provide structural protection for a free-floating non-native plant, European frogbit (*Hydrocharis morsus-ranae*). Cattail removal led to a fivefold decrease in frogbit cover because no other plants protected frogbit in otherwise open-water lakes (Monks *et al.*, 2019). The non-native floating water hyacinth (*Eichhornia crassipes*) in coastal wetlands in Texas, USA, also increased in abundance when the emergent invasive herb alligator weed (*Alternanthera philoxeroides*) was present (Wundrow *et al.*, 2012). In the coastal foredunes of the north-western USA, non-native invasive dune grasses (*Ammophila* spp.) alter the physical shape of sand dunes through differences in shoot and root growth relative to dominant native grasses (Zarnetske *et al.*, 2012). The biophysical alteration of sand dunes changes competitive relationships among native and non-native dune grasses and leads to competitive exclusion of native dune grasses and co-dominance of non-native dune grasses (Zarnetske *et al.*, 2013).

Finally, when non-native woody plants invade ecosystems primarily dominated by low-stature vegetation, such as trees invading early-successional habitats, they provide perching, nesting and protective structures for a variety of wildlife species. The non-native tree *Casuarina equisetifolia* is an aggressive invader of recent lava flows in La Réunion Island in the Indian Ocean (Potgieter *et al.*, 2014). Birds are attracted to *C. equisetifolia* branches as perch sites and deposit many seeds of other non-native plants underneath *C. equisetifolia*'s canopy (Potgieter *et al.*, 2014). Interestingly, one of the primary seed dispersers is the invasive red-whiskered bulbul (*P. jocosus*; see 'Non-native Plants and Seed Dispersers'), which has been implicated in invasional meltdown in the nearby island of Mauritius. Similarly, the non-native tree *Pyracantha angustifolia* in central Argentine shrublands harboured eight-times higher densities of other bird-dispersed non-native species than under a co-occurring native shrub or in areas without a shrub. The authors suggested this occurred because the fruiting phenology of non-native plants in these ecosystems were synchronous (Tecco *et al.*, 2006).

The accumulation of senesced plant litter can also change ecosystem structure. In California, USA grasslands, two herbaceous non-natives, medusa-head (*Elymus caput-medusae*) and wild oat (*Avena* spp.), form dense thatch layers. While native plant seed production and growth was negatively affected by deep thatch, wild oat growth and medusa-head fitness increased with thatch depth, promoting the persistence of both non-natives (Mariotte *et al.*, 2017). Finally, increasing amounts of senesced litter can increase the frequency or intensity of fires (Brooks *et al.*, 2004; Gaertner *et al.*, 2014). Non-native pine plantations increase fuel loads and the number of fires that spread from plantations into native vegetation. In New Caledonia in the south-west Pacific, more fires promote the fire-tolerant non-native fern *Pteridium aquilinum* in and around pine plantations, which then replaces native species and increases fire frequency (Simberloff *et al.*, 2010).

8.9 Herbivory and Invasional Meltdown

Interactions between non-native herbivores and plants could lead to a net positive effect for a plant if it responds positively to browsing, or if herbivory promotes dispersal (Fig. 8.2c). Non-native geese (*Branta canadensis*) in the Gulf Islands of British Columbia, Canada, predominantly browsed non-native grasses. Goose browse decreased grass height, increased tiller production and reduced accumulation of senesced grass litter, ultimately increasing non-native grass fitness relative to native forbs. The authors attributed this unexpected outcome as evidence for evolved tolerance to intense browsing of non-native grasses in their European native ranges (Best and Arcese, 2009). Herbivory may also promote the spread of non-native plants. The non-native rusty crayfish (*Orconectes rusticus*) feeds on the non-native submerged aquatic plant, Eurasian watermilfoil (*Myriophyllum spicatum*). At moderate densities, crayfish browse does not impact milfoil biomass but does increase milfoil dispersal rates by creating reproductive vegetative fragments that disperse through the water (Maezo *et al.*, 2010).

8.10 Indirect Interactions Among Non-native Species

Positive interactions may be the most conspicuous when considering direct interactions between pairs of non-natives. However, invasional meltdown is any instance where the *net effect* of all interactions is positive (Simberloff and Von Holle, 1999; Ricciardi, 2001; Simberloff, 2006; Fig. 8.1e). Thus, positive direct pairwise interactions are not the only pathway to meltdown. In communities comprising many interacting native and non-native species, the net effect of a non-native on another non-native will be the sum of all direct and indirect interactions (Miller, 1994; Ricciardi, 2001). Detecting indirect interactions in multispecies communities is difficult, controversial and frequently ignored (Strauss, 1991; Miller, 1994; Levine *et al.*, 2017; Wright *et al.*, 2017), including in invasion biology (White *et al.*, 2006; Godoy, 2019). In communities containing more than two species, indirect interactions may be highly influential (Miller, 1994; Ricciardi, 2001) but the paucity of studies that consider indirect effects in invasion limits our ability to fully assess the relative importance of indirect vs direct effects (see Chapter 17, this volume).

Below, I review a handful of case studies where indirect effects of multiple non-native species has promoted plant invasions. Many of these examples are interaction chains where non-native animals promote the growth or fitness of non-native plants by reducing herbivory on plants in a trophic cascade (Fig. 8.2d) or increasing herbivory on native competitors in apparent competition (Fig. 8.2e). Similarly, non-native plants may affect the fitness or growth of other non-native plants through indirect mutualisms mediated by shared mutualists or competitors (Fig. 8.2f).

8.11 Trophic Cascades

Non-native plants may indirectly benefit from non-native predators when the predator reduces the effects of herbivores on the non-native. Although I only found one example of non-native plants benefiting from trophic cascades, we may expect to find more examples in island ecosystems that tend to be invaded by multiple invasive species in a range of trophic positions (Russel and Kaiser-Bunbury, 2019). For example, on Christmas Island in the Indian Ocean, invasion by the non-native yellow crazy ant (*Anoplolepis gracilipes*) attacks and reduces the population of the red land crab (*Gecarcoidea natalis*), which consumes non-native plant seeds and seedlings (Green *et al.*, 2004, 2011). Thus, ants indirectly facilitate plant invasion by reducing crab herbivory and allowing non-native plants to establish in the island's forest interior (Green *et al.*, 2004, 2011).

8.12 Apparent Competition

Invasional meltdowns via apparent competition can occur when a non-native plant benefits relative to a native plant competitor through an asymmetrical relationship with a shared non-native herbivore, pest or pathogen (Fig. 8.2e; Chapter 17, this volume). In some cases, the non-native plant increases the population size of an herbivore, pest or pathogen that then reduces the population of a native plant competitor. The introduction of non-native biocontrol organisms is one avenue to apparent competition-mediated meltdowns (Pearson and Callaway, 2005, 2008). For example, in the midwestern USA, the non-native biocontrol weevil (*Rhinocyllus conicus*) had a three to five-fold increase in its attack rate on a native thistle (*Cirsium undulatum*) when the native thistle grew near the biocontrol's target non-native thistle (*Carduus nutans*; Rand and Louda, 2004). Non-native biocontrol gallflies (*Urophora* spp.) were introduced to control the invasive spotted knapweed (*Centaurea maculosa*) in western US grasslands. Populations of native deer mice (*Peromyscus maniculatus*) increased in abundance in knapweed patches because gallflies became a favoured food. Larger deer mouse populations increased predation rates on native plant seeds and decreased germination rates. This 'second order' apparent competition may indirectly benefit knapweed populations by reducing native forb competition (Pearson and Callaway, 2008).

Non-native plants may also harbour non-native pathogens that can 'spill over' to a native plant competitor (Strauss *et al.*, 2012; Flory and Clay, 2013), which may allow other non-native plants to invade. The invasive wild oat (*Avena fatua*) in western US grasslands is a reservoir of the non-native barley yellow dwarf virus and wild oat populations increase viral prevalence. Native grassland plants have reduced growth and fitness near virus-infected oat populations while other non-native species, like cheatgrass (*Bromus tectorum*), are unaffected by the virus (Power and Mitchell, 2004).

Apparent competition can also manifest when non-native herbivores preferentially browse native plants over co-occurring non-native plants (Parker *et al.*, 2006; de Villalobos *et al.*, 2011; Nuñez *et al.*, 2008; Relva *et al.*, 2010). Non-native herbivore-mediated apparent competition may be common between non-native and native plants; a meta-analysis of 63 published studies of 100 non-native plant species found that the presence of non-native herbivores enhanced the relative abundance of non-native plants by 65% while reducing the absolute abundance of native plants (Parker *et al.*, 2006). More specifically, in Patagonia, Argentina, a variety of large mammalian herbivores are altering the composition and abundance of native grassland and forest plant communities (Von Holle *et al.*, 2006; Relva *et al.*, 2010; de Villalobos *et al.*, 2011). In Patagonian grasslands where non-native feral horses preferentially grazed native plants, native plant cover and biomass decreased and non-native, unbrowsed *Pinus halepensis* seedlings were taller and had higher shoot mass relative to ungrazed locations (de Villalobos *et al.*, 2011). In Patagonian forests, non-native deer browse significantly reduced the growth of the native trees *Austrocedrus chilensis* and *Nothofagus dombeyi* but not the non-native trees *Pseudotsuga menziesii* and *Pinus ponderosa* (Nuñez *et al.*, 2008; Relva *et al.*, 2010). Non-native gypsy moth (*Lymantria dispar*) feeding preferences in eastern North America could indirectly promote the persistence of particular shrub species, including the non-native shrubs bush honeysuckle (*Lonicera maackii*) and Chinese privet (*L. sinense*) after large gypsy moth outbreaks (McEwan *et al.*, 2009). Similarly, the non-native snail *Pomacea maculata* intensively grazed native wetland plants in Texas, USA, which allowed the invasive herb alligatorweed *A. philoxeroides* to increase its abundance in the wetland (Meza-Lopes and Siemann, 2015).

Detecting apparent competition between native and non-native plants mediated through shared generalist herbivores may be empirically challenging. Ecologists typically use animal exclusion experiments to test the effects of herbivory, either using fences to exclude mammals or insecticides to exclude invertebrates. These coarse experimental treatments will exclude all native and non-native herbivores, making it difficult to distinguish the differences in the impacts of native vs non-native herbivores. This is likely why the few examples of these indirect interactions are either in areas where the predominant herbivores are all non-native – such as large mammals in Patagonia – or when the generalist non-native herbivore is the focus of the experimental study.

8.13 Indirect Mutualisms

Indirect mutualisms are defined as indirect positive effects of one species on another (Wootton, 1994; Fig. 8.2a). In the context of invasional meltdowns, two non-native plants may indirectly facilitate one another through a shared mutualist, herbivore or competitor. For example, non-native plants with large, showy floral displays may enhance pollination services of other non-native plants if they attract pollinators to their neighbour's flowers. This occurred in Chile for the pollen-limited *Carduus pycnocephalus* that has small, inconspicuous flowers. *Carduus* plants that grew near the large-flowered non-native *L. arboreus* had enhanced pollinator visitation and seed production relative when growing alone (Molina-Montenegro *et al.*, 2008). Non-native plants may also physically protect other non-native plants from herbivores. The invasive South African grass *Ehrharta calycina* escapes herbivory from a native herbivore, the black-tailed jackrabbit (*Lepus californicus*), in coastal foredunes in northern California, USA. When *E. calycina* is found near other non-native dune plants, *Ammophila arenaria* and *C. edulis*, it has lower herbivory, higher biomass and higher numbers of reproductive spikelets than in open areas of the dune (Cushman *et al.*, 2011).

Indirect mutualisms among non-native plants can also arise within competitive plant community networks if there is asymmetry in the strength of direct pairwise interactions (Stone and Roberts, 1991; Miller, 1994; White *et al.*, 2006; Levine *et al.*, 2017). Thus, non-native plants may indirectly facilitate one another if competitive interactions between native and non-native plants are stronger than competitive interactions between two or more non-native plants (Kuebbing and Nuñez, 2016). For example, the non-native herb garlic mustard (*Alliaria petiolata*) had higher biomass and more reproductive siliques in experimental plots with the non-native annual grass Japanese stiltgrass (*Microstegium vimineum*) relative to grass-free areas. The non-native grass indirectly facilitated the non-native herb by reducing native competitors within the plots (Flory and Bauer, 2014). In a mesocosm study of non-native plants common to the midwestern US tallgrass prairie, pairwise interactions among non-native plants were always competitive. However, the direct competitive effects of non-natives on one another were sometimes reduced when natives were present (Oschrin and Reynolds, 2019). Like many of these indirect mutualisms, it is impossible to assess how common or important asymmetric competition among co-occurring native and non-native plants is for promoting plant invasions. It would be premature to say that this mechanism is common or causing invasional meltdowns in plant communities, but it is rarely studied and warrants further research.

Importantly, indirect mutualisms among two or more non-native plants can lead to positive net effects even in the presence of *direct negative interactions* among non-native plants (Kuebbing and Nuñez, 2016). In other words, pairwise interactions among non-natives may be negative, but if the presence of native species alters those direct negative effects then the sum of all interactions in the community may be net positive and lead to non-native accumulation (Fig. 8.1e, Fig. 8.2d-f). Thus, just because studies of pairwise interactions among invaders find evidence of negative interactions (Kuebbing and Nuñez, 2015), these direct pairwise negative interactions may not manifest as non-native biotic resistance when those same two species are also interacting with other species (Ricciardi, 2001; Kuebbing and Nuñez, 2016). Ignoring the larger interaction web of co-occurring native and non-native species may inadvertently deemphasize the importance of indirect interactions and may potentially mask 'invasional meltdowns' occurring through positive indirect effects among non-natives. Thus, to truly assess the invasional meltdown paradigm, we need to consider how direct and indirect interactions in ecological communities comprised of many interacting non-native and native species shape invasion patterns and ecological impacts of non-native species.

8.14 Conclusions and Management Implications

The Invasional Meltdown Hypothesis was originally introduced as an alternative paradigm to a biotic resistance model of invasion (Simberloff and Von Holle, 1999). In a biotic resistance scenario, the probability of invasion by a non-native

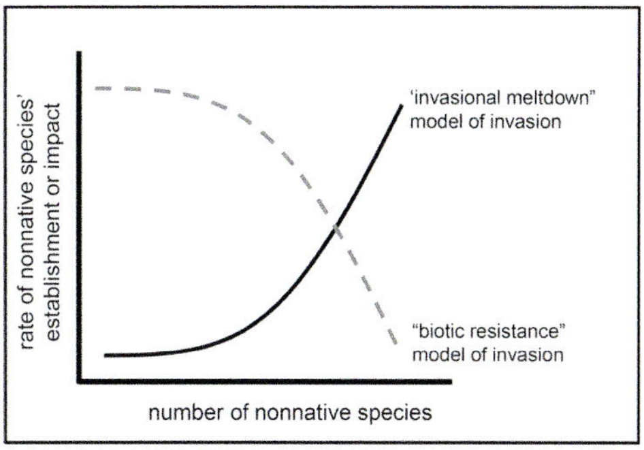

Fig. 8.3. The concept of invasional meltdown was originally proposed as an alternative to a biotic resistance model of invasion. As communities accumulate non-native species, resident non-native species could either facilitate or inhibit the establishment, population growth or spread of newly invading non-native species. In a biotic resistance invasion model, antagonistic interactions among non-natives would decrease the rate or probability of invasion by a new non-native (dashed line). In an invasional meltdown model, facilitative interactions among non-natives would increase the rate or probability of invasion by a new non-native (solid line).

species should become progressively *less* likely with an increasing number of non-native species in a community (Fig. 8.3). In an invasional meltdown scenario, the probability of invasion by a non-native species should become progressively *more* likely with an increasing number of invasive species in a community (Fig. 8.3). Differentiating between biotic resistance and invasional meltdown models is imperative for informing management decisions. For example, if non-native species are likely to repel other invaders, a manager may decide to allow a current non-native species to persist to decrease the likelihood of invasion by a potentially more damaging species. However, if non-natives are likely to promote the invasion of other non-natives, the removal of non-natives from the ecosystem may become a high priority for preventing a rapid accumulation of more non-natives.

Interestingly, current management of non-native plant species may provide evidence for determining whether plant invasions are hindered or facilitated by other non-natives. Frequently, the removal of a dominant non-native plant is followed by the invasion of another non-native plant and not the recovery of native plants (Kettenring and Adams, 2011; Pearson *et al.*, 2016a). This phenomenon has been termed a 'secondary invasion' (Pearson *et al.*, 2016b) or 'invasion treadmill' (Thomas and Reid, 2007) and suggests that non-native plants are suppressing or competitively excluding other non-native plants from that community. At face value, this supports the biotic resistance paradigm. However, there is abundant evidence that native plants are *also* competitively suppressed by non-native plants (Vilà and Weiner, 2004; Kuebbing and Nuñez, 2016), yet native plants typically do not increase in abundance after removal of dominant natives (Kettenring and Adams, 2011; Pearson *et al.*, 2016a). Native plants may be limited by the size of their soil seed banks, dispersal ability or other disturbances that exclude them from the ecosystem. Another plausible hypothesis is that dominant non-native invasive plants are competitively suppressing resident native and non-native plants, but the strength of this competitive suppression differs between non-natives and natives (Kuebbing and Nuñez, 2016). In other words, native plants may be indirectly facilitating the co-occurrence of multiple non-natives by mediating competitive interactions between non-natives (Stone and Roberts, 1991; Miller, 1994; Flory and Bauer, 2014; Northfield *et al.*, 2018). Under this scenario, you would still see the same pattern of secondary invasions and limited response of natives, but the net community effect would be promotion of multiple invasive plant species.

Preventing invasional meltdowns should be a top management priority. Yet, extracting management recommendations from this collection of examples is challenging because of the wide variety of ways non-native species may interact with one another. However, a few trends emerged from these examples that may reduce the likelihood of plant invasional meltdown:

- *Prevent the introduction of plant soil microbial mutualists:* a major pathway of plant microbes is the live-plant trade, which can introduce plants with their mutualist soil organisms that could facilitate their invasion or soil pathogens that could 'spill over' to infect native species (Liebhold *et al.*, 2012). It seems prudent to reduce, cautiously monitor or prohibit live-plant imports that vector soil microbes.

- *Prioritize prevention, early detection and control of 'repeat offenders':* there are some non-native taxa that are implicated in facilitative plant interactions in many different ecosystems around the globe. Some of the most frequently cited species in plant invasional meltdowns include the generalist pollinators European honeybee (*A. mellifera*) and bumblebee (*Bombus* spp.), ecosystem-modifying non-native earthworms and wild boar (*S. scrofa*), the widespread seed-dispersing songbirds the Japanese white-eye (*Zosterops japonicus*) and common bulbul bird (*P. jocosus*), as well as 'indirect mutualists' like the Argentine ant (*L. humile*).
- *Prioritize early detection of non-native plants that cannot spread because they lack their mutualists:* the vast majority of the global naturalized non-native flora are grown in domestic and botanical gardens (van Kleunen *et al.*, 2018). These cultivated plants provide an opportunity to test the likelihood a species may spread beyond their cultivated boundaries. If non-native plants are severely pollen-limited – suggesting they lack the necessary native pollinator mutualism – these plants could be prioritized for removal before a non-native pollinator is introduced (Bufford and Daehler, 2014; Moodley *et al.*, 2016).
- *Disrupt mutualisms to improve management efficacy:* there are many non-native plant mutualists, like introduced honeybees and bumblebees, that facilitate plant invasion. While complete eradication of these agriculturally important pollinators may be unrealistic, a reduction in their population sizes could facilitate non-native plant management. For example, in New Zealand, the efficacy of a seed-feeding biocontrol beetle (*Bruchidius villosus*) on the non-native Scotch broom (*C. scoparius*) increased when beehives were removed from fields near Scotch broom populations (Paynter *et al.*, 2010). Together, reduced mutualism and increased seed predation more effectively managed broom populations.
- *Manage more than one species at a time:* there is growing evidence that the removal of one non-native allows other non-natives to invade or increase in abundance (Courchamp *et al.*, 2011; Ballari *et al.*, 2016; Pearson *et al.*, 2016b). Unless interactions among multiple non-native species within a community are well-known, managers should be prepared for 'surprise effects', 'secondary invasions' or 'invasion treadmills' after management. When possible, managing multiple non-native species could reduce the likelihood of these adverse management outcomes.
- *Combine management and research more effectively:* removal experiments of non-native plants may be a potentially powerful means to test direct and indirect interactions within invaded communities. Removal experiments are used to understand species interactions in natural communities (Schmitz, 1997; Vázquez *et al.*, 2015; Kumschick *et al.*, 2015; Novak *et al.*, 2016) and may be particularly useful for invasion biology for a few of reasons. First, ethical issues surrounding introducing or manipulating non-native populations in natural communities is a limitation in invasion biology research. Removal experiments are the natural foil to introduction experiments and provide an ethical alternative to this problem. Second, non-natives are frequently removed by land managers concerned about non-native species in their conservation areas. Invasion biologists could capitalize on these removals by partnering with land managers to design experiments that both reduce the non-native abundance and teach us about non-native interactions.

References

Abe, T., Wada, K., Kato, Y., Makino, S. and Okochi, I. (2011) Alien pollinator promotes invasive mutualism in an insular pollination syndrome. *Biological Invasions* 13, 957–967.

Aizen, M.A., Morales, C.L. and Morales, J.M. (2008) Invasive mutualists erode native pollination webs. *PLoS Biology* 6, e31.

Ambrosini, A., de Souza, R. and Passaglia, L.M.P. (2016) Ecological role of bacterial inoculants and their potential impact on soil microbial diversity. *Plant and Soil* 400, 193–207.

Andrews, M., De Meyer, S., James, E.K., Stępkowski, T., Hodge, S. *et al.* (2018) Horizontal transfer of symbiosis genes within and between rhizobial genera: occurrence and importance. *Genes* 9, 321.

Andrus, A.D., Andam, C. and Parker, M.A. (2012) American origin of *Cupriavidus* bacteria associated with invasive *Mimosa* legumes in the Philippines. *FEMS Microbiology Ecology* 80, 747–750.

Baker, H.G. (1965) Characteristics and modes of origin of weeds. In: Baker, H.G. and Stebbins, G.L. (eds) *The Genetics of Colonizing Species*. Academic Press, New York, USA, pp. 147–172.

Baker, H.G. (1974) The evolution of weeds. *Annual Review of Systematics and Evolution* 5, 1–24.

Ballari, S.A., Kuebbing, S.E. and Nuñez, M.A. (2016) Potential problems of removing one invasive species at a time: a meta-analysis of the interactions between invasive vertebrates and unexpected effects of removal programs. *PeerJ* 4, e2029.

Barrios-Garcia, M.N. and Ballari, S.A. (2012) Impact of wild boar (*Sus scrofa*) in its introduced and native range: A review. *Biological Invasions* 14, 2283–2300.

Barrios-Garcia, M.N. and Simberloff, D. (2013) Linking the pattern to the mechanism: how an introduced mammal facilitates plant invasions. *Austral Ecology* 38, 884–890.

Bartomeus, I., Vil, M. and Santamaría, L. (2008) Contrasting effects of invasive plants in plant-pollinator networks. *Oecologia* 155, 761–770.

Bartuszevige, A.M. and Gorchov, D.L. (2006) Avian seed dispersal of an invasive shrub. *Biological Invasions* 8, 1013–1022.

Beavon, M.A. and Kelly, D. (2012) Invasional meltdown: Pollination of the invasive liana *Passiflora tripartita* var. *mollissima* (Passifloraceae) in New Zealand. *New Zealand Journal of Ecology* 36, 100–107.

Beavon, M.A. and Kelly, D. (2015) Dispersal of banana passionfruit (*Passiflora tripartita* var. *mollissima*) by exotic mammals in New Zealand facilitates plant invasiveness. *New Zealand Journal of Ecology* 39, 43–49.

Belote, R.T. and Jones, R.H. (2009) Tree leaf litter composition and non-native earthworms influence plant invasion in experimental forest floor mesocosms. *Biological Invasions* 11, 1045–1052.

Best, R.J. and Arcese, P. (2009) Exotic herbivores directly facilitate the exotic grasses they graze: mechanisms for an unexpected positive feedback between invaders. *Oecologia* 159, 139–150.

Birnbaum, C., Barrett, L.G., Thrall, P.H. and Leishman, M.R. (2012) Mutualisms are not constraining cross-continental invasion success of *Acacia* species within Australia. *Diversity and Distributions* 18, 962–976.

Birnbaum, C., Bissett, A., Thrall, P.H. and Leishman, M.R. (2014) Invasive legumes encounter similar soil fungal communities in their non-native and native ranges in Australia. *Soil Biology and Biochemistry* 76, 210–217. DOI: 10.1016/j.soilbio.2014.05.017.

Birnbaum, C., Bissett, A., Thrall, P.H. and Leishman, M.R. (2016) Nitrogen-Fixing bacterial communities in invasive legume nodules and associated soils are similar across introduced and native range populations in Australia. *Journal of Biogeography* 43, 1631–1644.

Bobadilla, S.Y., Benitez, V.V. and Guichón, M.L. (2016) Asiatic *Callosciurus* squirrels as seed dispersers of exotic plants in the Pampas. *Current Zoology* 62, 215–219.

Bourgeois, K., Suehs, C.M., Vidal, E. and Médail, F. (2005) Invasional meltdown potential: facilitation between introduced plants and mammals on French Mediterranean islands. *Ecoscience* 12(2), 248–256. DOI: 10.2980/i1195-6860-12-2-248.1.

Bourgeois, K., Vidal, E., Suehs, C.M. and Médail, F. (2004) Extreme invasional meltdown: multi-trophic interactions catalyse Mediterranean island invasions. In: Rhodes., Greece., Arianoutsou. and PApanastasis. (eds) *Proceedings 10th MEDECOS Conference, April 25- May 1*. Millpress, Rotterdam, UK, pp. 1–5.

Braga, R.R., Gómez-Aparicio, L., Heger, T., Vitule, J.R.S. and Jeschke, J.M. (2018a) Invasional meltdown hypothesis. In: Jeschke, J.M. and Heger, T. (eds) *Invasion Biology: Hypotheses and Evidence*. CABI, Wallingford, pp. 79–91.

Braga, R.R., Gómez-Aparicio, L., Heger, T., Vitule, J.R.S. and Jeschke, J.M. (2018b) Structuring evidence for invasional meltdown: broad support with biases and gaps. *Biological Invasions* 20, 923–936.

Brooks, M.L., D'Antonio, C.M., Richardson, D.M., Grace, J.B., Keeley, J.E. *et al.* (2004) Effects of invasive alien plants on fire regimes. *BioScience* 54, 677–688.

Bufford, J.L. and Daehler, C.C. (2014) Sterility and lack of pollinator services explain reproductive failure in non-invasive ornamental plants. *Diversity and Distributions* 20, 975–985.

Calvino-Cancela, M. (2011) Seed dispersal of alien and native plants by vertebrate herbivores. *Biological Invasions* 13, 895–904.
Cameron, E.K., Vil, M. and Cabeza, M. (2016) Global meta-analysis of the impacts of terrestrial invertebrate invaders on species, communities and ecosystems. *Global Ecology and Biogeography* 25, 596–606.
Carino, D.A. and Daehler, C.C. (2002) Can inconspicuous legumes facilitate alien grass invasions? Partridge peas and fountain grass in Hawai'i. *Ecography* 25, 33–41.
Catford, J.A., Jansson, R. and Nilsson, C. (2009) Reducing redundancy in invasion ecology by integrating hypotheses into a single theoretical framework. *Diversity and Distributions* 15, 22–40.
Chen, W.M., James, E.K., Chou, J.S., Sheu, S.Y., Yang, S.Z. *et al.* (2005) β-Rhizobia from *Mimosa pigra*, a newly discovered invasive plant in Taiwan. *New Phytologist* 168, 661–675.
Chimera, C.G. and Drake, D.R. (2010) Patterns of seed dispersal and dispersal failure in a Hawaiian dry forest having only introduced birds. *Biotropica* 42, 493–502.
Clause, J., Forey, E.L., C.J., Lambert, A.M. and Barot, S. (2015) Non-Native earthworms promote plant invasion by ingesting seeds and modifying soil properties. *Acta Oecologica* 64, 10–20.
Clergeau, P. and Mandon-Dalger, I. (2001) Fast colonization of an introduced bird: the case of *Pycnonotus jocosus* on the Mascarene Islands. *Biotropica* 33, 542–546.
Constible, J.M., Sweitzer, R.A., Van Vuren, D.H., Schuyler, P.T. and Knapp, D.A. (2005) Dispersal of non-native plants by introduced bison in an island ecosystem. *Biological Invasions* 7, 699–709.
Courchamp, F., Caut, S., Bonnaud, E., Bourgeois, K., Angulo, E. *et al.* (2011) Eradication of alien species: surprise effects and conservation successes. In: Veitch, C.R., Clout, M.N. and Towns, D.R. (eds) *Island Invasives: Eradication and Management. IUCN, Gland*. Switzerland, pp. 285–289.
Craven, D., Thakur, M.P., Cameron, E.K., Frelich, L.E.R., Beauséjour, R.B. *et al.* (2017) The unseen invaders: introduced earthworms as drivers of change in plant communities in North America forests (a meta-analysis). *Global Change Biology* 23, 1065–1074.
Crisóstomo, J.A., Rodríguez-Echeverría, S. and Freitas, H. (2013) Co-introduction of exotic rhizobia to the rhizosphere of the invasive legume *Acacia saligna*, an intercontinental study. *Applied Soil Ecology* 64, 118–126.
Cushman, J.H., Lortie, C.J. and Christian, C.E. (2011) Native herbivores and plant facilitation mediate the performance and distribution of an invasive exotic grass. *Journal of Ecology* 99, 524–531.
Dávalos, A., Nuzzo, V. and Blossey, B. (2015) Single and interactive effects of deer and earthworms on non-native plants. *Forest Ecology and Management* 351, 28–35.
Davis, N.E., Forsyth, D.M. and Coulson, G. (2010) Facilitative interactions between an exotic mammal and native and exotic plants: hog deer (*Axis porcinus*) as seed dispersers in south-eastern Australia. *Biological Invasions* 12, 1079–1092.
de Villalobos, A.E., Zalba, S.M. and Peláez, D.V. (2011) *Pinus halepensis* invasion in mountain pampean grassland: effects of feral horses grazing on seedling establishment. *Environmental Research* 111, 953–959.
Dickie, I.A., Bolstridge, N., Cooper, J.A. and Peltzer, D.A. (2010) Co-invasion by *Pinus* and its mycorrhizal fungi. *New Phytologist* 187, 475–484.
Ehrenfeld, J.G. (2010) Ecosystem consequences of biological invasions. *Annual Review of Ecology, Evolution and Systematics* 41, 59–80.
Ellison, A.M., Bank, M.S., Clinton, B.D., Colburn, E.A., Elliott, K. *et al.* (2005) Loss of foundation species: consequences for the structure and dynamics of forested ecosystems. *Frontiers in Ecology and the Environment* 3, 479–486.
Flory, S.L. and Bauer, J.T. (2014) Experimental evidence for indirect facilitation among invasive plants. *Journal of Ecology* 102(1), 12–18.
Flory, S.L. and Clay, K. (2013) Pathogen accumulation and long-term dynamics of plant invasions. *Journal of Ecology* 101, 607–613.
Foster, J.T. and Robinson, S.K. (2007) Introduced birds and the fate of Hawaiian rainforests. *Conservation Biology* 21, 1248–1257.
Gaertner, M., Biggs, R.T.B., Hui, M., Molofsky, J, C. and Richardson, D.M. (2014) Invasive plants as drivers of regime shifts: identifying high-priority invaders that alter feedback relationships. *Diversity and Distributions* 20, 733–744.
Godoy, O. (2019) Coexistence theory as a tool to understand biological invasions in species interaction networks: implications for the study of novel ecosystems. *Functional Ecology* 33, 1190–1201.

González-Muñoz, N., Costa-Tenorio, M. and Espigares, T. (2012) Invasion of alien *Acacia dealbata* on Spanish *Quercus robur* forests: Impact on soils and vegetation. *Forest Ecology and Management* 269, 214–221.

Goulson, D. (2003) Effects of introduced bees on native ecosystems. *Annual Review of Ecology, Evolution and Systematics* 34, 1–26.

Green, P.T., Lake, P.S. and O'Dowd, D.J. (2004) Resistance of island rainforest to invasion by alien plants: influence of microhabitat and herbivory on seedling performance. *Biological Invasions* 6, 1–9.

Green, P.T., O'Dowd, D.J., Abbott, K.L., Jeffery, M., Retallick, K. *et al.* (2011) Invader-invader mutualism facilitates a secondary invasion. *Ecology* 92, 1758–1768.

Gurevitch, J. (2006) Commentary on Simberloff (2006): Meltdowns, snowballs and positive feedbacks. *Ecology Letters* 9, 919–920.

Hayward, J., Horton, T.R. and Nuñez, M.A. (2015a) Ectomycorrhizal fungal communities coinvading with Pinaceae host plants in Argentina: Gringos bajo el bosque. *New Phytologist* 208, 497–506.

Hayward, J., Horton, T.R., Pauchard, A. and Nuñez, M.A. (2015b) A single ectomycorrhizal fungal species can enable a *Pinus* invasion. *Ecologist* 96, 1438–14441.

Hendrix, P.F. and Bohlen, P.J. (2002) Exotic earthworm invasions in North America: ecological and policy implications. *BioScience* 52, 801–811.

Heneghan, L., Steffen, J. and Fagen, K. (2007) Interactions of an introduced shrub and introduced earthworms in an Illinois urban woodland: impact on leaf litter decomposition. *Pedobiologia* 50, 543–551.

Horn, K., Parker, I.M., Malek, W., Rodríguez-Echeverría, S. and Parker, M.A. (2014) Disparate origin of *Bradyrhizobium* symbionts for invasive populations of *Cytisus scoparius* (Leguminosae) in North America. *FEMS Microbiology Ecology* 89, 89–98.

Hulme, P.E. (2009) Trade, transport and trouble: managing invasive species pathways in an era of globalization. *Journal of Applied Ecology* 46, 10–18.

Hynson, N.A., Merckx, V.S.F.T., Perry, B.A. and Treseder, K.K. (2013) Identities and distributions of the co-invading ectomycorrhizal fungal symbionts of exotic pines in the Hawaiian Islands. *Biological Invasions* 15, 2373–2385.

Jackson, M.C. (2015) Interactions among multiple invasive animals. *Ecology* 96, 2035–2041.

James, S.W. and Hendrix, P.F. (2004) Invasion of exotic earthworms into North America and other regions. In: Edwards, C.A. (ed.) *Earthworm Ecology*, 2nd edn. CRC Press, Boca Raton, Florida, USA, pp. 75–88.

Jones, C.G., Lawton, J.H. and Shachak, M. (1994) Organisms as ecosystem engineers. *Oikos* 69, 373–386.

Kawakami, K., Mizusawa, L. and Higuchi, H. (2009) Re-established mutualism in a seed-dispersal system consisting of native and introduced birds and plants on the Bonin Islands, Japan. *Ecological Research* 24, 741–748.

Kenis, M., Roques, A., Santini, A. and Liebhold, A.M. (2017) Impact of non-native invertebrates and pathogens on market forest tree resources. In: Vil, M. and Hulme, P.E. (eds) *Impact of Biological Invasions on Ecosystem Services*. Springer, Berlin, pp. 103–117.

Kettenring, K.M. and Adams, C.R. (2011) Lessons learned from invasive plant control experiments: a systematic review and meta-analysis. *Journal of Applied Ecology* 48, 970–979.

Klooster, W.S., Gandhi, K.J.K., Long, L.C., Perry, K.I., Rice, K.B. *et al.* (2018) Ecological impacts of emerald ash borer in forests at the epicenter of the invasion in North America. *Forests* 9, 250.

Kourtev, P.S., Huang, W.Z. and Ehrenfeld, J.G. (1999) Differences in earthworm densities and nitrogen dynamics in soils under exotic and native plant species. *Biological Invasions* 1, 237–245.

Kuebbing, S.E., Classen, A.T. and Simberloff, D. (2014) Two co-occurring invasive woody shrubs alter soil properties and promote subdominant invasive species. *Journal of Applied Ecology* 51, 124–133.

Kuebbing, S.E. and Nuñez, M.A. (2015) Negative, neutral, and positive interactions among nonnative plants: patterns, processes, and management implications. *Global Change Biology* 21, 926–934.

Kuebbing, S.E. and Nuñez, M.A. (2016) Invasive non-native plants have a greater effect on neighboring natives than other non-natives. *Nature Plants* 2, 16134.

Kuebbing, S.E., Nuñez, M.A. and Simberloff, D. (2013) Current mismatch between research and conservation efforts: the need to study co-occurring invasive plant species. *Biological Conservation* 160, 121–129. DOI: 10.1016/j.biocon.2013.01.009.

Kuebbing, S.E., Patterson, C.M., Classen, A.T. and Simberloff, D. (2016) Co-occurring nonnative woody shrubs have additive and non-additive soil legacies. *Ecological Applications* 26(6), 1896–1906. DOI: 10.1890/15-1931.1.

Kumschick, S., Gaertner, M., Vil, M., Essl, F., Pyšek, J.M. *et al.* (2015) Ecological impacts of alien species: quantification, scope, caveats, and recommendations. *BioScience* 65, 55–63.

La Pierre, K.J., Simms, E.L., Tariq, M., Zafar, M. and Porter, S.S. (2017) Invasive legumes can associate with many mutualists of native legumes, but usually do not. *Ecology and Evolution* 7, 8599–8611.

Lafleur, N.E., Rubega, M.A. and Elphick, C.S. (2007) Invasive fruits, novel foods, and choice: an investigation of European Starling and American Robin frugivory. *The Wilson Journal of Ornithology* 119, 429–438.

Lawrence, B., Fisk, M.C., Fahey, T.J. and Suárez, E.R. (2003) Influence of nonnative earthworms on mycorrhizal colonization of sugar maple (*Acer saccharum*). *New Phytologist* 157, 145–153.

Le Roux, J.J., Hui, C., Keet, J.H. and Ellis, A.G. (2017) Co-introduction vs ecological fitting as pathways to the establishment of effective mutualisms during biological invasions. *New Phytologist* 215, 1354–1360.

Levine, J.M., Bascompte, J., Adler, P.B. and Allesina, S. (2017) Beyond pairwise mechanisms of species coexistence in complex communities. *Nature* 546, 56–64.

Liao, C., Peng, R., Luo, Y., Zhou, X., Fang, C. *et al.* (2008) Altered ecosystem carbon and nitrogen cycles by plant invasion: a meta-analysis. *New Phytologist* 177, 706–714.

Liebhold, A.M., Brockerhoff, E.G., Garrett, L.J., Parke, J.L. and Britton, K.O. (2012) Live plant imports: the major pathway for forest insect and pathogen invasions of the US. *Frontiers in Ecology and the Environment* 10, 135–143.

Linnebjerg, J.F., Hansen, D.M., Bunbury, N. and Olesen, J.M. (2010) Diet composition of the invasive red-whiskered bulbul *Pycnonotus jocosus* in Mauritius. *Journal of Tropical Ecology* 26, 347–350.

Litchman, E. (2010) Invisible invaders: non-pathogenic invasive microbes in aquatic and terrestrial ecosystems. *Ecology Letters* 13, 1560–1572.

Liu, X.Y., Wei, S., Wang, F., James, E.K., Guo, X.Y. *et al.* (2012) *Burkholderia* and *Cupriavidus* spp. are the preferred symbionts of *Mimosa* spp. in Southern China. *FEMS Microbiology Ecology* 80, 417–426.

Lloyd, G., Mahon, M.B. and Crist, T.O. (2019) Invasive shrub cover and tree species composition influence exotic earthworms. *Forest Ecology and Management* 447, 53–59.

Lobe, J.W., Callaham Jr., M.A., Hendrix, P.F. and Hanula, J.L. (2014) Removal of an invasive shrub (Chinese privet: *Ligustrum sinense* Lour) reduces exotic earthworm abundance and promotes recovery of native North American earthworms. *Applied Soil Ecology* 83, 133–139.

Lovett, G.M., Weiss, M., Liebhold, A.M., Holmes, T.P., Leung, B. *et al.* (2016) Nonnative forest insects and pathogens in the United States: impacts and policy options. *Ecological Applications* 26, 1437–1455.

MacFarlane, A.E.T., Kelly, D. and Briskie, J.V. (2016) Introduced blackbirds and song thrushes: useful substitutes for lost mid-sized native frugivores. *or weed vectors? New Zealand Journal of Ecology* 40, 80–87.

Madritch, M.D. and Lindroth, R.L. (2009) Removal of invasive shrubs reduces exotic earthworm populations. *Biological Invasions* 11, 663–671.

Maezo, M.J., Fournier, H. and Beisner, B.E. (2010) Potential and realized interactions between two aquatic invasive species: Eurasian watermilfoil (*Myriophyllum spicatum*) and rusty crayfish (*Orconectes rusticus*). *Canadian Journal of Fisheries and Aquatic Science* 67, 684–700.

Mandon-Dalger, I., Clergeau, P., Tassin, J., Rivière, J.N. and Gatti, S. (2004) Relationships between alien plants and an alien bird species on reunion island. *Journal of Tropical Ecology* 20, 635–642.

Mariotte, P., Spotswood, E.N., Farrer, E.C. and Suding, K.N. (2017) Positive litter feedbacks of an introduced species reduce native diversity and promote invasion in California grasslands. *Applied Vegetation Science* 20, 28–39.

Martin, P.H., Canham, C.D. and Marks, P.L. (2008) Why forests appear resistant to exotic plant invasions: intentional introductions, stand dynamics, and the role of shade tolerance. *Frontiers in Ecology and the Environment* 7, 142–149.

Martin-Albarracin, V.L., Amico, G.C., Simberloff, D. and Nuñez, M.A. (2015) Impact of non-native birds on native ecosystems: a global analysis. *PLoS ONE* 10, e0143070.

McEwan, R.W., Rieske, L.K. and Arthur, M.A. (2009) Potential interactions between invasive woody shrubs and the gypsy moth (*Lymantria dispar*), an invasive insect herbivore. *Biological Invasions* 11, 1053–1058.

McGinn, K.J., van der Putten, W.H., Duncan, R.P., Shelby, N., Weser, C. *et al.* (2016) *Trifolium* species associate with a similar richness of soil-borne mutualists in their introduced and native ranges. *Journal of Biogeography* 43, 944–954.

McKey, D.B. and Kaufmann, S.C. (1991) Naturalization of exotic *Ficus* species (Moraceae) in South Florida. In: Center, T.D., Doren, R.F., Hofstetter, F.L., Myers, R.L. and Whiteaker, L.D. (eds) *Proceedings of the Symposium on Exotic Pest Plants*. Technical Report NPS/NREVER/NRTR-91/06.

Meza-Lopes, M.M. and Siemann, E. (2015) Experimental test of the Invasional Meltdown hypothesis: an exotic herbivore facilitates an exotic plant, but the plant does not reciprocally facilitate the herbivore. *Freshwater Biology* 60, 1475–1482.

Miller, T.E. (1994) Direct and indirect species interactions in an early old-field plant community. *American Naturalist* 143, 1007–1025.

Molina-Montenegro, M.A., Badano, E.I. and Cavieres, L.A. (2008) Positive interactions among plant species for pollinator service: assessing the 'magnet species' concept with invasive species. *Oikos* 117, 1833–1839.

Monks, A.M., Lishawa, S.C., Wellons, K.C., Albert, D.A., Mudrzynski, B. *et al.* (2019) European frogbit (*Hydrocharis morsus-ranae*) invasions facilitated by non-native cattails (*Typha*) in the Laurentian Great Lakes. *Journal of Great Lakes Research* 45, 912–920.

Moodley, D., Geerts, S., Richardson, D.M. and Wilson, J.R.U. (2016) The importance of pollinators and autonomous self-fertilisation in the early stages of plant invasions: Banksia and Hakea (Proteaceae) as case studies. *Plant Biology* 18(1), 124–131. DOI: 10.1111/plb.12334.

Morales, C.L. and Aizen, M.A. (2006) Invasive mutualisms and the structure of plant-pollinator interactions in the temperate forests of north-west Patagonia, Argentina. *Journal of Ecology* 94(1), 171–180. DOI: 10.1111/j.1365-2745.2005.01069.x.

Morales, C.L., Sáez, A., Garibaldi, L.A. and Aizen, M.A. (2017) Disruption of pollination services by invasive pollinator species. In: Vil, M. and Hulme, P.E. (eds) *Impact of Biological Invasions on Ecosystem Services*. Springer, Berlin, pp. 203–220.

Mueller, K.E., Lodge, A.G., Roth, A.M., Whitfield, T.J.S., Hobbie, S.E. *et al.* (2018) A tale of two studies: detection and attribution of the impacts of invasive plants in observational surveys. *Journal of Applied Ecology* 55, 1780–1789.

Ndlovu, J., Richardson, D.M., Wilson, J.R.U. and Le Roux, J.J. (2013) Co-invasion of South African ecosystems by and Australian legume and its rhizobial symbionts. *Journal of Biogeography* 40, 1240–1251.

Northfield, T.D., Laurance, S.G.W., Mayfield, M.M., Paini, D.R., Snyder, W.E. *et al.* (2018) Native turncoats and indirect facilitation of species invasions. *Proceedings of the Royal Society B* 285, 20171936.

Novak, M., Yeakel, J.D., Noble, A.E., Doak, D.F., Emmerson, M. *et al.* (2016) Characterizing species interactions to understand press perturbations: what is the community matrix? *Annual Review in Ecology, Evolution. and Systematics* 47, 409–432.

Nuñez, M.A. and Dickie, I.A. (2014) Invasive belowground mutualists of woody plants. *Biological Invasions* 16, 645–661.

Nuñez, M.A., Horton, T.R. and Simberloff, D. (2009) Lack of belowground mutualisms hinders Pinaceae invasions. *Ecology* 90(9), 2352–2359. DOI: 10.1890/08-2139.1.

Nuñez, M.A., Hayward, J., Horton, T.R., Amico, G.C., Dimarco, R.D. *et al.* (2013) Exotic mammals disperse exotic fungi that promote invasion by exotic trees. *PLoS ONE* 8(6), e66832. DOI: 10.1371/journal.pone.0066832.

Nuñez, M.A., Relva, M.A. and Simberloff, D. (2008) Enemy release or invasional meltdown? deer preference for exotic and native trees on Isla Victoria, Argentina. *Austral Ecology* 33(3), 317–323. DOI: 10.1111/j.1442-9993.2007.01819.x.

Nuzzo, V.A., Maerz, J.C. and Blossey, B. (2009) Earthworm invasion as the driving force behind plant invasion and community change in northeastern North American forests. *Conservation Biology* 23(4), 966–974. DOI: 10.1111/j.1523-1739.2009.01168.x.

Oschrin, E. and Reynolds, H.L. (2019) Co-Occurring invasive plant interactions do not predict the impacts of invasion in experimental tallgrass prairie. *Biological Invasions* 21, 2417–2430.

Padrón, B., Traveset, A., Biedenweg, T., Díaz, D., Nogales, M. *et al.* (2009) Impact of alien plant invaders on pollination networks in two archipelagos. *PLoS ONE* 4(7), e6275. DOI: 10.1371/journal.pone.0006275.

Parker, J.D., Burkepile, D.E. and Hay, M.E. (2006) Opposing effects of native and exotic herbivores on plant invasions. *Science* 311(5766), 1459–1461. DOI: 10.1126/science.1121407.

Paudel, S., Longcore, T., MacDonald, B., McCormick, M.K., Szlavecz, K. *et al.* (2016) Belowground interactions with aboveground consequences: invasive earthworms and arbuscular mycorrhizal fungi. *Ecology* 97(3), 605–614. DOI: 10.1890/15-1085.

Paynter, Q., Main, A., Gourlay, A.H., Peterson, P.G., Fowler, S.V. *et al.* (2010) Disruption of an exotic mutualism can improve management of an invasive plant: Varroa mite, honeybees and biological control of Scotch broom *Cytisus scoparius* in New Zealand. *Journal of Applied Ecology* 47, 309–317.

Pearson, D.E. and Callaway, R.M. (2005) Indirect nontarget effects of host-specific biological control agents: implications for biological control. *Biological Control* 35(3), 288–298.
Pearson, D.E. and Callaway, R.M. (2008) Weed-biocontrol insects reduce native-plant recruitment through second-order apparent competition. *Ecological Applications* 18(6), 1489–1500.
Pearson, D.E., Ortega, Y.K., Runyon, J.B. and Butler, J.L. (2016a) Secondary invasions: the bane of weed management. *Biological Conservation* 197, 8–17.
Pearson, D.E., Ortega, Y.K., Eren, Özkan. and Hierro, J.L. (2016b) Quantifying "apparent" impact and distinguishing impact from invasiveness in multispecies plant invasions. *Ecological Applications* 26(1), 162–173. DOI: 10.1890/14-2345.
Perez-Ramirez, N.O., Rogel, M.A., Wang, E.T., Castellanos, J.Z. and Martinez-Romero, E. (1998) Seeds of *Phaseolus vulgaris* bean carry *Rhizobium etli*. *FEMS Microbiology Ecology* 26, 289–296.
Potgieter, L.J., Wilson, J.R.U., Strasberg, D. and Richardson, D.M. (2014) *Casuarina* invasion alters primary succession on lava flows on La Réunion Island. *Biotropica* 46, 268–275.
Power, A.G. and Mitchell, C.E. (2004) Pathogen spillover in disease epidemics. *The American Naturalist* 164 Suppl 5, 79–89. DOI: 10.1086/424610.
Pringle, A., Bever, J.D., Gardes, M., Parrent, J.L., Rillig, M.C. *et al.* (2009) Mycorrhizal symbioses and plant invasions. *Annual Review of Ecology, Evolution, and Systematics* 40, 699–715.
Prior, K.M., Robinson, J.M., Meadley Dunphy, S.A. and Frederickson, M.E. (2015) Mutualism between co-introduced species facilitates invasion and alters plant community structure. *Proceedings. Biological sciences* 282(1800), 20142846. DOI: 10.1098/rspb.2014.2846.
Quackenbush, P.M., Butler, R.A., Emery, N.C., Jenkins, M.A., Kladivko, E.J. *et al.* (2012) *Lumbricus terrestris* prefers to consume garlic mustard (*Alliaria petiolata*) seeds. *Invasive Plant Science and Management* 5, 148–154.
Ramírez, B.W. and Montero, J.S. (1988) *Ficus microcarpa* L., *F. benjamina* L. and other species introduced in the New World, their pollinators (Agaonidae) and other fig wasps. *Revista de Biologia Tropical* 36, 441–446.
Rand, T.A. and Louda, S.M. (2004) Exotic weed invasion increases the susceptibility of native plants to attack by a biocontrol herbivore. *Ecology* 85(6), 1548–1554.
Relva, M,A., Nuñez, M.A. and Simberloff, D. (2010) Introduced deer reduce native plant cover and facilitate invasion of non-native tree species: evidence for invasional meltdown. *Biological Invasions* 12, 303–311.
Remigi, P., Zhu, J., Young, J.P.W. and Masson-Boivin, C. (2016) Symbiosis within symbiosis: evolving nitrogen-fixing legume symbionts. *Trends in Microbiology* 24(1), 63–75. DOI: 10.1016/j.tim.2015.10.007.
Ricciardi, A. (2001) Facilitative interactions among aquatic invaders: is an "invasional meltdown" occurring in the Great Lakes? *Canadian Journal of Fisheries and Aquatic Science* 58, 2513–2525.
Richardson, D.M., Allsopp, N., D'Antonio, C.M., Milton, S.J. and Rejmánek, M. (2000) Plant invasions--the role of mutualisms. *Biological Reviews* 75(1), 65–93. DOI: 10.1017/s0006323199005435.
Richardson, D.M. and Higgins, S.I. (1998) Pines as invaders in the southern hemisphere. In: Richardson, D.M. (ed.) *Ecology and Biogeography of Pinus*. Cambridge University Press, Cambridge, UK, pp. 450–474.
Richardson, D.M. and Rejmánek, M. (2011) Trees and shrubs as invasive alien species – a global review. *Diversity and Distributions* 17, 788–809.
Rodriguez-Cabal, M.A., Stuble, K.L., Nuñez, M.A. and Sanders, N.J. (2009) Quantitative analysis of the effects of the exotic Argentine ant on seed-dispersal mutualisms. *Biology Letters* 5(4), 499–502. DOI: 10.1098/rsbl.2009.0297.
Rodríguez-Echeverría, S. (2010) Rhizobial hitchhikers from down under: invasional meltdown in a plant-bacteria mutualism? *Journal of Biogeography* 37, 1611–1622.
Rodríguez-Echeverría, S., Fajardo, S., Ruiz-Díez, B. and Fernández-Pascual, M. (2012) Differential effectiveness of novel and old legume-rhizobia mutualisms: implications for invasion by exotic legumes. *Oecologia* 170(1), 253–261. DOI: 10.1007/s00442-012-2299-7.
Rodríguez-Echeverría, S., Le Roux, J.J., Crisóstomo, J.A. and Ndlovu, J. (2011) Jack-of-all-trades and master of many? How does associated rhizobial diversity influence the colonization success of Australian *Acacia* species? *Diversity and Distributions* 17, 946–957.
Rodríguez-Echeverría, S. and Traveset, A. (2015) Putative linkages between below- and aboveground mutualisms during alien plant invasions. *AoB PLANTS* 7, plv062. DOI: 10.1093/aobpla/plv062.

Roth, A.M., Whitfeld, T.J.S., Lodge, A.G., Eisenhauer, N., Frelich, L.E. *et al.* (2015) Invasive earthworms interact with abiotic conditions to influence the invasion of common buckthorn (*Rhamnus cathartica*). *Oecologia* 178(1), 219–230. DOI: 10.1007/s00442-014-3175-4.

Rowles, A.D. O'Dowd, D.J, O'Dowd, D.J. (2009) New mutualism for old: indirect disruption and direct facilitation of seed dispersal following Argentine ant invasion. *Oecologia* 158(4), 709–716. DOI: 10.1007/s00442-008-1171-2.

Russel, J.C. and Kaiser-Bunbury, C.N. (2019) Consequences of multispecies introductions on island ecosystems. *Annual Review of Ecology, Evolution and Systematics* 50, 169–190.

Russo, L. (2016) Positive and negative impacts of non-native bee species around the world. *Insects* 7(4), 69. DOI: 10.3390/insects7040069.

Schmitz, O.J. (1997) Press perturbations and the predictability of ecological interactions in a food web. *Ecology* 78, 55–69.

Shiels, A.B. (2011) Frugivory by introduced black rats (*Rattus rattus*) promotes dispersal of invasive plant seeds. *Biological Invasions* 13, 781–792.

Shiels, A.B. and Drake, D.R. (2011) Are introduced rats (*Rattus rattus*) both seed predators and dispersers in Hawaii? *Biological Invasions* 13, 883–894.

Siemann, E., Carrillo, J.A., Gabler, C.A., Zipp, R. and Rogers, W.E. (2009) Experimental test of the impacts of feral hogs on forest dynamics and processes in the southeastern US. *Forest Ecology and Management* 258, 546–553.

Simberloff, D. (2006) Invasional meltdown 6 years later: important phenomenon, unfortunate metaphor, or both? *Ecology Letters* 9(8), 912–919. DOI: 10.1111/j.1461-0248.2006.00939.x.

Simberloff, D. (2011) How common are invasion-induced ecosystem impacts? *Biological Invasions* 13(5), 1255–1268. DOI: 10.1007/s10530-011-9956-3.

Simberloff, D., Nuñez, M.A., Ledgard, N.J., Pauchard, A., Richardson, D.M. *et al.* (2010) Spread and impact of introduced conifers in South America: lessons from other southern hemisphere regions. *Austral Ecology* 35(5), 489–504. DOI: 10.1111/j.1442-9993.2009.02058.x.

Simberloff, D. and Von Holle, B. (1999) Positive interactions of nonindigenous species: invasional meltdown? *Biological Invasions* 1, 21–32. DOI: 10.1023/A:1010086329619.

Spotswood, E.N., Meyer, J.Y. and Bartolome, J.W. (2012) An invasive tree alters the structure of seed dispersal networks between birds and plants in French Polynesia. *Journal of Biogeography* 39, 2007–2020.

Stone, L. and Roberts, A. (1991) Conditions for a species to gain advantage from the presence of competitors. *Ecology* 72, 1964–1972.

Strauss, S.Y. (1991) Indirect effects in community ecology: their definition, study and importance. *Trends in Ecology & Evolution* 6(7), 206–210. DOI: 10.1016/0169-5347(91)90023-Q.

Stępkowski, T., Moulin, L., Krzyzańska, A., McInnes, A., Law, I.J. *et al.* (2005) European origin of *Bradyrhizobium* populations infecting lupins and serradella in soils of Western Australia and South Africa. *Applied and Environmental Microbiology* 71(11), 7041–7052. DOI: 10.1128/AEM.71.11.7041-7052.2005.

Strauss, A., White, A. and Boots, M. (2012) Invading with biological weapons: the importance of disease-mediated invasions. *Functional Ecology* 26(6), 1249–1261.

Tecco, P.A., Gurvich, D.E., Díaz, S., Pérez-Harguindeguy, N. and Cabido, M. (2006) Positive interaction between invasive plants: The influence of *Pyracantha angustifolia* on the recruitment of native and exotic woody species. *Austral Ecology* 31, 293–300.

Tekiela, D.R. and Barney, J.N. (2017) Co-invasion of similar invaders results in analogous ecological impact niches and no synergies. *Biological Invasions* 19, 147–159.

Thomas, M.B. and Reid, A.M. (2007) Are exotic natural enemies an effective way of controlling invasive plants? *Trends in Ecology & Evolution* 22(9), 447–453. DOI: 10.1016/j.tree.2007.03.003.

Traveset, A. and Richardson, D.M. (2014) Mutualistic interactions and biological invasions. *Annual Review of Ecology, Evolution and Systematics* 45, 89–113.

Turner, D.R. and Vitousek, P.M. (1987) Nodule biomass of the nitrogen-fixing alien *Myrica faya* Ait. in Hawaii Volcanoes National Park. *Pacific Science* 41, 1–4.

Ureclay, C., Longo, S., Geml, J., Tecco, P.A. and Nouhra, E. (2017) Co-invasive exotic pines and their ectomycorrhizal symbionts show capabilities for wide distance and altitudinal range expansion. *Fungal Ecology* 25, 50–58.

van Kleunen, M., Essl, F., Pergl, J., Brundu, G., Carboni, M. *et al.* (2018) The changing role of ornamental horticulture in alien plant invasions. *Biological reviews of the Cambridge Philosophical Society* 93(3), 1421–1437. DOI: 10.1111/brv.12402.

van Leeuwen, C.H.A. (2018) Internal and external dispersal of plants by animals: an aquatic perspective on alien interference. *Frontiers in Plant Science* 9, 153. DOI: 10.3389/fpls.2018.00153.

Vázquez, D.P., Ramos-Jiliberto, R., Urbani, P. and Valdovinos, F.S. (2015) A conceptual framework for studying the strength of plant-animal mutualistic interactions. *Ecology Letters* 18(4), 385–400. DOI: 10.1111/ele.12411.

Vellinga, E.C., Wolfe, B.E. and Pringle, A. (2009) Global patterns of ectomycorrhizal introductions. *The New Phytologist* 181(4), 960–973. DOI: 10.1111/j.1469-8137.2008.02728.x.

Vilà, M., Espinar, J.L., Hejda, M., Hulme, P.E., Jarošík, V. *et al.* (2011) Ecological impacts of invasive alien plants: a meta-analysis of their effects on species, communities and ecosystems. *Ecology Letters* 14(7), 702–708. DOI: 10.1111/j.1461-0248.2011.01628.x.

Vilà, M. and Weiner, J. (2004) Are invasive plant species better competitors than native plant species? - evidence from pair-wise experiments. *Oikos* 105(2), 229–238. DOI: 10.1111/j.0030-1299.2004.12682.x.

Von Holle, B., Joseph, K.A., Largay, E.F. and Lohnes, R.G. (2006) Facilitations between the introduced nitrogen-fixing tree, Robinia pseudoacacia, and nonnative plant species in the glacial outwash upland ecosystem of Cape Cod, MA. *Biodiversity and Conservation* 15(7), 2197–2215. DOI: 10.1007/s10531-004-6906-8.

Wei, G., Chen, W., Zhu, W., Chen, C., Young, J.P.W. *et al.* (2009) Invasive *Robinia* pseudoacacia in China is nodulated by *Mesorhizobium* and *Sinorhizobium* species that share similar nodulation genes with native American symbionts. *FEMS Microbiology Ecology* 68(3), 320–328. DOI: 10.1111/j.1574-6941.2009.00673.x.

Weir, B.S., Turner, S.J., Silvester, W.B., Park, D.-C. and Young, J.M. (2004) Unexpectedly diverse *Mesorhizobium* strains and *Rhizobium leguminosarum* nodulate native legume genera of New Zealand, while introduced legume weeds are nodulated by *Bradyrhizobium* species. *Applied and Environmental Microbiology* 70(10), 5980–5987. DOI: 10.1128/AEM.70.10.5980-5987.2004.

White, E.M., Wilson, J.C. and Clarke, A.R. (2006) Biotic indirect effects: a neglected concept in invasion biology. *Diversity and Distributions* 12, 443–455.

Widder, S., Allen, R.J., Pfeiffer, T., Curtis, T.P., Wiuf, C. *et al.* (2016) Challenges in microbial ecology: building predictive understanding of community function and dynamics. *The ISME Journal* 10(11), 2557–2568. DOI: 10.1038/ismej.2016.45.

Wood, J.R., Dickie, I.A., Moeller, H.V., Peltzer, D.A., Bonner, K.I. *et al.* (2015) Novel interactions between non-native mammals and fungi facilitate establishment of invasive pines. *Journal of Ecology* 103, 121–129.

Wootton, J.T. (1994) The nature and consequences of indirect effects in ecological communities. *Annual Review of Ecology and Systematics* 25, 443–466.

Wright, A.J., Wardle, D.A., Callaway, R. and Gaxiola, A. (2017) The overlooked role of facilitation in biodiversity experiments. *Trends in Ecology & Evolution* 32(5), 383–390. DOI: 10.1016/j.tree.2017.02.011.

Wundrow, E.J., Carrillo, J., Gabler, C.A., Horn, K.C. and Siemann, E. (2012) Facilitation and competition among invasive plants: a field experiment with alligatorweed and water hyacinth. *PLoS ONE* 7(10), e48444. DOI: 10.1371/journal.pone.0048444.

Zarnetske, P.L., Gouhier, T.C., Hacker, S.D., Seabloom, E.W. and Bokil, V.A. (2013) Indirect effects and facilitation among native and non-native species promote invasion success along an environmental stress gradient. *Journal of Ecology* 101, 905–915.

Zarnetske, P.L., Hacker, S.D., Seabloom, E.W., Ruggiero, P., Killian, J.R. *et al.* (2012) Biophysical feedback mediates effects of invasive grasses on coastal dune shape. *Ecology* 93(6), 1439–1450. DOI: 10.1890/11-1112.1.

Zenni, R.D. and Nuñez, M.A. (2013) The elephant in the room: the role of failed invasions in understanding invasion biology. *Oikos* 122, 801–815.

9 Biotic Resistance to Plant Invasions

John D. Parker[1]*, John L. Devaney[1,2] and Nathan P. Lemoine[3]

[1]Smithsonian Environmental Research Center, Edgewater, MD, USA; [2]Department of Biology, Maynooth University, Maynooth, Co. Kildare, Ireland; [3]Department of Biological Sciences, Marquette University, Milwaukee, WI, USA

Abstract

Biotic resistance to plant invasions takes many forms: consumption by native herbivores, competition with native plants and infection by native pathogens. But how often does biotic resistance prevent the damaging monocultures that typify the most problematic plant invaders, and how often is biotic resistance overwhelmed by the direct and indirect impacts of human activities? This chapter attempts to answer these questions, drawing on the long history of research into biotic resistance. We first briefly describe the major forms of biotic resistance to exotic plant invasions as an antecedent to other, more detailed chapters on competition, herbivory and pathogens. We then describe a new neutral model where variance in disturbance promotes invasions over the short term, but over longer timescales only propagule pressure drives invasions. These findings are a cautionary tale; pending increases in global trade and travel, particularly to the tropics, may provide the prerequisite disturbance and propagule pressure needed to ultimately stoke further invasions. Finally, we highlight case studies where invasions have been mitigated by restoration of biotic resistance from native herbivores and competitors. These studies provide strong empirical support that conservation of native biodiversity can be a nature-based solution to some invasions, although it remains to be seen if climate change will alter these effects over longer timescales.

9.1 Introduction

The Biotic Resistance (BR) Hypothesis, the idea that native species can limit or prevent invasions by introduced species, is a classic hypothesis in invasion biology (Chapman, 1931). In *On the Origin of Species*, Darwin remarked 'a prodigious number of plants ... never become naturalized for they cannot compete with our native plants nor resist destruction by our native animals' (Darwin, 1859). The concept of BR was later expanded by Charles Elton in *The Ecology of Invasions by Animals and Plants*, stating that many invasions fail because 'they meet with resistance', including competition, consumption and pathogens (Fig. 9.1, Elton, 1958). By arguing that failed invasions result from novel biotic interactions in a new range, Darwin and Elton implicitly acknowledged the role of evolutionary history in mediating biological invasions, a notion that underpins much of our modern understanding of BR (Lemoine *et al*., 2016; Pearson *et al*., 2018). However, the relative strength of BR in mediating plant invasion success, and especially how BR interacts with other forces including propagule pressure and disturbance, remain topics of considerable debate.

Empirical studies assessing the strength of BR exploded in the early 2000s (Fig. 9.2).

*Corresponding author: parkerj@si.edu

 Plant Invasions: The Role of Biotic Interactions (eds Anna Traveset and David M. Richardson)
DOI: 10.1079/9781789242171.0009

Fig. 9.1. (A) The native eastern cottontail rabbit *Sylvilagus floridanus* consuming dandelion *Taraxacum officinale,* one of the world's most widespread and economically costly introduced plants (photo by John Parker). (B) Lesions of the fungal pathogen *Bipolaris* infecting Japanese stiltgrass *Microstegium vimineum* (photo by Luke Flory). (C) Japanese stiltgrass invading the disturbed, high light areas surrounding a forest roadside (photo by John Parker).

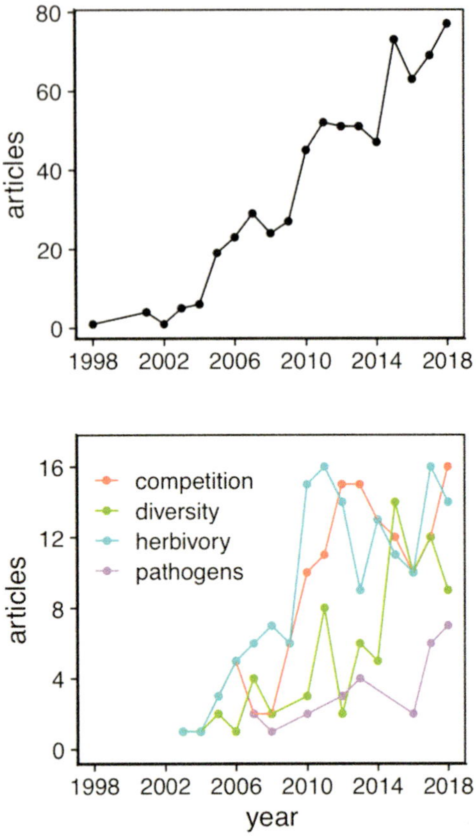

Fig. 9.2. Results from Web of Science search (19 May 2019) showing number of articles referencing biotic resistance and invasion from 1998 to 2018. Lower panel shows number of original empirical articles (i.e. excluding reviews and meta-analyses) examining different types of biotic resistance to plant invasions from 1998 to 2018.

Around this time, a seminal meta-analysis by Levine *et al.* (2004) showed negative effects of native competitors, herbivores and diverse plant assemblages on non-native plant establishment and performance, but little evidence of complete exclusion by BR.

This raised the question as to whether BR was a strong force in repelling plant invasions, or whether BR was relatively weak and commonly overwhelmed by other drivers. In this chapter, we will highlight how a number of studies have since examined BR in the context of multiple interacting forces, including anthropogenic disturbance and propagule pressure. These studies have improved our understanding of BR over the last two decades. Early studies addressed relatively straightforward questions like: *Do invaders meet with BR in their new ranges?* More recently, research has tackled increasingly complex questions like: *Does the strength of BR vary geographically? Do interactions between different drivers affect the strength of BR? Can we harness BR to manage invasions?* Our goal in this chapter is to survey these recent developments while pointing out areas of opportunity for future research.

9.2 Biotic Resistance from Competition

Competitive interactions among plants are the most basic and frequently cited component of BR. Although competitive interactions alone appear unlikely to completely inhibit invasions (Levine *et al.*, 2004), exotic species commonly suffer reduced establishment and growth rates in the presence of native competitors (Schmidt *et al.*, 2008; Manea and Leishman, 2011). Nevertheless, exotics can experience BR but still thrive by having asymmetrically negative effects on natives (Kuebbing *et al.*, 2013; Golivets and Wallin, 2018). Recently, Zhang and van Kleunen (2019) quantified competitive outcomes among 48 pairs of native and exotic annuals in Germany. Introduced plants outcompeted rare but not common natives, and interspecific interactions between natives and exotics were generally negative and not facilitative. Thus, competitive BR to invasions is potentially more common than facilitation, an intriguing result that warrants broader investigation across more species and habitats.

Importantly, native/non-native competitive outcomes in Zhang and van Kleunen (2019) were related to species' intrinsic growth rates, suggesting that species' traits are often better predictors of invasion success than origin alone. Indeed, van Kleunen *et al.* (2010) found that invasive species generally possess higher values of performance-related traits compared to non-invasive introduced species, including growth rate, size, fecundity and ecophysiology. Similarly, Ordonez *et al.* (2010) also found that when compared to native plants, invasive plants had faster growth rates, produced more seeds per unit reproductive effort and reached reproductive age faster. These results generally support the Limiting Similarity Hypothesis, whereby invading species are most successful when they are phenotypically dissimilar to the dominant resident natives (Wang *et al.*, 2013; Hulvey and Teller, 2018). In this model, increasing trait (and often phylogenetic) dissimilarity from the resident community enhances invader success by reducing niche overlap (e.g. Davies *et al.*, 2005, Davies *et al.*, 2011). Support can be mixed (Von Holle and Simberloff, 2004; Parker *et al.*, 2012; Price and Pärtel, 2013; Mason *et al.*, 2017), however, especially when traits are not phylogenetically conserved (Lemoine *et al.*, 2015; Yannelli *et al.*, 2017; Ng *et al.*, 2019). Moreover, when compared to equally common natives in the introduced range, invasive and common native plants can also possess similar trait values owing to environmental-biotic filtering that selects for phenotypically similar species regardless of origin (Lind and Parker, 2010; Lemoine *et al.*, 2015; Oduor *et al.*, 2016; van der Sande *et al.*, 2020). Importantly, the use of modern quantitative community assembly models can help disentangle whether plant success in a given habitat is driven most by plant traits, phylogeny or origin (Lemoine *et al.*, 2016).

9.3 Biotic Resistance from Species Diversity

Native species diversity might also impart BR because more speciose communities possess less available niche space to support incoming exotics. First proposed by Elton (1958), the Diversity-Invasibility Hypothesis predicts a negative relationship between native and exotic species richness (Davies *et al.*, 2005, Davies *et al.*, 2011). Initial studies supported this prediction (Case, 1990; Rejmánek, 1996; Knops *et al.*, 1999), but ecologists noticed that the direction and magnitude of diversity effects on invasion were often scale-dependent. Fine-scale surveys and experiments often reported negative relationships (Levine and D'Antonio, 1999; Levine, 2000), whereas larger-scale, mostly observational studies generally reported positive relationships (Stohlgren *et al.*, 2003), resulting in an invasion paradox (Fridley *et al.*, 2007). Covarying factors like productivity or disturbance partially explained some of these patterns (Parker *et al.*, 2010; Bennett *et al.*, 2016). In California grasslands, for example, BR is strongest when native species richness and functional diversity are low, and thus BR is more strongly related to native species biomass driven by underlying soil fertility (Grace *et al.*, 2017).

The invasion paradox remains controversial, but the burgeoning use of intensive statistical analyses with larger datasets is helping to provide clarity. For example, in a meta-analysis of 204 individual non-experimental studies, Peng *et al.* (2019) found positive native-exotic richness relationships across all spatial scales despite heterogeneity at smaller scales. Smith and Côté (2019) suggest that scale-dependent differences in diversity–invasibility relationships reflect systematic differences in the choice of invader success metric (e.g. richness vs performance) across different study types and scales. Small-scale experimental studies tend to focus on invader performance (e.g. growth or survival) and are more likely to uncover negative relationships between native richness and exotic performance (Smith and Côté, 2019). Larger-scale, observational studies generally focus on native and invader richness and mostly find positive relationships (e.g. Peng *et al.*, 2019). In contrast, results from Beaury *et al.* (2020) suggest that the positive relationships in larger scale studies are due to confounding effects of covarying factors like habitat quality, distance to human activities and climate. Once these factors were controlled for, they found universally negative relationships between native species richness and exotic plant occurrence across 24,456 plots of varying sizes (25 to 5400 m^2) in US National Parks. Thus, Beaury *et al.* (2020) appear to partially resolve the invasion paradox by demonstrating that biotic resistance by native species is common once underlying differences across sites are considered, providing strong evidence that conserving native plant biodiversity may help resist invasions.

9.4 Biotic Resistance from Herbivory

Herbivory-mediated establishment of introduced plants is a central tenet of invasion biology, with numerous examples of introduced plants supporting reduced enemy guilds in their new ranges, both compared to native plants in the new community, and when compared to conspecifics in their home range (Colautti *et al.*, 2004; Liu and Stiling, 2006). The evolutionary explanation behind these results contends that native herbivores are not adapted to consume novel plants with novel antiherbivore defences. Alternatively, introduced plants might be selectively eaten because they are defensively naïve against a novel suite of herbivores (Verhoeven *et al.*, 2009). The first meta-analysis addressing the prevalence of BR in plant introductions showed that resident herbivores had negative impacts on both performance and establishment of plant invaders (Levine *et al.*, 2004). Further, Parker *et al.* (2006) demonstrated that native generalist herbivores often suppressed plant invasions, an effect that was stronger for vertebrate than invertebrate herbivores and that differed by native vs non-native herbivores. This general result has since been supported in both the plant–herbivore and predator–prey literature (Parker and Hay, 2005; Morrison and Hay, 2011; Paolucci *et al.*, 2013).

Nevertheless, many individual exotic plant species appear to benefit from reduced herbivore pressure in their new ranges, supporting the Enemy Release Hypothesis. For example, white-tailed deer *Odocoileus virginianus* are one of the most dominant herbivores in eastern North America where they often decrease native plant species abundance and shift communities towards particular unpalatable exotic plants (Knight *et al.*, 2009; Kalisz *et al.*, 2014). However, in a larger analysis of 23 exclusion experiments across North America (Averill *et al.*, 2017), the general impact of white-tailed deer on 345 native vs 50 introduced species was equivocal. Such idiosyncratic impacts are consistent with a literature increasingly showing that BR from native herbivores is species-specific and hinges on the palatability and defensive traits of introduced plants relative to sympatric natives (Parker and Hay, 2005; Xiong *et al.*, 2008; Lind and Parker, 2010; Morrison and Hay, 2011).

Plant invasion ecology is still grappling to understand the context-dependency of herbivore control over plant invasions, and particularly whether herbivory is strong enough to repel invasions and thus be used as a management tool. For example, the reintroduction of large vertebrate herbivores has been suggested as a tool to restore native biodiversity and control invasions (Derham *et al.*, 2018), similar to the broader concept of rewilding to re-establish trophic diversity in ecosystems (Josh Donlan *et al.*, 2006). Early returns were equivocal, with reintroduced herbivores providing BR to some but not all invasive plants (Johnson and Cushman, 2007; Griffiths *et al.*, 2013; Ender *et al.*, 2017). In marine systems, different management scenarios dictate success of the invasive alga *Sargassum horneri*. Long-established marine protected areas (MPAs) and nearby heavily fished reefs are both relatively uninvaded due to the presence of strong sea urchin herbivory and competition from native seaweeds. Newly established MPAs, however, remain widely invaded due to insufficient time for recovery of herbivores and competition (Caselle *et al.*, 2018).

Recent results from Mozambique's Gorongosa National Park provide additional compelling evidence that trophic rewilding can reverse invasions. There, invasions of the shrub *Mimosa pigra*, one of the world's 100 worst invasive species (Lowe *et al.*, 2004), more than tripled during the war-induced collapse of large mammal populations from 1972 to 2015 (Guyton *et al.*, 2020). However, following concerted reintroduction efforts, large ungulate populations of waterbuck *Kobus ellipsiprymnus*, reedbuck *Redunca arundinum*, impala *Aepyceros melampus*, oribi *Ourebia ourebi*, and buffalo *Syncerus caffer* were restored to pre-war levels, after which *Mimosa* invasions in the park declined to pre-war levels. DNA meta-barcoding showed that *Mimosa* was a predominant diet item for these dominant herbivores, and a three-year exclusion experiment showed that herbivores essentially eliminated *Mimosa* reproduction and reduced biomass by over 99% (Guyton *et al.*, 2020). The last two decades of research have thus clearly demonstrated that native herbivore communities can cause substantial damage to introduced plants, with some compelling cases where herbivores have dramatically reduced invasions. Importantly, in Guyton *et al.* (2020), the effectiveness of biotic resistance by herbivores appears to hinge on herbivores escaping their own predators, suggesting that management of herbivore-mediated biotic resistance requires consideration of multiple trophic levels.

9.5 Biotic Resistance from Pathogens

Pathogen contributions to BR have been historically understudied relative to competition, diversity and herbivory (Fig. 9.2). However, bacteria, fungi and viruses can have large impacts on plant invasion success (Mitchell and Power, 2003; Callaway *et al.*, 2004; Inderjit and van der Putten, 2010; Faillace *et al.*, 2017). For example, soil fungal pathogens negatively affect growth and survival of seedlings of the Chinese tallow tree *Triadica sebifera*, a common tree invader of the American south-east, but have no effects on four sympatric native tree species (Nijjer *et al.*, 2007). Similarly, Chinese tallow trees are also more strongly susceptible to foliar fungal diseases than native hackberry trees *Celtis laevigata* (Siemann and Rogers, 2006). Interestingly, the effects of BR on tallow trees appear largely confined to tallow dominated habitats (Siemann and Rogers, 2003), suggesting that BR might be directly preventing the long-term persistence of invasive tallow tree monocultures. Other similar cases of BR from fungal pathogens have been demonstrated for European beachgrass *Ammophila arenaria* in California, USA (Beckstead and Parker, 2003), the vine *Mikania micrantha* in China (Gao *et al.*, 2013), and clovers *Trifolium* spp. invading New Zealand (McGinn *et al.*, 2018).

In contrast, studies indicating pathogen-mediated BR are counterbalanced by those showing facilitation of plant invasions by the native soil community, including for invasive black cherry trees *Prunus serotina* in Europe (Reinhart *et al.*, 2003), the grass *Cenchrus biflorus* in Botswana (van der Putten *et al.*, 2007), and knotweed *Fallopia x bohemica* in Switzerland (Parepa *et al.*, 2013). Thus, the overall effect of soil feedback on plant invasion remains largely neutral despite examples demonstrating both facilitation and BR (Inderjit and van der Putten, 2010; Suding *et al.*, 2013; Inderjit and Cahill, 2015). One caveat, however, is that the methodology involved in plant–soil feedback studies (e.g. the application of activated carbon and soil sterilization to remove pathogenic microbes, often conducted in greenhouses) is imprecise because it also eliminates beneficial microbes (Pernilla Brinkman *et al.*, 2010; Reinhart and

Rinella, 2016). Moreover, when field and greenhouse studies have been conducted simultaneously with similar methods, the effects observed in the greenhouse can be undetectable in the field (Schittko *et al.*, 2016).

In contrast to the 'black box' approach of plant soil feedback experiments, recent genetic evidence conclusively implicates foliar pathogens in BR to Japanese stiltgrass *Microstegium vimineum*. Stiltgrass is a prolific and problematic invader of disturbed forests in North America (Flory and Clay, 2010a, b), but over half of *Microstegium* populations across eastern North America have been recently diagnosed with infection by *Bipolaris* spp., a cosmopolitan fungal genus of grass pathogens (Flory *et al.*, 2011; Stricker *et al.*, 2016). Experimental suppression of *Bipolaris* reduces lesions by up to 81%, leading to a 39% increase in stiltgrass biomass and a 68% increase in seed set (Stricker *et al.*, 2016). However, given the ability of *Microstegium* to persist even after extreme reductions in biomass and seed set (Flory, 2010), it remains unclear if infection severity is having large-scale impacts on *Microstegium* persistence and spread (Flory *et al.*, 2018; Policelli *et al.*, 2018). Nevertheless, the probability of *Bipolaris* infection increases with population age (Stricker *et al.*, 2016), following a general pattern in invasion biology whereby many introduced species progressively accumulate new pathogens over time (Hawkes, 2007; Diez *et al.*, 2010; Mitchell *et al.*, 2010; Phillips *et al.*, 2010; Flory and Clay, 2013). One conclusion from these patterns is that stronger BR accrues over time, stressing the need for more chronosequence-style studies on invasion dynamics and BR.

9.6 Interactions Between Biotic Resistance and Other Drivers

Theory predicts that interactions between BR and other drivers should influence invasion success (Blumenthal, 2006; Mitchell *et al.*, 2006). In a particularly ambitious study, Heckman *et al.* (2017) factorially manipulated perennial plant diversity, soil nutrient supply and presence of above-ground foliar enemies (fungal pathogens and insect herbivores) and monitored plant communities for four years. Surprisingly, there were no interactions at any level, whereas increasing nutrients and excluding enemies independently increased exotic plant abundance. In contrast, Eschtruth and Battles (2009b) quantified the relative importance of canopy disturbance, propagule pressure, species diversity and herbivory in determining exotic plant invasion in ten eastern North America hemlock forests. White-tailed deer herbivory increased abundance of two of three plant invaders, consistent with simultaneous caging experiments (Eschtruth and Battles, 2009a). However, the primary predictor of invasions was an interaction between propagule pressure and canopy disturbance. Canopy defoliation caused by the invasive hemlock woolly adelgid *Adelges tsugae*, combined with high seed rain, increased invasions of garlic mustard *Alliaria petiolata*, Japanese barberry *Berberis thunbergii* and Japanese stiltgrass *M. vimineum*. In contrast, the combination of fire and rodents together provided stronger BR to cheatgrass *Bromus tectorum* invasions in the American west (St. Clair *et al.*, 2016). In fact, rodent granivory in particular and seed consumption in general appears to be a cryptic yet pervasive source of BR to many invasions (Connolly *et al.*, 2014; Maron *et al.*, 2014; Pearson *et al.*, 2014; Bogdziewicz *et al.*, 2019).

An increasing body of evidence suggests that biotic resistance is often overwhelmed by both disturbance and propagule pressure (Colautti *et al.*, 2006). In perhaps the seminal paper examining interactions between biotic resistance, disturbance and propagule pressure, Holle and Simberloff, 2005 reported that propagule supply was the only significant predictor of invader success through time in a temperate forest of the eastern USA. This pattern has been documented in multiple systems, including tropical forests (Edward *et al.*, 2009), temperate grasslands (Fensham *et al.*, 2013) and marine invertebrates (Hollebone and Hay, 2007; Clark and Johnston, 2009). Interactions between disturbance, biotic resistance and propagule pressure can still be important. In a sessile marine invertebrate community, non-native bryozoans could only invade when disturbance cleared open space and propagule pressure was high (Clark and Johnston, 2009). In scrubland communities, cheatgrass *B. tectorum* abundance is high when herbivorous rodents are absent, but high propagule pressure overwhelms this biotic resistance and allows cheatgrass to achieve abundances similar to those in the absence of herbivores (St. Clair and Bishop, 2019). To date, however, it is unclear which force, disturbance or propagule pressure is most important to invasion success over short and long timescales.

To test how disturbance and propagule pressure affect invasion success, we modified the neutral model of Herben *et al.* (2004) to simulate invasion success into native communities as a function of increasing disturbance and propagule pressure. We modelled a saturated community where the total number of individuals in the community, N, was held constant across time. Thus, an increasing proportion of non-native species in the community represents an erosion of biotic resistance due to disturbance or propagule pressure. The regional species pool consisted of P_n (native) and P_e (exotic) species. Communities were initially constructed by uniformly sampling N individuals from P_n with replacement. At each time step, a proportion of the community *d* was removed, simulating disturbance. Disturbance was stochastic and at each time step, *d* was drawn from a beta distribution with a known mean and variance. The mean represented average disturbance intensity, while variance depicted disturbance variability. After disturbance, the available spaces (l) were filled with recruits from the local community at a rate $l \times d \times N$. Immigrants from P_n and P_e filled the remaining spaces. Propagule pressure differed between native and exotic species. For native species, immigration rate was:

$I_n = m_n/(P_n + P_e)$

and the probability of exotic immigration was calculated as:

$I_e = m_e/(P_e + m_n P_n)$

where m_n and m_e are relative immigration rates. Setting $m_n = m_e = 1$ defines native and exotic species as having identical immigration rates; $m_n = 1$ and $m_e = 10$ defines immigration rates for exotic species as tenfold higher than native species. Immigrants were sampled from P_n and P_e according to the rates I_n and I_e. Initially, $N = 100$, $P_n = 20$, $P_e = 1$, and $l = 0.5$. Thus, a single exotic species invaded a native community of 100 individuals drawn from 20 native species. A complete model run consisted of 200 simulations per set of parameters, with output averaged at 5 and 500 time-steps.

In our model, disturbance variability rather than intensity dictated invasion success after five time-steps, with increasing propagule pressure enhancing overall invasion success but not swamping the disturbance signal (Fig. 9.3a,b). In contrast, neither disturbance intensity nor variability had an effect on long time scales; only propagule pressure dictated invasion success (Fig. 9.3c,d). These results have been broadly confirmed by recent field experiments. Carr *et al.* (2019) examined the invasion success of *Centaurea solstitialis* in Californian serpentine grasslands over almost two decades. Disturbance had no effect on *C. solstitialis* abundance in the long term, whereas propagule pressure was the only significant predictor of *C. solstitialis* abundance in all years except one early year. Thus, one conclusion from both simulated and empirical results is that if biotic resistance is to be successful in reducing or eliminating invasive species, it must overcome substantial and countervailing effects from stochastic disturbance in the short term and propagule pressure in the long term.

9.7 Are Latitudinal Gradients of Invasions Attributable to Biotic Resistance?

The prevailing pattern of relatively few invasions in the tropics (Rejmánek, 1996; Lonsdale, 1999; Sax, 2001; Hewitt, 2002) is thought to reflect stronger species interactions and thus stronger BR in those regions (Elton, 1958; Schemske *et al.*, 2009). However, the hypothesis that there is greater BR to invasions in the tropics is mostly correlative, has seen mixed support when examined in detail and is under-investigated in a mechanistic context. For example, global patterns of latitudinal gradients in invasions are largely weakened or eliminated once confounding factors like site area and introduction pressure are considered (Lonsdale, 1999; Blackburn and Duncan, 2001).

Mechanistically, lower numbers of naturalized species in the tropics might reflect stronger biotic interactions, including increased predation pressure and competition from more diverse assemblages. For example, invasive strains of the wetland plant *Phragmites australis* exhibit latitudinal clines in plant palatability suggestive of stronger enemy release in equatorward

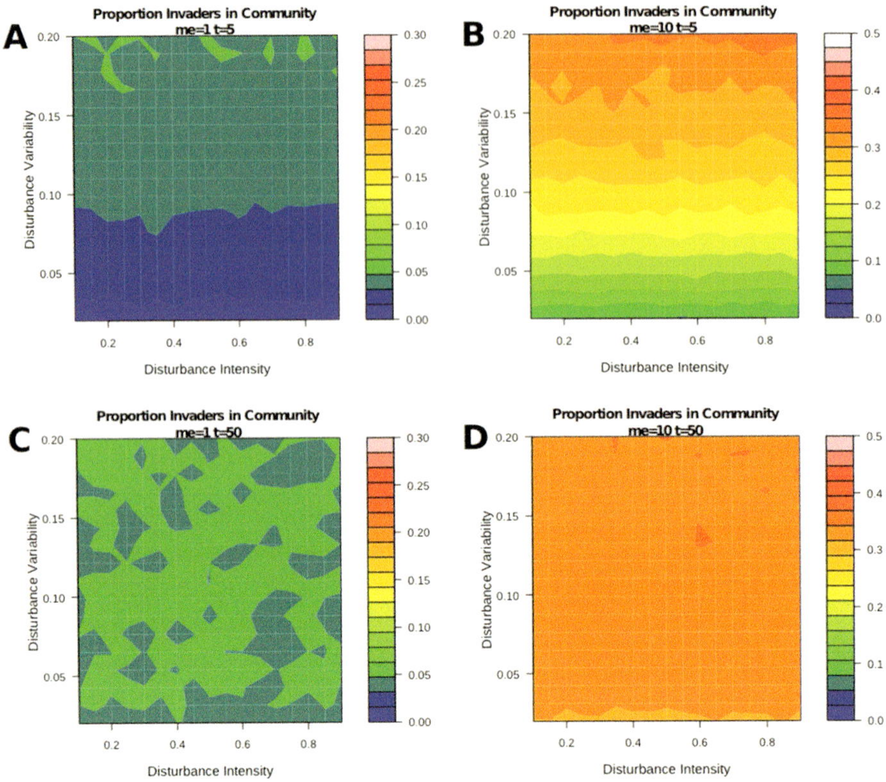

Fig. 9.3. (A) Over short time scales (t = 5 time steps) and equal propagule pressure from native and introduced species ($m_{n,e}$ = 1), invasion success is most strongly affected by disturbance variability. (B) After increasing propagule pressure of introduced species over native species by a factor of 10 (m_e = 10), invasion success was increased and disturbance variability still affected invasion success. Over longer time scales (t = 50 time steps), only enhanced propagule pressure affects invasion success (C, D).

populations (Cronin *et al.*, 2015; Allen *et al.*, 2017; Bhattarai *et al.*, 2017). Similarly, tropical invasive aquatic plants are less palatable than temperate plants to two herbivorous snails (Grutters *et al.*, 2017). Although these outcomes are consistent with stronger biotic resistance from consumers potentially selecting for greater plant defences in the tropics, the core question of whether herbivory and other forms of BR have stronger effects on invasive plants in the tropics has not been investigated to our knowledge. In benthic marine systems, predation resistance to invertebrate invasions can be stronger in the tropics compared to temperate sites (Freestone *et al.*, 2013; Kimbro *et al.*, 2013), but it can also be latitudinally invariant and overwhelmed by regional differences in propagule pressure (Cheng *et al.*, 2019). In freshwater systems, BR from consumers is stronger than that of competitors (similar to findings in marine systems, Papacostas *et al.*, 2017), but neither competition nor consumption has a latitudinal signal (Alofs and Jackson, 2014).

Thus, although the tropics generally appear less invaded than the temperate zone for many species, suggestive of stronger BR, the confounding effects of latitude, site area, productivity and invasion pressure are challenging, and the strength of biotic resistance may vary among different habitats (e.g. DeWalt *et al.*, 2004) and plant growth forms (e.g. Maron and Vila, 2001). Furthermore, we speculate that pending increases in trade and travel to the tropics will increase both anthropogenic disturbance and propagule supply of new invaders to these systems (Seebens *et al.*, 2015), potentially erasing these long-standing patterns. Here, we suggest that disentangling whether BR is stronger in the tropics is best accomplished with experiments manipulating multiple factors that could drive invasion success across habitats and latitudinal gradients, similar to insights gained from adopting home/away biogeographic comparisons (Hierro *et al.*, 2005).

9.8 Conclusions

The notion of biotic resistance was introduced by both Darwin and Elton to recognize that non-native plants would likely face stiff competition, herbivory and pathogen attack from native species in their new ranges. The effects of this BR, and their integration into an eco-evolutionary perspective on how plants successfully invade new communities, wouldn't be incorporated until almost a 100 years later, following an influential growth phase in invasion biology in the early 2000s. These early papers demonstrated that BR by native species significantly influenced exotic plant invasions (Levine *et al.*, 2004; Parker *et al.*, 2006). Since then, BR has been incorporated into modern community assembly models, whereby invasion success is modelled as a function of the total trait space of the recipient community, physiological matching to the environment, phylogenetic relatedness and various measures of habitat quality (Lemoine *et al.*, 2016). Overall, the picture that emerges is that BR to non-native plants is common and appears to get stronger over time. However, habitat quality, anthropogenic disturbance and particularly propagule pressure are often covarying drivers of invasive plant success that need to be incorporated before accurate assessments of the relative strength of BR can be drawn.

One of the key questions going forward is whether BR is strong enough to be effectively harnessed as a conservation tool to limit current and future invasions, which currently show no signs of slowing down and in fact may be increasing (Seebens *et al.*, 2017). Results from marine and terrestrial habitats alike suggest that restoration of trophic linkages can reduce exotic plant invasions (Caselle *et al.*, 2018; Guyton *et al.*, 2020), complementing the biotic resistance provided by native plant competitors (Beaury *et al.*, 2020) and pathogens (Stricker *et al.*, 2016). These studies provide promising benchmarks for studying how native biodiversity can be used as a conservation tool to resist and potentially even reverse some non-native plant invasions.

Finally, although BR can be both strengthened and weakened by interactive effects, too few studies have examined invasive plant population demography rather than more static measurements like plant cover, abundance or survival (Colautti *et al.*, 2014). While these metrics are informative, long-term population trajectories better elucidate the cumulative effects of BR over longer timescales (e.g. Kalisz *et al.*, 2014). The invasive thistle *Cirsium vulgare*, for example, has positive population growth rates in both low and intermediate levels of interspecific competition, but the combination of high competition and insect herbivory leads to negative population growth (Tenhumberg *et al.*, 2015). Thus, we propose that future studies incorporate quantitative aspects of plant population dynamics into modern community assembly models to test the relative importance of biotic vs abiotic drivers in determining invasion success. Moreover, given that climate warming can substantially alter the impacts of herbivores, competitors and pathogens (Yakob and Mumby, 2011; Altizer *et al.*, 2013; Vergés *et al.*, 2014; Lemoine *et al.*, 2017), a longer-term, demographic perspective will be needed to understand the effects of climate change on biotic resistance to invasions in the novel ecosystems of the future.

References

Allen, W.J., Meyerson, L.A., Cummings, D., Anderson, J., Bhattarai, G.P. *et al.* (2017) Biogeography of a plant invasion: drivers of latitudinal variation in enemy release. *Global Ecology and Biogeography* 26(4), 435–446. DOI: 10.1111/geb.12550.

Alofs, K.M. and Jackson, D.A. (2014) Meta-analysis suggests biotic resistance in freshwater environments is driven by consumption rather than competition. *Ecology* 95(12), 3259–3270. DOI: 10.1890/14-0060.1.

Altizer, S., Ostfeld, R.S., Johnson, P.T.J., Kutz, S. and Harvell, C.D. (2013) Climate change and infectious diseases: from evidence to a predictive framework. *Science* 341(6145), 514–519. DOI: 10.1126/science.1239401.

Averill, K.M., Mortensen, D.A., Smithwick, E.A.H., Kalisz, S., McShea, W.J. *et al.* (2017) A regional assessment of white-tailed deer effects on plant invasion. *AoB PLANTS* 10, 1–22. DOI: 10.1093/aobpla/plx047.

Beaury, E.M., Finn, J.T., Corbin, J.D., Barr, V. and Bradley, B.A. (2020) Biotic resistance to invasion is ubiquitous across ecosystems of the United States. *Ecology Letters* 23(3), 476–482. DOI: 10.1111/ele.13446.

Beckstead, J. and Parker, I.M. (2003) Invasiveness of *Ammophila arenaria*: Release from soil-borne pathogens? *Ecology* 84(11), 2824–2831. DOI: 10.1890/02-0517.

Bennett, J.A., Riibak, K., Kook, E., Reier, Ü., Tamme, R. *et al.* (2016) Species pools, community completeness and invasion: disentangling diversity effects on the establishment of native and alien species. *Ecology Letters* 19(12), 1496–1505. DOI: 10.1111/ele.12702.

Bhattarai, G.P., Meyerson, L.A., Anderson, J., Cummings, D., Allen, W.J. *et al.* (2017) Biogeography of a plant invasion: genetic variation and plasticity in latitudinal clines for traits related to herbivory. *Ecological Monographs* 87(1), 57–75. DOI: 10.1002/ecm.1233.

Blackburn, T.M. and Duncan, R.P. (2001) Determinants of establishment success in introduced birds. *Nature* 414(6860), 195–197. DOI: 10.1038/35102557.

Blumenthal, D.M. (2006) Interactions between resource availability and enemy release in plant invasion. *Ecology Letters* 9(7), 887–895. DOI: 10.1111/j.1461-0248.2006.00934.x.

Bogdziewicz, M., Lichti, N.I. and Zwolak, R. (2019) Consumer-mediated indirect interaction with a native plant lowers the fitness of an invasive competitor. *Journal of Ecology* 107(1), 12–22. DOI: 10.1111/1365-2745.13023.

Callaway, R.M., Thelen, G.C., Rodriguez, A. and Holben, W.E. (2004) Soil biota and exotic plant invasion. *Nature* 427(6976), 731–733. DOI: 10.1038/nature02322.

Carr, A.N., Hooper, D.U. and Dukes, J.S. (2019) Long-term propagule pressure overwhelms initial community determination of invader success. *Ecosphere* 10(8), e02826. DOI: 10.1002/ecs2.2826.

Case, T.J. (1990) Invasion resistance arises in strongly interacting species-rich model competition communities. *Proceedings of the National Academy of Sciences* 87(24), 9610–9614. DOI: 10.1073/pnas.87.24.9610.

Caselle, J.E., Davis, K. and Marks, L.M. (2018) Marine management affects the invasion success of a non-native species in a temperate reef system in California, USA. *Ecology Letters* 21(1), 43–53. DOI: 10.1111/ele.12869.

Chapman, R. (1931) *Animal Ecology*. McGraw-Hill, New York.

Cheng, B.S., Ruiz, G.M., Altieri, A.H. and Torchin, M.E. (2019) The biogeography of invasion in tropical and temperate seagrass beds: testing interactive effects of predation and propagule pressure. *Diversity and Distributions* 25, 285–297.

Clark, G.F. and Johnston, E.L. (2009) Propagule pressure and disturbance interact to overcome biotic resistance of marine invertebrate communities. *Oikos* 118(11), 1679–1686. DOI: 10.1111/j.1600-0706.2009.17564.x.

Colautti, R., Colautti, R., Parker, J.D., Cadotte, M.W., Pyšek, P. *et al.* (2014) Quantifying the invasiveness of species. *NeoBiota* 21, 7–27. DOI: 10.3897/neobiota.21.5310.

Colautti, R.I., Grigorovich, I.A. and MacIsaac, H.J. (2006) Propagule pressure: a null model for biological invasions. *Biological Invasions* 8(5), 1023–1037. DOI: 10.1007/s10530-005-3735-y.

Colautti, R.I., Ricciardi, A., Grigorovich, I.A. and MacIsaac, H.J. (2004) Is invasion success explained by the enemy release hypothesis? *Ecology Letters* 7(8), 721–733. DOI: 10.1111/j.1461-0248.2004.00616.x.

Connolly, B.M., Pearson, D.E. and Mack, R.N. (2014) Granivory of invasive, naturalized, and native plants in communities differentially susceptible to invasion. *Ecology* 95(7), 1759–1769. DOI: 10.1890/13-1774.1.

Cronin, J.T., Bhattarai, G.P., Allen, W.J. and Meyerson, L.A. (2015) Biogeography of a plant invasion: plant-herbivore interactions. *Ecology* 96(4), 1115–1127. DOI: 10.1890/14-1091.1.

Darwin, C. (1859) *The Origin of Species*. In: Literary Classics, Inc. (ed.), 1st.

Davies, K.F., Cavender-Bares, J. and Deacon, N. (2011) Native communities determine the identity of exotic invaders even at scales at which communities are unsaturated. *Diversity and Distributions* 17(1), 35–42. DOI: 10.1111/j.1472-4642.2010.00721.x.

Davies, K.F., Chesson, P., Harrison, S., Inouye, B.D., Melbourne, B.A. *et al.* (2005) Spatial heterogeneity explains the scale dependence of the native–exotic diversity relationship. *Ecology* 86(6), 1602–1610. DOI: 10.1890/04-1196.

Derham, T.T., Duncan, R.P., Johnson, C.N. and Jones, M.E. (2018) Hope and caution: rewilding to mitigate the impacts of biological invasions. *Philosophical Transactions of the Royal Society B: Biological Sciences* 373(1761), 20180127. DOI: 10.1098/rstb.2018.0127.

DeWalt, S.J., Denslow, J.S. and Ickes, K. (2004) Natural-enemy release facilitates habitat expansion of the invasive tropical shrub *Clidemia hirta*. *Ecology* 85(2), 471–483. DOI: 10.1890/02-0728.

Diez, J.M., Dickie, I., Edwards, G., Hulme, P.E., Sullivan, J.J. *et al.* (2010) Negative soil feedbacks accumulate over time for non-native plant species. *Ecology Letters* 13(7), 803–809. DOI: 10.1111/j.1461-0248.2010.01474.x.

Edward, E., Munishi, P.K.T. and Hulme, P.E. (2009) Relative roles of disturbance and propagule pressure on the invasion of humid tropical forest by *Cordia alliodora* (Boraginaceae) in Tanzania. *Biotropica* 41(2), 171–178. DOI: 10.1111/j.1744-7429.2008.00474.x.

Elton, C.S. (1958) *The Ecology of Invasions by Animals and Plants*. Methuen, London.

Ender, C.L., Christian, C.E. and Cushman, J.H. (2017) Native herbivores and environmental heterogeneity as mediators of an exotic grass invasion. *Ecology and Evolution* 7(5), 1561–1571. DOI: 10.1002/ece3.2727.

Eschtruth, A.K. and Battles, J.J. (2009a) Acceleration of exotic plant invasion in a forested ecosystem by a generalist herbivore. *Conservation Biology* 23(2), 388–399. DOI: 10.1111/j.1523-1739.2008.01122.x.

Eschtruth, A.K. and Battles, J.J. (2009b) Assessing the relative importance of disturbance, herbivory, diversity, and propagule pressure in exotic plant invasion. *Ecological Monographs* 79(2), 265–280. DOI: 10.1890/08-0221.1.

Faillace, C.A., Lorusso, N.S. and Duffy, S. (2017) Overlooking the smallest matter: viruses impact biological invasions. *Ecology Letters* 20(4), 524–538. DOI: 10.1111/ele.12742.

Fensham, R.J., Donald, S. and Dwyer, J.M. (2013) Propagule pressure, not fire or cattle grazing, promotes invasion of buffel grass *Cenchrus ciliaris*. *Journal of Applied Ecology* 50(1), 138–146. DOI: 10.1111/1365-2664.12009.

Flory, S.L. (2010) Management of *Microstegium vimineum* invasions and recovery of resident plant communities. *Restoration Ecology* 18(1), 103–112. DOI: 10.1111/j.1526-100X.2008.00425.x.

Flory, S.L. and Clay, K. (2010a) Non-native grass invasion alters native plant composition in experimental communities. *Biological Invasions* 12(5), 1285–1294. DOI: 10.1007/s10530-009-9546-9.

Flory, S.L. and Clay, K. (2010b) Non-native grass invasion suppresses forest succession. *Oecologia* 164(4), 1029–1038. DOI: 10.1007/s00442-010-1697-y.

Flory, S.L. and Clay, K. (2013) Pathogen accumulation and long-term dynamics of plant invasions. *Journal of Ecology* 101(3), 607–613. DOI: 10.1111/1365-2745.12078.

Flory, S.L., Kleczewski, N. and Clay, K. (2011) Ecological consequences of pathogen accumulation on an invasive grass. *Ecosphere* 2(10), art120. DOI: 10.1890/ES11-00191.1.

Flory, S.L., Alba, C., Clay, K., Holt, R.D. and Goss, E.M. (2018) Emerging pathogens can suppress invaders and promote native species recovery. *Biological Invasions* 20(1), 5–8. DOI: 10.1007/s10530-017-1438-9.

Freestone, A.L., Ruiz, G.M. and Torchin, M.E. (2013) Stronger biotic resistance in tropics relative to temperate zone: effects of predation on marine invasion dynamics. *Ecology* 94(6), 1370–1377. DOI: 10.1890/12-1382.1.

Fridley, J.D., Stachowicz, J.J., Naeem, S., Sax, D.F., Seabloom, E.W. *et al.* (2007) The invasion paradox: reconciling pattern and process in species invasions. *Ecology* 88(1), 3–17. DOI: 10.1890/0012-9658(2007)88[3:TIPRPA]2.0.CO;2.

Gao, L., Zan, Q., Li, M., Guo, Q., Hu, L. *et al.* (2013) Soil fungi of three native tree species inhibit biomass production and shift biomass allocation of invasive *Mikania micrantha* Kunth. *Chinese Science Bulletin* 58(7), 758–765. DOI: 10.1007/s11434-012-5394-6.

Golivets, M. and Wallin, K.F. (2018) Neighbour tolerance, not suppression, provides competitive advantage to non-native plants. *Ecology Letters* 21(5), 745–759. DOI: 10.1111/ele.12934.

Grace, J.B., Harrison, S. and Cornell, H. (2017) Is biotic resistance enhanced by natural variation in diversity? *Oikos* 126(10), 1484–1492. DOI: 10.1111/oik.04334.

Griffiths, C.J., Zuël, N., Jones, C.G., Ahamud, Z. and Harris, S. (2013) Assessing the potential to restore historic grazing ecosystems with tortoise ecological replacements. *Conservation Biology* 27(4), 690–700. DOI: 10.1111/cobi.12087.

Grutters, B.M.C., Roijendijk, Y.O.A., Verberk, W.C.E.P. and Bakker, E.S. (2017) Plant traits and plant biogeography control the biotic resistance provided by generalist herbivores. *Functional Ecology* 31(6), 1184–1192. DOI: 10.1111/1365-2435.12835.

Guyton, J.A., Pansu, J., Hutchinson, M.C., Kartzinel, T.R., Potter, A.B. *et al.* (2020) Trophic rewilding revives biotic resistance to shrub invasion. *Nature Ecology & Evolution* 4(5), 712–724. DOI: 10.1038/s41559-019-1068-y.

Hawkes, C. V. (2007) Are invaders moving targets? the generality and persistence of advantages in size, reproduction, and enemy release in invasive plant species with time since introduction. *The American Naturalist* 170(6), 832–843. DOI: 10.1086/522842.

Heckman, R.W., Halliday, F.W., Wilfahrt, P.A. and Mitchell, C.E. (2017) Effects of native diversity, soil nutrients, and natural enemies on exotic invasion in experimental plant communities. *Ecology* 98(5), 1409–1418. DOI: 10.1002/ecy.1796.

Herben, T., Mandák, B., Bímová, K. and Münzbergová, Z. (2004) Invasibility and species richness of a community: a neutral model and a survey of published data. *Ecology* 85(12), 3223–3233. DOI: 10.1890/03-0648.

Hewitt, C.L. (2002) Distribution and biodiversity of Australian tropical marine bioinvasions. *Pacific Science* 56(2), 213–222. DOI: 10.1353/psc.2002.0016.

Hierro, J.L., Maron, J.L. and Callaway, R.M. (2005) A biogeographical approach to plant invasions: the importance of studying exotics in their introduced and native range. *Journal of Ecology* 93(1), 5–15. DOI: 10.1111/j.0022-0477.2004.00953.x.

Holle, B.V. and Simberloff, D. (2005) Ecological resistance to biological invasion overwhelmed by propagule pressure. *Ecology* 86(12), 3212–3218.

Hollebone, A.L. and Hay, M.E. (2007) Propagule pressure of an invasive crab overwhelms native biotic resistance. *Marine Ecology Progress Series* 342, 191–196. DOI: 10.3354/meps342191.

Hulvey, K.B. and Teller, B.J. (2018) Site conditions determine a key native plant's contribution to invasion resistance in grasslands. *Ecology* 99(6), 1257–1264. DOI: 10.1002/ecy.2227.

Inderjit. and Cahill, J.F (2015) Linkages of plant-soil feedbacks and underlying invasion mechanisms. *AoB PLANTS* 7, plv022–plv022. DOI: 10.1093/aobpla/plv022.

Inderjit. and van der Putten, W.H (2010) Impacts of soil microbial communities on exotic plant invasions. *Trends in Ecology & Evolution* 25(9), 512–519. DOI: 10.1016/j.tree.2010.06.006.

Johnson, B.E. and Cushman, J.H. (2007) Influence of a large herbivore reintroduction on plant invasions and community composition in a California grassland. *Conservation Biology* 21(2), 515–526. DOI: 10.1111/j.1523-1739.2006.00610.x.

Josh Donlan, C., Berger, J., Bock, C.E., Bock, J.H., Burney, D.A. *et al.* (2006) Pleistocene rewilding: an optimistic agenda for twenty-first century conservation. *The American Naturalist* 168(5), 660–681. DOI: 10.1086/508027.

Kalisz, S., Spigler, R.B. and Horvitz, C.C. (2014) In a long-term experimental demography study, excluding ungulates reversed invader's explosive population growth rate and restored natives. *Proceedings of the National Academy of Sciences* 111(12), 4501–4506. DOI: 10.1073/pnas.1310121111.

Kimbro, D.L., Cheng, B.S. and Grosholz, E.D. (2013) Biotic resistance in marine environments. *Ecology Letters* 16(6), 821–833. DOI: 10.1111/ele.12106.

Knight, T.M., Dunn, J.L., Smith, L.A., Davis, J. and Kalisz, S. (2009) Deer facilitate invasive plant success in a Pennsylvania forest understory. *Natural Areas Journal* 29(2), 110–116. DOI: 10.3375/043.029.0202.

Knops, J.M.H., Tilman, D., Haddad, N.M., Naeem, S., Mitchell, C.E. *et al.* (1999) Effects of plant species richness on invasion dynamics, disease outbreaks, insect abundances and diversity. *Ecology Letters* 2(5), 286–293. DOI: 10.1046/j.1461-0248.1999.00083.x.

Kuebbing, S., Rodriguez-Cabal, M.A., Fowler, D., Breza, L., Schweitzer, J.A. *et al.* (2013) Resource availability and plant diversity explain patterns of invasion of an exotic grass. *Journal of Plant Ecology* 6(2), 141–149. DOI: 10.1093/jpe/rts018.

Lemoine, N.P., Burkepile, D.E. and Parker, J.D. (2016) Quantifying differences between native and introduced species. *Trends in Ecology & Evolution* 31(5), 372–381. DOI: 10.1016/j.tree.2016.02.008.

Lemoine, N.P., Doublet, D., Salminen, J.-P., Burkepile, D.E. and Parker, J.D. (2017) Responses of plant phenology, growth, defense, and reproduction to interactive effects of warming and insect herbivory. *Ecology* 98(7), 1817–1828. DOI: 10.1002/ecy.1855.

Lemoine, N.P., Shue, J., Verrico, B., Erickson, D., Kress, W.J. *et al.* (2015) Phylogenetic relatedness and leaf functional traits, not introduced status, influence community assembly. *Ecology* 96(10), 2605–2612. DOI: 10.1890/14-1883.1.

Levine, J.M. (2000) Species diversity and biological invasions: relating local process to community pattern. *Science* 288(5467), 852–854. DOI: 10.1126/science.288.5467.852.

Levine, J.M., Adler, P.B. and Yelenik, S.G. (2004) A meta-analysis of biotic resistance to exotic plant invasions. *Ecology Letters* 7(10), 975–989. DOI: 10.1111/j.1461-0248.2004.00657.x.

Levine, J.M. and D'Antonio, C.M. (1999) Elton revisited: a review of evidence linking diversity and invasibility. *Oikos* 87(1), 15–26. DOI: 10.2307/3546992.

Lind, E.M. and Parker, J.D. (2010) Novel weapons testing: are invasive plants more chemically defended than native plants? *PLoS ONE* 5(5), e10429. DOI: 10.1371/journal.pone.0010429.

Liu, H. and Stiling, P. (2006) Testing the enemy release hypothesis: a review and meta-analysis. *Biological Invasions* 8(7), 1535–1545. DOI: 10.1007/s10530-005-5845-y.

Lonsdale, W.M. (1999) Global patterns of plant invasions and the concept of invasibility. *Ecology* 80(5), 1522–1536. DOI: 10.1890/0012-9658(1999)080[1522:GPOPIA]2.0.CO;2.

Lowe, S., Browne, M., Boudjelas, S. and De Poorter, M. (2004) 100 of the world's worst invasive alien species: A selection from the Global Invasive Species Database. *The Invasive Species Specialist Group (ISSG) a specialist group of the Species Survival Commission (SSC) of the World Conservation Union.*

Manea, A. and Leishman, M.R. (2011) Competitive interactions between native and invasive exotic plant species are altered under elevated carbon dioxide. *Oecologia* 165(3), 735–744. DOI: 10.1007/s00442-010-1765-3.

Maron, J.L., Auge, H., Pearson, D.E., Korell, L., Hensen, I. *et al.* (2014) Staged invasions across disparate grasslands: effects of seed provenance, consumers and disturbance on productivity and species richness. *Ecology Letters* 17(4), 499–507. DOI: 10.1111/ele.12250.

Maron, J.L. and Vila, M. (2001) When do herbivores affect plant invasion? Evidence for the natural enemies and biotic resistance hypotheses. *Oikos* 95(3), 361–373. DOI: 10.1034/j.1600-0706.2001.950301.x.

Mason, T.J., French, K. and Jolley, D.F. (2017) Functional richness and identity do not strongly affect invasibility of constructed dune communities. *PLoS ONE* 12(1), e0169243. DOI: 10.1371/journal.pone.0169243.

McGinn, K.J., van der Putten, W.H., Hulme, P.E., Shelby, N., Weser, C. *et al.* (2018) The influence of residence time and geographic extent on the strength of plant-soil feedbacks for naturalised *Trifolium*. *Journal of Ecology* 106(1), 207–217. DOI: 10.1111/1365-2745.12864.

Mitchell, C.E., Agrawal, A.A., Bever, J.D., Gilbert, G.S., Hufbauer, R.A. *et al.* (2006) Biotic interactions and plant invasions. *Ecology Letters* 9(6), 726–740. DOI: 10.1111/j.1461-0248.2006.00908.x.

Mitchell, C.E., Blumenthal, D., Jarošík, V., Puckett, E.E. and Pyšek, P. (2010) Controls on pathogen species richness in plants' introduced and native ranges: roles of residence time, range size and host traits. *Ecology Letters* 13(12), 1525–1535. DOI: 10.1111/j.1461-0248.2010.01543.x.

Mitchell, C.E. and Power, A.G. (2003) Release of invasive plants from fungal and viral pathogens. *Nature* 421(6923), 625–627. DOI: 10.1038/nature01317.

Morrison, W.E. and Hay, M.E. (2011) Herbivore preference for native vs. exotic plants: generalist herbivores from multiple continents prefer exotic plants that are evolutionarily naïve. *PLoS ONE* 6(3), e17227. DOI: 10.1371/journal.pone.0017227.

Ng, J., Weaver, W.N. and Laport, R.G. (2019) Testing Darwin's Naturalization Conundrum using phylogenetic relationships: Generalizable patterns across disparate communities? *Diversity and Distributions* 25, 361–373.

Nijjer, S., Rogers, W.E. and Siemann, E. (2007) Negative plant–soil feedbacks may limit persistence of an invasive tree due to rapid accumulation of soil pathogens. *Proceedings of the Royal Society B* 274, 2621–2627.

Oduor, A.M.O., Leimu, R. and van Kleunen, M. (2016) Invasive plant species are locally adapted just as frequently and at least as strongly as native plant species. *Journal of Ecology* 104(4), 957–968. DOI: 10.1111/1365-2745.12578.

Ordonez, A., Wright, I.J. and Olff, H. (2010) Functional differences between native and alien species: a global-scale comparison. *Functional Ecology* 24(6), 1353–1361. DOI: 10.1111/j.1365-2435.2010.01739.x.

Paolucci, E.M., MacIsaac, H.J. and Ricciardi, A. (2013) Origin matters: alien consumers inflict greater damage on prey populations than do native consumers. *Diversity and Distributions* 19(8), 988–995. DOI: 10.1111/ddi.12073.

Papacostas, K.J., Rielly-Carroll, E.W., Georgian, S.E., Long, D.J., Princiotta, S.D. *et al.* (2017) Biological mechanisms of marine invasions. *Marine Ecology Progress Series* 565, 251–268. DOI: 10.3354/meps12001.

Parepa, M., Schaffner, U. and Bossdorf, O. (2013) Help from under ground: soil biota facilitate knotweed invasion. *Ecosphere* 4(2), art31. DOI: 10.1890/ES13-00011.1.
Parker, J.D., Burkepile, D.E. and Hay, M.E. (2006) Opposing effects of native and exotic herbivores on plant invasions. *Science* 311(5766), 1459–1461. DOI: 10.1126/science.1121407.
Parker, J.D., Burkepile, D.E., Lajeunesse, M.J. and Lind, E.M. (2012) Phylogenetic isolation increases plant success despite increasing susceptibility to generalist herbivores. *Diversity and Distributions* 18(1), 1–9. DOI: 10.1111/j.1472-4642.2011.00806.x.
Parker, J.D. and Hay, M.E. (2005) Biotic resistance to plant invasions? native herbivores prefer non-native plants. *Ecology Letters* 8(9), 959–967. DOI: 10.1111/j.1461-0248.2005.00799.x.
Parker, J.D., Richie, L.J., Lind, E.M. and Maloney, K.O. (2010) Land use history alters the relationship between native and exotic plants: the rich don't always get richer. *Biological Invasions* 12(6), 1557–1571. DOI: 10.1007/s10530-009-9568-3.
Pearson, D.E., Hierro, J.L., Chiuffo, M. and Villarreal, D. (2014) Rodent seed predation as a biotic filter influencing exotic plant abundance and distribution. *Biological Invasions* 16(5), 1185–1196. DOI: 10.1007/s10530-013-0573-1.
Pearson, D.E., Ortega, Y.K., Eren, Özkan. and Hierro, J.L. (2018) Community assembly theory as a framework for biological invasions. *Trends in Ecology & Evolution* 33(5), 313–325. DOI: 10.1016/j.tree.2018.03.002.
Peng, S., Kinlock, N.L., Gurevitch, J. and Peng, S. (2019) Correlation of native and exotic species richness: a global meta-analysis finds no invasion paradox across scales. *Ecology* 100(1), e02552. DOI: 10.1002/ecy.2552.
Pernilla Brinkman, E., Van der Putten, W.H., Bakker, E.-J. and Verhoeven, K.J.F. (2010) Plant-soil feedback: experimental approaches, statistical analyses and ecological interpretations. *Journal of Ecology* 98(5), 1063–1073. DOI: 10.1111/j.1365-2745.2010.01695.x.
Phillips, B.L., Kelehear, C., Pizzatto, L., Brown, G.P., Barton, D. *et al.* (2010) Parasites and pathogens lag behind their host during periods of host range advance. *Ecology* 91(3), 872–881. DOI: 10.1890/09-0530.1.
Policelli, N., Chiuffo, M.C., Moyano, J., Torres, A., Rodriguez-Cabal, M.A. *et al.* (2018) Pathogen accumulation cannot undo the impact of invasive species. *Biological Invasions* 20(1), 1–4. DOI: 10.1007/s10530-017-1439-8.
Price, J.N. and Pärtel, M. (2013) Can limiting similarity increase invasion resistance? A meta-analysis of experimental studies. *Oikos* 122(5), 649–656. DOI: 10.1111/j.1600-0706.2012.00121.x.
Reinhart, K.O., Packer, A., Van der Putten, W.H. and Clay, K. (2003) Plant-soil biota interactions and spatial distribution of black cherry in its native and invasive ranges. *Ecology Letters* 6(12), 1046–1050. DOI: 10.1046/j.1461-0248.2003.00539.x.
Reinhart, K.O. and Rinella, M.J. (2016) A common soil handling technique can generate incorrect estimates of soil biota effects on plants. *New Phytologist* 210(3), 786–789. DOI: 10.1111/nph.13822.
Rejmánek, M. (1996) Species richness and resistance to invasions. In: in Orians, G.H., Dirzo, R. and Cushman, J.H. (eds) *Biodiversity and Ecosystem Processes in Tropical Forests*. Springer Berlin Heidelberg (Ecological Studies), pp. 153–172.
Sax, D.F. (2001) Latitudinal gradients and geographic ranges of exotic species: implications for biogeography. *Journal of Biogeography* 28(1), 139–150. DOI: 10.1046/j.1365-2699.2001.00536.x.
Schemske, D.W., Mittelbach, G.G., Cornell, H.V., Sobel, J.M. and Roy, K. (2009) Is there a latitudinal gradient in the importance of biotic interactions? *Annual Review of Ecology, Evolution. and Systematics* 40, 245–269.
Schittko, C., Runge, C., Strupp, M., Wolff, S. and Wurst, S. (2016) No evidence that plant-soil feedback effects of native and invasive plant species under glasshouse conditions are reflected in the field. *Journal of Ecology* 104(5), 1243–1249. DOI: 10.1111/1365-2745.12603.
Schmidt, C.D., Hickman, K.R., Channell, R., Harmoney, K. and Stark, W. (2008) Competitive abilities of native grasses and non-native (*Bothriochloa* spp.) grasses. *Plant Ecology* 197(1), 69–80. DOI: 10.1007/s11258-007-9361-2.
Seebens, H., Blackburn, T.M., Dyer, E.E., Genovesi, P., Hulme, P.E. *et al.* (2017) No saturation in the accumulation of alien species worldwide. *Nature Communications* 8(1), 1–9. DOI: 10.1038/ncomms14435.
Seebens, H., Essl, F., Dawson, W., Fuentes, N., Moser, D. *et al.* (2015) Global trade will accelerate plant invasions in emerging economies under climate change. *Global Change Biology* 21(11), 4128–4140. DOI: 10.1111/gcb.13021.

Siemann, E. and Rogers, W.E. (2003) Herbivory, disease, recruitment limitation, and success of alien and native tree species. *Ecology* 84(6), 1489–1505. DOI: 10.1890/0012-9658(2003)084[1489:HDRLAS]2.0.CO;2.

Siemann, E. and Rogers, W.E. (2006) Recruitment limitation, seedling performance and persistence of exotic tree monocultures. *Biological Invasions* 8(5), 979–991. DOI: 10.1007/s10530-005-0825-9.

Smith, N.S. and Côté, I.M. (2019) Multiple drivers of contrasting diversity-invasibility relationships at fine spatial grains. *Ecology* 100(2), e02573. DOI: 10.1002/ecy.2573.

St. Clair, S.B. and Bishop, T.B.B. (2019) Loss of biotic resistance and high propagule pressure promote invasive grass-fire cycles. *Journal of Ecology* 107(4), 1995–2005. DOI: 10.1111/1365-2745.13156.

St. Clair, S.B., O'Connor, R., Gill, R. and McMillan, B. (2016) Biotic resistance and disturbance: rodent consumers regulate post-fire plant invasions and increase plant community diversity. *Ecology* 97(7), 1700–1711. DOI: 10.1002/ecy.1391.

Stohlgren, T.J., Barnett, D.T. and Kartesz, J.T. (2003) The rich get richer: patterns of plant invasions in the United States. *Frontiers in Ecology and the Environment* 1(1), 11–14. DOI: 10.1890/1540-9295(2003)001[0011:TRGRPO]2.0.CO;2.

Stricker, K.B., Harmon, P.F., Goss, E.M., Clay, K. Luke Flory, S, Flory, S.L. (2016) Emergence and accumulation of novel pathogens suppress an invasive species. *Ecology Letters* 19(4), 469–477. DOI: 10.1111/ele.12583.

Suding, K.N., Stanley Harpole, W., Fukami, T., Kulmatiski, A., MacDougall, A.S. *et al.* (2013) Consequences of plant-soil feedbacks in invasion. *Journal of Ecology* 101(2), 298–308. DOI: 10.1111/1365-2745.12057.

Tenhumberg, B., Suwa, T., Tyre, A.J., Russell, F.L. and Louda, S.M. (2015) Integral projection models show exotic thistle is more limited than native thistle by ambient competition and herbivory. *Ecosphere* 6(4), 1–18. DOI: 10.1890/ES14-00389.1.

van der Putten, W.H., Kowalchuk, G.A., Brinkman, E.P., Doodeman, G.T.A., van der Kaaij, R.M. *et al.* (2007) Soil feedback of exotic savanna grass relates to pathogen absence and mycorrhizal selectivity. *Ecology* 88(4), 978–988. DOI: 10.1890/06-1051.

van der Sande, M.T., Bruelheide, H., Dawson, W., Dengler, J., Essl, F. *et al.* (2020) Similar factors underlie tree abundance in forests in native and alien ranges. *Global Ecology and Biogeography* 29(2), 281–294. DOI: 10.1111/geb.13027.

van Kleunen, M., Weber, E. and Fischer, M. (2010) A meta-analysis of trait differences between invasive and non-invasive plant species. *Ecology Letters* 13(2), 235–245. DOI: 10.1111/j.1461-0248.2009.01418.x.

Vergés, A., Steinberg, P.D., Hay, M.E., Poore, A.G.B., Campbell, A.H. *et al.* (2014) The tropicalization of temperate marine ecosystems: climate-mediated changes in herbivory and community phase shifts. *Proceedings of the Royal Society B: Biological Sciences* 281(1789), 20140846. DOI: 10.1098/rspb.2014.0846.

Verhoeven, K.J.F., Biere, A., Harvey, J.A. van der Putten, W.H, Putten, W.H.V.D. (2009) Plant invaders and their novel natural enemies: who is naïve? *Ecology Letters* 12(2), 107–117. DOI: 10.1111/j.1461-0248.2008.01248.x.

Von Holle, B. and Simberloff, D. (2004) Testing Fox's assembly rule: does plant invasion depend on recipient community structure? *Oikos* 105(3), 551–563. DOI: 10.1111/j.0030-1299.2004.12597.x.

Wang, J., Ge, Y., Zhang, C.B., Bai, Y. and Du, Z.K. (2013) Dominant functional group effects on the invasion resistance at different resource levels. *PLoS ONE* 8(10), e77220. DOI: 10.1371/journal.pone.0077220.

Xiong, W.E.N., YU, D.A.N., Wang, Q., Liu, C. and Wang, L. (2008) A snail prefers native over exotic freshwater plants: implications for the enemy release hypotheses. *Freshwater Biology* 6, ???–2263. DOI: 10.1111/j.1365-2427.2008.02058.x.

Yakob, L. and Mumby, P.J. (2011) Climate change induces demographic resistance to disease in novel coral assemblages. *Proceedings of the National Academy of Sciences* 108(5), 1967–1969. DOI: 10.1073/pnas.1015443108.

Yannelli, F.A., Koch, C., Jeschke, J.M. and Kollmann, J. (2017) Limiting similarity and Darwin's naturalization hypothesis: understanding the drivers of biotic resistance against invasive plant species. *Oecologia* 183(3), 775–784. DOI: 10.1007/s00442-016-3798-8.

Zhang, Z. and van Kleunen, M. (2019) Common alien plants are more competitive than rare natives but not than common natives. *Ecology Letters* 22(9), 1378–1386. DOI: 10.1111/ele.13320.

10 EICA 2.0: A General Model of Enemy Release and Defence in Plant and Animal Invasions

Richard Honor and Robert I. Colautti*

Department of Biology, Queen's University, 116 Barrie St. Kingston, ON, K7L 3N6 Canada

Abstract

Plants and animals have evolved a variety of strategies to limit the negative fitness consequences of natural enemies (i.e. herbivores, predators, parasites and pathogens). Demographic bottlenecks occurring during the invasion process reduce the number of co-introduced natural enemies, providing opportunities to study rapid evolution in environments with different or reduced enemy loads. Enemy release theory provides a set of hypotheses and predictions about the role of natural enemies in the proliferation of invasive species. This body of theory includes the Enemy Release Hypothesis (ERH) and the related Evolution of Increased Competitive Ability Hypothesis (EICA), but there is often confusion about these hypotheses and the data needed to test them. We introduce a simple, general model of enemy release to identify and clarify some of the key assumptions and predictions implicit in enemy release theory and its impacts on invasion. Although introduced populations likely benefit from a reduction in the direct fitness impacts of natural enemies in the early stages of invasion, an evolutionary shift in resource allocation from defence to growth and reproduction is much less likely and depends on a delicate balance between the fitness costs and benefits of defence and the fitness impacts of natural enemies in both the native and introduced ranges. Even when the abundance of natural enemies is lower in the introduced range, the majority of scenarios do not favour evolution of less defended genotypes that are more competitive or more fecund, contrary to predictions of EICA. Perhaps surprisingly, we find that the level of damage by natural enemies in field surveys is not generally a good parameter for testing enemy release theory. Instead, common garden experiments characterizing fitness reaction norms of multiple genotypes from the native and introduced range are crucial to estimate the historic rate of adaptive evolution or predict it into the future. Incorporating spatial autocorrelation and methods from population genetics can further improve our understanding of the role of enemy release and evolution in the proliferation of invasive species.

10.1 Introduction

Human activity facilitates the movement of species across geographic barriers, homogenizing the world's biodiversity with negative impacts to human health and well-being (Olden *et al.*, 2004; Pejchar and Mooney, 2009; Pyšek and Richardson, 2010). Of all the species associated with human activity, relatively few establish in new regions and proliferate to the point where they are considered invasive (van Kleunen *et al.*, 2018). A long-standing priority for research on invasive species is to understand which ecological and genetic factors contribute most to proliferation and spread.

Research on invasive species can help inform strategies to mitigate their ecological and economic impacts, but biological invasions also provide opportunities to study ecology and evolution in action, which in turn can inform ecological and evolutionary theory (Cadotte *et al.*, 2006; Sax *et al.*, 2007; Lee, 2016; Sargent *et al.*, 2017; Lu-Irving *et al.*, 2018). Consider the invasive plant *Lythrum salicaria* (purple loosestrife),

*Corresponding author: robert.colautti@queensu.ca

 Plant Invasions: The Role of Biotic Interactions (eds Anna Traveset and David M. Richardson)
DOI: 10.1079/9781789242171.0010

which has a reputation as a problematic invader in North America, where it invaded some time in the late 1700s (Thompson *et al.*, 1987). From a few or perhaps a single region of origin, *L. salicaria* established a foothold near Boston, on the east coast of North America. There it must have experienced new biotic interactions and a different combination of abiotic characteristics (e.g. photoperiod, mean temperature, temperature range); changes in the biotic and abiotic environment represent changes in both natural selection and the environmental cues that signal favourable growing conditions. This beachhead population could have contained only a fraction of the available genetic variation maintained throughout its native range. Despite these bottlenecks and an initial lag phase of ~150 years, purple loosestrife spread rapidly throughout North America, colonizing gradients of temperature, moisture and seasonality. How is it that this one species was able to establish and thrive in such a wide range of environments so quickly?

One explanation for the proliferation of invasive species is that they escape regulation from their natural enemies – i.e. herbivores, pathogens, parasites or predators. This idea was formalized into the Enemy Release Hypothesis (ERH) (Keane and Crawley, 2002), which has two main predictions: first, invasive species benefit from a loss of natural enemies, particularly specialists. The loss of natural enemies is more likely for invasions occurring over large spatial scales (e.g. between continents) with severe demographic bottlenecks during transport and introduction. Second, new enemies acquired in the introduced range are more likely to be generalists that also affect native species. This reduction in the number of enemies and shift to a greater proportion of generalists may benefit an invasive species via direct increases in survival and reproduction or via indirect benefits of reduced competition if generalists do more damage to native competitors.

The ERH remains controversial, despite its intuitive appeal and a surge of empirical research beginning in the early 2000s. Indeed, several reviews and meta-analyses on this topic appear to reach contradictory conclusions, at least at a broad level. For example, one early review found that *community comparisons*, which are typically comparisons between invasive vs native species or invasive vs non-invasive exotic species in the invasive range, did not support a reduction in herbivores or herbivore damage as predicted by the ERH (Colautti *et al.*, 2004). A more recent common garden comparison of 61 species found a similar result – invasive species received more damage than did native species and non-invasive exotic species (Schultheis *et al.*, 2015). In contrast, there is more consistent support for the ERH in *biogeographical comparisons*, which compare native vs introduced populations of the same species (Mitchell and Power, 2003; Colautti *et al.*, 2004; Bossdorf *et al.*, 2005; Heger and Jeschke, 2014). In other words, introduced populations generally appear to have fewer natural enemies when compared to conspecific populations in the native range but do not consistently benefit from enemy release relative to their direct competitors. It is unclear whether the biogeographical release from natural enemies translates directly to population growth or abundance in the introduced range; a lack of good demographic field data has made it difficult to test this prediction directly (Parker *et al.*, 2013).

The original formulation of the ERH (Keane and Crawley, 2002) is fundamentally an ecological hypothesis, but the ecology and demography of a population are intimately linked to the genetics of its constituents. Natural selection operating during invasion favours traits that increase survival and reproduction in the new environment, and evolution of these traits can in turn alter rates of population growth and spread (Phillips *et al.*, 2006; Phillips, 2009; Colautti and Barrett, 2013; Pierson *et al.*, 2015). Therefore, it may be important to consider evolutionary changes associated with enemy release. In addition to direct and indirect effects of enemy release on host demography predicted by the ERH, a reduction in specialists and a shift to a higher proportion of generalists may trigger evolutionary changes in the amount and type of plant defences. For example, energy or nutrient limitations on defences that are effective against specialist enemies will create inherent trade-offs between the production of costly defences and the ability to compete and produce offspring when natural enemies are absent (Coley *et al.*, 1985; Simms and Triplett, 1994; Mauricio and Rausher, 1997; Strauss and Agrawal, 1999; Hahn and Maron, 2016). In other words, a reduction in specialist enemies should relax selection for costly defences, which could lead to the evolution of genotypes that are less defended but more competitive with higher rates of reproduction in the absence of natural enemies. This idea is the Evolution of Increased Competitive Ability Hypothesis (EICA) (Blossey and Nötzold, 1995).

In this chapter, we briefly review the enemy release theory of biological invasions with an emphasis on testable, quantitative predictions from a modern understanding of evolutionary ecology. In doing so, we attempt to clarify some of the challenges of comparative studies and identify promising avenues of future research on the role of enemy release in the spread of invasive species.

10.2 25 Years of the EICA

The common garden design is an elegant and powerful method for quantifying genetic variation underlying ecologically important traits, with a long history in the study of plant evolution. Just a few decades after Linnaeus proposed the original species concept (Linnaeus, 1735), and half a century before the rediscovery of Gregor Mendel's theory of heritability, the French Botanist Claude Thomas Alexis Jordan reported that phenotypic differences between populations of the common plant *Lobularia maritima* were maintained when grown in a common environment (Jordan, 1846). This work was initially ridiculed for proposing that the trait differences maintained in a common garden could be used as a basis for taxonomically splitting *L. maritima* into multiple species. It would be almost 100 years until the Modern Synthesis and a theory of population genetics could explain the evolution and maintenance of genetic differences among geographic populations of a single species. Despite the initial criticism, Jordan's approach set a precedent for later experiments testing the genetic basis of phenotypic differentiation by Gaston Bonnier (Bonnier, 1890), Göte Turesson (Turesson, 1922), and Jens Clausen and William Hiesey (Clausen and Hiesey, 1958). The common garden approach is now widely used in ecological genetics and remains an invaluable experimental tool for testing the genetic basis of ecologically relevant traits, even in the modern genomics era (Anderson *et al.*, 2011; Tiffin and Ross-Ibarra, 2014; Lowry *et al.*, 2019).

In the 1990s, a German researcher named Bernd Blossey began a simple common garden experiment that would influence hundreds of research papers on invasive species over the following decades. Working with Rolf Nötzold at the Christian-Albrects-University (CAU) in Kiel, Germany, seeds of *L. salicaria* were collected from a natural population near Ithaca, New York and another near Lucelle, Switzerland. In a common garden experiment the following year, they grew plants from each location to test for genetic differences in the growth and herbivory of *L. salicaria*.

Lythrum salicaria has a long history in the study of evolution. Darwin's classic pollination experiments with *L. salicaria* explored the reproductive consequences of tristyly and self-incompatibility in his opus on the evolution of flowering plants (Darwin, 1877). Ronald A. Fisher used *L. salicaria* to investigate the genetic basis of tristyly and polyploid inheritance (Fisher, 1949, 1941; Fisher and Martin, 1947; Fisher and Mather, 1943). In some sense, the study of Blossey and Nötzold (1995) is a continuation of this long and rich history of genetic studies involving *L. salicaria*, but a key difference is timescale. Whereas Fisher and Darwin assumed that evolutionary changes occurred on the scale of centuries to millennia, Blossey and Nötzold were curious to see whether evolution could have occurred rapidly enough in North American populations to be detectable ~200 years after its initial introduction. Given the early insights into fundamental genetic and evolutionary processes, what could modern studies of biological invasions reveal about adaptive evolution in novel environments on contemporary timescales?

Blossey and Nötzold reasoned that genotypes in North American populations would leave behind specialist natural enemies and therefore would have less need for costly defences, allowing them to invest more into competition, growth and reproduction. Growing the Lucelle and Ithaca genotypes in a common garden in Germany, they tested the hypothesis that North American populations evolved in the absence of *Galerucella pusilla* or *Hylobius transversovittatus* beetles for almost two centuries, resulting in more competitive genotypes with weaker herbivore defences. At the end of two separate growing seasons in 1991 and 1992, *L. salicaria* plants sampled from New York had roughly three times the biomass of plants from Switzerland, in support of the hypothesis. The root-feeding weevil (*H. transversovittatus*) had both lower survival and lower growth on the Swiss genotypes of *L. salicaria* compared to the plants from Ithaca, but the chrysomelid leaf-feeder (*G. pusilla*) did just as well on both genotypes. Blossey and Nötzold reasoned that the results of *H. transversovittatus* supported their hypothesis. They also argued that the lack of significant effects on the survival and growth of *G. pusilla* could reflect differences in defence investment, since leaf tissue is more easily replaced than roots, and therefore less defended.

In 1995 these results were published as the first test of Blossey and Nötzold's hypothesis, the EICA (Blossey and Nötzold, 1995). The study has been criticized for experimental flaws, including its small sample size, the use of bulked seed collections, rearing herbivores on leaves in petri dishes rather than whole plants, and a lack of fitness measurements. These are fair criticisms, but it is worth emphasizing how important EICA has been for the study of invasive plants and herbivory more generally. In 1995 most biologists did not realize how rapidly populations can evolve in nature, and EICA inspired more rigorous experiments from several dozen research groups, creating a rich source of data for meta-analysis (Bossdorf *et al.*, 2005; Colautti *et al.*, 2009; Felker-Quinn *et al.*, 2013; Colautti and Lau, 2015; Rotter and Holeski, 2018). However, EICA remains controversial despite this wealth of data. For example, a recent meta-analysis (N = 30 studies) found general support for EICA (Rotter and Holeski, 2018), in contrast to a similar study published five years earlier (N = 27), which found 'no support across invasive species for EICA's predictions' (Felker-Quinn *et al.*, 2013). Later in this chapter, we explore some of the reasons for this discrepancy.

A more recent variant of EICA is the Shifting Defence Hypothesis (SDH), which predicts a shift in the types of defence rather than a reduction in defences per se (Müller-Schärer *et al.*, 2004; Joshi and Vrieling, 2005). The rationale is that a shift in the community of natural enemies away from specialists toward generalists may select for different types of defence, rather than an overall reduction predicted by EICA. The SDH model assumes that plants use different defences for specialist and generalist herbivores. For example, a recent meta-analysis reported an increase in pyrrolizidine alkaloids but not digestibility-reducing defences in introduced genotypes relative to native genotypes (Doorduin and Vrieling, 2011). Chemical defences like pyrrolizidine alkaloids are known for toxic effects on generalists but can be beneficial to specialists that have evolved resistance mechanisms; these specialists can even sequester the plant chemicals as defences against their own enemies. However, even well-adapted specialists are negatively impacted by the chemical defences of their hosts and available data suggests that the specialist—generalist trade-off is only true for some chemicals and some species, and is not a general rule for plants (Koricheva *et al.*, 2004; Ali and Agrawal, 2012).

The ERH, EICA and SDH represent three of many overlapping formulations of what might be considered a general theory of enemy release. For example, the ERH posits that invasive species proliferate due to a reduced regulation from enemies that increases rates of survival and reproduction. In contrast, EICA and SDH imply that a loss of enemies has limited benefits initially, and it is the evolutionary reallocation of resources that gradually increases survival and reproduction. As we will see later, this distinction is crucial for understanding the evolution of defence and the role of trade-offs and enemy release in the proliferation of invasive species. Some authors have framed enemy release as a 'hierarchy of hypotheses' that breaks down larger hypotheses in ecology and evolution into smaller testable hypotheses that can be tackled by individual studies (Heger and Jeschke, 2014; Jeschke, 2014). Others have argued for a distinction between hypotheses and predictions (Farji-Brener and Amador-Vargas, 2014), though to our reading this distinction might be semantic. In this chapter, we introduce a general model of enemy release to provide a different perspective on the assumptions and predictions of enemy release theory, and the conditions under which ERH and EICA are likely to be found in nature.

10.3 A General Model of Enemy Release

Adaptive evolution is fundamentally an optimization algorithm. Genetic variants that improve survival and reproduction increase in frequency at the expense of alternative alleles with lower fitness. The resulting evolution of phenotypic traits is known as the evolutionary response to natural selection. Although natural selection favours genotypes with higher fitness, the evolution of a single, optimal genotype can be limited by genetic constraints that impose fitness trade-offs among traits. For example, an allele with a pleiotropic effect that delays reproduction but increases growth will limit adaptive evolution if natural selection favours both earlier flowering and larger size. Similarly, many chemical defences require nutrients like nitrogen or phosphorous that are often limited in natural populations (Lambers and Pons, 2008). Under nutrient limitation, an allele that upregulates production of a defence chemical would have an associated fitness cost by reducing allocation of limited resources to growth and development. These examples demonstrate how fitness trade-offs can limit the evolution of more adaptive genotypes.

Even without measuring specific traits or biochemical pathways, trade-offs can be quantified by measuring fitness of the same genotype reared in two or more environments. In cases where genotypes cannot be replicated, genetic families or populations can be a reasonable alternative. Ideally, these genetic groups would be assessed across a gradient of environments to compare their fitness reaction norms. A 'fitness reaction norm' is a graph showing how the fitness of one or more genotypes changes across environments. When the fitness reaction norms of two genotypes cross each other, then there is a genotype-by-environment (G × E) interaction for fitness with a trade-off that causes environment-dependent selection. That is, the genotype favoured by natural selection depends on the environment.

The fitness reaction norm is a useful tool for exploring adaptive evolution over contemporary timescales, but is rarely applied to studies of enemy release theory (e.g. ERH, EICA and SDH) or biological invasions generally. In this section we introduce a general model of enemy release and defence evolution based on a simple defence trade-off model. As a focal point for developing this model and its implications for fundamental research and the spread of invasive species, we focus on four main predictions that are key to testing the alternative hypotheses of enemy release theory: (i) A reduction in natural enemies in the introduced range, (ii) relaxes selection for costly defences, (iii) favouring genetic variants with weaker (or different) defences, (iv) increasing growth and reproduction.

10.3.1 A reduction in natural enemies

Herbivores can have substantial effects on the growth, development and reproduction of their plant hosts (Turcotte *et al.*, 2011). In response, plants have evolved a variety of physical and chemical defences to mitigate the fitness impacts of their natural enemies (Dangl and Jones, 2001). Animals have also evolved a variety of physical, chemical and behavioural defences against their natural enemies. Regardless of the type of organism and mechanism of defence, a natural enemy must, by definition, have negative effects on the fitness of its host. A release from the effects of natural enemies can occur in two ways: i) a *regulatory release* that acts immediately on population growth trajectories as individual plants benefit from increased rates of survival and reproduction; and ii) a *compensatory release*, which is a slower, evolutionary shift from well-defended genotypes to those that invest more into growth and reproduction. As noted in the introduction, empirical studies generally support a biogeographical release from natural enemies, which may lead to relaxed selection on costly defences. Both regulatory (ecological) and compensatory (evolutionary) release can increase fitness components in an invading species, which in turn increase vital rates of expanding populations.

10.3.2 Relaxes selection for costly defences

To understand why the strength of natural selection is of central importance to EICA, consider

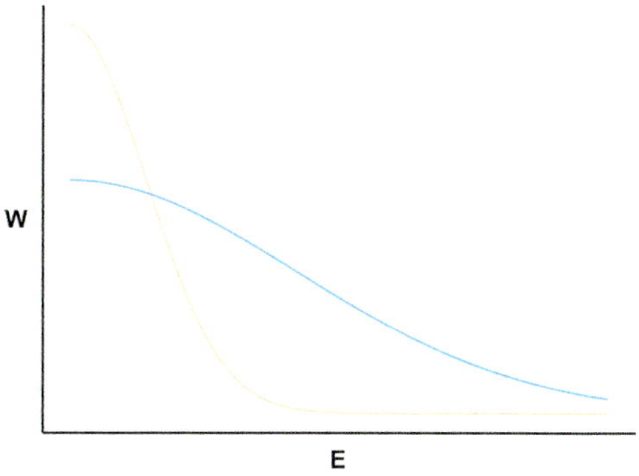

Fig. 10.1. Fitness reaction norms for a pair of genotypes with weak (susceptible, light orange) and strong defences (resistant/tolerant, blue). Fitness (*W*) is highest for both genotypes when herbivore abundance (*E*) is zero. The resistant/tolerant genotype maintains fitness at higher enemy abundance but pays a cost of lower fitness when natural enemies are rare or absent.

the relationship between the number or density (hereafter 'abundance') of enemies (*E*) in a plant population and the fitness of an average host plant (*W*). An important assumption of EICA is that genotypes differ in their defences, and the fitness costs of those defences change the relationship between enemy abundance and fitness. An example of this is shown in Fig. 10.1, in which two genotypes (or genetic families or populations) differ in their defence strategy (susceptible vs resistant). Resistant genotypes perform better when enemies are abundant but suffer a cost that reduces fitness relative to susceptible genotypes when enemies are rare.

In addition to chemical and physical defences that *resist* enemies, or behaviours that *avoid* enemies, individuals may differ in their capacity to *tolerate* damage from natural enemies. Tolerance can manifest through a number of mechanisms, including changes in growth form (e.g. secondary branching) or reallocation of resources from roots to shoots (Strauss and Agrawal, 1999). More generally, a plant is defined as tolerant to herbivory if its fitness is relatively unaffected by the damage it sustains from natural enemies. The general model we propose is therefore agnostic to the type of defence. For example, the two genotypes in Fig. 10.1 may receive the same amount of damage at any given level of *E*, however, the tolerant genotype exhibits a reduced slope across the reaction norm as its fitness is less affected by damage than the susceptible genotype. Similarly, the fitness of a plant that is highly resistant will not change much in the presence of enemies, as it experiences less damage than the susceptible genotype, resulting in a weaker slope in Fig. 10.1. Thus, despite the variety and complexity of defence mechanisms, the combined effects of resistance and tolerance traits can be captured in aggregate by the relationship between enemy abundance (*E*) and host fitness (*W*). By host fitness we mean absolute fitness (i.e. survival and reproduction). The relationship between host fitness and enemy abundance, as shown in Fig. 10.1, provides a convenient theoretical and empirical basis for testing enemy release theory without requiring knowledge of the specific traits or defence mechanisms underlying the relationship. Instead, we consider defence broadly as the combined benefit of resistance, avoidance and tolerance.

Plant populations are likely to contain many different types of enemies that vary in intensity and effects on host fitness. However, if the fitness effects of each type of enemy are roughly additive, then the abundance of each enemy type (E_j) can be weighted by its relative impact on fitness (w_j), yielding a weighted average metric of enemy abundance:

$$E = \sum w_j E_j$$

Thus, the total, weighted abundance of all enemies can be used as the enemy abundance parameter (*E*) in the model in Fig. 10.1. However, non-additive effects may also occur. For example, the presence of a pathogen may compound the damage incurred from a herbivore, or one type of enemy may reduce feeding by another. In cases like these, the relationship between the abundance of individual enemies (E_i) and host fitness (*W*) might have to be examined separately for different combinations of enemies. Nevertheless, the additive model will still be a good approximation in cases where interactive effects are weak relative to additive effects, or if at least one of a pair of strongly interacting enemies is at relatively low abundance in natural populations.

10.3.3 Favouring genetic variants with weaker (or different) defences

Fitness trade-offs are a fundamental feature of enemy release theory because they create environment-dependent selection in which the optimum genotype depends on the abundance of natural enemies (*E*). A highly defended genotype will become fixed by natural selection unless there is a cost to defence that favours less defended genotypes at low enemy abundance. Fig. 10.1 shows an example of the expected crossing reaction norms, where a genotype that is more resistant or tolerant to natural enemies has lower fitness when enemies are absent because of the cost of defence. The more resistant/tolerant genotype can, by definition, maintain fitness as enemy abundance increases but defences come at a cost. Together, the costs and benefits of defence create a crossing reaction norm for fitness with different genotypes favoured in different enemy environments.

One major benefit of reframing the ERH and EICA in terms of fitness reaction norms is that our model does not require knowledge of the specific mechanisms of defence, the taxonomy of the host plant and its natural enemies, or the genetic and ecological basis of defence trade-offs. Other fitness functions are possible in this model, but generally a defence trade-off model will include three main features: (i) the maximum fitness of each curve occurs when enemies are absent ($E=0$); (ii) enemies reduce fitness of all genotypes but defence ameliorates their effects; (iii) there is a fitness cost associated with defence that scales with its benefit. Although all defences have a theoretical cost, the cost—benefit ratio of a defence trait depends on many genetic and ecological factors (Strauss *et al.*, 2002). For example, if a chemical defence is energetically cheap to make but very effective against herbivores, then the cost—benefit ratio is low and genes for the defence should increase in frequency. If a chemical defence requires a great deal of energy or limiting nutrients and is required in abundance to confer resistance, then the cost—benefit ratio is high, and natural selection will tend to purge these defence genes. However, when the cost—benefit ratio is more balanced, then selection is more environment-dependent, maintaining genetic variation for defence traits. These three cost—benefit scenarios are shown in Fig. 10.2. Generally speaking, natural selection favours susceptible genotypes

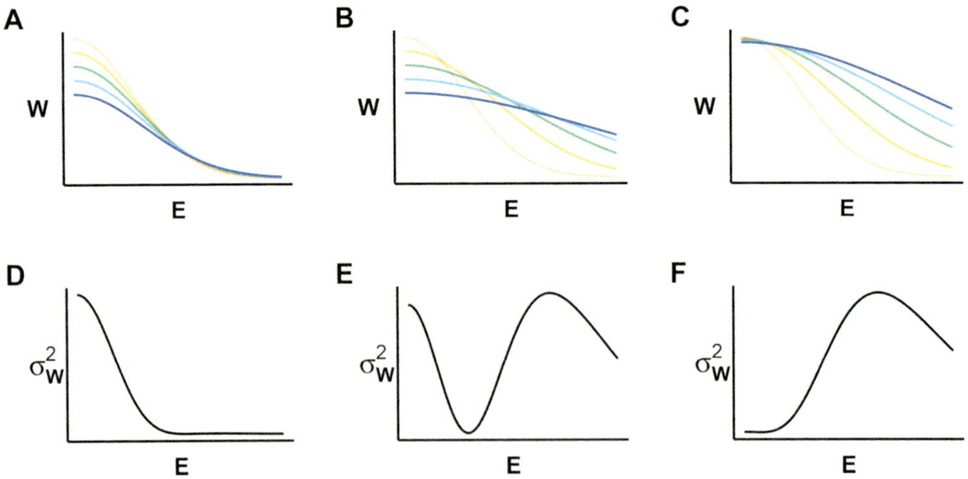

Fig. 10.2. A general defence trade-off model of defence traits along a gradient of enemy abundance. Fitness reaction norms (top row) show the relationship between enemy abundance (*E*) and fitness (*W*) for five different genotypes with defence allocation ranging from low (light red) to moderate (green) to high (dark blue). The three columns show different scenarios in which the defence cost—benefit ratios are high (A,D), intermediate (B,E) or low (C,F). The bottom row shows the variance in fitness for each of the three scenarios, with high variance corresponding to a rapid evolutionary response to selection.

when the cost—benefit ratio is high (Fig. 10.2a), and defended genotypes when the cost—benefit ratio is low (Fig. 10.2c). When costs and benefits are relatively balanced, then the outcome depends on the natural enemies in the environment (Fig. 10.2b).

The evolutionary response to selection on a defence trait is proportional to the additive genetic covariance between the trait and fitness (Robertson, 1966; Price, 1970). When the trait of interest is fitness (Fig. 10.2a,b,c), then the genetic variance for fitness (σ^2_w) provides an estimate of the expected rate of adaptive evolution (Fig. 10.2d,e,f). Selection favours genotypes with weak defences when costs are too high because the fitness benefits of defence are offset by their costs, resulting in very little variance in relative fitness when enemies are present (Fig. 10.2a,d). Conversely, when the costs of defence are too low, then the variance in relative fitness is low under enemy release and there is no evolutionary shift away from defence traits (Fig. 10.2c,f). A balance between the benefits and costs of defence is therefore necessary for the evolution of defence traits predicted by EICA and SDH.

When the costs and benefits of defence are balanced (Fig. 10.2b), then there are two peaks in the variance of relative fitness (Fig. 10.2e). In cases like this, well-defended genotypes are favoured when enemies are abundant and less defended genotypes are favoured when enemies are rare or absent. At moderate levels of enemy abundance, selection is weak because the fitness cost of resistance is balanced by the fitness benefits of defence (i.e. the 'trough' in Fig. 10.2e). At the highest levels of enemy abundance, genotypes are eventually overwhelmed and both fitness and the variance in fitness approach zero.

10.3.4 Increasing growth and reproduction

Enemy release in our model is simply a shift from high to low abundance of natural enemies (E), resulting in an increase in fitness (W). This increase in fitness is realized through higher rates of survival and reproduction associated directly with lower rates of herbivory. In addition to this initial increase in fitness owing to direct regulatory release, EICA predicts a compensatory increase in fitness as selection favours genotypes that are less defended but invest more into growth and reproduction. This is the usual formulation of EICA, but the scenarios described previously and shown in Fig. 10.2 reveal a crucial assumption that we believe has been largely overlooked: the cost of defence, relative to its benefit, must be (i) high enough for selection to favour weaker defences when enemies are rare, and (ii) low enough to favour defences when enemies are abundant (Fig. 10.2b). These costs are relative to the benefits of defence, which in our model determine the shape of the fitness reaction norm. The cost—benefit ratio of defence is perhaps the most important metric for EICA, yet empirical studies of EICA rarely measure costs or benefits as defined in our model, focusing instead on measurements of enemy abundance or host defence similar to those originally reported by Blossey and Nötzold (1995).

Another counterintuitive result of the general model in Fig. 10.2 is that a large reduction in the abundance of natural enemies between the native and introduced range may not often translate to a change in natural selection. Instead, we must consider the abundance of enemies in each range *relative to* the shapes of the fitness reaction norms. The scenario predicted by EICA depends on both the shape of the fitness reaction norm and the degree of release from enemies (Fig. 10.3). These scenarios are discussed in more detail in the next section and show that relatively unique circumstances must be met before the evolutionary shifts in defence as predicted by EICA can occur. For example, only one of the nine scenarios in Fig. 10.3 would favour a shift from higher to lower defences during invasion.

10.4 Predictions and Recommendations

So far in this chapter, we have reviewed enemy release theory and developed a simple but general model of enemy release. In this section we highlight predictions from the model to help identify new and less explored avenues of research, and to reinterpret some of the confusing or contradictory results of empirical studies.

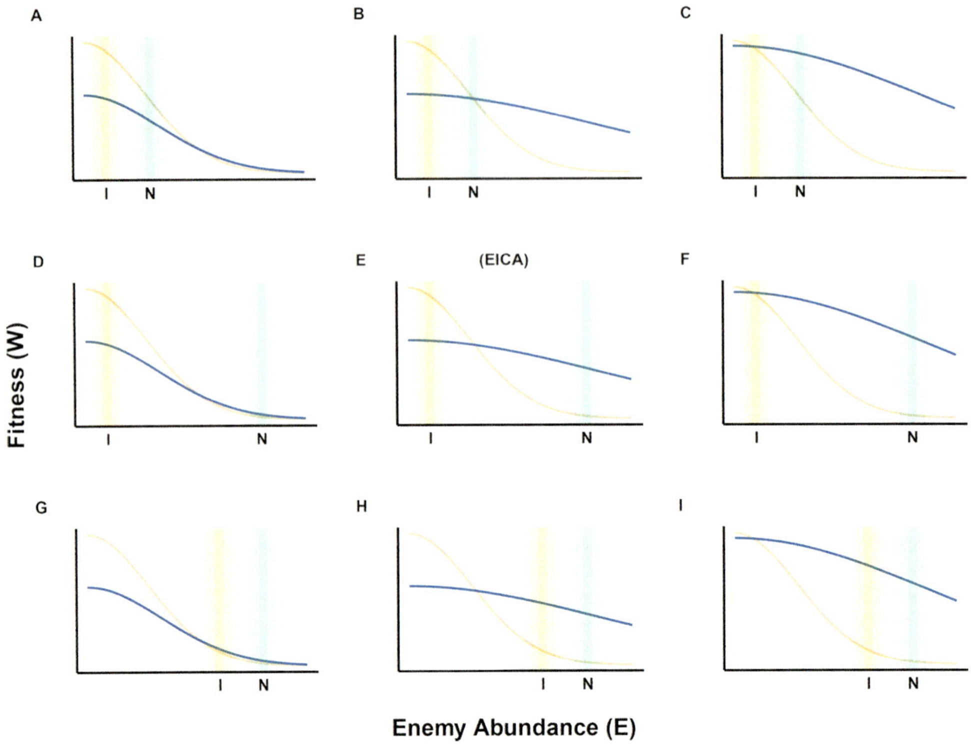

Fig. 10.3. Fitness reaction norms for the relationship between enemy abundance (*E*) and fitness (*W*) for a susceptible (light orange) and a resistant (blue) genotype. The three columns show different scenarios in which the defence cost–benefit ratios are high (A,D,G), moderate (B,E,H) or low (C,F,I). The three rows show different scenarios for variation in the abundance of enemies in the native (N, green) and introduced ranges (I, yellow). Of these nine scenarios, only one supports the evolution of increased competitive ability (EICA) in which natural selection for defence reverses between ranges (panel e).

10.4.1 Quantifying enemy release

A key premise of enemy release theory is that introduced populations experience a reduction in the abundance of natural enemies relative to the native range. The theoretical framework outlined in the previous section shows why biogeographical comparisons of the functional relationship between enemy abundance (*E*) and host fitness (*W*) are challenging but crucial to testing enemy release theory. In empirical studies of this sort, fitness can be measured on genotypes grown in a common environment as relative rates of survival and reproduction. Enemy abundance (*E*) can then be manipulated to characterize fitness reaction norms across levels that include enemy abundances that would naturally occur in the native and introduced ranges. Returning to the seminal research on *L. salicaria* by Blossey and Nötzold, we might be tempted to ignore fitness and enemy abundance in natural populations and instead measure the performance of natural enemies on native and invasive genotypes, such as *G. pusilla* or *H. transversovittatus* beetles on *L. salicaria*, or we might measure the damage inflicted by these herbivores on their host genotypes.

Notwithstanding the other insights that might be gained from herbivory assays, our analysis suggests that these are generally not strong metrics for testing enemy release theory because translating damage to reductions in host fitness is not straightforward. Resistance alone is not a good measure of defence because plants can vary considerably in their tolerance, both within and among species (Strauss and

Agrawal, 1999; Hahn and Maron, 2016). If the variance in tolerance is high relative to damage, then damage alone will be a weak predictor of fitness. Likewise, measurements of herbivore performance are unlikely to be good predictors of host fitness because herbivores are unaffected by tolerance mechanisms and because growth rates, fecundity and survival of herbivores in feeding assays do not directly predict their population growth trajectories or equilibrium abundance (*E*) in natural populations. As an alternative to measuring host damage or resistance, or herbivore performance, fitness reaction norms directly quantify the relationship between the herbivore community and their effect on host fitness.

10.4.2 Trade-offs

A fitness trade-off between growth and defence is a key assumption of EICA. Traditionally, a trade-off is a negative genetic correlation between two traits with antagonistic effects on fitness, yet these are rarely quantified. Additionally, a strong trade-off in growth and defence is detectable as a negative genetic correlation for fitness measured in environments with high vs low enemy abundance. However, our model of enemy release shows why a lack of genetic correlation does not rule out the existence of a trade-off: a negative genetic correlation for fitness would only be evident if a range of genotypes were assessed at enemy abundance levels (*E*) at or near both peaks of fitness variance in Fig. 10.3f. At other values of E, a negative genetic correlation (i.e. trade-off) might not be detectable above experimental error in fitness estimates.

Negative genetic correlations among genotypes are a likely indication of trade-offs, but genetic correlations among population means should be interpreted cautiously. This could be evidence for a cost of defence – a trade-off between growth and defence. Alternatively, natural selection could be acting independently on growth and defence traits, producing a spurious correlation. In this second scenario, the alleles for growth and defence traits are correlated across populations (i.e. in linkage disequilibrium), resulting in a correlation that is not indicative of a fitness trade-off. Instead of comparing population means, genetic correlations can be measured more directly on genotypes or genetic families segregating within populations. This removes the effect of linkage disequilibrium (i.e. correlated allele frequencies across populations) that results from correlated vectors of selection acting on different populations. However, some degree of linkage disequilibrium may still occur within populations as a result of assortative mating or correlational selection. Another approach is to create segregating F2 recombinant inbred lines (RILs) generated by crossing genotypes from genetically distinct populations.

10.4.3 Genotype-by-environment interactions

A common approach in studies testing EICA is to compare native and introduced genotypes of the same species growing in the same environment. These comparative common garden experiments can be more informative than field surveys if enemy abundance (*E*) is manipulated. Comparing damage and fitness of replicated native and introduced genotypes in a common garden study provides a more direct test for genetic variation in defence traits, as predicted by EICA and our model of enemy release (Fig. 10.2). In contrast, field surveys of damage in natural populations cannot parse genetic differences from environmental effects. One interesting prediction of our general model occurs when the cost—benefit ratio is consistent with predictions of EICA (Fig. 10.2b,e). In cases like these, the benefit of being defended against enemies is balanced by the cost of the defence trait at intermediate levels of enemy abundance. At these values of enemy abundance, defence traits are effectively neutral with respect to natural selection, and genetic variation for defence traits is expected to be the highest. Thus, while field surveys cannot distinguish enemy release and EICA from other genetic and environmental effects, they are important for characterizing variation in the abundance of natural enemies to generate fitness reaction norms.

Empirical studies of EICA should use as many different genotypes or populations across the invasive and native range as is technically feasible. Plant populations are not homogeneous throughout their native and introduced ranges and spatial autocorrelation in genotype and environment can confound comparisons between native and introduced ranges (Colautti *et al.*, 2009; Colautti and Lau, 2015). Genotypes in our model are represented by fitness reaction norms along a gradient of enemy abundance. Fitness reaction norms must cross to create two peaks in fitness variance in order for predictions of EICA to hold (Fig. 10.2b,e). Once these requisites are satisfied, enemy abundance in the native and introduced ranges must straddle the point at which reaction norms cross so that selection for defence traits reverses between the native and introduced environments (Fig. 10.3e). We have only explored a few scenarios, in all of which the variance in enemy abundance across populations is equal, with no overlap between the native and introduced range. In reality, populations may vary in enemy abundance more among populations than between the native and introduced ranges. This requires further attention as invasive species with high variability in enemy abundance (E) are less likely to evolve a shift from defence to growth and reproduction.

10.4.4 Natural selection

A key prediction of EICA and our general model of enemy release is that selection for defence scales with the abundance of natural enemies. Linear selection gradients provide a relatively simple but effective measure of natural selection (Lande and Arnold, 1983). In practice, the strength of selection for or against defence is a measure of the standardized covariance between the trait and fitness, which provides an empirical estimate of the fitness variance in Fig. 10.2. If natural selection favours a shift in defences as predicted by EICA, then selection gradients for defence traits should be positive in the native range but negative in the introduced range. Yet, we are not aware of any study that has tested this. Selection gradients can also help to test alternative hypotheses by comparing the agents of selection (e.g. herbivory vs climate), and their relative strengths. For example, latitudinal differences in the length of the growing season have selected for clines in growth and phenology of *L. salicaria* (Colautti *et al.*, 2010). The fitness consequences of flowering too late in a northern environment are strong and of similar magnitude to the fitness effects of the biocontrol beetles introduced to control *L. salicaria* (Colautti and Barrett, 2013). Selection gradients also provide a quantitative estimate of the relative strength of selection from herbivores, pathogens and competitors, in order to test alternative hypotheses such as the Novel Weapons Hypothesis (Callaway and Ridenour, 2004) or the Enemy of My Enemy Hypothesis (Colautti *et al.*, 2004, p. 200; Flory and Clay, 2013). Selection gradients are a well-established method in evolutionary biology that has been underutilized in studies of biological invasions.

Given the potential insights to be gained by directly measuring natural selection in invasive species, it is surprising how few studies have done so. In fact, a recent meta-analysis of selection gradients measured in natural populations found that just 9 of 113 studies involved non-native plants or animals (Colautti and Lau, 2015). The probability of randomly sampling only 9 of 113 is low ($P < 0.001$) in a binomial model assuming at least 20% of potential study species are non-native. This appears to be a research bias against measuring selection gradients in non-native species, and we hope that students and researchers interested in the evolution of invasive species will help to fill this research gap.

10.4.5 Enemy release equilibrium

Our enemy release model shows why a common garden approach with experimental manipulation of enemy abundance can be very effective for testing enemy release theory. Ideally, a broad range of enemy abundance (E) with replicated genotypes sampled throughout the native and introduced ranges would allow for a full characterization of fitness reaction norms and the variance in fitness, as shown in Fig. 10.2. Field surveys of the geographic origins of sampled genotypes could help to understand the spatial and temporal heterogeneity in enemy abundance that define the evolutionary outcomes summarized in Fig. 10.3. However, logistical difficulties with large field experiments will likely constrain the number of genotypes and the

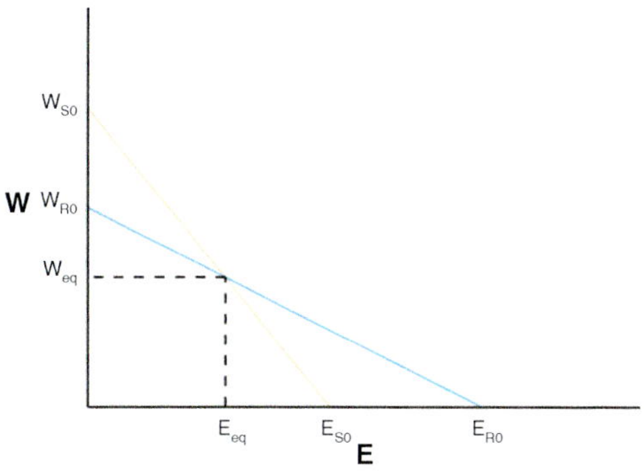

Fig. 10.4. Linear approximations of reaction norms for fitness (*W*) measured in a susceptible (*S*, light orange) and a resistant genotype (*R*, blue). The difference in y-intercepts ($W_{SO} - W_{RO}$) provides an estimate of the cost of resistance, while the slopes ($-W_0/E_0$) estimate of the benefits of defence.

levels of enemy abundance that can be included in a single experiment. Although fewer environments will be necessary, we suggest manipulating the abundance of natural enemies to more than two levels, in contrast to the majority of experiments testing EICA, which typically use only one or two rearing environments.

A full characterization of the fitness reaction norms is not possible when logistic and financial constraints limit the number of rearing environments. Instead, a subset of environments with linear models can provide a reasonable first-order approximation of enemy release. As shown in Fig. 10.4, simplifying the fitness reaction norms to linear equations allows estimation of four key parameters. First, the linear approximation of a fitness reaction norm for any genotype is defined by the x- and y- intercepts, thus:

$$W = W_0 - \frac{W_0}{E_0}E$$

Where E_0 and W_0 are the x- and y- intercepts, respectively. The y-intercept W_0 represents the maximum fitness of a genotype and should occur when enemies are absent. The difference in maximum fitness between susceptible (W_{SO}) and resistant (W_{RO}) genotypes provides an empirical estimate of the cost of resistance. The x-intercept (E_0) represents the abundance of enemies, which allows calculation of a slope (π) representing the benefit of defence:

$$\pi = -\frac{W_0}{E_0}$$

The theoretical maximum value of π is zero, which means that fitness is unaffected by enemy abundance. The point at which these reaction norms cross is the point at which the benefits of defence are offset by its costs, allowing us to define an equilibrium abundance of enemies E_{eq} (Fig. 10.4). For a pair of susceptible (S) and resistant (R) genotypes, the equilibrium abundance is:

$$E_{eq} = \frac{E_S E_R (W_S - W_R)}{W_S E_R - W_R E_S}$$

And the predicted fitness in this equilibrium environment is:

$$W_{eq} = \frac{W_S W_R (E_R - E_S)}{W_S E_R - W_R E_S}$$

Empirically, equilibrium estimates can help to test enemy release theory. For example, the predictions of EICA will only hold when the abundance of enemies in the introduced and native range are respectively below and above E_{eq}. This framework can also compare potential for evolution among populations or genotypes within the introduced range.

As articulated in the original study by Blossey and Nötzold (1995), introduced populations, having escaped some of their natural enemies, could evolve weaker defences in their introduced range and instead invest more into growth and reproduction. Perhaps the most surprising result of our enemy release model is that this original prediction is not likely to be true in many cases (Fig. 10.3). Instead, the evolution of defence depends on the cost—benefit ratio of the defence trait, which determines the shape of the fitness reaction norm (Fig. 10.2) and the levels of enemy abundance in both the native and introduced range (Fig. 10.3). There are many scenarios in which weaker defences will not evolve. Specifically, weaker defences will not evolve in our model under *any* of the following five conditions:

1. Genetic variation in cost is low relative to variation in defence, as shown in Fig. 10.3c,f and i.
2. Genetic variation in defence is low relative to its costs, as shown in Fig. 10.3a,d and g.
3. Native and introduced populations do not have large differences in enemy

abundance (E) relative to variation among populations within each range, as shown in Fig. 10.3a,b,c,g,h and i.

4. Native populations experience intermediate enemy abundances, such that selection does not favour high or low defences in the native range, as shown in Fig. 10.3b.
5. Introduced populations experience intermediate enemy abundances, such that selection does not favour high or low defences in the introduced range, as shown in Fig. 10.3h.

In scenarios 4 and 5, defence is essentially a neutral trait with respect to selection in one of the two ranges. This phenomenon is of particular interest given the stochastic nature of biological invasions in which founder effects and genetic drift can play a major role in evolution. When selection does not favour any level of defence, then repeated, independent invasions alone are sufficient to create heterogeneity in defence traits across introduced populations. In such cases, it is quite possible to observe shifts in defensive chemicals that have a quantifiable effect in experimental studies yet are neutral with respect to natural selection. This again highlights the importance of characterizing fitness reaction norms and measuring selection.

10.4.6 Increased performance

Of all the predictions relevant to EICA, an increase in growth and reproduction (or competitive ability) of introduced genotypes relative to natives is perhaps the most widely tested and the most difficult to interpret. Although dozens of common garden studies have compared physiology, growth and defence traits of native and introduced genotypes, spatial autocorrelation and non-random sampling have made it difficult to interpret or to draw generalities about evolution of native vs introduced genotypes (Colautti *et al.*, 2009; Colautti and Lau, 2015). Nevertheless, these studies demonstrate considerable variation among populations within the introduced range – a phenomenon worthy of further study in its own right. Newer approaches that reconstruct invasion history (i.e. demography) allow a more robust test of evolution during the invasion process (Agrawal *et al.*, 2015). Methods like these are crucial to disentangle evolutionary changes predicted by EICA from other drivers of evolutionary change (e.g. founder effects, genetic drift, local adaptation).

Much less is known about the relative performance of native and introduced populations in the wild. Absolute fitness is relatively easy to quantify by measuring rates of survival and reproduction, but translating fitness to realized abundance also requires consideration of inter- and intra-specific competition. An important but unanswered question is whether invasive species have higher population growth rates and reach higher abundances in their introduced range or whether they have similar levels of success and we just notice them more where they are non-native (Bossdorf *et al.*, 2005; Parker *et al.*, 2013; Colautti *et al.*, 2014). Comprehensive field surveys coupled with common garden experiments measuring fitness reaction norms are needed to test scenarios like those outlined in Fig. 10.3.

10.5 Conclusion

This chapter has reviewed the theoretical basis of enemy release theory, including the Enemy Release Hypothesis and the Evolution of Increased Competitive Ability Hypothesis. We introduced a simple but general model of enemy release in order to anchor ecological and evolutionary assumptions and predictions that can cause confusion. These include the relationship between enemy abundance and host fitness, the magnitude of fitness trade-offs and how to measure them, the strength and direction of natural selection, and heterogeneity in the native range (i.e. genetic and environmental). An analysis of a simple but general quantitative model of enemy release yielded several results that may not be intuitive. For example, we found that natural selection is a key parameter that has not received enough attention in empirical studies of enemy release. We also found that comparisons of damage by natural enemies in the native and introduced range are relatively uninformative for enemy release theory because of the potential for evolution of defences and for weak correlations between damage and defence. Finally, we found that evolutionary shifts in defence predicted by EICA are unlikely to hold under a wide range of scenarios. Twenty-five years after EICA was first proposed, there are still large gaps in our knowledge and exciting avenues for new research. Of these, studies examining natural selection, fitness reaction norms, genetic trade-offs and evolution among introduced populations may be the most promising. These new research areas will continue to demonstrate how experimental studies of invasive species can provide exciting new insights and help to address fundamental questions in ecology and evolution.

Acknowledgements

The authors are grateful for the invitation from Anna Traveset and Dave Richardson to contribute this chapter, for math support from C. Smith and E. Sit, and for discussions with S.E. Sultan, J. Anderson and S.B. Yakimowski. Funding for this research was provided by an NSERC Discovery grant to RIC.

References

Agrawal, A.A., Hastings, A.P., Bradburd, G.S., Woods, E.C., Züst, T. *et al.* (2015) Evolution of plant growth and defense in a continental introduction. *The American Naturalist* 186(1), E1–E15. DOI: 10.1086/681622.

Ali, J.G. and Agrawal, A.A. (2012) Specialist versus generalist insect herbivores and plant defense. *Trends in Plant Science* 17(5), 293–302. DOI: 10.1016/j.tplants.2012.02.006.

Anderson, J.T., Willis, J.H. and Mitchell-Olds, T. (2011) Evolutionary genetics of plant adaptation. *Trends in Genetics* 27(7), 258–266. DOI: 10.1016/j.tig.2011.04.001.

Blossey, B. and Nötzold, R. (1995) Evolution of increased competitive ability in invasive nonindigenous plants: a hypothesis. *The Journal of Ecology* 83(5), 887–889. DOI: 10.2307/2261425.

Bonnier, G. (1890) Cultures expérimentles dans les Alpes et les Pyrénées. *Revue Générale de Botanique* 2, 513–546.

Bossdorf, O., Auge, H., Lafuma, L., Rogers, W.E., Siemann, E. *et al.* (2005) Phenotypic and genetic differentiation between native and introduced plant populations. *Oecologia* 144(1), 1–11. DOI: 10.1007/s00442-005-0070-z.

Cadotte, M.W., McMahon, S.M. and Fukami, T. (eds) (2006) *Conceptual Ecology and Invasion Biology: Reciprocal Approaches to Nature*. Springer, Berlin.

Callaway, R.M. and Ridenour, W.M. (2004) Novel weapons: invasive success and the evolution of increased competitive ability. *Frontiers in Ecology and the Environment* 2(8), 436–443. DOI: 10.1890/1540-9295(2004)002[0436:NWISAT]2.0.CO;2.

Clausen, J. and Hiesey, W.M. (1958) *Experimental Studies on the Nature of Species. IV. Genetic Structure of Ecological Races*. Carnegie Institution of Washington Publication 615, Washington, DC.

Colautti, R.I. and Barrett, S.C.H. (2013) Rapid adaptation to climate facilitates range expansion of an invasive plant. *Science* 342(6156), 364–366. DOI: 10.1126/science.1242121.

Colautti, R., Colautti, R., Parker, J.D., Cadotte, M.W., Pyšek, P. *et al.* (2014) Quantifying the invasiveness of species. *NeoBiota* 21, 7–27. DOI: 10.3897/neobiota.21.5310.

Colautti, R.I., Eckert, C.G. and Barrett, S.C.H. (2010) Evolutionary constraints on adaptive evolution during range expansion in an invasive plant. *Proceedings of the Royal Society B: Biological Sciences* 277(1689), 1799–1806. DOI: 10.1098/rspb.2009.2231.

Colautti, R.I. and Lau, J.A. (2015) Contemporary evolution during invasion: evidence for differentiation, natural selection, and local adaptation. *Molecular Ecology* 24(9), 1999–2017. DOI: 10.1111/mec.13162.

Colautti, R.I., Maron, J.L. and Barrett, S.C.H. (2009) Common garden comparisons of native and introduced plant populations: latitudinal clines can obscure evolutionary inferences. *Evolutionary Applications* 2(2), 187–199. DOI: 10.1111/j.1752-4571.2008.00053.x.

Colautti, R.I., Ricciardi, A., Grigorovich, I.A. and MacIsaac, H.J. (2004) Is invasion success explained by the enemy release hypothesis? *Ecology Letters*. DOI: 10.1111/j.1461-0248.2004.00616.x.

Coley, P.D., Bryant, J.P. and Chapin, F.S. (1985) Resource availability and plant antiherbivore defense. *Science* 230(4728), 895–899. DOI: 10.1126/science.230.4728.895.
Dangl, J.L. and Jones, J.D. (2001) Plant pathogens and integrated defence responses to infection. *Nature* 411(6839), 826–833. DOI: 10.1038/35081161.
Darwin, C.R. (1877) *The Different Forms of Flowers on Plants of the Same Species*. John Murray, London.
Doorduin, L.J. and Vrieling, K. (2011) A review of the phytochemical support for the shifting defence hypothesis. *Phytochemistry Reviews* 10(1), 99–106. DOI: 10.1007/s11101-010-9195-8.
Farji-Brener, A.G. and Amador-Vargas, S. (2014) Hierarchy of hypotheses or cascade of predictions? A comment on Heger et al. (2013). *AMBIO* 43(8), 1112–1114. DOI: 10.1007/s13280-014-0549-0.
Felker-Quinn, E., Schweitzer, J.A. and Bailey, J.K. (2013) Meta-analysis reveals evolution in invasive plant species but little support for evolution of increased competitive ability (EICA). *Ecology and Evolution* 3(3), 739–751. DOI: 10.1002/ece3.488.
Fisher, R.A. (1941) The theoretical consequences of polyploid inheritance for the mid style form of *Lythrum salicaria*. *Annals of Eugenics* 11(1), 31–38. DOI: 10.1111/j.1469-1809.1941.tb02268.x.
Fisher, R.A. (1949) The linkage problem in a tetrasomic wild plant, *Lythrum salicaria*. *Hereditas* 35(S1), 225–233. DOI: 10.1111/j.1601-5223.1949.tb03335.x.
Fisher, R.A. and Martin, V.C. (1947) Spontaneous occurrence in *Lythrum salicaria* of plants duplex for the short-style gene. *Nature* 160(4068), 541–541. DOI: 10.1038/160541a0.
Fisher, R.A. and Mather, K. (1943) The inheritance of style length in *Lythrum salicaria*. *Annals of Eugenics* 12(1), 1–23. DOI: 10.1111/j.1469-1809.1943.tb02307.x.
Flory, S.L. and Clay, K. (2013) Pathogen accumulation and long-term dynamics of plant invasions. *Journal of Ecology* 101(3), 607–613. DOI: 10.1111/1365-2745.12078.
Hahn, P.G. and Maron, J.L. (2016) A framework for predicting intraspecific variation in plant defense. *Trends in Ecology & Evolution* 31(8), 646–656. DOI: 10.1016/j.tree.2016.05.007.
Heger, T. and Jeschke, J.M. (2014) The enemy release hypothesis as a hierarchy of hypotheses. *Oikos* 123(6), 741–750. DOI: 10.1111/j.1600-0706.2013.01263.x.
Jeschke, J.M. (2014) General hypotheses in invasion ecology. *Diversity and Distributions* 20(11), 1229–1234. DOI: 10.1111/ddi.12258.
Jordan, A. (1846) *Observations sur plusieurs plantes nouvelles, rares ou critiques de la France. Fragm. II.* Annales de la Société Linnéene de Lyon, Lyon, France.
Joshi, J. and Vrieling, K. (2005) The enemy release and EICA hypothesis revisited: incorporating the fundamental difference between specialist and generalist herbivores. *Ecology Letters* 8(7), 704–714. DOI: 10.1111/j.1461-0248.2005.00769.x.
Keane, R. and Crawley, M.J. (2002) Exotic plant invasions and the enemy release hypothesis. *Trends in Ecology & Evolution* 17(4), 164–170. DOI: 10.1016/S0169-5347(02)02499-0.
Koricheva, J., Nykänen, H. and Gianoli, E. (2004) Meta-analysis of trade-offs among plant antiherbivore defenses: are plants Jacks-of-All-Trades, masters of all? *The American Naturalist* 163(4), E64–E75. DOI: 10.1086/382601.
Lambers, H.F.C. and Pons, T.L. (2008) *Plant Physiological Ecology*. Springer, Berlin.
Lande, R. and Arnold, S.J. (1983) The measurement of selection on correlated characters. *Evolution* 37(6), 1210–1226. DOI: 10.1111/j.1558-5646.1983.tb00236.x.
Lee, C.E. (2016) Evolutionary mechanisms of habitat invasions, using the copepod *Eurytemora affinis* as a model system. *Evolutionary Applications* 9(1), 248–270. DOI: 10.1111/eva.12334.
Linnaeus, C. (1735) Systema naturæ, sive regna tria naturæ systematice proposita per classes, ordines, genera, & species. – pp. [1–12]. Lugduni Batavorum. (Haak).
Lowry, D.B., Lovell, J.T., Zhang, L., Bonnette, J., Fay, P.A. *et al.* (2019) QTL × environment interactions underlie adaptive divergence in switchgrass across a large latitudinal gradient. *Proceedings of the National Academy of Sciences* 116(26), 12933–12941. DOI: 10.1073/pnas.1821543116.
Lu-Irving, P., Marx, H.E. and Dlugosch, K.M. (2018) Leveraging contemporary species introductions to test phylogenetic hypotheses of trait evolution. *Current Opinion in Plant Biology* 42, 95–102. DOI: 10.1016/j.pbi.2018.04.011.
Mauricio, R. and Rausher, M.D. (1997) Experimental manipulation of putative selective agents provides evidence for the role of natural enemies in the evolution of plant defense. *Evolution* 51(5), 1435–1444. DOI: 10.1111/j.1558-5646.1997.tb01467.x.
Mitchell, C.E. and Power, A.G. (2003) Release of invasive plants from fungal and viral pathogens. *Nature* 421(6923), 625–627. DOI: 10.1038/nature01317.

Müller-Schärer, H., Schaffner, U. and Steinger, T. (2004) Evolution in invasive plants: implications for biological control. *Trends in Ecology & Evolution* 19(8), 417–422. DOI: 10.1016/j.tree.2004.05.010.

Olden, J.D., Leroy Poff, N., Douglas, M.R., Douglas, M.E. and Fausch, K.D. (2004) Ecological and evolutionary consequences of biotic homogenization. *Trends in Ecology & Evolution* 19(1), 18–24. DOI: 10.1016/j.tree.2003.09.010.

Parker, J.D., Torchin, M.E., Hufbauer, R.A., Lemoine, N.P., Alba, C. *et al.* (2013) Do invasive species perform better in their new ranges? *Ecology* 94(5), 985–994. DOI: 10.1890/12-1810.1.

Pejchar, L. and Mooney, H.A. (2009) Invasive species, ecosystem services and human well-being. *Trends in Ecology & Evolution* 24(9), 497–504. DOI: 10.1016/j.tree.2009.03.016.

Phillips, B.L. (2009) The evolution of growth rates on an expanding range edge. *Biology Letters* 5(6), 802–804. DOI: 10.1098/rsbl.2009.0367.

Phillips, B.L., Brown, G.P., Webb, J.K. and Shine, R. (2006) Invasion and the evolution of speed in toads. *Nature* 439(7078), 803. DOI: 10.1038/439803a.

Pierson, J.C., Beissinger, S.R., Bragg, J.G., Coates, D.J., Oostermeijer, J.G.B. *et al.* (2015) Incorporating evolutionary processes into population viability models. *Conservation Biology* 29(3), 755–764. DOI: 10.1111/cobi.12431.

Price, G.R. (1970) Selection and covariance. *Nature* 227(5257), 520–521. DOI: 10.1038/227520a0.

Pyšek, P. and Richardson, D.M. (2010) Invasive species, environmental change and management, and health. *Annual Review of Environment and Resources* 35(1), 25–55. DOI: 10.1146/annurev-environ-033009-095548.

Robertson, A. (1966) A mathematical model of the culling process in dairy cattle. *Animal Science* 8(1), 95–108. DOI: 10.1017/S0003356100037752.

Rotter, M.C. and Holeski, L.M. (2018) A meta-analysis of the evolution of increased competitive ability hypothesis: genetic-based trait variation and herbivory resistance trade-offs. *Biological Invasions* 20(9), 2647–2660. DOI: 10.1007/s10530-018-1724-1.

Sargent, R.D., Angert, A.L. and Williams, J.L. (2017) When are species invasions useful for addressing fundamental questions in plant biology? *American Journal of Botany* 104(6), 797–799. DOI: 10.3732/ajb.1700087.

Sax, D.F., Stachowicz, J.J., Brown, J.H., Bruno, J.F., Dawson, M.N. *et al.* (2007) Ecological and evolutionary insights from species invasions. *Trends in Ecology & Evolution* 22(9), 465–471. DOI: 10.1016/j.tree.2007.06.009.

Schultheis, E.H., Berardi, A.E. and Lau, J.A. (2015) No release for the wicked: enemy release is dynamic and not associated with invasiveness. *Ecology* 96(9), 2446–2457. DOI: 10.1890/14-2158.1.

Simms, E.L. and Triplett, J. (1994) Costs and benefits of plant responses to disease: resistance and tolerance. *Evolution* 48(6), 1973–1985. DOI: 10.1111/j.1558-5646.1994.tb02227.x.

Strauss, S.Y. and Agrawal, A.A. (1999) The ecology and evolution of plant tolerance to herbivory. *Trends in Ecology & Evolution* 14(5), 179–185. DOI: 10.1016/S0169-5347(98)01576-6.

Strauss, S.Y., Rudgers, J.A., Lau, J.A. and Irwin, R.E. (2002) Direct and ecological costs of resistance to herbivory. *Trends in Ecology & Evolution* 17(6), 278–285. DOI: 10.1016/S0169-5347(02)02483-7.

Thompson, D.Q., Thompson, E.B., Stuckey, R.L., Fish, U.S. and Service, W. (1987) *Spread, impact, and control of purple loosestrife (Lythrum salicaria) in North American wetlands, Fish and Wildlife Research 2*. U.S. Dept. of the Interior, Fish and Wildlife Service; National Technical Information Service, Washington, DC.

Tiffin, P. and Ross-Ibarra, J. (2014) Advances and limits of using population genetics to understand local adaptation. *Trends in Ecology & Evolution* 29(12), 673–680. DOI: 10.1016/j.tree.2014.10.004.

Turcotte, M.M., Reznick, D.N. and Hare, J.D. (2011) The impact of rapid evolution on population dynamics in the wild: experimental test of eco-evolutionary dynamics. *Ecology Letters* 14(11), 1084–1092. DOI: 10.1111/j.1461-0248.2011.01676.x.

Turesson, G. (1922) The genotypical response of the plant species to the habitat. *Hereditas* 3(3), 211–350. DOI: 10.1111/j.1601-5223.1922.tb02734.x.

van Kleunen, M., Bossdorf, O. and Dawson, W. (2018) The ecology and evolution of alien plants. *Annual Review of Ecology, Evolution, and Systematics* 49(1), 25–47. DOI: 10.1146/annurev-ecolsys-110617-062654.

11 The Role of Pathogens in Plant Invasions

Amy E. Kendig[1*], S. Luke Flory[1], Erica M. Goss[2], Robert D. Holt[3], Keith Clay[4], Philip F. Harmon[2], Brett R. Lane[2], Ashish Adhikari[2] and Christopher M. Wojan[5]

[1]Agronomy Department, University of Florida, Gainesville, USA; [2]Department of Plant Pathology, University of Florida, Gainesville, USA; [3]Department of Biology, University of Florida, Gainesville, USA; [4]Department of Ecology and Evolutionary Biology, Tulane University, New Orleans, USA; [5]Department of Biology, Indiana University, Bloomington, USA

Abstract

Plant–pathogen interactions occur throughout the process of plant invasion: pathogens can acutely influence plant survival and reproduction, while the large densities and spatial distributions of invasive plant species can influence pathogen communities. However, interactions between invasive plants and pathogens are often overlooked during the early stages of invasion. As with introductions of invasive plants, the introduction of agricultural crops to new areas can also generate novel host–pathogen interactions. The close monitoring of agricultural plants and resulting insights can inform hypotheses for invasive plants where research on pathogen interactions is lacking. This chapter reviews the known and hypothesized effects of pathogens on the invasion process and the effects of plant invasion on pathogens and infectious disease dynamics throughout the process of invasion. Initially, pathogens may inhibit the transport of potentially invasive plants. After arrival in a new range, pathogens can facilitate or inhibit establishment success of introduced plants depending on their relative impacts on the introduced plants and resident species. As invasive plants spread, they may encounter novel pathogens and alter the abundance and geographic range of pathogens. Pathogens can mediate interactions between invasive plants and resident species and may influence the long-term impacts of invasive plants on ecosystems. As invasive plants shift the composition of pathogen communities, resident species could be subject to higher disease risk. We highlight gaps in invasion biology research by providing examples from the agricultural literature and propose topics that have received little attention from either field.

11.1 Overview

Microbial communities of plants influence plant health and ecological interactions. The microbial communities of invasive non-native plants (invasive plants, hereafter) change as plants leave behind some microbes from their native ranges, including pathogens, and acquire new symbionts in their introduced ranges (Colautti *et al.*, 2004; Gilbert and Parker, 2006; Flory and Clay, 2013). It is well accepted that microbes associated with (or absent from) introduced plants, and soil microbial communities in particular, can influence initial plant establishment and subsequent impacts on resident species (van der Putten, 2010). For example, establishment of invasive pine species in Argentina was limited until mutualistic mycorrhizal fungi were introduced

*Corresponding author: aekendig@gmail.com

 Plant Invasions: The Role of Biotic Interactions (eds Anna Traveset and David M. Richardson)
DOI: 10.1079/9781789242171.0011

(Nuñez *et al.*, 2009), while the loss of soil pathogens may have promoted widespread and dense establishment of the black cherry tree (*Prunus serotina*) in Europe (Reinhart *et al.*, 2003) and spotted knapweed (*Centaurea maculosa*) in the USA (Callaway *et al.*, 2004). By contrast, garlic mustard (*Alliaria petiolata*) reduced native plant growth in US forests by suppressing microbial mutualists (Stinson *et al.*, 2006), which may have facilitated its own spread.

Although the ecological role of microbes in plant invasions is increasingly understood, microbial communities of invasive plants can change over time, altering the long-term dynamics and impacts of invaders (Lau and Suwa, 2016; Stricker *et al.*, 2016). Changes in microbial communities may be as straightforward as an invasive plant acquiring a pathogen from a resident species in its introduced range, or as complex as the evolution of microbes that are exposed to novel selection pressures during the introduction process (e.g. surviving transport and new environmental conditions, Fig. 11.1). Pathogens can be particularly influential microbes, affecting plant mortality, reproduction and competitive ability (Dobson and Crawley, 1994; Mordecai, 2011). Consequently, pathogens that infect crops have long received attention from researchers and practitioners. Like invasive plant species, agricultural plants are introduced to new geographical regions where new host–pathogen associations can arise and their interactions with pathogens can change over time. Therefore, case studies from agricultural plant pathology may provide important insights for understanding the relationships between invasive plants and their pathogens.

Using the stages of plant invasion as an organizational framework, we identify specific points where pathogens can exert influence (Fig. 11.1). First, pathogens may accompany and determine survival of hosts during transport to new geographic regions. During establishment and spread, introduced and native pathogens can alter plant survival and reproduction, as well as influence interactions with resident species. Invasive plants may facilitate the spread or emergence of plant pathogens as the plants reach uninvaded regions and interact with resident plant species. Pathogens can promote or attenuate the impact of invasive plant populations on resident species through altered ecological interactions or shifts in ecosystem functions. Finally, pathogen effects on invasive and resident species can change over time due to pathogen accumulation and evolution.

In this chapter, we review key research advances in understanding how pathogens affect and are affected by invasive plants. We use the stages of invasion (i.e. transport, establishment, spread, ecological impacts) to consider invasive plant–pathogen interactions and how they change over time (Fig. 11.1), highlight key examples from the invasive species and agricultural plant pathogen literature, and outline future research needs. The relationships between invasive plants and pathogens are relevant to multiple significant areas of invasion biology research that also consider plant–herbivore interactions (e.g. the Enemy Release Hypothesis, the Biotic Resistance Hypothesis and biological control, Gilbert and Parker, 2006). We cover such topics briefly and point readers to corresponding chapters in this book for more in-depth analyses. Given the burgeoning volume of global travel and commerce that provides continuous opportunities for transport of potential invaders into almost all habitats, elucidating how pathogens can modify interactions between invasive and resident species is more pressing than ever.

11.2 Transport

11.2.1 The role of pathogens in non-native plant transport

The deliberate or inadvertent transport of non-native plants to new ranges can include the movement of seeds, plant parts, juveniles and adult plants across geographic boundaries (Lehan *et al.*, 2013; Chapman *et al.*, 2017). Pathogens may affect the transport of non-native plants directly by affecting plant survival (Ragaert *et al.*, 2007; Kumar and Kalita, 2017), or indirectly, by facilitating the targeting of individual plants for quarantine or removal by human agents (Colautti *et al.*, 2004; Hulme *et al.*, 2008). Invasive plants are most frequently transported for ornamental trade (Lehan *et al.*, 2013), and often in the form of live plants (Chapman *et al.*, 2017). Invaders

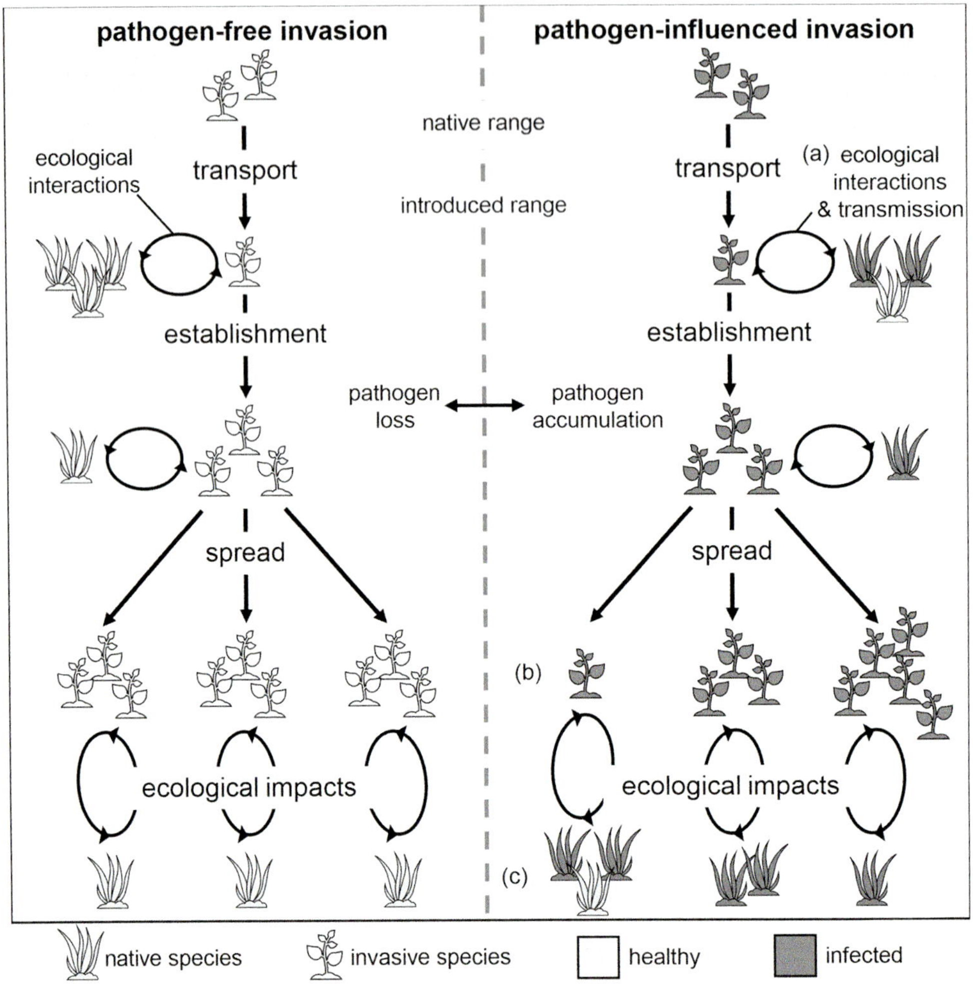

Fig. 11.1. Potential impacts of pathogens on the plant invasion process. Plants may complete the invasion process in the absence of pathogens (left panel), in the presence of pathogens (right panel), or by moving between these two states (pathogen loss and accumulation). Evidence from agricultural crops suggests that pathogens may affect survival of introduced plant populations throughout the invasion process. Examples from both invasive species and crops demonstrate pathogen loss following transport and pathogen accumulation following establishment (centre double-headed arrow). Curved arrows represent ecological interactions with resident plants, including competition and, when pathogens are present, transmission (a). Pathogens can be co-introduced with invasive species or transmitted from resident species. Pathogens may alter the size of established invasive plant populations through direct or indirect processes, including competition with native species (b). Studies on invasive plants have shown that pathogens can mediate competition with resident species through changes in invasive population size or disease transmission (c). Pathogen communities and infection prevalence may shift with the invasion process and exposure to new environments and hosts.

tend to be transported as seeds when movement is unintentional, contaminating soil or agricultural seeds (Lehan *et al.*, 2013). Pathogen infection of seeds and live plants during storage and transportation contributes to crop losses (Ragaert *et al.*, 2007; Kumar and Kalita, 2017) and may alter the survival of invasive plants during transport.

11.2.2 Transport of Plant Pathogens

The simultaneous or sequential introduction of two non-native species (i.e. a plant and a pathogen) is known as 'co-introduction', and 'co-invasion' if they both become invasive (Dickie *et al.*, 2017). Multiple methods exist to ensure introduced plants are disease-free, such as inspection, quarantine and sanitation of imported plants (Colautti *et al.*, 2004; Hulme *et al.*, 2008). Pathogens that cause obvious symptoms, kill plants or are detected with screening methods are more likely to be destroyed than those that go unnoticed by remaining dormant or causing symptoms later in the plant's life cycle (Liebhold *et al.*, 2012). In addition, plants transported during a specific life stage (e.g. seeds, juveniles) will lack pathogens that specialize on other life stages (Barrett and Heil, 2012). It is through these filters that co-introduction of certain classes of pathogens that have coevolved with invasive plants may be prevented (Keane and Crawley, 2002; Colautti *et al.*, 2004).

While screening methods can limit the transport of pathogens, there are numerous examples of pathogens being introduced despite prevention efforts. Some of the most devastating pathogens of crops and native species were transported on seeds, logs and live plants (Elmer, 2001; Brasier, 2008; Liebhold *et al.*, 2012). The origins and timing of introduction, however, are unclear for the majority of invasive plant pathogens (Desprez-Loustau *et al.*, 2007), making post hoc identification of co-introduction cases difficult. Nevertheless, some recently introduced invasive plant populations are infected with pathogens (Mitchell *et al.*, 2010; Stricker *et al.*, 2016), suggesting co-introduction or rapid acquisition, and there are numerous examples of co-introduced animals and parasites (Lymbery *et al.*, 2014).

Recurring transport or movement of individuals (i.e. propagule pressure) can help plants, and their microbes, overcome barriers to introduction (Lockwood *et al.*, 2005). For example, it is hypothesized that repeated importations of Japanese chestnut trees to North America resulted in the ultimate introduction of the chestnut blight fungus (*Cryphonectria parasitica*, Dutech *et al.*, 2012). Sudden oak death (*Phytophthora ramorum*), another virulent tree pathogen, was likely introduced to North America and Europe through repeated importations of nursery stock (Grünwald *et al.*, 2012).

11.3 Establishment: Survival and Reproduction

Following transport, introduced plants must survive and reproduce to become established (Blackburn *et al.*, 2011). Barriers such as small initial population size, limited resource availability, unfavourable climate, absence of a key mutualist and high herbivory can constrain plant establishment (Levine *et al.*, 2004; Lockwood *et al.*, 2005; Bradley *et al.*, 2010). It is hypothesized that introduced or resident pathogens can reduce survival and reproduction of introduced plants as well (i.e. biotic resistance, Elton, 1958), but evidence for such inhibitory interactions is limited (Levine *et al.*, 2004; Strauss *et al.*, 2012). For co-introduced pathogens to become established, they must be able to overcome multiple ecological and evolutionary barriers (Desprez-Loustau *et al.*, 2007). While the ecological impacts of invasive plants often occur well after the plant has established, the co-introduction of pathogens creates the opportunity for more immediate impacts on resident species, impacts that can spread spatially well beyond the geographic range of the introduced host (e.g. chestnut blight spread rapidly through essentially the entire range of the American chestnut, past the sites of introduction of the likely initial host, introduced Asian chestnuts, Dutech *et al.*, 2012).

11.3.1 The role of pathogens in non-native plant establishment

Co-introduced pathogens or those rapidly acquired from resident plants could prevent an introduced plant from becoming established (Elton, 1958), but these processes are rarely documented in ecological studies due to small population sizes and limited impacts of introduced plants during the establishment phase, which can keep these systems 'below the radar' of applied ecologists (Strauss *et al.*, 2012). Similar outcomes can occur, however, during monoculture production of non-native agricultural or forest crop plants, and novel disease pressures can hamper production (i.e. prevent 'invasion'). For example, the fungus *Dothistroma pini* spread through young plantations of the non-native pine *Pinus radiata* in Eastern and Central Africa between 1940 and 1965, damaging foliage, slowing tree growth and forcing foresters to substitute *P. radiata* for other species (Gibson, 1972). *D. pini* may have been introduced to the region through long-distance natural dispersal or air traffic (Gibson, 1972).

Resident pathogens can indirectly aid introduced plants by infecting resident species and reducing competition. The Enemy Release Hypothesis (Chapter 10, this volume) posits that the absence of pathogens and other types of enemies in the introduced range leads to greater performance or competitive ability of non-native species relative to the native range (Elton, 1958; Gillett, 1962; Keane and Crawley, 2002; Colautti *et al.*, 2004). Indeed, plants in introduced ranges tend to have fewer recorded pathogens than those in the native ranges (Mitchell and Power, 2003; Blumenthal *et al.*, 2009) and invasive plants can suffer less pathogen-induced damage and mortality than do resident species in the introduced range (Hawkes, 2007; Han *et al.*, 2008; Chun *et al.*, 2010; but see Blaney and Kotanen, 2001; Parker and Gilbert, 2007). In addition, crops tend to experience a 'honeymoon' period of limited pathogen pressure when initially planted in a new region (Bebber *et al.*, 2014a). However, few studies have connected lower pathogen pressure on invasive plants in their introduced range to higher population growth rates (Dunn *et al.*, 2012) or ultimate abundance or spatial extent of the invasion.

The nature and magnitude of disease impacts on introduced plant establishment are mediated by many factors, including pathogen specificity and phylogenetic relatedness of the introduced plant to resident species (Parker and Gilbert, 2004). Plant pathogen host ranges can vary widely, from a single plant species, such as the rust fungus *Puccinia chondrillina* used as a biocontrol for skeleton weed (*Chondrilla juncea* L.) in Australia (Cullen *et al.*, 1973), to rich arrays of phylogenetically distinct host species, such as the globally-distributed oomycete *Phytophthora cinnamomi*, which is native to Southeast Asia and can infect nearly 5000 plant species (Hardham and Blackman, 2018). Host range specificity depends on both coevolution between pathogen infection strategies and plant defences, and environmental conditions (Barrett and Heil, 2012). Introduced plants that fall within the host ranges of resident generalist pathogens or are closely related to hosts of more specific pathogens are more likely to experience disease during and after establishment (Parker *et al.*, 2015; Crous, 2017).

The invasion process is likely to influence the magnitude of disease impacts through changes in disease resistance in the introduced plant population. The genetic bottleneck often associated with non-native plant transport may lead to low genetic diversity in secondary metabolites or disease resistance genes of introduced plant species (Dlugosch and Parker, 2008), promoting eventual disease spread. In contrast, infection-induced mortality during transport or establishment may select for disease resistance (Blaney and Kotanen, 2001; Gos *et al.*, 2012). Multiple introductions can reintroduce genetic variation into the introduced plant population (Gladieux *et al.*, 2011), but each introduction may also bring novel pathogens and novel genotypes of established pathogens from the native range (Dutech *et al.*, 2012; Grünwald *et al.*, 2012).

11.3.2 Establishment of co-introduced plant pathogens

Because host population sizes can affect the ability of specialist pathogens to establish (Anderson and May, 1979), co-introduced pathogens may be limited by the initially small population size of their introduced plant host. However, pathogens may overcome this barrier through multiple mechanisms. Pathogens with broad host ranges can infect resident species while their introduced plant host is establishing and increasing in abundance (Strauss *et al.*, 2012), potentially with devastating effects. For example, the bacterium *Xylella fastidiosa* was introduced to southern Italy on ornamental plants from Central America and caused rapid mortality of historically important and quite old olive trees (Marcelletti and Scortichini, 2016). In addition, pathogens may remain dormant while the introduced plant species is establishing, only to emerge later in favourable conditions. For example, *P. ramorum* produces resting spores that can survive adverse environmental conditions (Tooley *et al.*, 2008).

Pathogens with specific requirements, such as insect vectors or narrow environmental tolerance, may face barriers to establishment. Viruses of crops have a more limited global distribution than their fungal counterparts, likely due to their dependence on insect vectors that are less widely distributed (Bebber *et al.*, 2014a). Environmental conditions may also differ greatly between the native and introduced ranges and could impact pathogen establishment. Honeysuckle leaf blight (*Insolibasidium deformans*) requires specific temperature and humidity conditions to infect its host, which may contribute to determining its geographic distribution and the timing of infection on invasive Amur honeysuckle (*Lonicera maackii*, Boyce *et al.*, 2014).

11.4 Spatial Spread: Dispersal and Repeated Establishment

The spread of invasive plants within an introduced range involves dispersal and repeated establishment events. Dispersal can occur through natural or human-assisted movement of seeds, plant parts and individuals (Reichard and White, 2001; Padayachee *et al.*, 2017), and establishment across space depends on complex interactions among plant traits, propagule pressure, landscape structure and environmental factors such as temperature and precipitation (Catford *et al.*, 2009). Because dispersal and repeated establishment are similar to transport and initial establishment, respectively, many of the plant–pathogen interactions that occur during those phases are relevant to spatial spread. However, pathogen impacts on short-distance dispersal can uniquely affect spatial spread.

11.4.1 The role of pathogens in invasive plant spread

Pathogens can influence the dispersal of invasive plants in ways similar to how they affect initial plant transport to the introduced range. For example, intentional movement of seeds or plants by humans is likely to select for visibly healthy individuals (Reichard and White, 2001). Movement within the introduced range differs from initial transport because it can involve more localized dispersal mechanisms, including animals, wind, water and machinery (Horvitz *et al.*, 2017). The infection status or defensive compounds of plants may affect animal foraging behaviour and resulting seed dispersal (Cazetta *et al.*, 2008). Pathogens also can affect plant size, seed size and seed production rates (Dobson and Crawley, 1994), which influence natural and unintentional dispersal (Thomson *et al.*, 2011; Tamme *et al.*, 2014).

Just as pathogens can induce mortality and reduce fecundity during initial plant establishment in a new range, pathogens also can inhibit establishment in new areas within the introduced range. Simple models of invasion suggest that in a homogeneous landscape, the wave of advance moves with a velocity proportional to $\sqrt{rD}$, where r is intrinsic growth rate, and D gauges the rate of diffusive dispersal (Andow *et al.*, 1990). A pathogen that reduces r is expected to reduce the rate at which an invasive plant moves through the landscape. This same qualitative expectation also holds in more complex models of invasion (Neubert and Parker, 2004). Moreover, environmental conditions mediate plant–pathogen interactions, such that the barrier to establishment caused by disease is likely to vary over space and time. For example, environmental variation drove differences in pathogen resistance between two ecotypes of native Australian flax (*Linum marginale*), leading to spatial differences in disease prevalence and host–pathogen coevolution (Laine *et al.*, 2014).

11.4.2 Pathogen community changes with spatial spread

As invasive plants spread, they may acquire new pathogens and maintain others through simultaneous or independent dispersal. The effectiveness of independent pathogen dispersal can vary by mechanism. For example, aphids, an important vector of plant viruses, can disperse up to hundreds of metres per day through wind-assisted movement, but frequently crawl among leaves on a single plant (Irwin *et al.*, 2007). Wind-dispersed fungal pathogens, however, may have even higher spread rates than those that depend on insects, as is the case for some invasive pathogens of US forests (Evans and Finkral, 2010). Plants may also lose some of their pathogens through the process of spatial spread (Chalkowski *et al.*, 2018), acquiring new pathogen communities as they encounter different species due to greater geographic range, habitat diversity and residence time (Clay, 1995; Mitchell *et al.*, 2010). Crop pathogens tend to accumulate as the intensity of cultivation increases (Bebber *et al.*, 2014b), while continuing to be limited by phylogenetic relatedness to resident species (Crous, 2017).

11.5 Ecological Impacts

Invasive plants can have significant impacts on resident communities and ecosystems. One of the most well-supported impacts of invasive plants is reduced diversity in the resident plant community (Vilà *et al.*, 2011; Pyšek *et al.*, 2012), which frequently arises through competition (Levine *et al.*, 2003). Pathogens can alleviate or enhance the negative effects of plant resource competition (Mordecai, 2011; Flory and Clay, 2013), which has been explored across a range of systems (Bradley *et al.*, 2008; Bagchi *et al.*, 2014; Spear and Mordecai, 2018). Invasive plants can also impact other aspects of the ecosystem, such as associated animal communities (Bateman and Ostoja, 2012), nutrient cycling (Allison and Vitousek, 2004) and fire regimes (Flory *et al.*, 2015), but the role of pathogens in mediating such impacts on communities and ecosystems has rarely been studied.

11.5.1 Pathogen-mediated plant competition

As invasive plants establish, resident plant communities can experience competition for shared resources (Brown *et al.*, 2002), allelopathy (Callaway and Ridenour, 2004) and damage from shared enemies (Dangremond *et al.*, 2010), ultimately leading to shifts in plant community diversity and structure. Pathogens can influence the ecological impact of invasive plants by altering the competitive abilities and niche overlap of native and invasive species (Mordecai, 2011). Theoretically, pathogens can push the outcome of competition between native and invasive plants in three directions: coexistence, competitive exclusion of the invasive species and competitive exclusion of the native species (Louda *et al.*, 1990; Mordecai, 2011; Flory and Clay, 2013).

Because the competitive exclusion of one species by another is a long-term process, a common approach to understanding the possible dynamics is to parameterize mathematical models with experimentally-derived values (Eppinga *et al.*, 2006; Borer *et al.*, 2007; Fabiszewski *et al.*, 2010; Mordecai, 2013). Such models have provided insights into pathogen-mediated competition. For example, the invasive marram grass (*Ammophila arenaria*) suffers from negative plant—soil feedbacks in its native and introduced ranges (Beckstead and Parker, 2003). Model predictions suggest that generalist pathogens that accumulate in *A. arenaria* soil can promote the exclusion of native species if native species incur more infection-induced fitness costs than *A. arenaria* (Eppinga *et al.*, 2006). In addition, invasive cheatgrass (*Bromus tectorum*) promotes infection of native squirreltail grass (*Elymus elymoides*) seeds (Beckstead *et al.*, 2010). However, modelling suggests that the pathogen does not promote competitive exclusion of the native squirreltail grass, partly due to its higher resistance to infection than the invasive species (Mordecai, 2013).

Short-term experiments often motivate the use of models to predict the outcome of pathogen-mediated competition (Malmstrom *et al.*, 2005; Beckstead *et al.*, 2010). Experiments with garlic mustard (*A. petiolata*, Cipollini and Enright, 2009) and stiltgrass (*Microstegium vimineum*, Stricker *et al.*, 2016) each suggest that foliar fungal pathogens could favour recovery of native species, while experiments with reed grass (White *et al.*, 2018) and Siam weed (*Chromoleana odorata*, Mangla and Callaway, 2008) suggest that soil pathogens contribute to the competitive exclusion of native species. Using data from such experiments to parameterize mathematical models can aid predictions about how pathogen impacts on native and invasive species can alter the outcome of competition, and how these outcomes may be modified by evolution, environmental conditions and other ecological interactions.

In addition to naturally occurring pathogens, intentionally introduced pathogens can modify interactions between native and invasive plants. Biological control (biocontrol) programmes introduce enemies that are specific to invasive plants and reduce their competitive effects on native species (Chapter 20, this volume). Fungal pathogen biocontrol agents can be highly effective, establishing quickly across large geographic areas and greatly reducing invasive plant abundance (Burdon *et al.*, 1981; Trujillo, 1985). Their efficacy can be sensitive to environmental conditions, such as precipitation (Cullen *et al.*, 1973) and the presence of competing plant species (Abu-Dieyeh, 2007; Mejri *et al.*, 2010). Successes and failures of biocontrol can signal potentially important factors to consider when evaluating the outcomes of natural pathogen accumulation.

11.5.2 Plant–pathogen interactions and broader effects

While it is clear that invasive plants can influence resident species through disease transmission, they may also indirectly promote disease through habitat modification (Chalkowski *et al.*, 2018). For example, dense plant invasions can change temperature, humidity and solar radiation, creating conditions favourable for fungal growth and sporulation (Burdon and Chilvers, 1982; Young *et al.*, 2012). High plant density can also reduce dispersal distances required for pathogens to infect new plants (Clay, 2008) and invasive plants can release chemicals into the soil that promote disease prevalence (Mangla and Callaway, 2008).

High invasive plant abundance also can affect other trophic levels and ecosystem functioning. Invasive plants can alter animal communities, trophic interactions, primary productivity, nutrient cycling and disturbance regimes (Levine *et al.*, 2003; Vilà *et al.*, 2011; Guiden and Orrock, 2017). Pathogen impacts on invasive plant fitness can mediate such effects. For example, reduced plant cover due to disease can affect soil formation and water regulation (Cheatham *et al.*, 2009). Disease-induced mortality of trees can increase the intensity of forest fires (Forrestel *et al.*, 2015) and alter nutrient cycling through food webs (Smock and MacGregor, 1988). Invasive species (Batten *et al.*, 2006) and pathogens (Essarioui *et al.*, 2019) can both affect the composition of soil microbial communities, but their combined effects are unclear.

11.6 Long-term Dynamics

The impacts of pathogen accumulation on invaded plant communities are likely to occur over long time periods because it can take years or decades for invasive plants to reach influential densities and experience epidemics (Enright and Cipollini, 2007; Stricker *et al.*, 2016). It is common for the enemy communities of invasive plants, consisting of herbivores and pathogens, to increase in diversity and damage over time (Table 11.1, Hawkes, 2007; Schultheis *et al.*, 2015). In many cases, the ideal long-term outcome (from a management perspective) of pathogen accumulation on an invasive plant species would be decline of the invasive species and recovery of the native species. However, there are both ecological and evolutionary causes for lags in pathogen effects on invaders and their interactions with native species (D'Antonio and Flory, 2017).

11.6.1 Long-term ecological dynamics of invasive plants and their pathogens

The accumulation of host-specific pathogens on an invasive species could lead to its decline and the recovery of native species (Flory and Clay, 2013). While there are few recorded examples of natural pathogen-driven population crashes of invasive plants (Simberloff and Gibbons, 2004), similar dynamics have occurred in agricultural systems, where crop failures can be driven by epidemics (Desprez-Loustau *et al.*, 2007).

Legacy effects, including alterations to the soil microbial community or chemical composition, may prevent native species recovery following pathogen-induced invasive plant decline (Corbin and D'Antonio, 2012; D'Antonio and Flory, 2017). Indeed, changes in soil microbial

Table 11.1. Examples of natural pathogen accumulation (i.e. not as the result of the intentional introduction of agents for biocontrol) and impacts in invaded systems.

Invasive plant species	Pathogen species	Pathogen in native range?	Accumulation time	Disease impacts on invader	Disease impacts on community	References
Ailanthus altissima (tree-of-heaven)	*Verticillium albo-atrum and Verticillium nonalfalfae*	No (Asia)	~200 years	85% mortality	Negligible	Harris *et al.*, 2013; Maschek and Halmschlager, 2018
Alliaria petiolata (garlic mustard)	*Erysiphe cruciferarum* (powdery mildew)	Yes (Europe)	~140 years	50% reduction in seed production	Eliminates the competitive impacts (reduced seed production) of garlic mustard on a native species	Enright and Cipollini, 2007; Cipollini and Enright, 2009
Ammophila arenaria (marram grass)	Soil fungi (e.g. *Fusarium* spp., *Alternaria* spp.)	Yes (Europe)	Unknown	16% reduction in germination, 13% reduction in seedling survival, and 81% reduction in biomass	Competitive exclusion of native possible if it is more sensitive to soil pathogens than *A. arenaria*	Van der Putten *et al.*, 1993, an Van der Putten and Peters, 1997; Beckstead and Parker, 2003; Eppinga *et al.*, 2006
Avena fatua, *Bromus hordeaceus*, and other annual grasses	Barley and cereal yellow dwarf viruses	Yes (Europe)	Unknown, but viruses are thought to have been established prior to plant introduction	89% reduction in biomass, 88% reduction in seed production	Disease can allow invasion by competitively inferior invasive species	Malmstrom *et al.*, 2005, Malmstrom *et al.*, 2007; Borer *et al.*, 2007
Bromus tectorum (cheatgrass)	*Pyrenophora semeniperda* (black fingers of death)	Yes (Europe)	May have been simultaneously introduced	Can kill more than 80% of dormant seeds	Promotes coexistence between invasive and native grass	Beckstead *et al.*, 2010; Stewart *et al.*, 2009; Boose *et al.*, 2011; Mordecai, 2013
Cirsium arvense (Canada thistle)	*Puccinia punctiformis* (rust fungus)	Yes (SE Europe, N Africa)	May have been simultaneously introduced	30–100% reduction in density	Unknown	Berner *et al.*, 2013, 2015

Continued

Table 11.1. Continued

Invasive plant species	Pathogen species	Pathogen in native range?	Accumulation time	Disease impacts on invader	Disease impacts on community	References
Lonicera maackii (Amur honeysuckle)	*Insolibasidium deformans* (honeysuckle leaf blight)	No (W Asia)	Unknown	Stem dieback	Unknown	Riffle and Peterson, 1986; Boyce *et al*., 2014; Farr and Rossman, 2015
Microstegium vimineum (stiltgrass)	*Bipolaris* spp. (foliar fungi)	Some species (S/E/SE Asia)	0–100 years	40–81% reduction in seed production	Increased native biomass through decreased invader biomass	Flory *et al*., 2011; Stricker *et al*., 2016
Pueraria montana var. *lobata* (kudzu)	*Phakospora pachyrhizi* (soybean rust)	Yes (E/SE Asia and Pacific)	~100 years	Negligible	Kudzu serves as an overwintering host for soybean rust, which kills soybeans	Harmon *et al*., 2005; Fabiszewski *et al*., 2010
Rhododendron ponticum (common rhododendron)	*Phytophthora ramorum* (sudden oak death)	Yes (Europe)	~240 years	Can be asymptomatic or cause foliar necrosis	Rhododendron acts as a reservoir host for the pathogen, which increases mortality of beech and other native woody species	Denman *et al*., 2005, 2009; Purse *et al*., 2013

communities and chemistry induced by invasive plants can negatively impact native plant performance (Callaway and Ridenour, 2004; Stinson *et al.*, 2006; Mangla and Callaway, 2008). In addition, invasive plant performance can be lower when grown in soil collected from conspecifics in the introduced range compared to sterilized soil or conspecifics in the native range, providing evidence for soil pathogen accumulation (Beckstead and Parker, 2003; Nijjer *et al.*, 2007; Diez *et al.*, 2010; McGinn *et al.*, 2018).

If pathogens suppress populations of invasive plants, extinction debt can still prevent the recovery of native species (Tilman *et al.*, 1994). A native plant population may be reduced to the point that its trajectory leads to extinction and it is unable to recover after the invasive plant no longer exerts strong competitive effects (Gilbert and Levine, 2013). However, native species richness tends to stay the same or increase following invasive plant removal (Andreu and Vilà, 2011; Abella, 2014), suggesting that extinction debt may be the exception rather than the rule.

11.6.2 Long-term evolutionary dynamics of introduced plants and their pathogens

While invasive plants and agricultural crops are ecologically similar in their initial release and accumulation of pathogens, invasive plants differ from most crops because they are free to evolve in response to accumulating pathogens. Indeed, invasive animals have evolved resistance to parasitoid and pathogen biological controls (Dwyer *et al.*, 1990; Tomasetto *et al.*, 2017). In addition, invaders may evolve tolerance to disease, as was demonstrated for European clover species introduced to California (Gilbert and Parker, 2010), and serve as a reservoirs for transmission to other plant species. In agricultural settings, wild plants can serve as pathogen reservoirs for crops (Wisler and Norris, 2005), including maintaining genetic variation in pathogen populations through diversifying selection (Meaden and Koskella, 2017). Coevolutionary dynamics between invasive plants and accumulating pathogens may result in more long-term stability in the interaction than typically observed in the boom-and-bust cycles of disease in agricultural systems (McDonald and Linde, 2002; but see Simberloff and Gibbons, 2004), but this issue requires more systematic theoretical and empirical study. Evolutionary shifts in host range due to selection on host susceptibility, tolerance and pathogen traits is likely to lead to unpredictability in the ultimate abundance and geographical ranges of introduced species in their novel ranges.

11.7 Conclusions

The impacts of plant invasions and how they change over time can be mediated by their pathogen communities, which are themselves dynamic due to exposure to new environments, host species and microbial communities. While our understanding of invasive plant–pathogen interactions is driven by studies conducted after the invasive plant has established, insights from agricultural and natural systems suggest that biotic resistance to establishment, initial pathogen accumulation and long-term plant–pathogen coevolution can have important consequences for plant populations. Studies that evaluate the relative importance of pathogens (or absence of pathogens) and other factors on the process and impacts of invasion are needed to improve context and generality for the processes reviewed in this chapter. For example, experimental invasions combined with different types of 'enemy removal' treatments could allow for comparisons of the relative effects of soil-borne pathogens, foliar pathogens, insects and larger herbivores on the impacts of invasive plants (Borer *et al.*, 2015). In addition, studies on interactions between pathogens and other factors, such as herbivores (Swope and Parker, 2010) and resource availability (Heckman *et al.*, 2016), as well as anthropogenic drivers such as land use change, are needed to generate more realistic predictions about the long-term dynamics of invasive species.

References

Abella, S.R. (2014) Effectiveness of exotic plant treatments on National Park service lands in the United States. *Invasive Plant Science and Management* 7(1), 147–163. DOI: 10.1614/IPSM-D-13-00058.1.

Abu-Dieyeh, M.H. (2007) Sclerotinia minor for dandelion control: effect of dandelion accession, age and grass competition. *Weed Research* 82, 1068–1072.

Allison, S.D. and Vitousek, P.M. (2004) Rapid nutrient cycling in leaf litter from invasive plants in Hawai'i. *Oecologia* 141(4), 612–619. DOI: 10.1007/s00442-004-1679-z.

Anderson, R.M. and May, R.M. (1979) Population biology of infectious diseases: Part I. *Nature* 280(5721), 361–367. DOI: 10.1038/280361a0.

Andow, D.A., Kareiva, P.M., Levin, S.A. and Okubo, A. (1990) Spread of invading organisms. *Landscape Ecology* 4(2-3), 177–188. DOI: 10.1007/BF00132860.

Andreu, J. and Vilà, M. (2011) Native plant community response to alien plant invasion and removal. *Management of Biological Invasions* 2(1), 81–94. DOI: 10.3391/mbi.2011.2.1.08.

Bagchi, R., Gallery, R.E., Gripenberg, S., Gurr, S.J., Narayan, L. *et al.* (2014) Pathogens and insect herbivores drive rainforest plant diversity and composition. *Nature* 506(7486), 85–88. DOI: 10.1038/nature12911.

Barrett, L.G. and Heil, M. (2012) Unifying concepts and mechanisms in the specificity of plant–enemy interactions. *Trends in Plant Science* 17(5), 282–292. DOI: 10.1016/j.tplants.2012.02.009.

Bateman, H.L. and Ostoja, S.M. (2012) Invasive woody plants affect the composition of native lizard and small mammal communities in riparian woodlands. *Animal Conservation* 15(3), 294–304. DOI: 10.1111/j.1469-1795.2011.00517.x.

Batten, K.M., Scow, K.M., Davies, K.F. and Harrison, S.P. (2006) Two invasive plants alter soil microbial community composition in serpentine grasslands. *Biological Invasions* 8(2), 217–230. DOI: 10.1007/s10530-004-3856-8.

Bebber, D.P., Holmes, T. and Gurr, S.J. (2014a) The global spread of crop pests and pathogens. *Global Ecology and Biogeography* 23(12), 1398–1407. DOI: 10.1111/geb.12214.

Bebber, D.P., Holmes, T., Smith, D. and Gurr, S.J. (2014b) Economic and physical determinants of the global distributions of crop pests and pathogens. *New Phytologist* 202(3), 901–910. DOI: 10.1111/nph.12722.

Beckstead, J., Meyer, S.E., Connolly, B.M., Huck, M.B., Street, L.E. *et al.* (2010) Cheatgrass facilitates spillover of a seed bank pathogen onto native grass species. *Journal of Ecology* 98(1), 168–177. DOI: 10.1111/j.1365-2745.2009.01599.x.

Beckstead, J. and Parker, I.M. (2003) Invasiveness of *Ammophila arenaria:* Release from soil-borne pathogens? *Ecology* 84(11), 2824–2831. DOI: 10.1890/02-0517.

Berner, D., Smallwood, E., Cavin, C., Lagopodi, A., Kashefi, J. *et al.* (2013) Successful establishment of epiphytotics of Puccinia punctiformis for biological control of *Cirsium arvense*. *Biological Control* 67(3), 350–360. DOI: 10.1016/j.biocontrol.2013.09.010.

Berner, D.K., Smallwood, E.L., Cavin, C.A., McMahon, M.B., Thomas, K.M. *et al.* (2015) Asymptomatic systemic disease of Canada thistle (Cirsium arvense) caused by Puccinia punctiformis and changes in shoot density following inoculation. *Biological Control* 86, 28–35. DOI: 10.1016/j.biocontrol.2015.02.006.

Blackburn, T.M., Pyšek, P., Bacher, S., Carlton, J.T., Duncan, R.P. *et al.* (2011) A proposed unified framework for biological invasions. *Trends in Ecology and Evolution* 26(7), 333–339. DOI: 10.1016/j.tree.2011.03.023.

Blaney, C.S. and Kotanen, P.M. (2001) Effects of fungal pathogens on seeds of native and exotic plants: a test using congeneric pairs. *Journal of Applied Ecology* 38(5), 1104–1113. DOI: 10.1046/j.1365-2664.2001.00663.x.

Blumenthal, D., Mitchell, C.E., Pysek, P. and Jarosik, V. (2009) Synergy between pathogen release and resource availability in plant invasion. *Proceedings of the National Academy of Sciences USA* 106(19), 7899–7904. DOI: 10.1073/pnas.0812607106.

Boose, D., Harrison, S., Clement, S. and Meyer, S. (2011) Population genetic structure of the seed pathogen *Pyrenophora semeniperda* on *Bromus tectorum* in western North America. *Mycologia* 103(1), 85–93. DOI: 10.3852/09-310.

Borer, E.T., Hosseini, P.R., Seabloom, E.W. and Dobson, A.P. (2007) Pathogen-induced reversal of native dominance in a grassland community. *Proceedings of the National Academy of Sciences USA* 104(13), 5473–5478. DOI: 10.1073/pnas.0608573104.
Borer, E.T., Lind, E.M., Ogdahl, E.J., Seabloom, E.W., Tilman, D. *et al.* (2015) Food-web composition and plant diversity control foliar nutrient content and stoichiometry. *Journal of Ecology* 103(6), 1432–1441. DOI: 10.1111/1365-2745.12461.
Boyce, R.L., Brossart, S.N., Bryant, L.A., Fehrenbach, L.A., Hetzer, R. *et al.* (2014) The beginning of the end? extensive dieback of an open-grown Amur honeysuckle stand in northern Kentucky, USA. *Biological Invasions* 16(10), 2017–2023. DOI: 10.1007/s10530-014-0656-7.
Bradley, B.A., Blumenthal, D.M., Wilcove, D.S. and Ziska, L.H. (2010) Predicting plant invasions in an era of global change. *Trends in Ecology & Evolution* 25(5), 310–318. DOI: 10.1016/j.tree.2009.12.003.
Bradley, D.J., Gilbert, G.S. and Martiny, J.B.H. (2008) Pathogens promote plant diversity through a compensatory response. *Ecology Letters* 11(5), 461–469. DOI: 10.1111/j.1461-0248.2008.01162.x.
Brasier, C.M. (2008) The biosecurity threat to the UK and global environment from international trade in plants. *Plant Pathology* 57(5), 792–808. DOI: 10.1111/j.1365-3059.2008.01886.x.
Brown, B.J., Mitchell, R.J. and Graham, S.A. (2002) Competition for pollination between an invasive species (purple loosestrife) and a native congener. *Ecology* 83(8), 2328–2336. DOI: 10.1890/0012-9658(2002)083[2328:CFPBAI]2.0.CO;2.
Burdon, J.J. and Chilvers, G.A. (1982) Host density as a factor in plant disease ecology. *Annual Review of Phytopathology* 20(1), 143–166. DOI: 10.1146/annurev.py.20.090182.001043.
Burdon, J.J., Groves, R.H. and Cullen, J.M. (1981) The impact of biological control on the distribution and abundance of *Chondrilla juncea* in south-eastern Australia. *The Journal of Applied Ecology* 18(3), 957–966. DOI: 10.2307/2402385.
Callaway, R.M. and Ridenour, W.M. (2004) Novel weapons: invasive success and the evolution of increased competitive ability. *Frontiers in Ecology and the Environment* 2(8), 436–443. DOI: 10.1890/1540-9295(2004)002[0436:NWISAT]2.0.CO;2.
Callaway, R.M., Thelen, G.C., Rodriguez, A. and Holben, W.E. (2004) Soil biota and exotic plant invasion. *Nature* 427(6976), 731–733. DOI: 10.1038/nature02322.
Catford, J.A., Jansson, R. and Nilsson, C. (2009) Reducing redundancy in invasion ecology by integrating hypotheses into a single theoretical framework. *Diversity and Distributions* 15(1), 22–40. DOI: 10.1111/j.1472-4642.2008.00521.x.
Cazetta, E., Schaefer, H.M. and Galetti, M. (2008) Does attraction to frugivores or defense against pathogens shape fruit pulp composition? *Oecologia* 155(2), 277–286. DOI: 10.1007/s00442-007-0917-6.
Chalkowski, K., Lepczyk, C.A. and Zohdy, S. (2018) Parasite ecology of invasive species: conceptual framework and new hypotheses. *Trends in Parasitology* 34(8), 655–663. DOI: 10.1016/j.pt.2018.05.008.
Chapman, D., Purse, B.V., Roy, H.E. and Bullock, J.M. (2017) Global trade networks determine the distribution of invasive non-native species. *Global Ecology and Biogeography* 26(8), 907–917. DOI: 10.1111/geb.12599.
Cheatham, M.R., Rouse, M.N., Esker, P.D., Ignacio, S., Pradel, W. *et al.* (2009) Beyond yield: plant disease in the context of ecosystem services. *Phytopathology* 99(11), 1228–1236. DOI: 10.1094/PHYTO-99-11-1228.
Chun, Y.J., Van Kleunen, M. and Dawson, W. (2010) The role of enemy release, tolerance and resistance in plant invasions: linking damage to performance. *Ecology Letters* 13, 937–946. DOI: 10.1111/j.1461-0248.2010.01498.x.
Cipollini, D. and Enright, S. (2009) A powdery mildew fungus levels the playing field for garlic mustard (*Alliaria petiolata*) and a North American native plant. *Invasive Plant Science and Management* 2(3), 253–259. DOI: 10.1614/IPSM-08-144.1.
Clay, K. (1995) Correlates of pathogen species richness in the grass family. *Canadian Journal of Botany* 73(S1), 42–49. DOI: 10.1139/b95-223.
Clay, K*et al.* (2008) Red Queen communities. In: Ostfeld, R.S., Keesing, F. and Eviner, V.T. (eds) *Infectious Disease Ecology: Effects of Ecosystems on Disease and of Disease on Ecosystems*. Princeton University Press, Princeton, pp. 145–178.
Colautti, R.I., Ricciardi, A., Grigorovich, I.A. and MacIsaac, H.J. (2004) Is invasion success explained by the enemy release hypothesis? *Ecology Letters* 7(8), 721–733. DOI: 10.1111/j.1461-0248.2004.00616.x.
Corbin, J.D. and D'Antonio, C.M. (2012) Gone but not forgotten? invasive plants' legacies on community and ecosystem properties. *Invasive Plant Science and Management* 5(1), 117–124. DOI: 10.1614/IPSM-D-11-00005.1.

Crous, C.J., Burgess, T.I., Le Roux, J.J., Richardson, D.M., Slippers, B. *et al.* (2017) Ecological disequilibrium drives insect pest and pathogen accumulation in non-native trees. *AoB PLANTS* 9, 1–16.

Cullen, J.M., Kable, P.F. and Catt, M. (1973) Epidemic spread of a rust imported for biological control. *Nature* 244(5416), 462–464. DOI: 10.1038/244462a0.

D'Antonio, C. and Flory, S.L. (2017) Long-term dynamics and impacts of plant invasions. *Journal of Ecology* 105(6), 1459–1461. DOI: 10.1111/1365-2745.12879.

Dangremond, E.M., Pardini, E.A. and Knight, T.M. (2010) Apparent competition with an invasive plant hastens the extinction of an endangered lupine. *Ecology* 91(8), 2261–2271. DOI: 10.1890/09-0418.1.

Denman, S., Kirk, S.A., Brasier, C.M. and Webber, J.F. (2005) In vitro leaf inoculation studies as an indication of tree foliage susceptibility to *Phytophthora ramorum* in the UK. *Plant Pathology* 54(4), 512–521. DOI: 10.1111/j.1365-3059.2005.01243.x.

Denman, S., Kirk, S.A., Moralejo, E. and Webber, J.F. (2009) *Phytophthora ramorum* and *Phytophthora kernoviae* on naturally infected asymptomatic foliage. *EPPO Bulletin* 39(1), 105–111. DOI: 10.1111/j.1365-2338.2009.02243.x.

Desprez-Loustau, M.-L., Robin, C., Buée, M., Courtecuisse, R., Garbaye, J. *et al.* (2007) The fungal dimension of biological invasions. *Trends in Ecology & Evolution* 22(9), 472–480. DOI: 10.1016/j.tree.2007.04.005.

Dickie, I.A., Bufford, J.L., Cobb, R.C., Desprez-Loustau, M.-L., Grelet, G. *et al.* (2017) The emerging science of linked plant-fungal invasions. *New Phytologist* 215(4), 1314–1332. DOI: 10.1111/nph.14657.

Diez, J.M., Dickie, I., Edwards, G., Hulme, P.E., Sullivan, J.J. *et al.* (2010) Negative soil feedbacks accumulate over time for non-native plant species. *Ecology Letters* 13(7), 803–809. DOI: 10.1111/j.1461-0248.2010.01474.x.

Dlugosch, K.M. and Parker, I.M. (2008) Founding events in species invasions: genetic variation, adaptive evolution, and the role of multiple introductions. *Molecular Ecology* 17(1), 431–449. DOI: 10.1111/j.1365-294X.2007.03538.x.

Dobson, A. and Crawley, M. (1994) Pathogens and the structure of plant communities. *Trends in Ecology & Evolution* 9(10), 393–398. DOI: 10.1016/0169-5347(94)90062-0.

Dunn, A.M., Torchin, M.E., Hatcher, M.J., Kotanen, P.M., Blumenthal, D.M. *et al.* (2012) Indirect effects of parasites in invasions. *Functional Ecology* 26(6), 1262–1274. DOI: 10.1111/j.1365-2435.2012.02041.x.

Dutech, C., Barrès, B., Bridier, J., Robin, C., Milgroom, M.G. *et al.* (2012) The chestnut blight fungus world tour: successive introduction events from diverse origins in an invasive plant fungal pathogen. *Molecular Ecology* 21(16), 3931–3946. DOI: 10.1111/j.1365-294X.2012.05575.x.

Dwyer, G., Levin, S.A. and Buttel, L. (1990) A simulation model of the population dynamics and evolution of myxomatosis. *Ecological Monographs* 60(4), 423–447. DOI: 10.2307/1943014.

Elmer, W.H. (2001) Seeds as vehicles for pathogen importation. *Biological Invasions* 3(3), 263–271. DOI: 10.1023/A:1015217308477.

Elton, C.S. (1958) *The Ecology of Invasions by Animals and Plants*. Methuen, London.

Enright, S.M. and Cipollini, D. (2007) Infection by powdery mildew *Erysiphe cruciferarum* (Erysiphaceae) strongly affects growth and fitness of *Alliaria petiolata* (Brassicaceae). *American Journal of Botany* 94(11), 1813–1820. DOI: 10.3732/ajb.94.11.1813.

Eppinga, M.B., Rietkerk, M., Dekker, S.C., De Ruiter, P.C., Van der Putten, W.H. *et al.* (2006) Accumulation of local pathogens: a new hypothesis to explain exotic plant invasions. *Oikos* 114(1)pp., 168–176. DOI: 10.1111/j.2006.0030-1299.14625.x.

Essarioui, A., LeBlanc, N., Otto-Hanson, L., Schlatter, D.C., Kistler, H.C. *et al.* (2019) Inhibitory and nutrient use phenotypes among co-existing *Fusarium* and *Streptomyces* populations suggest local coevolutionary interactions in soil. *Environmental Microbiology* 1462-2920(14782).

Evans, A.M. and Finkral, A.J. (2010) A new look at spread rates of exotic diseases in North American forests. *Forest Science* 56, 453–459.

Fabiszewski, A.M., Umbanhowar, J. and Mitchell, C.E. *et al.* (2010) Modeling landscape-scale pathogen spillover between domesticated and wild hosts: Asian soybean rust and kudzu. *Ecological Applications* 20(2), 582–592. DOI: 10.1890/08-0820.1.

Farr, D.F. and Rossman, A.Y. (2015) *Fungal Databases. U.S. National Fungus Collections, ARS, USDA.* Available at: https://nt.ars-grin.gov/fungaldatabases/ (accessed 15 March 2019).

Flory, S.L. and Clay, K. (2013) Pathogen accumulation and long-term dynamics of plant invasions. *Journal of Ecology* 101(3), 607–613. DOI: 10.1111/1365-2745.12078.

Flory, S.L., Clay, K., Emery, S.M., Robb, J.R., Winters, B. *et al.* (2015) Fire and non-native grass invasion interact to suppress tree regeneration in temperate deciduous forests. *Journal of Applied Ecology* 52(4), 992–1000. DOI: 10.1111/1365-2664.12437.

Flory, S.L., Kleczewski, N. and Clay, K. (2011) Ecological consequences of pathogen accumulation on an invasive grass. *Ecosphere* 2(10), art120–12. DOI: 10.1890/ES11-00191.1.

Forrestel, A.B., Ramage, B.S., Moody, T., Moritz, M.A., Stephens, S.L. *et al.* (2015) Disease, fuels and potential fire behavior: impacts of sudden oak death in two coastal California forest types. *Forest Ecology and Management* 348, 23–30. DOI: 10.1016/j.foreco.2015.03.024.

Gibson, I.A.S. (1972) Dothistroma blight of Pinus radiata. *Annual Review of Phytopathology* 10(1), 51–72. DOI: 10.1146/annurev.py.10.090172.000411.

Gilbert, B. and Levine, J.M. (2013) Plant invasions and extinction debts. *Proceedings of the National Academy of Sciences* 110(5), 1744–1749. DOI: 10.1073/pnas.1212375110.

Gilbert, G.S. and Parker, I.M. (2006) Invasions and the regulation of plant populations by pathogens. In: Cadotte, M.W., McMahon, S.M. and Fukami, T. (eds) *Conceptual Ecology and Invasion Biology: Reciprocal Approaches to Nature. Springer, Dordrecht*, pp. 289–305.

Gilbert, G.S. and Parker, I.M. (2010) Original article: rapid evolution in a plant-pathogen interaction and the consequences for introduced host species. *Evolutionary Applications* 3(2), 144–156. DOI: 10.1111/j.1752-4571.2009.00107.x.

Gillett, J.B. (1962) Pest pressure, an underestimated factor in evolution. *Systematics Association Publication* 4, 37–46.

Gladieux, P., Giraud, T., Kiss, L., Genton, B.J., Jonot, O. *et al.* (2011) Distinct invasion sources of common ragweed (*Ambrosia artemisiifolia*) in Eastern and Western Europe. *Biological Invasions* 13(4), 933–944. DOI: 10.1007/s10530-010-9880-y.

Gos, G., Slotte, T. and Wright, S.I. (2012) Signatures of balancing selection are maintained at disease resistance loci following mating system evolution and a population bottleneck in the genus Capsella. *BMC Evolutionary Biology* 12(1), 152. DOI: 10.1186/1471-2148-12-152.

Grünwald, N.J., Garbelotto, M., Goss, E.M., Heungens, K., Prospero, S. *et al.* (2012) Emergence of the sudden oak death pathogen *Phytophthora ramorum*. *Trends in Microbiology* 20(3), 131–138. DOI: 10.1016/j.tim.2011.12.006.

Guiden, P.W. and Orrock, J.L. (2017) Invasive exotic shrub modifies a classic animal-habitat relationship and alters patterns of vertebrate seed predation. *Ecology* 98(2), 321–327. DOI: 10.1002/ecy.1678.

Han, X., Dendy, S.P., Garrett, K.A., Fang, L., Smith, M.D. *et al.* (2008) Comparison of damage to native and exotic tallgrass prairie plants by natural enemies. *Plant Ecology* 198(2), 197–210. DOI: 10.1007/s11258-008-9395-0.

Hardham, A.R. and Blackman, L.M. (2018) Phytophthora cinnamomi. *Molecular Plant Pathology* 19(2), 260–285. DOI: 10.1111/mpp.12568.

Harmon, P.F., Momol, M.T., Marois, J.J., Dankers, H., Harmon, C.L. *et al.* (2008) Asian soybean rust caused by *Phakopsora pachyrhizi* on Soybean and Kudzu in Florida. *Plant Health Progress* 6(1), 9. DOI: 10.1094/PHP-2005-0613-01-RS.

Harris, P.T., Cannon, G.H., Smith, N.E. and Muth, N.Z. (2013) Assessment of plant community restoration following Tree-of-Heaven (*Ailanthus altissima*) control by *Verticillium albo-atrum*. *Biological Invasions* 15(9), 1887–1893. DOI: 10.1007/s10530-013-0430-2.

Hawkes, C. V. (2007) Are invaders moving targets? the generality and persistence of advantages in size, reproduction, and enemy release in invasive plant species with time since introduction. *The American Naturalist* 170(6), 832–843. DOI: 10.1086/522842.

Heckman, R.W., Wright, J.P. and Mitchell, C.E. (2016) Joint effects of nutrient addition and enemy exclusion on exotic plant success. *Ecology* 97(12), 3337–3345. DOI: 10.1002/ecy.1585.

Horvitz, N., Wang, R., Wan, F.-H. and Nathan, R. (2017) Pervasive human-mediated large-scale invasion: analysis of spread patterns and their underlying mechanisms in 17 of China's worst invasive plants. *Journal of Ecology* 105(1), 85–94. DOI: 10.1111/1365-2745.12692.

Hulme, P.E., Bacher, S., Kenis, M., Klotz, S., Kühn, I. *et al.* (2008) Grasping at the routes of biological invasions: a framework for integrating pathways into policy. *Journal of Applied Ecology* 45(2), 403–414. DOI: 10.1111/j.1365-2664.2007.01442.x.

Irwin, M.E., Kampmeier, G.E. and Weisser, W.W. (2007) Aphid movement: process and consequences. In: van Emden, H.F. and Harrington, R. (eds) *Aphids as Crop Pests*. CAB International, Wallingford, UK, pp. 153–186.

Keane, R. and Crawley, M.J. (2002) Exotic plant invasions and the enemy release hypothesis. *Trends in Ecology & Evolution* 17(4), 164–170. DOI: 10.1016/S0169-5347(02)02499-0.

Kumar, D. and Kalita, P. (2017) Reducing postharvest losses during storage of grain crops to strengthen food security in developing countries. *Foods* 6(1), 8. DOI: 10.3390/foods6010008.

Laine, A.-L., Burdon, J.J., Nemri, A. and Thrall, P.H. (2014) Host ecotype generates evolutionary and epidemiological divergence across a pathogen metapopulation. *Proceedings of the Royal Society B* 281(1787), 20140522. DOI: 10.1098/rspb.2014.0522.

Lau, J.A. and Suwa, T. (2016) The changing nature of plant–microbe interactions during a biological invasion. *Biological Invasions* 18(12), 3527–3534. DOI: 10.1007/s10530-016-1245-8.

Lehan, N.E., Murphy, J.R., Thorburn, L.P. and Bradley, B.A. (2013) Accidental introductions are an important source of invasive plants in the continental United States. *American Journal of Botany* 100(7), 1287–1293. DOI: 10.3732/ajb.1300061.

Levine, J.M., Adler, P.B. and Yelenik, S.G. (2004) A meta-analysis of biotic resistance to exotic plant invasions. *Ecology Letters* 7(10), 975–989. DOI: 10.1111/j.1461-0248.2004.00657.x.

Levine, J.M., Vilà, M., Antonio, C.M.D., Dukes, J.S., Grigulis, K. *et al.* (2003) Mechanisms underlying the impacts of exotic plant invasions. *Proceedings of the Royal Society of London. Series B: Biological Sciences* 270(1517), 775–781. DOI: 10.1098/rspb.2003.2327.

Liebhold, A.M., Brockerhoff, E.G., Garrett, L.J., Parke, J.L., Britton, K.O. *et al.* (2012) Live plant imports: the major pathway for forest insect and pathogen invasions of the US. *Frontiers in Ecology and the Environment* 10(3), 135–143. DOI: 10.1890/110198.

Lockwood, J.L., Cassey, P. and Blackburn, T. (2005) The role of propagule pressure in explaining species invasions. *Trends in Ecology & Evolution* 20(5), 223–228. DOI: 10.1016/j.tree.2005.02.004.

Louda, S.M., Keeler, K.H. and Holt, R.D. (1990) Herbivore influences on plant performance and competitive interactions. In: Grace, J.B. and Tilman, D. (eds) *Perspectives in Plant Competition*. Academic Press, New York, pp. 413–444.

Lymbery, A.J., Morine, M., Kanani, H.G., Beatty, S.J., Morgan, D.L. *et al.* (2014) Co-invaders: the effects of alien parasites on native hosts. *International Journal for Parasitology: Parasites and Wildlife* 3, 171–177.

Malmstrom, C.M., Hughes, C.C., Newton, L.A. and Stoner, C.J. (2005) Virus infection in remnant native bunchgrasses from invaded California grasslands. *New Phytologist* 168(1), 217–230. DOI: 10.1111/j.1469-8137.2005.01479.x.

Malmstrom, C.M., Shu, R., Linton, E.W., Newton, L.A., Cook, M.A. *et al.* (2007) Barley yellow dwarf viruses (BYDVs) preserved in herbarium specimens illuminate historical disease ecology of invasive and native grasses. *Journal of Ecology* 95(6), 1153–1166. DOI: 10.1111/j.1365-2745.2007.01307.x.

Mangla, S., Inderjit. and Callaway, R.M. (2008) Exotic invasive plant accumulates native soil pathogens which inhibit native plants. *Journal of Ecology* 96, 58–67.

Marcelletti, S. and Scortichini, M. (2016) *Xylella fastidiosa* CoDiRO strain associated with the olive quick decline syndrome in southern Italy belongs to a clonal complex of the subspecies pauca that evolved in central America. *Microbiology* 162(12), 2087–2098. DOI: 10.1099/mic.0.000388.

Maschek, O. and Halmschlager, E. (2018) Effects of *Verticillium nonalfalfae* on *Ailanthus altissima* and associated indigenous and invasive tree species in eastern Austria. *European Journal of Forest Research* 137(2), 197–209. DOI: 10.1007/s10342-018-1099-y.

McDonald, B.A. and Linde, C. (2002) The population genetics of plant pathogens and breeding strategies for durable resistance. *Euphytica* 124(2), 163–180. DOI: 10.1023/A:1015678432355.

McGinn, K.J., van der Putten, W.H., Hulme, P.E., Shelby, N., Weser, C. *et al.* (2018) The influence of residence time and geographic extent on the strength of plant-soil feedbacks for naturalised *Trifolium*. *Journal of Ecology* 106(1), 207–217. DOI: 10.1111/1365-2745.12864.

Meaden, S. and Koskella, B. (2017) Adaptation of the pathogen, *Pseudomonas syringae*, during experimental evolution on a native vs. alternative host plant. *Molecular Ecology* 26(7), 1790–1801. DOI: 10.1111/mec.14060.

Mejri, D., Gamalero, E., Tombolini, R., Musso, C., Massa, N. *et al.* (2010) Biological control of great brome (*Bromus diandrus*) in durum wheat (*Triticum durum*): specificity, physiological traits and impact on plant growth and root architecture of the fluorescent pseudomonad strain X33d. *BioControl* 55(4), 561–572. DOI: 10.1007/s10526-010-9285-y.

Mitchell, C.E., Blumenthal, D., Jarošík, V., Puckett, E.E., Pyšek, P. *et al.* (2010) Controls on pathogen species richness in plants' introduced and native ranges: roles of residence time, range size and host traits. *Ecology Letters* 13(12), 1525–1535. DOI: 10.1111/j.1461-0248.2010.01543.x.

Mitchell, C.E. and Power, A.G. (2003) Release of invasive plants from fungal and viral pathogens. *Nature* 421(6923), 625–627. DOI: 10.1038/nature01317.

Mordecai, E.A. (2011) Pathogen impacts on plant communities: unifying theory, concepts, and empirical work. *Ecological Monographs* 81(3), 429–441. DOI: 10.1890/10-2241.1.

Mordecai, E.A. (2013) Despite spillover, a shared pathogen promotes native plant persistence in a cheatgrass-invaded grassland. *Ecology* 94(12), 2744–2753. DOI: 10.1890/13-0086.1.

Neubert, M.G. and Parker, I.M. (2004) Projecting rates of spread for invasive species. *Risk Analysis* 24(4), 817–831. DOI: 10.1111/j.0272-4332.2004.00481.x.

Nijjer, S., Rogers, W.E. and Siemann, E. (2007) Negative plant-soil feedbacks may limit persistence of an invasive tree due to rapid accumulation of soil pathogens. *Proceedings of the Royal Society B: Biological Sciences* 274(1625), 2621–2627. DOI: 10.1098/rspb.2007.0804.

Nuñez, M.A., Horton, T.R. and Simberloff, D. (2009) Lack of belowground mutualisms hinders Pinaceae invasions. *Ecology* 90(9), 2352–2359. DOI: 10.1890/08-2139.1.

Padayachee, A.L., Irlich, U.M., Faulkner, K.T., Gaertner, M., Procheş, Şerban. *et al.* (2017) How do invasive species travel to and through urban environments? *Biological Invasions* 19(12), 3557–3570. DOI: 10.1007/s10530-017-1596-9.

Parker, I.M. and Gilbert, G.S. (2004) The evolutionary ecology of novel plant-pathogen interactions. *Annual Review of Ecology, Evolution, and Systematics* 35(1), 675–700. DOI: 10.1146/annurev.ecolsys.34.011802.132339.

Parker, I.M. and Gilbert, G.S. (2007) When there is no escape: the effects of natural enemies on native, invasive, and noninvasive plants. *Ecology* 88(5), 1210–1224. DOI: 10.1890/06-1377.

Parker, I.M., Saunders, M., Bontrager, M., Weitz, A.P., Hendricks, R. *et al.* (2015) Phylogenetic structure and host abundance drive disease pressure in communities. *Nature* 520(7548), 542–544. DOI: 10.1038/nature14372.

Purse, B.V., Graeser, P., Searle, K., Edwards, C., Harris, C. *et al.* (2013) Challenges in predicting invasive reservoir hosts of emerging pathogens: mapping *Rhododendron ponticum* as a foliar host for *Phytophthora ramorum* and *Phytophthora kernoviae* in the UK. *Biological Invasions* 15(3), 529–545. DOI: 10.1007/s10530-012-0305-y.

Pyšek, P., Jarošík, V., Hulme, P.E., Pergl, J., Hejda, M. *et al.* (2012) A global assessment of invasive plant impacts on resident species, communities and ecosystems: the interaction of impact measures, invading species' traits and environment. *Global Change Biology* 18(5), 1725–1737. DOI: 10.1111/j.1365-2486.2011.02636.x.

Ragaert, P., Devlieghere, F. and Debevere, J. (2007) Role of microbiological and physiological spoilage mechanisms during storage of minimally processed vegetables. *Postharvest Biology and Technology* 44(3), 185–194. DOI: 10.1016/j.postharvbio.2007.01.001.

Reichard, S.H. and White, P. (2001) Horticulture as a pathway of invasive plant introductions in the United States. *BioScience* 51(2), 103. DOI: 10.1641/0006-3568(2001)051[0103:HAAPOI]2.0.CO;2.

Reinhart, K.O., Packer, A., Van der Putten, W.H. and Clay, K. (2003) Plant-soil biota interactions and spatial distribution of black cherry in its native and invasive ranges. *Ecology Letters* 6(12), 1046–1050. DOI: 10.1046/j.1461-0248.2003.00539.x.

Riffle, J.W. and Peterson, G.W. (1986) *Diseases of trees in the Great Plains. Gun. Tech. Rep. RM-129. Fort Collins, CO: U.S.* Department of Agriculture, Forest Service, Rocky Mountain Forest and Range Experiment Station.

Schultheis, E.H., Berardi, A.E. and Lau, J.A. (2015) No release for the wicked: enemy release is dynamic and not associated with invasiveness. *Ecology* 96(9), 2446–2457. DOI: 10.1890/14-2158.1.

Simberloff, D. and Gibbons, L. (2004) Now you See them, Now you don't! – Population Crashes of Established Introduced Species. *Biological Invasions* 6(2), 161–172. DOI: 10.1023/B:BINV.0000022133.49752.46.

Smock, L.A. and MacGregor, C.M. (1988) Impact of the American chestnut blight on aquatic shredding Macroinvertebrates. *Journal of the North American Benthological Society* 7(3), 212–221. DOI: 10.2307/1467421.

Spear, E.R. and Mordecai, E.A. (2018) Foliar pathogens are unlikely to stabilize coexistence of competing species in a California grassland. *Ecology* 99(10), 2250–2259. DOI: 10.1002/ecy.2427.

Stewart, T.E., Allen, P.S. and Meyer, S.E. (2009) First report of *Pyrenophora seminiperda* in turkey and Greece. *Plant Disease* 93(12), 1351. DOI: 10.1094/PDIS-93-12-1351B.

Stinson, K.A., Campbell, S.A., Powell, J.R., Wolfe, B.E., Callaway, R.M. *et al.* (2006) Invasive plant suppresses the growth of native tree seedlings by disrupting belowground mutualisms. *PLoS Biology* 4(5), e140–731. DOI: 10.1371/journal.pbio.0040140.

Strauss, A., White, A. and Boots, M. (2012) Invading with biological weapons: the importance of disease-mediated invasions. *Functional Ecology* 26(6), 1249–1261. DOI: 10.1111/1365-2435.12011.

Stricker, K.B., Harmon, P.F., Goss, E.M., Clay, K., Luke Flory, S. *et al.* (2016) Emergence and accumulation of novel pathogens suppress an invasive species. *Ecology Letters* 19(4), 469–477. DOI: 10.1111/ele.12583.

Swope, S.M. and Parker, I.M. (2010) Trait-mediated interactions and lifetime fitness of the invasive plant *Centaurea solstitialis*. *Ecology* 91(8), 2284–2293. DOI: 10.1890/09-0855.1.

Tamme, R., Götzenberger, L., Zobel, M., Bullock, J.M., Hooftman, D.A.P., Kaasik, A. *et al.* (2014) Predicting species' maximum dispersal distances from simple plant traits. *Ecology* 95(2), 505–513. DOI: 10.1890/13-1000.1.

Thomson, F.J., Moles, A.T., Auld, T.D. and Kingsford, R.T. (2011) Seed dispersal distance is more strongly correlated with plant height than with seed mass. *Journal of Ecology* 99(6), 1299–1307. DOI: 10.1111/j.1365-2745.2011.01867.x.

Tilman, D., May, R.M., Lehman, C.L. and Nowak, M.A. (1994) Habitat destruction and the extinction debt. *Nature* 371(6492), 65–66. DOI: 10.1038/371065a0.

Tomasetto, F., Tylianakis, J.M., Reale, M., Wratten, S., Goldson, S.L. *et al.* (2017) Intensified agriculture favors evolved resistance to biological control. *Proceedings of the National Academy of Sciences* 114(15), 3885–3890. DOI: 10.1073/pnas.1618416114.

Tooley, P.W., Browning, M. and Berner, D. (2008) Recovery of *Phytophthora ramorum* following exposure to temperature extremes. *Plant Disease* 92(3), 431–437. DOI: 10.1094/PDIS-92-3-0431.

Trujillo, E.E. (1985) Biological control of Hamakua pa-makani with *Cercosporella* sp. in Hawaii. In: Zwölfer, H. and Delfosse, E.S. (eds) *Proceedings of the VI International Symposium on Biological Control of Weeds*, pp. 661–671.

van der Putten, W.H (2010) Impacts of soil microbial communities on exotic plant invasions. *Trends in ecology & evolution* 25(9), 512–519. DOI: 10.1016/j.tree.2010.06.006.

Van der Putten, W.H., Van Dijk, C. and Peters, B.A.M. (1993) Plant-Specific soil-borne diseases contribute to succession in foredune vegetation. *Nature* 362(6415), 53–56. DOI: 10.1038/362053a0.

Van der Putten, W.H. and Peters, B.A.M. (1997) How soil-borne pathogens may affect plant competition. *Ecology* 78(6), 1785–1795. DOI: 10.1890/0012-9658(1997)078[1785:HSBPMA]2.0.CO;2.

Vilà, M., Espinar, J.L., Hejda, M., Hulme, P.E., Jarošík, V. *et al.* (2011) Ecological impacts of invasive alien plants: a meta-analysis of their effects on species, communities and ecosystems. *Ecology Letters* 14(7), 702–708. DOI: 10.1111/j.1461-0248.2011.01628.x.

White, J.F., Kingsley, K.I., Kowalski, K.P., Irizarry, I., Micci, A. *et al.* (2018) Disease protection and allelopathic interactions of seed-transmitted endophytic pseudomonads of invasive reed grass (*Phragmites australis*). *Plant and Soil* 422(1-2), 195–208. DOI: 10.1007/s11104-016-3169-6.

Wisler, G.C. and Norris, R.F. (2005) Interactions between weeds and cultivated plants as related to management of plant pathogens. *Weed Science* 53(6), 914–917. DOI: 10.1614/WS-04-051R.1.

Young, H.M., George, S., Narváez, D.F., Srivastava, P., Schuerger, A.C. *et al.* (2012) Effect of solar radiation on severity of soybean rust. *Phytopathology* 102(8), 794–803. DOI: 10.1094/PHYTO-10-11-0294.

12 Direct and Indirect Effects of Herbivores Influencing Plant Invasions

Peter M. Kotanen*

Department of Ecology and Evolutionary Biology, University of Toronto Mississauga, Mississauga, Ontario, Canada

Abstract

Non-native plants rarely escape damage by herbivores. Instead, upon arrival in a new region, they begin to acquire new enemies, replacing those they have lost during their migration. These herbivores can include both natives to the new region and species that have themselves been accidentally or deliberately introduced from elsewhere, potentially including examples originating from the invader's original range. Shifts of new enemies from other hosts can occur over a range of timescales, depending in part on whether evolutionary change is required, but are likely to be faster for plants that are widespread and phylogenetically related to a herbivore's original host, and faster for generalist herbivores than for specialists. The occurrence of herbivores is not necessarily uniform across an invader's range; instead, they may be less diverse or abundant in host populations that are geographically or ecologically marginal, though existing evidence is mixed. Collectively, these new suites of herbivores can affect the growth and fitness of invaders, both directly by damaging them and indirectly by attacking their competitors. Studies comparing the demographic consequences of herbivory for successful vs unsuccessful invaders may help to clarify how often such impacts limit invasiveness. The view that an invader enters 'enemy-free' space is inaccurate; instead, persistence and spread of non-native plants often may be affected by the novel and changing assemblage of herbivores that they acquire within their new distribution.

12.1 Introduction

Non-native plants lose many of their original herbivores during the invasion process (Liu and Stiling, 2006; Bezemer *et al.*, 2014; Chapter 10, this volume), yet this often does not translate to a reduction in herbivore damage, at least compared to native plants (Colautti *et al.*, 2004; Chun *et al.*, 2010; Bezemer *et al.*, 2014; Heger and Jeschke, 2014). One reason is that invaders rarely enter an empty environment: instead, they are likely to encounter a novel set of new herbivores, some of which may be capable of inflicting significant damage (Levine *et al.*, 2004; Mitchell *et al.*, 2006; Chapter 9, this volume). In this chapter, I discuss the sources of these new enemies and their importance for the success and impacts of non-native plants. My focus will be mostly on insect and vertebrate herbivores; other chapters in this volume discuss pathogens (Chapter 11, this volume) and feedbacks involving microherbivores such as nematodes (Chapter 3, this volume).

12.2 Acquisition of New Enemies

Although it long has been recognized that non-native plants acquire new enemies in the regions that they invade (Elton, 1958), more attention has often been paid to the loss of enemies, as predicted by the Enemy Release Hypothesis (Keane and Crawley, 2002; Chapter 10, this volume). Consequently, it is often assumed that invaders

*peter.kotanen@utoronto.ca

DOI: 10.1079/9781789242171.0012

typically experience little herbivore damage. Instead, it is not unusual for non-native plants to experience high levels of herbivory. For instance, in southern Canadian populations of common burdock, *Arctium minus*, my laboratory has documented losses of 90% of seeds to larvae of a single gelechiid moth, *Metzneria lappella*, while leaves are heavily attacked by insects including various chewers and mining tephritid flies (Kambo and Kotanen, 2014). Similarly, we have measured losses of almost 50% of the capitula of the invasive thistle *Cirsium arvense* to a weevil, *Larinus planus*, as well as 12% of this plant's leaf area to insects including the tortoise beetle *Cassida rubiginosa* (Nunes *et al.*, 2016). Where do such enemies come from? All these examples involve introduced insects, but this is often not the case. Instead, there are several distinct sources of such enemies, each with different dynamics and implications.

12.2.1 Co-invaders

Herbivores of non-native plants often have themselves been introduced from elsewhere: many food webs contain numerous non-native insects, while introduced mammals such as feral goats, European rabbits and brushtail possums (as well as livestock) can eclipse the effects of native vertebrates. From the perspective of the host plant, these herbivores may not be new at all, but instead may be old enemies originating in the non-native plant's native range; examples include the reassociation of the specialist parsnip webworm (*Depressaria pastinacella*) with European wild parsnip (*Pastinaca sativa*) in North America (Zangerl and Berenbaum, 2005), and Chinese tallow tree (*Triadica sebifera*) with the Asian beetle *Adoretus sinicus* in its introduced range in Hawaii (Siemann and Rogers, 2003). Alternatively, herbivores may be species from different biogeographic regions encountering non-native hosts for the first time: population growth of the invasive Brazilian shrub *Eugenia uniflora* in Florida is significantly reduced by *Myllocerus undatus*, a weevil itself introduced from Sri Lanka (Stricker and Stiling, 2012).

Many non-native invertebrate herbivores were likely introduced accidentally, as contaminants of plants, soil or wood products; however, this is not always the case. For instance, some of the most prominent North American herbivores of non-native thistles were introduced as biocontrol organisms (Cripps *et al.*, 2011), while many vertebrates have been repeatedly translocated for economic reasons such as agriculture, hunting or fur production. Whether deliberate or accidental, introductions of non-native herbivores aren't necessarily coincident with the introduction of their non-native hosts. Generalists, including many insects and most vertebrates, may have been introduced well before the arrival of particular hosts. In contrast, specialists (including most biocontrol organisms) are likely to follow only after their food plants are established. One exception may be specialist herbivores intimately associated with their hosts, which in some cases may be able to disperse with them: the moth *M. lappella* develops inside the seed heads (capitula) of burdock, which are also the units of dispersal (by adhering to fur or clothing). Overall, it is likely that colonization of non-native plants by co-invaders occurs over a spectrum of timescales from decades after their host's arrival (as for *Depressaria* and many biocontrol agents) to nearly simultaneous with initial introduction (as may be true for some predispersal seed predators), even to preceding their future hosts (as may be the case for deliberately introduced game or domesticated mammals).

12.2.2 Ecological host shifts

Native herbivores frequently use non-native plants (Maron and Vila, 2001; Bezemer *et al.*, 2014); for instance, 34% of California's 236 butterfly species feed or oviposit on introduced taxa (Graves and Shapiro, 2003). These herbivores include species which already possessed traits enabling them to colonize new hosts by 'ecological fitting' without a need for genetic changes. In particular, native generalists may readily attack newly-arrived plant species, as long as they are not too heavily defended (Parker *et al.*, 2006; Bertheau *et al.*, 2010; Heard and Sax, 2013). For example, fall webworm (*Hyphantria cunia*) has a huge host range, including both native and non-native plants (Hinman *et al.*, 2019); white-tailed deer in eastern North America likewise consume a wide range of natives and non-natives (Averill *et al.*, 2016). Even oligophagous herbivores may be able to attack a newly-arrived non-native if it is chemically and physically similar to their usual hosts. For instance, a diverse set of native insects associated with a native North American thistle, *Cirsium altissimum*, have significant impacts on the invasive bull thistle, *Cirsium vulgare* (Eckberg *et al.*, 2012, 2014).

Most mammals are generalists (compared to insects), suggesting mammalian impacts on new invaders often may be immediate and significant. For invertebrates, it might be expected that more specific feeders should be less likely to use new arrivals; for example, in a study of plants in a botanical garden, Kirichenko *et al.* (2013) provided evidence that internal feeders (leaf miners and gall makers) were slower to colonize new hosts than external feeders. Similarly, the probability that a species of caterpillar used introduced *Piper* spp. in New Guinea increased with that caterpillar's host range, from 3% for oligophagous species to 92% for polyphagous species (Novotny *et al.*, 2003).

12.2.3 Adaptive change

An invading plant that initially escapes damage may gradually acquire new enemies as these herbivores evolve the behavioural or physiological adaptations necessary to exploit these new and often abundant food sources (Strong *et al.*, 1984; Carroll, 2007). Since evolution is generally slow compared to invasion, non-native plants often may not have coexisted with native herbivores long enough for significant adaptation to have occurred. Nonetheless, instances have been documented of such adaptive shifts (Cox, 2004; Carroll, 2007). For instance, soapberry bugs, *Jadera haematoloma*, in Florida originally fed on the native balloon vine *Cardiospermum corindum*, but within 50 years of the introduction of Chinese flametree, *Koelrueteria elegans*, some populations evolved shorter beaks and other physical and developmental changes that allowed them to exploit this new host (Carroll *et al.*, 1998; Carroll, 2007). These changes went beyond a simple expansion of dietary range. Instead, *J. haematoloma* has now evolved a host race specialized on *K. elegans*, at the cost of reduced performance on its original host. Shifts onto other hosts have been documented elsewhere in North America (Carroll, 2007), while changes in beak length and other traits have allowed a different native bug, *Leptocoris tagalicus*, to colonize another introduced balloon vine *Cardiospermum grandiflorum* in Australia (Carroll *et al.*, 2005). As another example, native *Rhagoletis* tephritid flies colonized fruits of Asian honeysuckles, *Lonicera* spp., within about 250 years of their introduction to North America; these flies are hybrids of two native species, apparently representing a new, reproductively isolated species (Schwarz *et al.*, 2005). Finally, in Nevada, a population of the butterfly *Euphydryas editha* evolved a preference for the introduced *Plantago lanceolata* over about a decade, nearly abandoning its native hosts (Singer *et al.*, 1993). Whether cases such as these are typical is unclear; will most of the hundreds of non-native plants present in many floras eventually acquire adapted herbivores? If so, a long-term result of plant invasions might be a radiation of new insect species adapted to non-native hosts.

12.3 Variation in Herbivore Communities

Impacts of natural enemies on non-native plants are far from uniform, but instead vary among species and populations with important implications for invasions. There are many sources of such variation; I will discuss just a few examples.

12.3.1 Time since invasion

Accumulation of new enemies can be a protracted process. As discussed above, new enemies may colonize an invader almost immediately upon its arrival in a new location; in other cases, this process may take many years, resulting in a gradual increase in the diversity of a non-native plant's herbivore community (Strong *et al.*, 1984). As well, as invaders spread geographically, they may be more likely to encounter potential enemies; as a rule, plants with broader distributions

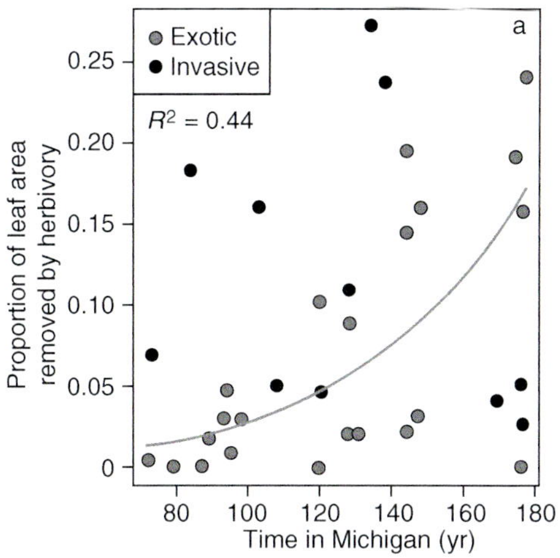

Fig. 12.1. Evidence that herbivores gradually colonize non-native plants following their arrival in a new region. Points indicate damage by invertebrate herbivores to non-invasive and invasive non-native plants vs years since their arrival in Michigan, USA. Insect damage increased (p = 0.02) with residence time for non-native (but not invasive) plants. Adapted from Fig. 12.3 from Schultheis *et al*. (2015) No release for the wicked: enemy release is dynamic and not associated with invasiveness *Ecology* 96, 2446–2457. Used with permission. © 2015 by the Ecological Society of America.

harbour a greater diversity of herbivores (Strong *et al.*, 1984; Lewinsohn *et al.*, 2005; Branco *et al.*, 2015). Again, the result can be a gradual accumulation of enemies over time. Although a larger herbivore community does not necessarily translate directly to more damage, the end result nonetheless can be that a plant species that has been present in its new region for a longer period of time experiences more damage. For instance, Schultheis *et al.* (2015) found insect (though not mammal) damage increased with the length of time a non-invasive non-native species had occurred in Michigan (Fig. 12.1). Interestingly, this pattern was absent in invasive non-natives, which instead tended to experience high levels of herbivory, suggesting interspecific variation in damage does not necessarily correlate with invasiveness (see below).

Accumulation of new enemies over time may mean that a species which initially benefits from enemy release might gradually become less invasive as it is incorporated into native food webs, at least unless it evolves stronger defences in response to this increasing herbivore pressure (see below). However, the accumulation of new enemies can be difficult to document, since the long timescale means that it is usually not possible to observe directly. Instead, some studies have attempted to correlate damage experienced by non-native plants with the time since their first recorded arrival. Results have been mixed: some studies have failed to detect the expected temporal trend (Carpenter and Cappuccino, 2005), while others found evidence of enemy accumulation, though often with low explanatory power (Hawkes, 2007; Brändle *et al.*, 2008; Schultheis *et al.*, 2015). One reason may be that many other factors also affect rates of accumulation, including those mentioned below. Another impediment is that dates of arrival are often not known with confidence; records in the literature and herbaria generally represent latest possible dates. Finally, 'date of arrival' is frustratingly ambiguous: arrival in a new continent, in a particular region, or at a particular study site can differ by many decades; it is often not obvious which date should be used as a starting point.

One alternative for quantifying accumulation of enemies involves substituting space for time: comparing populations near the invasion front with older populations nearer the original site of introduction. Some results of studies using this approach support the gradual accumulation of enemies; for example, for *Senecio madagascariensis* in Australia, abundance and diversity of invertebrate herbivores (but not damage) decreased from the point of introduction to the invasion front (Harvey *et al.*, 2013). Other studies have documented more complicated patterns or even increased damage near the invasion front (Wan and Bonser, 2016; Tabassum and Leishman, 2018). One reason is that comparisons over large geographic ranges may confound time with spatial variation in climate, isolation, enemy distributions or other factors (see below). Still, if herbivory is at least partly related to an invader's date of arrival at each site, this may suggest that time in a new continent is less important than time at a particular location.

12.3.2 Phylogenetic isolation

The rate of enemy accumulation may depend in part on the phylogenetic relationship of a non-native with the invaded flora. Close relatives generally share similar traits, notably defensive chemistry; as a result, plants that are more closely related are more likely to share herbivores, particularly insects (*e.g.* Odegaard *et al.*, 2005; Brändle and Brandl, 2006; Weiblen *et al.*, 2006; Bertheau *et al.*, 2010). This suggests that escape from enemies is less likely to occur, and more likely to quickly decay, if an invading plant co-occurs with close relatives from which it may acquire enemies.

A number of studies have related herbivore damage to phylogenetic isolation of non-native species. For example, damage to plants in an arboretum in Tanzania was greater for species with relatives in the local native flora (Dawson *et al.*, 2009; Fig. 12.2); other arboretum and common garden studies have found similar results (Hill and Kotanen, 2009; Pearse and Hipp, 2009; Kirichenko and Kenis, 2016). Support also has come from wild populations (Hill and Kotanen, 2009; Harvey *et al.*, 2012), though patterns may be more difficult to detect against natural background variation (Hill and Kotanen, 2009). Even in the best-controlled experiments, important differences may still occur between close relatives because of unique chemical defences or other species-specific traits; as a result, correlations in phylogenetic studies often are weak, even if the overall trend is clear.

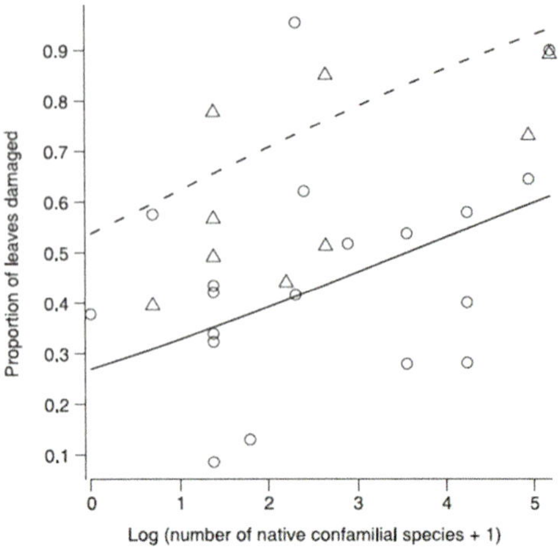

Fig. 12.2. Evidence that the presence of native relatives at both the generic and familial level increase the risk of herbivory for non-native plants. Symbols indicate the proportion of leaves damaged by herbivores in a botanical garden in Tanzania vs the number of related species in the local native flora. Triangles and dashed lines indicate results for species with native congeners; circles and solid line indicate results for species lacking native congeners. Damage increases with the number of confamilial relatives ($p = 0.01$); species with local native congeners experience more damage than those without ($p = 0.01$). From Dawson *et al.* (2009) Herbivory is related to taxonomic isolation, but not to invasiveness of tropical alien plants. *Diversity and Distributions* 15, 141–147. Used with permission. © 2008 The Authors. Journal compilation © 2008 Blackwell Publishing Ltd.

Phylogenetically conserved trends in damage may shed light on some other patterns. For instance, they suggest that phylogenetic outliers may be especially problematic invaders because they are more likely to escape enemies (but see Parker *et al.*, 2012). A classic example would be the invasion of Australia by *Opuntia* spp.; these cacti belong to a family entirely absent from the native Australian flora, and as such escaped significant insect herbivory until the moth *Cactoblastis cactorum* was deliberately introduced, resulting in a dramatic reduction in abundance (Moran, 1980). Such phylogenetic isolation may be continent-dependent: the probability that a non-native species escapes its enemies may be greater in regions where relatives are poorly represented. Some of the best evidence of phylogenetic trends in both invasiveness and damage comes from regions like Australia, New Zealand and East Africa that are very floristically distinct from the primarily European sources of invaders: it may be more difficult for European plants to escape in regions like North America, which shares most plant families and many genera with western Europe.

12.3.3 Geographic variation

Herbivory is not necessarily uniform throughout an invader's new range; instead, there can be pronounced geographic gradients in damage. These differences can be very strong: in *A. minus*,

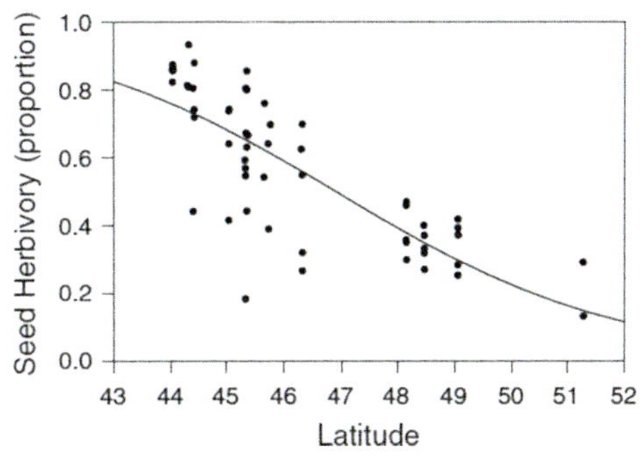

Fig. 12.3. Example of a geographic trend in herbivory. Predispersal seed predation of the invader common burdock, *Arctium minus*, by the moth *Metzneria lappella* declines with increasing latitude in Ontario, Canada ($p < 0.001$). From Kambo and Kotanen (2014) Latitudinal trends in herbivory and performance of an invasive species, common burdock (*Arctium minus*). *Biological Invasions* 16, 101–112. Used with permission. © Springer Science + Business Media Dordrecht 2013.

we found losses of seeds to predispersal predators declined over a latitudinal gradient from 80–90% in southern Ontario to only 20% near this invader's northern range limit (Kambo and Kotanen, 2014; Fig. 12.3). Similarly, Lu *et al.* (2013) found beetle damage to the semi-aquatic invader *Alternanthera philoxeroides* declined with increasing latitude in China; these patterns mirror the latitudinal decline in herbivore damage often reported for native plants (Anstett *et al.*, 2016). Other studies also have found that geographically marginal populations experience reduced herbivore damage (e.g. Vaupel and Matthies, 2012; Woods *et al.*, 2012), but exceptions exist (Harvey *et al.*, 2013; Wan and Bonser, 2016; Tabassum and Leishman, 2018).

There are numerous non-exclusive factors that can contribute to geographic trends in herbivory, including temperature, growing season duration and water availability. In addition, marginal populations are likely to be newer, smaller and more isolated than central populations, and thus less likely to be found or stably occupied by herbivores. We have found that newly-created experimental populations of the invaders *Ambrosia artemisiifolia* (MacKay and Kotanen, 2008) and *Cirsium arvense* (Nunes and Kotanen, 2018) initially experience reduced damage even when they are only tens of metres away from potential sources of herbivores. Such short-lived escape may be especially important to disturbance-loving invaders such as *Ambrosia* that repeatedly found equally short-lived populations.

Spatial variation in herbivory implies that factors limiting populations of non-native plants may vary regionally: in some areas populations have escaped most of their enemies, while in others they have not. Such variation may have significant consequences for the spread of invaders; for example, if damage is reduced in marginal populations, this may improve their performance, potentially favouring further range extensions (Kambo and Kotanen, 2014). In particular, as ranges of non-native plants change in response to changing climate (Walther *et al.*, 2009; Bradley *et al.*, 2010) such range-edge interactions may influence the rates at which invaders advance into formerly climatically unsuitable areas, or retreat from increasingly hostile sites.

12.3.4 Habitat variation

As well as varying across geographic gradients, herbivory also can be strongly influenced by local habitat variation (Louda *et al.*, 1990; Mitchell *et al.*, 2006). Herbivores may be more abundant in habitats that provide alternative resources, suitable physical conditions or refuge from predators, while plant defences and recovery from damage also may be affected by the local environment. This sort of local variation can have important consequences for non-native plants. For example, in one of the most iconic examples of biological control, populations of invasive Klamath weed (*Hypericum perforatum*) in California were greatly reduced by the introduction of a suite of chrysomelid beetles. However, these beetles are only effective in sunny, warm sites: shady, wet sites constitute refuges where this plant can survive (Huffaker and Kennett, 1959). *Arctium minus* is an example of an invader exhibiting the opposite pattern: in both its native and introduced ranges, plants in open sites experience less damage from insect herbivores than do shaded plants (Straw, 1991; Lee and

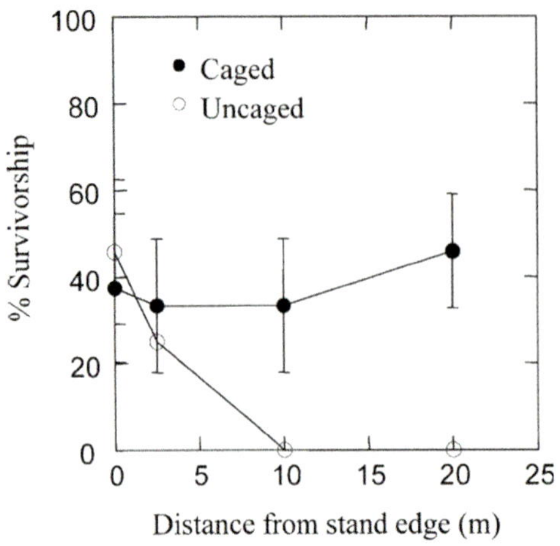

Fig. 12.4. Example of local variation in herbivory. Survival of seedlings of the invasive grass *Cortaderia jubata* transplanted into a stand of chaparral in California. Filled circles indicate plants protected by cages; open circles indicate unprotected plants. Survival is higher for uncaged plants near the stand edge, but does not vary with distance into the stand for protected plants, suggesting vertebrate herbivory increases deeper into the stand. Interaction $p = 0.03$. From Lambrinos, J.G. (2006) Spatially variable propagule pressure and herbivory influence invasion of chaparral shrubland by an exotic grass. *Oecologia* 147, 327–334. Used with permission. © Springer-Verlag 2005.

Kotanen, 2017). Habitat variation may provide protection from vertebrate herbivores as well as invertebrates. For instance, risk-averse foragers may be more likely to consume plants growing in refuges than exposed plants: herbivory by rabbits decreases recruitment of the invasive grass *Cortaderia jubata* in the interior of patches of Californian chaparral vegetation far more than near their edges (Lambrinos, 2006; Fig. 12.4). Patterns of habitat-dependent herbivory may themselves change during the invasion process. In its native range in Costa Rica, *Clidemia hirta* is largely excluded from understorey sites by insect herbivores and pathogens; instead, it is restricted to open areas (DeWalt *et al.*, 2004a). In contrast, in Hawaii, where this species has been introduced without its principal enemies, it occupies both open and understorey sites (DeWalt *et al.*, 2004a). In the long term, occupancy of forested habitats may lead to evolutionary changes in physiology or other traits, but the results of a common garden study suggest that this is not yet the case (DeWalt *et al.*, 2004b).

12.4 Effects of Herbivores on Non-native Hosts

12.4.1 Linking herbivore damage and plant performance

As these examples illustrate, herbivore damage to non-native plants is common, but does it matter? To be ecologically important, damage must affect host performance: growth, reproduction and ultimately abundance or distribution. However, it is often not clear that herbivores reduce the performance of the non-native plants they attack. Many non-natives remain abundant despite high levels of damage, and it has been argued that herbivory may rarely limit their success (Maron and Vila, 2001; Schultheis *et al.*, 2015). This argument parallels the debate for plants in general: many species are highly tolerant of damage, particularly to leaves, and can maintain growth and reproduction despite it. However, this view is likely to be pessimistic (Levine *et al.*, 2004). First, rigorously documenting the consequences of damage usually requires controlled experiments rather than just observation. Studies that count holes in leaves are common, but studies that determine the effects of that damage are much less frequent. Second, even when herbivores fail to prevent the establishment or spread of an invader, they may reduce abundance, growth rate, reproduction or other important demographic parameters. This may be especially difficult to observe in interspecific studies where between-species differences in herbivore attack obscure within-species demographic responses. Third, most research tends to focus on abundant or problematic species, yet these are the ones least likely to be constrained by herbivore damage; conversely, it is rarely possible to study failed invasions, yet these may include examples where herbivory played a key role. Finally, an important effect of herbivory can be reducing the ability of plants to withstand competition (Louda *et al.*,

1990). For instance, Suwa and Louda (2012) found that herbivory reduced the ability of the non-native thistle *C. vulgare* to invade tallgrass prairie by reducing its ability to compete with native vegetation. Such competition-mediated effects suggest the influence of herbivory on performance of non-natives can be seriously underestimated in studies lacking a competitive environment.

12.4.2 Effects on performance

Despite these caveats, many examples exist of herbivores significantly affecting the performance of non-native plants. Some of the most dramatic are provided by biological control (Chapter 20, this volume); for instance the cases of *Opuntia* and *Hypericum*. These examples are atypical, both because they were deliberately designed to have a large effect, and because many biocontrol insects nonetheless fail to produce such large reductions of their hosts (Thomas and Reid, 2007); however, they make the point that some herbivores can have strong impacts on non-natives.

Non-native plants also can suffer significant damage from unplanned interactions with herbivores. When herbivore exclusion experiments are conducted, they frequently reveal impacts on non-native species; for example, an insect exclusion experiment found biomass and probability of flowering of the introduced thistle *Cirsium vulgare* were reduced by a diverse set of native and non-native insects (Suwa and Louda, 2012). In some cases, impacts of herbivores on non-natives may be even greater than those on natives. For example, some studies have found non-native plants are more palatable to native herbivores than natives (Parker and Hay, 2005; Morrison and Hay, 2011), and are damaged equally (Chun *et al.*, 2010; Dostál *et al.*, 2013) or more severely (Heard and Sax, 2013; Schultheis *et al.*, 2015) than native plants.

The evolution of new defences in invaded regions also implies that the selection pressure exerted by herbivores must be consistently strong enough to produce these responses. Invaders initially may lose defences (Blossey and Nötzold, 1995), at least to the specialist enemies they have escaped (Joshi and Vrieling, 2005; Zhang *et al.*, 2018), but in the longer term, selection pressure from enemies in the invaded range may reverse this trend (Müller-Schärer *et al.*, 2004). For instance, Liao *et al.* (2014) found evidence that introduced populations of the invader *Chromolaena odorata* in China have evolved increased resistance to local generalist herbivores. Resistance may also be selected for by specialist herbivores, when present. For instance, Stastny and Sargent (2017) found evidence that a specialist beetle (*Neogalerucella calmariensis*) introduced for biocontrol in Canada selected for increased defences in populations of the wetland invader Purple Loosestrife (*Lythrum salicaria*). Wan *et al.* (2019) found similar results for invasive Chinese populations of *Ambrosia artemisiifolia* attacked by the accidentally introduced specialist beetle *Ophraella communa*.

12.4.3 Tolerance

Whether damage by herbivores translates to effects on the performance of their host plants depends in part on that plant's tolerance to herbivory: the ability of a plant to compensate for damage, maintaining growth and reproduction (Rosenthal and Kotanen, 1994; Strauss and Agrawal, 1999). When tolerance is high, damage to a plant may be only weakly correlated with its performance; a classic example is provided by pasture grasses, which can survive intense grazing. Tolerance can allow grazing-adapted natives to resist the invasion of more sensitive non-natives. For example, cattle in regions of North America formerly used by bison can exclude non-native plants from native grassland, while 'protecting' areas with fencing promotes invasion by non-natives (Milchunas *et al.*, 1992); in regions without a history of grazing, cattle can have the opposite effect (Mack and Thompson, 1982; Seastedt and Pyšek, 2011).

Invaders may spread and persist if they are highly tolerant, even if they experience substantial herbivory; for example, *A. artemisiifolia* is highly tolerant of leaf damage (MacDonald and Kotanen, 2010), likely contributing to its success as an invasive weed. Similarly, high tolerance of *A. philoxeroides* to intense damage by a native beetle may help to explain the success of this invader in China (Lu *et al.*, 2014).

Increased tolerance may be selected for in invading populations: in a recent meta-analysis, Zhang *et al.* (2018) found tolerance to generalist insect herbivores (but not specialists) was higher in introduced plant populations than in native populations. As an example, introduced populations of Chinese tallow tree (*T. sebifera*) are more herbivore-tolerant than native populations (Rogers and Siemann, 2004; Zou *et al.*, 2008; Huang *et al.*, 2010; Wang *et al.*, 2011), suggesting evolution of tolerance may have contributed to the invasiveness of this species. Increased tolerance also has been found in other invaders including *Solidago* (Liao *et al.*, 2016) and *Lythrum* (Stastny and Sargent, 2017), but see Bossdorf *et al.* (2004), Beckmann *et al.* (2016), and Chun *et al.* (2010) for counterexamples.

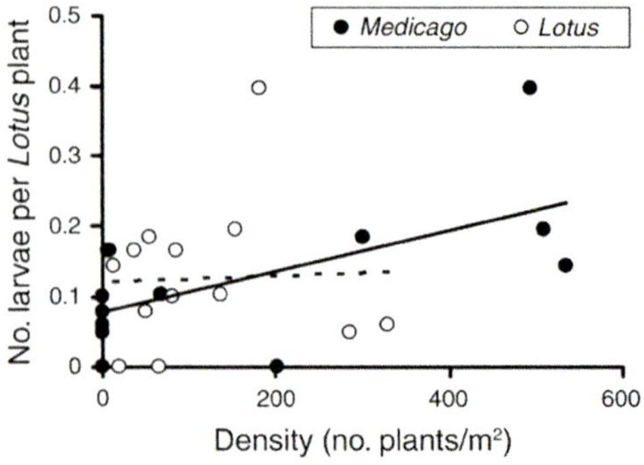

Fig. 12.5. Example of spillover. The number of introduced weevils, *Hypera brunneipennis*, feeding on a native plant, *Lotus wrangelianus*, in California vs the density of *Lotus* (dashed line) and the density of a non-native alternative host, *Medicago polymorpha* (solid line). Weevils on *Lotus* increased with increasing *Medicago* density ($p = 0.03$), but were unaffected by *Lotus* density. From Lau and Strauss (2005) Insect herbivores drive important indirect effects of exotic plants on native communities. *Ecology* 86, 2990–2997. Used with permission.

12.4.4 Indirect effects of herbivores

As well as direct effects on non-native host plants, herbivores may affect non-natives by a variety of indirect pathways (Dunn *et al.*, 2012). Indirect interactions also can have important implications for native plants (Pearson and Callaway, 2003; Chapters 8 and 17, this volume), but here I focus on the consequences of herbivory for the invaders themselves.

Herbivores can benefit non-native plants by suppressing their competitors, an idea which has been referred to as the Enemy of My Enemy Hypothesis (Colautti *et al.*, 2004). For instance, native natural enemies may damage their usual native hosts more than competing non-natives, as predicted by the Enemy Release Hypothesis (Chapter 10, this volume). As an example, native rodents may favour invasion by the grass *Bromus tectorum* because they prefer seeds of natives over those of this aggressive invader (Lucero, 2018; Lucero and Callaway, 2018). Non-natives may also benefit when particularly effective introduced herbivores physically alter an otherwise resistant ecosystem by removing native dominants. For example, introduced insects such as hemlock woolly adelgid (*Adelges tsugae*) and emerald ash borer (*Agrilus planipennis*) may create opportunities for non-native plants in North American forests by killing canopy trees and increasing light levels at the forest floor (Gandhi and Herms, 2010). Similarly, vertebrate grazers (including livestock) can provide opportunities for non-natives by creating space and disturbing or enriching soil; for instance, in eastern North America, overabundant white-tailed deer (*Odocoileus virginianus*) are implicated in the success of invasive garlic mustard (*Alliaria petiolata*), likely because they create open space on the forest floor and/or contribute to increasing earthworm populations (Kalisz *et al.*, 2014; Dávalos *et al.*, 2015).

Non-native plants can benefit from a herbivore that attacks their competitors even if they experience significant damage themselves (Chapter 17, this volume). For instance, non-native and native plants can interact through apparent competition (Holt, 1977; Mitchell *et al.*, 2006): an increase in one plant species has negative consequences for the other because it hosts enemies that attack that plant. In one example, the non-native legume *Medicago polymorpha* increased damage to the native *Lotus wrangelianus* through the shared, introduced weevil *Hypera brunneipennis* (Lau and Strauss, 2005; Fig. 12.5). Similarly, impacts on native thistles of seed-eating weevils introduced to control the invasive thistle *Carduus nutans* were greater where *C. nutans* was more abundant (Rand and Louda, 2004). Even more complex interactions may occur when herbivores act as disease vectors; for instance, aphid-vectored barley and cereal dwarf viruses shared between native and non-native grasses may have contributed to the replacement of perennial California grasslands with invasive annual grasses (Malmstrom *et al.*, 2005, 2006). Such effects can easily be mistaken for resource-based competition in field experiments, though the mechanism is entirely different.

Apparent competition can act through spillover or spillback, depending on the identity of the host acting as the main reservoir for the herbivore. In spillover, a herbivore 'spills over' from its original host to damage co-occurring plants, as in the case for *Carduus* and *Medicago*. In spillback, herbivores multiply on a novel host and then 'spill back' to attack their original hosts. For example, in South Africa, the native fruit fly *Ceratitis cosyra* spills back from non-native commercial mango (*Mangifera indica*) fields to attack wild marula (*Sclerocarya birrea*) fruits (Moxley *et al.*, 2017). Both spillover and spillback from a non-native plant potentially could benefit it by suppressing its competitors.

12.5 Conclusions

An invading plant may initially escape many of its native-range herbivores, but most soon begin to accumulate a diverse new set of natural enemies. As a result, herbivore damage to invaders is frequent and widespread – but not universal or uniform. Instead, damage to an invader varies with factors including the interval since its arrival, its evolutionary history and its location. This invaded-range herbivory has the potential to strongly affect populations of both non-native plants and their native competitors, but it is still unclear how often this is the case. Rigorous studies documenting the effects of herbivores on the demography of non-native plants are badly needed to determine how often damage translates to population-level consequences. Such studies need to consider failed or scarce non-natives as well as abundant invasives if they are to clarify the role herbivory plays in determining an invader's success. Regardless, interactions with herbivores don't end when a plant invades a new region. Instead, as Elton (1958) wrote, invaders exchange 'new food-chains for old'.

Acknowledgements

This work was supported by an NSERC Discovery Grant to PMK.

References

Anstett, D.N., Nunes, K.A., Baskett, C. and Kotanen, P.M. (2016) Sources of controversy surrounding latitudinal patterns in herbivory and defence. *Trends in Ecology & Evolution* 10, 789–802.

Averill, K.M., Mortensen, D.A., Smithwick, E.A.H. and Post, E. (2016) Deer feeding selectivity for invasive plants. *Biological Invasions* 18(5), 1247–1263. DOI: 10.1007/s10530-016-1063-z.

Beckmann, M., Bruelheide, H. and Erfmeier, A. (2016) Reduced tolerance to simulated herbivory on clonal organs in alien genotypes: a multi-species experiment with native and introduced origins. *Biological Invasions* 18(2), 549–563. DOI: 10.1007/s10530-015-1027-8.

Bertheau, C., Brockerhoff, E.G., Roux-Morabito, G., Lieutier, F. and Jactel, H. (2010) Novel insect-tree associations resulting from accidental and intentional biological 'invasions': a meta-analysis of effects on insect fitness. *Ecology Letters* 13(4), 506–515. DOI: 10.1111/j.1461-0248.2010.01445.x.

Bezemer, T.M., Harvey, J.A. and Cronin, J.T. (2014) Response of native insect communities to invasive plants. *Annual Review of Entomology* 59(1), 119–141. DOI: 10.1146/annurev-ento-011613-162104.

Blossey, B. and Nötzold, R. (1995) Evolution of increased competitive ability in invasive nonindigenous plants: a hypothesis. *The Journal of Ecology* 83(5), 887–889. DOI: 10.2307/2261425.

Bossdorf, O., Schröder, S., Prati, D. and Auge, H. (2004) Palatability and tolerance to simulated herbivory in native and introduced populations of *Alliaria petiolata* (Brassicaceae). *American Journal of Botany* 91(6), 856–862. DOI: 10.3732/ajb.91.6.856.

Bradley, B.A., Blumenthal, D.M., Wilcove, D.S. and Ziska, L.H. (2010) Predicting plant invasions in an era of global change. *Trends in Ecology & Evolution* 25(5), 310–318. DOI: 10.1016/j.tree.2009.12.003.

Branco, M., Brockerhoff, E.G., Castagneyrol, B., Orazio, C. and Jactel, H. (2015) Host range expansion of native insects to exotic trees increases with area of introduction and the presence of congeneric native trees. *Journal of Applied Ecology* 52(1), 69–77. DOI: 10.1111/1365-2664.12362.

Brändle, M. and Brandl, R. (2006) Is the composition of phytophagous insects and parasitic fungi among trees predictable? *Oikos* 113(2), 296–304. DOI: 10.1111/j.2006.0030-1299.14418.x.

Brändle, M., Kühn, I., Klotz, S., Belle, C. and Brandl, R. (2008) Species richness of herbivores on exotic host plants increases with time since introduction of the host. *Diversity and Distributions* 14(6), 905–912. DOI: 10.1111/j.1472-4642.2008.00511.x.

Carpenter, D. and Cappuccino, N. (2005) Herbivory, time since introduction and the invasiveness of exotic plants. *Journal of Ecology* 93(2), 315–321. DOI: 10.1111/j.1365-2745.2005.00973.x.

Carroll, S.P. (2007) Natives adapting to invasive species: ecology, genes, and the sustainability of conservation. *Ecological Research* 22(6), 892–901. DOI: 10.1007/s11284-007-0352-5.

Carroll, S.P., Klassen, S.P. and Dingle, H. (1998) Rapidly evolving adaptations to host ecology and nutrition in the soapberry bug. *Evolutionary Ecology* 12(8), 955–968. DOI: 10.1023/A:1006568206413.

Carroll, S.P., Loye, J.E., Dingle, H., Mathieson, M., Famula, T.R. *et al.* (2005) And the beak shall inherit - evolution in response to invasion. *Ecology Letters* 8(9), 944–951. DOI: 10.1111/j.1461-0248.2005.00800.x.

Chun, Y.J., Van Kleunen, M. and Dawson, W. (2010) The role of enemy release, tolerance and resistance in plant invasions: linking damage to performance. *Ecology Letters* 6, no–946. DOI: 10.1111/j.1461-0248.2010.01498.x.

Colautti, R.I., Ricciardi, A., Grigorovich, I.A. and MacIsaac, H.J. (2004) Is invasion success explained by the enemy release hypothesis? *Ecology Letters* 7(8), 721–733. DOI: 10.1111/j.1461-0248.2004.00616.x.

Cox, G.W. (2004) *Alien Species and Evolution*. Island Press, Washington DC.

Cripps, M.G., Gassmann, A., Fowler, S.V., Bourdôt, G.W., McClay, A.S. *et al.* (2011) Classical biological control of *Cirsium arvense*: Lessons from the past. *Biological Control* 57(3), 165–174. DOI: 10.1016/j.biocontrol.2011.03.011.

Dávalos, A., Nuzzo, V. and Blossey, B. (2015) Single and interactive effects of deer and earthworms on non-native plants. *Forest Ecology and Management* 351, 28–35. DOI: 10.1016/j.foreco.2015.04.026.

Dawson, W., Burslem, D.F.R.P. and Hulme, P.E. (2009) Herbivory is related to taxonomic isolation, but not to invasiveness of tropical alien plants. *Diversity and Distributions* 15(1), 141–147. DOI: 10.1111/j.1472-4642.2008.00527.x.

DeWalt, S.J., Denslow, J.S. and Ickes, K. (2004a) Natural-enemy release facilitates habitat expansion of the invasive tropical shrub *Clidemia hirta*. *Ecology* 85(2), 471–483. DOI: 10.1890/02-0728.

DeWalt, S.J., Denslow, J.S. and Hamrick, J.L. (2004b) Biomass allocation, growth, and photosynthesis of genotypes from native and introduced ranges of the tropical shrub *Clidemia hirta*. *Oecologia* 138(4), 521–531. DOI: 10.1007/s00442-003-1462-6.

Dostál, P., Allan, E., Dawson, W., van Kleunen, M., Bartish, I. *et al.* (2013) Enemy damage of exotic plant species is similar to that of natives and increases with productivity. *Journal of Ecology* 101(2), 388–399. DOI: 10.1111/1365-2745.12037.

Dunn, A.M., Torchin, M.E., Hatcher, M.J., Kotanen, P.M., Blumenthal, D.M. *et al.* (2012) Indirect effects of parasites in invasions. *Functional Ecology* 26(6), 1262–1274. DOI: 10.1111/j.1365-2435.2012.02041.x.

Eckberg, J.O., Tenhumberg, B. and Louda, S.M. (2012) Insect herbivory and propagule pressure influence *Cirsium vulgare* invasiveness across the landscape. *Ecology* 93(8), 1787–1794. DOI: 10.1890/11-1583.1.

Eckberg, J.O., Tenhumberg, B. and Louda, S.M. (2014) Native insect herbivory limits population growth rate of a non-native thistle. *Oecologia* 175(1), 129–138. DOI: 10.1007/s00442-013-2876-4.

Elton, C.S. (1958) *The Ecology of Invasions by Animals and Plants*. The University of Chicago Press, Chicago, IL.

Gandhi, K.J.K. and Herms, D.A. (2010) Direct and indirect effects of alien insect herbivores on ecological processes and interactions in forests of eastern North America. *Biological Invasions* 12(2), 389–405. DOI: 10.1007/s10530-009-9627-9.

Graves, S.D. and Shapiro, A.M. (2003) Exotics as host plants of the California butterfly fauna. *Biological Conservation* 110(3), 413–433. DOI: 10.1016/S0006-3207(02)00233-1.

Harvey, K.J., Nipperess, D.A., Britton, D.R. and Hughes, L. (2012) Australian family ties: does a lack of relatives help invasive plants escape natural enemies? *Biological Invasions* 14(11), 2423–2434. DOI: 10.1007/s10530-012-0239-4.

Harvey, K.J., Nipperess, D.A., Britton, D.R. and Hughes, L. (2013) Does time since introduction influence enemy release of an invasive weed? *Oecologia* 173(2), 493–506. DOI: 10.1007/s00442-013-2633-8.

Hawkes, C. V. (2007) Are invaders moving targets? the generality and persistence of advantages in size, reproduction, and enemy release in invasive plant species with time since introduction. *The American Naturalist* 170(6), 832–843. DOI: 10.1086/522842.

Heard, M.J. and Sax, D.F. (2013) Coexistence between native and exotic species is facilitated by asymmetries in competitive ability and susceptibility to herbivores. *Ecology Letters* 16(2), 206–213. DOI: 10.1111/ele.12030.

Heger, T. and Jeschke, J.M. (2014) The enemy release hypothesis as a hierarchy of hypotheses. *Oikos* 123(6), 741–750. DOI: 10.1111/j.1600-0706.2013.01263.x.

Hill, S.B. and Kotanen, P.M. (2009) Evidence that phylogenetically novel non-Indigenous plants experience less herbivory. *Oecologia* 161(3), 581–590. DOI: 10.1007/s00442-009-1403-0.

Hinman, E.D., Fridley, J.D. and Parry, D. (2019) Plant defense against generalist herbivores in the forest understory: a phylogenetic comparison of native and invasive species. *Biological Invasions* 21(4), 1269–1281. DOI: 10.1007/s10530-018-1898-6.

Holt, R.D. (1977) Predation, apparent competition, and the structure of prey communities. *Theoretical Population Biology* 12(2), 197–229. DOI: 10.1016/0040-5809(77)90042-9.

Huang, W., Siemann, E., Wheeler, G.S., Zou, J., Carrillo, J. *et al.* (2010) Resource allocation to defence and growth are driven by different responses to generalist and specialist herbivory in an invasive plant. *Journal of Ecology* 98(5), 1157–1167. DOI: 10.1111/j.1365-2745.2010.01704.x.

Huffaker, C.B. and Kennett, C.E. (1959) A ten-year study of vegetational changes associated with biological control of Klamath weed. *Journal of Range Management* 12(2), 69–82. DOI: 10.2307/3894934.

Joshi, J. and Vrieling, K. (2005) The enemy release and EICA hypothesis revisited: incorporating the fundamental difference between specialist and generalist herbivores. *Ecology Letters* 8(7), 704–714. DOI: 10.1111/j.1461-0248.2005.00769.x.

Kalisz, S., Spigler, R.B. and Horvitz, C.C. (2014) In a long-term experimental demography study, excluding ungulates reversed invader's explosive population growth rate and restored natives. *Proceedings of the National Academy of Sciences* 111(12), 4501–4506. DOI: 10.1073/pnas.1310121111.

Kambo, D. and Kotanen, P.M. (2014) Latitudinal trends in herbivory and performance of an invasive species, common burdock (*Arctium minus*). *Biological Invasions* 16(1), 101–112. DOI: 10.1007/s10530-013-0506-z.

Keane, R. and Crawley, M,J. (2002) Exotic plant invasions and the enemy release hypothesis. *Trends in Ecology & Evolution* 17(4), 164–170. DOI: 10.1016/S0169-5347(02)02499-0.

Kirichenko, N. and Kenis, M. (2016) Using a botanical garden to assess factors influencing the colonization of exotic woody plants by phyllophagous insects. *Oecologia* 182(1), 243–252. DOI: 10.1007/s00442-016-3645-y.

Kirichenko, N., Péré, C., Baranchikov, Y., Schaffner, U. and Kenis, M. (2013) Do alien plants escape from natural enemies of congeneric residents? Yes but not from all. *Biological Invasions* 15(9), 2105–2113. DOI: 10.1007/s10530-013-0436-9.

Lambrinos, J.G. (2006) Spatially variable propagule pressure and herbivory influence invasion of chaparral shrubland by an exotic grass. *Oecologia* 147(2), 327–334. DOI: 10.1007/s00442-005-0259-1.

Lau, J.A. and Strauss, S.Y. (2005) Insect herbivores drive important indirect effects of exotic plants on native communities. *Ecology* 86(11), 2990–2997. DOI: 10.1890/04-1779.

Lee, Y. and Kotanen, P.M. (2017) Differences in herbivore damage to *Arctium minus* in open and forest habitats in its non-native range. *Botany* 95(8), 841–845. DOI: 10.1139/cjb-2017-0055.

Levine, J.M., Adler, P.B. and Yelenik, S.G. (2004) A meta-analysis of biotic resistance to exotic plant invasions. *Ecology Letters* 7(10), 975–989. DOI: 10.1111/j.1461-0248.2004.00657.x.

Lewinsohn, T.M., Novotny, V. and Basset, Y. (2005) Insects on plants: diversity of herbivore assemblages revisited. *Annual Review of Ecology, Evolution, and Systematics* 36(1), 597–620. DOI: 10.1146/annurev.ecolsys.36.091704.175520.

Liao, H., Gurgel, P.C.S., Pal, R.W., Hooper, D. and Callaway, R.M. (2016) *Solidago gigantea* plants from nonnative ranges compensate more in response to damage than plants from the native range. *Ecology* 97(9), 2355–2363. DOI: 10.1002/ecy.1481.

Liao, Z.-Y., Zheng, Y.-L., Lei, Y.-B. and Feng, Y.-L. (2014) Evolutionary increases in defense during a biological invasion. *Oecologia* 174(4), 1205–1214. DOI: 10.1007/s00442-013-2852-z.
Liu, H. and Stiling, P. (2006) Testing the enemy release hypothesis: a review and meta-analysis. *Biological Invasions* 8(7), 1535–1545. DOI: 10.1007/s10530-005-5845-y.
Louda, S.M., Keeler, K.H. and Holt, R.D. (1990) Herbivore influences on plant performance and competitive interactions. In: Grace, J.B. and Tilman, D. (eds) *Perspectives on Plant Competition*. Academic Press, San Diego, CA, pp. 413–444.
Lu, X., Shao, X. and Ding, J. (2014) No impact of a native beetle on exotic plant performance and competitive ability due to plant compensation. *Plant Ecology* 215(3), 275–284. DOI: 10.1007/s11258-014-0296-0.
Lu, X., Siemann, E., Shao, X., Wei, H., Ding, J. *et al.* (2013) Climate warming affects biological invasions by shifting interactions of plants and herbivores. *Global Change Biology* 19(8), 2339–2347. DOI: 10.1111/gcb.12244.
Lucero, J.E. (2018) Do seeds from invasive bromes experience less granivory than seeds from native congeners in the great Basin desert? *Plant Ecology* 219(9), 1053–1061. DOI: 10.1007/s11258-018-0858-7.
Lucero, J.E. and Callaway, R.M. (2018) Native granivores reduce the establishment of native grasses but not invasive *Bromus tectorum*. *Biological Invasions* 20(12), 3491–3497. DOI: 10.1007/s10530-018-1789-x.
MacDonald, A.A.M. and Kotanen, P.M. (2010) Leaf damage has weak effects on growth and fecundity of common ragweed (*Ambrosia artemisiifolia*). *Botany* 88(2), 158–164. DOI: 10.1139/B09-110.
Mack, R.N. and Thompson, J.N. (1982) Evolution in steppe with few large, hooved mammals. *The American Naturalist* 119(6), 757–773. DOI: 10.1086/283953.
MacKay, J. and Kotanen, P.M. (2008) Local escape of an invasive plant, common ragweed (*Ambrosia artemisiifolia* L.), from above-ground and below-ground enemies in its native area. *Journal of Ecology* 96(6), 1152–1161. DOI: 10.1111/j.1365-2745.2008.01426.x.
Malmstrom, C.M., McCullough, A.J., Johnson, H.A., Newton, L.A. and Borer, E.T. (2005) Invasive annual grasses indirectly increase virus incidence in California native perennial bunchgrasses. *Oecologia* 145(1), 153–164. DOI: 10.1007/s00442-005-0099-z.
Malmstrom, C.M., Stoner, C.J., Brandenberg, S. and Newton, L.A. (2006) Virus infection and grazing exert counteracting influences on survivorship of native bunchgrass seedlings competing with invasive exotics. *Journal of Ecology* 94(2), 264–275. DOI: 10.1111/j.1365-2745.2006.01101.x.
Maron, J.L. and Vila, M. (2001) When do herbivores affect plant invasion? Evidence for the natural enemies and biotic resistance hypotheses. *Oikos* 95(3), 361–373. DOI: 10.1034/j.1600-0706.2001.950301.x.
Milchunas, D.G., Lauenroth, W.K. and Chapman, P.L. (1992) Plant competition, abiotic, and long- and short-term effects of large herbivores on demography of opportunistic species in a semiarid grassland. *Oecologia* 92(4), 520–531. DOI: 10.1007/BF00317844.
Mitchell, C.E., Agrawal, A.A., Bever, J.D., Gilbert, G.S., Hufbauer, R.A. *et al.* (2006) Biotic interactions and plant invasions. *Ecology Letters* 9(6), 726–740. DOI: 10.1111/j.1461-0248.2006.00908.x.
Moran, V.C. (1980) Interactions between phytophagous insects and their *Opuntia* hosts. *Ecological Entomology* 5(2), 153–164. DOI: 10.1111/j.1365-2311.1980.tb01136.x.
Morrison, W.E. and Hay, M.E. (2011) Herbivore preference for native vs. exotic plants: generalist herbivores from multiple continents prefer exotic plants that are evolutionarily naïve. *PLoS ONE* 6(3), e17227. DOI: 10.1371/journal.pone.0017227.
Moxley, C., Lammers, W., van Veen, F.J.F., Thébault, E., Esler, K.J. *et al.* (2017) A major subtropical fruit pest accumulates in crop fields and spills over to a wild host. *Agriculture, Ecosystems & Environment* 242, 102–109. DOI: 10.1016/j.agee.2017.03.024.
Müller-Schärer, H., Schaffner, U. and Steinger, T. (2004) Evolution in invasive plants: implications for biological control. *Trends in ecology & evolution* 19, 4167–4422.
Novotny, V., Miller, S.E., Cizek, L., Leps, J., Janda, M. *et al.* (2003) Colonising aliens: caterpillars (Lepidoptera) feeding on *Piper aduncum* and *P. umbellatum* in rainforests of Papua New Guinea. *Ecological Entomology* 28(6), 704–716. DOI: 10.1111/j.1365-2311.2003.00558.x.
Nunes, K.A., Cassin, C.M. and Kotanen, P.M. (2016) Variation in herbivory along a latitudinal gradient for native and exotic Asteraceae. *Plant Ecology* 217(5), 481–493. DOI: 10.1007/s11258-016-0593-x.
Nunes, K.A. and Kotanen, P.M. (2018) Does local isolation allow an invasive thistle to escape enemy pressure? *Oecologia* 188(1), 139–147. DOI: 10.1007/s00442-018-4175-6.
Odegaard, F., Diserud, O.H. and Østbye, K. (2005) The importance of plant relatedness for host utilization among phytophagous insects. *Ecology Letters* 8(6), 612–617. DOI: 10.1111/j.1461-0248.2005.00758.x.

Parker, J.D., Burkepile, D.E. and Hay, M.E. (2006) Opposing effects of native and exotic herbivores on plant invasions. *Science* 311(5766), 1459–1461. DOI: 10.1126/science.1121407.
Parker, J.D., Burkepile, D.E., Lajeunesse, M.J. and Lind, E.M. (2012) Phylogenetic isolation increases plant success despite increasing susceptibility to generalist herbivores. *Diversity and Distributions* 18(1), 1–9. DOI: 10.1111/j.1472-4642.2011.00806.x.
Parker, J.D. and Hay, M.E. (2005) Biotic resistance to plant invasions? native herbivores prefer non-native plants. *Ecology Letters* 8(9), 959–967. DOI: 10.1111/j.1461-0248.2005.00799.x.
Pearse, I.S. and Hipp, A.L. (2009) Phylogenetic and trait similarity to a native species predict herbivory on non-native oaks. *Proceedings of the National Academy of Sciences* 106(43), 18097–18102. DOI: 10.1073/pnas.0904867106.
Pearson, D.E. and Callaway, R.M. (2003) Indirect effects of host-specific biological control agents. *Trends in Ecology & Evolution* 18(9), 456–461. DOI: 10.1016/S0169-5347(03)00188-5.
Rand, T.A. and Louda, S.M. (2004) Exotic weed invasion increases the susceptibility of native plants to attack by a biocontrol herbivore. *Ecology* 85(6), 1548–1554. DOI: 10.1890/03-3067.
Rogers, W.E. and Siemann, E. (2004) Invasive ecotypes tolerate herbivory more effectively than native ecotypes of the Chinese tallow tree *Sapium sebiferum*. *Journal of Applied Ecology* 41(3), 561–570. DOI: 10.1111/j.0021-8901.2004.00914.x.
Rosenthal, J.P. and Kotanen, P.M. (1994) Terrestrial plant tolerance to herbivory. *Trends in Ecology & Evolution* 9(4), 145–148. DOI: 10.1016/0169-5347(94)90180-5.
Schultheis, E.H., Berardi, A.E. and Lau, J.A. (2015) No release for the wicked: enemy release is dynamic and not associated with invasiveness. *Ecology* 96(9), 2446–2457. DOI: 10.1890/14-2158.1.
Schwarz, D., Matta, B.M., Shakir-Botteri, N.L. and McPheron, B.A. (2005) Host shift to an invasive plant triggers rapid animal hybrid speciation. *Nature* 436(7050), 546–549. DOI: 10.1038/nature03800.
Seastedt, T.R. and Pyšek, P. (2011) Mechanisms of plant invasions of North American and European grasslands. *Annual Review of Ecology, Evolution, and Systematics* 42(1), 133–153. DOI: 10.1146/annurev-ecolsys-102710-145057.
Siemann, E. and Rogers, W.E. (2003) Increased competitive ability of an invasive tree may be limited by an invasive beetle. *Ecological Applications* 13(6), 1503–1507. DOI: 10.1890/03-5022.
Singer, M.C., Thomas, C.D. and Parmesan, C. (1993) Rapid human-induced evolution of insect–host associations. *Nature* 366(6456), 681–683. DOI: 10.1038/366681a0.
Stastny, M. and Sargent, R.D. (2017) Evidence for rapid evolutionary change in an invasive plant in response to biological control. *Journal of Evolutionary Biology* 30(5), 1042–1052. DOI: 10.1111/jeb.13078.
Strauss, S.Y. and Agrawal, A.A. (1999) The ecology and evolution of plant tolerance to herbivory. *Trends in Ecology & Evolution* 14(5), 179–185. DOI: 10.1016/S0169-5347(98)01576-6.
Straw, N.A. (1991) Resource limitation of tephritid flies on lesser burdock, *Arctium minus* (Hill) Bernh. (Compositae). *Oecologia* 86(4), 492–502. DOI: 10.1007/BF00318315.
Stricker, K.B. and Stiling, P. (2012) Herbivory by an introduced Asian weevil negatively affects population growth of an invasive Brazilian shrub in Florida. *Ecology* 93(8), 1902–1911. DOI: 10.1890/11-1328.1.
Strong, D.R., Lawton, J.H. and Southwood, R. (1984) *Insects on Plants*. Harvard University Press, Cambridge, MA.
Suwa, T. and Louda, S.M. (2012) Combined effects of plant competition and insect herbivory hinder invasiveness of an introduced thistle. *Oecologia* 169(2), 467–476. DOI: 10.1007/s00442-011-2207-6.
Tabassum, S. and Leishman, M.R. (2018) Does enemy damage vary across the range of exotic plant species? Evidence from two coastal dune plant species in eastern Australia. *Oecologia* 186(2), 303–309. DOI: 10.1007/s00442-017-4008-z.
Thomas, M. and Reid, A. (2007) Are exotic natural enemies an effective way of controlling invasive plants? *Trends in Ecology & Evolution* 22(9), 447–453. DOI: 10.1016/j.tree.2007.03.003.
Vaupel, A. and Matthies, D. (2012) Abundance, reproduction, and seed predation of an alpine plant decrease from the center toward the range limit. *Ecology* 93(10), 2253–2262. DOI: 10.1890/11-2026.1.
Walther, G.-R., Roques, A., Hulme, P.E., Sykes, M.T., Pyšek, P. *et al.* (2009) Alien species in a warmer world: risks and opportunities. *Trends in Ecology & Evolution* 24(12), 686–693. DOI: 10.1016/j.tree.2009.06.008.
Wan, J., Huang, B., Yu, H. and Peng, S. (2019) Reassociation of an invasive plant with its specialist herbivore provides a test of the shifting defence hypothesis. *Journal of Ecology* 107(1), 361–371. DOI: 10.1111/1365-2745.13019.

Wan, J.S.H. and Bonser, S.P. (2016) Enemy release at range edges: do invasive species escape their herbivores as they expand into new areas? *Journal of Plant Ecology* 9(5), 636–647. DOI: 10.1093/jpe/rtw003.

Wang, Y., Huang, W., Siemann, E., Zou, J., Wheeler, G.S. *et al.* (2011) Lower resistance and higher tolerance of invasive host plants: biocontrol agents reach high densities but exert weak control. *Ecological Applications* 21(3), 729–738. DOI: 10.1890/09-2406.1.

Weiblen, G.D., Webb, C.O., Novotny, V., Basset, Y. and Miller, S.E. (2006) Phylogenetic dispersion of host use in a tropical insect herbivore community. *Ecology* 87(sp7), S62–S75. DOI: 10.1890/0012-9658(2006)87[62:PDOHUI]2.0.CO;2.

Woods, E.C., Hastings, A.P., Turley, N.E., Heard, S.B. and Agrawal, A.A. (2012) Adaptive geographical clines in the growth and defense of a native plant. *Ecological Monographs* 82(2), 149–168. DOI: 10.1890/11-1446.1.

Zangerl, A.R. and Berenbaum, M.R. (2005) Increase in toxicity of an invasive weed after reassociation with its coevolved herbivore. *Proceedings of the National Academy of Sciences* 102(43), 15529–15532. DOI: 10.1073/pnas.0507805102.

Zhang, Z., Pan, X., Blumenthal, D., van Kleunen, M., Liu, M. *et al.* (2018) Contrasting effects of specialist and generalist herbivores on resistance evolution in invasive plants. *Ecology* 99(4), 866875–875. DOI: 10.1002/ecy.2155.

Zou, J., Rogers, W.E. and Siemann, E. (2008) Increased competitive ability and herbivory tolerance in the invasive plant *Sapium sebiferum*. *Biological Invasions* 10(3), 291–302. DOI: 10.1007/s10530-007-9130-0.

13 Impacts of Non-native Plants on Plant–Pollinator Interactions

Marcelo A. Aizen* and Carolina L. Morales

Instituto de Investigaciones en Biodiversidad y Medio Ambiente (INIBIOMA), Universidad Nacional del Comahue-CONICET, 8400 San Carlos de Bariloche, Rio Negro, Argentina

Abstract

There has been growing interest in the consequences of invasive non-native plants for the plant–pollinator mutualism, most likely because of its relevance for the maintenance of terrestrial biodiversity and food production. However, the development of this research field has been thematically uneven and the overall evidence inconclusive. Many studies have focused on how non-native plants interact with native plants via pollinator sharing, which have allowed meta-analytical syntheses, whereas several others have looked at how frequently non-native plants integrate into native plant–pollinator webs and how they affect network structure. However, relatively few studies have addressed the consequences of invasive plants for pollinators. Overall, the research approach in this area has been predominantly phenomenological rather than mechanistic, which has hindered our understanding of apparently contradictory evidence. One key characteristic of invasive non-native plants that seems to mediate negative effects on the pollination mutualism is the high relative abundance that they reach at late stages of invasion. This high dominance is apparently the main trigger of all the disruptive direct and indirect effects that are discussed in this chapter. Finally, we identify several intriguing questions on the ecological and evolutionary consequences of invasive plants for the plant–pollinator mutualism waiting to be answered.

13.1 Introduction

Although the independence of seed production from pollinators can facilitate the colonization and spread of non-native plants (Grossenbacher *et al.*, 2015, 2017; Razanajatovo *et al.*, 2016), many non-native plants rely on animal pollinators to produce seeds in their native and introduced ranges (e.g. Parker, 1997; Larson *et al.*, 2002; Geerts and Pauw, 2009; Goodell *et al.*, 2010; Issaly *et al.*, 2020). Several highly pollinator-dependent plants can become aggressive invaders, reaching high local abundances and spreading across extensive geographical areas outside their native ranges (Fig. 13.1). This pollinator dependence implies that, upon arrival in a new region, non-native plants rely on resident pollinators to establish and spread. In fact, non-native plants are commonly visited and pollinated by both native and non-native pollinators (Chapter 4, this volume). Usurpation of the native pollinator fauna and the use of non-native pollinators are characteristics of non-native plants that can increase their invasion potential with consequences for both native pollinators and plants, and their interactions (Morales and Aizen, 2002; Aizen *et al.*, 2008). In particular, because of their high abundance, often associated with high rewards and large floral displays (Fig. 13.1), non-native plants can cause the alteration and disruption of interactions between native plants and pollinators, modify the diet and assemblage structure of pollinators and affect the reproduction of native plants via pollinator sharing (Vanbergen *et al.*, 2018).

*Corresponding author: maizen@comahue-conicet.gob.ar

DOI: 10.1079/9781789242171.0013

Fig. 13.1. Invasive Scotch broom (*Cytisus scoparius*) can form dense stands (left) that during flowering are clearly distinguished in 1 m pixel aerial photographs (right) (Cavallero *et al.*, 2018). This European species has invaded not only NW Patagonia, Argentina, where these pictures were taken, but also large parts of the USA, Canada, Australia and New Zealand (Potter *et al.*, 2009).

The pollination of the non-native plants themselves and/or of the native plants that non-native plants interact with have been important topics in the plant invasion literature. A Google Scholar search encompassing the last five decades (1970–2019) shows that about one fifth of the publications in the growing research topic of plant invasions mention or study pollination (Fig. 13.2), a proportion that has been relatively constant over time. However, the topic of plant invasions does not feature so prominently in the wider literature on pollination, with an average of just 1.5% of all articles involving plant–pollinator interactions mentioning the term 'invasive species'. However, this proportion has grown to about 4% in the last 15 years (Fig. 13.2). This increase in both the absolute and relative number of published articles in this field reflects a growing interest in the study of the consequences of invasive non-native plants for the plant–pollinator mutualism. This trend is probably related to the increasing awareness of the relevance of this interaction for the maintenance of terrestrial biodiversity and for food production, as well as an increasing perception that plant invasions could be an important driver of change. In absolute terms, nowadays about 800 publications per year focus on the link between invasive plants and pollination. Despite the relevance of pollination in the plant invasion literature, with several recent studies focused on how invasive plants affect the pollination of native plants and the structure of plant–pollinator interaction networks, we are still lacking a clear understanding of why the impacts of invasive plants on plant–pollinator interactions can greatly differ in sign and magnitude. In fact, a previous consensus on a pervasive negative effect of non-native plants on flower visitation and reproduction of native plants (Morales and Traveset, 2009), two relatively well-studied impacts, has been recently confronted (Charlebois and Sargent, 2017). Therefore, a unifying conceptual framework is still needed to accommodate disparate results and predict the impacts of invasive species on native mutualists.

Impacts of invasive non-native plants on native pollinators are expected to be a consequence of the spatial and temporal availability of floral resources (i.e. nectar and pollen), whereas impacts on native plants are expected to be mediated by pollinator sharing. In this chapter, we go over these different impacts and explore the mechanisms involved. In addition to providing evidence of the impact of non-native plants on native pollinators and plants, we discuss the role of invasive non-native plants in the context of plant–pollinator interaction webs. This network context is important because the effect of non-native plants

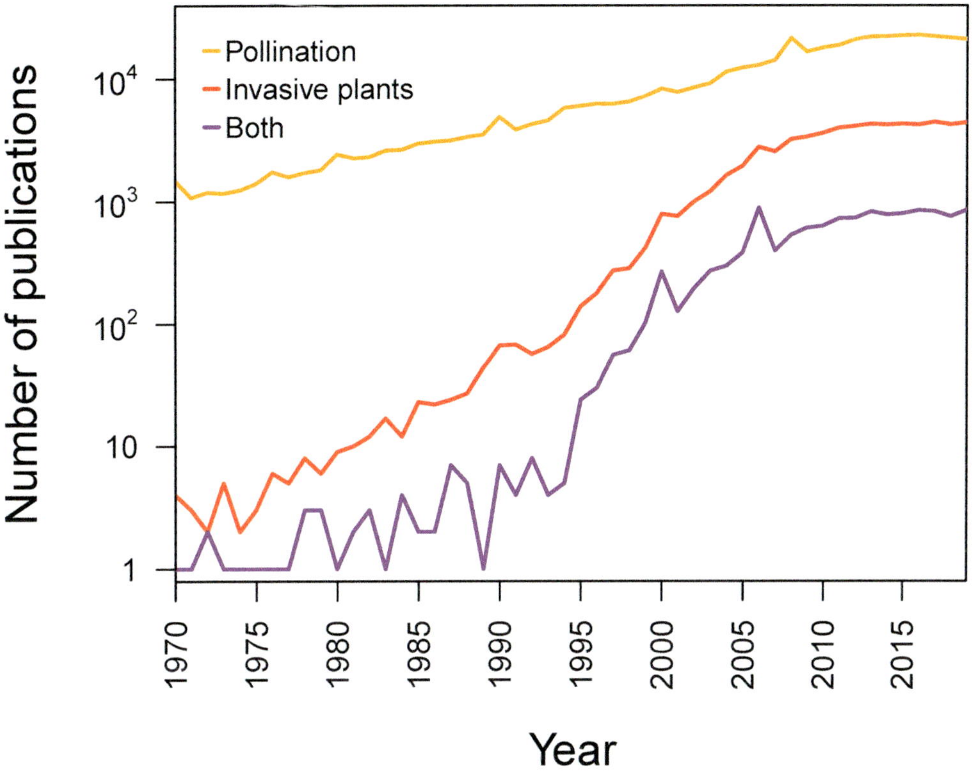

Fig. 13.2. Temporal trend (1970–2019) in the number of publications including 'pollination', 'invasive plants' and both terms (i.e. 'pollination' AND 'invasive plants') according to Google Scholar (search on 18 August 2019). The total number of publications for the year 2019 was estimated assuming a constant rate of publications during the remaining 136 days until the end of the year after 18 August.

on particular plant–pollinator interactions can affect the stability and resilience of whole mutualistic assemblages through multiple indirect pathways (Guimarães *et al.*, 2017; Frost *et al.*, 2019; Simmons *et al.*, 2019a, b; see also Chapter 18, this volume). As a unifying theme, we propose that one characteristic of invasive species, their high abundance, is key to understanding their negative effects on native pollinators and plants via pollinator sharing.

13.2 Impacts on Native Pollinators

Non-native plants can have manifold effects on native pollinator abundance and diversity. These effects are mainly mediated by the high abundance reached by non-native plants when they become invasive. First, the local extinction or the persistence at extremely low density of many native plant species outcompeted by the invaders could cause the loss of many pollinator species. Particularly susceptible are those pollinators specialized in these extinct or rare plants (Aizen *et al.*, 2012). Second, monospecific flower stands have been associated with pollinator nutritional stress. In particular, bees fed on monotypic pollen diets may suffer a higher incidence of sublethal effects, impairing their immunocompetence and thus increasing susceptibility to pathogens (Branchiccela *et al.*, 2019). Third, even though large, dense flower stands of invasive plants (Fig. 13.1) represent a food cornucopia for pollinators, either the nectar or pollen produced by these non-native species may only be available for a short time period. Shortening of the flowering season should lead to more marked cycles of feast and starvation that may further limit pollinator diversity (Dorado and Vázquez, 2016). Also, depending on the spatial extent of the invasions, pollinator responses might be behavioural only, if invasions are local, or can be translated into more severe demographical changes in larger invaded areas, when changing foraging behaviour or moving to another areas within flying distances is not an option. Therefore, because of their dominance, invasive non-native plants can be an important force affecting pollinator populations and communities.

In a context of a plant invasion, specialized pollinators are those most prompt to decline when their food plants are outcompeted and eventually vanish, at least locally, from the plant community (Powell *et al.*, 2011). This could be particularly important for many solitary, oligolectic bees that limit pollen harvesting to the pollen produced by one or a few related plant species. These specialized bees often fail to develop on pollen of unrelated plant species (Praz *et al.*, 2008) and may avoid harvesting pollen from non-native species (Palladini and Maron, 2014). In addition, the effect of the loss of their pollen plants, even though occurring at local spatial scales, could become further exacerbated by the small foraging range of these bees (Gathmann and Tscharntke, 2002). Therefore, the decline and eventual extinction of native plant species outcompeted by non-native plant species can have an indirect negative impact on the pollinators that feed on them. Surprisingly, empirical evidence on this potentially important factor structuring pollinator assemblages is scarce and still non confirmatory. For instance, in Poland meadows invaded by goldenrod (*Solidago* spp.) sustain much less diverse pollinator assemblages than non-invaded meadows. However, reductions in pollinator diversity were not associated with either food or nesting specialization (Moroń *et al.*, 2009; see also Fründ *et al.*, 2010).

Floral resource dominance by non-native plant species can not only affect specialist but also generalist pollinators due to the low nutrition quality associated with monotypic diets and its consequence for animal health. For instance, honeybee colonies feeding on a plantation of *Eucalyptus grandis*, a species that produces pollen with low protein and lipid content, had a higher incidence of *Nosema* spp., a microsporidian pathogen, and lower brood and adult bee population compared to colonies supplemented with polyfloral pollen throughout the flowering season (Branchiccela *et al.*, 2019). Even though the nectar or pollen of non-native plants can be of high quality, many generalist pollinators need a diversity of macro- and micronutrients for growth, survival and reproduction (Paoli *et al.*, 2014a, b; Vaudo *et al.*, 2016, 2018); nutritional requirements that might not be met in a community dominated by a single or a few plant species (Harmon-Threatt and Kremen, 2015).

Finally, the temporal flowering dimension is also a key aspect that might explain reductions in the diversity of pollinator communities. In general, higher flower diversity should stabilize the availability of floral resources along the flowering season (Dorado and Vázquez, 2016). More specifically, low diverse plant communities dominated by invasive plants are expected to also have more shortened community-wide flowering phenologies. In turn, a restricted flowering should filter out those pollinators whose activity period overlaps the flowering of the outcompeted native plants, and also reduce temporal food availability for generalist pollinators active during time periods exceeding the flowering phenology of one or a few plant species (Dorado and Vázquez, 2016). This phenological curtailment has been proposed as a cause of decline in pollinator populations and in overall pollinator diversity in agricultural landscapes (Mandelik *et al.*, 2016). However, to our knowledge no study has assessed the importance of flowering phenological restriction to explain reductions in pollinator diversity in the context of plant invasions.

Pollinators surviving in a landscape dominated by one or a few invasive non-native flowering plants are also subjected to novel selection forces. A reduced number of pollen and nectar sources can select for convergent pollinator adaptations to increase efficiency in the use of those resources (Nuismer *et al.*, 2018). These novel selection pressures can be expected to be highly effective whenever the invasive plant species occupy large areas and persist in the landscape over time (cf. Endler, 1986). These conditions could induce consistent selection pressures on morphological, physiological and behavioural traits to maximize exploitation of those resources, as well as select for changes in pollinator activity time to increase overlap with the flowering non-native plants. However, whether or not these selection pressures are consistent enough to result in evolutionary change in the pollinators is an open question. Unlike selection on floral traits, very few microevolutionary studies have looked at selection by flowers on pollinator traits (e.g. Anderson and Johnson, 2009; Pauw *et al.*, 2009), and we are aware of only one that relates to this issue in a context of a plant invasion. This is the case of a study on *Nicotiana glauca*, a South American hummingbird-pollinated invasive plant, which invaded South Africa where it is visited and pollinated by sunbirds (Issaly *et al.*, 2020). Interestingly, the lack of a floral perch in the tubular flowers or inflorescences of this plant species induces hovering in sunbirds, an unusual behaviour in these birds (Geerts and Pauw, 2009). However, whether this novel behaviour results in selection for more efficient hovering that translates into higher nectar intake is unknown.

13.3 Impacts on Native Plants

Non-native plants can affect seed production of native plants when sharing pollinators. By flowering together, showy stands of flowering non-native species may increase pollinator attraction, facilitating in some instances the pollination of native plants by enhancing visitation frequency. However, most commonly, native plants can experience decreasing visitation when flowering in proximity to non-native plants (Morales and Traveset, 2009). Following Ratchke's model (Rathcke, 1983; see also Feinsinger, 1987), it is expected that the effect of non-native plant species on the visitation of native plants is density dependent. Whereas a facilitative effect is likely to increase pollination of native plants via increasing pollinator attraction at low density of the non-native species (Bjerknes *et al.*, 2007) or due to the presence of the non-native species at the landscape scale (Albrecht *et al.*, 2016), this effect can shift to competition at high density when large stands of non-native species saturate the local pollinator pool and the maximum number of visits get diluted among an increasing number of flowers (Cavallero *et al.*, 2018). It is likely, however, that the shift from the facilitative to the competitive phase of these pollinator-mediated interactions, as well as the intensity of density-dependent facilitation and competition, will depend on intrinsic features

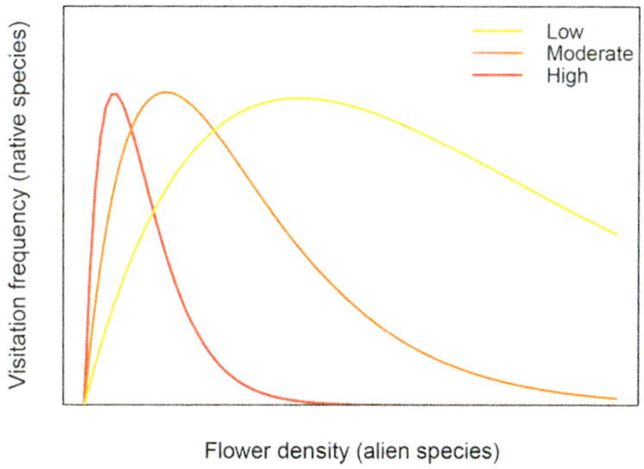

Fig. 13.3. Graphical representation of the visit frequency to flowers of a native plant species as a function of the flower density of a non-native plant species (modified from Rathcke, 1983). The non-native plant species will facilitate visitation to the flowers of the native species to the left of the maximum of each curve while decreasing visitation to the flowers of the native species to the right of the maximum. The range of flower densities in which facilitation occurs will narrow as the attractiveness and/or rewardness of the non-native species increase. The attractiveness and/or rewardness of non-native species has been classified in three categories (i.e. low, moderate and high), represented in different colours.

of the interacting non-native and native species. More specifically, the shift between these phases could occur at lower densities and the intensity of these interactions be stronger when the attractiveness and/or rewardness of the non-native species increases in relation to the native species (Fig. 13.3). For instance, the presence of Himalayan balsam, *Impatiens glandulifera*, a high-nectar producing species, greatly depresses visitation to the flowers of *Stachys palustris* in invaded areas in Europe, even when the non-native is at low density (Chittka and Schürkens, 2001). In any case, the model depicted in Fig. 13.3 awaits formal testing.

Independent of changes in visitation frequency, two other negative effects are associated with mixed flower visitation and the ensuing improper pollen transfer (Morales and Traveset, 2008). The first is conspecific pollen loss, which occurs when pollen of one species is lost on the stigmas or other flower parts of a second species (e.g. Murcia and Feinsinger, 1996). Beyond any negative effect on plant female fitness (i.e. seed production) due to increasing pollen limitation, conspecific pollen loss can affect male fitness through limiting siring opportunities (Aizen and Rovere, 2010). The second effect is heterospecific pollen deposition; foreign pollen can occupy space on the stigmas of a target species, thus diminishing the stigmatic area available for the deposition of conspecific pollen (e.g. Galen and Gregory, 1989). Heterospecific pollen on stigmas may also have allelopathic effects inhibiting the germination of conspecific pollen (e.g. Murphy and Aarssen, 1995). These, and other less common effects of heterospecific pollen transfer, e.g. the closure of sensitive stigmas and ovule usurpation (reviewed in Morales and Traveset, 2008), could be unavoidable negative side impacts of non-native plants on native plants sharing pollinators. All these effects are expected to be density dependent, being aggravated in a context of plant invasion because of the high abundance and dominance that non-native species can reach. With few exceptions (e.g. Martyniuk *et al.*, 2015), the role of improper pollen transfer in mediating pollination-mediated effects of non-native plants on native plants has been rarely studied but inferred (Morales and Traveset, 2009). For instance, a decrease in pollinator visitation frequency to the flowers of native *Dipterocarpus obtusifolia*, as well as a decrease and an increase in the deposition of conspecific and heterospecific pollen, respectively, were found along a gradient of understorey flower density of invasive *Chromolaena odorata* associated with logging intensity in Southeast Asia. However, effects on seed production were not evident, probably because the reproduction of *D. obtusifolia* is resource limited (Ghazoul, 2002). Yet, effects of invasive plants on the male fitness of native plants could still be widespread (Aizen and Rovere, 2010).

Results of a seminal meta-analysis showed that non-native plant species commonly have negative effects on flower visitation frequency and reproductive success of native plants (Morales and Traveset, 2009). Pollinator-mediated negative impacts of non-native on native plant species seem to become stronger with the increasing relative abundance of non-native plants, showing that non-native plant dominance increases the magnitude of these negative effects. Intrinsic plant factors, like similar flower symmetry and colour, and overall phylogenetic relatedness, could also exacerbate these negative effects (Morales and Traveset, 2009). However, a more recent expanded meta-analysis has challenged these findings (Charlebois and Sargent, 2017). Indeed, this latter study reveals that the overall effect of non-native plants on native flower visitation and reproductive success can range from facilitation to competition (Fig. 13.3), with no apparent effect of similarity in flower symmetry or colour, or phylogenetic relatedness between non-native and native species on this outcome. However, these authors found that greater spatial interspersion of native and non-native plants seems to negatively affect the reproductive success of the natives, without having a sizable effect on flower visitation (Charlebois and Sargent, 2017). This effect of increasing spatial intermixing, which occurs along an invasion process, could point to a role for interspecific pollen transfer as an important factor determining a negative effect of non-native plants on the pollination of native plants (e.g. Martyniuk *et al.*, 2015).

Plant traits that confer greater fitness in native plants in the presence of an invasive plant could be selected for, opening the possibility that plant invasions could lead to evolutionary change. Research focused on invasive plant species has found adaptative changes in their pollination and mating systems in invaded areas occurring in a few generations (see Chapter 4, this volume), and thus rapid evolutionary changes in reproductive traits could be expected for native plant species in the invaded communities as well. Also, empirical evidence supports the view that native plants that have coexisted with invasive plants for several generations have evolved vegetative traits that confer greater growth and/or reproductive success than conspecific individuals that experience interactions with these non-native plants for first time (Oduor, 2013). In the case of pollinator-mediated interactions, there could be selection for convergence or divergence in flower morphology and/or flowering phenology depending on whether these interactions are of a facilitative or competitive nature, respectively (Johnson and Schiestl, 2016; Moreira-Hernández and Muchhala, 2019). Improper pollen transfer and, particularly, conspecific pollen loss could be strong selection forces for divergent flower traits, including flowering phenology, as this mechanism could affect plant fitness through both male and female functions (Moreira-Hernández and Muchhala, 2019). Indeed, a negative effect of non-native species on native reproductive success can increase with similarity in flower morphology, colour and overall phylogenetic relatedness (Morales and Traveset, 2009). At least one study showed increased phenotypic selection for flowers with short corollas in North American *Impatiens capensis* growing in mixed stands with the invasive Asian *I. glandulifera*, a congeneric species producing flowers with long corollas. Although the mechanisms behind this pattern could not be disentangled, competition via limiting flower visitation and stigmatic interference by the non-native pollen could be behind selection for floral divergence (Beans and Roach, 2015). Yet, studies reporting differences in floral traits in native plant species between invaded and non-invaded plant populations are missing.

13.4 Impacts on Plant–Pollinator Networks

By 'usurping' pollinators, non-native plants can cause the alteration or disruption of pre-existing interactions between native mutualists. When non-native plants are accompanied by non-native pollinators, such as honeybees and European bumblebees introduced in many regions of the world, they can form complexes that can boost each other's invasive potential (Morales and Aizen, 2002) and lead to an 'invasional meltdown' (Simberloff and Von Holle, 1999; see also Chapter 8, this volume). Boosting of plant invasions might also occur in areas where honeybees or bumblebees are considered native, and their abundances are managed and thus increased artificially (Geslin *et al.*, 2017). In the case of non-native pollinators, the formation of these invasion complexes can be facilitated by a second driver, most frequently anthropogenic habitat disturbance (Morales and Aizen, 2002; Tylianakis *et al.*, 2008). The new established interactions between non-native plants and native pollinators can not only change the ecological and evolutionary dynamics of the native pollinators they directly interact with, but also of the many other native plants and pollinators through a cascade of indirect interactions established through pollinator and plant sharing, respectively (Guimarães *et al.*, 2017; Frost *et al.*, 2019; Simmons *et al.*, 2019a, b).

Non-native plants can easily integrate into native pollination webs (Memmott and Waser, 2002). Several studies have shown that when non-native plants are introduced outside their native geographic region, they will be readily visited by native pollinators. This could be expected because of the generalized nature of most plant–pollinator mutualisms (Waser *et al.*, 1996). For instance, generalized pollinators seem to have facilitated the integration of Himalayan balsam, *I. glandulifera*, into European plant–pollinator networks (Lopezaraiza-Mikel *et al.*, 2007). However, non-native plant species can integrate into relatively specialized plant–pollinator webs, even if they do not share a common evolutionary history with their new interaction partners or with a phylogenetically related group. Although this pre-adaptation has been suggested as a prerequisite for integration into native pollination webs (Richardson *et al.*, 2000; Ollerton *et al.*, 2012), there are frequent examples in which non-native plants can acquire a completely different set of pollinators, including pollinators of different orders in their new geographical range. For instance, non-native plants can integrate into plant–hummingbird pollination networks in the New World, despite not showing previous association with bird pollinators in their native ranges (Maruyama *et al.*, 2016). However, the lack of highly specialized pollinators might limit the invasive potential of other non-native plants. This could be the case of hummingbird-pollinated *N. glauca* that integrates into plant–pollinator networks in invaded regions hosting bird pollinators, like sunbirds in Africa, while relying exclusively on autogamous pollination in regions lacking native nectarivorous birds (Ollerton *et al.*, 2012). Also, extremely few examples of invasive species exist in highly diverse groups like figs and orchids (Richardson *et al.*, 2000). In fact, the few cases of known invasive species in these groups are explained by co-introduction of their specialized pollinator in the case of figs (McKey, 1989) or autogamy in the case of orchids (Ackerman *et al.*, 2014; see also Chapter 4, this volume).

Once integrated, both non-native plants and pollinators can alter the structure of plant–pollinator networks, and these changes intensify along invasion gradients (e.g. non-native flower abundance). Interestingly, the results of studies looking at the effects of non-native plants on the structure of plant–pollinator networks are mixed (Aizen *et al.*, 2008; Bartomeus *et al.*, 2008; Padrón *et al.*, 2009; Vilà *et al.*, 2009; Kaiser-Bunbury *et al.*, 2011; Traveset *et al.*, 2015; Parra-Tabla *et al.*, 2019). However, these discrepancies can stem from focusing on communities encompassing different ranges of invasion intensity, and also on parameters describing the architecture of the whole network rather than parameters describing the architecture of the sub-web formed by the native species and their interactions. For instance, although the overall connectivity and total interaction frequency of a network may not change along invasion, the number of distinctive interactions and the frequency of individual interaction links between native species can decrease. This implies an increase in the number and interaction frequency of novel links established between native and non-native species (Aizen *et al.*, 2008; Kaiser-Bunbury *et al.*, 2011). Also, changes in the identity of interaction partners may not necessarily change other parameters of the whole network associated with community stability, such as nestedness (Bastolla *et al.*, 2009; Thébault and Fontaine, 2010), but they may have indirect consequences for the fitness of the native plants, through for instance, pollinator-mediated competition (see above).

Without neglecting the effect of other factors, such as flower morphology, species abundance seems to be key in determining the degree of generalization in species interactions. One common observation in network studies is that the most abundant species tend to be those with the highest number of interaction links (Vázquez *et al.*, 2009). Indeed, recent analyses suggest that generalization in species interactions in mutualistic webs is a consequence of a species being abundant rather than its cause (Fort *et al.*, 2016; Simmons *et al.*, 2019b). Therefore, the increasing abundance of either non-native plant or pollinator species along an invasion process seems to be critical to understand changes in their increasing role as central nodes in network structure. When non-native mutualists, in this case plants, colonize a new site they are found typically at low abundance, being visited and pollinated by only the most generalized pollinators (Aizen *et al.*, 2008; Tylianakis *et al.*, 2008). However, while increasing in abundance along the invasion process, non-native plants do not only interact with generalized but also with more specialized pollinators, a consequence of the asymmetrical nature of plant–pollinator webs and of the fact that abundance increases generalization (see above). The presence of non-native pollinators may also contribute to boost each other's abundances (Fig. 13.4), because of the preferential interactions that can be established between them (Morales and Aizen, 2002). In the advanced stages of invasion, non-native plants and pollinators can become 'super-generalists' as they monopolize a disproportionate number of interactions and total interaction frequency (Olesen *et al.*, 2002). As non-natives increase in abundance, native plants and pollinators interact differentially and in an increasingly non-reciprocal fashion (i.e. asymmetrically) with non-native species, which can determine their ecological and evolutionary fate (Aizen *et al.*, 2008; Tylianakis *et al.*, 2008).

From an ecological perspective, networks highly invaded with non-native plants can become dominated by the transfer of non-native pollen, which may be associated with a reduction in the transfer of conspecific pollen of native species (Lopezaraiza-Mikel *et al.*, 2007). Also, high interaction asymmetry has been linked to more robust community configurations (Bascompte *et al.*, 2006), which can resist efforts to restore native interactions without a severe control of the abundances of non-native species (Tylianakis, 2008). From an evolutionary perspective, because of disparate abundances, selection regimes can become less reciprocal and the presence of non-natives dictate the future evolutionary trajectories of both native plants and pollinators (Nuismer *et al.*, 2018). In the case of native plants, there could be, for instance, selection for displaced flowering phenologies to decrease overlap with the flowering of the dominant non-natives and decrease pollinator sharing (see above). In terms of network architecture, over evolutionary time this could translate into distinctive time-structured modules coexisting in space but with low phenological matching among them, some of which could recreate at least part of the original native network (Valdovinos *et al.*, 2009). This displacement, however, could be limited by pollinator availability, climatic constraints,

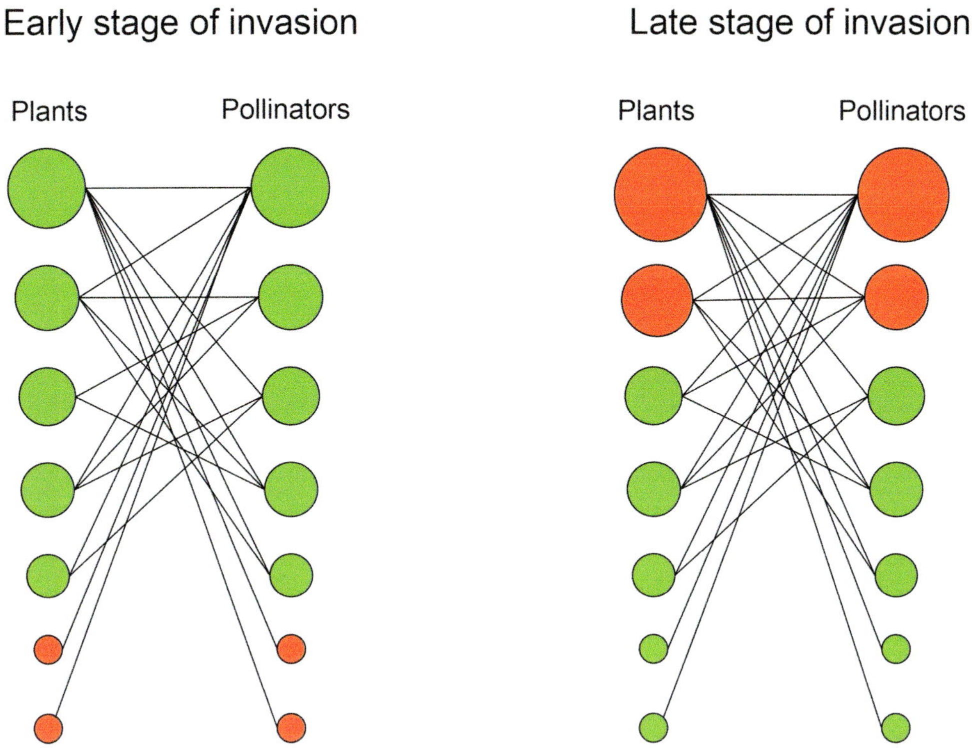

Fig. 13.4. Plant—pollinator networks that include native (green) and non-native (red) plant and pollinator species. The size of the circle is proportional to a species abundance, which in turn will influence total interaction frequency and the number of interaction partners. Early in the invasion, non-native species are at low abundances and interact mostly with native generalists. Late in the invasion, non-native species are at very high abundances, becoming 'super generalists' and interacting with both native and non-native species (modified from Tylianakis, 2008).

pollinator-mediated competition, other biotic pressures acting on flowering time, or even species persistence in the community (Elzinga *et al.*, 2007). In any event, questions about the consequences of the disruption of interaction links between native species due to invasion of non-native species for species persistence and evolution remain unexplored. Network studies along invasion gradients that include phylogenetic perspective can be helpful to predict some of these consequences (e.g. Aizen *et al.*, 2016).

13.5 Research Opportunities

Biological invasions are one of the most important causes of biodiversity loss (Sala *et al.*, 2000) through their manifold effects on direct and indirect interspecific interactions (Valiente-Banuet *et al.*, 2015). The very existence of this chapter points to the consideration that, in the case of non-native plant invasions, part of this biodiversity loss could be occurring through the diverse effects of plant invasions on the plant–pollinator milieu. However, without pretending to make an exhaustive examination of the existing literature, throughout this chapter we have emphasized the many existing knowledge gaps and conflicting evidence in this area of research.

The most direct and important effects on biodiversity could be occurring through the local extinction of pollinators that feed on the flowers, or overlap their foraging period with the flowering of outcompeted native plants. Despite a handful of reports comparing pollinator diversity between invaded and non-invaded sites (e.g. Moroń *et al.*, 2009; Nienhuis *et al.*, 2009; Kaiser-Bunbury *et al.*, 2017), we are far from being able to explain these differences. Even more, we need to understand how much pollinator decline can be explained by invasive non-native plants and how much by associated drivers like human-related habitat disturbance (e.g. Morales and Aizen, 2002). It is interesting that despite its feasibility, few existing studies address the mechanisms and potential causes underpinning pollinator losses driven by plant invasions. Some more progress has been made to assess the nutritional consequences of monotypic or low-quality diets for bee health. However, most of these essays have focused on resilient, social bees, like *Apis mellifera* (Branchiccela *et al.*, 2019) and *Bombus terrestris* (Arnold *et al.*, 2014), which can provide a biased picture of potential nutritional effects. As a consequence, we need research focused on how the pollen or nectar of invasive non-native plants can affect the health and performance of different native pollinator species.

Unlike the impact of non-native plants on pollinators, much more research has been conducted on the effects of these plants on native plant reproduction via pollinator sharing (Morales and Traveset, 2009; Charlebois and Sargent, 2017). However, despite the high number of studies, no consensus yet exists on the sign and magnitude of the effects on non-native plants on pollinator visitation and reproductive success of native plants (Morales and Traveset, 2009; Charlebois and Sargent, 2017). Part of this disagreement could arise from not being able to distinguish between non-native plants that have reached (or not) invasive status based, for instance, on their dominance in the community. We propose that further progress in this area could be achieved by designing studies on pollinator-mediated interactions between non-native and native plant species within the conceptual framework depicted in Fig. 13.3. More generally, we are in need of a more mechanistic approach to understand how invasive non-native plant species, or any other type of human-driven disturbance, affect plant reproduction via pollination (Aizen, 2007). This is particularly important because, beyond monitoring flower visitation, the different antagonistic mechanisms associated with improper pollen transfer (see above) remain mostly unaddressed either experimentally or observationally.

There is another process that could affect the pollination and reproduction of native plants that has seldom been considered in the context of plant invasion: the indirect consequences of shifts in the diversity and composition of the pollinator assemblage (Aizen *et al.*, 2014; Morales *et al.*, 2017). To our knowledge, the first and only attempt so far to address this issue was the study by Kaiser-Bunbury *et al.* (2017), who compared pollinator diversity and native plant reproduction in unrestored habitats and restored habitats (i.e. with non-native shrubs removed) in the Seychelles. These authors showed that plant invasion was associated with lower interaction diversity and reduced sexual reproduction in native plants, although they did not explore the mechanistic link between these two consequences of the presence of non-natives (Kaiser-Bunbury *et al.*, 2017). Mechanistically, the pollination and reproduction of many native plants can be affected to the extent that invasive plant species favour the increasing dominance of some pollinators, like the common honeybee or non-native bumblebees (e.g. Morales and Aizen, 2002). For instance, when pollinator assemblages become dominated by pollen-collecting bees, the efficiency of nectarivorous pollinators (such as hummingbirds) could decrease because of bees depressing overall pollen availability (Hargreaves *et al.*, 2009; Chalcoff *et al.*, 2012). This context-dependent nature of a pollinator's efficiency is a particularly intriguing and unstudied topic in the context of plant invasions.

Research has also been conducted to assess and understand the impacts of non-native plant species on the structure and stability of plant–pollinator networks (see above). However, there is still no consensus on these impacts, in part because of evaluation of different network parameters and of comparing networks representing different extents of invasion. Therefore, future comparative studies need to consider more explicitly these methodological issues. Besides this, the answer to several conceptual questions could benefit from a network approach. First, to what extent co-optation of native pollinators and differential association with non-native pollinators (Morales and Aizen, 2002) facilitate the transition from low abundance of non-native plants (as is the case for all non-native species recently arrived in a region) to a high abundance and dominance of non-native plants (typical of non-native species when they reach invasive status) (Fig. 13.4). Second, to what extent the richness and evenness of the pre-existing plant and pollinator community facilitate the integration of non-native plants upon arrival in plant–pollinator webs and their posterior impact on native plants and pollinators; an important question in the context of hypotheses relating to enemy-free space and biotic resistance (Chapters 9 and 10, this volume). Third, to what extent functional trait dissimilarity and/or phylogenetic relatedness predict the integration and impact of non-native plants on plant–pollinator networks; a relevant question in the context of the Empty Niche Hypothesis (MacDougall *et al.*, 2009). Overall, a network approach could be useful to understand the consequences of invasive plants for mutualistic communities that could suffer sudden collapses because of crossing critical transition thresholds and reaching new stable states (Lever *et al.*, 2014; see Chapter 18, this volume).

Finally, the incorporation of an evolutionary dimension into studies of the effects on invasive non-native plants on the plant–pollinator mutualism offers many opportunities for novel research, much needed to predict the long-term pollination-related consequences of plant invasions. Pollinator and plant evolution in the context of non-native plant invasions are topics that have barely been investigated. This includes from molecular work examining, for instance, gene flow in invaded plant communities to basic comparisons of pollinator and plant traits between invaded and non-invaded communities. Given the fragmentation of native plant populations in advanced stages of invasion, evolutionary change generated by both natural selection and genetic drift should be considered (Vanbergen *et al.*, 2018).

13.6 Concluding Remarks

In this chapter, we have reviewed the different mechanisms whereby invasive non-native plants can affect native pollinators, native plants and their interactions. It is apparent that plant origin per se (i.e. native or non-native) does not seem to be a major predictor of these effects (Carvalheiro *et al.*, 2014; Charlebois and Sargent, 2017). Instead, there seems to be some consensus that effects of non-native plants on the plant–pollinator mutualism are basically density-dependent, particularly associated with the high relative abundance these plants and their floral displays can reach when they become invasive (Morales and Traveset, 2009). Several direct and indirect density-dependence effects that are key to understand the impact of non-native plants on native plants, pollinators and their interactions can be framed within the view of the pollination mutualism as a benefit–cost interaction (Morris *et al.*, 2010; Aizen *et al.*, 2014). In any event, more mechanistic-oriented studies are needed, not only to improve our understanding of the processes involved, but also to predict the effects of invasive non-native plants on pollinators, plant reproduction and community persistence.

Acknowledgments

We thank A. Traveset and two reviewers for helpful comments and suggestions. Grants from the Argentine Fund for Science and Technology, FONCYT, (PICT 2012–3015 and PICT 2015–2333) are acknowledged. This work is part of the “Safeguarding Pollination Services in a Changing World: theory into practice” (SURPASS2), an international collaboration funded by the Newton Fund Latin American Biodiversity Programme, awarded through the UKRI Natural Environment Research Council (NERC) NE/S011870/1 and the National Council of Scientific and Technological Research of Argentina (CONICET) RD 1984-19.

References

Ackerman, J.D., Falcón, W., Molinari, J., Vega, C., Espino, I. *et al.* (2014) Biotic resistance and invasional meltdown: consequences of acquired interspecific interactions for an invasive orchid, Spathoglottis plicata in Puerto Rico. *Biological Invasions* 16, 2435–2447.

Aizen, M.A. (2007) Enfoques en El estudio de la reproducción sexual de las plantas en ambientes alterados: Limitaciones Y perspectivas. *Ecologia Austral* 17, 7–19.

Aizen, M.A., Morales, C.L. and Morales, J.M. (2008) Invasive mutualists erode native pollination webs. *PLoS Biology* 6, e31.

Aizen, M.A., Morales, C.L., Vázquez, D.P., Garibaldi, L.A., Sáez, A. *et al.* (2014) When mutualism goes bad: density-dependent impacts of introduced bees on plant reproduction. *New Phytologist* 204, 322–328.

Aizen, M.A.G., Sabatino, G., Gilarranz, M., Bascompte, J, L.J. and Verdú, M. (2016) The phylogenetic structure of plant-pollinator networks increases with habitat size and isolation. *Ecology Letters* 19, 29–36.

Aizen, M.A., Sabatino, M. and Tylianakis, J.M. (2012) Specialization and rarity predict nonrandom loss of interactions from mutualist networks. *Science* 335, 1486–1489.

Aizen, M.A. and Rovere, A.E. (2010) Reproductive interactions mediated by flowering overlap in a temperate hummingbird-plant assemblage. *Oikos* 119, 696–706.

Albrecht, M., Ramis, M.R. and Traveset, A. (2016) Pollinator-Mediated impacts of alien invasive plants on the pollination of native plants: the role of spatial scale and distinct behaviour among pollinator guilds. *Biological Invasions* 18, 1801–1812.

Anderson, B. and Johnson, S.D. (2009) Geographical covariation and local convergence of flower depth in a guild of fly-pollinated plants. *New Phytologist* 182, 533–540.

Arnold, S.E.J., Idrovo, M.E.P., Arias, L.J.L., Belmain, S.R. and Stevenson, P.C. (2014) Herbivore defence compounds occur in pollen and reduce bumblebee colony fitness. *Journal of Chemical Ecology* 40, 878–881.

Bartomeus, I., Vil, M. and Santamaría, L. (2008) Contrasting effects of invasive plants in plant-pollinator networks. *Oecologia* 155, 761–770.

Bascompte, J., Jordano, P. and Olesen, J.M. (2006) Asymmetric coevolutionary networks facilitate biodiversity maintenance. *Science* 312, 431–433.

Bastolla, U., Fortuna, M.A., Pascual-García, A., Ferrera, A., Luque, B. *et al.* (2009) The architecture of mutualistic networks minimizes competition and increases biodiversity. *Nature* 458, 1018–1020.

Beans, C.M. and Roach, D.A. (2015) An invasive plant alters pollinator-mediated phenotypic selection on a native congener. *American Journal of Botany* 102, 50–57.

Bjerknes, A.L., Totland, Ø., Hegland, S.J. and Nielsen, A. (2007) Do alien plant invasions really affect pollination success in native plant species. *Biological Conservation* 138, 1–12.

Branchiccela, B.C., Corona, L., Díaz-Cetti, M., Invernizzi, C, S., Martínez de la Escalera, G. *et al.* (2019) Impact of nutritional stress on the honeybee colony health. *Scientific Reports* 9, 10156.

Carvalheiro, L.G., Biesmeijer, J.C., Benadi, G., Fründ, J., Stang, M. *et al.* (2014) The potential for indirect effects between co-flowering plants via shared pollinators depends on resource abundance, accessibility and relatedness. *Ecology Letters* 17, 1389–1399.

Cavallero, L., Morales, C.L., Montero-Castaño, A., Gowda, J.H. and Aizen, M.A. (2018) Scale-Dependent effects of conspecific flower availability on pollination quantity and quality in an invasive shrub. *Oecologia* 188, 501–513.

Chalcoff, V.R., Aizen, M.A. and Ezcurra, C. (2012) Erosion of a pollination mutualism along an environmental gradient in a south Andean treelet, *Embothrium coccineum* (Proteaceae). *Oikos* 121, 471–480.

Charlebois, J.A. and Sargent, R.D. (2017) No consistent pollinator-mediated impacts of alien plants on natives. *Ecology Letters* 20, 1479–1490.

Chittka, L. and Schürkens, S. (2001) Successful invasion of a floral market. *Nature* 411, 653.

Dorado, J. and Vázquez, D.P. (2016) Flower diversity and bee reproduction in an arid ecosystem. *PeerJ* 4, e2250.

Elzinga, J.A., Atlan, A., Biere, A., Gigord, L., Weis, A.E. *et al.* (2007) Time after time: flowering phenology and biotic interactions. *Trends in Ecology & Evolution* 22, 432–439.

Endler, J.A. (1986) *Natural Selection in the Wild*. Princeton University Press, Princeton.

Feinsinger, P. (1987) Effects of plant species on each other's pollination: Is community structure influenced? *Trends in Ecology & Evolution* 2, 123–126.

Fort, H., Vázquez, D.P. and Lan, B.L. (2016) Abundance and generalisation in mutualistic networks: solving the chicken-and-egg dilemma. *Ecology Letters* 19, 4–11.

Frost, C.M., Allen, W.J., Courchamp, F., Jeschke, J.M., Saul, W.C. *et al.* (2019) Using network theory to understand and predict biological invasions. *Trends in Ecology & Evolution* 34, 831–843.

Fründ, J., Linsenmair, K.E. and Blüthgen, N. (2010) Pollinator diversity and specialization in relation to flower diversity. *Oikos* 119, 1581–1590.

Galen, C. and Gregory, T. (1989) Interspecific pollen transfer as a mechanism of competition: Consequences of foreign pollen contamination for seed set in the alpine wildflower, *Polemonium viscosum*. *Oecologia* 81, 120–123.

Gathmann, A. and Tscharntke, T. (2002) Foraging ranges of solitary bees. *Journal of Animal Ecology* 71, 757–764.

Geerts, S. and Pauw, A. (2009) African sunbirds hover to pollinate an invasive hummingbird-pollinated plant. *Oikos* 118, 573–579.

Geslin, B., Gauzens, B., Baude, M., Dajoz, I., Fontaine, C. *et al.* (2017) Massively introduced managed species and their consequences for plant–pollinator interactions. *Advances in Ecological Research* 57, 147–199.

Ghazoul, J. (2002) Flowers at the front line of invasion? *Ecological Entomology* 27, 638–640.

Goodell, K., McKinney, A.M. and Lin, C.H. (2010) Pollen limitation and local habitat-dependent pollinator interactions in the invasive shrub *Lonicera maackii*. *International Journal of Plant Sciences* 171, 63–72.

Grossenbacher, D.L., Brandvain, Y., Auld, J.R., Burd, M., Cheptou, P.O. *et al.* (2017) Self-compatibility is over-represented on islands. *New Phytologist* 215, 469–478.

Grossenbacher, D., Briscoe Runquist, R., Goldberg, E.E. and Brandvain, Y. (2015) Geographic range size is predicted by plant mating system. *Ecology Letters* 18, 706–713.

Guimarães, P.R., Pires, M.M., Jordano, P., Bascompte, J. and Thompson, J.N. (2017) Indirect effects drive coevolution in mutualistic networks. *Nature* 550, 511–514.

Hargreaves, A.L., Harder, L.D. and Johnson, S.D. (2009) Consumptive emasculation: the ecological and evolutionary consequences of pollen theft. *Biological Reviews* 84, 259–276.

Harmon-Threatt, A.N. and Kremen, C. (2015) Bumble bees selectively use native and exotic species to maintain nutritional intake across highly variable and invaded local floral resource pools. *Ecological Entomology* 40, 471–478.

Issaly, E.A., Sérsic, A.N., Pauw, A., Cocucci, A.A., Traveset, A. *et al.* (2020) Reproductive ecology of the bird-pollinated Nicotiana glauca across native and introduced ranges with contrasting pollination environments. *Biological Invasions* 22, 485–498.

Johnson, S.D. and Schiestl, F.P. (2016) *Floral Mimicry*. Oxford University Press, Oxford.

Kaiser-Bunbury, C.N., Valentin, T., Mougal, J., Matatiken, D. and Ghazoul, J. (2011) The tolerance of island plant-pollinator networks to alien plants. *Journal of Ecology* 99(1), 202–213. DOI: 10.1111/j.1365-2745.2010.01732.x.

Kaiser-Bunbury, C.N., Mougal, J., Whittington, A.E., Valentin, T., Gabriel, R. *et al.* (2017) Ecosystem restoration strengthens pollination network resilience and function. *Nature* 542(7640), 223–227. DOI: 10.1038/nature21071.

Larson, K.C., Fowler, S.P. and Walker, J.C. (2002) Lack of pollinators limits fruit set in the exotic *Lonicera japonica*. *American Midland Naturalist* 148, 54–60.

Lever, J., van Nes, E.H., Scheffer, M. and Bascompte, J. (2014) The sudden collapse of pollinator communities. *Ecology Letters* 17, 350–359.

Lopezaraiza-Mikel, M.E., Hayes, R.B., Whalley, M.R. and Memmott, J. (2007) The impact of an alien plant on a native plant-pollinator network: an experimental approach. *Ecology Letters* 10, 539–550.

MacDougall, A.S., Gilbert, B. and Levine, J.M. (2009) Plant invasions and the niche. *Journal of Ecology* 97, 609–615.

Mandelik, Y.W., Neeson, T, R. and Kremen, C. (2016) Complementary habitat use by wild bees in agro-natural landscapes. *Ecological Applications* 22, 1535–1546.

Martyniuk, N.A., Morales, C.L. and Aizen, M.A. (2015) Invasive conifers reduce seed set of a native Andean cedar through heterospecific pollination competition. *Biological Invasions* 17, 1055–1067.

Maruyama, P.K., Vizentin-Bugoni, J., Sonne, J., Martín González, A.M., Schleuning, M. *et al.* (2016) The integration of alien plants in mutualistic plant-hummingbird networks across the Americas: the importance of species traits and insularity. *Diversity and Distributions* 22, 672–681.

McKey, D. (1989) Population biology of figs: applications for conservation. *Experientia* 45, 661–673.

Memmott, J. and Waser, N.M. (2002) Integration of alien plants into a native flower-pollinator visitation web. *Proceedings of the Royal Society B* 269, 395–399.

Morales, C.L. and Aizen, M.A. (2002) Does invasion of exotic plants promote invasion of exotic flower visitors? A case study from the temperate forests of the southern Andes. *Biological Invasions* 4, 87–100.

Morales, C.L., Sáez, A., Garibaldi, L.A. and Aizen, M.A. (2017) Disruption of pollination services by invasive pollinator species. In: Vila, M., Hulme, P. and Ruiz, G. (eds) *Impact of Biological Invasions on Ecosystem Services*. Springer, Berlin, pp. 203–220.

Morales, C.L. and Traveset, A. (2008) Interspecific pollen transfer: magnitude, prevalence and consequences for plant fitness. *Critical Reviews in Plant Sciences* 27, 221–238.

Morales, C. and Traveset, A. (2009) A meta-analysis of impacts of alien vs. native plants on pollinator visitation and reproductive success of co-flowering native plants. *Ecology Letters* 12, 716–728.

Moreira-Hernández, J.I. and Muchhala, N. (2019) Importance of pollinator-mediated interspecific pollen transfer for angiosperm evolution. *Annual Review of Ecology, Evolution, and Systematics* 50, 1–27.

Moroń, D., Lenda, M., Skórka, P., Szentgyörgyi, H., Settele, J. *et al.* (2009) Wild pollinator communities are negatively affected by invasion of alien goldenrods in grassland landscapes. *Biological Conservation* 142, 1322–1332.

Morris, W.F., Vázquez, D.P. and Chacoff, N.P. (2010) Benefit and cost curves for typical pollination mutualisms. *Ecology* 91(5), 1276–1285. DOI: 10.1890/08-2278.1.

Murcia, C. and Feinsinger, E. (1996) Interspecific pollen loss by hummingbirds visiting flower mixtures: effects of floral architecture. *Ecology* 77, 550–560.

Murphy, S.D. and Aarssen, L.W. (1995) Reduced seed set in *Elytrigia repens* caused by allelopathic pollen from *Phleum pratense*. *Canadian Journal of Botany* 73, 1417–1422.

Nienhuis, C.M., Dietzsch, A.C. and Stout, J.C. (2009) The impacts of an invasive alien plant and its removal on native bees. *Apidologie* 40, 450–463.

Nuismer, S.L., Week, B. and Aizen, M.A. (2018) Coevolution slows the disassembly of mutualistic networks. *American Naturalist* 192, 490–502.

Oduor, A.M.O. (2013) Evolutionary responses of native plant species to invasive plants: a review. *New Phytologist* 200, 986–992.

Olesen, J.M., Eskildsen, L.I. and Venkatasamy, S. (2002) Invasion of pollination networks on oceanic islands: importance of invader complexes and endemic super generalists. *Diversity Distributions* 8(3), 181–192. DOI: 10.1046/j.1472-4642.2002.00148.x.

Ollerton, J., Watts, S., Connerty, S., Lock, J., Parker, L. *et al.* (2012) Pollination ecology of the invasive tree tobacco *Nicotiana glauca*: comparisons across native and non-native ranges. *Journal of Pollination Ecology* 9, 85–95.

Padrón, B., Traveset, A., Biedenweg, T., Díaz, D., Nogales, M. *et al.* (2009) Impact of alien plant invaders on pollination networks in two archipelagos. *PLoS ONE* 4, e6275.

Palladini, J.D. and Maron, J.L. (2014) Reproduction and survival of a solitary bee along native and exotic floral resource gradients. *Oecologia* 176, 789–798.

Paoli, P.P., Donley, D., Stabler, D., Saseendranath, A., Nicolson, S.W. *et al.* (2014a) Nutritional balance of essential amino acids and carbohydrates of the adult worker honeybee depends on age. *Amino Acids* 46, 1449–1458.

Paoli, P.P., Wakeling, L.A., Wright, G. and Ford, D. (2014b) The dietary proportion of essential amino acids and Sir2 influence lifespan in the honeybee. *Age* 36, 1239–1247.

Parker, I.M. (1997) Pollinator limitation of *Cytisus scoparius* (Scotch broom), an invasive exotic shrub. *Ecology* 78, 1457–1470.

Parra-Tabla, V., Angulo-Pérez, D., Albor, C., Campos-Navarrete, M.J., Tun-Garrido, J. *et al.* (2019) The effect of alien species on plant-pollinator network structure across a gradient of plant species invasion. *bioRxiv*.

Pauw, A., Stofberg, J. and Waterman, R.J. (2009) Flies and flowers in Darwin's race. *Evolution* 63, 268–279.

Potter, K.J.B., Kriticos, D.J., Watt, M.S. and Leriche, A. (2009) The current and future potential distribution of *Cytisus scoparius*: A weed of pastoral systems, natural ecosystems and plantation forestry. *Weed Research* 49, 271–282.

Powell, K.I., Chase, J.M. and Knight, T.M. (2011) A synthesis of plant invasion effects on biodiversity across spatial scales. *American Journal of Botany* 98, 539–548.
Praz, C.J., Müller, A. and Dorn, S. (2008) Specialized bees fail to develop on non-host pollen: do plants chemically protect their pollen? *Ecology* 89, 795–804.
Rathcke, B. (1983) Competition and facilitation among plants for pollination. In: Leslie, L.A. (ed.) *Pollination Biology. Academic Press*. Orlando, pp. 305–329.
Razanajatovo, M., Maurel, N., Dawson, W., Essl, F., Kreft, H. *et al.* (2016) Plants capable of selfing are more likely to become naturalized. *Nature Communications* 7, 1–9.
Richardson, D.M., Allsopp, N., D'Antonio, C.M., Milton, S.J. and Rejmanek, M. (2000) Plant invasions - The role of mutualisms. *Biological Reviews* 75, 65–93.
Sala, O., Stuart Chapin, F., Armesto, J.J., Berlow, E.L., Bloomfield, J.P. *et al.* (2000) Biodiversity-Global biodiversity scenarios for the year 2100. *Science* 287, 1770–1774.
Simberloff, D. and Von Holle, B. (1999) Positive interactions of nonindigenous species: Invasional meltdown? *Biological Invasions* 1, 21–32.
Simmons, B.I., Cirtwill, A.R., Baker, N.J., Wauchope, H.S., Dicks, L.V. *et al.* (2019a) Motifs in bipartite ecological networks: uncovering indirect interactions. *Oikos* 128, 154–170.
Simmons, B.I., Vizentin-Bugoni, J., Maruyama, P.K., Cotton, P.A., Marín-Gómez, O.H. *et al.* (2019b) Abundance drives broad patterns of generalisation in plant–hummingbird pollination networks. *Oikos*, 1287–1295.
Thébault, E. and Fontaine, C. (2010) Stability of ecological communities and the architecture of mutualistic and trophic networks. *Science* 329, 853–856.
Traveset, A., Chamorro, S., Olesen, J.M. and Heleno, R. (2015) Space, time and aliens: Charting the dynamic structure of Galápagos pollination networks. *AoB PLANTS* 7, plv068.
Tylianakis, J.M. (2008) Understanding the web of life: the birds, the bees, and sex with aliens. *PLoS Biology* 6, e47.
Tylianakis, J.M., Didham, R.K., Bascompte, J. and Wardle, D. (2008) Global change and species interactions in terrestrial ecosystems. *Ecology Letters* 11, 1351–1363.
Valdovinos, F.S.R.-J., Flores, R., Espinoza, C, J.D. and López, G. (2009) Structure and dynamics of pollination networks: the role of alien plants. *Oikos* 118, 1190–1200.
Valiente-Banuet, A., Aizen, M.A., Alcántara, J.M., Arroyo, J., Cocucci, A. *et al.* (2015) Beyond species loss: the extinction of ecological interactions in a changing world. *Functional Ecology* 29, 299–307.
Vanbergen, A.J., Espíndola, A. and Aizen, M.A. (2018) Risks to pollinators and pollination from invasive alien species. *Nature Ecology and Evolution* 2, 16–25.
Vaudo, A.D., Farrell, L.M., Patch, H.M., Grozinger, C.M. and Tooker, J.F. (2018) Consistent pollen nutritional intake drives bumble bee (*Bombus impatiens*) colony growth and reproduction across different habitats. *Ecology and Evolution* 8, 5765–5776.
Vaudo, A.D., Patch, H.M., Mortensen, D.A., Tooker, J.F. and Grozinger, C.M. (2016) Macronutrient ratios in pollen shape bumble bee (*Bombus impatiens*) foraging strategies and floral preferences. *Proceedings of the National Academy of Sciences USA* 113, e4035–e4042.
Vázquez, D.P.B., Cagnolo, L, N. and Chacoff, N.P. (2009) Uniting pattern and process in plant-animal mutualistic networks: a review. *Annals of Botany* 103, 1445–1457.
Vilà, M., Bartomeus, I., Dietzsch, A.C., Petanidou, T., Steffan-Dewenter, I. *et al.* (2009) Invasive plant integration into native plant-pollinator networks across Europe. *Proceedings of the Royal Society B: Biological Sciences* 276(1674), 3887–3893. DOI: 10.1098/rspb.2009.1076.
Waser, N.M., Chittka, L., Price, M.V., Williams, N.M. and Ollerton, J. (1996) Generalization in pollination systems, and why it matters. *Ecology* 77, 1043–1060.

14 The Effect of Non-native Plant Invasions on the Dispersal of Native Seeds

Ruben H. Heleno*

Centre for Functional Ecology, Department of Life Sciences, University of Coimbra, Coimbra, Portugal

Abstract

Non-native plants change the communities they integrate in multiple ways, including direct and indirect effects on co-occurring native vegetation. While direct effects are more obvious, indirect effects, i.e. those mediated by biotic interactions with other trophic levels, can also have pervasive consequences for long-term community persistence. Seed dispersal is a critical stage during the life cycle of most plants, as it lays the foundations for plant recruitment patterns and long-term vegetation dynamics. By interacting with seed-dispersing animals, primarily frugivorous birds and mammals, plants can positively or negatively affect the dispersal of co-occurring native seeds. In an increasingly invaded world, it is thus critically important to identify general trends on the direction and magnitude of these effects. This chapter reviews the empirical evidence supporting such changes and the potential underlying mechanisms driving them. While the direct impacts of plant invasions are relatively easy to document, indirect effects are much harder to detect. Nevertheless, the most important consequence of the incorporation of new fruiting plants into native communities seems to be a direct competition for the services provided by the local dispersers, negatively affecting native seed dispersal rates. However, another key message emerging from the literature is that responses are highly idiosyncratic, and usually habitat- and species-specific, and therefore resistant to broad generalizations. Fruiting phenology, and in particular the synchrony/asynchrony between the availability of native and non-native fruits, seems to be a particularly important driver of the direction of the responses (i.e. towards facilitation or competition). However, most evidence is still derived from anecdotal observations and formal community level assessments are largely missing. Similarly, how invasive plants change the emergent structure of seed dispersal networks remains uncertain, with early evidence suggesting that novel seed dispersal networks might be structurally very similar to native ones. Bringing together classic experimental designs and new technical and analytical tools to provide broad synthesis will be vital in the near future to clarify the direction, magnitude and generality of these effects.

14.1 Introduction

Due to their sessile nature, most adult plants are confronted with two serious limitations: they cannot freely run away from local threats – such as wildfires or herbivores; and they cannot actively colonize new available grounds. It's only during a relatively short stage during their life cycle that most plants can overcome these limitations, namely by enclosing copies of their genomes inside relatively small 'shipping vessels' we call seeds. Protected inside seeds, plants can expand their distribution range, colonize new territories, escape high mortality rates near the parent plant and survive during periods that would be lethal to adult plants (Howe and Smallwood, 1982). Accordingly, seed dispersal is a critical process with direct implications for plant recruitment and long-term vegetation dynamics (Traveset *et al.*, 2014). Seeds can be dispersed by abiotic vectors, such as wind and water currents, or by animals: either passively

*rheleno@uc.pt

 Plant Invasions: The Role of Biotic Interactions (eds Anna Traveset and David M. Richardson)
DOI: 10.1079/9781789242171.0014

attaching to their body surface or, more frequently, by being dispersed in the animals' gut after ingestion (van der Pijl, 1982; Costa *et al.*, 2014). The internal dispersal of seeds by frugivores has several important advantages over other dispersal mechanisms (Costa *et al.*, 2014) and can be particularly relevant in assisting the spread of non-native plants (Simberloff, 2009; Traveset and Richardson, 2011; Muñoz and Ackerman, 2013). Such concerns initially created a strong research interest in identifying the dispersers of some of the most aggressive invasive plants, such as *Acacia cyclops* in South Africa (Glyphis *et al.*, 1981), *Asparagus asparagoides* in Australia (Stansbury, 2001), *Triadica sebifera* (previously *Sapium sebiferum*) in the USA (Renne *et al.*, 2002) or *Maesopsis eminii* in Tanzania (Cordeiro *et al.*, 2004), reviewed in Chapter 5 (this volume). A second group of studies explored the problem from the opposite angle, namely by evaluating the role of invasive animals as potential seed dispersers in the communities they infiltrate (e.g. García *et al.*, 2014). Both perspectives confirmed the generalist nature of seed dispersal interactions, reflected in the rapid incorporation of native and non-native fruits in the diets of native and non-native dispersers whenever new species came into contact (Mandon-Dalger *et al.*, 2004; Bartuszevige and Gorchov, 2006; García *et al.*, 2014).

As the field of biological invasions developed, researchers realized that invasive plants often change the recipient communities in multiple ways and became increasingly interested in understanding the full range of changes perpetrated by non-native plants at the level on the entire communities (e.g. Heleno *et al.*, 2009). The most immediate impacts are those associated with direct competition for resources such as light, nutrients and space, negatively affecting the native vegetation (Hejda *et al.*, 2009). However, a second order of effects has also been documented, concerned with changes in the myriad of ecological interactions that bind species into functional communities (Heleno *et al.*, 2009; Traveset and Richardson, 2014; López-Núñez *et al.*, 2017). These interactions, such as those established between plants and their seed dispersers, herbivores, parasites or pollinators, are key for the reproduction and survival of plants and animals and are thus ultimately responsible for their long-term persistence (Bond, 1994; McCann, 2007). For example, the simple presence of non-native flowers can shift the attention of local pollinators and change the pollination services available to native plants even before any other effect related to direct competition for space takes place (Traveset and Richardson, 2006). Similarly, it is reasonable to postulate that the presence of non-native fruits can change the seed dispersal services available to native plants, either hampering or boosting their dispersal patterns, independently of other impacts imposed by invasive species (Knight, 1986). If so, understanding exactly how seed dispersal services are affected in a world that is increasingly dominated by non-native species is urgently needed (Traveset and Richardson, 2014). Here, I review the existing evidence for changes on the dispersal of native plants as a result of the incorporation of non-native plants into their local communities, and explore the generality of the findings, their underlying mechanisms, and current challenges and opportunities to advance the field.

14.2 Direct Effects: Competitive Exclusion of Native Plants

The first consequence of the arrival of a new species in an area where it didn't naturally occur is inevitably an increase, even if short-term, in the diversity of the local community. In the specific case of seed dispersal, this means that either new resources become available to local frugivores or more animals can potentially disperse the seeds of the local plants. However, we now know that a small proportion of such introduced plants can, at least under some circumstances, outcompete the native vegetation, rapidly increasing in density and eventually becoming invasive (Richardson *et al.*, 2000b; Daehler, 2003). The direct competition for resources such as space, light, water and nutrients often reduces the resources available to native plants, reducing their diversity and relative abundance and eventually leading to a simplification of the vegetation (McKinney and Lockwood, 1999; Hejda *et al.*, 2009; Heleno *et al.*, 2009). The competitive exclusion of native plants from highly invaded sites implies that their fruits will tend to become locally rare and eventually become extinct from the local community, being no longer available for dispersal and thus extirpated from the local seed rain – i.e. the seeds that reach the ground. Such direct impacts – where native plants are no longer dispersed simply because they have been physically replaced by non-native plants – are likely the main mechanism by which plant invasions affect native plants, including their dispersal. However, in this review I focus on the indirect effects of invasive plants, i.e. those that are mediated by changes on the activity of seed dispersers.

14.3 Indirect Effects: Fruit Neighbourhood and Seed Dispersers' Preferences

The replacement of diverse native floras by one or a few dominant invasive plants naturally affects the composition of the plant community, and concomitantly the fruit neighbourhood (Carlo, 2005) and the composition of the seed dispersers' fauna (Williams and Karl, 2002; Ceia *et al.*, 2009). These shifts in the diversity, density and identity of plants and animals affect frugivorous interactions and seed dispersal due to changes in species encounter probabilities, and in the potential interactions due to morphological constraints and feeding preferences (Vázquez *et al.*, 2009; Muñoz and Ackerman, 2013; Dormann *et al.*, 2017). Fig. 14.1 represent a typical plant invasion scenario, where the increase in the abundance of an invasive plant species slowly boosts encounter probabilities with local frugivores, indirectly affecting the dispersal opportunities for co-occurring native plants.

14.3.1 Changes in the abundance, distribution and diversity of seed dispersers

Reducing the density of seed dispersers is a powerful way by which invasive plants can negatively impact the dispersal of native plants. For example, Cohan *et al.* (1978) reported that the non-native *Tamarix chinensis* displaces native vegetation in the USA, negatively affecting frugivores' abundance. Similarly, the invasion of *Pittosporum undulatum* in the Azores has been found to truncate the density of several important seed dispersers (Ceia *et al.*, 2009). Logically, decreases in the abundance and/or diversity of local seed dispersers will in turn

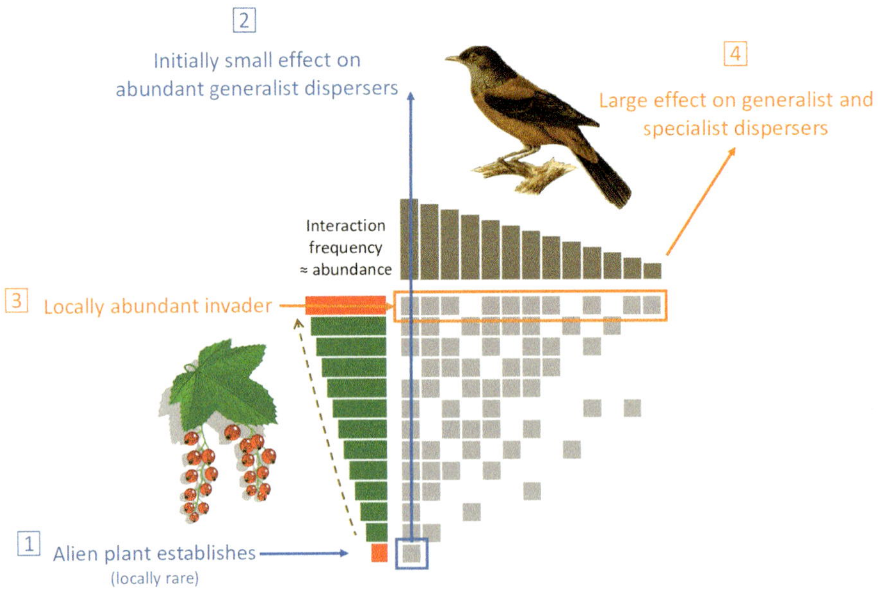

Fig. 14.1. Simplified diagram representing the progression of the links (dark grey cells) established by a fleshy-fruited invasive plant species since it integrates (1) and spreads (3) into a hypothetical native recipient community.

hamper the dispersal services available to native seeds (Rumeu *et al.*, 2017).

However, non-native plants can also attract and support higher densities of seed dispersers, potentially facilitating the dispersal of native plants. This mechanism has been reported on several occasions, usually with non-native plants subsidizing range expansions of native frugivores. For example, native rainforest pigeons in Australia benefit from non-native berries as a winter food source, allowing them to occupy areas where they couldn't otherwise persist (Date *et al.*, 1995). Non-native resources, and particularly fruits, might have also facilitated the expansion of mockingbirds in the USA. (Stiles 1982 *in* Reichard *et al.*, 2001), and the pied barbet in South Africa (Macdonald, 1986), two important seed dispersers. However, whether these reinforcements of the community of seed dispersers had positive cascading effects on the dispersal of native plants remains to be unequivocally shown.

A particular mechanism by which non-native plants might affect the distribution of dispersers is by altering their migration patterns. Animal migrations, which involve many frugivores, are largely triggered by the need to guarantee access to resources for survival. By providing new resources, namely abundant fruit, in areas or periods of natural food shortage, non-native plants can reduce the need of animals to migrate, which will truncate seed dispersal distances. This has been proposed on a number of occasions, including for migrant birds in Europe and North America (Debussche and Isenmann, 1990; White and Stiles, 1992), and for flying foxes in Australia (Parry-Jones and Augee, 2001); however, the exact consequences for the dispersal of native plants are still to be quantified.

14.3.2 Neighbourhood effects, facilitation and competition for dispersers

Invasive plants can also affect the dispersal of co-occurring native plants by influencing frugivores' choices in a new biological context. So far, very few studies have rigorously measured the importance of such fruit neighbourhood effects, and those that did produced somehow conflicting evidence. Two studies, Carlo (2005) and Gleditsch and Carlo (2011), showed that the dispersal of *Solanum americanum* fruits increased in the presence of *Cestrum diurnum* and Lonicera spp. fruits, respectively, following an increase in the number of bird visits. On the other hand, Saracco *et al.* (2005) found that frugivore visits were negatively affected by the presence of conspecific fruits in the neighbourhood, suggesting that individual trees compete for a limited seed disperser fauna. Interestingly, the identity of the species in the neighbourhood seems to have a specific effect on the direction and magnitude of the interference (Saracco *et al.*, 2005), as well as dispersers' behavioural changes across years (Martínez *et al.*, 2014). More recently, Rumeu *et al.* (2019) also showed that hawthorn fruits are less frequently dispersed and by fewer disperser species in the presence of other lipid-rich fruits in the neighbourhood due to the competition for dispersers. Although other studies reveal a less clear importance of fruit neighbourhood effects (Blendinger and Villegas, 2011), the general rule seems to be that seed dispersal of endozoochorous plants is dependent on the broader biological context in which the interaction takes place, but that the direction and magnitude of these effects is in itself highly context-dependent.

14.3.3 Disperser preferences

Across the world, native plants are now integrated into novel communities where they coexist with different groups of non-native and often invasive plants (Richardson and Rejmánek, 2011). In this context, whether birds show any intrinsic preference for native or non-native fruit might have disproportionately relevant consequences for future community composition, and therefore the topic has raised considerable interest (e.g. Daehler, 2003; Drummond, 2005; Gosper *et al.*, 2005; Aslan and Rejmanek, 2012; Muñoz and Ackerman, 2013). Obviously, frugivores do not use plant species' origin as a decision rule for selecting their fruit choices, so the previous question should be separated into two previous assessments. First: do frugivores consistently prefer certain types of fruit? And secondly: are non-native fruits consistently different than native fruits? The first question has received slightly more consideration in ecological literature, however neither has yet received unequivocal evidence to support robust generalizations (Aslan and Rejmanek, 2012; Muñoz and Ackerman, 2013). Daehler's (2003) review suggests that invasive plant species have at least a marginal advantage in seed dispersal when compared to natives. Some studies have also documented particular cases where dispersers show a preference for non-native congeneric fruits (Sallabanks, 1993; Vilà and D'Antonio, 1998), although most studies did not find a clear overall preference for invasive and non-invasive fruits (Reichard *et al.*, 2001), or between native and non-native fruits (Heleno *et al.*, 2013b). More recent reviews showed that frugivores' preferences seem highly context-specific (Muñoz and Ackerman, 2013), depending for example on the resemblance of the non-native fruit morphology with that of the regional native fruits (Aslan and Rejmanek, 2012). However, even in the absence of any specific preferences for native or non-native fruits, experimental work showed that the presence of the invasive *Leycesteria formosa* strongly limited the number of native *Vaccinium cylindraceum* fruits dispersed in the Azores (Table 14.1). While on a relatively small scale, this study shows that non-native plants can dispute the services from the limited disperser fauna, negatively affecting performance of co-occurring natives, which supports the prevalence of competition over facilitation in seed dispersal (Heleno *et al.*, 2013b). Such effects might not be consistent across species though, which raises further difficulties in identifying general patterns. For example, some bird species change from dispersing mostly native seeds in undisturbed Bonin Islands to mostly dispersing non-native seeds in heavily invaded islands, however not all species change in the same way, hindering the detection of an overall signal (Kawakami *et al.*, 2009). Similarly, the presence of the fleshy-fruited invasive *Chrysanthemoides monilifera* did not affect the number of native fruits removed by flying foxes from feeding stations in eastern Australia (Gosper *et al.*, 2006). However, this study points to the importance of fruiting synchrony/asynchrony in the intensity of the effects on native species dispersal.

14.3.4 Importance of fruiting phenology

Despite differences in disperser preferences or fruit morphology and nutritive content, the time of ripening and the duration of the fruiting period can be determinant for the fate of native

Table 14.1. Results from studies reporting the indirect effects of the presence of non-native plant species on the dispersal of co-occurring native plants.

Non-native plant(s)	Native plant(s)	Location	Variable	Effect	Reference
Chrysanthemoides monilifera	*Monotoca elliptica* *Elaeocarpus spp.*	Eastern Australia	Native fruit removal from experimental feeding stations	No effect	Gosper *et al.* (2006)
Bischofia javanica *Morus australis* *Ficus microcarpa* and others	*Solanum nigrum* *Ficus nishimurae* *Trema orientalis* *Wikstroemia pseudoretusa* and others	Bonin Islands, Japan	Proportion of droppings with native seeds	No overall effect, but significant for some disperser species	Kawakami *et al.* (2009)
Lonicera maackii *Lonicera morrowii*	*Solanum americanum*	Pennsylvania, USA	Native fruit removal from potted plants	Positive effect	Gleditsch and Carlo (2011)
Leycesteria formosa	*Vaccinium cylindraceum*	Azores, Portugal	Native fruit removal from plants	Strong negative effect	Heleno *et al.* (2013b)

and non-native seeds in the landscape (Williams and Karl, 1996; Corlett, 2005). Indeed, it has been suggested that some bird-dispersed non-native plant species may succeed because they have a prolonged fruiting display strategy that allows plants to present fruit when native species have completed fruiting (Reichard *et al.*, 2001). For example, this has been indicated as a likely important driver facilitating the spread of invasive species in the USA (White and Stiles, 1992), France (Debussche and Isenmann, 1990), New Zealand (Williams and Karl, 1996) and Galapagos (Heleno *et al.*, 2013a). Although there are obviously many exceptions, it seems that invasive plants tend to have longer fruiting seasons and/or produce fruits outside the main native fruiting periods – i.e. presenting a gap-filling phenology (Gosper and Vivian-Smith, 2009; Heleno *et al.*, 2013a; Muñoz and Ackerman, 2013), that might be particularly relevant in temperate regions, where fruit production is highly concentrated.

14.3.5 It's complicated: more complex indirect effects

While ecologists strive to find general patterns, they are often confronted with enormous sources of intra- and inter-specific variability associated with the very different biological contexts in which organisms interact (Lawton, 1999). This is even more problematic when trying to detect general patterns that are mediated by indirect interactions such as those involving frugivores. Accordingly, several studies have reported impacts of plant invasions on seed dispersal processes whose generality is still hard to evaluate. Such impacts include, for example, the effect of introduced trees as facilitators of nucleation processes, particularly in dry environments. In South African savannas, isolated *Prosopis* trees concentrate bird-dispersed seeds under their canopies, facilitating subsequent invasions by fleshy-fruited plant species (Milton *et al.*, 2007, see also Chapter 7, this volume). However, whether the outcome of nucleation processes is positive or negative will largely depend on the identity of the seeds dispersed into these new recruitment sites (Hervías *et al.*, 2014). Contrastingly, Erskine Ogden and Rejmánek (2005) reported that native plant recruitment was greater in areas invaded by *Foeniculum vulgare* than in cleared areas due to the attraction of avian dispersers. Such effects are also likely important for the reclamation of abandoned fields, with non-native plants potentially accelerating the ecological succession via increased seed dispersal (Neilan *et al.*, 2006). Although generalizations are extremely hard, native plant recruitment under non-native canopies is probably less common than the opposite, as expected due to the intrinsic population dynamics of invasive species.

Even harder to predict are consequences that are mediated by multiple cascading effects. For example, Schmidt and Whelan (1999) suggested that frugivorous birds that build their nests in non-native plants experienced higher nest predation, which might indirectly hinder the dispersal of native plant species. These and other indirect effects are inherently hard to predict, particularly when they affect the fate of the few, yet disproportionately important, seeds that are dispersed over long distances (Nathan, 2006), for example by truncating animal migrations (Debussche and Isenmann, 1990).

Another order of effects that is still poorly understood is related to the composition of seed mixes in animal droppings. The spread of non-native plants throughout the globe will likely translate in the increase of the proportion of animal droppings containing both native and non-native seed mixes, independent of the sheer number of seeds dispersed. Exploring interference effects on germination and seedling survival, including direct competition and potential allelopathic effects between native and non-native plants, is thus particularly relevant (Thiébaut *et al.*, 2019).

Finally, de novo interactions that might be established in species invasive ranges, but that do not reflect standard dispersal mechanisms observed in any of the native ranges, are also particularly hard to predict, such as the dispersal of invasive pine seeds by glossy black cockatoos in Australia (Richardson *et al.*, 2000a).

14.3.6 Changes to the entire seed-dispersal network structure

As shown above, most existing evidence on the effects of plant invasions on native seed dispersal emerged from observations or experimentation involving a few co-occurring plant species. Typically, studies compared the fruit removal frequency of focal species presented either in isolation or in the presence of competing fruit resources (e.g. Vilà and D'Antonio, 1998; Carlo, 2005; Heleno *et al.*, 2013b). The growing recognition that species–interactions networks are particularly important for the persistence of biological communities prompted ecologists to document all plant species dispersed by a given disperser (e.g. Williams, 2006; Milton *et al.*, 2007; Kawakami *et al.*, 2009; Linnebjerg *et al.*, 2010) or all potential dispersers of a given plant species (e.g. Cordeiro *et al.*, 2004; Bartuszevige and Gorchov, 2006). However, the quantification of the entire seed dispersal interactions of complete species assemblages, including all native and non-native plants and animals, is still rare (Heleno *et al.*, 2013b). Such community-level studies are highly challenging but are particularly important to identify changes in community structure and in the performance of ecosystem functions such as the consequences of plant invasions on seed dispersal. These studies took at least three alternative approaches: the first approach is to describe the entire seed dispersal network of one area and look for consistent differences in the dispersal patterns of native and non-native seeds (Table 14.2). This approach revealed that two thirds of the seeds dispersed in one of the last native forest remnants in the Azores are actually non-native, and that these tended to be served by a greater diversity of dispersers (Heleno *et al.*, 2013b). However, such an approach is mostly descriptive and has limited power to infer about potential changes in community dynamics. A second approach is to contrast the structure of seed dispersal networks in one invaded site with that of other native-dominated networks available in the literature. This approach was adopted by Vizentin-Bugoni *et al.* (2019) to show that interactions between native plants and animals are now virtually absent in the highly invaded island of Oahu, Hawaii, where less than 7% of the dispersed seeds are native. Surprisingly, and despite this incredible transformation, the emergent structural properties of the seed dispersal network did not significantly differ from most native networks reported in the literature (Vizentin-Bugoni *et al.*, 2019). This study highlights the extreme plasticity of seed dispersal interactions, however the statistical power to detect differences in such broad datasets is naturally hindered by the high intrinsic variability of heterogeneous datasets assembled for multiple purposes (e.g. Traveset *et al.*, 2016). A third approach consists of the implementation of classic experimental designs, with replicated communities exposed to different pressures from invasive plants. This approach has been implemented in Galapagos but also hasn't found significant changes in the structure of invaded and uninvaded communities, likely due to the high heterogeneity within the few sampled sites (Heleno *et al.*, 2013a).

Given these contingencies, and the lack of sufficient data for broad synthesis, it is still hard to pinpoint what changes in the structure of seed dispersal networks occur following plant invasions. However, the provisional consensus seems to indicate that although most non-native fleshy fruits are readily dispersed by many frugivores, increasing generalization levels, the overall community structure might not change much due to the high variability and idiosyncrasy of the inter-specific interactions (Heleno *et al.*, 2013a; Vizentin-Bugoni *et al.*, 2019), as observed for example in plant—pollinator networks (Timóteo *et al.*, 2018; Frost *et al.*, 2019).

14.4 Current Limitations and Future Prospects

14.4.1 More and better data

As evident from Tables 14.1 and 14.2, very few studies have explored the consequences of plant invasions on the dispersal of native plants. This paucity is particularly evident when compared with the literature focusing on pollination – the other key plant reproductive mutualism (Traveset and Richardson, 2006; Frost *et al.*, 2019). Consequently, there is still a great need for detailed studies that measure the efficiency of native seed dispersal in the absence and presence of non-native and invasive plant species. Both types of species- and community-based studies can still be highly relevant as they offer complementary results. Namely, species-centred studies allow easier manipulative experiments and fine fitness measures, while community-based

Table 14.2. Reported effects of plant invasions on the emergent structure of seed dispersal networks.

Site/Reference	Experimental design	Parameter	Effect
Galapagos Heleno *et al.* (2013a)	4 invaded vs 4 uninvaded forest patches	Network size	No effect
		Plant species richness	No effect
		Disperser species richness	No effect
		Number of links	No effect
		Connectance	No effect
		Nestedness (weighted)	No effect
		Robustness to dispersers extinctions	No effect
		Plant niche overlap	No effect
		Dispersers generality	No effect
		Interaction evenness	No effect
		Plant linkage level	No effect
		Plant specialization (d')	No effect
		Plant species strength	No effect
Azores Heleno *et al.* (2013b)	Relative deviations of native and non-native subnetwork parameters in relation to randomized networks	Connectance	Moderate decrease
		Linkage density	Moderate increase
		Plant linkage level	Moderate increase
		Interaction evenness	Moderate increase
		Robustness to dispersers extinctions	Moderate increase
		Plant species strength	No difference
Hawai'i Vizentin-Bugoni *et al.* (2019)	One invaded network vs many native-dominated networks from the literature	Nestedness (binary)	No difference
		Nestedness (weighted)	No difference
		Modularity (binary)	No difference
		Modularity (weighted)	No difference[a]
		Complementary specialization (H2')	No difference[a]
		Robustness to dispersers extinctions	No difference
		Robustness to plant extinctions	No difference
French Polynesia Spotswood *et al.* (2013)	Bootstrapped networks from one heavily invaded island and a moderately invaded island	Number of plant species	Decrease
		Number of links	Decrease
		Nestedness (binary)	Decrease
		Nestedness (weighted)	Decrease
		Linkage density	Decrease
		Interaction evenness	Decrease
		Centrality of native plants	Decrease

[a]Some partial differences found due to differences in latitude.

studies, usually taking advantage of natural gradients resulting from plant invasions or plant restoration projects, can elucidate about emergent community structure, and on the generality of the effects across species. Recent advances in the use of molecular techniques to identify both the disperser and the dispersed seeds from environmental DNA are likely to be an important complementary tool towards that goal (González-Varo *et al.*, 2014).

The need for further research is even more evident for seed dispersal mechanisms other than the dispersal of fleshy-fruited plants by frugivorous animals. Plant invasions might also affect the dispersal of native plants by affecting seed-caching birds and scatter-hoarding rodents, ant-dispersal mechanisms, epizoochory and even abiotic dispersal mechanisms. However, such effects are still largely unexplored.

14.4.2 Moving from a frequentist to a seed dispersal effectiveness landscape

Plant—frugivore interactions are not always mutualistic, but span across a continuum from highly mutualistic to highly antagonistic interactions, passing through interactions with a net neutral effect on plant fitness (Heleno *et al.*, 2011; Muñoz and Ackerman, 2013). Seed dispersal effectiveness – i.e. the contribution of seed dispersal to plant fitness via changes in recruitment – depends on multiple aspects, including the number of seeds dispersed, the treatment conferred to the seeds and the specific conditions of their deposition site (Schupp *et al.*, 2010). Accordingly, while most studies to date have simply reported the number of fruits removed, or the number of seeds dispersed, such measures provide only an incomplete proxy for the actual effectiveness of seed dispersal (Howe, 2016; Simmons *et al.*, 2018; Fricke *et al.*, 2019). Therefore, unless we successfully integrate the quantitative and qualitative effects of seed dispersal on seed recruitment probabilities, we cannot fully evaluate whether the effect of a given frugivore on the recruitment of a given plant is actually strong, weak, positive or negative. For example, while it has been shown that granivorous birds, large herbivores and seed-catching birds and rodents can also disperse a non-negligible proportion of viable seeds, the net effect of these interactions on plant fitness is still hard to quantify (Feldman *et al.*, 1999; Heleno *et al.*, 2011; Albert *et al.*, 2015, see also Chapter 6, this volume). Recent studies have already begun to incorporate the effect of seed passage through the dispersers gut on germination to refine the measures of interaction outcome in seed dispersal networks (Fricke *et al.*, 2019), and also to estimate seedling survival (González-Castro *et al.*, 2015). These are very important steps towards effectively closing the seed dispersal loop (see Wang and Smith, 2002) in community level studies, i.e. to estimate the seed dispersal effectiveness of all pairwise interactions in a given community, and to fully understand the impacts of plant invasions on native seed dispersal.

14.4.3 Generating a new synthesis

As more studies document the impacts of plant invasions on the dispersal of native plants in different contexts, opportunities for broader synthesis will naturally emerge. Such synthesis, based on large datasets and modelling approaches, will be critical to identify general patterns and allow us to move from a descriptive to a predictive framework. In particular, it is expected that the collation of a sufficient number of high-quality community-level seed dispersal networks will allow the detection of structural changes on species interaction patterns (e.g. Dugger *et al.*, 2019), for example as the proportion of non-native and invasive plant species increases in the studied communities. However, the spatial representation, and the quality of such datasets, chiefly in terms of sampling effort and on the completeness of the plant and animal groups included, is still highly heterogeneous.

14.5 Concluding Remarks

Despite the relevance of biotic invasions as a major driver of current global changes, the specific impacts of plant invasions on the dispersal of co-occurring native plants are still poorly understood (Traveset and Richardson, 2006; Muñoz and Ackerman, 2013). Although a handful of studies in the last decade have provided valuable insights into such impacts, generalizations remain difficult. This is due both to high intrinsic variation in species-specific effects and ecosystem-level responses – which are by themselves relevant results – but also to limited research and methodological shortcomings that need to be addressed. Some studies have quantified changes in the dispersal of one or a few focal native plants in the presence of non-native plants, but very few have explored changes in the overall assemblages of plants and their dispersers in the context of plant invasions (but see Heleno *et al.*, 2013a; Vizentin-Bugoni *et al.*, 2019). Moreover, it is often difficult to disentangle the independent effects of potential seed dispersal failure from other mechanisms that might limit native plant recruitment in an invasion scenario, such as direct competition for space, light and nutrients. Accordingly, it is still remarkably difficult to know how the magnitude and direction of indirect effects mediated by seed dispersers compares with the direct effects of plant invasions.

Observing how species interact, survive and reproduce in new biological contexts of biological invasions has been very important for refining ecological theory (Richardson, 2011). On the other hand, the urgency posed by the rapid homogenization of the world's biota and the accelerated rate of species extinctions calls for the urgent prioritization of conservation targets and a greater efficiency in the restoration of degraded ecosystem functions (e.g. Kaiser-Bunbury and Blüthgen, 2015, see also Chapter 22, this volume). Several authors have explored ways to manipulate natural seed rain as a cost-effective way to assist conservation, either by controlling non-native fruit production, managing the seed disperser's fauna, or actively managing seed deposition patterns, for example using bird perches (Wenny, 2001; Gosper *et al.*, 2005; Hervías *et al.*, 2014). However, while knowledge of the seed dispersal patterns of native and non-native plants has been particularly useful for establishing conservation priorities and evaluating restoration success (Heleno *et al.*, 2010; Ribeiro da Silva *et al.*, 2015; Correia *et al.*, 2017), its applied potential as an active conservation tool remains unfulfilled.

The generalism and plasticity of seed dispersal interactions implies that most plant invaders will readily encounter suitable dispersers in their introduced ranges (Traveset and Richardson, 2014). Positive feedbacks on invaders' performance and secondary effects on native and other introduced plants are likely to involve a wide range of species, and to be complex, variable and context-specific (Gosper *et al.*, 2005; Muñoz and Ackerman, 2013). Nevertheless, understanding such effects has both a large theoretical and applied potential for ecology and conservation biology.

Acknowledgements

This work was funded by the Portuguese Fundação para a Ciência e a Tecnologia (FCT) through grants UID/BIA/04004/2020 and CEECIND/00092/2017.

References

Albert, A., Auffret, A.G., Cosyns, E., Cousins, S.A.O., D'hondt, B. *et al.* (2015) Seed dispersal by ungulates as an ecological filter: a trait-based meta-analysis. *Oikos* 124(9), 1109–1120. DOI: 10.1111/oik.02512.

Aslan, C. and Rejmanek, M. (2012) Native fruit traits may mediate dispersal competition between native and non-native plants. *NeoBiota* 12(1), 1–24. DOI: 10.3897/neobiota.12.2357.

Bartuszevige, A.M. and Gorchov, D.L. (2006) Avian seed dispersal of an invasive shrub. *Biological Invasions* 8(5), 1013–1022. DOI: 10.1007/s10530-005-3634-2.

Blendinger, P.G. and Villegas, M. (2011) Crop size is more important than neighborhood fruit availability for fruit removal of Eugenia uniflora (Myrtaceae) by bird seed dispersers. *Plant Ecology* 212(5), 889–899. DOI: 10.1007/s11258-010-9873-z.

Bond, W.J. (1994) Do mutualisms matter - assessing the impact of pollinator and disperser disruption on plant extinction. *Philosophical Transactions of the Royal Society B-Biological Sciences* 344, 83–90.

Carlo, T.A. (2005) Interspecific neighbors change seed dispersal pattern of an avian-dispersed plant. *Ecology* 86(9), 2440–2449. DOI: 10.1890/04-1479.

Ceia, R., Heleno, R. and Ramos, J. (2009) Summer abundance and ecological distribution of passerines in native and exotic forests in São Miguel, Azores. *Ardeola* 56, 25–39.

Cohan, D.R., Anderson, B.W. and Ohmart, R.D. (1978) Avian population responses to salt cedar along the lower Colorado river. *USDA Forest Service General Technical Report* WO-12, 371–382.

Cordeiro, N.J., Patrick, D.A.G., Munisi, B. and Gupta, V. (2004) Role of dispersal in the invasion of an exotic tree in an East African submontane forest. *Journal of Tropical Ecology* 20(4), 449–457. DOI: 10.1017/S026646740400152X.

Corlett, R.T. (2005) Interactions between birds, fruit bats and exotic plants in urban Hong Kong, South China. *Urban Ecosystems* 8(3-4), 275–283. DOI: 10.1007/s11252-005-3260-x.

Correia, M., Timóteo, S., Rodríguez-Echeverría, S., Mazars-Simon, A. and Heleno, R. (2017) Refaunation and the reinstatement of the seed-dispersal function in Gorongosa National Park. *Conservation Biology* 31(1), 76–85. DOI: 10.1111/cobi.12782.

Costa, J.M., Ramos, J.A., da Silva, L.P., Timoteo, S., Araújo, P.M. *et al.* (2014) Endozoochory largely outweighs epizoochory in migrating passerines. *Journal of Avian Biology* 45(1), 59–64. DOI: 10.1111/j.1600-048X.2013.00271.x.

Daehler, C.C. (2003) Performance comparisons of co-occurring native and alien invasive plants: implications for conservation and restoration. *Annual Review of Ecology, Evolution, and Systematics* 34(1), 183–211. DOI: 10.1146/annurev.ecolsys.34.011802.132403.

Date, E., Recher, H., Ford, H. and Stewart, D. (1996) The conservation and ecology of rainforest pigeons in northeastern New South Wales. *Pacific Conservation Biology* 2(3), 299–308. DOI: 10.1071/PC960299.

Debussche, M. and Isenmann, P. (1990) Introduced and cultivated fleshy-fruited plants: consequences of a mutualistic Mediterranean plant-bird system. *Biological invasions in Europe and the Mediterranean Basin* Springer, 399–416.

Dormann, C.F., Fründ, J. and Schaefer, H.M. (2017) Identifying causes of patterns in ecological networks: opportunities and limitations. *Annual Review of Ecology, Evolution, and Systematics* 48(1), 559–584. DOI: 10.1146/annurev-ecolsys-110316-022928.

Drummond, B.A. (2005) The selection of native and invasive plants by frugivorous birds in Maine. *Northeastern Naturalist* 12(1), 33–44. DOI: 10.1656/1092-6194(2005)012[0033:TSONAI]2.0.CO;2.

Dugger, P.J., Blendinger, P.G., Böhning-Gaese, K., Chama, L., Correia, M. *et al.* (2019) Seed-dispersal networks are more specialized in the Neotropics than in the Afrotropics. *Global Ecology and Biogeography* 28(2), 248–261. DOI: 10.1111/geb.12833.

Erskine Ogden, J.A. and Rejmánek, M. (2005) Recovery of native plant communities after the control of a dominant invasive plant species, Foeniculum vulgare: implications for management. *Biological Conservation* 125(4), 427–439. DOI: 10.1016/j.biocon.2005.03.025.

Feldman, R., Tomback, D.F. and Koehler, J. (1999) Cost of mutualism: competition, tree morphology, and pollen production in limber pine clusters. *Ecology* 80(1), 324–329. DOI: 10.1890/0012-9658(1999)080[0324:COMCTM]2.0.CO;2.

Fricke, E.C., Bender, J., Rehm, E.M. and Rogers, H.S. (2019) Functional outcomes of mutualistic network interactions: a community-scale study of frugivore gut passage on germination. *Journal of Ecology* 107, 757–767.

Frost, C.M., Allen, W.J., Courchamp, F., Jeschke, J.M., Saul, W.-C. *et al.* (2019) Using network theory to understand and predict biological invasions. *Trends in Ecology & Evolution* 34(9), 831–843. DOI: 10.1016/j.tree.2019.04.012.

García, D., Martínez, D., Stouffer, D.B. and Tylianakis, J.M. (2014) Exotic birds increase generalization and compensate for native bird decline in plant-frugivore assemblages. *Journal of Animal Ecology* 83(6), 1441–1450. DOI: 10.1111/1365-2656.12237.

Gleditsch, J.M. and Carlo, T.A. (2011) Fruit quantity of invasive shrubs predicts the abundance of common native avian frugivores in central Pennsylvania. *Diversity and Distributions* 17(2), 244–253. DOI: 10.1111/j.1472-4642.2010.00733.x.

Glyphis, J.P., Milton, S.J. and Siegfried, W.R. (1981) Dispersal of *Acacia cyclops* by birds. *Oecologia* 48(1), 138–141. DOI: 10.1007/BF00347002.

González-Castro, A., Calviño-Cancela, M. and Nogales, M. (2015) Comparing seed dispersal effectiveness by frugivores at the community level. *Ecology* 96(3), 808–818. DOI: 10.1890/14-0655.1.

González-Varo, J.P., Arroyo, J.M. and Jordano, P. (2014) Who dispersed the seeds? the use of DNA barcoding in frugivory and seed dispersal studies. *Methods in Ecology and Evolution* 5(8), 806–814. DOI: 10.1111/2041-210X.12212.

Gosper, C.R. and Vivian-Smith, G. (2009) The role of fruit traits of bird-dispersed plants in invasiveness and weed risk assessment. *Diversity and Distributions* 15(6), 1037–1046. DOI: 10.1111/j.1472-4642.2009.00599.x.

Gosper, C.R., Stansbury, C.D. and Vivian-Smith, G. (2005) Seed dispersal of fleshy-fruited invasive plants by birds: contributing factors and management options. *Diversity and Distributions* 11(6), 549–558. DOI: 10.1111/j.1366-9516.2005.00195.x.

Gosper, C.R., Whelan, R.J. and French, K. (2006) The effect of invasive plant management on the rate of removal of vertebrate-dispersed fruits. *Plant Ecology* 184(2), 351–363. DOI: 10.1007/s11258-005-9078-z.

Hejda, M., Pyšek, P. and Jarošík, V. (2009) Impact of invasive plants on the species richness, diversity and composition of invaded communities. *Journal of Ecology* 97(3), 393–403. DOI: 10.1111/j.1365-2745.2009.01480.x.

Heleno, R.H., Ceia, R.S., Ramos, J.A. and Memmott, J. (2009) Effects of alien plants on insect abundance and biomass: a food-web approach. *Conservation Biology* 23(2), 410–419. DOI: 10.1111/j.1523-1739.2008.01129.x.

Heleno, R.H., Lacerda, I., Ramos, J.A. and Memmott, J. (2010) Evaluation of restoration effectiveness: community response to the removal of alien plants. *Ecological Applications* 20(5), 1191–1203. DOI: 10.1890/09-1384.1.

Heleno, R.H., Ross, G., Everard, A.M.Y., Memmott, J. and Ramos, J.A. (2011) The role of avian 'seed predators' as seed dispersers. *Ibis* 153(1), 199–203. DOI: 10.1111/j.1474-919X.2010.01088.x.

Heleno, R.H., Olesen, J.M., Nogales, M., Vargas, P. and Traveset, A. (2013a) Seed dispersal networks in the Galápagos and the consequences of alien plant invasions. *Proceedings of the Royal Society B: Biological Sciences* 280(1750), 20122112. DOI: 10.1098/rspb.2012.2112.

Heleno, R.H., Ramos, J.A. and Memmott, J. (2013b) Integration of exotic seeds into an Azorean seed dispersal network. *Biological Invasions* 15(5), 1143–1154. DOI: 10.1007/s10530-012-0357-z.

Hervías, S., Ceia, R., Ramos, J., Sampaio, H. and Heleno, R. (2014) Tiptoeing between restoration and invasion: seed rain into natural gaps within a highly invaded relic forest in the Azores. *European Journal of Forest Research* 133, 383–390.

Howe, H.F. (2016) Making dispersal syndromes and networks useful in tropical conservation and restoration. *Global Ecology and Conservation* 6, 152–178. DOI: 10.1016/j.gecco.2016.03.002.

Howe, H.F. and Smallwood, J. (1982) Ecology of seed dispersal. *Annual Review of Ecology and Systematics* 13(1), 201–228. DOI: 10.1146/annurev.es.13.110182.001221.

Kaiser-Bunbury, C.N. and Blüthgen, N. (2015) Integrating network ecology with applied conservation: a synthesis and guide to implementation. *AoB Plants* 7, plv076. DOI: 10.1093/aobpla/plv076.

Kawakami, K., Mizusawa, L. and Higuchi, H. (2009) Re-established mutualism in a seed-dispersal system consisting of native and introduced birds and plants on the Bonin Islands, Japan. *Ecological Research* 24(4), 741–748. DOI: 10.1007/s11284-008-0543-8.

Knight, R.S. (1986) Fruit displays of indigenous and invasive alien plants in the south-western Cape. *South African Journal of Botany* 52(3), 249–255. DOI: 10.1016/S0254-6299(16)31559-9.

Lawton, J.H. (1999) Are there general laws in ecology? *Oikos* 84(2), 177–192. DOI: 10.2307/3546712.

Linnebjerg, J.F., Hansen, D.M., Bunbury, N. and Olesen, J.M. (2010) Diet composition of the invasive red-whiskered bulbul *Pycnonotus jocosus* in Mauritius. *Journal of Tropical Ecology* 26(3), 347–350. DOI: 10.1017/S0266467409990617.

López-Núñez, F.A., Heleno, R.H., Ribeiro, S., Marchante, H. and Marchante, E. (2017) Four-trophic level food webs reveal the cascading impacts of an invasive plant targeted for biocontrol. *Ecology* 98(3), 782–793. DOI: 10.1002/ecy.1701.

Macdonald, I.A.W. (1986) Range expansion in the Pied Barbet and the spread of alien tree species in southern Africa. *Ostrich* 57(2), 75–94. DOI: 10.1080/00306525.1986.9634130.

Mandon-Dalger, I., Clergeau, P., Tassin, J., Rivière, J.-N. and Gatti, S. (2004) Relationships between alien plants and an alien bird species on reunion island. *Journal of Tropical Ecology* 20(6), 635–642. DOI: 10.1017/S0266467404001774.

Martínez, D., García, D. and Herrera, J.M. (2014) Consistency and reciprocity of indirect interactions between tree species mediated by frugivorous birds. *Oikos* 123(4), 414–422. DOI: 10.1111/j.1600-0706.2013.00558.x.

McCann, K. (2007) Protecting biostructure. *Nature* 446(7131), 29–29. DOI: 10.1038/446029a.

McKinney, M.L. and Lockwood, J.L. (1999) Biotic homogenization: a few winners replacing many losers in the next mass extinction. *Trends in Ecology & Evolution* 14(11), 450–453. DOI: 10.1016/S0169-5347(99)01679-1.

Milton, S.J., Wilson, J.R.U., Richardson, D.M., Seymour, C.L., Dean, W.R.J. *et al.* (2007) Invasive alien plants infiltrate bird-mediated shrub nucleation processes in arid savanna. *Journal of Ecology* 95(4), 648–661. DOI: 10.1111/j.1365-2745.2007.01247.x.

Muñoz, M. and Ackerman, J. (2013) Invasive plants and mutualistic interactions between fleshy fruits and frugivorous animals. *Invasive Plant Ecology*, 121–136.

Nathan, R. (2006) Long-distance dispersal of plants. *Science* 313(5788), 786–788. DOI: 10.1126/science.1124975.

Neilan, W., Catterall, C.P., Kanowski, J. and McKenna, S. (2006) Do frugivorous birds assist rainforest succession in weed dominated oldfield regrowth of subtropical Australia? *Biological Conservation* 129(3), 393–407. DOI: 10.1016/j.biocon.2005.11.007.

Parry-Jones, K. and Augee, M. (2001) Factors affecting the occupation of a colony site in Sydney, New South Wales by the Grey-headed Flying-fox Pteropus poliocephalus (Pteropodidae). *Austral Ecology* 26, 47–55.

Reichard, S.H., Chalker-Scott, L. and Buchanan, S. (2001) Interactions among non-native plants and birds. In: Marzluff, J.M., Bowman, R. and Donnelly, R. (eds) *Avian Ecology and Conservation in an Urbanizing World*. Springer, Boston, pp. 179–223.

Renne, I.J., Barrow, W.C., Johnson Randall, L.A. and Bridges, W.C. (2002) Generalized avian dispersal syndrome contributes to Chinese tallow tree (Sapium sebiferum, Euphorbiaceae) invasiveness. *Diversity and Distributions* 8(5), 285–295. DOI: 10.1046/j.1472-4642.2002.00150.x.

Ribeiro da Silva, F., Montoya, D., Furtado, R., Memmott, J., Pizo, M.A. *et al.* (2015) The restoration of tropical seed dispersal networks. *Restoration Ecology* 23(6), 852–860. DOI: 10.1111/rec.12244.

Richardson, D.M. (ed.) (2011) *Fifty years of invasion ecology: the legacy of Charles Elton*. Wiley-Blackwell, Oxford.

Richardson, D.M. and Rejmánek, M. (2011) Trees and shrubs as invasive alien species - a global review. *Diversity and Distributions* 17(5), 788–809. DOI: 10.1111/j.1472-4642.2011.00782.x.

Richardson, D.M., Allsopp, N., D'Antonio, C.M., Milton, S.J. and Rejmanek, M. (2000a) Plant invasions – the role of mutualisms. *Biological Reviews of the Cambridge Philosophical Society* 75(1), 65–93. DOI: 10.1017/S0006323199005435.

Richardson, D.M., Pyšek, P., Rejmanek, M., Barbour, M.G., Panetta, F.D. *et al.* (2000b) Naturalization and invasion of alien plants: concepts and definitions. *Diversity and Distributions* 6(2), 93–107. DOI: 10.1046/j.1472-4642.2000.00083.x.

Rumeu, B., Devoto, M., Traveset, A., Olesen, J.M., Vargas, P. *et al.* (2017) Predicting the consequences of disperser extinction: richness matters the most when abundance is low. *Functional Ecology* 31(10), 1910–1920. DOI: 10.1111/1365-2435.12897.

Rumeu, B., Álvarez-Villanueva, M., Arroyo, J.M. and González-Varo, J.P. (2019) Interspecific competition for frugivores: population-level seed dispersal in contrasting fruiting communities. *Oecologia* 190(3), 605–617. DOI: 10.1007/s00442-019-04434-9.

Sallabanks, R. (1993) Fruiting plant attractiveness to avian seed dispersers: native vs invasive Crataegus in Western Oregon. *Madrono* 40, 108–116.

Saracco, J.F., Collazo, J.A., Groom, M.J. and Carlo, T.A. (2005) Crop size and fruit neighborhood effects on bird visitation to fruiting Schefflera morototoni trees in Puerto Rico1. *Biotropica* 37(1), 81–87. DOI: 10.1111/j.1744-7429.2005.04040.x.

Schmidt, K.A. and Whelan, C.J. (1999) Effects of exotic Lonicera and Rhamnus on songbird nest predation. *Conservation Biology* 13(6), 1502–1506. DOI: 10.1046/j.1523-1739.1999.99050.x.

Schupp, E.W., Jordano, P. and Gómez, J.M. (2010) Seed dispersal effectiveness revisited: a conceptual review. *New Phytologist* 188(2), 333–353. DOI: 10.1111/j.1469-8137.2010.03402.x.

Simberloff, D. (2009) The role of propagule pressure in biological invasions. *Annual Review of Ecology, Evolution, and Systematics* 40(1)pp., 81–102. DOI: 10.1146/annurev.ecolsys.110308.120304.

Simmons, B.I., Sutherland, W.J., Dicks, L.V., Albrecht, J., Farwig, N. *et al.* (2018) Moving from frugivory to seed dispersal: incorporating the functional outcomes of interactions in plant-frugivore networks. *Journal of Animal Ecology* 87(4), 995–1007. DOI: 10.1111/1365-2656.12831.

Spotswood, E.N., Meyer, J.-Y. and Bartolome, J.W. (2013) Preference for an invasive fruit trumps fruit abundance in selection by an introduced bird in the Society Islands, French Polynesia. *Biological Invasions* 15(10), 2147–2156.

Stansbury, C.D. (2001) Dispersal of the environmental weed Bridal Creeper, *Asparagus asparagoides*, by Silvereyes, *Zosterops lateralis*, in south-western Australia. *Emu - Austral Ornithology* 101(1), 39–45. DOI: 10.1071/MU00069.

Thiébaut, G., Tarayre, M. and Rodríguez-Pérez, H. (2019) Allelopathic effects of native versus invasive plants on one major invader. *Frontiers in Plant Science* 10. DOI: 10.3389/fpls.2019.00854.

Timóteo, S., O'Connor, C.J., López-Núñez, F.A., Costa, J.M., Gouveia, A.C. *et al.* (2018) Pollination networks from natural and anthropogenic-novel communities show high structural similarity. *Oecologia* 188(4), 1155–1165. DOI: 10.1007/s00442-018-4281-5.

Traveset, A. and Richardson, D.M. (2006) Biological invasions as disruptors of plant reproductive mutualisms. *Trends in Ecology & Evolution* 21(4), 208–216. DOI: 10.1016/j.tree.2006.01.006.

Traveset, A. and Richardson, D.M. (2011) *Mutualisms: Key Drivers of Invasions. Key Casualties of Invasions*. Wiley-Blackwell, Oxford, UK.

Traveset, A. and Richardson, D.M. (2014) Mutualistic interactions of invasive species. *Annual Review of Ecology, Evolution and Systematics* 45, 89–113.

Traveset, A., Heleno, R.H. and Nogales, M. (2014) The ecology of seed dispersal. In: Gallagher, R.S. (ed.) *Seeds: The ecology of regeneration in plant communities*. CABI, Oxfordshire, UK, pp. 62–93.

Traveset, A., Tur, C., Trøjelsgaard, K., Heleno, R., Castro-Urgal, R. *et al.* (2016) Global patterns of mainland and insular pollination networks. *Global Ecology and Biogeography* 25(7), 880–890. DOI: 10.1111/geb.12362.

van der Pijl, L. (1982) *Principles of dispersal in higher plants*. Springer-Verlag, Berlin.

Vilà, M. and D'Antonio, C.M. (1998) Fruit choice and seed dispersal of invasive vs. noninvasive carpobrotus (Aizoaceae) in coastal California. *Ecology* 79(3), 1053–1060. DOI: 10.1890/0012-9658(1998)079[1053:FCASDO]2.0.CO;2.

Vázquez, D.P., Chacoff, N.P. and Cagnolo, L. (2009) Evaluating multiple determinants of the structure of plant–animal mutualistic networks. *Ecology* 90(8), 2039–2046. DOI: 10.1890/08-1837.1.

Vizentin-Bugoni, J., Tarwater, C.E., Foster, J.T., Drake, D.R., Gleditsch, J.M. *et al.* (2019) Structure, spatial dynamics, and stability of novel seed dispersal mutualistic networks in Hawai'i. *Science* 364(6435), 78–82. DOI: 10.1126/science.aau8751.

Wang, B.C. and Smith, T.B. (2002) Closing the seed dispersal loop. *Trends in Ecology & Evolution* 17(8), 379–386. DOI: 10.1016/S0169-5347(02)02541-7.

Wenny, D.G. (2001) Advantages of seed dispersal: a re-evaluation of directed dispersal. *Evolutionary Ecology Research* 3, 51–74.

White, D.W. and Stiles, E.W. (1992) Bird dispersal of fruits of species introduced into eastern North America. *Canadian Journal of Botany* 70(8), 1689–1696. DOI: 10.1139/b92-208.

Williams, P.A. (2006) The role of blackbirds (Turdus merula) in weed invasion in New Zealand. *New Zealand Journal of Ecology* 30, 285–291.

Williams, P.A. and Karl, B.J. (1996) Fleshy fruits of indigenous and adventive plants in the diet of birds in forest remnants, Nelson, New Zealand. *New Zealand Journal of Ecology* 20, 127–145.

Williams, P.A. and Karl, B.J. (2002) Birds and small mammals in kanuka (Kunzea ericoides) and gorse (*Ulex europaeus*) scrub and the resulting seed rain and seedling dynamics. *New Zealand Journal of Ecology*, 3ndn1–41.

15 Allelopathic Disruptions of Biotic Interactions Due to Non-native Plants

Lauren M. Smith-Ramesh*

National Institute for Mathematical and Biological Synthesis, Knoxville, TN USA

Abstract

Allelopathy, or the process by which plants influence the growth and performance of their neighbours through the release of chemicals, may play a key role in mediating the impacts of non-native invasive species on their neighbours. The Novel Weapons Hypothesis purports that non-native invasive species are in part successful because they produce harmful allelochemicals to which resident species are particularly susceptible because residents lack a shared evolutionary history with the invader. While allelopathic non-native invaders may reduce the growth and performance of neighbours through direct phytotoxicity, they may more often exert negative impacts through disruption of biotic interactions among resident species. Allelopathy by non-native plants may disrupt mutualisms between resident plants and microbes, plant–herbivore interactions or existing competitive and facilitative interactions among resident plants. For example, several non-native plants are known to disrupt the mutualism between resident plants and mycorrhizal fungi, reducing resident plant fitness to the benefit of the invader. Allelopathic non-natives may also disrupt interactions among resident plants and their herbivores when allelochemicals also influence herbivore behaviour or fitness. Alternatively, biotic interactions can also be protective for resident species, which may be less susceptible to the impacts of non-native species when their mutualisms are intact. As we advance our understanding of allelopathy and its role in mediating the impacts of invasive plant species, we may gain new insights by viewing invasions within a network context rather than focusing on pairwise interactions.

15.1 Introduction

Allelopathy, the process by which plants exhibit positive or negative effects on the growth and performance of their neighbours through the release of chemical compounds, is often invoked to explain the success of non-native invasive plant species. The term 'allelopathy', derived from *allelon* (mutual) and *pathos* (harm or affection), was originally coined to describe the effects of ethylene gas on fruit ripening (Molisch, 1937; Inderjit and Duke, 2003). While the word allelopathy may have initially described mutual benefit or harm imposed by plants on one another, as expected in coevolved plant communities characterized by interactions such as facilitation or interference competition through chemical 'arms races', allelopathic effects of invasive plant species are often hypothesized to be more one-sided. This phenomenon, in which the invader gains a unilateral advantage through allelochemicals that exert strong negative effects on naïve native species, is known as the Novel Weapons Hypothesis (Callaway and Ridenour, 2004). While the role of allelopathy via direct

*laurensmithramesh@gmail.com

 Plant Invasions: The Role of Biotic Interactions (eds Anna Traveset and David M. Richardson)
DOI: 10.1079/9781789242171.0015

phytotoxicity has been called into question in recent decades (Uren, 2007; Willis, 2007), a growing body of evidence suggests that allelopathy mediated by indirect biotic interactions, particularly via fungal mutualists, may be an important driver of several plant invasions. Here, I briefly discuss a few key ways that allelopathic invaders are known to disrupt biotic interactions among plants and other organisms. I then discuss the future of research on allelopathy and invasion, with an emphasis on how allelopathic invaders that disrupt biotic interactions can alter the structure of species interaction networks in their invaded range.

15.2 Defining Allelopathy

Allelopathy in plants is typically defined as the release of compounds that enter the rhizosphere as root exudates or leachates from aboveground tissue and influence the fitness of co-occurring plants (Willis, 2007). Allelopathy can also include chemical compounds that travel through the air (i.e. volatile organic compounds) to influence neighbouring plants (Souza-Alonso *et al.*, 2018). Allelochemicals can be directly phytotoxic (i.e. +/- catechin) or inhibit plant fitness through indirect interactions mediated by microbes (i.e. glucosinolates), or both (Cipollini *et al.*, 2012).

Allelopathy is commonly suggested as a mechanism driving plant invasion (for some leading examples see Callaway and Aschehoug, 2000; Roberts and Anderson, 2001; Dorning and Cipollini, 2006; Abhilasha *et al.*, 2008), but invoking this factor has been controversial due to questions about whether allelochemicals can travel through soils and persist long enough to impose harm on neighbouring plants (Uren, 2007; Willis, 2007; Duke, 2010). Furthermore, allelopathy research has been plagued by methodological issues in the past, including using unrealistically high concentrations of allelochemicals in experiments, forcing the release of chemicals, and using substrates such as filter paper, gels or artificial soils rather than natural soils (Duke, 2010; Cipollini *et al.*, 2012). However, in recent decades, improved methodological approaches as well as consideration of edaphic properties and interactions with rhizosphere microbes when assessing allelopathy as an invasion mechanism have improved confidence in the ecological importance of allelopathy (Cipollini *et al.*, 2012; Parepa and Bossdorf, 2016). Extensive reviews of allelopathy in ecological systems have been published elsewhere (for a few examples, see Hierro and Callaway, 2003; Bais *et al.*, 2006; Cipollini *et al.*, 2012).

15.3 Allelopathy and Invasion: The Novel Weapons Hypothesis

While allelopathy has been long studied as a form of interference competition between plants, microbes and even marine benthic invertebrates, its role as a mechanism driving plant invasions came about more recently (Hierro and Callaway, 2003). The Novel Weapons Hypothesis (NWH) describes invasion by allelopathic invaders that gain an advantage they would not normally experience in their native range because recipient communities are 'naïve' and are unable to tolerate allelochemicals from plants with which they share no evolutionary history (Callaway and Aschehoug, 2000, Hierro and Callaway, 2003; Callaway and Ridenour, 2004; Chapter 3, this volume). Several biogeographic studies have backed up the NWH, demonstrating that naïve communities are more susceptible to invader allelochemicals than co-occurring plants in the allelopath's native range (Callaway and Aschehoug, 2000; Becerra *et al.*, 2018). While the potential of allelopathy to play a key role in invasion for at least some introduced plant species is well established (Table 15.1), its impact on native species interaction networks, evolutionary responses to allelopathic effects of invaders on natives (and vice versa), and impacts on ecosystem function are all ripe areas for future research.

15.4 Allelochemical Disruptions of Biotic Interactions

Allelopathic invaders can disrupt biotic interactions among native species in several ways. Most research to date focuses on plant–plant interactions (interference competition), plant–microbe

Table 15.1. A sampling of allelopathic invasive plant species with key references. Allelopathic invaders occur globally and include representatives from diverse plant families that produce many different classes of allelochemical.

Species	Invaded range	Origin	Allelochemical	References
Acacia dealbata	Widespread (Africa, Europe, South America)	Australia	Unknown, several candidates	(Lorenzo *et al.*, 2008, Lorenzo *et al.*, 2011; Aguilera *et al.*, 2015)
Ageratina altissima	East Asia	Eastern North America	Phenolics (scopoletin)	(Kim and Lee, 2011)
Ageratina adenophora	East and South Asia	North America (Mexico)	Volatile organic compounds (VOCs)	(Inderjit *et al.*, 2011)
Alliaria petiolata	Eastern North America	Eurasia	Glucosinolates (sinigrin)	(Vaughn and Berhow, 1999; Cipollini and Cipollini, 2016)
Centaurea spp.	North America	Europe	Flavonoids (+/- catechin)	(Callaway and Aschehoug, 2000)
Carduus nutans	North America, Australia, New Zealand	Eurasia	Aplotaxene	(Wardle *et al.*, 1994; Silva *et al.*, 2014)
Erigeron canadensis	Widespread (Eurasia, Australia)	North America	Phenolics	(Djurdjević *et al.*, 2012)
Solidago canadensis *Solidago gigantea*	East Asia, Europe	North America	Polyacetylene, diterpene lactone derivatives	(Abhilasha *et al.*, 2008) (Pal *et al.*, 2015)

interactions and plant–herbivore interactions (Fig. 15.1). Each of these categories is detailed below.

15.5 Plant–Plant Interactions: Direct Phytotoxicity

The most straightforward way in which allelochemicals can influence biotic interactions involving invasive plants is through direct plant–plant interactions. By releasing phytotoxic chemicals into soils via root exudation, litter decomposition, or leachates, invasive plants may reduce competitor fitness and gain an advantage in resource competition against native species. For allelopathy to yield a net competitive advantage to the invader, the compound of interest must have a stronger negative effect on heterospecific competitors than on conspecifics. This appears to be the case for a number of invasive plants examined, such as bush honeysuckle (*Lonicera maackii*), whose root exudates are toxic to commonly co-occurring native species but are not autotoxic (Dorning and Cipollini, 2006). Allelopathy may also be a driving mechanism behind the invasion of diffuse knapweed (*Centaurea diffusa*) and spotted knapweed (*Centaurea stoebe*, formerly *Centaurea maculosa*), potentially via direct phytotoxicity of 8-hydroxyquinoline and (+/-)-catechin, respectively (Callaway and Aschehoug, 2000; Callaway and Ridenour, 2004; Vivanco *et al.*, 2004).

Studies that conclude that direct phytotoxicity of allelochemicals drive invader impacts have often been criticized on the basis that many purported allelochemicals cannot travel through soil and are often degraded by microbes so quickly that a direct phytotoxic effect is unlikely (Uren, 2007; Willis, 2007). However, allelochemicals that are processed by microbes in the rhizosphere may indeed have significant effects on plant fitness if plant–microbe interactions are disrupted by allelochemicals (Inderjit, 2005; Cantor *et al.*, 2011; Cipollini *et al.*, 2012).

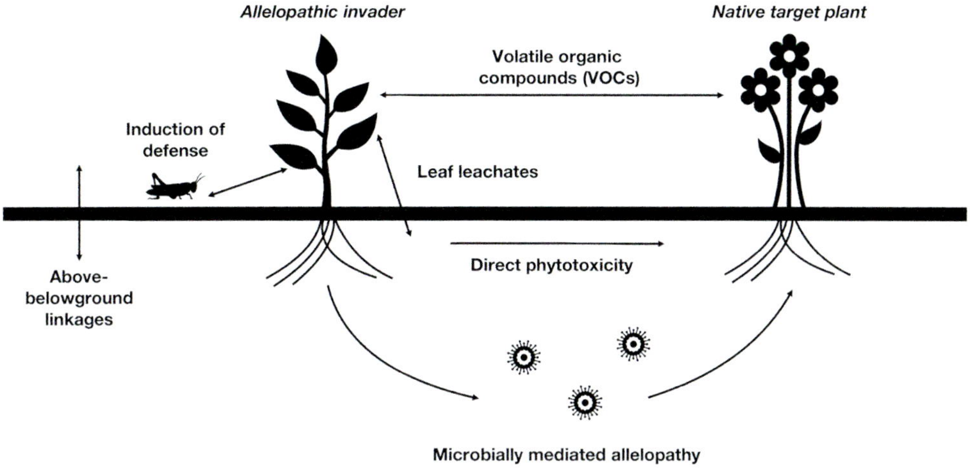

Fig. 15.1. Allelopathic invasive plants can disrupt biotic interactions through several pathways. Allelochemicals reach the soil through direct root exudation or as leachates from live or decomposing leaves, or travel between plants aboveground in the form of volatile organic compounds (VOCs). Allelopathy can alter the outcome of direct competition between plants via phytotoxicity, where allelochemicals directly reduce the ability of a native target plant to grow and take up nutrients. Allelopathy can also indirectly inhibit target plant growth through a microbe-mediated pathway, for example by disrupting the interaction between native plants and mutualists such as mycorrhizal fungi. Because many allelochemicals that are phytotoxic or antimicrobial may also act as antiherbivore defence, allelopathic invaders may influence herbivore behaviour or survival. In turn, herbivores both above- and below-ground may influence allelopathy through induction of chemical defence, resulting in increased inputs of allelochemicals through litter inputs from aboveground tissue or increased root exudation resulting from above- and below-ground linkages in plant physiology. Other intrinsic and extrinsic factors may also influence allelochemical production, such as plant phenological stage or abiotic conditions (e.g. light and nutrient availability).

15.6 Plant–Microbe Interactions

Plant–microbe interactions have emerged as key factors structuring plant communities (Reynolds *et al.*, 2003; Mangan *et al.*, 2010; Schnitzer *et al.*, 2011) and appear to play key roles in driving invasions and mediating native community responses to invasion (Klironomos, 2002; Pringle *et al.*, 2009; also see Chapter 3, this volume). Microbial mutualists, such as the rhizobia that enable nitrogen fixation in association with legumes, or the mycorrhizal fungi that aid in nutrient uptake, defence and plant–water relations, are key drivers of plant community structure and ecosystem function (van der Heijden *et al.*, 1998; van der Heijden and Sanders, 2002; Rillig, 2004; Pringle *et al.*, 2009; Delavaux *et al.*, 2017). Microbial pathogens also act as key determinants of plant community structure (Packer and Clay, 2000; Reinhart *et al.*, 2003; Clay *et al.*, 2008).

Leguminous plants depend upon associated nitrogen-fixing rhizobia for access to nitrogen, and act as an important nitrogen flux into terrestrial ecosystems (Palacios and Verma, 1988). When invaders disrupt the biotic interaction between legumes and rhizobia, they can not only inhibit the success of these plant species, but also potentially alter nitrogen availability on an ecosystem scale. Such a phenomenon has been documented for the allelopathic invader *C. nutans*, whose exudates inhibit nodulation of white clover (*Trifolium repens*) by rhizobia. The disruption of this mutualism significantly reduces the performance of legumes in areas invaded by *Carduus nutans*, indicating that this disrupted biotic interaction is likely to alter plant community structure as well as ecosystem functioning

(Wardle *et al.*, 1994). Recent evidence suggests that root exudates of garlic mustard (*Alliaria petiolata*) can inhibit nodule formation in legumes (Portales-Reyes *et al.*, 2015). Other weedy species have been observed to reduce nodulation of legumes by rhizobia, although most of these observations come from agricultural systems (Cipollini *et al.*, 2012).

In addition to disrupting the mutualism between legumes and rhizobia, several invaders are known to disrupt the mutualism between native plants and mycorrhizal fungi. The vast majority of vascular plants globally participate in mutualisms with mycorrhizal fungi, including arbuscular mycorrhizal fungi (AMF), ectomycorrhizal fungi (EMF) and ericoid mycorrhizal fungi (van der Heijden and Sanders, 2002). Because these mutualisms are both ubiquitous and important to plant growth and performance, invaders that release allelochemicals that disrupt plant–fungal mutualisms may have powerful impacts on native communities (Reinhart and Callaway, 2006; Müller, 2009; van der Putten, 2010). Members of the family Brassicaceae, which are some of the relatively small proportion of terrestrial plants that generally do not associate with fungal mutualists, often produce a class of antifungal compounds called glucosinolates that can disrupt the plant–fungal mutualism. Well known allelopathic invaders from the Brassicaceae include *Brassica nigra*, which invades grasslands in western North America, and *A. petiolata*, which invades deciduous forests in eastern North America (Müller, 2009). The role of allelopathy in invasion by *A. petiolata* (garlic mustard) is detailed in Box 15.1.

Allelochemicals that disrupt plant–microbe interactions will alter plant community structure, ecosystem functioning and the evolutionary trajectory of native plants, microbes and invaders. As naïve native communities are exposed to new allelochemicals from invasive plants, we may see the replacement of native species that depend upon microbial mutualisms with less dependent species. We may see changes in resource availability as microbes involved in nutritive mutualisms are negatively affected by allelochemicals (Wardle *et al.*, 1994). Finally, we may see native plants and microbes adapting to invasive allelochemicals, either when native plants become less dependent upon microbial mutualists, or native microbes become more tolerant to allelochemicals (Callaway *et al.*, 2005; Lankau and Nodurft, 2013). A complete understanding of the impact of these allelopathic invaders will require a multifaceted approach to understand how each of these factors interact to determine plant community structure and ecosystem function.

15.7 Plant–Herbivore Interactions

While we generally do not define antiherbivore defence as allelopathy, many secondary metabolites that have allelopathic effects can also deter herbivores both above- and below-ground. This antiherbivore effect can grant invasive species an advantage through enemy release when naïve native herbivores avoid chemically defended invaders and instead consume native forage. This phenomenon is well documented for garlic mustard (*A. petiolata*), where the glucosinolates that act as allelochemicals to disrupt the mutualism between mycorrhizal fungi and native plants also act as strong chemical defences against deer, mollusc and insect herbivory (Knight *et al.*, 2009; Müller, 2009; Hahn and Dornbush, 2012; Dornbush and Hahn, 2013). The herbivore-mediated advantage granted by garlic mustard's allelochemicals has been shown to be important in its ability to invade native forests, to the extent that the invader has a negative population growth rate in the absence of the herbivore and positive population growth when the herbivore is present (Kalisz *et al.*, 2014).

Because of the dual role of many secondary metabolites as allelochemicals and antiherbivore compounds, the production and release of allelochemicals can be highly influenced by interaction with herbivores. Leaf and root herbivory can induce increased production of allelochemicals, which can subsequently influence plants and microbes when those chemicals are exuded into the soil or leaf litter decomposes and releases chemicals (Thelen *et al.*, 2005). Due to above- and below-ground linkages in plant physiology, induction of chemical defence aboveground can influence chemical production belowground as well, either increasing or decreasing root chemical content depending on the physiology of secondary metabolite production (Bardgett *et al.*,

Box 15.1. Disrupted plant–fungal interactions: garlic mustard as a case study.

Garlic mustard (*Alliaria petiolata*) invades deciduous forest ecosystems in eastern North America. As its name suggests, garlic mustard is a member of the Brassicaceae (mustard) family, and produces a class of sulphur-containing compounds known as glucosinolates (Rodgers *et al.*, 2008; Cipollini and Cipollini, 2016). Glucosinolates give mustards, such as the cruciferous vegetables we consume, their distinct flavour, and serve as defence against herbivory on both above- and below-ground tissue (Knight *et al.*, 2009; Hahn and Dornbush, 2012; Dornbush and Hahn, 2013). Glucosinolates also act as allelochemicals, inhibiting germination (Prati and Bossdorf, 2004) and exhibiting antifungal properties that disrupt mutualisms between mycorrhizal fungi and native plants (Vaughn and Berhow, 1999; Roberts and Anderson, 2001; Stinson *et al.*, 2006; Cantor *et al.*, 2011; Wolfe *et al.*, 2008; Cantor *et al.*, 2011). Fungi in garltoric mustard's invaded range appear to be more susceptible to its allelochemicals than fungi in its native Eurasia, making garlic mustard's invasion consistent with the Novel Weapons Hypothesis (Callaway *et al.*, 2008; Barto *et al.*, 2010).

Glucosinolates, when broken down by the enzyme myrosinase, yield antifungal compounds known as isothiocyanates that inhibit colonization of roots by mycorrhizal fungi (Schreiner and Koide, 1993). Detailed microbial community analyses show that garlic mustard reduces the relative abundance of ectomycorrhizal fungi in the soil relative to saprotrophic and pathophytic fungi, and reduces beta diversity of soil fungi (Anthony *et al.*, 2017). Garlic mustard also reduces colonization of plant roots by arbuscular mycorrhizal fungi (Roberts and Anderson, 2001; Barto *et al.*, 2011). Studies have demonstrated that disruption of the plant–fungal mutualism reduces native tree growth (Stinson *et al.*, 2006), inhibits physiological function in native understorey plants (Hale *et al.*, 2011, Hale *et al.*, 2016; Brouwer *et al.*, 2015) and results in legacy effects that inhibit forest understorey restoration following invader removal (Anderson *et al.*, 2014).

By disrupting the plant–fungal mutualism, invaders such as garlic mustard may shift native communities in favour of plant species that are less dependent upon fungal mutualists, or depend upon classes of fungal mutualists that are less susceptible to the effects of glucosinolates. Given that many native plant species that are highly dependent upon mutualisms with mycorrhizal fungi are conservative, slow-growing perennial species, these impacts are worrisome from a conservation perspective. In addition to changing community composition, both the invader and native communities are expected to evolve in response to changing selection pressures over the course of invasion. Glucosinolate content varies in natural populations of garlic mustard (Smith and Reynolds, 2015), and selection appears to favour reduced allelochemical production over the course of an invasion (Lankau *et al.*, 2009). Researchers have documented the evolution of reduced mycorrhizal dependence among native plant species that have been exposed to garlic mustard invasion (Lankau, 2012, 2013; Lankau and Nodurft, 2013), as well as variation in the susceptibility of arbuscular mycorrhizal fungi to garlic mustard's allelochemicals (Barto *et al.*, 2011). As research on this focal species continues, our challenge will be to unravel how all of these changes – to the native plant community, the native microbial community, and the genetics of the invader and natives – ultimately work together to determine the impact of invasion on native biodiversity and community structure.

1998). Therefore, the presence and intensity of herbivory on invasive species may influence the role that allelopathy plays in driving invasion.

15.8 Biotic Interactions as Defence Against Allelopathic Invaders

While I focus here on how allelochemicals can disrupt biotic interactions, we should also note that there are well-documented cases where biotic interactions among native species can be protective against the effects of allelopathic plants. For example, grasses associated with mutualistic endophytic fungi may be more resistant to allelopathy than grasses that are not infected with fungal associates (Bao *et al.*, 2015). Rhizosphere fungi and bacteria can break down and detoxify allelochemicals, thereby protecting associated plants (Cipollini *et al.*, 2012). Native or exotic enemies that interact with invaders directly can also modulate allelopathy. For example, if an allelopathic invader is subject to intense herbivory, it may have reduced resources to put towards allelochemical production and release, or may have reduced biomass from which allelochemicals can leach from above-ground tissue into soils (Smith-Ramesh and Reynolds, 2017; Heinze *et al.*, 2019; Bennett *et al.*, 2020).

15.9 Viewing Allelopathic Invader Impacts Through the Lens of Network Theory

Recent calls to action have emphasized the need to examine biological invasions through the lens of network theory (Hui and Richardson, 2017, 2019; Frost *et al.*, 2019). Species interaction networks can provide powerful ways to view the causes and consequences of biological invasion in a broader community-level context. Furthermore, examining invasion within a complex adaptive network framework allows consideration of evolution and environmental change (Hui and Richardson, 2017). Many past examinations of invaders in a network context have focused on food webs and trophic interactions (Romanuk *et al.*, 2009; Baiser *et al.*, 2010; Lurgi *et al.*, 2014; Smith-Ramesh *et al.*, 2017). However, allelopathic invaders that disrupt biotic interactions will also be particularly interesting to study within a network context because they can alter the pattern of mutualistic interactions within a community.

Because allelopathic invaders are chemically defended, they may interact less with pathogens, herbivores and mutualists compared to non-allelopathic exotics, thus having unique impacts on network structure. We might predict that allelopathic invaders may reduce network connectance by disrupting interactions that previously existed, even if they do not alter biodiversity. For example, if allelopathic invaders alter community structure in such a way that native plants that depend on mycorrhizal mutualisms are replaced with plant species that perform well without fungal partners, those plant–microbe links may be lost from the interaction network. Future work should examine whether allelopathic invaders do indeed reduce food web complexity (Fig. 15.2 for a simplified example), as well as how they influence network properties such as mean interaction strength, modularity and nestedness. By determining how allelopathic invaders uniquely alter species interaction networks in their invaded range, we can better predict whether they will weaken the robustness and stability of native species interaction networks, potentially leading to species losses, subsequent invasions or changes to ecosystem function.

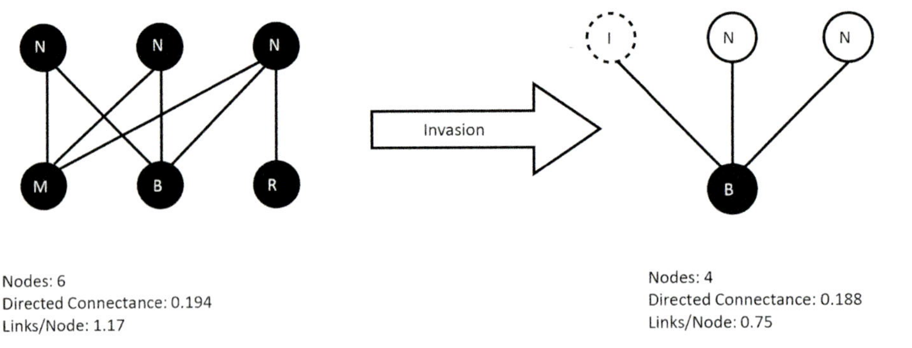

Fig. 15.2. Allelopathic invaders that disrupt biotic interactions will alter the structure of species interaction networks within invaded communities. This simplified example shows a mutualistic network prior to invasion, where three example native plant species (N) (top row of filled nodes), interact with mycorrhizal fungi (M), rhizosphere bacteria (B) and rhizobia (R). After the introduction of an allelopathic invader (I, unfilled node with dashed border), disrupted biotic interactions result in the loss of several microbial mutualists and the replacement of native plant species that depend upon microbial mutualisms with different species that are not dependent upon mutualists (N, unfilled nodes). This alteration of community structure results in a new mutualistic network with fewer species, lower directed connectance (L/S^2), and fewer links per species (L/S).

References

Abhilasha, D., Quintana, N., Vivanco, J. and Joshi, J. (2008) Do allelopathic compounds in invasive Solidago canadensis s.l. restrain the native European flora? *Journal of Ecology* 96(5), 993–1001. DOI: 10.1111/j.1365-2745.2008.01413.x.

Aguilera, N., Becerra, J., Villaseñor-Parada, C., Lorenzo, P., González, L. *et al*. (2015) Effects and identification of chemical compounds released from the invasive *Acacia dealbata* Link. *Chemistry and Ecology* 31(6), 479–493. DOI: 10.1080/02757540.2015.1050004.

Anderson, R.C., Anderson, M.R., Bauer, J.T., Slater, M., Herold, J. *et al*. (2014) Effect of removal of garlic mustard (Alliaria petiolata, Brassicaeae) on arbuscular mycorrhizal fungi inoculum potential in forest soils. *The Open Ecology Journal* 3(1), 41–47. DOI: 10.2174/1874213001003010041.

Anthony, M.A., Frey, S.D. and Stinson, K.A. (2017) Fungal community homogenization, shift in dominant trophic guild, and appearance of novel taxa with biotic invasion. *Ecosphere* 8(9), e01951. DOI: 10.1002/ecs2.1951.

Bais, H.P., Weir, T.L., Perry, L.G., Gilroy, S. and Vivanco, J.M. (2006) The role of root exudates in rhizosphere interactions with plants and other organisms. *Annual Review of Plant Biology* 57, 233–266. DOI: 10.1146/annurev.arplant.57.032905.105159.

Baiser, B., Russell, G.J. and Lockwood, J.L. (2010) Connectance determines invasion success via trophic interactions in model food webs. *Oikos* 119(12), 1970–1976. DOI: 10.1111/j.1600-0706.2010.18557.x.

Bao, G., Saikkonen, K., Wang, H., Zhou, L., Chen, S. *et al*. (2015) Does endophyte symbiosis resist allelopathic effects of an invasive plant in degraded grassland? *Fungal Ecology* 17, 114–125. DOI: 10.1016/j.funeco.2015.05.006.

Bardgett, R.D., Wardle, D.A. and Yeates, G.W. (1998) Linking above-ground and below-ground interactions: how plant responses to foliar herbivory influence soil organisms. *Soil Biology and Biochemistry* 30(14), 1867–1878. DOI: 10.1016/S0038-0717(98)00069-8.

Barto, E.K., Antunes, P.M., Stinson, K., Koch, A.M., Klironomos, J.N. *et al*. (2011) Differences in arbuscular mycorrhizal fungal communities associated with sugar maple seedlings in and outside of invaded garlic mustard forest patches. *Biological Invasions* 13(12), 2755–2762. DOI: 10.1007/s10530-011-9945-6.

Barto, E.K., Powell, J.R. and Cipollini, D. (2010) How novel are the chemical weapons of garlic mustard in North American forest understories? *Biological Invasions* 12(10), 3465–3471. DOI: 10.1007/s10530-010-9744-5.

Becerra, P.I., Catford, J.A., Luce McLeod, M., Andonian, K., Aschehoug, E.T. *et al*. (2018) Inhibitory effects of *Eucalyptus globulus* on understorey plant growth and species richness are greater in non-native regions. *Global Ecology and Biogeography* 27(1), 68–76. DOI: 10.1111/geb.12676.

Bennett, S.I., Howard, C., Albrecht, R., Smith-Ramesh, L.M. and Reynolds, H. (2020) Simulated herbivory weakens Plant-Soil feedbacks in competitive mixtures of native and invasive woodland plants. *Frontiers in Ecology and Evolution* 7, 497. DOI: 10.3389/fevo.2019.00497.

Brouwer, N.L., Hale, A.N. and Kalisz, S. (2015) Mutualism-disrupting allelopathic invader drives carbon stress and vital rate decline in a forest perennial herb. *AoB PLANTS* 7, pii:plv014. DOI: 10.1093/aobpla/plv014.

Callaway, R.M. and Aschehoug, E.T. (2000) Invasive plants versus their new and old neighbors: a mechanism for exotic invasion. *Science* 290(5491), 521–523. DOI: 10.1126/science.290.5491.521.

Callaway, R.M., Cipollini, D., Barto, K., Thelen, G.C., Hallett, S.G. *et al*. (2008) Novel weapons: invasive plant suppresses fungal mutualists in America but not in its native Europe. *Ecology* 89(4), 1043–1055. DOI: 10.1890/07-0370.1.

Callaway, R.M. and Ridenour, W.M. (2004) Novel weapons: invasive success and the evolution of increased competitive ability. *Frontiers in Ecology and the Environment* 2(8), 436–443. DOI: 10.1890/1540-9295(2004)002[0436:NWISAT]2.0.CO;2.

Callaway, R.M., Ridenour, W.M., Laboski, T., Weir, T. and Vivanco, J.M. (2005) Natural selection for resistance to the allelopathic effects of invasive plants. *Journal of Ecology* 93(3), 576–583. DOI: 10.1111/j.1365-2745.2005.00994.x.

Cantor, A., Hale, A., Aaron, J., Traw, M.B. and Kalisz, S. (2011) Low allelochemical concentrations detected in garlic mustard-invaded forest soils inhibit fungal growth and AMF spore germination. *Biological Invasions* 13(12), 3015–3025. DOI: 10.1007/s10530-011-9986-x.

Cipollini, D. and Cipollini, K. (2016) A review of garlic mustard (*Alliaria petiolata*, Brassicaceae) as an allelopathic plant. *The Journal of the Torrey Botanical Society* 143(4), 339–348. DOI: 10.3159/TORREY-D-15-00059.

Cipollini, D., Rigsby, C.M. and Barto, E.K. (2012) Microbes as targets and mediators of allelopathy in plants. *Journal of Chemical Ecology* 38(6), 714–727. DOI: 10.1007/s10886-012-0133-7.

Clay, K., Reinhart, K., Rudgers, J., Tintjer, T., Koslow, J. *et al.* (2008) Red Queen communities. In: Ostfeld, R., Keesing, F. and Eviner, V. (eds) *Infectious Disease Ecology: Effects of Ecosystems on Disease and of Disease on Ecosystems*. Princeton University Press, Princeton, NJ, pp. 145–178.

Delavaux, C.S., Smith-Ramesh, L.M. and Kuebbing, S.E. (2017) Beyond nutrients: a meta-analysis of the diverse effects of arbuscular mycorrhizal fungi on plants and soils. *Ecology* 98(8), 2111–2119. DOI: 10.1002/ecy.1892.

Djurdjević, L., Gajić, G., Kostić, O., Jarić, S., Pavlović, M. *et al.* (2012) Seasonal dynamics of allelopathically significant phenolic compounds in globally successful invader Conyza canadensis L. plants and associated sandy soil. *Flora - Morphology, Distribution, Functional Ecology of Plants* 207(11), 812–820. DOI: 10.1016/j.flora.2012.09.006.

Dornbush, M.E. and Hahn, P.G. (2013) Consumers and establishment limitations contribute more than competitive interactions in sustaining dominance of the exotic herb garlic mustard in a Wisconsin, USA forest. *Biological Invasions* 15(12), 2691–2706. DOI: 10.1007/s10530-013-0484-1.

Dorning, M. and Cipollini, D. (2006) Leaf and root extracts of the invasive shrub, Lonicera maackii, inhibit seed germination of three herbs with no autotoxic effects. *Plant Ecology* 184(2), 287–296. DOI: 10.1007/s11258-005-9073-4.

Duke, S.O. (2010) Allelopathy: current status of research and future of the discipline: a commentary. *Allelopathy Journal* 25, 17–29.

Evans, H., Crocoll, C., Bajpai, D., Kaur, R., Feng, R. *et al.* (2011) Volatile chemicals from leaf litter are associated with invasiveness of a Neotropical weed in Asia. *Ecology* 92(2), 316–324. DOI: 10.1890/10-0400.1.

Frost, C.M., Allen, W.J., Courchamp, F., Jeschke, J.M., Saul, W.-C. *et al.* (2019) Using network theory to understand and predict biological invasions. *Trends in Ecology & Evolution* 34(9), 831–843. DOI: 10.1016/j.tree.2019.04.012.

Hahn, P.G. and Dornbush, M.E. (2012) Exotic consumers interact with exotic plants to mediate native plant survival in a midwestern forest herb layer. *Biological Invasions* 14(2), 449–460. DOI: 10.1007/s10530-011-0089-5.

Hale, A., Lapointe, L. and Kalisz, S. (2016) Invader disruption of a mycorrhizal-plant mutualism reduces carbon acquisition, allocation and growth in a native forest herb. *The New phytologist* 209, 542–549.

Hale, A.N., Tonsor, S.J. and Kalisz, S. (2011) Testing the mutualism disruption hypothesis: physiological mechanisms for invasion of intact perennial plant communities. *Ecosphere* 2(10), art110. DOI: 10.1890/ES11-00136.1.

Heinze, J., Simons, N.K., Seibold, S., Wacker, A., Weithoff, G. *et al.* (2019) The relative importance of plant-soil feedbacks for plant-species performance increases with decreasing intensity of herbivory. *Oecologia* 3, 651–664.

Hierro, J.L. and Callaway, R.M. (2003) Allelopathy and exotic plant invasion. *Plant and Soil* 256(1), 29–39. DOI: 10.1023/A:1026208327014.

Hui, C. and Richardson, D.M. (2017) *Invasion Dynamics*. Oxford University Press, Oxford.

Hui, C. and Richardson, D.M. (2019) How to invade an ecological network. *Trends in Ecology & Evolution* 34(2), 121–131. DOI: 10.1016/j.tree.2018.11.003.

Inderjit. (2005) Soil microorganisms: an important determinant of allelopathic activity. *Plant and Soil* 274(1-2), 227–236. DOI: 10.1007/s11104-004-0159-x.

Inderjit. and Duke, S.O. (2003) Ecophysiological aspects of allelopathy. *Planta* 217(4), 529–539. DOI: 10.1007/s00425-003-1054-z.

Kalisz, S., Spigler, R.B. and Horvitz, C.C. (2014) In a long-term experimental demography study, excluding ungulates reversed invader's explosive population growth rate and restored natives. *Proceedings of the National Academy of Sciences* 111(12), 4501–4506. DOI: 10.1073/pnas.1310121111.

Kim, Y.O. and Lee, E.J. (2011) Comparison of phenolic compounds and the effects of invasive and native species in East Asia: support for the novel weapons hypothesis. *Ecological Research* 26(1), 87–94. DOI: 10.1007/s11284-010-0762-7.

Klironomos, J.N. (2002) Feedback with soil biota contributes to plant rarity and invasiveness in communities. *Nature* 417(6884), 67–70. DOI: 10.1038/417067a.

Knight, T.M., Dunn, J.L., Smith, L.A., Davis, J. and Kalisz, S. (2009) Deer facilitate invasive plant success in a Pennsylvania forest Understory. *Natural Areas Journal* 29(2), 110–116. DOI: 10.3375/043.029.0202.

Lankau, R.A. (2012) Coevolution between invasive and native plants driven by chemical competition and soil biota. *Proceedings of the National Academy of Sciences* 109(28), 11240–11245. DOI: 10.1073/pnas.1201343109.

Lankau, R.A. (2013) Species invasion alters local adaptation to soil communities in a native plant. *Ecology* 94(1), 32–40. DOI: 10.1890/12-0675.1.

Lankau, R.A. and Nodurft, R.N. (2013) An exotic invader drives the evolution of plant traits that determine mycorrhizal fungal diversity in a native competitor. *Molecular Ecology* 22(21), 5472–5485. DOI: 10.1111/mec.12484.

Lankau, R.A., Nuzzo, V., Spyreas, G. and Davis, A.S. (2009) Evolutionary limits ameliorate the negative impact of an invasive plant. *Proceedings of the National Academy of Sciences* 106(36), 15362–15367. DOI: 10.1073/pnas.0905446106.

Lorenzo, P., Pazos-Malvido, E., González, L. and Reigosa, M.J. (2008) Allelopathic interference of invasive *Acacia dealbata*: physiological effects. *Allelopathy Journal* 22, 425–462.

Lorenzo, P., Palomera-Pérez, A., Reigosa, M.J. and González, L. (2011) Allelopathic interference of invasive Acacia dealbata link on the physiological parameters of native understory species. *Plant Ecology* 212(3), 403–412. DOI: 10.1007/s11258-010-9831-9.

Lurgi, M., Galiana, N., Lopez, B.C., Joppa, L.N. and Montoya, J.M. (2014) Network complexity and species traits mediate the effects of biological invasions on dynamic food webs. *Frontiers in Ecology and Evolution*.

Mangan, S.A., Schnitzer, S.A., Herre, E.A., Mack, K.M.L., Valencia, M.C. *et al.* (2010) Negative plant-soil feedback predicts tree-species relative abundance in a tropical forest. *Nature* 466(7307), 752–755. DOI: 10.1038/nature09273.

Molisch, H. (1937) *Der Einfluss einer Pflanze auf die andere-Allelopathie*. Fischer.

Müller, C. (2009) Role of glucosinolates in plant invasiveness. *Phytochemistry Reviews* 8(1), 227–242. DOI: 10.1007/s11101-008-9116-2.

Packer, A. and Clay, K. (2000) Soil pathogens and spatial patterns of seedling mortality in a temperate tree. *Nature* 404(6775), 278–281. DOI: 10.1038/35005072.

Pal, R.W., Chen, S., Nagy, D.U. and Callaway, R.M. (2015) Impacts of Solidago gigantea on other species at home and away. *Biological Invasions* 17(11), 3317–3325. DOI: 10.1007/s10530-015-0955-7.

Palacios, R. and Verma, D.P.S. (1988) *Molecular Genetics of Plant-Microbe Interactions*. APS Press, St Paul, Minnesota.

Parepa, M. and Bossdorf, O. (2016) Testing for allelopathy in invasive plants: it all depends on the substrate! *Biological Invasions* 18(10), 2975–2982. DOI: 10.1007/s10530-016-1189-z.

Portales-Reyes, C., Van Doornik, T., Schultheis, E.H. and Suwa, T. (2015) A novel impact of a novel weapon: allelochemicals in Alliaria petiolata disrupt the legume-rhizobia mutualism. *Biological Invasions* 17(9), 2779–2791. DOI: 10.1007/s10530-015-0913-4.

Prati, D. and Bossdorf, O. (2004) Allelopathic inhibition of germination by Alliaria petiolata (Brassicaceae). *American Journal of Botany* 91(2), 285–288. DOI: 10.3732/ajb.91.2.285.

Pringle, A., Bever, J.D., Gardes, M., Parrent, J.L., Rillig, M.C. *et al.* (2009) Mycorrhizal symbioses and plant invasions. *Annual Review of Ecology, Evolution, and Systematics* 40(1), 699–715. DOI: 10.1146/annurev.ecolsys.39.110707.173454.

Reinhart, K.O. and Callaway, R.M. (2006) Soil biota and invasive plants. *New Phytologist* 170(3), 445–457. DOI: 10.1111/j.1469-8137.2006.01715.x.

Reinhart, K.O., Packer, A., Van der Putten, W.H. and Clay, K. (2003) Plant-soil biota interactions and spatial distribution of black cherry in its native and invasive ranges. *Ecology Letters* 6(12), 1046–1050. DOI: 10.1046/j.1461-0248.2003.00539.x.

Reynolds, H.L., Packer, A., Bever, J.D. and Clay, K. (2003) Grassroots ecology: Plant-microbe-soil interactions as drivers of plant community structure and dynamics. *Ecology* 84(9), 2281–2291. DOI: 10.1890/02-0298.

Rillig, M.C. (2004) Arbuscular mycorrhizae and terrestrial ecosystem processes. *Ecology Letters* 7(8), 740–754. DOI: 10.1111/j.1461-0248.2004.00620.x.

Roberts, K.J. and Anderson, R.C. (2001) Effect of garlic mustard [Alliaria petiolata (Beib. Cavara & Grande)] extracts on plants and Arbuscular Mycorrhizal (AM) fungi. *The American Midland Naturalist* 146(1), 146–152. DOI: 10.1674/0003-0031(2001)146[0146:EOGMAP]2.0.CO;2.

Rodgers, V.L., Stinson, K.A. and Finzi, A.C. (2008) Ready or not, garlic mustard is moving in: Alliaria petiolata as a member of eastern North American forests. *BioScience* 58(5), 426–436. DOI: 10.1641/B580510.

Romanuk, T.N., Zhou, Y., Brose, U., Berlow, E.L., Williams, R.J. *et al.* (2009) Predicting invasion success in complex ecological networks. *Philosophical Transactions of the Royal Society B: Biological Sciences* 364(1524), 1743–1754. DOI: 10.1098/rstb.2008.0286.

Schnitzer, S.A., Klironomos, J.N., Hillerislambers, J., Kinkel, L.L., Reich, P.B. *et al.* (2011) Soil microbes drive the classic plant diversity-productivity pattern. *Ecology* 92(2), 296–303. DOI: 10.1890/10-0773.1.

Schreiner, R. and Koide, R.T. (1993) Mustards, mustard oils and mycorrhizas. *New Phytologist* 123(1), 107–113. DOI: 10.1111/j.1469-8137.1993.tb04536.x.

Silva, F.M.L., Donega, M.A., Cerdeira, A.L., Corniani, N., Velini, E.D. *et al.* (2014) Roots of the invasive species Carduus nutans L. and C. acanthoides L. produce large amounts of aplotaxene, a possible allelochemical. *Journal of Chemical Ecology* 40(3), 276–284. DOI: 10.1007/s10886-014-0390-8.

Smith, L.M. and Reynolds, H.L. (2015) Extended leaf phenology, allelopathy, and inter-population variation influence invasion success of an understory forest herb. *Biological Invasions* 17(8), 2299–2313. DOI: 10.1007/s10530-015-0876-5.

Smith-Ramesh, L.M. and Reynolds, H.L. (2017) The next frontier of plant-soil feedback research: unraveling context dependence across biotic and abiotic gradients. *Journal of Vegetation Science* 28(3), 484–494. DOI: 10.1111/jvs.12519.

Smith-Ramesh, L.M., Moore, A.C. and Schmitz, O.J. (2017) Global synthesis suggests that food web connectance correlates to invasion resistance. *Global Change Biology* 23(2), 465–473. DOI: 10.1111/gcb.13460.

Souza-Alonso, P., González, L., López-Nogueira, A., Cavaleiro, C. and Pedrol, N. (2018) Volatile organic compounds of *Acacia longifolia* and their effects on germination and early growth of species from invaded habitats. *Chemistry and Ecology* 34(2), 126–145. DOI: 10.1080/02757540.2017.1404584.

Stinson, K.A., Campbell, S.A., Powell, J.R., Wolfe, B.E., Callaway, R.M. *et al.* (2006) Invasive plant suppresses the growth of native tree seedlings by disrupting belowground mutualisms. *PLoS Biology* 4(5), e140–731. DOI: 10.1371/journal.pbio.0040140.

Thelen, G.C., Vivanco, J.M., Newingham, B., Good, W., Bais, H.P. *et al.* (2005) Insect herbivory stimulates allelopathic exudation by an invasive plant and the suppression of natives. *Ecology Letters* 8(2), 209–217. DOI: 10.1111/j.1461-0248.2004.00713.x.

Uren, N.C. (2007) Types, amounts, and possible functions of compounds released into the rhizosphere by soil-grown plants. In: Pinton, R., Varanini, Z. and Nannipieri, P. (eds) *The Rhizosphere: Biochemistry and Organic Substances at the Soil-Plant Interface*. Taylor & Francis Group, Boca Raton, FL, pp. 1–21.

van der Heijden, M.G.A., Klironomos, J.N., Ursic, M., Moutoglis, P., Streitwolf-Engel, R. *et al.* (1998) Mycorrhizal fungal diversity determines plant biodiversity, ecosystem variability and productivity. *Nature* 396(6706), 69–72. DOI: 10.1038/23932.

van der Heijden, M.G.A. and Sanders, I.R. (2002) *Mycorrhizal Ecology*. Springer, Berlin.

van der Putten, W.H. (2010) Impacts of soil microbial communities on exotic plant invasions. *Trends in ecology & evolution* 25(9), 512–519. DOI: 10.1016/j.tree.2010.06.006.

Vaughn, S.F. and Berhow, M.A. (1999) Allelochemicals isolated from tissues of the invasive weed garlic mustard (*Alliaria petiolata*). *Journal of Chemical Ecology* 25(11), 2495–2504. DOI: 10.1023/A:1020874124645.

Vivanco, J.M., Bais, H.P., Stermitz, F.R., Thelen, G.C. and Callaway, R.M. (2004) Biogeographical variation in community response to root allelochemistry: novel weapons and exotic invasion. *Ecology Letters* 7(4), 285–292. DOI: 10.1111/j.1461-0248.2004.00576.x.

Wardle, D.A., Nicholson, K.S., Ahmed, M. and Rahman, A. (1994) Interference effects of the invasive plant Carduus nutans L. against the nitrogen fixation ability of Trifolium repens L. *Plant and Soil* 163(2), 287–297. DOI: 10.1007/BF00007978.

Willis, R.J. (2007) *The History of Allelopathy*. Springer, Berlin.

Wolfe, B.E., Rodgers, V.L., Stinson, K.A. and Pringle, A. (2008) The invasive plant *Alliaria petiolata* (garlic mustard) inhibits ectomycorrhizal fungi in its introduced range. *Journal of Ecology* 96(4), 777–783. DOI: 10.1111/j.1365-2745.2008.01389.x.

16 Competition Between Native and Non-native Plants

Elizabeth M. Wandrag[1,2]* and Jane A. Catford[3,4]

[1]School of Environmental and Rural Science, University of New England, Armidale, NSW 2350, Australia; [2]Institute for Applied Ecology, University of Canberra, Canberra, ACT 2617, Australia; [3]Department of Geography, King's College, London, UK; [4]School of BioSciences, University of Melbourne, Vic 3010, Australia

Abstract

The introduction of species to new locations leads to novel competitive interactions between resident native and newly-arriving non-native species. The nature of these competitive interactions can influence the suitability of the environment for the survival, reproduction and spread of non-native plant species, and the impact those species have on native plant communities. Indeed, the large literature on competition among plants reflects its importance in shaping the composition of plant communities, including the invasion success of non-native species. While competition and invasion theory have historically developed in parallel, the increasing recognition of the synergism between the two themes has led to new insights into how non-native plant species invade native plant communities, and the impacts they have on those plant communities.

This chapter provides an entry point into the aspects of competition theory that can help explain the success, dominance and impacts of invasive species. It focuses on resource competition, which arises wherever the resources necessary for establishment, survival, reproduction and spread are in limited supply. It highlights key hypotheses developed in invasion biology that relate to ideas of competition, outlines biotic and abiotic factors that influence the strength of competition and species' relative competitive abilities, and describes when and how competition between non-native and native plant species can influence invasion outcomes. Understanding the processes that influence the strength of competition between non-native and native plant species is a necessary step towards understanding the causes and consequences of biological invasions.

16.1 Introduction

Competition among plant species can influence the suitability of local environmental conditions for species' survival and reproduction. Consequently, it can determine the distribution and abundance of plant species. Competition is broadly defined as the negative influence of neighbours on an individual's performance (Keddy, 2001), and can include interference competition, apparent competition (Holt, 1977) and resource (or exploitative) competition (Table 16.1; Park, 1954). This chapter focuses on resource competition (though see Box 16.1 for a brief description of interference and apparent competition), and considers how it affects and is affected by biological invasions.

Resource competition arises wherever the resources necessary for establishment, survival, reproduction and spread are in limited supply. It includes competition between individuals for a shared resource, as well as situations where one individual depletes a resource such that it is no longer available to another. Individuals can compete for biotic and abiotic resources. Biotic resources include pollinators and seed dispersers. For example, the density of individuals with the same pollination strategy

*Corresponding author: elizabethwandrag@gmail.com

DOI: 10.1079/9781789242171.0016

Table 16.1. Key competition theories referenced in this chapter.

Theoretical basis	Summary	Key references
Exploitative vs interference competition	Species can be superior competitors by preempting use of shared resources leaving them unavailable to other species, or by directly preventing other species from accessing resources.	Park, 1954
Intra- vs interspecific competition	The strength of intraspecific (within species) competition influences interspecific (between species) competition; if the former is sufficiently strong, it can prevent species that are superior competitors from achieving competitive dominance.	Connell, 1983
Effect and response	Competitive ability is divided into an effect and response component, where a species' competitive ability can either be measured as its ability to suppress another (effect) or avoid being suppressed by another (response). There is increasing evidence suggesting that competitive response is most important.	Goldberg and Landa, 1991; Goldberg, 1996
Interspecific tradeoffs	Differences among species in life-history strategies mean that no single species can outperform all other species under all conditions.	Macarthur and Levins, 1967; Tilman, 2011
Niche vs fitness differences	Niche differences refer to differences among species in their use of both abiotic and biotic resources. They arise when species limit themselves more than others: intraspecific competition constrains population growth such that species have an advantage when rare. Fitness differences focus on species' differences that drive dominance but will only operate where niches overlap such that species are competing for at least some of the same limiting resources.	Chesson, 2000
Intransitive competition	Rock–paper–scissors: communities are not characterised by strict competitive hierarchies and no species is a universally superior or inferior competitor.	Gilpin, 1975; May and Leonard, 1975

(either of the same or different species) will affect floral visitation rates and thus reproductive success (Rathcke, 1983). Abiotic resources include light, nutrients, carbon dioxide, water and space. In this chapter, we focus on competition for these resources. We first outline some key concepts in competition theory and community ecology that are relevant to understanding many concepts in the literature on non-native plant invasions. We then describe factors that influence whether non-native and native plant species compete for shared

Box 16.1. Overview of aspects of competition not covered in this chapter.

Interference competition

Interference competition refers to the direct effects of one species on another, such that the ability of the recipient to exploit a resource is decreased. While there are a variety of ways in which interference competition can occur in animals, in plants it is largely restricted to allelopathy: the release of phytotoxins by some plant species that inhibits the performance of others. Evidence suggests that allelopathy can influence the impact of invasive plant species on native plant species (Callaway and Aschehoug, 2000; Becerra *et al.*, 2018). For example, *Centaurea maculosa* (spotted knapweed), a species native to Europe but invasive in the USA, releases the phytotoxin (-)-catechin, which inhibits germination and growth of native plant species in the invaded range (Ridenour and Callaway, 2001; Bais *et al.*, 2003). Allelopathy may also indirectly influence competitive interactions between native and non-native plant species. *Alliaria petiolata*, also native to Europe but invasive in the USA, achieves competitive dominance in the invaded range by releasing allelopathic chemicals that affect mycorrhizal fungi and indirectly reduce the performance of mycorrhizae-dependent native plant species but not itself (it does not rely on mycorrhizae).

Apparent competition

Plant species can also indirectly reduce the abundance of other plant species through their effects on the abundance of shared consumers (e.g. seed predators or herbivores), a phenomenon known as apparent competition (Holt, 1977). Apparent competition is implicated in the invasive success and impact of non-native plant species. Impacts are thought to arise through alterations to interactions between consumers (either native or non-native) and native plant species. For example, the presence of the invasive annual plant *Brassica nigra* in Californian grasslands leads to a nine-fold reduction in the establishment of the native perennial grass species *Nassella pulchra*. This reduction arises due to increases in densities of native seed consumers (e.g. the harvest mouse *Reithrodontomys megalotis*), most likely because the presence of *B. nigra* increases the availability of habitat refugia for these species, which then increase consumer pressure on the native grass (Orrock *et al.*, 2008).

resources, the strength of that competition and its implications for the invasion success and impact of non-native plant species.

16.2 The Process of Competition: General overview of competition theory

16.2.1 Intra- versus interspecific competition

Individuals of the same species have the same resource requirements and adaptations for acquiring resources. Consequently, the number of conspecific individuals that can occur in a given location is limited. The more conspecific individuals present, the more competition for shared resources intensifies, reducing individual performance and slowing population growth. Understanding factors that influence the strength of competition within species (intraspecific competition) is important because intraspecific competition can also influence the degree to which individuals of different species compete (interspecific competition, Table 16.1). Strong intraspecific competition among individuals of an otherwise superior competitor can prevent it from reaching densities high enough to exclude or even significantly reduce the abundance of competitively inferior species. Indeed, the relative balance of intra- vs interspecific competition is thought to influence the ability of plant species to coexist (Connell, 1983). For example, according to Chesson's (2000) theory of coexistence, stronger intra- than interspecific competition is a requirement for species coexistence.

16.2.2 Competitive effect and response

Quantifying a species' ability to compete for shared resources is central to predicting what species will ultimately win in competition to become dominant (Hart *et al.*, 2018). Competitive ability can be measured as the ability of one species to suppress the abundance of another (competitive effect), or the ability of one species to avoid being suppressed by another (competitive repsonse; Goldberg and Landa, 1991; Goldberg, 1996). These two aspects of competitive ability, sometimes referred to as competitive suppression vs tolerance (MacDougall and Turkington, 2004), are not mutually exclusive (Hart *et al.*, 2018), and can apply to both intra- and interspecific competition. Often, studies focus on a species' competitive effect because there is strong evidence that a specific set of traits can predict competitive effect (Wang *et al.*, 2010), meaning that it can seem straightforward to predict the superior competitor. In particular, traits associated with attaining a large size (in height or biomass) have frequently been found to confer a competitive advantage (Gaudet and Keddy, 1988), because the ability to grow large is generally associated with the ability to preempt resources such that the larger individuals in any system should be the most competitive (Peart, 1989; Hautier *et al.*, 2009).

However, the potential to achieve dominance is not only linked to the ability of one species to suppress the performance of another. An individual of one species could supress another by simply being larger, but it may not achieve dominance if it also suppresses conspecific individuals (i.e. intraspecific competition is strong). Recent theoretical work has demonstrated that competitive ability and hence dominance arises from differences between species in their competitive response (tolerance), rather than competitive effect (suppression) (Hart *et al.*, 2018). Because competitive effects act on both intra- and interspecific competition, species that are most able to grow and reproduce in the face of competition from larger individuals should always win in competition. In other words, species with higher growth may often be the superior competitors, but the ability to tolerate low levels of a given resource will ultimately determine whether they achieve competitive dominance (see also 'Resource use and interspecific tradeoffs'). The importance of competitive response in determining overall competitive ability may be one reason that identifying traits associated with good competitors has not always been straightforward (Wang *et al.*, 2010), since species may each develop different strategies to tolerate resource limitation, with resource limitation in turn varying among locations.

16.2.3 Resource use and interspecific tradeoffs

Another way of identifying which species will likely win in competition is to consider species' resource requirements and use. Under this model, the dominant competitor is the species that can maintain positive population growth at the lowest level of the most limiting resource (Tilman, 1982). This resource level is denoted R^*: the species with the lowest R^* is the dominant competitor for that given resource. This means that the superior competitor is not only the species whose population can deplete the resource to a lower level at equilibrium, but also the species with seedlings that are then able to establish at those low resource levels, i.e. on the resources leftover and unused by resident plants. If a single resource is limiting, or becomes limiting, then a monoculture made up of the most competitive species should result. However, if species differ in their resource requirements and their ability to compete for different limiting resources (e.g. competitive abilities for water, W^*, and light, L^*, as well as nutrients), and there are interspecific tradeoffs (see below) in these competitive abilities, habitat heterogeneity should promote the coexistence of multiple species (Tilman, 1982).

Interspecific tradeoffs rest on the assumption that a single species cannot be good at everything; species trade being good at one thing for being good at another, and no species can thus outperform all others under all conditions (Box 16.2). An interspecific tradeoff that is considered important for invasion outcomes is the competition–colonization tradeoff explained in Box 16.2, but the theory could apply to any strategy that allows a species to outperform others under some conditions. Key outcomes that arise from this theory are that: i) the more dissimilar two species are, the more likely they are to avoid competing with each other (Macarthur and Levins, 1967); and ii) ecologically similar species can coexist in heterogeneous habitats because each species

Box 16.2. Interspecific tradeoffs and their implications for competition between natives and non-natives.

Interspecific (or life-history) tradeoffs offer a leading explanation for species coexistence (Tilman, 1994; Harpole *et al.*, 2016). If a species occupies a unique position along a tradeoff surface, it cannot be competitively displaced because no single species outperforms all others under all conditions.

The competition–colonization tradeoff

A well-known tradeoff is the competition–colonization tradeoff, where species that are good colonizers are poor competitors and vice versa. This tradeoff happens because organisms that allocate more biomass towards reproduction and dispersal have less biomass to allocate towards competing for a limiting resource (Tilman, 1994). This is nicely demonstrated by species found in North American grasslands. At Cedar Creek Ecosystem Reserve in Minnesota, where soil nitrogen is the key limiting resource, dominant competitors will allocate up to 70% of their biomass to roots and <10% to reproductive structures, whereas good colonizers will allocate as much as 50% towards reproduction and <10% towards roots (Gleeson and Tilman, 1990, 1994). While two-dimensional tradeoffs are usually examined (for conceptual, mathematical and empirical ease), tradeoffs occur across multiple dimensions. It is these multidimensional tradeoffs that would enable such a high diversity of species to exist in heterogeneous landscapes because each species will likely have the best combination of traits under a specific set of conditions, somewhere in space or time.

The Universal Tradeoff Hypothesis

Understanding interspecific tradeoffs may help predict the outcome of invasions. In 2011, Tilman proposed the Universal Tradeoff Hypothesis (UTH), which posits that traits of ecologically similar species are bound to the same interspecific tradeoff surfaces, regardless of their biogeographic or phylogenetic origins (Tilman, 2011). Evidence from mass species incursions in the geological past support this hypothesis: the fossil record shows that, with few exceptions, ecologically similar species have coexisted for at least a million years following past biogeographic migrations. However, applying the UTH to modern, human-mediated invasions and focusing on the competition–colonization tradeoff, Catford *et al.* (2018) noted that if all species are bound by the same rules regardless of their origin, then any conditions that allow species to overcome such tradeoffs could disrupt species coexistence and enable competitive exclusion.

Competition–colonization tradeoffs and non-native plant invasion

There are several reasons why human-mediated invasions could allow species to overcome interspecific tradeoffs. Like Catford *et al.* (2018), we describe two but note that there are more.

First, by planting and dispersing non-native species in high numbers, humans artificially increase the colonization ability of some species. Historical dispersal and migration of invaders in the fossil record was reliant on invaders' own colonization abilities, which were limited through tradeoffs. Second, by rapidly and often selectively moving species outside their historical range, humans can release introduced species from their enemies (Joshi and Vrieling, 2005). Catford *et al.* (2018) demonstrated that these two processes could move non-native species off the universal tradeoff curve (Fig. 16.1) by enabling them to become better colonizers, better competitors or longer-lived than their native counterparts – at no cost to the invaders themselves (Catford *et al.*, 2018). Consequently, human-mediated invasions could break the rules of coexistence, providing non-natives with an advantage over resident natives and potentially resulting in native species extinctions (Fig. 16.1c).

is a superior competitor over some portion of a heterogeneous habitat (Catford *et al.*, 2018). When heterogeneity in resource limitation declines, for example through the addition of multiple limiting nutrients, a collapse in the number of species that can coexist is predicted (Harpole *et al.*, 2016).

16.2.4 Niche versus fitness differences

Modern coexistence theory developed by Chesson (2000) describes how species' niche vs fitness differences can determine the outcome of competition. Niche differences focus on the difference

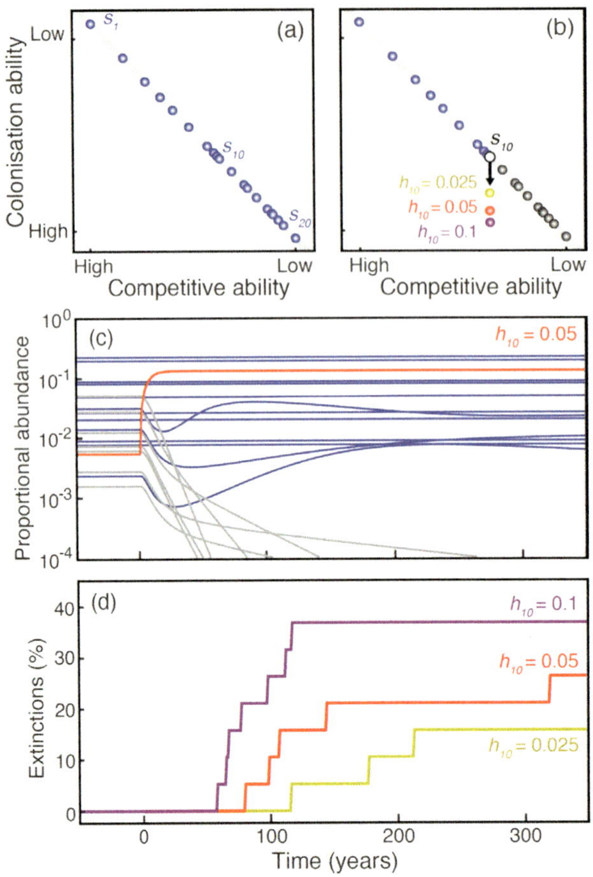

Fig. 16.1. Simulated species extinctions resulting from human-mediated species invasions and disruption of competition–colonization interspecific tradeoffs. (a) In a coexistence model based on a competition–colonization tradeoff (grey line), species are ranked from the best to the poorest competitor (s_1 to s_{20}) in a 20-species metacommunity (Tilman, 1994). Species can coexist if their colonization abilities are inversely related to their competitive rank (this is the tradeoff). This assumes a perfect competitive hierarchy where superior competitors can displace inferior competitors regardless of time of arrival or establishment, and the weakest competitors only colonize vacant sites. Regardless of positions on the *X*-axis, differences in species' competitive abilities are determined by rank only, such that the competitive difference between s_1 and s_2 is the same as the competitive difference between s_{18} and s_{19}. (b) Schematic drawing of a disrupted tradeoff curve showing the shifted positions of a designated 'invader' (s_{10}), an intermediate competitor in the community, given differences in colonization ability that arise from increasing levels of propagule input from external sources (i.e. via human-assisted dispersal), indicated by the term *h*. White circle indicates $h_{10} = 0$, where no externally-sourced propagules are added; coloured circles indicate increasing levels of external propagule input from species 10 [h_{10} increases from 0 to 0.025 (yellow), 0.05 (red) and 0.1 (violet)]; blue circles indicate 'native' species that are superior competitors and thus unaffected by the invasion; grey circles indicate the ten natives that could be displaced by s_{10}. (c) Relative abundance (log scale) of each of the 20 species over time. Each line represents a species: the red line indicates the invader after elevating h_{10} to 0.05; blue lines indicate persistent native species; grey lines indicate native species driven to extinction (i.e. relative abundance drops below 0.0001 of species' original equilibrium abundance) as a consequence of increasing h_{10}. (d) A timeline of percentage of the 19 native species driven to extinction as a consequence of elevating h_{10} to 0.025, 0.05 (same scenario shown in c) and 0.1. Reprinted from Catford *et al*. (2018).

between intra- and interspecific competition outlined in Section 16.2.1.1. They arise when species limit themselves (strong intraspecific competition) more than they limit others (weak interspecific competition). For example, where two species have different adaptations for acquiring a limiting resource, individuals within a species are more likely to compete for that resource than individuals of the different species. The intraspecific competition component of niche differences can arise from density-dependent interactions with biotic partners as well as from density-dependent competition for shared resources. The larger the niche difference between two species, the more likely that intraspecific competition will outweigh interspecific competition.

While niche differences require that species have an advantage when rare, fitness differences do not. According to modern coexistence theory, fitness differences include anything that can favour one competitor over the other to drive dominance and should not be confused with Darwinian fitness. Differences among species in their ability to draw down (and tolerate lower levels of) limiting resources determine which species can dominate, meaning that both the competitive response ability described previously (in 16.2.1.2 —Competitive effect and response) and Tilman's (1982) R* theory (described in 16.2.1.3 — Resource use and interspecific tradeoffs) could be considered as fitness differences. The ability of two species to coexist could be influenced by either processes that influence average niche differences (often referred to as stabilizing processes), or those that influence average fitness differences (equalizing processes).

Modern coexistence theory predicts that species can coexist by avoiding competition due to occupying distinct niches, or that one species can exclude the other if species have overlapping niches and strong fitness differences such that one species is always the superior competitor for shared resources. This prediction has proved helpful in understanding non-native plant invasion and impact (see 16.2.1.6 — Applying competition theory to invasions; MacDougall *et al.*, 2009).

16.2.5 Intransitive competition

Intransitive competition can occur when there is no clear competitive hierarchy, such that no species is a universally superior or inferior competitor. It is likened to the game of rock–paper–scissors: species A excludes species B, species B excludes C, and C excludes A (Gilpin, 1975; May and Leonard, 1975). Although intransitive competition often appears to promote species coexistence and prevent competitive dominance, few studies have directly sought to identify communities structured by intransitive competition (Gallien, 2017). Current research suggests that intransitivity may be more common in heterogeneous environments, in species-rich plant communities, and among common rather than rare species (Allesina and Levine, 2011; Soliveres *et al.*, 2015). However, research on intransitive competition remains limited, particularly in the context of invasion. Recent evidence that plant communities characterized by intransitive dynamics may be more vulnerable to species introductions and losses (Mack *et al.*, 2019) – both consequences of invasions – suggests that further research is warranted.

16.2.6 Applying competition theory to invasions

Many hypotheses developed in invasion ecology to explain the success, dominance and impacts of invasive species relate to the ideas of competition outlined above (Catford *et al.*, 2009; Enders *et al.*, 2018). While the specific predictions of each of these hypotheses differ somewhat, they generally fall into some combination of two categories.

First, non-native plant species can invade native plant communities by outcompeting native species for shared resources (Fig. 16.2). Non-natives may outcompete native species if they are inherently superior competitors relative to native species (Gurevitch, 2011). This could arise by chance because the large number of non-native plant species introduced is likely to include a highly competitive species (global competition, Table 16.2; Alpert, 2006), or if non-native species were selectively introduced due to traits associated with high competitive ability, such as the ability to attain a large size (Richardson and Pyšek, 2012). Non-natives may also outcompete native species if factors associated with introduction to new locations

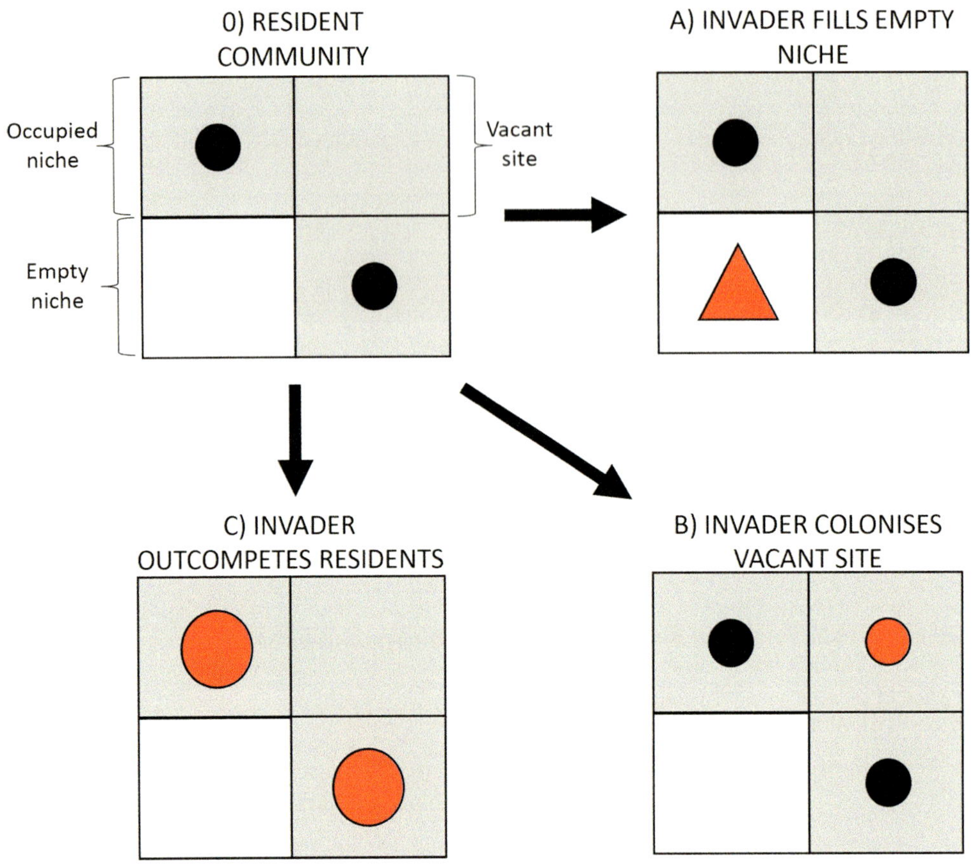

Fig. 16.2. Non-native (red) plant species could invade into a native plant community if they A) exploit under- or unused resources due to occupying a different niche to resident native (black) species; B) native species do not occupy all locations they could (i.e. due to propagule limitation; or C) occupy similar niche space to native species but are superior competitors for shared resources.

give them a competitive advantage over natives (Gurevitch, 2011). Such factors could include the loss of natural enemies that regulate population growth in native ranges, or local environmental changes that often go hand in hand with the introduction of non-native species.

Second, non-native plant species can invade native plant communities simply by exploiting unused or underused resources in the introduced location and avoiding competition (Fig. 16.2). This could occur if non-native and native species have different fundamental niches and non-natives occupy empty niches, or if native species are propagule limited (due either to reproductive constraints or dispersal limitation) and do not occupy all locations that are suitable. An example of the former is invaders that belong to novel functional groups. For example, the invasive success of non-native tree species into treeless South African fynbos is attributed to their novel functional group relative to resident natives (Rundel *et al.*, 2014).

Identifying whether non-native plant species invade because they occupy distinct niches or because they outcompete native species for shared niche space can help predict the impact of non-native plant species. MacDougall *et al.* (2009) highlighted how viewing hypotheses developed in invasion ecology in terms of niche vs fitness differences (see Chesson, 2000; see Section 2.1.4) could help understand the success and impact of non-native plant species.

Table 16.2. Competition-based invasion hypotheses (adapted from Catford *et al*., 2009; Enders and Jeschke, 2018).

Hypothesis	Premise	Mechanism	Competitive outcome	Key references
Adaptation	Non-native may either be pre-adapted to local conditions or may adapt following introduction.	Life-history strategies and traits of the invading species, potentially combined with the ability to avoid enemies, allows invading species to preempt or better exploit resources than native species, making them better competitors.	Increased competitive ability relative to resident natives.	Elton, 1958; Rejmánek and Richardson, 1996; Alpert, 2006; MacDougall *et al*., 2018
Biotic Indirect Effects	Natives indirectly mediate non-native success through, e.g. their influence on soil biotic communities.	Native species may cultivate unique biotic communities that enhance or limit non-native plant performance. A notable example is plant–soil feedbacks whereby native-cultivated soil microbes benefit non-natives.	Increased or decreased competitive ability relative to resident natives, depending on influence of biota associated with natives.	Callaway *et al*., 2004
Biotic Resistance	Competitors and enemies in the non-native range limit colonization of invader.	Either resident native species are inherently better competitors than the invading species, or invading species lack defences against natural enemies in the invaded range such that natural enemies limit their competitive ability relative to resident natives.	Decreased competitive ability relative to resident natives.	Levine *et al*., 2004
Darwin's Naturalization	The success of non-native species is linked to their relatedness with resident natives.	Non-natives that are closely related to natives may be better adapted to local conditions but more likely to compete for shared resources.	Stronger competitive effects of the resident community for closely related non-natives.	Darwin, 1859
Dynamic Equilibrium Model	Properties of the physical environment in the non-native range influence species interactions.	Natural fluctuations of native ecosystems that affect species growth and mortality will influence competitive interactions among species. Links with hypotheses centred on resource availability and disturbance.	Environment dependent: could increase or decrease the strength of competition among species.	Huston, 1979

Continued

Table 16.2. Continued

Hypothesis	Premise	Mechanism	Competitive outcome	Key references
Evolution of Increased Competitive Ability (EICA)	Following introduction to a new location, natural enemies are lost or decline in abundance.	The loss or reduction of natural enemies allows the invader to allocate resources previously used for defence into traits that make them better competitors against native species.	Increased competitive ability relative to resident natives.	Blossey and Notzold, 1995; Joshi and Vrieling, 2005
Evolutionary Imbalance Hypothesis (EIH)	Species from different biogeographic regions vary in their relative competitive abilities.	Environmental conditions and the intensity of local competition should influence the strength of selection for competitive ability, leading to differences in the competitive ability of species from different locations.	Depends on the conditions in the native relative to the introduced range.	Fridley and Sax, 2014
Empty Niche	Like limiting similarity but focuses on characteristics of the environment rather than the species.	Non-natives avoid competition with natives by occupying distinct niches.	Indirect benefit to invaders of avoiding competition.	Godoy and Levine, 2014; Rundel *et al*., 2014
Enemy Release / Enemy Reduction	Following introduction to a new location, natural enemies are lost or decline in abundance.	The loss or reduction of natural enemies either: i) allows species to more rapidly increase population size than in their native range; or ii) reallocate resources previously used for defence to traits that make them better competitors (e.g. resource acquisition or fast growth). Like EICA but does not require evolution.	Increased competitive ability relative to resident natives.	Keane and Crawley, 2002; Colautti and MacIsaac, 2004; Joshi and Vrieling, 2005
Enemy of My Enemy	Non-native plant species accumulate local pathogens that are more detrimental to native species.	Also based on plant–soil feedback models, this posits that greater sensitivity of natives to pathogens accumulated by non-natives influences their competitive ability.	Increased competitive ability of non-native relative to native.	Eppinga *et al*., 2006
Enemy Inversion	Natural enemies have reduced or opposite effects for non-native plants in non-native ranges due to changed abiotic conditions and/or other novel biotic interactions.	Reduced or opposite effect of natural enemies could affect non-natives as described for the Enemy Release Hypothesis, above.	Increased competitive ability relative to resident natives.	Colautti *et al*., 2004

Continued

Table 16.2. Continued

Hypothesis	Premise	Mechanism	Competitive outcome	Key references
Environmental Heterogeneity	Variation in resource availability creates distinct niches that are suitable to both native and invasive species.	Native and invasive species avoid competition by occupying distinct niches.	Indirect benefit to invaders of avoiding competition.	Melbourne *et al.*, 2007
Evolutionary Reduced Competitive Ability (ERCA)	Following introduction to a new location, species may encounter fewer or weaker competitors.	If competitive ability involves traits that have a fitness cost, then any reduction in interspecific competition in the introduced range could lead to selection against competitive ability.	Should not influence the outcome of competition between natives and non-natives, but should lead to some fitness benefit to non-natives.	Bossdorf *et al.*, 2004
Global Competition / Sampling Effect	As the number of species introduced increases, the likelihood that a competitive species will be in the invading species pool increases	Life-history strategies and traits of the invading species, potentially combined with the ability to avoid enemies, allows invading species to preempt or better exploit resources than native species, making them better competitors for shared resources.	Increased competitive ability relative to resident natives.	Elton, 1958; Rejmánek and Richardson, 1996; Alpert, 2006; MacDougall *et al.*, 2018
Habitat Filtering	Non-natives may be pre-adapted to local conditions	As with Darwin's Naturalization Hypothesis, species pre-adapted to local conditions may also be more likely to compete with resident natives for shared resources.	Increased strength of competition with native, no predictions as to the outcome.	Darwin, 1859
Ideal Weed / Selection Bias	Species may be selected for traits that pre-adapt them to the conditions of the invaded system (selection bias), which can result in the preferential introduction of 'ideal weeds'.	As for global competition/sampling effect above: life-history strategies and traits of the invading species may make them better competitors.	Increased competitive ability relative to resident natives.	Elton, 1958; Rejmánek and Richardson, 1996; Alpert, 2006; MacDougall *et al.*, 2018
Increased Resource Availability / Fluctuating Resources / Invasion Windows / Disturbance	Resource supply increases or resource use decreases, providing opportunities for invasion.	Resources are fully utilized under normal conditions but an increase in their availability reduces competitive constraints on either the invader or resident native (or both).	Indirect benefit to invaders of avoiding competition.	Davis *et al.*, 2000

Continued

Table 16.2. Continued

Hypothesis	Premise	Mechanism	Competitive outcome	Key references
Increased Susceptibility	Genetic bottlenecks of non-natives increase their susceptibility to enemies.	Increased susceptibility to enemies reduces competitive ability.	Decreased competitive ability relative to resident natives.	Colautti *et al.*, 2004
Invasional Meltdown	Counter to novel associations. Either the invader brings coevolved mutualists with them, or mutualists are themselves introduced. Can also include the introduction of other non-native species that facilitate the focal plant species.	Either through facilitation or mutualism, the presence of another non-native species makes the environment more suitable for the focal invader.	Increased competitive ability relative to resident natives.	Simberloff and Von Holle, 1999
Limiting Similarity	Invading species are functionally distinct from resident natives.	Functionally distinct invaders occupy distinct niches to native species, thereby avoiding competition.	Indirect benefit to invaders of avoiding competition.	Darwin, 1859; Macarthur and Levins, 1967
Loss of Mutualists	Counter to enemy release, invaders lose mutualistic partners in the new range.	The loss of coevolved mutualists from species' native ranges leads to a decrease in competitive ability relative to resident natives.	Decreased competitive ability relative to resident natives.	Alpert, 2006; Mitchell *et al.*, 2006
Novel Associations	Invading species form novel associations with mutualists in the invaded range.	If species form new associations with mutualists in the invaded range (e.g. mutualistic microbes) their competitive ability may be enhanced relative to resident natives.	Increased competitive ability relative to resident natives.	Callaway *et al.*, 2004; Mitchell *et al.*, 2006
Novel Weapons / Inference Competition	Invading species directly inhibit the performance of resident natives, usually through the release of allelopathic chemicals.	Native species lack defences to novel allelopathic chemicals and experience reduced performance. Does not necessarily involve competition for shared resources, but the two can be difficult to disentangle.	Indirect benefit to non-natives from lowered competitive ability of natives.	Callaway and Aschehoug, 2000; Hierro *et al.*, 2005
Plasticity Hypothesis	Non-natives are more phenotypically plastic than natives	Phenotypic plasticity can allow species to increase competitive ability (like EICA but without the requirement for evolutionary change)	Increased competitive ability relative to resident natives.	Richards *et al.*, 2006

Continued

Table 16.2. Continued

Hypothesis	Premise	Mechanism	Competitive outcome	Key references
Propagule Pressure	High propagule pressure may allow species to overcome competition–colonization tradeoffs	Overcoming competition–colonization tradeoffs allows strongly competitive species to also become widely dispersed and hence more likely to compete with widely dispersed but weakly competitive native species	Increased competitive ability relative to resident natives.	Catford *et al.*, 2018
Reckless Invader	Selective introduction for traits that allow species to rapidly exploit resources to invest in fast growth, but at the expense of defence/stress tolerance.	Species success arises through their ability to exploit disturbances (e.g. rapid growth and high fecundity), allowing them to increase in abundance. However, those traits represent a tradeoff with stress tolerance such that native species may eventually outcompete them.	Able to exploit available resources in low competitive environments, but likely lower competitive ability relative to natives.	Simberloff and Gibbons, 2004; Alpert, 2006
Sampling	A large number of non-native species is more likely to become invasive than a small number.	As the number of non-native species increases, the strength of interspecific competition for native species should also increase.	Increased competitive ability relative to resident natives.	Marler *et al.*, 1999

Niche differences should facilitate the establishment of non-native plant species irrespective of their fitness relative to native species, but they should also promote coexistence between non-native and native plants, since species avoid direct competition. In contrast, where species' niches overlap and fitness differences favour non-natives, non-natives should not only establish but also exclude native species.

16.3 Factors That Influence Competitive Ability: Does Origin Matter?

The universal interspecific tradeoffs described in Box 16.2 mean that native and non-native plant species are not expected to be fundamentally different (Tilman, 2011). Indeed, under controlled conditions, non-native and native species tend to respond similarly to the same biotic and abiotic variables, including competition (Dawson *et al.*, 2012). However, some non-native plant species appear to be more competitive in their introduced than native range (Ni *et al.*, 2010), or relative to resident native species. Consequently, the process of moving a species outside its native range must, for some species, alter their relative competitive abilities. In this section, we discuss four potential explanations for increases in the relative competitive abilities of non-native plant species following introduction (Buckley and Catford, 2016; see also Table 16.2).

16.3.1 Human association

The distribution of non-native plant species is overwhelmingly associated with human activities (Pyšek *et al.*, 2010). Three aspects of this association could alter the relative competitive abilities of non-native plant species. First, humans influence the identity of species that are introduced. Non-native plant species are rarely a random sample from a general community of plant species, but rather have been selected for traits (whether deliberately or not) that predispose them to succeed in introduced locations. For example, plants used in gardens are often selected for traits such as fast growth, ease of propagation, long and early-onset flowering periods and low susceptibility to disease. These are all traits associated with invasion success, and might explain why up to 93% of naturalised non-native plant species likely originate from the ornamental plant trade (van Kleunen *et al.*, 2018). Second, humans influence the locations to which species are introduced (van Kleunen *et al.*, 2015). This influence includes selective climate matching of non-native plant species with introduced locations (McGregor *et al.*, 2012), and human-mediated modification of the environment, which may favour non-native relative to native plant species (Catford *et al.*, 2014). Third, humans tend to repeatedly introduce propagules of non-native species and to do so at high densities (Driscoll *et al.*, 2014). This can lead to niche preemption and other benefits associated with high propagule pressure (Simberloff, 2009). It could also allow species to overcome competition–colonization tradeoffs (Box 16.2) such that strongly competitive species also become widely dispersed and hence more likely to compete with widely dispersed but weakly competitive native species (Catford *et al.*, 2018).

A notable example of all three aspects of human association favouring non-native relative to native plant species is the invasion success of European plant species in temperate grasslands. This invasion is generally attributed to: i) the long-association of these species with human cultivation; ii) their consequent pre-adaptation to agricultural landscapes; and iii) the preferential introduction of these species, at high rates, into agricultural landscapes worldwide (van Kleunen *et al.*, 2015; MacDougall *et al.*, 2018). While these factors do not always confer competitive advantage, there appear to be at least some situations in which non-native species are competitively superior to native species (HilleRisLambers *et al.*, 2010; O'Reilly-Nugent *et al.*, 2019), an advantage that environmental change, including climate change, seems to be exacerbating (Catford and Jones, 2019).

16.3.2 Evolutionary history

16.3.2.1 Relatedness

Darwin (1859) hypothesized that evolutionary history should influence the ability of non-native plant species to establish and spread. Non-native plant species that are closely related to resident natives may gain an advantage through being pre-adapted to the conditions of the introduced location. However, greater overlap in resource requirements between closely related native and non-native species could increase competition for shared resources. Indeed, a corollary of the interspecific tradeoffs that are predicted to ensure that no single species can outperform all others under all conditions is that the more similar two species are, the more likely they will be to compete for shared resources (Macarthur and Levins, 1967).

In contrast, non-native plant species that are only distantly related to native plant species should be more likely to use a different set of resources to resident natives. This should be a disadvantage where specialist adaptations to local conditions are required, but an advantage in locations where empty niche space is present (Table 16.2; Elton, 1958) and resources are underutilized by resident natives (Stachowicz and Tilman, 2005). In such situations, non-native species may succeed by avoiding competition with resident natives. While there are examples of relatedness both facilitating and hindering the establishment success of non-native plant species, there is increasing evidence that there are simply some traits that confer high competitive ability, independently of species' origin (Kunstler *et al.*, 2012; O'Reilly-Nugent *et al.*, 2019). For example, although functional dissimilarity of non-native to native species in Germany appears to increase survival, this effect is largely confined to taller individuals with smaller specific leaf area, both traits associated with high competitive ability more generally (Conti *et al.*, 2018).

16.3.2.2 The Evolutionary Imbalance Hypothesis

In contrast to the Universal Tradeoff Hypothesis (Tilman, 2011), the Evolutionary Imbalance Hypothesis (EIH) posits that differences in the evolutionary histories of species can lead to variation in the competitive abilities of species from different biogeographic regions (Fridley and Sax, 2014). Central to EIH is the idea that evolution acts to produce species adapted to a given set of environmental constraints. These adaptations should be optimized by repeated exposure to the same set of environmental conditions and to strong competition from multiple competitors. Consequently, species from regions with high phylogenetic diversity (many potential competitors acting to select for increased competitive ability) and stable environments are predicted to be competitively superior to species from regions of low phylogenetic diversity (Fridley and Sax, 2014). Analyses using data on non-native plants in eastern North America, the Czech Republic and New Zealand support this hypothesis, suggesting that non-native species from areas with high phylogenetic diversity have a higher probability of naturalization (Fridley and Sax, 2014).

16.3.3 Changes to other biotic interactions affecting competitive ability

16.3.3.1 Natural enemy release

The Enemy Release Hypothesis (Keane and Crawley, 2002) posits that non-native plant species that lose native range natural enemies (predators, herbivores, pathogens, parasites) can experience increased success in introduced locations. This increased success can arise, in part, due to increased competitive ability. Two mechanisms for increased competitive ability are recognized (Keane and Crawley, 2002; Catford *et al.*, 2009). First, release from enemies that directly limit native population sizes (*regulatory release*) can allow rapid population growth, such that species experience reduced intraspecific competition and hence increased ability to competitively supress native species. Second, release from enemies can allow species to reallocate resources previously used for defence into growth and reproduction (Keane and Crawley, 2002; Catford *et al.*, 2009). Greater allocation to growth and reproduction (also known as *compensatory release*) should increase species' competitive abilities relative to resident natives that are constrained by their own natural enemies. Evidence for increased competitive ability arising from enemy release is equivocal (Daehler, 2003), and the effects are likely transient (Hawkes, 2007; Diez *et al.*, 2010). Nonetheless, this hypothesis remains central to many studies that examine the putative competitive superiority of non-native plant species.

16.3.3.2 Changing mutualistic associations

There are three ways in which mutualistic interactions involving non-native plant species can change in non-native locations to alter their competitive ability relative to native species. First, non-natives may leave behind key mutualists from their native range. This could either reduce the competitive ability of non-natives relative to native species with intact mutualisms, or enhance the competitive ability of non-natives if they can reallocate resources previously allocated to establishing mutualistic associations towards faster growth or reproduction. Second, non-natives may encounter novel or co-introduced mutualists in non-native ranges, which could enhance their competitive ability relative to native species, particularly if non-natives concurrently lose natural enemies. Finally, non-natives may disrupt native mutualisms (Pinzone *et al.*, 2018) to indirectly enhance their competitive ability relative to resident native species.

One mutualism that appears to strongly influence species' relative competitive abilities is that between plants and mutualistic soil microbes, such as mycorrhizal fungi (Bever, 2003). A positive effect on growth rate due to the presence of soil mutualists, or any decrease in growth rate due to their absence, can sometimes allow one species to generate stronger interspecific competition on another than it would otherwise. A notable example of this is the non-native *Centaurea maculosa* in the USA, where the mycorrhizae present enhance its ability to compete with resident native grass species (Marler *et al.*, 1999). Indeed, soil microbes appear to be particularly important in mediating competition between non-native and native species. The loss of soil mutualists for the non-native *Hypericum perforatum* in North America may enhance its competitive ability due to greater allocation towards reproduction (Seifert *et al.*, 2009), and there is mounting evidence that non-native plant species disrupt native plant–soil mutualist interactions (Le Roux *et al.*, 2018).

16.4 Post-introduction Evolution

Introduction to new environments changes the selection pressures acting on plant species and may induce rapid evolution that alters their competitive ability. We highlight two key mechanisms through which this is expected to alter the relative competitive abilities of non-native and native plant species.

16.4.1 The Evolution of Increased Competitive Ability hypothesis

The Evolution of Increased Competitive Ability (EICA) hypothesis focuses on changes to non-native competitive ability that result from the loss of natural enemies. Although similar to the compensatory release aspect of natural enemy release described earlier, EICA involves a long-term evolutionary change to species competitive abilities, while natural enemy release could arise due to phenotypic plasticity. Key predictions of EICA are that non-native plants species will be less well defended and grow larger than conspecifics in their native range (Blair and Wolfe, 2004). While there is considerable evidence for the loss of defence mechanisms by non-native species, evidence for increased growth has been mixed (Siemann and Rogers, 2001; Bossdorf *et al.*, 2004). For example, while invasive genotypes of *Lespedeza cuneata* in North America had lower defences but greater competitive ability than conspecifics from the native Japanese range (Beaton *et al.*, 2011), invasive individuals of *Alliaria petiolata* did not grow larger than conspecifics from the native European range when grown alone and, in competition, were outperformed by native range conspecifics (Bossdorf *et al.*, 2004). Additionally, growth differences do not always translate to increased competitive ability. Several studies have reported growth increases for non-natives in the absence of interspecific competition, but failed to find differences when the same species are grown together with competitors (Leger and Rice, 2003; Blumenthal and Hufbauer, 2007). Further, some research suggests that non-native plant species may evolve reduced competitive ability in introduced locations if the interspecific competition in the new environment is weaker (Evolutionary Reduced Competitive Ability; Bossdorf *et al.*, 2004).

16.4.1.1 Coevolution between native and non-native plant species

Any increases in competition strength arising from novel combinations of plant species could lead to selection for competitive genotypes of both non-native and native plant species (Strauss *et al.*, 2006; Oduor, 2013). Recent work on the invasive *A. petiolata* and native *Pilea pumila* in the USA provides an elegant demonstration of this (Huang *et al.*, 2018). The allelopathic influence of *A. petiolata* can cause declines in native performance via its effects on mycorrhizal fungi (Box 16.1). However, *P. pumila* appears to evolve tolerance to these effects by both reduced dependence on and greater ability to associate with residual mycorrhizal fungi. By quantifying competitive interactions between and within populations of each species that differed in their history of co-occurrence, the authors demonstrated that both species evolved greater tolerance of interspecific competition. For the non-native *A. petiolata*, greater competitive tolerance came at the cost of lower growth in the absence of competition, further highlighting that the faster growth prediction of EICA may not always hold.

16.5 How Does Context Mediate the Outcome of Native–Non-native Competition?

Intra- and interspecific competition, and their relative importance, vary in response to local conditions, particularly resource availability. When resources are limiting, species-specific adaptations for resource acquisition should promote niche partitioning and weaken competition. In contrast, higher resource availability should provide fewer opportunities for niche partitioning and strengthen interspecific competition (Gause, 1934; Grime, 1979; Tilman, 1982). Resource availability can be affected by both resource supply and resource uptake (Davis *et al.*, 2000). For example, eutrophication will lead to an increase in resource supply, while the loss of resident vegetation will lead to a decrease in resource uptake, with both processes increasing resource availability. In the following sections, we outline how changes to resource supply and disturbance can influence competitive outcomes. Both factors are closely associated with human activities and thus non-native introductions, and we note that they often interactively affect competitive outcomes.

16.5.1 Resource supply

Native and non-native species often come into contact under conditions that generate strong interspecific interactions. This is partly because human activities are often associated with resource-rich environments (Dukes and Mooney, 1999), and partly because humans directly increase resource supply through processes such as fertilization and irrigation. The observation that the dominance of non-native plant species is correlated with a decline in the abundance of native species and is more marked when resource supply is high (Seabloom *et al.*, 2015) has resulted in a large literature dedicated to the role of human-mediated changes to resource supply in mediating competition between natives and non-natives.

Water, nutrients and carbon dioxide are key resources that tend to increase due to human activity. These have served as focal points for much research, with considerable evidence that they modify competitive interactions between non-native and native species. For example, although two shrub species native to North America (*Encelia californica* and *Eriogonum fasciculatum*) showed increased survival and biomass in response to increasing water availability, this effect was reversed in competition with the non-native grass species *Avena fatua*: competition with *A. fatua* under high water availability led to a decline in survival and a more than tenfold decrease in biomass of native shrub species (Fig. 16.3; Esch *et al.*, 2018). However, shifts in the relative competitive abilities of non-native and native species in relation to shifts in resource supply will depend on the species involved. In another North American experiment examining species' response to water availability, all non-native plant species examined competitively suppressed native species regardless of water availability (Maron and Marler, 2008). Moreover, resident native plant species showed some ability to competitively exclude invading non-native plant species, an effect that was moderated by water availability for only one species.

16.5.2 Disturbance

Disturbance includes processes that involve complete biomass removal, such as soil tilling or herbicide use, and those that only involve only partial biomass removal, such as grazing or fire. There are at least two ways in which a reduction

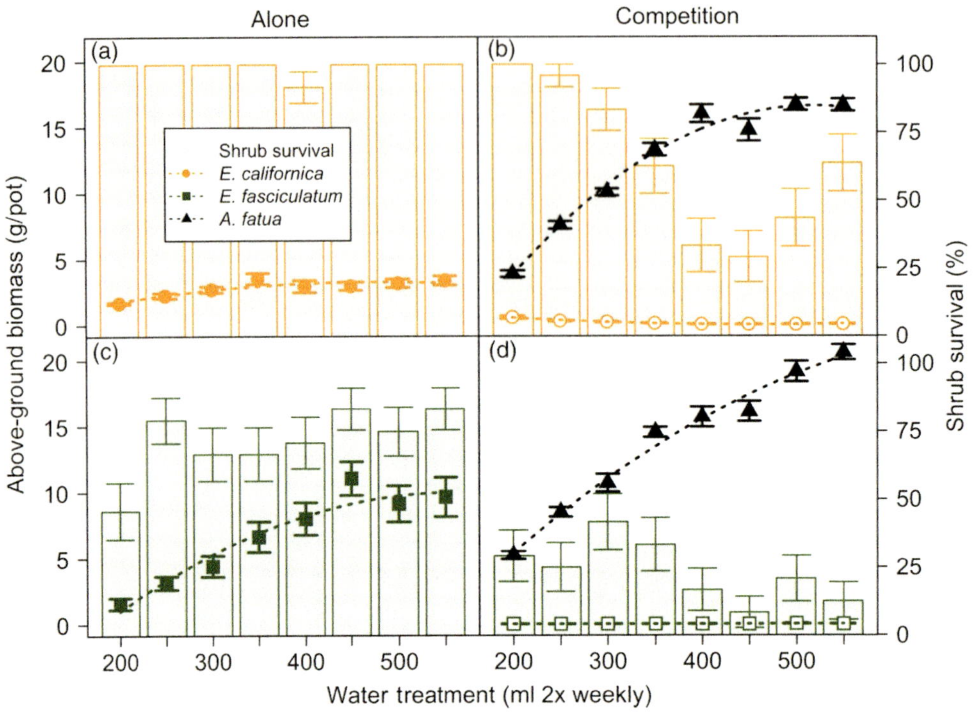

Fig. 16.3. Aboveground biomass responses (points) and survival (grey bars) of two native shrub species in response to increased water supply when grown alone (a, c) and when grown in competition (b, d) with the non-native grass species, *Avena fatua* (black points in b and d). Lines represent the best-fit linear or quadratic model for the above-ground responses of each species. Reprinted from Esch *et al*., 2018.

in the density or biomass of neighbours due to disturbance can influence competition: i) by directly limiting the growth of an otherwise dominant competitor; and ii) by freeing resources that would otherwise be used by resident vegetation (i.e. increasing resource availability by decreasing resource uptake). The two are often interrelated and difficult to disentangle. For example, the suppression of native grass species due to grazing by large herbivores is implicated in the invasion success of many grass and herb species in grasslands worldwide (Seabloom *et al*., 2015), and is a major driver of *Pinus* invasions into South African fynbos and globally (Richardson and Bond, 1991; Higgins and Richardson, 1998). However, benefits to non-natives likely arise due to different mechanisms in each case. For non-native grasses and herbs, biomass removal of resident natives is thought to allow species that are good competitors for light (e.g. taller species with high biomass; O'Reilly-Nugent *et al*., 2019) to competitively exclude resident natives. For *Pinus* species, benefits likely arise from increases in light availability due to the suppression of species that would otherwise be superior competitors for light (Richardson and Bond, 1991).

The type and frequency of the disturbance is often also an important determinant of the benefit derived by non-natives from the loss of native species (Catford *et al*., 2011). Many non-natives benefit from disturbances by being able to rapidly exploit available resources. However, traits associated with the ability to exploit disturbances do not always confer a competitive advantage. Consequently, some non-native plant species do not persist through time due to displacement by species that are better able to tolerate competition in undisturbed conditions. For example, in the USA, MacDougall and Turkington (2004) showed that resource preemption and fast growth in the non-native *Dactylis glomerata* enabled it to competitively supress native grass species following disturbance by fire, but it was ultimately the ability to tolerate competition that determined the long-term distribution and abundance of grasses, including the non-native *Poa pratensis*. Nonetheless, by reducing competition from native species, disturbance is widely held to provide windows of opportunity for the invasion of non-native plant species that can rapidly exploit available resources (Seabloom *et al.*, 2003b).

16.6 What Are the Implications of Competition Between Natives and Non-natives?

The previous sections have described how species-specific variation and local context influence the relative competitive abilities of native and non-native species. This is important because it can determine both the invasion success of non-natives and the impact they have on native plant communities (Fig. 16.2). Specifically: the ability of native plant species to competitively suppress non-native species will determine whether non-natives are able to establish and spread in native plant communities, and the ability of non-native plant species to suppress native species will determine the impact of non-natives on native plant communities. In this section, we elaborate on these two alternatives, but emphasize that a continuum between the two likely exists.

16.6.1 Biotic resistance

Competition with native plants is a major biotic interaction influencing non-native plant establishment and spread. Consequently, competition can act as a form of biotic resistance of the native plant community to non-native invasion (Levine *et al.*, 2004): high competitive ability of native relative to non-native plant species should limit invasion. Biotic resistance is often identified by comparing non-native plant performance in plots with and without resident native plant species. In general, there is strong evidence that resident plant communities reduce the performance of invading non-native plant species. For example, in South Africa, belowground competition from native shrub species halves the survival and growth for the non-native tree *Acacia cyclops* (Morris *et al.*, 2015), while competition from native grasses leads to a similar reduction in the growth of the non-native shrub *Chromolaena odorata* (te Beest *et al.*, 2018). However, while many studies document some competitive effect of the resident plant community, few show that it completely prevents non-native establishment (Levine *et al.*, 2004). Indeed, there is increasing evidence that fine-scale variation in microsite conditions and species-specific variation in resource requirements (i.e. niche differentiation) often mean that local establishment of non-native plant species is possible (Wainwright *et al.*, 2019; Wandrag *et al.*, 2019). Nonetheless, the potential for native plant species to competitively suppress non-native plant species is of ongoing interest due to its potential to aid in the management of non-native plant species (see Section 6).

16.6.2 Impact of invasive species

Increases in non-native plant dominance are often associated with decreases in the abundance or species richness of native species (Seabloom *et al.*, 2013). Historically, this dominance was thought to arise from their ability to competitively suppress native species. However, as understanding of both invasions and interspecific competition has evolved, it has become increasingly clear that many species successfully invade and dominate by occupying a different niche to native species. With this has come recognition of the fact that increasing rates of non-native plant invasion may be more strongly linked to the association of non-native plants with human-modified landscapes than to their ability to competitively suppress natives.

The phenomenon of non-native plant species invading due to niche differentiation is not restricted to landscapes that have been heavily modified by humans, but is one that has received considerable attention. The passenger–driver model of non-native plant invasion

(MacDougall and Turkington, 2006) focuses on human-mediated global change, and proposes two mechanisms through which non-native plant dominance could correlate with native decline. First, non-native plant species may be passengers of change. Their strong association with humans and pre-adaptation to human modified systems means they are better suited to human-modified landscapes than resident native species such that native species decline due to land use change rather than direct competition with non-native species. Second, non-native plant species may be drivers of change. Any decline in native species richness or abundance associated with invasion is due to the direct competitive suppression of native by non-native plant species.

An example of non-native plant species as passengers of change is the invasion success of European plant species in California. Although over 9 million hectares have experienced native plant losses with concurrent increases in non-native plant species (Seabloom *et al.*, 2003a), there is little evidence of the competitive suppression of native by non-native plant species. Instead, changing land uses and associated disruptions to native disturbance regimes and nutrient availability appear to independently favour non-native over native plant species (Seabloom *et al.*, 2003a; HilleRisLambers *et al.*, 2010). Nonetheless, there are examples of non-native plant species as drivers of change. In Australian grasslands, the non-native grass *A. fatua* appears to competitively suppress native species under some conditions (O'Reilly-Nugent *et al.*, 2019). It is likely that human-mediated land use and species inherent competitive abilities interactively influence non-native plant dominance in some situations (HilleRisLambers *et al.*, 2010), allowing non-native plant species to become 'back-seat drivers' of change by both benefiting from and exacerbating the effects of global change (Bauer, 2012).

16.7 Future Directions and Conclusions

Understanding the processes that influence where and how non-native and native plant species compete is crucial to preventing the establishment and spread of non-native species, and managing their impact on native plant communities. The challenge of doing so is particularly pressing as global change factors, such as climate change and agricultural intensification, increasingly create new combinations of species, and alter the conditions under which they interact. We identify four key avenues for future work.

16.7.1 Measuring competition

The myriad ways to examine competition among species (see e.g. Hart *et al.*, 2018) highlight both the importance of competition in shaping the distribution of species, and the difficulties associated with identifying and quantifying competitive interactions. A considerable effort has been dedicated to developing experimental (Alexander *et al.*, 2015, 2016; Wandrag *et al.*, 2019) and modelling (O'Reilly-Nugent *et al.*, 2019) approaches to examining competition to advance our understanding of its role in shaping species distributions, including non-native plant invasions. Novel insights that have emerged from these efforts include recognition that: i) facilitative interactions (rather than competitive) may be more important than currently recognized (Bimler *et al.*, 2018; Wandrag *et al.*, 2019); ii) intransitive dynamics are more common than previously thought (Soliveres *et al.*, 2015); iii) niche differences often override competition in shaping species distributions (O'Reilly-Nugent *et al.*, 2019; Wainwright *et al.*, 2019; Wandrag *et al.*, 2019); and iv) some species are simply superior competitors, regardless of biotic and abiotic conditions (Kunstler *et al.*, 2012; Seabloom *et al.*, 2015; O'Reilly-Nugent *et al.*, 2019). These emerging insights highlight the crucial need to continue to both develop approaches to measuring competition, and develop our theoretical understanding of when and how it matters.

16.7.2 Understanding the implications of global change

The introduction and invasion success of non-native plant species are typically associated with environmental change and human activities (Pyšek *et al.*, 2010), processes that can also influence the strength and direction of competition. Consequently, research that simultaneously examines how shifts in biotic (i.e. community composition) and abiotic conditions alter competitive outcomes is needed. Recognizing this need, researchers in Europe have recently used manipulative experiments along elevational gradients to demonstrate that differences in the rate at which competing species migrate have implications for their response to climate change (Alexander *et al.*, 2015). Applying a similar approach to invasion, considering aspects of the process of introduction to new locations that may pre-adapt non-natives to change, is an important avenue for future experimental work. Specifically, improving understanding of when and how species respond to changing biotic interactions and environmental conditions will be central to predicting the outcome of novel non-native–native interactions, as well as how existing invaded communities will respond to changing environmental conditions.

16.7.3 Applying competition theory to managing invasions

Competition theory has considerable promise as a tool to prevent and mitigate non-native plant invasions. For example, maintaining community diversity may help prevent non-native plant species from establishing (Funk *et al.*, 2008), while sowing desirable species known to be competitive under local conditions, or identified based on a specific set of traits known to be important, may help restore communities following invasion (Funk *et al.*, 2008; Laughlin, 2014; Schuster *et al.*, 2018). However, the most appropriate action is likely highly context dependent (Funk *et al.*, 2008), and would be difficult to identify in locations where multiple non-native plant species have successfully invaded, since no single strategy may apply universally. This important application of competition theory to plant invasions highlights the need for research that bridges that gap between the two disciplines.

16.7.4 Filling in geographic and taxonomic gaps

There are considerable geographic and taxonomic biases in the current competition literature. In particular, little research has examined competitive interactions between plants (regardless of origin) in aquatic systems or remote locations, and there is a strong bias towards grassland systems and species. For example, much of competition theory has been developed from research in North American and European grasslands where competition has historically been considered an important process structuring community composition. However, recent work in less productive Australian grasslands has repeatedly failed to find much evidence of competitive interactions among species (Bimler *et al.*, 2018; O'Reilly-Nugent *et al.*, 2019; Wainwright *et al.*, 2019; Wandrag *et al.*, 2019). This suggests that expanding research to fill current geographic and taxonomic gaps may generate new insights into the processes that affect and are affected by competition.

16.8 Conclusions

It is increasingly clear that community ecology and invasion biology each have much to offer the other. While the two disciplines have historically developed in parallel, attempts to bridge the gap between them have already provided valuable insights into how plant communities assemble, and what enables some plant species to invade and impact resident native vegetation. Competitive interactions between native and non-native plants are a strong example of this. Understanding the processes that influence the strength of competition between non-native and native plant species will help predict which non-native plant species could invade, and where they are most likely to displace native plant species. Consequently, developing a predictive framework for identifying and managing non-native plant invasions will necessarily involve a deeper understanding of the rules governing the influence of neighbours on an individual's performance.

Acknowledgements

We thank Dave Richardson and Anna Traveset for inviting us to write this chapter, and two anonymous reviewers for helpful feedback. Discussions with numerous colleagues have helped develop our understanding of many ideas presented in this chapter. We would especially like to acknowledge useful discussions with Richard Duncan on the application of coexistence theory to understanding non-native plant invasions.

References

Alexander, J.M., Diez, J.M. and Levine, J.M. (2015) Novel competitors shape species' responses to climate change. *Nature* 525(7570), 515–518. DOI: 10.1038/nature14952.

Alexander, J.M., Diez, J.M., Hart, S.P. and Levine, J.M. (2016) When climate reshuffles competitors: a call for experimental macroecology. *Trends in Ecology & Evolution* 31(11), 831–841. DOI: 10.1016/j.tree.2016.08.003.

Allesina, S. and Levine, J.M. (2011) A competitive network theory of species diversity. *Proceedings of the National Academy of Sciences* 108(14), 5638–5642. DOI: 10.1073/pnas.1014428108.

Alpert, P. (2006) The advantages and disadvantages of being introduced. *Biological Invasions* 8(7), 1523–1534. DOI: 10.1007/s10530-005-5844-z.

Bais, H.P., Vepachedu, R., Gilroy, S., Callaway, R.M., Vivanco, J.M. *et al.* (2003) Allelopathy and exotic plant invasion: from molecules and genes to species interactions. *Science* 301(5638), 1377–1380. DOI: 10.1126/science.1083245.

Bauer, J.T. (2012) Invasive species: "back-seat drivers" of ecosystem change? *Biological Invasions* 14(7), 1295–1304. DOI: 10.1007/s10530-011-0165-x.

Beaton, L.L., Van Zandt, P.A., Esselman, E.J. and Knight, T.M. (2011) Comparison of the herbivore defense and competitive ability of ancestral and modern genotypes of an invasive plant, *Lespedeza cuneata*. *Oikos* 120(9), 1413–1419. DOI: 10.1111/j.1600-0706.2011.18893.x.

Becerra, P.I., Catford, J.A., Inderjit., Luce McLeod, M., Andonian, K. *et al.* (2018) Inhibitory effects of *Eucalyptus globulus* on understorey plant growth and species richness are greater in non-native regions. *Global Ecology and Biogeography* 27(1), 68–76. DOI: 10.1111/geb.12676.

Bever, J.D. (2003) Soil community feedback and the coexistence of competitors: conceptual frameworks and empirical tests. *New Phytologist* 157(3), 465–473. DOI: 10.1046/j.1469-8137.2003.00714.x.

Bimler, M.D., Stouffer, D.B., Lai, H.R. and Mayfield, M.M. (2018) Accurate predictions of coexistence in natural systems require the inclusion of facilitative interactions and environmental dependency. *Journal of Ecology* 106(5), 1839–1852. DOI: 10.1111/1365-2745.13030.

Blair, A.C. and Wolfe, L.M. (2004) The evolution of an invasive plant: An experimental study with *Silene latifolia*. *Ecology* 85(11), 3035–3042. DOI: 10.1890/04-0341.

Blossey, B. and Notzold, R. (1995) Evolution of increased competitive ability in invasive nonindigenous plants: a hypothesis. *The Journal of Ecology* 83(5), 887–889. DOI: 10.2307/2261425.

Blumenthal, D.M. and Hufbauer, R.A. (2007) Increased plant size in exotic populations: a common-garden test with 14 invasive species. *Ecology* 88(11), 2758–2765. DOI: 10.1890/06-2115.1.

Bossdorf, O., Prati, D., Auge, H. and Schmid, B. (2004) Reduced competitive ability in an invasive plant. *Ecology Letters* 7(4), 346–353. DOI: 10.1111/j.1461-0248.2004.00583.x.

Buckley, Y.M. and Catford, J. (2016) Does the biogeographic origin of species matter? Ecological effects of native and non-native species and the use of origin to guide management. *Journal of Ecology* 104(1), 4–17. DOI: 10.1111/1365-2745.12501.

Callaway, R.M. and Aschehoug, E.T. (2000) Invasive plants versus their new and old neighbors: a mechanism for exotic invasion. *Science* 290(5491), 521–523. DOI: 10.1126/science.290.5491.521.

Callaway, R.M., Thelen, G.C., Barth, S., Ramsey, P.W., Gannon, J.E. *et al.* (2004) Soil fungi alter interactions between the invader *Centaurea maculosa* and North American natives. *Ecology* 85(4), 1062–1071. DOI: 10.1890/02-0775.

Catford, J.A., Bode, M. and Tilman, D. (2018) Introduced species that overcome life history tradeoffs can cause native extinctions. *Nature Communications* 9(1), 2131. DOI: 10.1038/s41467-018-04491-3.

Catford, J.A., Downes, B.J., Gippel, C.J. and Vesk, P.A. (2011) Flow regulation reduces native plant cover and facilitates exotic invasion in riparian wetlands. *Journal of Applied Ecology* 48(2), 432–442. DOI: 10.1111/j.1365-2664.2010.01945.x.

Catford, J.A. and Jones, L.P. (2019) Grassland invasion in a changing climate. In: Gibson, D.J. and Newman, J. (eds) *Grasslands and Climate Change*. Cambridge University Press, Cambridge, pp. 149–171.

Catford, J.A., Jansson, R. and Nilsson, C. (2009) Reducing redundancy in invasion ecology by integrating hypotheses into a single theoretical framework. *Diversity and Distributions* 15(1), 22–40. DOI: 10.1111/j.1472-4642.2008.00521.x.

Catford, J.A., Morris, W.K., Vesk, P.A., Gippel, C.J., Downes, B.J. *et al.* (2014) Species and environmental characteristics point to flow regulation and drought as drivers of riparian plant invasion. *Diversity and Distributions* 20(9), 1084–1096. DOI: 10.1111/ddi.12225.

Chesson, P. (2000) Mechanisms of maintenance of species diversity. *Annual Review of Ecology and Systematics* 31(1), 343–366. DOI: 10.1146/annurev.ecolsys.31.1.343.

Colautti, R.I. and MacIsaac, H.J. (2004) A neutral terminology to define 'invasive' species. *Diversity and Distributions* 10(2), 135–141. DOI: 10.1111/j.1366-9516.2004.00061.x.

Colautti, R.I., Ricciardi, A., Grigorovich, I.A. and MacIsaac, H.J. (2004) Is invasion success explained by the enemy release hypothesis? *Ecology Letters* 7(8), 721–733. DOI: 10.1111/j.1461-0248.2004.00616.x.

Connell, J.H. (1983) On the prevalence and relative importance of interspecific competition: evidence from field experiments. *The American Naturalist* 122(5), 661–696. DOI: 10.1086/284165.

Conti, L., Block, S., Parepa, M., Münkemüller, T., Thuiller, W. *et al.* (2018) Functional trait differences and trait plasticity mediate biotic resistance to potential plant invaders. *Journal of Ecology* 106(4), 1607–1620. DOI: 10.1111/1365-2745.12928.

Daehler, C.C. (2003) Performance comparisons of co-occurring native and alien invasive plants: implications for conservation and restoration. *Annual Review of Ecology, Evolution, and Systematics* 34(1), 183–211. DOI: 10.1146/annurev.ecolsys.34.011802.132403.

Darwin, C. (1859) *On the Origin of Species*. Murray, London.

Davis, M.A., Grime, J.P. and Thompson, K. (2000) Fluctuating resources in plant communities: a general theory of invasibility. *Journal of Ecology* 88(3), 528–534. DOI: 10.1046/j.1365-2745.2000.00473.x.

Dawson, W., Fischer, M. and van Kleunen, M. (2012) Common and rare plant species respond differently to fertilisation and competition, whether they are alien or native. *Ecology Letters* 15(8), 873–880. DOI: 10.1111/j.1461-0248.2012.01811.x.

Diez, J.M., Dickie, I., Edwards, G., Hulme, P.E., Sullivan, J.J. *et al.* (2010) Negative soil feedbacks accumulate over time for non-native plant species. *Ecology Letters* 13(7), 803–809. DOI: 10.1111/j.1461-0248.2010.01474.x.

Driscoll, D.A., Catford, J.A., Barney, J.N., Hulme, P.E., Inderjit. *et al.* (2014) New pasture plants intensify invasive species risk. *Proceedings of the National Academy of Sciences* 111(46), 16622–16627. DOI: 10.1073/pnas.1409347111.

Dukes, J.S. and Mooney, H.A. (1999) Does global change increase the success of biological invaders? *Trends in Ecology & Evolution* 14(4), 135–139. DOI: 10.1016/S0169-5347(98)01554-7.

Elton, C.S. (1958) *The Ecology of Invasions by Animals and Plants*. Methuen, London.

Enders, M., Hütt, M.-T. and Jeschke, J.M. (2018) Drawing a map of invasion biology based on a network of hypotheses. *Ecosphere* 9(3), e02146. DOI: 10.1002/ecs2.2146.

Enders, M. and Jeschke, J.M. (2018) A network of invasion hypotheses. In: Jeschke, J.M. and Heger, T. (eds) *Invasion Biology: Hypotheses and Evidence*. CABI, Wallingford, UK, pp. 49–59.

Eppinga, M.B., Rietkerk, M., Dekker, S.C., De Ruiter, P.C., Van der Putten, W.H. *et al.* (2006) Accumulation of local pathogens: a new hypothesis to explain exotic plant invasions. *Oikos* 114(1), 168–176. DOI: 10.1111/j.2006.0030-1299.14625.x.

Esch, E.H., Ashbacher, A.C., Kopp, C.W. and Cleland, E.E. (2018) Competition reverses the response of shrub seedling mortality and growth along a soil moisture gradient. *Journal of Ecology* 106(5), 2096–2108. DOI: 10.1111/1365-2745.12964.

Fridley, J.D. and Sax, D.F. (2014) The imbalance of nature: revisiting a Darwinian framework for invasion biology. *Global Ecology and Biogeography* 23(11), 1157–1166. DOI: 10.1111/geb.12221.

Funk, J.L., Cleland, E.E., Suding, K.N. and Zavaleta, E.S. (2008) Restoration through reassembly: plant traits and invasion resistance. *Trends in Ecology & Evolution* 23(12), 695–703. DOI: 10.1016/j.tree.2008.07.013.

Gallien, L. (2017) Intransitive competition and its effects on community functional diversity. *Oikos* 126(5), 615–623. DOI: 10.1111/oik.04033.
Gaudet, C.L. and Keddy, P.A. (1988) A comparative approach to predicting competitive ability from plant traits. *Nature* 334(6179), 242–243. DOI: 10.1038/334242a0.
Gause, G.F. (1934) *The Struggle for Existence*. Williams and Wilkins, Baltimore, Maryland.
Gilpin, M.E. (1975) Limit cycles in competition communities. *The American Naturalist* 109(965), 51–60. DOI: 10.1086/282973.
Gleeson, S.K. and Tilman, D. (1990) Allocation and the transient dynamics of succession on poor soils. *Ecology* 71(3), 1144–1155. DOI: 10.2307/1937382.
Gleeson, S.K. and Tilman, D. (1994) Plant allocation, growth rate and successional status. *Functional Ecology* 8(4), 543–550. DOI: 10.2307/2390080.
Godoy, O. and Levine, J.M. (2014) Phenology effects on invasion success: insights from coupling field experiments to coexistence theory. *Ecology* 95(3), 726–736. DOI: 10.1890/13-1157.1.
Goldberg, D.E. (1996) Competitive ability: definitions, contingency and correlated traits. philosophical transactions of the Royal Society of London. *Series B* 351, 1377–1385.
Goldberg, D.E. and Landa, K. (1991) Competitive effect and response: hierarchies and correlated traits in the early stages of competition. *The Journal of Ecology* 79(4), 1013. DOI: 10.2307/2261095.
Grime, J.P. (1979) *Plant Strategies and Vegetation Processes*. Wiley, London.
Gurevitch, J. (2011) Competition, plant. In: Simberloff, D. and Rejmánek, M. (eds) *Encyclopedia of Biological Invasions*. University of California Press, Berkeley, pp. 122–125.
Harpole, W.S., Sullivan, L.L., Lind, E.M., Firn, J., Adler, P.B. *et al.* (2016) Addition of multiple limiting resources reduces grassland diversity. *Nature* 537(7618), 93–96. DOI: 10.1038/nature19324.
Hart, S.P., Freckleton, R.P. and Levine, J.M. (2018) How to quantify competitive ability. *Journal of Ecology* 106(5), 1902–1909. DOI: 10.1111/1365-2745.12954.
Hautier, Y., Niklaus, P.A. and Hector, A. (2009) Competition for light causes plant biodiversity loss after eutrophication. *Science* 324(5927), 636–638. DOI: 10.1126/science.1169640.
Hawkes, C. V. (2007) Are invaders moving targets? the generality and persistence of advantages in size, reproduction, and enemy release in invasive plant species with time since introduction. *The American Naturalist* 170(6), 832–843. DOI: 10.1086/522842.
Hierro, J.L., Maron, J.L. and Callaway, R.M. (2005) A biogeographical approach to plant invasions: the importance of studying exotics in their introduced and native range. *Journal of Ecology* 93(1), 5–15.
Higgins, S.I. and Richardson, D.M. (1998) Pine invasions in the southern hemisphere: modelling interactions between organism, environment and disturbance. *Plant Ecology* 135(1), 79–93. DOI: 10.1023/A:1009760512895.
HilleRisLambers, J., Yelenik, S.G., Colman, B.P. and Levine, J.M. (2010) California annual grass invaders: the drivers or passengers of change? *Journal of Ecology* 98(5), 1147–1156. DOI: 10.1111/j.1365-2745.2010.01706.x.
Holt, R.D. (1977) Predation, apparent competition, and the structure of prey communities. *Theoretical Population Biology* 12(2), 197–229. DOI: 10.1016/0040-5809(77)90042-9.
Huang, F., Lankau, R. and Peng, S. (2018) Coexistence via coevolution driven by reduced allelochemical effects and increased tolerance to competition between invasive and native plants. *New Phytologist* 218(1), 357–369. DOI: 10.1111/nph.14937.
Huston, M. (1979) A general hypothesis of species diversity. *The American Naturalist* 113(1), 81–101. DOI: 10.1086/283366.
Joshi, J. and Vrieling, K. (2005) The enemy release and EICA hypothesis revisited: incorporating the fundamental difference between specialist and generalist herbivores. *Ecology Letters* 8(7), 704–714. DOI: 10.1111/j.1461-0248.2005.00769.x.
Keane, R. and Crawley, M.J. (2002) Exotic plant invasions and the enemy release hypothesis. *Trends in Ecology & Evolution* 17(4), 164–170. DOI: 10.1016/S0169-5347(02)02499-0.
Keddy, P.A. (2001) *Competition*. Kluwer Academic Publishers, Dordrecht.
Kunstler, G., Lavergne, S., Courbaud, B., Thuiller, W., Vieilledent, G. *et al.* (2012) Competitive interactions between forest trees are driven by species' trait hierarchy, not phylogenetic or functional similarity: implications for forest community assembly. *Ecology Letters* 15(8), 831–840. DOI: 10.1111/j.1461-0248.2012.01803.x.
Laughlin, D.C. (2014) Applying trait-based models to achieve functional targets for theory-driven ecological restoration. *Ecology Letters* 17(7), 771–784. DOI: 10.1111/ele.12288.

Le Roux, J.J., Ellis, A.G., van Zyl, L.-M., Hosking, N.D., Keet, J.-H. *et al.* (2018) Importance of soil legacy effects and successful mutualistic interactions during Australian Acacia invasions in nutrient-poor environments. *Journal of Ecology* 106(5), 2071–2081. DOI: 10.1111/1365-2745.12965.
Leger, E.A. and Rice, K.J. (2003) Invasive California poppies (*Eschscholzia californica* Cham.) grow larger than native individuals under reduced competition. *Ecology Letters* 6(3), 257–264. DOI: 10.1046/j.1461-0248.2003.00423.x.
Levine, J.M., Adler, P.B. and Yelenik, S.G. (2004) A meta-analysis of biotic resistance to exotic plant invasions. *Ecology Letters* 7(10), 975–989. DOI: 10.1111/j.1461-0248.2004.00657.x.
MacDougall, A.S., Gilbert, B. and Levine, J.M. (2009) Plant invasions and the niche. *Journal of Ecology* 97(4), 609–615. DOI: 10.1111/j.1365-2745.2009.01514.x.
MacDougall, A.S., McCune, J.L., Eriksson, O., Cousins, S.A.O., Pärtel, M. *et al.* (2018) The Neolithic plant invasion hypothesis: the role of preadaptation and disturbance in grassland invasion. *New Phytologist* 220(1), 94–103. DOI: 10.1111/nph.15285.
MacDougall, A.S. and Turkington, R.O.Y. (2004) Relative importance of suppression-based and tolerance-based competition in an invaded oak savanna. *Journal of Ecology* 92(3), 422–434. DOI: 10.1111/j.0022-0477.2004.00886.x.
MacDougall, A.S. and Turkington, R. (2006) Dispersal, competition, and shifting patterns of diversity in a degraded oak savanna. *Ecology* 87(7), 1831–1843. DOI: 10.1890/0012-9658(2006)87[1831:DCASPO]2.0.CO;2.
Macarthur, R. and Levins, R. (1967) The limiting similarity, convergence, and divergence of coexisting species. *The American Naturalist* 101(921), 377–385. DOI: 10.1086/282505.
Mack, K.M.L., Eppinga, M.B. and Bever, J.D. (2019) Plant-soil feedbacks promote coexistence and resilience in multi-species communities. *PLoS ONE* 14(2), e0211572–20. DOI: 10.1371/journal.pone.0211572.
Marler, M.J., Zabinski, C.A. and Callaway, R.M. (1999) Mycorrhizae indirectly enhance competitive effects of an invasive forb on a native bunchgrass. *Ecology* 80(4), 1180–1186. DOI: 10.1890/0012-9658(1999)080[1180:MIECEO]2.0.CO;2.
Maron, J.L. and Marler, M. (2008) Field-based competitive impacts between invaders and natives at varying resource supply. *Journal of Ecology* 96(6), 1187–1197. DOI: 10.1111/j.1365-2745.2008.01440.x.
May, R.M. and Leonard, W.J. (1975) Nonlinear aspects of competition between three species. *SIAM Journal on Applied Mathematics* 29(2), 243–253. DOI: 10.1137/0129022.
McGregor, K.F., Watt, M.S., Hulme, P.E. and Duncan, R.P. (2012) What determines pine naturalization: species traits. *climate suitability or forestry use? Diversity and Distributions* 18, 1013–1023.
Melbourne, B.A., Cornell, H.V., Davies, K.F. and Dugaw, C.J. *et al.* (2007) Invasion in a heterogeneous world: resistance. *coexistence or hostile takeover? Ecology Letters* 10, 77–94.
Mitchell, C.E., Agrawal, A.A., Bever, J.D., Gilbert, G.S., Hufbauer, R.A. *et al.* (2006) Biotic interactions and plant invasions. *Ecology Letters* 9(6), 726–740. DOI: 10.1111/j.1461-0248.2006.00908.x.
Morris, T.L., Barger, N.N. and Cramer, M.D. (2015) Competitive resistance of a native shrubland to invasion by the alien invasive tree species, *Acacia cyclops*. *Biological Invasions* 17(12), 3563–3577. DOI: 10.1007/s10530-015-0978-0.
Ni, G.-Y., Schaffner, U., Peng, S.-L. and Callaway, R.M. (2010) *Acroptilon repens*, an Asian invader, has stronger competitive effects on species from America than species from its native range. *Biological Invasions* 12(11), 3653–3663. DOI: 10.1007/s10530-010-9759-y.
Oduor, A.M.O. (2013) Evolutionary responses of native plant species to invasive plants: a review. *New Phytologist* 200(4), 986–992. DOI: 10.1111/nph.12429.
Orrock, J.L., Witter, M.S. and Reichman, O.J. (2008) Apparent competition with an exotic plant reduces native plant establishment. *Ecology* 89(4), 1168–1174. DOI: 10.1890/07-0223.1.
O'Reilly-Nugent, A., Wandrag, E.M., Catford, J.A. and Gruber, B. *et al.* (2019) Measuring competitive impact: Joint-species modelling of invaded plant communities. *Journal of Ecology*.
Park, T. (1954) Experimental studies of interspecies competition II. temperature, humidity, and competition in two species of *Tribolium*. *Physiological Zoology* 27(3), 177–238. DOI: 10.1086/physzool.27.3.30152164.
Peart, D.R. (1989) Species interactions in a successional grassland. II. colonization of vegetated sites. *Journal of Ecology* 77, 236–251.
Pinzone, P., Potts, D., Pettibone, G. and Warren, R. (2018) Do novel weapons that degrade mycorrhizal mutualisms promote species invasion? *Plant Ecology* 219(5), 539–548. DOI: 10.1007/s11258-018-0816-4.

Pyšek, P., Jarošík, V., Hulme, P.E., Kuhn, I., Wild, J. *et al.* (2010) Disentangling the role of environmental and human pressures on biological invasions across Europe. *Proceedings of the National Academy of Sciences* 107(27), 12157–12162. DOI: 10.1073/pnas.1002314107.

Rathcke, B. (1983) Competition and facilitation among plants for pollination. In: Real, L. (ed.) *Pollination Biology*. Academic Press, New York, United States, pp. 305–329.

Rejmánek, M. and Richardson, D.M. (1996) What attributes make some plant species more invasive? *Ecology* 77(6), 1655–1661. DOI: 10.2307/2265768.

Richards, C.L., Bossdorf, O., Muth, N.Z., Gurevitch, J., Pigliucci, M. *et al.* (2006) Jack of all trades, master of some? on the role of phenotypic plasticity in plant invasions. *Ecology Letters* 9(8), 981–993. DOI: 10.1111/j.1461-0248.2006.00950.x.

Richardson, D.M. and Bond, W.J. (1991) Determinants of plant distribution: evidence from pine invasions. *The American Naturalist* 137(5), 639–668. DOI: 10.1086/285186.

Richardson, D.M. and Pyšek, P. (2012) Naturalization of introduced plants: ecological drivers of biogeographical patterns. *New Phytologist* 196(2), 383–396. DOI: 10.1111/j.1469-8137.2012.04292.x.

Ridenour, W.M. and Callaway, R.M. (2001) The relative importance of allelopathy in interference: the effects of an invasive weed on a native bunchgrass. *Oecologia* 126(3), 444–450. DOI: 10.1007/s004420000533.

Rundel, P.W., Dickie, I.A. and Richardson, D.M. (2014) Tree invasions into treeless areas: mechanisms and ecosystem processes. *Biological Invasions* 16(3), 663–675. DOI: 10.1007/s10530-013-0614-9.

Schuster, M.J., Wragg, P.D. and Reich, P.B. (2018) Using revegetation to suppress invasive plants in grasslands and forests. *Journal of Applied Ecology* 55(5), 2362–2373. DOI: 10.1111/1365-2664.13195.

Seabloom, E.W., Harpole, W.S., Reichman, O.J. and Tilman, D. (2003a) Invasion, competitive dominance, and resource use by exotic and native California grassland species. *Proceedings of the National Academy of Sciences* 100(23), 13384–13389. DOI: 10.1073/pnas.1835728100.

Seabloom, E.W., Borer, E.T., Boucher, V.L., Burton, R.S., Cottingham, K.L. *et al.* (2003b) Competition, seed limitation, disturbance, and reestablishment of California native annual forbs. *Ecological Applications* 13(3), 575–592. DOI: 10.1890/1051-0761(2003)013[0575:CSLDAR]2.0.CO;2.

Seabloom, E.W., Borer, E.T., Buckley, Y., Cleland, E.E., Davies, K. *et al.* (2013) Predicting invasion in grassland ecosystems: is exotic dominance the real embarrassment of richness? *Global Change Biology* 19(12), 3677–3687. DOI: 10.1111/gcb.12370.

Seabloom, E.W., Borer, E.T., Buckley, Y.M., Cleland, E.E., Davies, K.F. *et al.* (2015) Plant species' origin predicts dominance and response to nutrient enrichment and herbivores in global grasslands. *Nature Communications* 6(1), 7710. DOI: 10.1038/ncomms8710.

Seifert, E.K., Bever, J.D. and Maron, J.L. (2009) Evidence for the evolution of reduced mycorrhizal dependence during plant invasion. *Ecology* 90(4), 1055–1062. DOI: 10.1890/08-0419.1.

Siemann, E. and Rogers, W.E. (2001) Genetic differences in competitive abilities of an invasive tree species. *Ecology Letters* 4, 514–518.

Simberloff, D. (2009) The role of propagule pressure in biological invasions. *Annual Review of Ecology, Evolution, and Systematics* 40(1), 81–102. DOI: 10.1146/annurev.ecolsys.110308.120304.

Simberloff, D. and Gibbons, L. (2004) Now you see them, now you don't! – Population crashes of established introduced species. *Biological Invasions* 6(2), 161–172. DOI: 10.1023/B:BINV.0000022133.49752.46.

Simberloff, D. and Von Holle, B. (1999) Positive interactions of nonindigenous species: invasional meltdown? *Biological Invasions* 1(1), 21–32. DOI: 10.1023/A:1010086329619.

Soliveres, S., Maestre, F.T., Ulrich, W., Manning, P., Boch, S. *et al.* (2015) Intransitive competition is widespread in plant communities and maintains their species richness. *Ecology Letters* 18(8), 790–798. DOI: 10.1111/ele.12456.

Stachowicz, J.J. and Tilman, D. (2005) Species invasions and the relationships between species diversity, community saturation, and ecosystem functioning. In: Sax, D.F., Stachowicz, J.J. and Gaines, S.D. (eds) *Species Invasions: Insights into Ecology, Evolution, and Biogeography*. Sinauer Sunderland, Massachusetts, pp. 41–64.

Strauss, S.Y., Lau, J.A. and Carroll, S.P. (2006) Evolutionary responses of natives to introduced species: what do introductions tell us about natural communities? *Ecology Letters* 9(3), 357–374. DOI: 10.1111/j.1461-0248.2005.00874.x.

te Beest, M., Mpandza, N.J. and Olff, H. (2018) Biotic resistance affects growth and reproduction, but not survival of a high-impact woody invader in African savannas. *Journal of Vegetation Science* 29(3), 532–540. DOI: 10.1111/jvs.12633.

Tilman, D. (1982) *Resource Competition and Community Structure*. Princeton University Press, Princeton.

Tilman, D. (1994) Competition and biodiversity in spatially structured habitats. *Ecology* 75(1), 2–16. DOI: 10.2307/1939377.

Tilman, D. (2011) Diversification, biotic interchange, and the universal trade-off hypothesis. *The American Naturalist* 178(3), 355–371. DOI: 10.1086/661245.

van Kleunen, M., Bossdorf, O. and Dawson, W. (2018) The ecology and evolution of alien plants. *Annual Review of Ecology, Evolution, and Systematics* 49(1), 25–47. DOI: 10.1146/annurev-ecolsys-110617-062654.

van Kleunen, M., Dawson, W., Essl, F., Pergl, J., Winter, M. *et al*. (2015) Global exchange and accumulation of non-native plants. *Nature* 525(7567), 100–103. DOI: 10.1038/nature14910.

Wainwright, C.E., HilleRisLambers, J., Lai, H.R., Loy, X., Mayfield, M.M. *et al*. (2019) Distinct responses of niche and fitness differences to water availability underlie variable coexistence outcomes in semi-arid annual plant communities. *Journal of Ecology* 107(1), 293–306. DOI: 10.1111/1365-2745.13056.

Wandrag, E.M., Catford, J.A. and Duncan, R.P. (2019) Quantifying niche availability, niche overlap and competition for recruitment sites in plant populations without explicit knowledge of niche axes. *Journal of Ecology* 107(4), 1791–1803. DOI: 10.1111/1365-2745.13137.

Wang, P., Stieglitz, T., Zhou, D.W. and Cahill, J.F. (2010) Are competitive effect and response two sides of the same coin. *or fundamentally different? Functional Ecology* 24, 196–207.

17 Indirect Biotic Interactions of Plant Invasions with Native Plants and Animals

Warwick J. Allen*

The Bio-Protection Research Centre, School of Biological Sciences, University of Canterbury, Christchurch, New Zealand

Abstract

Invasive plants often occur at high densities and tend to be highly generalist in their interactions with herbivores, pathogens, mycorrhiza, endophytes and pollinators. These characteristics mean that invasive plants should frequently participate in diverse indirect biotic interactions with the surrounding community, mediated by their direct interaction partners (e.g. antagonists and mutualists). Indirect interactions play an important role in many ecological processes, yet we still lack a systematic understanding of the circumstances under which they influence the success and impacts of invasive species. In this chapter, I first describe several of the indirect interaction pathways that are commonly encountered in invasion biology and review their contribution to the impacts of plant invasions on co-occurring species. The literature review revealed that there are now many case studies describing various indirect impacts of invasive plants. However, identical interaction motifs (e.g. plant–enemy–plant, plant–mutualist–plant) can bring about several possible outcomes, depending upon each species' provenance, relative abundances and interaction strengths, abiotic resource availability, spatial and temporal scale and the influence of other species. Moreover, knowledge gaps identified include a lack of studies of indirect facilitation outside of plant—pollinator systems, limited consideration of indirect invader impacts on other non-native species, and the scarcity of generalizable results to date. Second, I integrate the literature with some trending research areas in invasion biology (interaction networks, biogeography, invasion dynamics) and identify some potential future research directions. Finally, I discuss how knowledge about indirect biotic interactions could be incorporated into the management of invasive plants.

17.1 Introduction

Invasive species have a diverse range of impacts that often occur through their direct interactions with other species in the community (Mitchell *et al.*, 2006; Tylianakis *et al.*, 2008). Whereas these direct impacts can usually be observed, measured and predicted with relative ease, the indirect impacts of invaders remain less well understood (White *et al.*, 2006). Indirect biotic interactions are defined as impacts of one species on the growth, fitness or population dynamics of another through changes in the population or behaviour of a third species (indirect interactions mediated through abiotic factors were not considered for this review). Indirect interactions play an important role in the structure, dynamics and evolution of ecological communities (Strauss, 1991; Wootton, 1994; Walsh, 2013), which can be equal to that of direct interactions (Menge, 1995). Given their disproportionately high abundance (van Kleunen *et al.*, 2010) and generalist associations (Aizen *et al.*, 2008; Moora *et al.*, 2011; Albrecht *et al.*, 2014; Klock *et al.*, 2015), invasive species may be expected to engage in strong and diverse indirect interactions with the surrounding community. Yet, we still lack a systematic understanding of the role

*warwick.j.allen@gmail.com

 Plant Invasions: The Role of Biotic Interactions (eds Anna Traveset and David M. Richardson)
DOI: 10.1079/9781789242171.0017

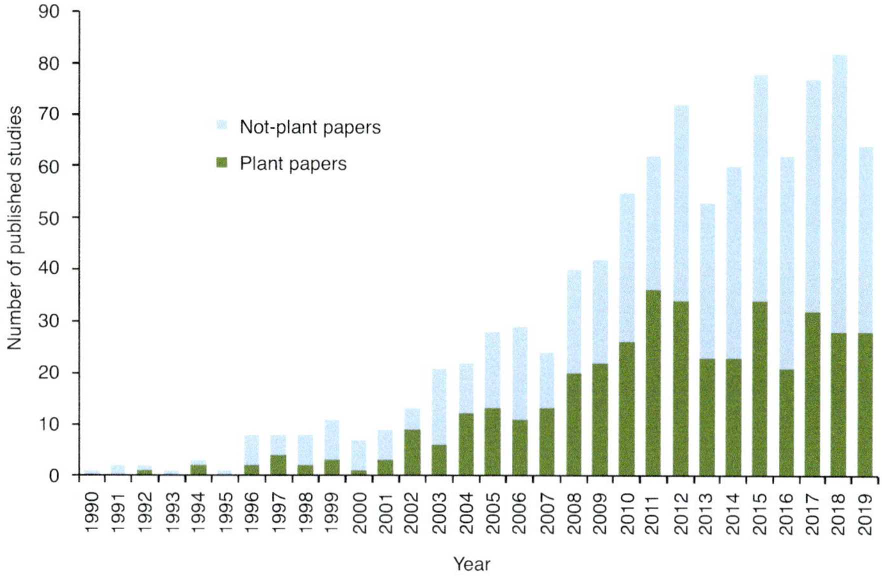

Fig. 17.1. Results of a Web of Science search showing the number of studies of indirect interactions involving non-native species between 1990 and 2019, separated into those investigating plants vs other taxa (total $n = 945$). See Appendix 17.1 for the search terms.

of indirect interactions in mediating the success and impacts of invasive species.

The last major review of indirect interactions in invasion ecology was by White *et al.* (2006), who identified and synthesized the findings of 16 published examples of apparent competition involving at least one non-native species, 11 examples of indirect mutualism/facilitation (via either a third competitor or shared natural enemy) and eight examples each of trophic cascades and exploitative competition (for a biotic resource). Since then, the literature has grown substantially (Fig. 17.1), especially through case studies describing the myriad ways that indirect interactions can contribute to the success or impact of specific invasive species. Alongside this growth in the literature, it has become apparent that indirect species interactions are highly context-dependent and species-specific, so far limiting our ability to make general inferences and predictions about their importance to the success, impacts and management of invasive plants. In this chapter, I aim to integrate the idiosyncratic literature with some emergent themes in invasion ecology (i.e. interaction networks, biogeography, invasion dynamics) and suggest fruitful avenues for future research. First, I describe the indirect interaction pathways that are commonly encountered in invasion biology (Box 17.1 and Fig. 17.2) and review the current state of knowledge on their role in plant invasion impacts. Second, I outline potential future research directions and outstanding questions (Box 17.2), revolving around a common theme of considering the community, spatial and temporal scale in studies of indirect interactions and invasions. Finally, I discuss ways in which indirect interactions may be integrated into the management of invasive plants.

17.2 State of Knowledge on Indirect Interactions and Invasions

Invasive plants participate in a wide variety of indirect interactions that become increasingly complex as more species are added to the system. As species richness in a community increases, the number of potential three-species indirect interaction pathways also grows at a disproportionate rate to that of direct interactions

Box 17.1. Indirect interactions commonly studied in invasion ecology.

Apparent competition

Apparent competition describes negative interactions between two or more species mediated by changes in the population or behaviour of shared natural enemies (Fig. 17.2, Ai) (Holt, 1977; Holt and Bonsall, 2017). In theory, this occurs when one competitor (e.g. a plant species) increases in density, resulting in an increase in the density of its natural enemies (e.g. herbivores or pathogens), which in turn negatively impacts other competitors (e.g. other plant species). Both theoretical (Holt, 1977) and empirical (Bonsall and Hassell, 1997; Morris *et al.*, 2004; Frost *et al.*, 2016) evidence supports the importance of apparent competition to the structure and dynamics of natural communities. Apparent competitive impacts may help explain invasion by plant species that are inferior competitors for abiotic resources, and are likely underreported because they are difficult to observe and easily mistaken for resource competition. Nonetheless, evidence for apparent competition as a mechanism of invasion success and impacts has accumulated in several plant systems, mediated by both pathogens (Power and Mitchell, 2004; Borer *et al.*, 2007) and herbivores (Sessions and Kelly, 2002; Dangremond *et al.*, 2010; Orrock *et al.*, 2015; see Appendix S2 of Bhattarai *et al.*, 2017a, for a complete list of studies and their findings).

Apparent facilitation

Apparent facilitation can be described using a similar framework to apparent competition, except that the species mediating the indirect interaction between plants is a mutualist, such as pollinators, rhizobia, mycorrhizal fungi or dispersal vectors (Fig. 17.2, Bii and Cii). Despite long being recognized as a potential driver of community structure and dynamics (Levine, 1980), this type of indirect interaction was not considered in the earlier review by White *et al.* (2006). Apparent facilitation has largely been demonstrated in the invasion literature through plant–pollinator interactions, with mixed support for their importance (Aizen *et al.*, 2008; Bartomeus *et al.*, 2008; Morales and Traveset, 2009; Stouffer *et al.*, 2014; Charlebois and Sargent, 2017), whereas other plant mutualists have received surprisingly little consideration (but see Rodríguez-Echeverría, 2010; Dickie *et al.*, 2014; Grove *et al.*, 2017; Allen *et al.*, 2020).

Competition chains

One form of indirect plant–plant facilitation occurs through competition chains, where one plant species indirectly benefits another through suppression of a shared competitor (Fig. 17.2, Biii and Ciii), as long as this indirect effect is stronger than the direct competition between the two focal species. Plant–plant facilitation has been widely studied in the community assembly literature (Levine, 1999; Aschehoug and Callaway, 2015) and shown to facilitate plant invasions (i.e. native turncoats; Northfield *et al.*, 2018; Chapter 7, this volume); yet, there have been surprisingly few empirical tests of its role in driving invasion impacts (but see Flory and Bauer, 2014). However, the recent meta-analysis by Kuebbing and Nuñez (2016) showed that negative competitive impacts of non-native plants on native plants are two times stronger than their impacts on other non-native plants, suggesting that indirect interactions among multiple native and non-native species may favour non-native dominance of plant communities.

Trophic cascades

The need for invasion ecology to include a multitrophic perspective has long been acknowledged (White *et al.*, 2006; Harvey *et al.*, 2010), but relatively few studies have investigated the role of invasive plants in trophic cascades. A trophic cascade typically describes how a plant indirectly benefits through top-down predation of its consumers by higher trophic level species (Fig. 17.2, Biv and Civ) (Strong, 1992), an interaction that has been suggested as a mechanism for enemy release of invasive plants (Engelkes *et al.*, 2012; Allen *et al.*, 2015). However, trophic cascades may also occur from the bottom up, a concept often neglected in invasion ecology, although studies conducted in this area have invariably demonstrated cascading effects spanning up to four trophic levels (Heleno *et al.*, 2009; Lau, 2013; McCary *et al.*, 2016; López-Núñez *et al.*, 2017).

Bottom-up impacts

Although invasive plants benefit organisms that directly utilize them as a resource (e.g. consumers, parasites or mutualists), these organisms may also experience negative indirect impacts if invaders displace plant species that are higher quality resources (Fig. 17.2, Aiii), acting as a kind of

Continued

ecological trap (Schlaepfer *et al*., 2002). For example, when compared to the native shrubs they displace, some invasive plants in North American forests provide birds with fruit of lower nutritional value (Greenberg and Walter, 2010) and poorer nesting habitat due to high predation rates (Rodewald, 2012). Alternatively, positive bottom-up indirect impacts could arise if the invasive plant facilitates higher quality host plant species (e.g. through providing shelter) (Fig. 17.2, Bvi and Cv).

Other indirect interactions

There are many other indirect interactions involving invasive plants, such as competition for a biotic resource (e.g. plants competing for mutualists or two herbivores competing for an invasive plant) (Fig. 17.2, Aii and Bv), belowground interactions with whole communities (e.g. plant–soil feedbacks and soil legacies), and increasingly complex indirect interactions including four or more species. These interactions are covered in detail in Chapters 3, 13 and 18, this volume.

(Northfield *et al*., 2018). Regardless of the species involved or the complexity of these interactions, indirect impacts of invasive plants on other species in the community have three fundamental outcomes: facilitation, antagonism or no impact. These outcomes can be achieved through a variety of indirect pathways, with some of the most frequently studied three-species indirect interactions in the invasion biology literature described in Box 17.1 and Fig. 17.2. For the rest of the section, I synthesize the accumulated evidence on the role of these indirect interactions in plant invasions and summarize the strengths and weaknesses of the literature to date.

All three potential outcomes of indirect interactions (facilitation, antagonism or no impact) can occur through various interaction pathways (Fig. 17.2). Moreover, identical motifs of interacting species can lead to several possible outcomes, depending upon each species' provenance,

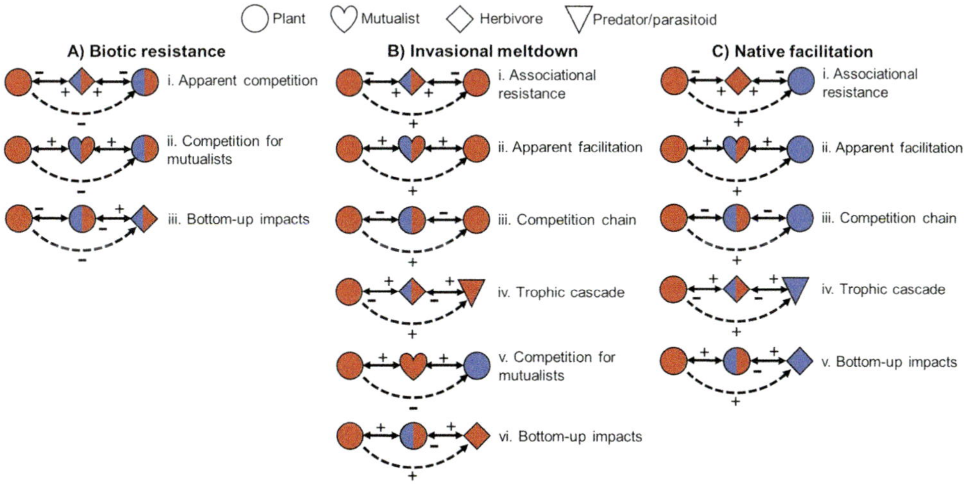

Fig. 17.2. Three-species interaction pathways demonstrating a subset of the possible indirect interactions involving invasive plants and their potential outcomes, depending upon the provenance of the other species: a) biotic resistance of non-native and native species; b) invasional meltdown of non-native species; and c) native facilitation. The focal invasive plant driving the indirect impact is always on the left-hand side of each motif and the recipient of the indirect impact on the right. Solid and dashed arrows represent direct and indirect interactions, respectively. Red and blue symbols represent non-native and native species, respectively, with symbols that are half of each colour indicating that the outcome of the indirect interaction is the same regardless of species provenance.

Box 17.2. Unanswered research questions regarding indirect interactions and plant invasions.

1. What controls the relative importance of direct and indirect interactions in the invasion process?
2. To what extent do observed indirect interactions impact individual fitness and population and community dynamics?
3. Do indirect impacts of invasive plants on native and non-native plant species differ systematically?
4. What types of indirect interactions contribute most strongly to invasion success and impacts?
5. How can we predict indirect interactions involving a species that has not yet arrived?
6. How far do indirect impacts of invasive plants propagate through communities?
7. How do indirect interactions involving invasive plants vary across spatial and temporal scales of invasion?

relative abundances and interaction strengths, availability of abiotic resources, spatial and temporal scale, and the influence of other species outside of the simplified motif. For example, apparent competition between invasive plants and other co-occurring plant species (Box 17.1) can generate both negative and positive impacts (Fig. 17.2, Ai, Bi and Ci). Negative impacts can occur through several mechanisms, including greater resistance (Rand *et al.*, 2004; Bhattarai *et al.*, 2017a) and tolerance (Ashton and Lerdau, 2008; Zhang *et al.*, 2018) of invaders, non-trophic effects (Orrock *et al.*, 2010; Enge *et al.*, 2013), and extended leaf phenology (Smith and Hall, 2016; Martinrod and Gorchov, 2017). On the other hand, susceptible invaders may attract herbivores or pathogens away from co-occurring species that share them, often known as associational resistance (Atwater *et al.*, 2011). Likewise, indirect interactions between plants mediated by shared mutualists can also lead to several possible outcomes (Dickie *et al.*, 2017), including native facilitation (Fig. 17.2, Cii), biotic resistance (Fig. 17.2, Aii) and invasional meltdown (Fig. 17.2, Biii). For example, the seminal study by Lopezaraiza-Mikel *et al.* (2007) found that the invasive plant *Impatiens glandulifera* attracted pollinators on to co-occurring species (a positive indirect effect, Fig. 17.2, Cii and Biii), but that the pollinators were covered in invader pollen, potentially inhibiting conspecific pollination (a negative indirect effect). Moreover, invasive plants can also indirectly harm co-occurring species in the short term via competition for shared mutualists (Aizen *et al.*, 2008; Bartomeus *et al.*, 2008; Morales and Traveset, 2009; but see Charlebois and Sargent, 2017; Chapter 13, this volume), yet this negative impact may be offset in the long term because both plants promote the mutualist (Tylianakis and Morris, 2017). Finally, another interesting and relatively unexplored system concerns common mycelial networks, which can mediate positive and negative indirect interactions among plants (Newman, 1988; van der Heijden and Horton, 2009; Horton, 2015) through several mechanisms, including increasing mycorrhizal inoculum (Dickie *et al.*, 2014), altering competitive interactions via shared mycorrhizae (Waller *et al.*, 2016) and alerting neighbours to the presence of natural enemies (Johnson and Gilbert, 2015). However, common mycelial networks are yet to be investigated in an invasion context. Based on these examples, we can conclude that there are many possible outcomes of indirect interactions involving three species, and that we still do not have a clear idea about the interactions that are most common or important to the success and impacts of invasive plants. However, some promising recent results have provided some early insight into this question, with Simmons *et al.* (2019a,b) showing that some interaction motifs are more common than others and that native and non-native species may fill very different motif roles.

17.2.1 Literature limitations and knowledge gaps

Several limitations and knowledge gaps are apparent in the literature on the role of indirect interaction in plant invasions. First, the literature to date has largely focused on negative indirect impacts of invasive plants (i.e. apparent competition), with indirect facilitation primarily studied through plant—pollinator systems (Bartomeus *et al.*, 2008; Padrón *et al.*, 2009; Charlebois and Sargent, 2017) overlooking important belowground interactions (e.g. mycorrhizal fungi and rhizobia bacteria) and trophic cascades. Second, nearly all studies to date have been performed between invasive and native species (Bhattarai *et al.*, 2017a), with limited consideration of invader impacts on other non-native species (but see Aizen *et al.*, 2008; Dickie *et al.*, 2014; Allen *et al.*, 2020). The provenance of interaction partners has important implications for interpreting the outcome of indirect interactions (Fig. 17.2) and planning a management programme. For example, if indirect facilitation occurs between two invasive plants, this may represent an indirect 'invasional meltdown' (Simberloff and Von Holle, 1999; Chapter 8, this volume), a phenomenon that has received little attention relative to direct facilitation (Braga *et al.*, 2018). Moreover, if the mediating species (e.g. pollinator or mycorrhizae) is native, then management approaches that target this partner may be limited. On the other hand, apparent competition between two invasive species would represent the opposite of invasional meltdown (i.e. invasional interference; Yang *et al.*, 2011). A final limitation is the prevalence of idiosyncratic case studies of single pairwise indirect interactions. These represent valuable examples of indirect impacts, but limit general understanding or predictive ability, leaving it unclear whether indirect impacts of invasive species are systematically positive or negative and favour native or non-native species. In light of these limitations, a recent study compared indirect impacts of Scotch broom (*Cytisus scoparius*) on 21 native and non-native plant species via multiple interaction partners, finding that indirect interactions tended to favour non-native species belowground (i.e. invasional meltdown via soil fungi) and native species aboveground (i.e. biotic resistance via reduced herbivory) (Allen *et al.*, 2020). However, more studies incorporating multiple native and non-native species and types of interactions are urgently needed. Moreover, the three-species examples explored in Box 17.1 and Fig. 17.2 do not include complex interactions involving four or more species. Because most community-level studies of indirect impacts to date have been conducted through observation and modelling (e.g. O'Dowd *et al.*, 2003; Bumbeer *et al.*, 2018; but see Aschehoug and Callaway, 2015), there is an urgent need for manipulative experiments of whole communities.

For invasion ecologists to develop useful management tools, it is highly desirable to predict the presence, direction and magnitude of indirect interactions among species. However, this has remained a major challenge, including for indirect impacts of invasive species (Allen *et al.*, 2020). One promising approach to this problem revolves around using interaction network analyses, phylogeny and traits to predict the outcomes of indirect interactions at the community level. This may be broken down into two key steps: i) predicting the invader's role in the non-native range, informed by their home range role (Emer *et al.*, 2016) or their phylogenetic similarity to native species (Pearse and Altermatt, 2013); and ii) predicting the indirect impacts of the invader based on the structure of the recipient network (Tack *et al.*, 2011; Frost *et al.*, 2016). However, these methods have had mixed success to date, possibly due to the confounding influence of other important variables such as species' abundances (Fort *et al.*, 2016) and host quality (Carvalheiro *et al.*, 2014). Furthermore, they are yet to be used in an invasion context. A related direction also receiving research attention involves comparing an invader's phylogenetic and functional traits to those of the recipient community, with the aim of predicting which species interactions may drive invasion success and impacts, as well as their strength. A recent review by Gallien and Carboni (2017) concluded that phylogenetic and functional similarity of invaders to the recipient community can predict their success at large and local spatial scales, respectively. However, current studies are limited by competing outcomes of high trait similarity for different types of interactions (e.g. similarity is expected to enhance mutualisms but increase competition) and the lack of a multitrophic, temporal (i.e. invasion stage) and biogeographic (i.e. comparisons within and between the native and invaded ranges) perspective. The following section outlines a systematic approach towards improving our understanding and prediction of indirect interactions.

17.3 Future Directions

17.3.1 Scaling up the community using large-scale experiments and interaction networks

As studies have accumulated, the staggering complexity of biotic interactions involving invasive species has become increasingly clear. These complex indirect interactions can occur over multiple trophic levels and also involve abiotic factors. For example, an experiment by Smith-Ramesh, 2017 showed that spiders living on siliques (dried flower structures) of invasive *Alliaria petiolata* mediated positive indirect effects on native plants via reduced herbivore abundance and increased phosphorous deposition from herbivore carcasses. Further examples include the cascading impacts of an invasive plant (*Acacia longifolia*) over four trophic levels (López-Núñez *et al.*, 2017) or the indirect facilitation of invasive tree species in New Zealand and Argentina through the dispersal of mycorrhizal fungi by non-native mammals (Nuñez *et al.*, 2013; Wood *et al.*, 2015). Even these complex, multi-species indirect interactions rarely occur in isolation, and as more species and interaction pathways are considered, the potential for diffuse indirect impacts increases rapidly (Northfield *et al.*, 2018). Incorporating more comprehensive communities will enable broader assessment of the balance between antagonistic and mutualistic indirect interactions, which ultimately determines the net indirect impact of invaders. However, as the simultaneous consideration of multiple species and types of interactions expands, our ability to grasp the intricacies and importance of all these direct and indirect interactions becomes increasingly strained. Therefore, to make useful and general predictions about which, where and when indirect interactions are likely to be important, we need to carefully consider how to best incorporate studies of entire communities and ecosystems into invasion ecology.

Scaling up the study of indirect interactions and invasions to the community level may be achieved through two main complementary approaches. First, there is an urgent need for large-scale experiments that manipulate and measure species interactions in entire communities and ecosystems, with data preferably collected that can be used to predict population dynamics, coexistence and ecosystem function. Examining plant invasions in this broader community context will advance our understanding of invasions through the inclusion of both direct and indirect interactions across multiple trophic levels and taxonomic groups. This ambitious goal will require large-scale, long-term and collaborative research efforts, such as the famous Jena experiment that has produced many insights into the diversity—ecosystem function relationship (Weisser *et al.*, 2017), the MEND (Maintenance of Exotic vs Native Diversity) experiment that compares common garden communities dominated by native vs non-native plants (Wilsey *et al.*, 2011), or mesocosm experiments that manipulate community composition and species interactions (Waller *et al.*, 2020). Second, analytical approaches to species interaction networks present an ever-expanding toolkit that can be used to interpret the data from such large-scale experiments, due to their capacity to describe and disentangle the complexity of natural systems (reviewed by Delmas *et al.*, 2019). Indeed, the incorporation of species interaction networks into invasion ecology is a topic of increasing interest (see Chapter 18, this volume). A surprising result to emerge from early reviews on the topic is that invaders do not usually disrupt the structure of the species interaction networks they invade (David *et al.*, 2017; Frost *et al.*, 2019), suggesting that invader impacts may be limited in how far they spread through recipient networks. Moreover, species interaction networks represent a promising framework that may be used to predict invasion success, impacts and even management outcomes (e.g. establishment of biological control agents or unforeseen consequences of eradication). This potential is supported by early theoretical studies suggesting that network structure may influence invasibility (e.g. Romanuk *et al.*, 2009; Minoarivelo and Hui, 2016), and recent work outlining how adaptive network models based on community stability theory may be used to predict the invasibility of a recipient network and impacts of invaders (Hui and Richardson, 2017, 2019). However, manipulative experiments remain scarce (but see Macfadyen *et al.*, 2009; Wei *et al.*, 2015).

17.3.2 Spatial scale and biogeography of indirect interactions and invasions

Many invasive plants are widespread in their novel range, distributed over large, even continental, spatial scales. Yet, studies examining the biotic mechanisms underlying the success and impacts of invasive species have generally focused on small scales, ignoring the potential influence of large-scale ecological patterns such as latitudinal gradients in biodiversity (Hillebrand, 2004) and species interactions (Schemske *et al.*, 2009; Moles *et al.*, 2011). Because of their different ecological and evolutionary histories, native and invasive species may experience non-parallel relationships along such large-scale gradients, with the potential to generate biogeographical heterogeneity in the role of species interactions in invasions (i.e. latitudinal variation in enemy release or biotic resistance; Bezemer *et al.*, 2014). This hypothesis has been tested almost exclusively using plant—herbivore interactions, with varying levels of support (Cronin *et al.*, 2015; Allen *et al.*, 2017; Bhattarai *et al.*, 2017b; Lu *et al.*, 2019). However, only two studies to date have explored biogeographic variation in indirect interactions and invasions. Bhattarai *et al.* (2017a) found stronger apparent competitive impacts at higher latitudes for an invasive lineage of the wetland grass *Phragmites australis* on a co-occurring native lineage, mediated by stem galling flies (*Lipara* spp.). Furthermore, Lu *et al.* (2018) demonstrated non-parallel latitudinal gradients in plant growth for native and invasive species of *Alternanthera* grown with soil biota collected from different latitudes in China, altering the performance of the herbivorous beetle *Agasicles hygrophila*. Like latitude, other large-scale gradients that may modify indirect interactions should also be considered. For example, outcomes of indirect interactions are expected to vary along productivity gradients (Oksanen, 1990), yet empirical tests are virtually nonexistent, despite the recognition that understanding the biotic and abiotic context dependency of species interactions is a central problem in community ecology (Agrawal *et al.*, 2007). Moreover, species interactions are also thought to increase in strength towards a species' range centre (Alexander *et al.*, 2007), suggesting that invaders may generate stronger indirect impacts as invasion proceeds (see below). Taken together, these results suggest that conclusions regarding the role of species interactions in invasions may be misleading when only considering a narrow selection of study locations in the invaded range. The asymmetry of native—invasive indirect interactions may vary over broad spatial scales, highlighting the need to integrate this complexity into models describing heterogeneity in the success and impacts of invasive plants.

17.3.3 Temporal scale of indirect interactions and invasions

Temporal variation is an underappreciated aspect of species interactions in plant invasions (Strayer *et al.*, 2006). Temporal variation may be important because the strength of different types of indirect interactions could vary throughout the invasion process, driving a range of invasion dynamics, spanning introduction failure, naturalization, lag phases, exponential growth, boom and bust cycles and severe impacts on native communities. One theoretical expectation that underpins these predictions is the idea that the strength of indirect interactions is related to the density of the species involved (Holt, 1977). In that way, invaders at low densities (i.e. the early establishment phase of invasions) may be unlikely to exert strong negative indirect impacts on co-occurring native species, suggesting that apparent competition on to native species may not explain establishment success of invasive plants. Instead, apparent competition at early stages of invasion should take the form of biotic resistance from generalist native enemies that are shared with much of the recipient community (Parker and Hay, 2005; Parker *et al.*, 2006), potentially contributing to invasion failure or limiting spread. Moreover, once an invader becomes established and increases in density, indirect biotic resistance against it may further increase in strength due to increased connectivity and pathogen spread via a more contiguous matrix of hosts. However, if the invader is tolerant to generalist enemies (Ashton and Lerdau, 2008; Zhang *et al.*, 2018), then indirect negative impacts on co-occurring species (i.e. apparent competition) should also increase with invader density. To date, no studies have empirically tested how indirect interactions change over the course of an invasion, with the exception of Grove *et al.* (2017) who found no change in the indirect impact of Scotch broom (*C. scoparius*) on the ectomycorrhizal mutualism of Douglas fir (*Pseudotsuga menziesii*) over a 31-year invasion chronosequence. Finally, apparent competition from primary invaders may also provide an ecosystem service through resisting secondary invasions, a concept that has been largely unexplored.

In contrast to apparent competition, indirect facilitation via shared mutualists should be expected to benefit invasions in the early establishment phase, when invader density is low. These mutualisms could involve native species (i.e. the Enhanced Mutualist Hypothesis; Richardson *et al.*, 2000) or at least one other non-native species (i.e. invasional meltdown, Simberloff and Von Holle, 1999). For established invasions, indirect facilitation may continue to assist the growth and spread of the invader in a positive invasion feedback; however, invasive plants may also indirectly facilitate other species, due to their high density and generality of interactions, potentially resulting in 'boom and bust' cycles of invasion if they facilitate superior competitors.

Other temporal aspects of biological invasions and species interactions may also influence invasion dynamics. As non-native species continue to be transported around the world (Seebens *et al.*, 2017), natural enemies from the home range will accumulate on non-native plants over the course of an invasion, demonstrated through higher pathogen diversity or prevalence (Mitchell *et al.*, 2010; Flory and Clay, 2013; Flory *et al.*, 2018), increased herbivory (Siemann *et al.*, 2006; Brändle *et al.*, 2008) and more negative plant–soil feedbacks (Diez *et al.*, 2010). Similar patterns may also be expected for plant—mutualist interactions as invasion progresses, and these have been proposed as a potential mechanism to explain lag phases in plant invasions via asynchronous co-invasion of mutualists (Dickie *et al.*, 2017). However, to my knowledge the accumulation of mutualists by invasive plants remains to be tested. Furthermore, native pathogens or mutualists can also incorporate novel hosts into their interactions (i.e. host switch or host range expansion; Carroll and Boyd, 1992; Flory and Clay, 2013), resulting in more shared enemies or mutualists as the invasion progresses. Finally, soil legacies also represent temporal variation in indirect interactions of invasive species, whereby invaders can indirectly influence subsequent residents even after removal, through leaving behind an altered soil biota (Corbin and D'Antonio, 2012). Many questions remain to be answered around soil legacies, including how long they last, the influence of other community members and how to predict their direction and magnitude. Incorporating these different sources of temporal variation into studies of indirect interactions and invasions will help clarify how their influence changes over the course of an invasion, informing potential management approaches.

17.4 Integrating Indirect Interactions into Invasive Species Management

There are several ways that indirect interactions involving invasive plants could be integrated into their management. First, the inclusion of indirect interactions can extend the concept of biological control, which has been traditionally predicated on direct trophic interactions (i.e. the introduction of herbivores or pathogens to control an invasive plant). Because of the importance of indirect interactions, management practices that target indirect interaction partners may provide novel avenues to control invaders and mitigate invasion impacts. Although not a plant invasion, Spencer *et al.* (2017) present a great example of exploiting indirect interactions for the management of the invasive venomous redback spider (*Latrodectus hasseltii*) in Otago, New Zealand. The spider is facilitated by burrows of another invader, the European rabbit (*Oryctolagus cuniculus*), and preys upon the critically endangered endemic Cromwell chafer beetle (*Prodontria lewisii*). By filling in rabbit burrows, the researchers were able to observe a near total decrease in spider density and capture of the chafer beetle (Spencer *et al.*, 2017). Second, knowledge of indirect interactions could inform the potential indirect impacts of desirable non-native organisms, such as agricultural and horticultural crops, pollinators and biological control agents. For example, biological control agents of invasive plants are predicted to have the potential for indirect impacts on closely related native plant species (Kiviat *et al.*, 2019) or higher trophic levels (Pearson and Callaway, 2003), concerns exemplified by classical studies that demonstrated strong negative impacts of the invasive thistle *Carduus nutans* on a native thistle *Cirsium undulatum*, via a biological control agent, the weevil *Rhinocyllus conicus* (Rand and Louda, 2004; Rand *et al.*, 2004). Alternatively, the presence of low-quality alternate hosts for biological control agents may improve long-term efficacy when primary hosts become scarce. Finally, knowledge of indirect interactions can also be useful to predict unanticipated impacts of the removal of invaders (e.g. Bergstrom *et al.*, 2009). This is especially important due to the central role that many successful invaders take in invaded communities (Aizen *et al.*, 2008; Heleno *et al.*, 2013; Albrecht *et al.*, 2014).

17.5 Concluding Remarks

This review has described the common indirect interactions studied by invasion ecologists and outlined the accumulated evidence for these interactions to date. There are now many case studies that demonstrate indirect impacts of invasive plants through various interaction pathways, but a comprehensive understanding of the factors contributing to the context dependency of these interactions has remained elusive, hampering our ability to predict the indirect impacts of invaders. To this end, I have identified several potential future research directions and unanswered questions (Box 17.2) that, if answered, could help move the field towards a systematic understanding of the role of indirect interactions in mediating the success and impacts of invasive species.

Acknowledgements

I thank Ian Dickie, Jason Tylianakis, Anna Traveset and two anonymous reviewers for constructive discussions and comments on earlier versions of the manuscript. This contribution was supported by Centre of Research Excellence funding to the Bio-Protection Research Centre from the Tertiary Education Commission.

References

Agrawal, A.A., Ackerly, D.D., Adler, F., Arnold, A.E., Cáceres, C. *et al.* (2007) Filling key gaps in population and community ecology. *Frontiers in Ecology and the Environment* 5(3), 145–152. DOI: 10.1890/1540-9295(2007)5[145:FKGIPA]2.0.CO;2.

Aizen, M.A., Morales, C.L. and Morales, J.M. (2008) Invasive mutualists erode native pollination webs. *PLoS Biology* 6(2), e31. DOI: 10.1371/journal.pbio.0060031.

Albrecht, M., Padrón, B., Bartomeus, I. and Traveset, A. (2014) Consequences of plant invasions on compartmentalization and species' roles in plant–pollinator networks. *Proceedings of the Royal Society B: Biological Sciences* 281(1788), 20140773. DOI: 10.1098/rspb.2014.0773.

Alexander, H.M., Price, S., Houser, R., Finch, D. and Tourtellot, M. (2007) Is there reduction in disease and pre-dispersal seed predation at the border of a host plant's range? Field and herbarium studies of *Carex blanda*. *Journal of Ecology* 95(3), 446–457. DOI: 10.1111/j.1365-2745.2007.01228.x.

Allen, W.J., Meyerson, L.A., Cummings, D., Anderson, J., Bhattarai, G.P. *et al.* (2017) Biogeography of a plant invasion: drivers of latitudinal variation in enemy release. *Global Ecology and Biogeography* 26(4), 435–446. DOI: 10.1111/geb.12550.

Allen, W.J., Young, R.E., Bhattarai, G.P., Croy, J.R., Lambert, A.M. *et al.* (2015) Multitrophic enemy escape of invasive *Phragmites australis* and its introduced herbivores in North America. *Biological Invasions* 17(12), 3419–3432. DOI: 10.1007/s10530-015-0968-2.

Allen, W.J., Wainer, R., Tylianakis, J.M., Barratt, B.I.P., Shadbolt, M.-R.. *et al.* (2020) Community-level direct and indirect impacts of an invasive plant favour exotic over native species. *Journal of Ecology* 39. DOI: 10.1111/1365-2745.13433.

Aschehoug, E.T. and Callaway, R.M. (2015) Diversity increases indirect interactions, attenuates the intensity of competition, and promotes coexistence. *The American Naturalist* 186(4), 452–459. DOI: 10.1086/682901.

Ashton, I.W. and Lerdau, M.T. (2008) Tolerance to herbivory, and not resistance, may explain differential success of invasive, naturalized, and native North American temperate vines. *Diversity and Distributions* 14(2), 169–178. DOI: 10.1111/j.1472-4642.2007.00425.x.

Atwater, D.Z., Bauer, C.M. and Callaway, R.M. (2011) Indirect positive effects ameliorate strong negative effects of *Euphorbia esula* on a native plant. *Plant Ecology* 212(10), 1655–1662. DOI: 10.1007/s11258-011-9938-7.

Bartomeus, I., Vilà, M. and Santamaría, L. (2008) Contrasting effects of invasive plants in plant–pollinator networks. *Oecologia* 155(4), 761–770. DOI: 10.1007/s00442-007-0946-1.

Bergstrom, D.M., Lucieer, A., Kiefer, K., Wasley, J., Belbin, L. *et al.* (2009) Indirect effects of invasive species removal devastate world heritage island. *Journal of Applied Ecology* 46(1), 73–81. DOI: 10.1111/j.1365-2664.2008.01601.x.

Bezemer, T.M., Harvey, J.A. and Cronin, J.T. (2014) Response of native insect communities to invasive plants. *Annual Review of Entomology* 59(1), 119–141. DOI: 10.1146/annurev-ento-011613-162104.

Bhattarai, G.P., Meyerson, L.A. and Cronin, J.T. (2017a) Geographic variation in apparent competition between native and invasive *Phragmites australis*. *Ecology* 98(2), 349–358. DOI: 10.1002/ecy.1646.

Bhattarai, G.P., Meyerson, L.A., Anderson, J., Cummings, D., Allen, W.J. *et al.* (2017b) Biogeography of a plant invasion: genetic variation and plasticity in latitudinal clines for traits related to herbivory. *Ecological Monographs* 87(1), 57–75. DOI: 10.1002/ecm.1233.

Bonsall, M.B. and Hassell, M.P. (1997) Apparent competition structures ecological assemblages. *Nature* 388(6640), 371–373. DOI: 10.1038/41084.

Borer, E.T., Hosseini, P.R., Seabloom, E.W. and Dobson, A.P. (2007) Pathogen-induced reversal of native dominance in a grassland community. *Proceedings of the National Academy of Sciences* 104(13), 5473–5478. DOI: 10.1073/pnas.0608573104.

Braga, R.R., Gómez-Aparicio, L., Heger, T., Vitule, J.R.S. and Jeschke, J.M. (2018) Structuring evidence for invasional meltdown: broad support but with biases and gaps. *Biological Invasions* 20(4), 923–936. DOI: 10.1007/s10530-017-1582-2.

Brändle, M., Kühn, I., Klotz, S., Belle, C. and Brandl, R. (2008) Species richness of herbivores on exotic host plants increases with time since introduction of the host. *Diversity and Distributions* 14(6), 905–912. DOI: 10.1111/j.1472-4642.2008.00511.x.

Bumbeer, J., da Rocha, R.M., Bornatowski, H., Robert, M. de C. and Ainsworth, C. (2018) Predicting impacts of lionfish (*Pterois volitans*) invasion in a coastal ecosystem of southern Brazil. *Biological Invasions* 20(5), 1257–1274. DOI: 10.1007/s10530-017-1625-8.

Carroll, S.P. and Boyd, C. (1992) Host race radiation in the soapberry bug: natural history with the history. *Evolution* 46(4), 1052–1069. DOI: 10.1111/j.1558-5646.1992.tb00619.x.

Carvalheiro, L.G., Biesmeijer, J.C., Benadi, G., Fründ, J., Stang, M. *et al.* (2014) The potential for indirect effects between co-flowering plants via shared pollinators depends on resource abundance, accessibility and relatedness. *Ecology Letters* 17(11), 1389–1399. DOI: 10.1111/ele.12342.

Charlebois, J.A. and Sargent, R.D. (2017) No consistent pollinator-mediated impacts of alien plants on natives. *Ecology Letters* 20(11), 1479–1490. DOI: 10.1111/ele.12831.

Corbin, J.D. and D'Antonio, C.M. (2012) Gone but not forgotten? invasive plants' legacies on community and ecosystem properties. *Invasive Plant Science and Management* 5(1), 117–124. DOI: 10.1614/IPSM-D-11-00005.1.

Cronin, J.T., Bhattarai, G.P., Allen, W.J. and Meyerson, L.A. (2015) Biogeography of a plant invasion: plant–herbivore interactions. *Ecology* 96(4), 1115–1127. DOI: 10.1890/14-1091.1.

Dangremond, E.M., Pardini, E.A. and Knight, T.M. (2010) Apparent competition with an invasive plant hastens the extinction of an endangered lupine. *Ecology* 91(8), 2261–2271. DOI: 10.1890/09-0418.1.

David, P., Thébault, E., Anneville, O., Duyck, P.-F., Chapuis, E. *et al.* (2017) Impacts of invasive species on food webs: a review of empirical data. *Advances in Ecological Research* 56, 1–60. DOI: 10.1016/bs.aecr.2016.10.001.

Delmas, E., Besson, M., Brice, M.-H., Burkle, L.A., Dalla Riva, G.V. *et al.* (2019) Analysing ecological networks of species interactions. *Biological Reviews* 94(1), 16–36. DOI: 10.1111/brv.12433.

Dickie, I.A., Bufford, J.L., Cobb, R.C., Desprez-Loustau, M.-L., Grelet, G. *et al.* (2017) The emerging science of linked plant-fungal invasions. *New Phytologist* 215(4), 1314–1332. DOI: 10.1111/nph.14657.

Dickie, I.A., St John, M.G., Yeates, G.W., Morse, C.W., Bonner, K.I. *et al.* (2014) Belowground legacies of *Pinus contorta* invasion and removal result in multiple mechanisms of invasional meltdown. *AoB PLANTS* 6, plu056. DOI: 10.1093/aobpla/plu056.

Diez, J.M., Dickie, I., Edwards, G., Hulme, P.E., Sullivan, J.J. *et al.* (2010) Negative soil feedbacks accumulate over time for non-native plant species. *Ecology Letters* 13(7), 803–809. DOI: 10.1111/j.1461-0248.2010.01474.x.

Emer, C., Memmott, J., Vaughan, I.P., Montoya, D. and Tylianakis, J.M. (2016) Species roles in plant-pollinator communities are conserved across native and alien ranges. *Diversity and Distributions* 22(8), 841–852. DOI: 10.1111/ddi.12458.

Enge, S., Nylund, G.M. and Pavia, H. (2013) Native generalist herbivores promote invasion of a chemically defended seaweed via refuge-mediated apparent competition. *Ecology Letters* 16(4), 487–492. DOI: 10.1111/ele.12072.

Engelkes, T., Wouters, B., Bezemer, T.M., Harvey, J.A. and van der Putten, W.H. (2012) Contrasting patterns of herbivore and predator pressure on invasive and native plants. *Basic and Applied Ecology* 13(8), 725–734. DOI: 10.1016/j.baae.2012.10.005.

Flory, S.L., Alba, C., Clay, K., Holt, R.D. and Goss, E.M. (2018) Emerging pathogens can suppress invaders and promote native species recovery. *Biological invasions* 20(1), 5–8. DOI: 10.1007/s10530-017-1438-9.

Flory, S.L. and Bauer, J.T. (2014) Experimental evidence for indirect facilitation among invasive plants. *Journal of Ecology* 102(1), 12–18. DOI: 10.1111/1365-2745.12186.

Flory, S.L. and Clay, K. (2013) Pathogen accumulation and long-term dynamics of plant invasions. *Journal of Ecology* 101(3), 607–613. DOI: 10.1111/1365-2745.12078.

Fort, H., Vázquez, D.P. and Lan, B.L. (2016) Abundance and generalisation in mutualistic networks: solving the chicken-and-egg dilemma. *Ecology Letters* 19(1), 4–11. DOI: 10.1111/ele.12535.

Frost, C.M., Allen, W.J., Courchamp, F., Jeschke, J.M., Saul, W.-C. *et al.* (2019) Using ecological network theory to predict biological invasions. *Trends in Ecology & Evolution* 34, 831–843.

Frost, C.M., Peralta, G., Rand, T.A., Didham, R.K., Varsani, A. *et al.* (2016) Apparent competition drives community-wide parasitism rates and changes in host abundance across ecosystem boundaries. *Nature Communications* 7(1), 12644. DOI: 10.1038/ncomms12644.

Gallien, L. and Carboni, M. (2017) The community ecology of invasive species: where are we and what's next? *Ecography* 40(2), 335–352. DOI: 10.1111/ecog.02446.

Greenberg, C.H. and Walter, S.T. (2010) Fleshy fruit removal and nutritional composition of winter-fruiting plants: a comparison of non-native invasive and native species. *Natural Areas Journal* 30(3), 312–321. DOI: 10.3375/043.030.0306.

Grove, S., Parker, I.M. and Haubensak, K.A. (2017) Do impacts of an invasive nitrogen-fixing shrub on Douglas-fir and its ectomycorrhizal mutualism change over time following invasion? *Journal of Ecology* 105(6), 1687–1697. DOI: 10.1111/1365-2745.12764.

Harvey, J.A., Bukovinszky, T. and van der Putten, W.H. (2010) Interactions between invasive plants and insect herbivores: a plea for a multitrophic perspective. *Biological Conservation* 143(10), 2251–2259. DOI: 10.1016/j.biocon.2010.03.004.

Heleno, R.H., Ceia, R.S., Ramos, J.A. and Memmott, J. (2009) Effects of alien plants on insect abundance and biomass: a food-web approach. *Conservation Biology* 23(2), 410–419. DOI: 10.1111/j.1523-1739.2008.01129.x.

Heleno, R.H., Ramos, J.A. and Memmott, J. (2013) Integration of exotic seeds into an Azorean seed dispersal network. *Biological Invasions* 15(5), 1143–1154. DOI: 10.1007/s10530-012-0357-z.

Hillebrand, H. (2004) On the generality of the latitudinal diversity gradient. *The American Naturalist* 163(2), 192–211. DOI: 10.1086/381004.

Holt, R.D. (1977) Predation, apparent competition, and the structure of prey communities. *Theoretical Population Biology* 12(2), 197–229. DOI: 10.1016/0040-5809(77)90042-9.

Holt, R.D. and Bonsall, M.B. (2017) Apparent competition. *Annual Review of Ecology, Evolution, and Systematics* 48(1), 447–471. DOI: 10.1146/annurev-ecolsys-110316-022628.

Horton, T.R. (2015) *Mycorrhizal Networks*. Springer, Dordrecht, the Netherlands.

Hui, C. and Richardson, D.M. (2017) *Invasion Dynamics*. Oxford University Press, Oxford.

Hui, C. and Richardson, D.M. (2019) How to invade an ecological network. *Trends in Ecology & Evolution* 34(2), 121–131. DOI: 10.1016/j.tree.2018.11.003.

Johnson, D. and Gilbert, L. (2015) Interplant signalling through hyphal networks. *New Phytologist* 205(4), 1448–1453. DOI: 10.1111/nph.13115.

Kiviat, E., Meyerson, L.A., Mozdzer, T.J., Allen, W.J., Baldwin, A.H. *et al.* (2019) Evidence does not support the targeting of cryptic invaders at the subspecies level using classical biological control: the example of *Phragmites*. *Biological Invasions* 21(8), 2529–2541. DOI: 10.1007/s10530-019-02014-9.

Klock, M.M., Barrett, L.G., Thrall, P.H. and Harms, K.E. (2015) Host promiscuity in symbiont associations can influence exotic legume establishment and colonization of novel ranges. *Diversity and Distributions* 21(10), 1193–1203. DOI: 10.1111/ddi.12363.

Kuebbing, S.E. and Nuñez, M.A. (2016) Invasive non-native plants have a greater effect on neighbouring natives than other non-natives. *Nature Plants* 2(10), 16134. DOI: 10.1038/nplants.2016.134.

Lau, J.A. (2013) Trophic consequences of a biological invasion: do plant invasions increase predator abundance? *Oikos* 122(3), 474–480. DOI: 10.1111/j.1600-0706.2012.20774.x.

Levine, S. (1980) Indirect mutualism: variations on a theme. *American Naturalist* 116, 441–448.

Levine, J.M. (1999) Indirect facilitation: evidence and predictions from a riparian community. *Ecology* 80(5), 1762–1769. DOI: 10.1890/0012-9658(1999)080[1762:IFEAPF]2.0.CO;2.

Lu, X., He, M., Ding, J. and Siemann, E. (2018) Latitudinal variation in soil biota: testing the biotic interaction hypothesis with an invasive plant and a native congener. *The ISME Journal* 12(12), 2811–2822. DOI: 10.1038/s41396-018-0219-5.

Lu, X., He, M., Tang, S., Wu, Y., Shao, X. *et al.* (2019) Herbivory may promote a non-native plant invasion at low but not high latitudes. *Annals of Botany* 124(5), 819–827. DOI: 10.1093/aob/mcz121.

López-Núñez, F.A., Heleno, R.H., Ribeiro, S., Marchante, H. and Marchante, E. (2017) Four-trophic level food webs reveal the cascading impacts of an invasive plant targeted for biocontrol. *Ecology* 98(3), 782–793. DOI: 10.1002/ecy.1701.

Lopezaraiza-Mikel, M.E., Hayes, R.B., Whalley, M.R. and Memmott, J. (2007) The impact of an alien plant on a native plant–pollinator network: an experimental approach. *Ecology Letters* 10(7), 539–550. DOI: 10.1111/j.1461-0248.2007.01055.x.

Macfadyen, S., Gibson, R., Polaszek, A., Morris, R.J., Craze, P.G. *et al.* (2009) Do differences in food web structure between organic and conventional farms affect the ecosystem service of pest control? *Ecology Letters* 12(3), 229–238. DOI: 10.1111/j.1461-0248.2008.01279.x.

Martinrod, K.L. and Gorchov, D.L. (2017) White-Tailed deer browse on an invasive shrub with extended leaf phenology meets assumptions of an apparent competition hypothesis. *AoB PLANTS* 9, plx006. DOI: 10.1093/aobpla/plx006..

McCary, M.A., Mores, R., Farfan, M.A. and Wise, D.H. (2016) Invasive plants have different effects on trophic structure of green and brown food webs in terrestrial ecosystems: a meta-analysis. *Ecology Letters* 19(3), 328–335. DOI: 10.1111/ele.12562.

Menge, B.A. (1995) Indirect effects in marine rocky intertidal interaction webs: patterns and importance. *Ecological Monographs* 65(1), 21–74. DOI: 10.2307/2937158.

Minoarivelo, H.O. and Hui, C. (2016) Invading a mutualistic network: to be or not to be similar. *Ecology and Evolution* 6(14), 4981–4996. DOI: 10.1002/ece3.2263.

Mitchell, C.E., Agrawal, A.A., Bever, J.D., Gilbert, G.S., Hufbauer, R.A. *et al.* (2006) Biotic interactions and plant invasions. *Ecology Letters* 9(6), 726–740. DOI: 10.1111/j.1461-0248.2006.00908.x.

Mitchell, C.E., Blumenthal, D., Jarošík, V., Puckett, E.E. and Pyšek, P. (2010) Controls on pathogen species richness in plants' introduced and native ranges: roles of residence time, range size and host traits. *Ecology Letters* 13(12), 1525–1535. DOI: 10.1111/j.1461-0248.2010.01543.x.

Moles, A.T., Bonser, S.P., Poore, A.G.B., Wallis, I.R. and Foley, W.J. (2011) Assessing the evidence for latitudinal gradients in plant defence and herbivory. *Functional Ecology* 25(2), 380–388. DOI: 10.1111/j.1365-2435.2010.01814.x.

Moora, M., Berger, S., Davison, J., Öpik, M., Bommarco, R. *et al.* (2011) Alien plants associate with widespread generalist arbuscular mycorrhizal fungal taxa: evidence from a continental-scale study using massively parallel 454 sequencing. *Journal of Biogeography* 38(7), 1305–1317. DOI: 10.1111/j.1365-2699.2011.02478.x.

Morales, C.L. and Traveset, A. (2009) A meta-analysis of impacts of alien vs. native plants on pollinator visitation and reproductive success of co-flowering native plants. *Ecology Letters* 12(7), 716–728. DOI: 10.1111/j.1461-0248.2009.01319.x.

Morris, R.J., Lewis, O.T. and Godfray, H.C.J. (2004) Experimental evidence for apparent competition in a tropical forest food web. *Nature* 428(6980), 310–313. DOI: 10.1038/nature02394.

Newman, E.I. (1988) Mycorrhizal links between plants: their functioning and ecological significance. *Advances in Ecological Research* 18, 243–270.

Northfield, T.D., Laurance, S.G.W., Mayfield, M.M., Paini, D.R., Snyder, W.E. *et al.* (2018) Native turncoats and indirect facilitation of species invasions. *Proceedings of the Royal Society B: Biological Sciences* 285(1871), 20171936. DOI: 10.1098/rspb.2017.1936.

Nuñez, M.A., Hayward, J., Horton, T.R., Amico, G.C., Dimarco, R.D. *et al.* (2013) Exotic mammals disperse exotic fungi that promote invasion by exotic trees. *PLoS ONE* 8(6), e66832. DOI: 10.1371/journal.pone.0066832.

O'Dowd, D.J., Green, P.T. and Lake, P.S. (2003) Invasional 'meltdown' on an oceanic island. *Ecology Letters* 6(9), 812–817. DOI: 10.1046/j.1461-0248.2003.00512.x.

Oksanen, L. (1990) Predation, herbivory, and plant strategies along gradients of primary productivity. In: Grace, J.B. and Tilman, D. (eds) *Perspectives on Plant Competition*. Academic Press, San Diego, pp. 445–474.

Orrock, J.L., Dutra, H.P., Marquis, R.J. and Barber, N. (2015) Apparent competition and native consumers exacerbate the strong competitive effect of an exotic plant species. *Ecology* 96(4), 1052–1061. DOI: 10.1890/14-0732.1.

Orrock, J.L., Holt, R.D. and Baskett, M.L. (2010) Refuge-mediated apparent competition in plant-consumer interactions. *Ecology Letters* 13(1), 11–20. DOI: 10.1111/j.1461-0248.2009.01412.x.

Padrón, B., Traveset, A., Biedenweg, T., Díaz, D., Nogales, M. *et al.* (2009) Impact of alien plant invaders on pollination networks in two archipelagos. *PLoS ONE* 4(7), e6275. DOI: 10.1371/journal.pone.0006275.

Parker, J.D., Burkepile, D.E. and Hay, M.E. (2006) Opposing effects of native and exotic herbivores on plant invasions. *Science* 311(5766), 1459–1461. DOI: 10.1126/science.1121407.

Parker, J.D. and Hay, M.E. (2005) Biotic resistance to plant invasions? native herbivores prefer non-native plants. *Ecology Letters* 8(9), 959–967. DOI: 10.1111/j.1461-0248.2005.00799.x.

Pearse, I.S. and Altermatt, F. (2013) Predicting novel trophic interactions in a non-native world. *Ecology Letters* 16(8), 1088–1094. DOI: 10.1111/ele.12143.

Pearson, D.E. and Callaway, R.M. (2003) Indirect effects of host-specific biological control agents. *Trends in Ecology & Evolution* 18(9), 456–461. DOI: 10.1016/S0169-5347(03)00188-5.

Power, A.G. and Mitchell, C.E. (2004) Pathogen spillover in disease epidemics. *The American Naturalist* 164(S5), S79–S89. DOI: 10.1086/424610.

Rand, T.A. and Louda, S.M. (2004) Exotic weed invasion increases the susceptibility of native plants to attack by a biocontrol herbivore. *Ecology* 85(6), 1548–1554. DOI: 10.1890/03-3067.

Rand, T.A., Russell, F.L. and Louda, S.M. (2004) Local- vs landscape-scale indirect effects of an invasive weed on native plants. *Weed Technology* 18(sp1), 1250–1254. DOI: 10.1614/0890-037X(2004)018[1250:LVLIEO]2.0.CO;2.

Richardson, D.M., Allsopp, N., D'Antonio, C.M., Milton, S.J. and Rejmánek, M. (2000) Plant invasions – the role of mutualisms. *Biological Reviews of the Cambridge Philosophical Society* 75(1), 65–93. DOI: 10.1017/S0006323199005435.

Rodewald, A.D. (2012) Spreading messages about invasives. *Diversity and Distributions* 18(1), 97–99. DOI: 10.1111/j.1472-4642.2011.00817.x.

Rodríguez-Echeverría, S. (2010) Rhizobial hitchhikers from down under: invasional meltdown in a plant-bacteria mutualism? *Journal of Biogeography* 25, 1611–1622. DOI: 10.1111/j.1365-2699.2010.02284.x.

Romanuk, T.N., Zhou, Y., Brose, U., Berlow, E.L., Williams, R.J. *et al.* (2009) Predicting invasion success in complex ecological networks. *Philosophical Transactions of the Royal Society B: Biological Sciences* 364(1524), 1743–1754. DOI: 10.1098/rstb.2008.0286.

Schemske, D.W., Mittelbach, G.G., Cornell, H.V., Sobel, J.M. and Roy, K. (2009) Is there a latitudinal gradient in the importance of biotic interactions? *Annual Review of Ecology, Evolution, and Systematics* 40(1), 245–269. DOI: 10.1146/annurev.ecolsys.39.110707.173430.

Schlaepfer, M.A., Runge, M.C. and Sherman, P.W. (2002) Ecological and evolutionary traps. *Trends in Ecology & Evolution* 17(10), 474–480. DOI: 10.1016/S0169-5347(02)02580-6.

Seebens, H., Blackburn, T.M., Dyer, E.E., Genovesi, P., Hulme, P.E. *et al.* (2017) No saturation in the accumulation of alien species worldwide. *Nature Communications* 8(1), 14435. DOI: 10.1038/ncomms14435.

Sessions, L. and Kelly, D. (2002) Predator-mediated apparent competition between an introduced grass, *Agrostis capillaris*, and a native fern, *Botrychium australe* (Ophioglossaceae), in New Zealand. *Oikos* 96(1), 102–109. DOI: 10.1034/j.1600-0706.2002.960111.x.

Siemann, E., Rogers, W.E. and Dewalt, S.J. (2006) Rapid adaptation of insect herbivores to an invasive plant. *Proceedings of the Royal Society B: Biological Sciences* 273(1602), 2763–2769. DOI: 10.1098/rspb.2006.3644.

Simberloff, D. and Von Holle, B. (1999) Positive interactions of nonindigenous species: invasional meltdown? *Biological Invasions* 1(1), 21–32. DOI: 10.1023/A:1010086329619.

Smith, L.M. and Hall, S. (2016) Extended leaf phenology may drive plant invasion through direct and apparent competition. *Oikos* 125(6), 839–848. DOI: 10.1111/oik.02529.

Simmons, B.I., Cirtwill, A.R., Baker, N.J., Wauchope, H.S., Dicks, L.V. *et al.* (2019a) Motifs in bipartite ecological networks: uncovering indirect interactions. *Oikos* 128(2), 154–170. DOI: 10.1111/oik.05670.

Simmons, B.I., Sweering, M.J.M., Schillinger, M., Dicks, L.V., Sutherland, W.J. *et al.* (2019b) bmotif: A package for motif analyses of bipartite networks. *Methods in Ecology and Evolution* 10(5), 695–701. DOI: 10.1111/2041-210X.13149.

Smith-Ramesh, L.M. (2017) Invasive plant alters community and ecosystem dynamics by promoting native predators. *Ecology* 98(3), 751–761. DOI: 10.1002/ecy.1688.

Spencer, J., van Heezik, Y., Seddon, P.J. and Barratt, B.I.P. (2017) Synergy between two invasive species, redback spiders and rabbits, threaten the endangered Cromwell chafer beetle. *Biological Invasions* 19(5), 1379–1387. DOI: 10.1007/s10530-016-1352-6.

Stouffer, D.B., Cirtwill, A.R., Bascompte, J. and Bartomeus, I. (2014) How exotic plants integrate into pollination networks. *Journal of Ecology* 102(6), 1442–1450. DOI: 10.1111/1365-2745.12310.

Strauss, S.Y. (1991) Indirect effects in community ecology: their definition, study and importance. *Trends in Ecology & Evolution* 6(7), 206–210. DOI: 10.1016/0169-5347(91)90023-Q.

Strayer, D.L., Eviner, V.T., Jeschke, J.M. and Pace, M.L. (2006) Understanding the long-term effects of species invasions. *Trends in Ecology & Evolution* 21(11), 645–651. DOI: 10.1016/j.tree.2006.07.007.

Strong, D.R. (1992) Are trophic cascades all wet? differentiation and donor-control in speciose ecosystems. *Ecology* 73(3), 747–754. DOI: 10.2307/1940154.

Tack, A.J.M., Gripenberg, S. and Roslin, T. (2011) Can we predict indirect interactions from quantitative food webs? - an experimental approach. *Journal of Animal Ecology* 80(1), 108–118. DOI: 10.1111/j.1365-2656.2010.01744.x.

Tylianakis, J.M., Didham, R.K., Bascompte, J. and Wardle, D.A. (2008) Global change and species interactions in terrestrial ecosystems. *Ecology Letters* 11(12), 1351–1363. DOI: 10.1111/j.1461-0248.2008.01250.x.

Tylianakis, J.M. and Morris, R.J. (2017) Ecological networks across environmental gradients. *Annual Review of Ecology, Evolution, and Systematics* 48(1), 25–48. DOI: 10.1146/annurev-ecolsys-110316-022821.

van Kleunen, M., Weber, E. and Fischer, M. (2010) A meta-analysis of trait differences between invasive and non-invasive plant species. *Ecology Letters* 13(2), 235–245. DOI: 10.1111/j.1461-0248.2009.01418.x.

van der Heijden, M.G.A. and Horton, T.R. (2009) Socialism in soil? the importance of mycorrhizal fungal networks for facilitation in natural ecosystems. *Journal of Ecology* 97(6), 1139–1150. DOI: 10.1111/j.1365-2745.2009.01570.x.

Waller, L.P., Allen, W.J., Barratt, B.I.P., Condron, L.M., França, F.M. *et al.* (2020) Biotic interactions drive ecosystem responses to exotic plant invaders. *Science* 368(6494), 967–972. DOI: 10.1126/science.aba2225.

Waller, L.P., Callaway, R.M., Klironomos, J.N., Ortega, Y.K. and Maron, J.L. (2016) Reduced mycorrhizal responsiveness leads to increased competitive tolerance in an invasive exotic plant. *Journal of Ecology* 104(6), 1599–1607. DOI: 10.1111/1365-2745.12641.

Walsh, M.R. (2013) The evolutionary consequences of indirect effects. *Trends in Ecology & Evolution* 28(1), 23–29. DOI: 10.1016/j.tree.2012.08.006.

Wei, Z., Yang, T., Friman, V.-P., Xu, Y., Shen, Q. *et al.* (2015) Trophic network architecture of root-associated bacterial communities determines pathogen invasion and plant health. *Nature Communications* 6, 8413. DOI: 10.1038/ncomms9413.

Weisser, W.W., Roscher, C., Meyer, S.T., Ebeling, A., Luo, G. *et al.* (2017) Biodiversity effects on ecosystem functioning in a 15-year grassland experiment: patterns, mechanisms, and open questions. *Basic and Applied Ecology* 23, 1–73. DOI: 10.1016/j.baae.2017.06.002.

White, E.M., Wilson, J.C. and Clarke, A.R. (2006) Biotic indirect effects: a neglected concept in invasion biology. *Diversity and Distributions* 12(4), 443–455. DOI: 10.1111/j.1366-9516.2006.00265.x.

Wilsey, B.J., Daneshgar, P.P. and Polley, H.W. (2011) Biodiversity, phenology and temporal niche differences between native- and novel exotic-dominated grasslands. *Perspectives in Plant Ecology, Evolution and Systematics* 13(4), 265–276. DOI: 10.1016/j.ppees.2011.07.002.

Wood, J.R., Dickie, I.A., Moeller, H.V., Peltzer, D.A., Bonner, K.I. *et al.* (2015) Novel interactions between non-native mammals and fungi facilitate establishment of invasive pines. *Journal of Ecology* 103(1), 121–129. DOI: 10.1111/1365-2745.12345.
Wootton, J.T. (1994) Predicting direct and indirect effects: an integrated approach using experiments and path analysis. *Ecology* 75(1), 151–165. DOI: 10.2307/1939391.
Yang, S., Ferrari, M.J. and Shea, K. (2011) Pollinator behavior mediates negative interactions between two congeneric invasive plant species. *The American Naturalist* 177(1), 110–118. DOI: 10.1086/657433.
Zhang, Z., Pan, X., Blumenthal, D., van Kleunen, M., Liu, M. *et al.* (2018) Contrasting effects of specialist and generalist herbivores on resistance evolution in invasive plants. *Ecology* 99(4), 866–875. DOI: 10.1002/ecy.2155.

18 How a Network Approach Has Advanced the Field of Plant Invasion Ecology

Carine Emer[1,2]* and Sérgio Timóteo[3]

[1]Department of Ecology, Institute of Biosciences, São Paulo State University, Rio Claro, São Paulo, Brazil; [2]National Centre for Wild Birds Research and Conservation, Cabedelo, Paraíba, Brazil; [3]University of Coimbra, Centre for Functional Ecology, Department of Life Sciences, Calçada Martim de Freitas, 3000-456, Coimbra, Portugal

Abstract

Every organism on Earth, whether in natural or anthropogenic environments, is connected to a complex web of life, the famous 'entangled bank' coined by Darwin in 1859. Non-native species can integrate into local 'banks' by establishing novel associations with the resident species. In that context, network ecology has been an important tool to study the interactions of non-native species and the effects on recipient communities due to its ability to simultaneously investigate the assembly and disassembly of species interactions as well as their functional roles. Its visually appealing tools and relatively simple metrics gained momentum among scientists and are increasingly applied in different areas of ecology, from the more theoretical grounds to applied research on restoration and conservation. A network approach helps us to understand how plant invasions may or may not form novel species associations, how they change the structure of invaded communities, the outcomes for ecosystem functionality and, ultimately, the implications for the conservation of ecological interactions. Networks have been widely used on pollination studies, especially from temperate zones, unveiling their nested patterns and the mechanisms by which non-native plants integrate into local communities. Yet, very few papers have used network approaches to assess plant invasion effects in other systems such as plant–herbivore, plant–pathogen or seed-dispersal processes. Here we describe how joining network ecology with plant invasion biology started and how it has developed over the last few decades. We show the extent of its contribution, despite contradictory results and biases, to a better understanding of the role of non-native plant species in shaping community structure. Finally, we explore how it can be further improved to answer emerging questions.

18.1 Bringing Network Theory to Ecology

Over the last decades ecologists have embraced network theory to unveil Darwin's entangled bank, in which species coexist, interacting with each other through complex ecological and evolutionary processes. Network theory has become useful to understand the structure of complex systems; it has evolved from, and expanded to, unrelated fields of research, from physics to social sciences, medicine, engineering and conservation (e.g. Barabási and Albert, 1999; Bassett *et al.*, 2011). This transdisciplinarity has resulted in an interchange of techniques, models and terms between disciplines, and ecologists have been borrowing many of them to understand patterns of species interactions. Nonetheless, theoretical aspects have also been developed by ecologists, creating a

*Corresponding author: c.emer09@gmail.com

 Plant Invasions: The Role of Biotic Interactions (eds Anna Traveset and David M. Richardson)
DOI: 10.1079/9781789242171.0018

Box 18.1. A network glossary for ecologists.

Centrality: species position within a network structure; usually related to species roles and functions.
Connectance: proportion of all possible links between species that are actually realized.
Degree: number of interactions of a given species.
Evenness: uniformity in the distribution of interactions across species in the network.
Generalist: species that establish links with many partners, i.e. the opposite of specialist.
Interaction strength: proportion of interactions between two species, also called dependence.
Linkage density: average number of links per species.
Modularity: a pattern in which groups of tightly connected species (modules) interact more strongly than expected among themselves than with species in other modules.
Nestedness: a pattern in which specialist species interact with a subset of the species with whom the more generalist species interact, which in turn form a core of species that connect the whole network.
Robustness: the ability of a network to withstand a cascade of secondary extinctions; the proportion of extinct species and interactions resulting in the network collapse.
Specialization H′2: measures complementary specialization, reflecting the community-wide selectiveness of the observed interactions.
Specialization d′: measures the selectivity within the range of partners per species.
Species strength: sum of a species' dependencies, reflecting its importance for its partners at the other level of the network.

novel field called 'Ecological Networks' (Delmas *et al.*, 2019). Within this framework, communities are viewed as a network of interactions in which species (sometimes individuals or higher taxa, e.g. genus, families, guilds) are regarded as nodes, while the interactions between them are the links connecting those nodes (Box 18.1 shows the most common network terminology).

Interest in the relationships between species goes back a long time, and researchers have depicted them as diagrams at least since the late 19th century (Camerano, 1880, cited in Bersier, 2007). Throughout the 20th century, many studies have focused on these interspecific relationships, especially on trophic interactions (Summerhayes and Elton, 1923), with the first quantitative, comparative studies aiming to find generalities about the structure of food webs emerging during the 1970s (Cohen, 1978). More recently, mutualistic interactions have received much attention and the use of a network approach to study them has become common. Since the early studies of Jordano (1987) and Memmott (1999), the use of ecological networks has increased substantially due to their ability to simultaneously investigate the assembly and disassembly of ecological communities (Bascompte *et al.*, 2003; Lewinsohn *et al.*, 2006; Montoya *et al.*, 2015) as well as the functional roles of their components (Olesen *et al.*, 2007; Martín González *et al.*, 2010; Stouffer *et al.*, 2014). Nowadays, ecological network studies have gone beyond theoretical and experimental science and have been used in more applied scenarios (Memmott, 2009; Tylianakis and Morris, 2017; Frost *et al.*, 2019). For instance, ecologists have applied network theory to understand the restoration of ecological processes (Montoya *et al.*, 2012; Kaiser-Bunbury *et al.*, 2017), the consequences of habitat loss and fragmentation (Tylianakis *et al.*, 2007; Emer *et al.*, 2018), the mechanisms of biological control (Henneman and Memmott, 2001; Macfadyen *et al.*, 2011), urban ecology (Cruz *et al.*, 2013) and biological invasions (Padrón *et al.*, 2009; Vil *et al.*, 2009; Traveset *et al.*, 2015a; Vizentin-Bugoni *et al.*, 2019). Moreover, networks help us to understand the interplay of structural and functional roles between non-native and native species as invasions occur (Dupont *et al.*, 2009; Arroyo-Correa *et al.*, 2020). Overall and despite contradictory results reported in the literature, network ecology provides an efficient way to visualize and study how non-native species are accommodated in their recipient communities, how they may change the structure of invaded communities, the outcomes for the maintenance of ecological functions and, ultimately, for the conservation of ecological interactions (Tylianakis, 2008; Valiente-Banuet *et al.*, 2015).

Bipartite networks are by far the most studied type of ecological network, also in the context

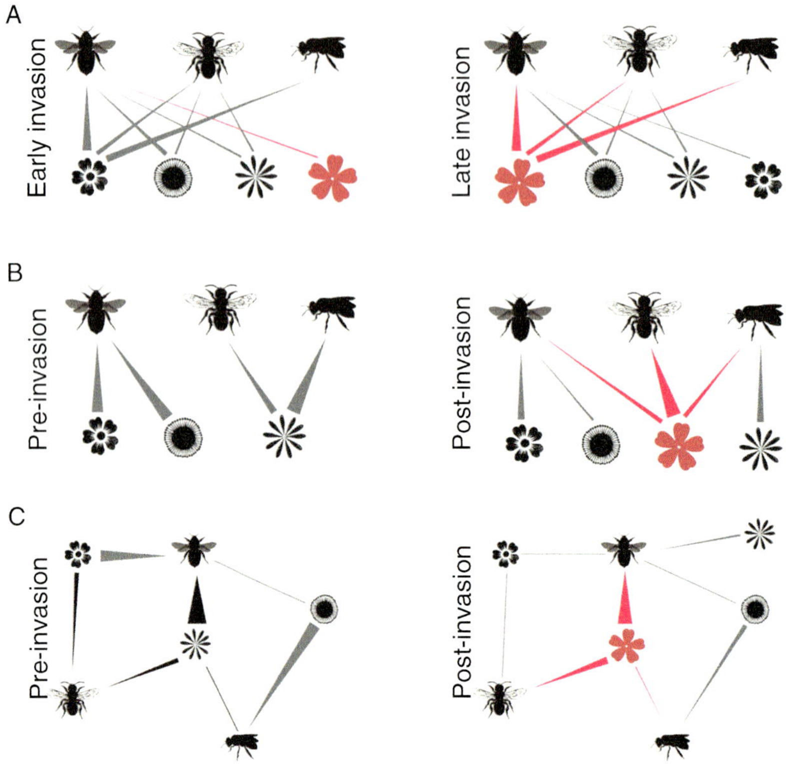

Fig. 18.1. Example of how invasive plant species integrate and reorganize interactions in animal–plant networks (see Box 18.1 for definition of concepts). A and B represent bipartite mutualistic networks between plants and their flower visitors; pink flowers and links picture the invasive plant species and the interactions it establishes in the community; the width of the links is the strength of that interaction. (A) Nested networks showing a simplified gradient of invasion in which invasive plants initially enter the community by interacting with highly generalist species; as invasion progress, they displace the links of some native species and become themselves super-generalists. (B) Networks picturing pre-invaded communities with only a few interactions organized in two distinct modules, and post-invaded communities in which the invasive plants establish novel links with the resident species, connect the previously isolated modules and change the strength of the interactions among native species. (C) An alternative representation of networks which emphasizes the structural roles of species in the network topology. From pre- to post-invaded networks, the invasive plant has displaced the native species by usurping some of its links and becoming itself highly central, thus functioning as a network hub, which concentrates most of the interactions of the local pollinators, thus performing an important role in maintaining network cohesiveness. Animal silhouettes reproduced from PhyloPic.org under the Public Domain Dedication 1.0 license. Bombus terretris reproduced with no changes from Melissa Broussard (https://creativecommons.org/licenses/by/3.0/).

of biological invasions. This type of network is composed of two distinct interacting groups of species that do not interact within themselves (Fig. 18.1). Here, ecological processes are studied in isolation, such as the mutualistic seed-dispersal interactions in which an assemblage of fleshy-fruited plant species are dispersed by an assemblage of frugivore species, or the antagonistic plant—herbivore interactions, in which an assemblage of herbivore species feed upon an assemblage of plant species.

Likewise, the large body of work on ecological networks so far accumulated has sought to describe the emergent structural patterns that characterize these networks and that are related to community stability (Montoya *et al.*, 2006;

Fortuna *et al.*, 2010; Thébault and Fontaine, 2010). Among many network metrics available, two descriptors have received much attention by ecologists: nestedness and modularity. These network descriptors capture how species interact at the community level. For instance, regarding ecological functions and niche partitioning (Schleuning *et al.*, 2015), dietary specialization (Mello *et al.*, 2015) or geographical variation (Dügger *et al.*, 2018), while also providing insightful information about the evolutionary aspects that may have shaped ecological communities (Rezende *et al.*, 2007; Schleuning *et al.*, 2014; Peralta, 2016). Therefore, changes in those patterns can be good predictors of changes in community stability due to anthropogenic disturbance, such as the invasion by non-native plants.

Interestingly, nestedness and modularity may have opposing effects on the stability of mutualistic and antagonistic networks (Thébault and Fontaine, 2010). Many ecological networks, particularly those involving mutualistic interactions, are structured around a core of very well-connected species interacting with less connected ones, therefore configuring a 'nested' pattern (Bascompte *et al.*, 2003; Cantor *et al.*, 2017). In mutualistic networks, nestedness tends to stabilize communities by reducing interspecific competition while increasing biodiversity by enhancing the number of co-occurring species (Bastolla *et al.*, 2009), with the opposite happening in antagonistic networks (Thébault and Fontaine, 2010). Moreover, nestedness makes it difficult for species to become isolated when the system is subject to perturbations, thus conferring cohesion to mutualistic networks (Bascompte *et al.*, 2003). On the other hand, a nested pattern in bipartite networks may facilitate the integration of non-native species into the community through the interactions established by highly generalist species (Aizen *et al.*, 2008, Fig. 18.1a). Another recurring structure of ecological networks, modularity, allows the analysis of the interaction patterns within and among modules (subgroups of species that interact strongly among themselves) and the identification of species' structural roles. Modularity in antagonistic networks is associated with increasing stability because it may retain perturbation impacts within modules (Thébault and Fontaine, 2010). Therefore, one can expect that a modular structure make it difficult for non-native species to fully integrate into the local communities because novel interactions established would be restricted to the invaded modules (Krause *et al.*, 2003). On the other hand, a highly invasive species could change modularity by establishing interactions with resident species from previously isolated modules (Albrecht *et al.*, 2014, Fig. 18.1b). Nestedness and modularity are not mutually exclusive though, and several studies have identified an important interplay between them that may be determinant for the stability of different networks (e.g. Lewinsohn *et al.*, 2006; Fortuna *et al.*, 2010; Thébault and Fontaine, 2010).

18.2 Integrating Ecological Networks and Plant Invasion

In a world dominated by human activities, invasive species are one of the major threats to global biodiversity (IPBES, 2019). Although much progress has been made in understanding biological invasions, there is still no quantitative and/or predictive theory of biological invasions because of the complexity involved, and often the lack of sufficient data to predict whether and how non-native species can successfully integrate into local communities and/or become a hazard (Frost *et al.*, 2019, Valdovinos, 2019; but see Hui and Richardson, 2017, 2019 for recent proposals on increasing predictability of invasions through ecological networks). Early on, ecologists perceived the potential of a network approach to help solve such a conundrum. Among the first papers to apply network theory to understand plant invasion stands the work by Memmott and Waser (2002), which reanalysed early 20th-century records of flower visitation by pollinators from Illinois, USA, in a community with ca. 12% non-native plant species (Fig. 18.2). This study showed that non-native plants received fewer visits than native species, most pollinators had no preference for non-native plant species (rather, native plant species showed higher preference by pollinators) and connectance was lower in the invaded communities. The authors hypothesized that the integration of non-natives may have occurred through i) competitive interactions between non-native and native species

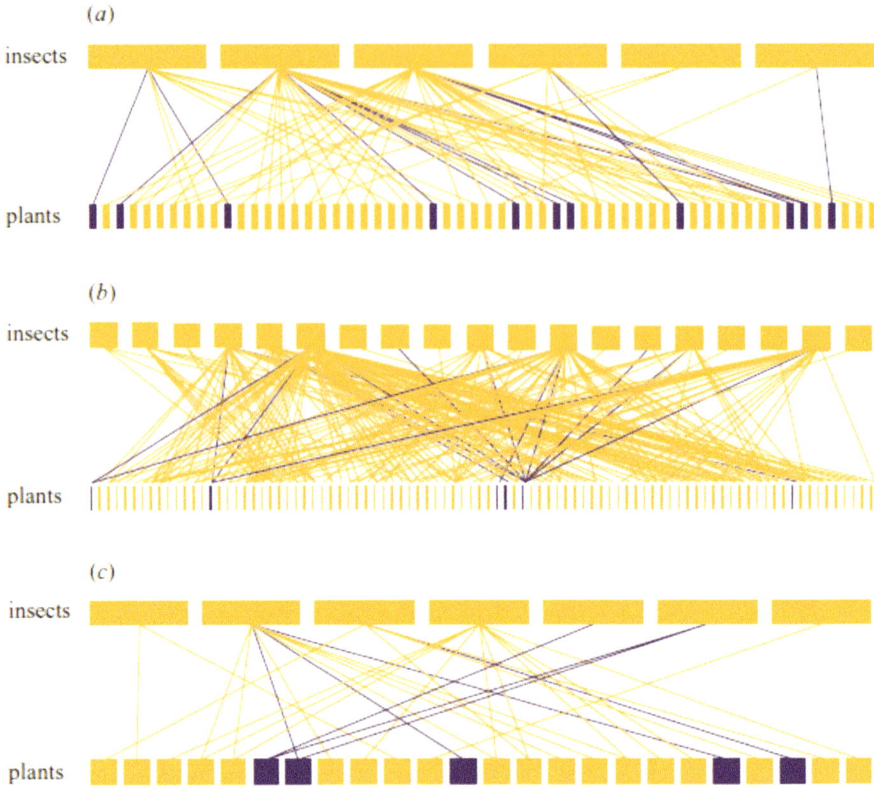

Fig. 18.2. Invaded pollination networks from the seminal work by Memmott and Waser (2002) showing the integration of non-native plants into native communities using a network approach. The three sections represent interactions involving Anthophoridae (a), Sarcophagidae (b) and Sphingidae (c). Bottom boxes represent plant species, top boxes their flower visitors and the lines connecting both levels represent the interactions among species. The width of boxes and lines is proportional to the frequency of interaction. Non-native plant species and their interactions are shown in blue, whereas native species and interactions are shown in yellow. Reproduced with permission of Royal Society.

(attracting pollinators that would otherwise visit native species), ii) facilitation (attracting pollinators for co-flowering non-native and native species), or iii) indirect mutualisms between native and non-native species (maintaining a continuous high density of pollinators when flowering did not overlap). Concomitantly, a study conducted in oceanic islands (Olesen *et al.*, 2002) reported that the generalism of non-native plant species did not differ significantly from the native species, or was even lower. Similar to the first study, non-native plants were well integrated into local communities, apparently not outcompeting native species. These studies demonstrated new avenues to understand the effects of non-native species at the community level, in contrast to the classic 'species-centred' focus. They also provided a new perspective on the integration of non-native plants in recipient communities, and their potential effects on native species.

These early network-based studies (Memmott and Waser, 2002; Olesen *et al.*, 2002) contrasted with previous literature, in which non-native species were mostly reported as causing drastic damage to biodiversity across the globe. Negative effects of non-native species included modifications to the availability and quality of resources, reduction of species diversity and community functioning, and disruption of interspecific interactions (Richardson *et al.*, 2000a; Traveset and Richardson, 2006). It was noticed, both in network and non-network studies, that non-native plants had higher invasion success when they had high taxonomic affinity with the local community, or their partners were phylogenetically close to species in their natural ranges (Memmott and Waser, 2002; Razanajatovo *et al.*, 2015), but also when non-natives and natives exhibited similar morphological characteristics that allowed them to interact (Morales and Traveset, 2009). Moreover, at higher invasion levels, interactions were established between non-native partners, forming 'invader complexes' (Richardson *et al.*, 2000a; Kaiser-Bunbury *et al.*, 2011), especially in insular ecosystems (Traveset and Richardson, 2006; Arroyo-Correa *et al.*, 2020). Evidence was mounting that facilitative interactions were allowing the successful establishment and eventual dominance by non-native plant species through mutualistic relationships with either native or non-native organisms (Richardson *et al.*, 2000a; Traveset and Richardson, 2011; Richardson and Pyšek, 2012). It also became apparent that this came with a cost: interaction partners were being diverted away from native to non-native species and the strength of their interactions changing, compromising native species (Chittka and Schürkens, 2001; Ghazoul, 2004; Aizen *et al.*, 2008; Goodell and Parker, 2017). As network-based studies became common, a higher number of negative impacts of invasive plant species also started to emerge within this field. Non-native effects can propagate and be detected across different levels of networks, often resulting in the simplification of invaded communities (Heleno *et al.*, 2009; Carvalheiro *et al.*, 2010; López-Núñez *et al.*, 2017).

18.3 Overall Patterns – The Case of Pollination Networks

Most of our understanding of the effects of plant invasions on networks are based on plant pollination systems from temperate zones. Despite this clear bias, plus the contrasting effects of non-native plants found in the literature (reviewed in Frost *et al.*, 2019), pollination networks have helped us to elucidate important aspects of plant invasions. They revealed that invaded plant—pollinator communities are mostly nested (Olesen *et al.*, 2007; Traveset *et al.*, 2015a): non-native species can assume highly generalist positions while displacing native species from central to peripheral roles (Olesen *et al.*, 2002; Aizen *et al.*, 2008; Fig. 18.1a,c). Hence transference of visits from native to non-native plant species is commonly detected when researchers look through the lenses of ecological networks (Aizen *et al.*, 2008; Traveset *et al.*, 2015b), yet the functional consequences of such changes are still not fully understood (Morales and Traveset, 2009, but see Kaiser-Bunbury *et al.*, 2011). The high generalization of pollination systems detected in network-based studies contradicted the previous literature, which stated that interactions are highly specialized between plants and pollinators (Waser *et al.*, 1996; Waser and Ollerton, 2006). Such contradictions may be due to the fact that the vast majority of data concerning pollination networks are based on observations of flower visitations, which is only the first stage of pollination and may not adequately represent non-native pollen transfer to native species (King *et al.*, 2013; Emer *et al.*, 2015). Networks based on pollen transport and pollen transfer look at the exchange of pollen grains among co-occurring flowering species and prove to better describe the impacts of non-native plants on pollination at the community level (Lopezaraiza–Mikel *et al.*, 2007; Tur *et al.*, 2014). Indeed, these networks tend to be more specialized and less nested than flower visitation networks (Tur *et al.*, 2014; Emer *et al.*, 2015), a finding that reopens early discussions on the generalist–specialist paradigm of pollination systems (Waser and Ollerton, 2006).

Because invasive plants can establish many interactions with resident species, they can alter modular structures by connecting different network areas and/or increasing the size of the modules they invade (Fig. 18.1b). Invasive plants may occupy central roles in the invaded networks by either functioning as network 'connectors' and increasing overall network connectivity, or by increasing the number of interactions within their own module or among modules, thus acting as module and/or network 'hubs', respectively. Invasive hubs were detected in bird pollination networks of Galápagos, where the invasive plant *Psidium guajava* was the most connected species in the network during the hot season (Traveset *et al.*, 2015a). Using a data set of 40 networks with different levels of invasion, Albrecht *et al.* (2014) also found that invasive plants attracted more pollinators to the modules they had invaded and exhibited higher generalization in respect to their pollinators than did native species, thus occupying more central roles in the network. Hence, after invasion, approximately 50% of the native species switched from occupying central roles to more peripheral positions in the network, performing secondary roles that neither contributed much to modules' connectivity nor affected the overall network structure. The use of a network approach helped to disclose the usurpation of pollinators from resident species while attracting even more flower visitors to the community, and even though the overall network structure did not change, native species were being relegated to marginal roles in the community.

Although invasive plants can occupy central positions in networks and modify the role of native species, the asymmetric relationship among mutualistic partners (Bascompte *et al.*, 2006) seems to provide resistance of the network structure to plant invasion, at least for pollination systems. Asymmetries are found in networks when the level of dependency between partner species is not reciprocal, i.e. one of the partners depends significantly more on the other. In pollination, plants tend to be more specialized than pollinators thus are more dependent on pollinators than the other way around. Hence, a generalist pollinator can switch its visitation from a native to a non-native plant species according to its needs, or to the resources available (Arroyo-Correa *et al.*, 2020). If a non-native plant overcompetes a native species in terms of floral resources, pollinators may prefer to visit them instead of natives, therefore switching partner species but maintaining the same network structure. To test this, Vil *et al.* (2009) studied the effects of five highly invasive plants across Europe and found that none affected network structure in terms of connectance, linkage density or nestedness. Even when changes to properties of invaded networks were detected, such as increasing nestedness or decreasing connectance, they were mainly driven by fluctuations in the number and abundance of species in the local community as a consequence of the arrival of the newcomers, rather than the displacement of native species (Stouffer *et al.*, 2014). Thus, changes to network structure promoted by invasive non-native plants happen mostly by rewiring local interactions, i.e. reorganizing plant–pollinator associations in the invaded communities (Aizen *et al.*, 2008; Padrón *et al.*, 2009; García *et al.*, 2014).

The functional consequences of rewiring for both native and non-native species' fitness may be quite distinct. Nonetheless, native pollinators can show preference for non-native plants (Stouffer *et al.*, 2014), especially when they exhibit highly attractive floral resources (Chittka and Schürkens, 2001). Because of this, shared pollinators can transport high amounts of non-native pollen grains throughout the community (Lopezaraiza–Mikel *et al.*, 2007; Tur *et al.*, 2014). Yet, this does not necessarily translate in non-native pollen grain deposition on the native stigmas (Emer *et al.*, 2015), which can influence native pollen germination and reduce plant fitness. Indeed, in the Seychelles islands, fruit setting of native plants was mostly not affected by the establishment of non-native plants, even though the number and the uniformity of distribution of interactions were slightly lower in the invaded networks (Kaiser-Bunbury *et al.*, 2011). It has been suggested that whether a non-native plant affects or not ecological functions can be predicted by its network role in the native range (Emer *et al.*, 2016). Accordingly, a non-native plant species with a low number of interactions in its native distribution will perform likewise when it establishes outside its natural range, therefore with lower chances of negatively affecting native plant fitness. On the other hand, highly central plants in their home range are predicted to have a high number of interactions in the communities they invade. Therefore, highly central non-native plants have greater probability of promoting the rewiring of interactions between partner species, potentially affecting pollination functions for native plants.

18.4 Beyond Pollination Networks

The impacts of non-native plants on networks other than plant–pollinator interactions are much less understood; there is a lack of studies on the effect of invasive plants on plant–herbivore, plant–pathogen or plant–mycorrhizal networks, for example, and as described for pollination networks, existing studies show conflicting results. Very few papers have assessed plant invasion effects on seed-dispersal or plant–frugivore networks, and most of them have been restricted to islands (e.g. Heleno *et al.*, 2012; Vizentin-Bugoni *et al.*, 2019). In the Azores, birds seem to prefer fruits of non-native plant species, which were involved in approximately two thirds of the interactions, despite native plants constituting the most common fruit resource (Heleno *et al.*, 2013). In that archipelago, non-native plants attracted more bird species than native ones, increasing linkage density, interaction evenness and network robustness. Contrarily, the proportion of interactions involving non-native plants was lower than the proportion of native–native interactions in seed-dispersal networks of two Galápagos islands (Heleno *et al.*, 2012). Even though both islands showed different invasion histories, overall network descriptors did not change with invasion time and the proportion of non-native species. A perfect example of an extreme case of a community dominated by non-native species comes from Hawaii, where non-native plant and frugivore birds dominate the networks (Vizentin-Bugoni *et al.*, 2019). This study showed that non-native plants are also very well integrated in the seed dispersal networks, and that the structure (nestedness, modularity and network specialization) and stability (robustness) of the networks is very similar to that of native-dominated networks from around the globe. Nonetheless, slight decreases in modularity and specialization were detected when the Hawaiian novel community was compared with non-tropical networks. These findings suggest that network properties may be evolutionarily conserved (Rezende *et al.*, 2007; Kéfi *et al.*, 2016), yet the coevolutionary trajectories of species are not a requirement for the emergence of network patterns given that the features that allow the interaction to occur in seed dispersal, e.g. trait matching between seed size and beak length (González-Castro *et al.*, 2015), are independent of partner species being native or non-natives.

Even though poorly explored, networks have also been applied to elucidate the consequences of non-native plant invasions on different food webs. For example, a study performed in Portugal (López-Núñez *et al.*, 2017) examined the cascading effects of the highly invasive *Acacia longifolia* through several levels of the network. In this case, a strong indirect effect of *A. longifolia* was found on plant–galler–parasitoid–inquiline interactions mediated by apparent competition. Another study from the Ogasawara oceanic islands of Japan unveiled that non-native plants were easily integrated into plant–insect communities by performing interactions with both native and non-native cerambycid beetle species that feed on their living or dead wood (Sugiura, 2010). Non-native plants tend to be used by a higher number of non-native beetles, suggesting a facilitative effect between non-native species and the formation of 'invader complexes'. The plant–beetle networks were highly nested, but this structure was not affected by the presence of non-native species. The outcomes of these studies suggest that the effects of non-native plant species on trophic networks are context-dependent, and are likely to vary according to the traits of the invader and the biotic and abiotic characteristics of the recipient community.

18.5 Networks Along Gradients of Invasion

An important aspect of plant invasions, and of their impact on the structure of ecological networks, is the extent to which they dominate recipient communities (Aizen *et al.*, 2008; Heleno *et al.*, 2009; Kaiser-Bunbury *et al.*, 2011). Changes in network properties, at least in mutualistic systems, are expected to occur at high levels of invasion (Kaiser-Bunbury *et al.*, 2011); depending on whether there are a few highly dominant, or many non-dominant non-native species, such changes will deliver different results (Timóteo *et al.*, 2018a; Vizentin-Bugoni *et al.*, 2019). Many studies so far have simply compared non-invaded against invaded plots, usually focusing on the impact of a single species or a few species that become highly abundant, e.g. *Impatiens glandulifera* (in the UK; Emer *et al.*, 2015), *Opuntia* spp. (Canaries and Balearics; Padrón *et al.*, 2009) or *Ailanthus altissima*, *Carpobrotus* spp. and *Oxalis pes-caprae* (Balearics;

Lapiedra *et al.*, 2015). However, invasion is actually a progression starting with the arrival of a species in a community outside its natural range, through its establishment and naturalization, and eventual dominance over other species in the recipient community (Richardson *et al.*, 2000b, Fig. 18.1a). To understand this entire process, researchers can work on gradients of invasion by plant species. For instance, empirical temporal networks can be sampled to follow how invasion advances through time. Also, spatial networks built at different locations can be used to generate a chrono-sequence that illustrates the structural response of networks to the gradient of invasion. Using a gradient approach helps to clarify the inconsistent results to date for the effect of plant invasions on networks, the existence of thresholds from which network effects are detectable, and how species-level metrics and species roles change as networks become more invaded (Tylianakis and Morris, 2017; Frost *et al.*, 2019).

A few studies have pointed out that beyond a certain level of invasion, non-native plants promote detectable changes in network structure that result in simpler and more heterogeneous networks, even though ecological functions are mostly not affected (Aizen *et al.*, 2008; Kaiser-Bunbury *et al.*, 2011). As networks become invaded, the integration of non-native plants results in a gradual decrease in mutualism strength, increase in interaction asymmetry, and decrease in connectivity between native plants and pollinators (Aizen *et al.*, 2008). This results from the increased interaction generalization of non-native plants, to the point of becoming super-generalists and highly central in the invaded network, then becoming important players in the structural cohesion of their networks (Aizen *et al.*, 2008; Kaiser-Bunbury *et al.*, 2011; Albrecht *et al.*, 2014). Changes in the strength of non-native plant species create highly asymmetrical interactions that in turn increase network stability and robustness (Bascompte *et al.*, 2006), which promotes the persistence of non-native plants and reinforces the invasion process (Traveset and Richardson, 2014). By investigating the changes produced by increasing non-native floral resources in the Seychelles, Kaiser-Bunbury *et al.* (2011) detected that plant linkage and interaction evenness significantly decreased. Interaction evenness, particularly, showed a split behaviour with no changes being observed up to a point beyond which it decreased, implying an initial tolerance by the native community to invasion. Once this invasion threshold is crossed, links established by abundant generalist pollinators become stronger and/or they interact more with different sets of plants, which results in a decrease in interaction evenness. Native plants become dependent on fewer pollinators, participate in fewer interactions and seed quality may be reduced, thus making the native plant community more vulnerable.

A different approach was used by Timóteo *et al.* (2018a) who quantified the magnitude of invasion as the number of increased non-native species in their recipient communities. In that study, the authors asked whether the network structure of novel communities, i.e. communities formed by species without a shared evolutionary history (Hobbs *et al.*, 2006), differed from that of 50 native-dominated networks from across the globe (0% to 38% non-native plant species in the community). Contrary to expectations, no significant changes were detected in connectance, specialization or robustness across networks. Moreover, these descriptors did not differ between those 50 networks and a highly anthropogenic novel community (with 85% of non-invasive, non-native plant species), although a slight tendency towards higher specialization of the latter was apparent (Timóteo *et al.*, 2018a). No differences were further detected in interaction patterns (linkage level, species specialization and strength) between the native and non-native sub-communities of the highly anthropogenic community. This lack of differences may result from the contrasting effect of invasive plants that promote generalism (Aizen *et al.*, 2008; Albrecht *et al.*, 2014) and non-invasive, non-native species that appear to promote network specialization. Previous studies suggested that species' phylogenetic relationships are an important determinant of the structure of mutualistic networks (Rezende *et al.*, 2007; Peralta, 2016). However, both Timóteo *et al.* (2018a) and Vizentin-Bugoni *et al.* (2019) show that interaction patterns in native-dominated mutualistic networks can also emerge in novel communities dominated by non-native plant species, in which interacting partners have none or little shared evolutionary history.

18.6 Removing Non-native Plants From Ecological Networks

One of the many applications of a network approach lies in understanding the effects of disturbance on the community, and many studies have assessed the robustness of networks to cascades of secondary species extinctions (Dunne *et al.*, 2002; Valdovinos *et al.*, 2009). To counter the effect of non-native species, conservation and restoration programmes often undertake ambitious plans to control or eradicate non-native plants. Ecological networks allow ecologists and practitioners to evaluate the cascading effects that such removals may have on the remaining community (Carvalheiro *et al.*, 2008; Heleno *et al.*, 2009; Kaiser-Bunbury *et al.*, 2017). Although somewhat counter-intuitive, the fact that non-native plants become deeply integrated in the architecture of their recipient network confers higher robustness against extinction cascades, as Valdovinos *et al.* (2009) demonstrated. These authors applied a dynamic approach to follow the disassembly of pollination networks to understand the structural role of non-native plants in their networks. Non-native plant species increased network connectivity, occupying upper positions on the species degree distribution and, for that reason, had an important role in network temporal persistence. As the asymmetry in interaction strength increases, networks dominated by invasive plants eventually become more stable and difficult to restore (Tylianakis, 2008). While modelling the removal of non-native plants, Valdovinos *et al.* (2009) found that it decreased the chances of species persistence because the remaining community becomes more dependent on highly generalist non-natives therefore more susceptible to local extinctions. The result is a restored native community that is more loosely connected, more fragile and susceptible to the loss of species if these generalist non-natives are suddenly removed, i.e. invaded communities are expected to be sensitive to the removal of the most connected species, which is then the case of invasive plants (Valdovinos *et al.*, 2009). However, in a field experiment of non-native plant species removal, Heleno *et al.* (2010) found somewhat different results: the removal of an abundant non-native plant species did not result in changes in network descriptors. The native community was able to re-establish both trophic and mutualistic interactions, which highlights the important role of native species in restructuring the network of interactions and maintaining its robustness to the removal of non-native species (Heleno *et al.*, 2010; Kaiser-Bunbury *et al.*, 2017). This is encouraging for habitat restoration from plant invasions because it shows signs of recovery by the native vegetation. Nonetheless, the community response to the removal of non-native plant species can be very context-dependent. The importance of non-native plants as a resource in the recipient community calls for some caution when designing restoration activities, because its eradication can create a resource gap that may become detrimental for the native partner species (Ceia *et al.*, 2010). The quality of the remaining resources and the ability of the animal community to use them will also play an essential role in the overall community response to removal of non-native plants (Carvalheiro *et al.*, 2008).

18.7 Future Directions

Ecologists are facing an exciting time with the advances recently made in the field of network ecology and are further exploring a variety of factors that influence network structure and stability. The study of invasions by non-native plant species will undoubtedly benefit from this development and, simultaneously, contribute to the advancement of community ecology. Ecological systems are not static, and changes promoted by biological invasions result from complex dynamics that may not be captured by simply studying network topology. The process of invasion by plant species (and indeed other organisms) will certainly contribute to provide invaluable information regarding the importance of interspecific interactions on community assembly and the factors that may determine the establishment of species in a community (Sargent and Ackerly,

2008; Pantel *et al.*, 2017). These include the temporal dynamics (Dupont *et al.*, 2009; Arroyo-Correa *et al.*, 2020) and spatial connectivity of interactions and network architecture (Pilosof *et al.*, 2017; Emer *et al.*, 2018; Timóteo *et al.*, 2018b), the interdependence of different types of trophic and non-trophic interactions (Fontaine *et al.*, 2011; Kéfi *et al.*, 2016) and the phylogenetic relationship between interacting species (Rezende *et al.*, 2007; Peralta, 2016; Bastazani *et al.*, 2017). Emerging fields are those that increase our understanding of the role of species traits (Maglianesi *et al.*, 2014; Stouffer *et al.*, 2014), functional diversity (Montoya *et al.*, 2015; Dehling and Stouffer, 2018) and intraspecific variation in shaping network structure (Dupont *et al.*, 2011; Tinker *et al.*, 2012; Tur *et al.*, 2014). Species traits, and their effect on functional redundancy and complementarity, may allow us to predict the potential invasion ability of a species and the susceptibility of a network to be invaded by non-native plants (Hui *et al.*, 2016; Rojas *et al.*, 2019). Finally, biological invasion closely relates to community (in)stability and phase transitions, and often exhibits nonlinear and adaptative behaviours that contribute to the difficulty in predicting invasions. A 'wind vane' to reveal the performance and response of species to invasions was proposed based on the statistical properties of the interaction matrix of a local invaded network (Hui and Richardson, 2019). The use of adaptive ecological network analysis, that allows for a continuous update of the interaction matrix as networks rewire during invasion processes, can further contribute to a better understanding and forecasts of the functional consequences of biological invasions (Hui and Richardson, 2017).

18.8 Concluding Remarks

Network theory is already a powerful tool in ecology; it is a fast-moving field that is generating a vast body of theoretical and applied knowledge, furthering our understanding of the mechanisms of plant invasions, how to better predict invaders and how to deal with them. Yet, networks also revealed how complex are the outcomes of plant invasion in the interplay of biotic interactions in native communities. Despite the contradictory results mentioned throughout this chapter, and the existing biases in the literature regarding the type of interactions studied and their locations, the use of a network approach to study plant invasions allows us to make some broad generalizations: i) non-native plants mostly integrate into native communities by interacting initially with native, generalist species; ii) most non-native plant species that become invasive are often generalists in their habitat requirements and mutualistic partners, which allows them to establish novel interactions with the native species; iii) invasive plants can usurp interaction visits from natives, displacing them from very important central positions to peripheral roles within the network; iv) in most cases, invasive plants do not change the overall structure of the local networks up to a threshold, but rearrange the interaction links among partner species within the recipient communities.

Most important is the conclusion that the devil is in the detail. As often mentioned, network structure is not dramatically changed by plant invasions, at least up to a given point. However, if we have a closer look at what each individual species (native or non-native) is doing, and how they relate to each other within the structure of invaded networks, we will find that invasive plants are usurping interactions and displacing native species from highly central and important roles to marginal positions, with unknown consequences for ecosystem functionality. As network analysis techniques develop (e.g. use of multilayer networks that include more than a single ecological process at a time), the potential for a broader understanding of plant invasions, and invasions in general, increases because researchers will be looking at the multiple pathways that allow a non-native plant to become invasive. Such work will yield important insights into community assembly and disassembly, but should also provide conservationists with a powerful tool to design effective restoration strategies that account for the sometimes hidden knock-on effects of the removal of non-native species.

Acknowledgements

CE was funding by the São Paulo Research Foundation (FAPESP 2105/15172-7, 2016/18355-8). ST was funded by the Portuguese Fundação para a Ciência e a Tecnologia (FCT - UID/BIA/04004/2020, CEECIND/00135/2017)

References

Aizen, M.A., Morales, C.L. and Morales, J.M. (2008) Invasive mutualists erode native pollination webs. *PLoS Biology* 6, e31.

Albrecht, M., Padron, B., Bartomeus, I. and Traveset, A. (2014) Consequences of plant invasions on compartmentalization and species' roles in plant-pollinator networks. *Proceedings of the Royal Society B* 281, 20140773–20140773.

Arroyo-Correa, B., Burkle, L. and Emer, C. (2020) Alien plants and flower visitors disrupt the seasonal dynamics of mutualistic networks. *Journal of Ecology* 108, 1475–1486.

Barabási, A.L. and Albert, R. (1999) Emergence of scaling in random networks. *Science* 286, 509–512.

Bascompte, J., Jordano, P., Melian, C.J. and Olesen, J.M. (2003) The nested assembly of plant-animal mutualistic networks. *Proceedings of the National Academy of Sciences USA* 100, 9383–9387.

Bascompte, J., Jordano, P. and Olesen, J.M. (2006) Asymmetric coevolutionary networks facilitate biodiversity maintenance. *Science* 312, 431–433.

Bassett, D.S., Wymbs, N.F., Porter, M.A., Mucha, P.J., Carlson, J.M. *et al.* (2011) Dynamic reconfiguration of human brain networks during learning. *Proceedings of the National Academy of Sciences USA* 108, 7641–7646.

Bastazani, V.A.G., Ferreira, P.M.A., Azambuja, B.O., Casas, G., Debastiani, V.J. *et al.* (2017) Untangling the tangled bank: a novel method for partitioning the effects of phylogenies and traits on ecological networks. *Evolutionary Biology* 4, 312–324.

Bastolla, U., Fortuna, M., Pascual-García, A., Ferrera, A., Luque, B. *et al.* (2009) The architecture of mutualistic networks minimizes competition and increases biodiversity. *Nature* 458, 1018–1020.

Bersier, L.F. (2007) A history of the study of ecological networks. In: Képès, F. (ed.) *Biological Networks. Complex Systems and Interdisciplinary Science*. 3. World Scientific, Singapore, pp. 365–421.

Camerano, L. (1880) Dell'equilibrio dei viventi merc`e la reciproca distruzione. *Atti della Reale Accademia della Scienze di Torino* 15, 393–414.

Cantor, M., Pires, M.M., Marquitti, F.M.D., Raimundo, R.L.G., Sebastián-González, E. *et al.* (2017) Nestedness across biological scales. *PLoS ONE* 12, e0171691.

Carvalheiro, L.G., Barbosa, E.R.M. and Memmott, J. (2008) Pollinator networks, alien species and the conservation of rare plants: Trinia glauca as a case study. *Journal of Applied Ecology* 45, 1419–1427.

Carvalheiro, L.G., Buckley, Y.M. and Memmott, J. (2010) Diet breadth influences how the impact of invasive plants is propagated through food webs. *Ecology* 91, 1063–1074.

Ceia, R.S., Sampaio, H.L., Parejo, S.H., Heleno, R.H., Arosa, M.L. *et al.* (2010) Throwing the baby out with the bathwater: does laurel forest restoration remove a critical winter food supply for the critically endangered Azores bullfinch? *Biological Invasions* 13, 93–104.

Chittka, L. and Schürkens, S. (2001) Successful invasion of a floral market. *Nature* 411, 653–653.

Cohen, J.E. (1978) *Food Webs and Niche Space*. Princeton University Press, Princeton.

Cruz, J.C., Ramos, J.A., Silva, L.P., Tenreiro, P.Q. and Heleno, R.H. (2013) Seed dispersal networks in an urban novel ecosystem. *European Journal of Forest Research* 132, 887–897.

Dehling, D.M. and Stouffer, D.B. (2018) Bringing the Eltonian niche into functional diversity. *Oikos* 127, 1711–1723.

Delmas, E., Besson, M., Brice, M.H., Burkle, L.A., Dalla Riva, G. V. *et al.* (2019) Analysing ecological networks of species interactions. *Biological Reviews* 94, 16–36.

Dunne, J.A., Williams, R.J. and Martinez, N.D. (2002) Network structure and biodiversity loss in food webs: robustness increases with connectance. *Ecology Letters* 5, 558–567.

Dupont, Y.L., Padrón, B., Olesen, J.M. and Petanidou, T. (2009) Spatio-Temporal variation in the structure of pollination networks. *Oikos* 118, 1261–1269.

Dupont, Y.L., Trøjelsgaard, K. and Olesen, J.M. (2011) Scaling down from species to individuals: a flower-visitation network between individual honeybees and thistle plants. *Oikos* 120, 170–177.
Dügger, P.B., Bohning-Gaese, P.G., Chama, K. and Correia, M, L. *et al.* (2018) Seed-dispersal networks are more specialized in the Neotropics than in the Afrotropics. *Global Ecology and Biogeography* 28, 248–261.
Emer, C., Galetti, M., Pizo, M.A., Guimarães, P.R., Moraes, S. *et al.* (2018) Seed-dispersal interactions in fragmented landscapes - a metanetwork approach. *Ecology Letters* 21, 484–493.
Emer, C., Memmott, J., Vaughan, I.P., Montoya, D. and Tylianakis, J.M. (2016) Species roles in plant–pollinator communities are conserved across native and alien ranges. *Diversity and Distributions* 22, 841–852.
Emer, C., Vaughan, I.P., Hiscock, S. and Memmott, J. (2015) The impact of the invasive alien plant, *Impatiens glandulifera*, on pollen transfer networks. *PLoS ONE* 10, e0143532.
Fontaine, C., Guimarães, P.R., Kéfi, S., Loeuille, N., Memmott, J. *et al.* (2011) The ecological and evolutionary implications of merging different types of networks. *Ecology Letters* 14, 1170–1181.
Fortuna, M.A., Stouffer, D.B., Olesen, J.M., Jordano, P., Mouillot, D. *et al.* (2010) Nestedness versus modularity in ecological networks: two sides of the same coin? *Journal of Animal Ecology* 79, 811–817.
Frost, C.M., Allen, W.J., Courchamp, F., Jeschke, J.M., Saul, W.C. *et al.* (2019) Using network theory to understand and predict biological invasions. *Trends in Ecology & Evolution* 34, 831–843.
García, D., Martínez, D., Stouffer, D.B. and Tylianakis, J.M. (2014) Exotic birds increase generalization and compensate for native bird decline in plant-frugivore assemblages. *Journal of Animal Ecology* 83, 1441–1450.
Ghazoul, J. (2004) Alien abduction: disruption of native plant-pollinator interactions by invasive species. *Biotropica* 36, 156.
González-Castro, A., Yang, S., Nogales, M. and Carlo, T.A. (2015) Relative importance of phenotypic trait matching and species' abundances in determining plant - avian seed dispersal interactions in a small insular community. *AoB PLANTS* 7, 1–10.
Goodell, K. and Parker, I.M. (2017) Invasion of a dominant floral resource: effects on the floral community and pollination of native plants. *Ecology* 98, 57–69.
Heleno, R.H., Ceia, R.S., Ramos, J.A. and Memmott, J. (2009) Effects of alien plants on insect abundance and biomass: a food-web approach. *Conservation Biology* 23, 410–419.
Heleno, R., Lacerda, I., Ramos, J. and Memmott, J. (2010) Evaluation of restoration effectiveness: community response to the removal of alien plants. *Ecological Applications* 20, 1191–1203.
Heleno, R.H., Olesen, J.M., Nogales, M., Vargas, P. and Traveset, A. (2012) Seed dispersal networks in the Galapagos and the consequences of alien plant invasions. *Proceedings of the Royal Society B* 280, 20122112–20122112.
Heleno, R.H., Ramos, J.A. and Memmott, J. (2013) Integration of exotic seeds into an Azorean seed dispersal network. *Biological Invasions* 15, 1143–1154.
Henneman, M.L. and Memmott, J. (2001) Infiltration of a Hawaiian community by introduced biological control agents. *Science* 293, 1314–1316.
Hobbs, R.J., Arico, S., Aronson, J., Baron, J.S., Bridgewater, P. *et al.* (2006) Novel ecosystems: theoretical and management aspects of the new ecological world order. *Global Ecology and Biogeography* 15, 1–7.
Hui, C. and Richardson, D.M. (2017) *Invasion dynamics*. Oxford University Press, Oxford.
Hui, C. and Richardson, D.M. (2019) How to invade an ecological network. *Trends in Ecology and Evolution* 34, 121–131.
Hui, C., Richardson, D.M., Landi, P., Minoarivelo, H.O., Garnas, J. *et al.* (2016) Defining invasiveness and invasibility in ecological networks. *Biological Invasions* 18(4), 971–983. DOI: 10.1007/s10530-016-1076-7.
IPBES (2019) Brondizio, E.S. (ed.) *Global assessment report on biodiversity and ecosystem services of the Intergovernmental Science — Policy Platform on Biodiversity and Ecosystem Services*. IPBES Secretariat, Bonn, Germany.
Jordano, P. (1987) Patterns of mutualistic interactions in pollination and seed dispersal: connectance, dependence asymmetries, and coevolution. *American Naturalist* 129, 657–677.
Kaiser-Bunbury, C.N., Mougal, J., Whittington, A.E., Valentin, T., Gabriel, R. *et al.* (2017) Ecosystem restoration strengthens pollination network resilience and function. *Nature* 542, 223–227.
Kaiser-Bunbury, C.N., Valentin, T., Mougal, J., Matatiken, D. and Ghazoul, J. (2011) The tolerance of island plant-pollinator networks to alien plants. *Journal of Ecology* 99, 202–213.

King, C., Ballantyne, G. and Willmer, P.G. (2013) Why flower visitation is a poor proxy for pollination: measuring single-visit pollen deposition, with implications for pollination networks and conservation. *Methods in Ecology and Evolution* 4, 811–818.

Krause, A.E., Frank, K.A., Mason, D.M., Ulanowicz, R.E. and Taylor, W.W. (2003) Compartments revealed in food-web structure. *Nature* 426, 282–285.

Kéfi, S., Miele, V., Wieters, E.A., Navarrete, S.A. and Berlow, E.L. (2016) How structured is the entangled bank? the surprisingly simple organization of multiplex ecological networks leads to increased persistence and resilience. *PLoS Biology* 14, e1002527.

Lapiedra, O., Sol, D., Traveset, A. and Vil, M. (2015) Random processes and phylogenetic loss caused by plant invasions. *Global Ecology and Biogeography* 24, 774–785.

Lewinsohn, T.M., Inácio Prado, P., Jordano, P., Bascompte, J. and Olesen, J, M. (2006) Structure in plant-animal interaction assemblages. *Oikos* 113, 174–184.

Lopezaraiza-Mikel, M.E., Hayes, R.B., Whalley, M.R. and Memmott, J. (2007) The impact of an alien plant on a native plant–pollinator network: an experimental approach. *Ecology Letters* 10, 539–550.

López-Núñez, F.A., Heleno, R.H., Ribeiro, S., Marchante, H. and Marchante, E. (2017) Four-trophic level food webs reveal the cascading impacts of an invasive plant targeted for biocontrol. *Ecology* 98, 782–793.

Macfadyen, S., Gibson, R.H., Symondson, W.O.C. and Memmott, J. (2011) Landscape structure influences modularity patterns in farm food webs: consequences for pest control. *Ecological Applications* 21, 516–524.

Maglianesi, M.A., Blüthgen, N., Böhning-Gaese, K. and Schleuning, M. (2014) Morphological traits determine specialization and resource use in plant–hummingbird networks in the neotropics. *Ecology* 95, 3325–3334.

Martín González, A.M., Dalsgaard, B. and Olesen, J.M. (2010) Centrality measures and the importance of generalist species in pollination networks. *Ecological Complexity* 7, 36–43.

Mello, M.A.R., Rodrigues, F.A., Costa, L.D.F., Kissling, W.D., Sekercioglu, Ç. H. *et al*.., ¸ (2015) Keystone species in seed dispersal networks are mainly determined by dietary specialization. *Oikos* 124, 1031–1039.

Memmott, J. (1999) The structure of a plant-pollinator food web. *Ecology Letters* 2, 276–280.

Memmott, J. (2009) Food webs: a ladder for Picking strawberries or a practical tool for practical problems? *Philosophical Transactions of the Royal Society B* 364, 1693–1699.

Memmott, J. and Waser, N.M. (2002) Integration of alien plants into a native flower-pollinator visitation web. *Proceedings of the Royal Society B* 269, 2395–2399.

Montoya, J.M., Pimm, S.L. and Sole, R.V. (2006) Ecological networks and their fragility. *Nature* 442, 259–264.

Montoya, D., Rogers, L. and Memmott, J. (2012) Emerging perspectives in the restoration of biodiversity-based ecosystem services. *Trends in Ecology & Evolution* 27, 666–672.

Montoya, D., Yallop, M.L. and Memmott, J. (2015) Functional group diversity increases with modularity in complex food webs. *Nature Communications* 6, 7379.

Morales, C.L. and Traveset, A. (2009) A meta-analysis of impacts of alien vs. native plants on pollinator visitation and reproductive success of co-flowering native plants. *Ecology Letters* 12, 716–728.

Olesen, J.M., Bascompte, J., Dupont, Y.L. and Jordano, P. (2007) The modularity of pollination networks. *Proceedings of the National Academy of Sciences USA* 104, 19891–19896.

Olesen, J.M., Eskildsen, L.I. and Venkatasamy, S. (2002) Invasion of pollination networks on oceanic islands: importance of invader complexes and endemic super generalists. *Diversity and Distributions* 8, 181–192.

Padrón, B., Traveset, A., Biedenweg, T., Díaz, D., Nogales, M. *et al*. (2009) Impact of alien plant invaders on pollination networks I two archipelagos. *PLoS ONE* 4.

Pantel, J.H., Bohan, D.A., Calcagno, V., David, P., Duyck, P.F. *et al*. (2017) 14 questions for invasion in ecological networks. *Advances in Ecological Research* 56, 293–340.

Peralta, G. (2016) Merging evolutionary history into species interaction networks. *Functional Ecology* 30, 1917–1925.

Pilosof, S., Porter, M.A., Pascual, M. and Kéfi, S. (2017) The multilayer nature of ecological networks. *Nature Ecology & Evolution* 1, 0101.

Razanajatovo, M., Föhr, C., Fischer, M., Prati, D. and van Kleunen, M. (2015) Non-naturalized alien plants receive fewer flower visits than naturalized and native plants in a Swiss botanical garden. *Biological Conservation* 182, 109–116.

Rezende, E.L., Lavabre, J.E., Guimarães, P.R., Jordano, P. and Bascompte, J. (2007) Non-random coextinctions in phylogenetically structured mutualistic networks. *Nature* 448, 925–928.

Richardson, D.M. and Pyšek, P. (2012) Naturalization of introduced plants: ecological drivers of biogeographical patterns. *New Phytologist* 196(2), 383–396. DOI: 10.1111/j.1469-8137.2012.04292.x.

Richardson, D.M., Allsopp, N., D'Antonio, C.M., Milton, S.J. and Rejmánek, M. (2000a) Plant invasions – the role of mutualisms. *Biological Review* 75, 65–93.

Richardson, D.M., Pyšek, P., Rejmanek, M., Barbour, M.G., Panetta, F.D. *et al.* (2000b) Naturalization and invasion of alien plants: concepts and definitions. *Diversity and Distributions* 6(2), 93–107. DOI: 10.1046/j.1472-4642.2000.00083.x.

Rojas, T.N., Gallo, M.C.F., Vergara-Tabares, D.L., Nazaro, M.G., Zampini, I.C. *et al.* (2019) Being popular or freak: how alien plants integrate into native plant-frugivore networks. *Biological Invasions* 21, 2589–2598.

Sargent, R.D. and Ackerly, D.D. (2008) Plant-pollinator interactions and the assembly of plant communities. *Trends in Ecology & Evolution* 23, 123–130.

Schleuning, M., Fründ, J. and García, D. (2015) Predicting ecosystem functions from biodiversity and mutualistic networks: an extension of trait-based concepts to plant-animal interactions. *Ecography* 38, 001–013.

Schleuning, M., Ingmann, L., Strauß, R., Fritz, S.A., Dalsgaard, B. *et al.* (2014) Ecological, historical and evolutionary determinants of modularity in weighted seed-dispersal networks. *Ecology Letters* 17, 454–463.

Stouffer, D.B., Cirtwill, A.R. and Bascompte, J. (2014) How exotic plants integrate into pollination networks. *Journal of Ecology* 102, 1442–1450.

Sugiura, S. (2010) Species interactions-area relationships: biological invasions and network structure in relation to island area. *Proceedings of the Royal Society B* 277, 1807–1815.

Summerhayes, V.S. and Elton, C. (1923) Contributions to the ecology of Spitzbergen and bear island. *Journal of Ecology* 11, 214–216.

Thébault, E. and Fontaine, C. (2010) Stability of ecological communities and the architecture of mutualistic and trophic networks. *Science* 329, 853–856.

Timóteo, S., O'Connor, C.J., López-Núñez, F.A., Costa, J.M., Gouveia, A.C. *et al.* (2018a) Pollination networks from natural and anthropogenic-novel communities show high structural similarity. *Oecologia* 188(4), 1155–1165. DOI: 10.1007/s00442-018-4281-5.

Timóteo, S., Correia, M., Rodríguez-Echeverría, S., Freitas, H. and Heleno, R. (2018b) Multilayer networks reveal the spatial structure of seed-dispersal interactions across the great Rift landscapes. *Nature Communications* 9(1), 140. DOI: 10.1038/s41467-017-02658-y.

Tinker, T.M., Guimarães, P.R., Novak, M., Marquitti, F.M.D., Bodkin, J.L. *et al.* (2012) Structure and mechanism of diet specialisation: testing models of individual variation in resource use with sea otters. *Ecology Letters* 15, 475–483.

Traveset, A. and Richardson, D.M. (2011) Mutualisms: key drivers of invasions... key casualties of invasions. In: Richardson, D.M. (ed.) *Fifty Years of Invasion Ecology: The Legacy of Charles Elton*. Wiley-Blackwell, Oxford, pp. 143–160.

Traveset, A., Olesen, J.M., Nogales, M., Vargas, P., Jaramillo, P. *et al.* (2015a) Bird–flower visitation networks in the Galápagos unveil a widespread interaction release. *Nature Communications* 6, 6376.

Traveset, A., Chamorro, S., Olesen, J.M. and Heleno, R. (2015b) Space, time and aliens: Charting the dynamic structure of Galápagos pollination networks. *AoB PLANTS* 7, plv068.

Traveset, A. and Richardson, D.M. (2006) Biological invasions as disruptors of plant reproductive mutualisms. *Trends in Ecology & Evolution* 21, 208–216.

Traveset, A. and Richardson, D.M. (2014) Mutualistic interactions and biological invasions. *Annual Review of Ecology, Evolution, and Systematics* 45, 89–113.

Tur, C., Vigalondo, B., Trøjelsgaard, K., Olesen, J.M. and Traveset, A. (2014) Downscaling pollen-transport networks to the level of individuals. *Journal of Animal Ecology* 83, 306–317.

Tylianakis, J.M. (2008) Understanding the web of life: the birds, the bees, and sex with aliens. *PLoS Biology* 6, e47.

Tylianakis, J.M. and Morris, R.J. (2017) Ecological networks across environmental gradients. *Annual Review of Ecology, Evolution, and Systematics* 48, 25–48.

Tylianakis, J.M., Tscharntke, T. and Lewis, O.T. (2007) Habitat modification alters the structure of tropical host–parasitoid food webs. *Nature* 445, 202–205.

Valdovinos, F.S. (2019) Mutualistic networks: moving closer to a predictive theory. *Ecology Letters* 22, 1517–1534.

Valdovinos, F.S., Ramos-Jiliberto, R., Flores, J.D., Espinoza, C. and López, G. (2009) Structure and dynamics of pollination networks: the role of alien plants. *Oikos* 118, 1190–1200.

Valiente-Banuet, A., Aizen, M.A., Alcántara, J.M., Arroyo, J., Cocucci, A. *et al.* (2015) Beyond species loss: the extinction of ecological interactions in a changing world. *Functional Ecology* 29, 299–307.

Vil, M., Bartomeus, I., Dietzsch, A.C., Petanidou, T., Steffan-Dewenter, I. *et al.* (2009) Invasive plant integration into native plant–pollinator networks across Europe. *Proceedings of the Royal Society B* 276, 3887–3893.

Waser, N.M., Chittka, L., Price, M.V., Williams, N.M. and Ollerton, J. (1996) Generalization in pollination systems, and why it matters. *Ecology* 77(4), 1043–1060. DOI: 10.2307/2265575.

Vizentin-Bugoni, J., Tarwater, C.E., Foster, J.T., Drake, D.R., Gleditsch, J.M. *et al.* (2019) Structure, spatial dynamics, and stability of novel seed dispersal mutualistic networks in Hawai'i. *Science* 364(6435), 78–82. DOI: 10.1126/science.aau8751.

Waser, N.M. and Ollerton, J. (2006) *Plant-Pollinator Interactions: From Specialization to Generalization*. University of Chicago Press, Chicago.

19 Molecular Ecology of Plant–Microbial Interactions During Invasions: Progress and Challenges

Johannes J. Le Roux*

Department of Biological Sciences, Macquarie University, Sydney, 2109NSW, Australia

Abstract

Microbes are omnipresent, yet their interactions with invasive plants remain understudied. This is surprising, given the importance of microbes in plant community ecology and their influence on plant performance in new environments. Recent advances in molecular genetic approaches have opened the door to studying this unseen majority in great detail and to understand how they fit into ecological interaction networks. Molecular approaches allow rapid assessments of microbial diversity at reasonable cost while providing both taxonomic and evolutionary information. Here I discuss how these approaches have contributed to a better understanding of plant–microbial interactions in the context of biological invasions. By drawing insights from various case studies, I illustrate how next-generation sequencing (DNA barcoding) has revolutionized the way we understand such interactions. Tight-knit and coevolved mutualist (e.g. mycorrhizal) and antagonist (e.g. pathogen) interactions appear particularly promising to understand the structure and function of invasive plant–microbial interaction networks, the impacts of invasive plants on native networks and the vulnerability of native networks to infiltration by non-native species. I also discuss novel ways in which molecular data can aid the study of invasive plant–microbial interactions, such as incorporating phylogenetic data into network analyses to better understand the role of evolutionary history in network dynamics and how such dynamics respond to plant invasions. DNA barcoding of microbes also presents unique challenges to the study of network ecology, such as uncertainty in the legitimacy and efficiency of interactions. Future research should incorporate overall plant-associated microbial communities (microbiomes) into interaction networks to better understand the role microbes play during plant invasions.

19.1 Introduction

Upon arrival in new environments, non-native plants represent near 'clean slates' of biotic interactions, typically being liberated from the majority of the organisms they historically interacted with in their native ranges. Establishing new interactions has obvious and important implications for plant success under novel environmental conditions. On the one hand, liberation from natural enemies may enhance colonization success; on the other, lost mutualists may impede establishment success (Fig. 19.1). This may be particularly true when interactions relate to key processes associated with the completion of the plant's life cycle, e.g. pollination or seed predation (Traveset and Richardson, 2014).

Ecological interaction networks have provided us with many insights of how non-native and invasive plants reassemble biotic interactions under novel ecological contexts (Frost *et al.*, 2019). They have shed light on how non-native species infiltrate resident networks (e.g. Heleno *et al.*, 2013), the dynamics of networks

*jaco.leroux@mq.edu.au

 Plant Invasions: The Role of Biotic Interactions (eds Anna Traveset and David M. Richardson)
DOI: 10.1079/9781789242171.0019

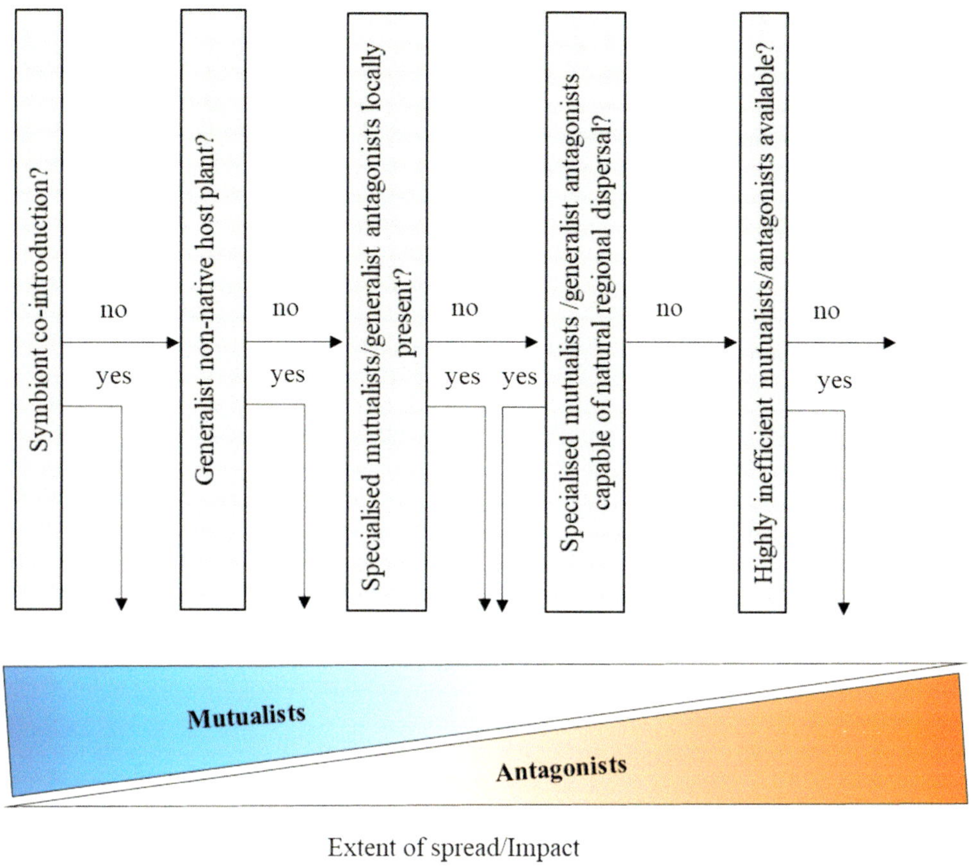

Fig. 19.1. Conceptual framework to illustrate how levels of historical specialization of non-native plant–microbial interactions (mutualisms and antagonisms), along with introduction history, makeup of resident communities and dispersibility of symbionts, are predicted to influence the establishment success and invasiveness of non-native plants (i.e. extent of spread and impact) in different ways. The blue triangle illustrates how these demographic and biogeographic processes are expected to influence the extent of spread/impact during plant–mutualist interaction reassembly, while the orange triangle illustrates how the reassembly of non-native plant-antagonist interactions may impact the extent of spread/impact. Dark and light shaded areas in both triangles indicate high and low extent of spread/impact, respectively.

during invasion (e.g. Spotswood *et al.*, 2012) and the impacts of invasive species on native species (e.g. Aizen *et al.*, 2008). From these (mostly above-ground) interaction networks, e.g. plant–pollinator and plant–herbivore networks, a few generalizations have emerged. First, non-native plants often infiltrate networks because they have highly generalist interaction requirements, allowing them to rapidly re-establish interactions with (usually generalist too) resident species (Richardson *et al.*, 2000). Indeed, it appears that the generalist nature of many plant–pollinator and plant–seed disperser networks makes them vulnerable to infiltration and disruption by invasive plants or animals (Olesen *et al.*, 2002). Secondly, introduced plants often encounter phylogenetically or functionally similar partners in their new ranges (Traveset and Richardson, 2014). Lastly, specialized interactions can be maintained if plants are co-introduced into the new environment with their interacting specialist partners, a phenomenon that appears to be commonplace (e.g. Mitchell and Power, 2003; Díez, 2005; Dickie *et al.*, 2010; Moeller *et al.*, 2015; Le Roux *et al.*, 2017; Warrington *et al.*, 2019). It follows that, in the absence of co-introduction, non-native plants with generalist mutualistic and specialist antagonist interaction requirements may have higher establishment success and performance compared to plants with specialized mutualist and general antagonist interactions (Fig. 19.1).

The importance of co-introduction during plant invasion also becomes evident when one considers the broad-scale biogeographic signatures that underlie many *native* plant–symbiont interactions. For instance, Steidinger *et al.* (2019) recently provided a global analysis on the biogeography of three major plant–mutualist interactions: those involving ectomycorrhizal fungi, arbuscular mycorrhizal fungi and nitrogen-fixers. This study found different mutualisms to have remarkably strong and disjunct biogeographic structures. Trees native to low-latitudinal areas are predominantly associated with arbuscular mycorrhizae, becoming largely nitrogen-fixing in mid-latitudinal ranges and ectomycorrhizal in higher latitudes. These biogeographic patterns may have important implications for plant invasion. For instance, the success of ectomycorrhizal trees native to high latitudinal areas, in the absence of co-introduction with their mutualist fungi to low latitudinal regions, will likely be limited by the availability and compatibility of resident symbionts.

Compared to interaction networks between plants and their above-ground macrofauna, far less is known about plant–microbial networks, especially in the context of biological invasions. This is surprising given the usually strong link between the success of plants and their interactions with microbes. For example, pine trees (family Pinaceae) have been extensively planted around the globe as a valuable source of wood and pulp. These trees have also become invasive in many of the areas to which they were introduced (Richardson and Rejmánek, 2011). This is perhaps unsurprising given that pine introductions tick all the right boxes for any species to become invasive: their old and new ranges often have good climatic matches, they have been introduced in high numbers (i.e. high propagule pressure), they often lack specialist natural enemies in their new ranges and they have been artificially selected and cared for by humans (Richardson, 2006). What is surprising though is that many pine introductions actually failed to establish self-sustaining populations (Richardson *et al.*, 1994). This was mainly because of the absence of coevolved mutualists in their new ranges (Nuñez *et al.*, 2017), specifically ectomycorrhizal fungi (EMF) (Pringle *et al.*, 2009; Policelli *et al.*, 2019). EMF, and mycorrhizae in general, form integral components of soil biodiversity and their association with plants significantly affects the absorption of nutrients and protection against diseases, and therefore growth of plants. EMF are not free-living and often show some degree of host specificity. For non-native pines this meant that they were unable to form successful symbioses with resident EMF, especially when floras in their new ranges lacked phylogenetically closely-related species (Molina *et al.*, 1992; Vellinga *et al.*, 2009). It was only after the introduction of pine-specific EMF to these regions that pines successfully established and started to become widespread (Vellinga *et al.*, 2009). Interestingly, most pine invasions have been facilitated by a single EMF species (Hayward *et al.*, 2015). This illustrates that a single mutualist co-introduction can facilitate invasion of multiple plant species, as was also recently found for various invasive Australian acacias and their associated rhizobia in New Zealand and South Africa (Warrington *et al.*, 2019).

It is only recently that we have started to explore invasive plant–microbial interactions from an ecological interaction network viewpoint (e.g. Dickie *et al.*, 2017a, b; Le Roux *et al.*, 2016). Unlike their above-ground counterparts, microbes pose unique challenges that make the 'traditional' observational approaches used for characterizing aboveground interaction networks difficult (hereafter I use the term 'traditional' to contrast molecular approaches for network construction with field-based observational studies). By virtue of their size, most, if not all, microbes are difficult to identify without the use of molecular approaches. Moreover, many microbes are biotrophic (i.e. symbionts that need a host to stay alive) meaning that standard microbiological approaches of isolation and culturing cannot be used to detect, identify and quantify them.

Like all interaction networks, plant–microbial networks comprise interactions that vary in their specificity. It is thus expected that these networks will display nestedness, i.e. specialist plants/microbes in a community will only interact with a subset of microbes/plants that generalist species interact with (Bascompte, 2009; but also see Roy-Bolduc *et al.*, 2016). Specialization can also mean that some plant–microbial pairs may not have the same probability of interacting; and if groups of plants only interact with a given group of coevolved microbes for example, or vice versa, the network will become modular (Bascompte, 2009). These expectations suggest that evolutionary history can affect interaction network architecture. As I will discuss later in the chapter, this has important implications for our understanding of how plant invasions occur and the impacts they cause. For introduced non-native plants with highly specialized microbial interactions, co-introduction with coevolved microbes may be the only way to maintain effective associations, and can either enhance (i.e. mutualist interactions) or impede (i.e. antagonist interactions) establishment success. While co-introduced enemies can sometimes affect resident native species more negatively than their original hosts (e.g. the Enemy of My Enemy Hypothesis), this is rarely the case for highly specialized interactions as evidenced from successful biological control programmes where such spillovers have been rarely observed. On the other hand, non-native plant performance is expected to be negatively affected in the absence of specialist mutualist co-introduction. These scenarios are also anticipated to influence network structures in predictable ways. For example, co-introduction of highly specialized interacting species pairs is expected to manifest as separate, largely unconnected, modules in resident networks (Fig. 19.2; Le Roux *et al.*, 2017). While such co-introductions may not directly disrupt the structure of resident network structures, they may do so indirectly through their impacts on abiotic conditions. For example, as co-introduced plants and their mutualists become dominant, they may dramatically alter soil nutrient cycling through positive feedbacks (e.g. Le Roux *et al.*, 2018), with native plants and their mutualists being replaced/excluded through changes in abiotic conditions mediated by co-invading plant–mutualist pairs. On the other hand, it is expected that when plants are co-introduced with coevolved, but generalist, microbes, that non-native plant performance will only be moderate based on both mutualist and antagonist interactions (Le Roux *et al.*, 2017). Under these circumstances, introduced plants can also establish interactions with resident microbes, potentially affecting neighbouring resident native plants by competing with them with for available microbes. Moreover, co-introduced generalist microbes could switch hosts, thereby affecting resident plants. Lastly, generalist introduced plants could establish interactions with existing generalist (native or non-native) microbes they have never encountered before (Dickie *et al.*, 2017a; Le Roux *et al.*, 2017).

In this chapter, I summarize the use of molecular approaches to better understand plant–microbial interactions and the networks they form. I discuss advances in molecular genetics and draw on recent studies based on plant–mutualist and plant–antagonist interactions to illustrate how these approaches can aid a better understanding of how networks are structured and what their responses to biological invasions are. Lastly, I offer some ideas on priorities for future research in this emerging and exiting field.

19.2 Advances in Molecular Approaches to Understand Microbial Diversity and Their Interactions with Plants

Microbes have not yet been fully incorporated into network analyses, partly because of difficulties associated with identifying them and understanding their roles in ecosystems, and partly because determining the legitimacy of their interactions with each other and/or with macrobes is not easy. Even when interactions can be established, it is often challenging to ascertain their nature, i.e. whether they are antagonistic, mutualistic, etc. Moreover, the nature of many microbial interactions can be context-specific, i.e. being either neutral, beneficial or harmful depending on the local environmental conditions under which they operate. To this day, most new microbial species are described using a tedious combination of isolation in pure culture, Sanger sequencing of diagnostic gene regions (e.g. 16S rRNA genes for prokaryotes),

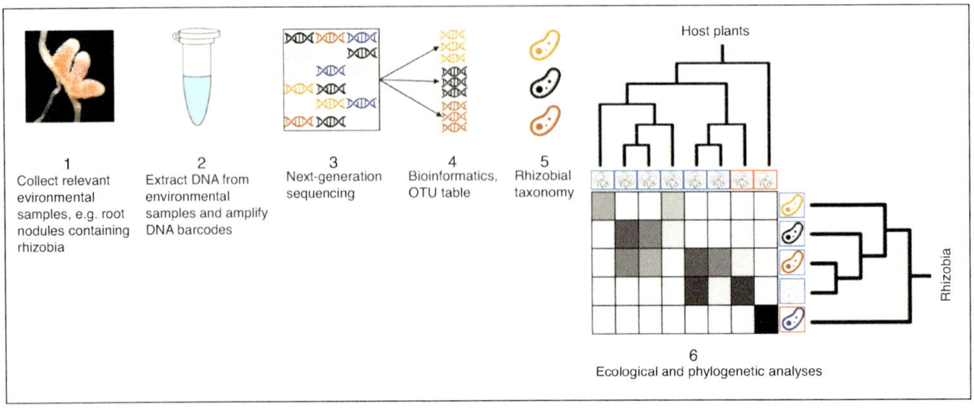

Fig. 19.2. Flow chart for incorporating evolutionary history (phylogeny) into ecological interaction networks based on NGS data. 1) Relevant material (e.g. root nodules housing rhizobia on legume roots) are collected. 2) DNA is extracted from root nodules and from host plant material and 3) specific DNA barcodes for each amplified. Rhizobial DNA barcodes are 'indexed' according to host plant identity. 4) Amplified DNA fragments are sequenced and sorted into Operational Taxonomic Units (OTUs) based and DNA sequence similarity resulting in a legume host x OTU count table which can be used as a proxy for interaction strength. 5) OTUs are compared (e.g. blasted) against reference databases for classification. DNA barcodes can be used to reconstruct host plant and rhizobial phylogenies (also see section 19.3.4). Numerous inferences can be made from such phylogenetically-informed networks. For instance, if phylogenetic relatedness between host plant species is positively correlated with their interaction similarities, i.e. closely related plants tend to interact with the same microbial partners, then we can predict whether non-native plants (indicated by red borders in 6) will successfully integrate into native plant–mutualist networks. On the other hand, non-native plants that are distantly related to resident plants (e.g. legumes) with specialist mutualist requirements may only be able to form successful interactions with their own, co-introduced, microbial (e.g. rhizobial) partners. The latter will lead to strongly modular and possibly disconnected co-introduced legume x rhizobium interaction modules. Note that cell colours in 6 represent OTU counts where darker shades indicate high abundances and lighter shades lower abundances.

whole genome approaches (e.g. DNA–DNA hybridization), morphological characteristics and physiological data. This so-called 'polyphasic' taxonomic approach remains the best (and in some cases the only) way to classify microorganisms and to understand their ecological functions and evolutionary placement (Gillis *et al.*, 2001). Sanger sequencing of taxonomically-informative loci, like the 16S rRNA gene, also remains valuable for identifying microbes isolated from various hosts and such data have been successfully incorporated into network analyses (e.g. Le Roux *et al.*, 2016). However, most bacteria and archaea are unculturable (Rappé and Giovannoni, 2003), and therefore our appreciation and understanding of the true diversity of plant–microbe interactions remain severely limited. Given that networks of plants and animals are embedded in whole systems that comprise extraordinary microbial diversity, there is an urgent need to incorporate this unseen majority into ecological networks. Recent developments in massively parallel sequencing technologies (i.e. next-generation sequencing; NGS) enable us, for the first time, to overcome some of the limitations associated with identifying and quantifying microbes, and provide access to cost-effective sequencing platforms and high-quality 'big data'. This, together with conventional microbiological approaches, has opened the door for unprecedented ways to explore microbial diversity and function, not just generally, but also in the context of ecological networks (Toju, 2015).

NGS DNA 'metabarcoding' allows for the typing of microbial diversity with high levels of precision and detail, often obtained as bulk samples (for a recent review see Derocles *et al.*,

2018). Utilizing short, but informative, parts of DNA barcodes specific to certain taxa not only provides information about diversity, structure and composition of microbial communities but also, to some extent, function. For example, DNA barcodes based on partial 16S rRNA sequences provide one of the most important culture-independent methods to identify and quantify abundances of prokaryotes from complex environmental samples like plant tissues, animal gut samples and rhizosphere soils, among others (Derocles *et al.*, 2018). Such estimates of abundances are not possible for higher organisms since tissue biomass cannot be accurately estimated from sequence data (Clare, 2014).

Microbial NGS data are usually clustered into so-called OTUs (Operational Taxonomic Units) based on the DNA sequence similarity of individual barcodes (Evans *et al.*, 2016). Like species identities, microbial OTUs can provide information on various community-level metrics, like (OTU) richness, abundance, evenness, diversity turnover and community composition, and at various spatial and temporal scales (e.g. see Keet *et al.*, 2019). Moreover, extensive taxonomic and genetic repositories allow for identification of OTUs against reference data (Fig. 19.2; Claesson *et al.*, 2010). It should be noted, however, that NGS typically generates short DNA sequence reads (≤600 bp) which can limit the level to which taxonomic assignments can be confidently made, often resulting in numerous 'unclassified OTUs' in data sets. While unclassified OTUs are still distinguishable within analyses, they are difficult to reconcile between different studies (Evans *et al.*, 2016). Despite these shortcomings, NGS-derived taxonomic information can in some instances provide useful information on the functional roles of microbial communities. For example, high abundances of *Bacillus* OTUs in the rhizospheres of plants may suggest a role of these bacteria in promoting plant growth, due to ability of most *Bacillus* species to solubilize mineral phosphorus and promote its availability to plants (Bulgarelli *et al.*, 2013). However, many functional genes do not follow the phylogenetic histories of the genes frequently used to classify microbes and used in metabarcoding studies. While functional inferences are difficult to incorporate reliably in these studies, promising advances are helping us to overcome these limitations. For example, the development of so-called epicPCR (emulsion, paired isolation, and concatenation PCR), allows for the physical concatenation of two target genes from a single microbial genome (Spencer *et al.*, 2016). This implies that it is now possible to amplify a diagnostic gene (e.g. bacterial 16S rRNA) and a functional gene (e.g. nitrogenase *nifH*) from the same bacterial genome (Spencer *et al.*, 2016).

NGS technologies are also able to detect remnant DNA shed into the environment by organisms, in what is has been coined environmental DNA (e-DNA) approaches, potentially allowing for the typing of organisms even when they are not physically present (e.g. Chariton *et al.*, 2010). eDNA is usually degraded and therefore shorter barcodes are generally targeted which may further restrict taxonomic assignments of, and phylogeny reconstructions between, taxa. For obvious reasons, eDNA approaches are applied to detect multicellular organisms rather than unicellular organisms like bacteria.

Incorporating NGS data into ecological interaction network analyses is in its infancy, but holds great promise to advance our understanding of how plant–microbial interactions respond to anthropic perturbations such as climate change and biological invasions. A major reason behind the limited use of NGS data to date in the construction of ecological networks, is the absence of actual interaction data (Evans *et al.*, 2016), i.e. how do we know for sure that sequenced OTUs are actually interacting with putative host plants? How to treat this complexity of information remains problematic and challenging, consequently many researchers have been 'filtering' out NGS data when interactions are unknown (Derocles *et al.*, 2018). For example, if we consider coevolved and tight-knit plant–microbial associations, such as endosymbiotic interactions, then realized links between organisms can be assumed. Similarly, putative plant–herbivore interactions can be inferred using eDNA gut-content analyses, but again it may be difficult to disentangle contamination from actual consumption when only including a limited number of replicates. Analogous to gut-content analysis, organs or tissue of plants, together with symbiont-specific DNA barcodes, can be used to construct conventional networks based on molecular data (e.g. see Sepp *et al.*,

2019). A further advantage of network analyses based on molecular data is that, compared to traditional approaches, more independent replicates can be incorporated into a single experimental design. Compared to traditional observational studies, numerous interactions can be easily sampled over short time periods when using molecular approaches. However, contamination, the often non-linear relationship between biomass and barcode abundance (for macrobes), and sequencing errors, all complicate the use of NGS approaches in network analyses. These issues can strongly bias the overall interpretation of species detectability and their abundances (Lahoz-Monfort *et al.*, 2016) and to overcome this limitation requires high levels of replication (Mata *et al.*, 2019). For instance, for plant–microbial networks a large number of individual host plant species are needed to confidently disentangle interactions. A drawback of such high replication remains the high costs associated with NGS analyses.

Microbial communities are often depicted as co-occurrence networks (e.g. Toju *et al.*, 2014). It is important to not confuse these types of networks with traditional bipartite interaction networks. Co-occurrence networks only provide information on which species co-occur in a particular environment or can be informative on how taxon co-occurrence change in response to different environmental conditions or perturbations. In other words, these networks usually only indicate covariance, which does not necessarily translate into actual interactions (e.g. Widder *et al.*, 2014).

Recent research, albeit not always directly related to network ecology, provides insights into how NGS data can be incorporated into plant–endosymbiont networks and how this can inform a better understanding of biological invasions. For example, Keet *et al.* (2017) tested the hypothesis that highly invasive and geographically widespread Australian acacias (genus *Acacia*) in South Africa are less specialized in their associations with mutualistic rhizobia than other introduced, but locally restricted and less invasive, acacias. These authors hypothesized that the smaller ranges of localized acacias may indicate that their specialized rhizobium mutualist requirements are not being met in their new range. To test this, Keet *et al.* (2017) sampled root nodules from acacias (within which rhizobia live and fix atmospheric nitrogen) and used NGS sequencing approaches to analyse a rhizobium-specific DNA barcode targeting a so-called 'nodulation' gene. Importantly, rhizobial nodulation genes are involved in legume–rhizobia communication and are important determinants of symbiotic specificity, i.e. the molecular 'talk' between legumes and rhizobia that facilitate root nodule development between compatible legumes and rhizobia. Keet *et al.* (2017) rejected their initial hypothesis and found widespread and localized acacias to share a few and highly abundant *Bradyrhizobium* strains (also see Box 19.1). Interaction efficiency (i.e. nitrogen fixation) mirrored these results, with nitrogen isotope analyses indicating that widespread and localized acacias do not differ in their ability to fix atmospheric nitrogen. While this study did not explicitly construct interaction networks (but see following section) it illustrates how NGS data can inform ecological network analyses. Specifically, Keet *et al.* (2017) showed that it is not only possible to get a reasonable estimate of interactions (i.e. networks links) and their strengths (i.e. link weights) between interacting partners (i.e. rhizobial OTU identities and abundances within acacia root nodules), but that these estimates can also be correlated with interaction efficiency (i.e. nitrogen fixation efficiency measured by isotope analysis).

19.3 Molecular Ecology of Plant–Microbial Interactions During Invasion: Insights From Plant–Mutualist and Plant–Antagonist Networks

19.3.1 Molecular ecology of plant–mutualist interactions: insights from rhizobium and mycorrhizal mutualisms

As discussed above, coevolved endosymbiotic associations provide ideal systems to study plant–microbial interaction networks using molecular approaches. This is because NGS approaches utilizing taxon-specific DNA barcodes can quickly resolve taxa (i.e. OTUs = network nodes), interactions (i.e. network links) and interaction strength (i.e. link weight based on OTU and plant abundance data). Such coevolved interactions are also expected to play prominent roles in mediating plant establishment success and invasiveness (i.e. the Missed Mutualisms and Enemy Release Hypotheses; Traveset and Richardson, 2014). For example, associations with rhizobia are thought to play a key role in the success of legumes under novel environmental conditions (Parker, 2001). As mentioned, rhizobia form specialized structures (root, or less-frequently stem, nodules) on their legume hosts within which they can fix atmospheric nitrogen into forms that can be utilized by host plants. In exchange, legumes provide rhizobia with photosynthate. The interaction between legumes and rhizobia involves complex and intricate molecular communication (Le Roux *et al.*, 2017). This communication is initiated when legumes release signalling chemicals such as (iso)flavonoids into the rhizosphere to which rhizobia are attracted. Rhizobia then colonize the root hairs of compatible legume hosts through the activation of their own nodulation genes (Perret *et al.*, 2000), the latter being important determinants of bacterial control over interaction specificity (Rogel *et al.*, 2011).

Le Roux *et al.* (2016) not only reported the first ecological interaction networks between native legumes and rhizobia, but also provided evidence for how invasive legumes impact these networks (also see Box 19.1). In this study the authors constructed networks across three sites in South Africa: pristine and uninvaded communities, communities that have been moderately invaded by Australian *Acacia* species (1–10 acacia trees/m^2) and communities that have been heavily invaded by acacias (>10 acacia trees/m^2). Pure rhizobial cultures were isolated from root nodules of all legume hosts present at each site and their 16S rRNA gene sequenced using Sanger sequencing technologies. Rhizobial

Box. 19.1

Legume–rhizobium mutualist interactions as a model system to study the molecular ecology of plant–microbial interactions: insights from Australian acacias.

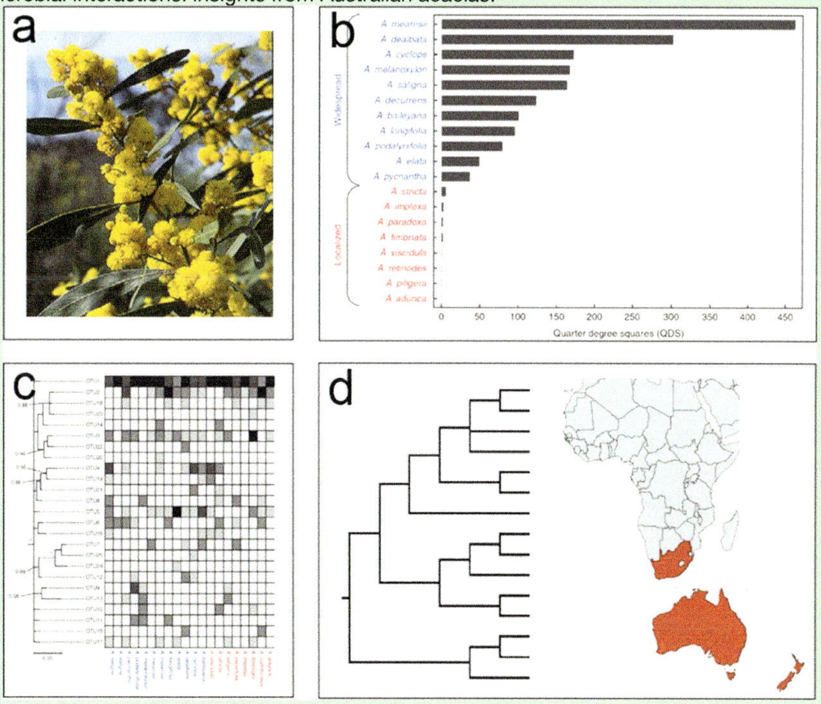

Continued

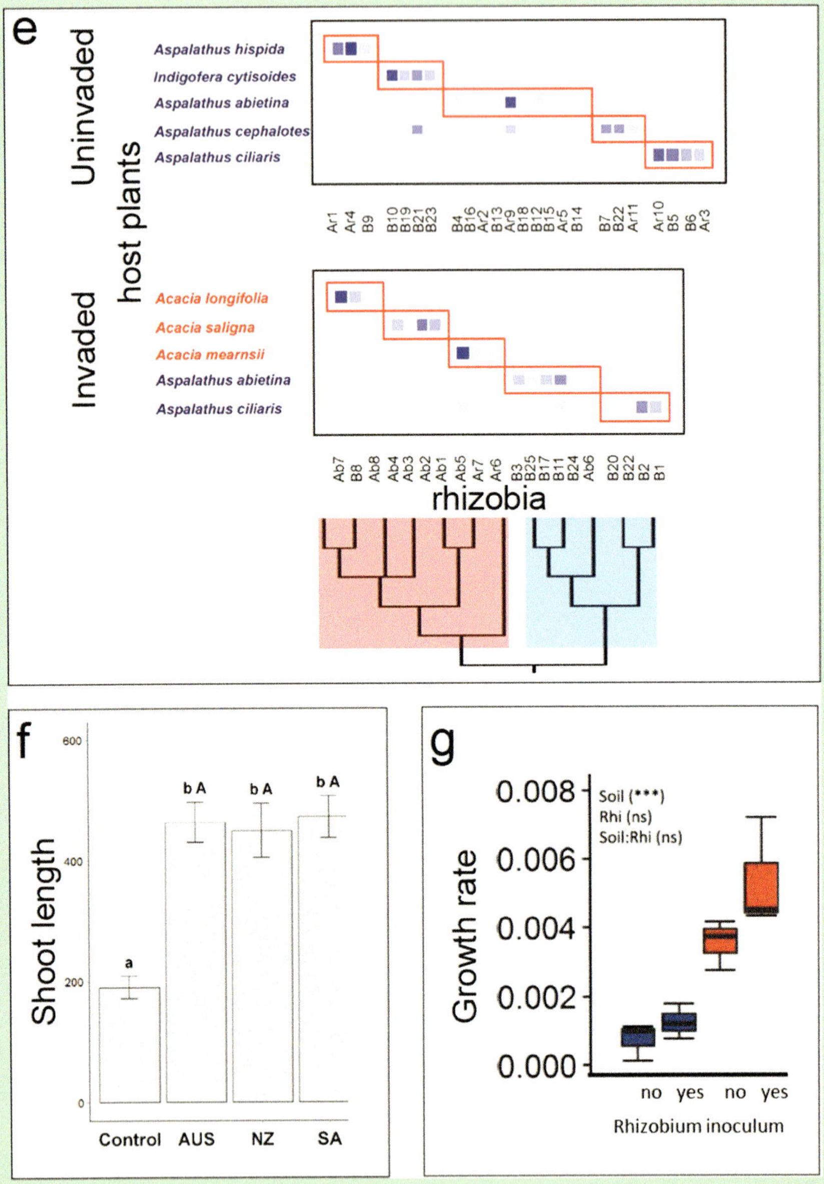

Australian acacias (a) have become widespread invasive species in many areas of the world (Richardson *et al.*, 2011). However, in places like South Africa, acacias show variable invasiveness (b) based on geographical spread (number of quarter degree squares occupied; Keet *et al.*, 2017). Despite this, highly invasive and less invasive acacias seem to share the same rhizobia (Keet *et al.*, 2017; i.e. OTU1 in (c). Based on phylogenetic information, Warrington *et al.* (2019) recently confirmed that rhizobia have been co-introduced with their acacia hosts from Australia (d). Network analyses in South Africa found native and invasive legumes and their associated rhizobia to form distinct modules (Le Roux *et al.*, 2016; red boxes in e). Moreover, co-introduced Australian rhizobia in association with invasive acacias are phylogenetically distinct from native rhizobia found in association with South

Continued

African legumes. Unsurprisingly, co-introduction results in equal performances of acacias when inoculated with Australian (AUS in pane f) and co-introduced rhizobia (NZ and SA in pane f; data from Warrington *et al.*, 2019). Furthermore, Le Roux *et al.* (2018) found that invasive acacias homogenize soil rhizobial communities and that co-invading acacia–rhizobium pairs may lead to positive soil feedbacks in the invasive range (g), i.e. acacias performed better when grown invaded soils (red plots in g) compared to when grown in pristine soils (blue plots in g). Images modified from Le Roux *et al.* (2016), Le Roux *et al.*, 2018; panes e and g), Keet *et al.* (2017); panes b and c) and Warrington *et al.* (2019); pane f).

isolates were then grouped into distinct strains (OTUs) based on two levels of DNA sequence similarity, at 95 and 97%. Networks based on both OTU classification schemes indicated that invasive acacias did not infiltrate resident native networks, but rather formed distinct interaction modules. This observation is consistent with the prediction of co-introduction of acacias and their rhizobia from Australia (Le Roux *et al.*, 2017), which was recently confirmed by a phylogeographic study that found various acacias to have been co-introduced with Australian *Bradyrhizobium* strains to South Africa (Warrington *et al.*, 2019).

So, what are the consequences of strong and disconnected interaction modules resulting from co-introduced invasive acacias and rhizobia? Le Roux *et al.* (2018) compared soil rhizobium communities in paired acacia-invaded and uninvaded sites in South Africa. They found that acacias not only affect the diversity, but also the structure, of rhizobial communities. Specifically, soils beneath invasive acacias had lower diversity and homogenised rhizobial communities compared to soils sampled from neighbouring uninvaded areas. These authors also found overall acacia-induced soil changes to benefit the performance of acacias, i.e. creating positive plant–soil feedbacks (also see Box 19.1). It is conceivable that these abiotic and biotic soil changes may facilitate the establishment other invasive acacias, whereby host-switching between co-introduced rhizobia and different acacia species may occur. Indeed, localized and widespread acacias in South Africa share a few abundant *Bradyrhizobium* strains (Keet *et al.*, 2017) of Australian origin (Warrington *et al.*, 2019).

Interestingly, only few native South African legumes associate naturally with rhizobia from the genus *Bradyrhizobium* (Lemaire *et al.*, 2015). Networks sampled in acacia-invaded and uninvaded areas corroborated this, indicating that native legumes and invasive acacias interact with phylogenetically unrelated rhizobia (Le Roux *et al.*, 2016, Box 19.1). Moreover, native legumes with specialized rhizobial requirements were not present in acacia-invaded sites (and thus networks), while generalist natives persisted, but only in association with compositionally different rhizobia. This suggests that specialist native legumes may be more severely impacted by co-invading acacia–rhizobium pairs than generalist native legumes (Le Roux *et al.*, 2016). Whether compositional changes in rhizobia associated with generalist native legumes under acacia invasion translate into lowered symbiotic effectiveness (i.e. nitrogen fixation) remains unknown. Evidence from elsewhere suggests that this is certainly possible. In Portugal, the performance of invasive *Acacia longifolia* is greatly enhanced in association with co-introduced Australian *Bradyrhizobium* strains compared to when the species is in association with Portuguese bradyrhizobia, while the opposite is true for native Portuguese legumes (Rodríguez-Echeverría, 2010; Rodríguez-Echeverría *et al.*, 2012).

The legume–rhizobium examples discussed above confirm the expectations that plant–mutualist networks usually display some degree of nestedness and modularity, properties that have been confirmed for other plant–microbial mutualist networks. For example, a comprehensive network analysis based on *native* plant–arbuscular mycorrhizal fungal (AMF) interactions also found networks to be nested and modular (Montesinos-Navarro *et al.*, 2012). These authors identified AMF based on Sanger DNA sequencing data of the fungal DNA barcode, the internal transcribed (*ITS*) region, directly sequenced from the root material of 25 different host plant species. More recently, Sepp *et al.* (2019) sampled root material from the whole plant community (35 species) found in a 1,000-m^2 grassland area in Estonia. Sampling was repeated across two seasons and NGS data from 100 AMF taxa found in root material was used to construct networks. These networks were again found to be nested and modular, but importantly, also showed that plant traits may be important determinants of network structure. Specifically, while AMF interactions were dependent on plant species identity, plant functional groups (forb vs grass) and their abundances had the most significant impact on how plant–AMF interactions were structured, with limited seasonal variation in these interactions (Sepp *et al.*, 2019). While the plant–AMF networks discussed above did not include any invasive species, they do illustrate the utility of molecular approaches in studying plant–microbial mutualist interaction networks.

Dickie *et al.* (2017b) provided one of the first plant–mycorrhizal networks (based on sporocarp observational data) to understand how non-native species fit into resident networks. This study included EMF species in association with native and non-native trees in New Zealand and the UK. Dickie *et al.* (2017b) found non-native trees in both regions to be mostly linked into large non-specific modules via their generalist EMF interactions. In New Zealand, this large module was exclusively dominated by non-native trees, while in the UK modules included both native and non-native species. In both regions, network infiltration by generalist non-native trees appears to reduce network modularity and increase nestedness (Dickie *et al.*, 2017b; also see Santos *et al.*, 2012; Traveset *et al.*, 2013). Dickie *et al.* (2017b) also showed that EMF interacting with native trees in both the UK and New Zealand had lower host breadth (i.e. higher host specificity) than the EMF associated with non-native trees. The authors concluded that most non-native tree–EMF associations likely represent novel associations. However, some non-native trees in their data set, such as pines, are known to have been co-introduced with their EMF to New Zealand (Dickie *et al.*, 2010). The lack of a strong pine–EMF module could point to the fact that other non-native trees are forming associations with pine-associated EMF. Such novel interactions where both partners are non-native, but lack previous interaction history, have been termed co-xenic associations, a phenomenon thought to be common during the reassembly of plant–fungal interactions during invasion (Dickie *et al.*, 2017a). A major shortcoming of the work by Dickie *et al.* (2017b), acknowledged by the authors themselves, is the fact that observational data only provide information on a fraction of the overall fungal community associated with host plants and are unlikely to be correlated with the true belowground diversity and abundance of fungi. Moreover, this study was conducted national level, whereas biological interactions during invasion must be understood at local spatial scales.

Molecular NGS approaches provide ways to circumvent the issues discussed above and to obtain high-resolution data on species identities and their abundances (e.g. see Sepp *et al.*, 2019). While the use of NGS data in plant–mycorrhizal networks has, to the best of my knowledge, never been explicitly incorporated into analyses in an invasion ecology context, this approach intuitively has much potential. For example, Gazol *et al.* (2016) used such data to study the impact of co-invading pines and their EMF on native AMF diversity associated with native plants in two southern hemisphere countries; Argentina and South Africa. Importantly, almost no indigenous EMF is found in South Africa while they are native and naturally abundant in Argentinean soils. These authors showed that pine invasions impacted local AMF communities differently in these two countries. In South Africa, the presence of pines led to decreased local AMF richness, significantly affecting AMF community composition, while in Argentina no such effects were detected. Invasive pines did cause some changes in phylogenetic structure of AMF communities in both countries. These data contain measures of taxonomic identity, interaction and abundance and could therefore be easily incorporated into constructing ecological networks, and illustrate the exciting and fruitful opportunities for future research using such approaches.

19.3.2 Molecular ecology of plant–antagonist interactions: insights from fungal pathogens

Less is generally known about the dynamics and structure of plant–antagonist microbial networks compared to plant–mutualist interactions (but see e.g. Bufford *et al*., 2020). Available evidence from above-ground networks suggests that antagonist network structures may differ, in some respects, from other interaction types (e.g. see Rohr and Bascompte, 2014). For example, analyses of four different trophic levels of antagonistic interactions (plants, aphids, primary parasitoids and secondary parasitoids), which included phylogenetic data, found evolutionary relatedness among resource species to be correlated with the overlap of their consumers, but that evolutionary relatedness among consumers was not at all, or negatively, correlated with overlap in their resources (Elias *et al*., 2013). This was true across all network levels, suggesting that both phylogenetic constraints on resource breadth, and exploitative competition among phylogenetically closely related consumers that limits resource overlap, may influence the structure of antagonistic networks (Elias *et al*., 2013).

Almost nothing is known about the nature and dynamics of plant–microbial antagonist networks, especially in the context of biological invasions. Certainly, I know of no research that has utilized molecular approaches to better understand plant–microbial antagonist networks during invasion. This is surprising, given the perceived importance of pathogens in determining the dynamics of species introductions and invasiveness (i.e. release from pathogenic bacteria, fungi and viruses, Mitchell and Power, 2003). The often-inconspicuous nature of some pathogens also renders traditional network approaches inadequate to fully capture interactions. Similar to the discussion on plant–mutualist interactions above I will draw on insights form available research to illustrate the potential utility of molecular approaches to better understand the dynamics of plant–microbial antagonist networks during invasion. All molecular approaches discussed up until now are obviously transferable to pathogens, including bacteria, fungi and viruses, and have been successfully applied outside of invasion science. However, there are important limitations to keep in mind when constructing molecular plant–pathogen networks. First, pathogenicity genes are often carried on plasmids, and hence only knowing which 'species' are involved in interactions is insufficient to infer true pathogenic interactions. Second, pathogen interactions can only be reliably invoked in conjunction with actual disease symptoms, so again, identifying species alone is insufficient to characterize effective interactions. Despite these shortcomings some progress has been made. For example, Toju *et al*. (2014) reported one of the first overall network structures for highly diverse and *native* plant and fungal communities (including all fungi, i.e. beneficial, antagonistic and neutral interactions). The overall network, based on interactions between 33 plant species and 387 fungal taxa, was significantly modular, but only weakly so (Toju *et al*., 2014). Contrary to most ecological networks, this network lacked appreciable nestedness (Toju *et al*., 2014). What underlies this unique property remains unknown, but it may reflect the fact that all possible plant–fungal interaction types, i.e. mutualistic, antagonistic, amensalistic, etc., were captured in this single network.

Some networks between plant and pathogens have been constructed based on traditional observational data. For example, forest health records collected throughout France enabled Vacher *et al*. (2008) to construct a bipartite network between 51 tree taxa and 157 associated parasitic fungal species. Interactions were defined as fungal observations that could unambiguously be linked to plant disease symptoms. The resulting network revealed a fully connected and significantly modular structure. Moreover, crude estimates of phylogenetic relatedness for both trees and fungi revealed that network modules largely mirrored deep divergences in plant phylogeny, i.e. phylogenetically closely related trees were more likely to share the same fungal species than phylogenetically distantly related trees. Vacher *et al*.'s (2008) network included both native and non-native tree species and, in a follow-up study, these authors asked whether the interactions of non-native trees are different from those of native trees or whether network infiltration by non-native trees have changed any of the network's properties (Vacher *et al*., 2010). This study revealed that phylogenetically closely-related non-native and native trees with comparable life-histories had similar types and numbers of interactions, and that network modularity did not change when networks were infiltrated by non-native species (Vacher *et al*., 2010). However, it has been suggested that a lack of quantitative data may preclude detection of structural changes to networks (Tylianakis *et al.*, 2007). Incorporating NGS data may be useful in addressing these issues (e.g. Toju *et al.*, 2014), not only by providing semi-quantitative data for interactions, but also by providing a more complete picture, i.e. being able to identify 'missing links' or rare interactions (Evans *et al.*, 2016) not easily captured by observational approaches.

19.3.3 Molecular network approaches provide novel solutions to old problems: understanding interaction efficiency

A major difference between traditional and molecular network approaches is the potential for fine-scale evolutionary and functional information that can be gained from the latter. For example, pollination networks based on observational data typically only provide crude estimates of interaction frequency and often lack insight into the efficiency of interactions (i.e. pollen transfer and reproductive output, but see Ballantyne *et al.*, 2015). The time and cost constraints associated with capturing such data at the community level are difficult to overcome, e.g. measuring successful pollen transfer of all possible interactions. Similarly, inferences of network properties such as levels of interaction specialization and strength are limited by the typical spatial scales included in observational studies, effort required to capture most interactions, dispersibility of interacting partners, reliable taxonomic assignment of species and temporal variation in interactions, among others (Evans *et al.*, 2016). All these constraints are particularly problematic when dealing with networks in species-rich ecosystems. NGS metabarcoding provides a means to overcome some of these issues. For example, NGS data for network analyses have been found to be taxonomically more comprehensive, and more time-efficient to produce, than traditional observational data (Ji *et al.*, 2013). NGS-based species identifications are less reliant upon taxonomic expertise, making them especially valuable for sampling poorly known taxa and ecosystems. Genetic approaches also allow for the quantification of networks over short time intervals and/or across larger spatial scales (Evans *et al.*, 2016).

So, how can the advantages associated with molecular approaches help us to better understand the efficiency of ecological interactions? NGS data can provide some information on the strength of individual interactions (e.g. OTU abundance), essential data for determining interaction efficiency. It should be noted, however, that NGS approaches are probably, at best, semi-quantitative, as sequence-read abundance may be affected when multiple copies of the target region exist. While it is expected that DNA barcode abundance will be related to the benefits derived from mutualists, i.e. more efficient mutualists will be more abundant, it is important to remember that beneficial and 'cheater' symbionts can often coexist. NGS data are not capable of differentiating between these symbiont types. However, non-genetic techniques, such stable isotope analysis, provide useful ways to confirm interaction efficiency and have recently gained traction in invasion ecology (Bodey *et al.*, 2011; McCue *et al.*, 2020). For example, for legume–rhizobium interactions, efficiency can be measured as the amount of nitrogen gained through biological nitrogen fixation (from the host plant's perspective) or, of photosynthetically-derived sugars received (from the bacterium's perspective). Let us consider interaction efficiency from the plant's viewpoint. Nitrogen is the major macronutrient required by all plants and, within legume root nodules, rhizobia fix it through nitrogenase activity. This enzyme reduces atmospheric N_2 to ammonia by preferentially incorporating the lighter isotope of atmospheric nitrogen (i.e. ^{14}N, Unkovich, 2013). Measures of change in the heavier nitrogen isotope ($\delta^{15}N$) can therefore be used to determine the contribution of biological nitrogen fixation to the nitrogen budget of legumes. This, in turn, can be correlated with interaction strength, i.e. rhizobial OTU abundances within root nodules.

Similar to legume–rhizobium interactions, aspects of the efficiency of plant–mycorrhizal interactions may be inferred using stable isotope analysis. For example, there is mounting evidence to suggest that AMF associations increase plant water-use efficiency (i.e. the ratio of photosynthesis to transpiration), but this relationship may be dependent on which plant and fungal species interact (see Augé, 2001 for review). Stable isotope analysis can measure water use efficiency because, Rubisco, the key CO_2-fixing enzyme during photosynthesis, has a higher affinity for the abundantly available and lighter $^{12}CO_2$ than for the less abundant and heavier $^{13}CO_2$. Water stress conditions usually causes a reduction in stomatal conductance, leading to reduced internal substomatal CO_2 concentrations in leaves. Under these conditions ^{12}C:^{13}C ratios are expected to decline as ^{12}C gets depleted and Rubisco discriminates less against the available ^{13}C. Lower discrimination by Rubisco can be driven by stomatal closure and/or a higher photosynthetic water-use efficiency (Richards, 1996). Using $\delta^{13}C$ isotope analysis may therefore be useful to interpret the water relations of plants in association with different mycorrhizae, e.g. native resident species found in neighbouring invaded and uninvaded areas or invasive species forming novel associations in their new ranges. Such analyses may also pave the way for addressing interesting and pressing questions in network ecology, e.g. whether a link exists between levels of interaction specialization and cheating by partners, or whether interaction efficiency trade-offs exist between generalist and specialist interactions.

19.3.4 Incorporating evolutionary history and plant–microbial interaction networks

In addition to reasonable taxonomic assignment and semi-quantitative estimates of abundances, another major advantage of incorporating molecular data into networks is that relatedness of taxa, i.e. phylogenies based on DNA sequence data, can be reconstructed. This is important, since species traits are key modulators of network properties (Rezende *et al.*, 2007a) and this, together with their often-shared evolutionary history, suggests that phylogenetic constraints may influence species interactions (Minoarivelo *et al.*, 2014). For example, due to high coevolutionary history, reciprocally specialized plant–endosymbiont interactions are expected to a show strong phylogenetic signal. This also implies that network infiltration by non-native plants may be phylogenetically constrained. Indeed, phylogenetic constraints on species interactions may partially explain the modular nature of most plant–mutualist networks (Bascompte, 2009). These constraints may also affect plant–antagonist networks, e.g. herbivorous insects may have limited host ranges so that phylogenetically closely related species often interact with the same suite of plants or vice versa (Rezende *et al.*, 2007b). Although a phylogenetic signal may be present in all ecological networks, its strength may vary according to interaction type, interacting species pairs and specific components of network structure (Rohr and Bascompte, 2014). Empirical support for the role of evolutionary history in shaping network structure remains modest (e.g. Minoarivelo *et al.*, 2014) and some network properties often show little or no phylogenetic signal. This may reflect the fact that coevolution is also expected to lead to diffuse diversification, making coevolutionary history less tightly linked to interaction strength.

Inferring phylogenetically-structured ecological interaction networks is a rapidly growing field of interest in network ecology and will greatly enhance our understanding of the processes that underlie network dynamics (e.g. Elias *et al.*, 2013). Networks based on Sanger DNA sequencing data, and to a limited extent NGS data, allow opportunities for phylogeny reconstructions for host plants, interacting microbes, or both (e.g. see Rohr and Bascompte, 2014; Le Roux *et al.*, 2016; Keet *et al.*, 2017), providing us with the prospect of incorporating evolutionary relatedness into network analyses and better understanding how it influences invasiveness and invasive species impacts (Fig. 19.2; e.g. Charlery de la Masselière *et al.*, 2017). The short DNA sequence reads typical of NGS metabarcoding approaches make phylogenetic inferences less reliable than those based on the longer DNA sequence reads obtained through Sanger sequencing. However, it is possible to use longer sequencing reads from closely-related taxa obtained from databases like Genbank to reconstruct more robust phylogenies for taxa initially identified using NGS DNA barcodes (Evans *et al.*, 2016).

Although, to my knowledge, phylogenetic estimates have not been fully incorporated into plant–microbial networks in invasion ecology, some studies have suggested ways this can be achieved. For example, the modular legume–rhizobium networks reported by Le Roux *et al.* (2016), including distinct co-invading acacia–*Bradyrhizobium* modules, largely reflect evolutionary history, at least based on rhizobium phylogenies (Ndlovu *et al.*, 2013; Le Roux *et al.*, 2016; Warrington *et al.*, 2019, also see Box 19.1, Fig. 19.2). That is, different and distinct genera of native South African and invasive Australian legumes were found to interact with phylogenetically distinct rhizobial genera, echoing the distinct native and invasive legume–rhizobium modules identified in the network topology reported by Le Roux *et al.* (2016). By mapping associated host plants on to rhizobium phylogenies, Le Roux *et al.* (2016) showed that the observed modularity had strong phylogenetic signal, that is, phylogenetically distinct invasive and native legume genera interacted (almost exclusively) with unique and phylogenetically deeply-diverged rhizobial strains. Together with metrics like interaction specificity, such phylogenetic inferences may provide valuable insights into the susceptibility of networks to infiltration by invasive species in the absence of co-introduction with their symbionts, or how the removal of invasive species and subsequent restoration may impact resident network dynamics over time (Fig. 19.2; Raimundo *et al.*, 2018). As discussed below, incorporating phylogenetic relatedness in multi-level interactions networks, e.g. between rhizobia, plants and herbivores, provides unique opportunities to test whether phylogenetic signal underlies complex network structures and variation within and between interaction levels (e.g. see Elias *et al.*, 2013).

Associations between plants and their microbial symbionts are largely modulated via gene-by-gene interactions, akin to trait-by-trait interactions between plants and their macrofauna, e.g. floral tube and pollinator mouthpart morphologies. In some instances, interacting plant and microbial genes are well characterized and their functions known. For example, as mentioned, molecular communication during the initiation of root nodule formation by rhizobia represents a crucial component of legume–rhizobium interaction specificity (Lira *et al.*, 2015). The excretion of (iso)flavonoids by legume roots initiates the attraction of compatible rhizobia, starting the process of nodule development (Lira *et al.*, 2015). The role of these exudates in symbiotic specificity is illustrated by the fact that their number and concentration tend to increase when compatible rhizobia are detected by host legumes (Lira *et al.*, 2015 and references therein). Recently, Ren *et al.*, 2019 showed that rhizobial transfer RNA (tRNA)-derived fragments (tRFs) are also critically important for successful nodulation. These tRFs are enriched *in planta* and target various host legume genes involved in plant development, including in root hair formation (the entry point for many rhizobia). This link suggests a role for tRF signalling root hair deformation and the creation of infection threads for rhizobia to reach legume root epidermal cells. Initial signal exchange by legume exudates also leads to the activation of a range of so-called nodulation genes, some of which are involved in the rhizobial regulation of host-specificity. These genes are responsible for the expression of so-called nodulation factors (Nod factors), which in turn are recognized by legume Nod factor receptors (Lira *et al.*, 2015; Le Roux *et al.*, 2017). These receptors correspond to the structures of Nod factors and may therefore act as legume determinants for symbiotic specificity. The genes encoding Nod factors (i.e. rhizobial nodulation genes), and to some extent their receptors (i.e. legume genes), have been characterized (e.g. Spaink, 2000; Endre *et al.*, 2002). Using sequencing data from these genes to construct networks may provide valuable insights into the precise coevolutionary mechanisms that underpin the emergence and dynamics of network structures and how invasive legumes infiltrate and change resident networks. For example, *a priori* information on rhizobium–legume phylogenetic co-diversification based on these genes may inform us on the invasibility of networks (i.e. network infiltration risk, Fig. 19.2) based on the availability of compatible symbiont genes in the new range or, alternatively, can provide information on the impact of plant–rhizobium co-introduction on resident networks (also see Raimundo *et al.*, 2018).

Phylogenetic information may also inform us about how invasion may impact on interacting native species pairs, such as the risk of co-extinction cascades due to phylogenetic relatedness (Rezende *et al.*, 2007b), e.g. shared susceptibility of closely related native plants to co-introduced plant pathogens (Evans *et al.*, 2016). Such data can provide a better understanding of how native plants and microbes with differing evolutionary histories respond to novel ecological conditions, including invasion

(Raimundo *et al.*, 2018) and the role of other evolutionary processes, e.g. convergence, in shaping the structure of ecological networks for different types of interactions. The pace and mode of network rewiring may vary between different interaction types, species and habitats. Hence, the similarities and dissimilarities of networks based on coevolutionary history can be informative about how, and at what pace, network changes may occur in response to invasion (e.g. Pearse and Altermatt, 2013). For instance, Rezende *et al.* (2007a) found that in 50% of the 36 plant–pollinator and 23 plant–frugivore mutualistic networks they examined, phylogenetic relatedness between paired species was positively correlated with their interaction similarities, i.e. closely related taxa often interact with the same partners. In such cases, phylogenetic information retrieved using molecular genetic approaches may be useful to predict which interactions are likely or unlikely to occur/be impacted following invasion by particular species (Raimundo *et al.*, 2018). Some of these theoretical concepts have been used in plant–insect networks, illustrating that coevolutionary history, and its associated phylogenetic constraints, can predict invasive species interactions (Pearse and Altermatt, 2013; Charlery de la Masselière *et al.*, 2017).

19.3.5 Linking above-ground and below-ground networks between unicellular and higher organisms

Traditional and molecular-based networks have been very informative about the processes that shape the structure and function of biological communities. They have, however, focused mainly on individual interaction levels (e.g. plant–mycorrhizal, plant–pollinator; but see e.g. Elias *et al.*, 2013; Clare *et al.*, 2019). Capturing all possible community interactions using observational data, even within such individual subnetworks, is impossible due to logistical challenges such as the availability of time, funding and limitations relating to basic organismal biology and taxonomy (Evans *et al.*, 2016).

NGS approaches have greatly advanced our ability to construct high-quality multi-level interaction networks (e.g. see Morriën *et al.*, 2017 for a multi-level network that included both traditional and molecular approaches). Molecular approaches are also faster than traditional approaches (Evans *et al.*, 2016) and, because the molecular systematics of macrobes has enjoyed much attention over the last decade (CBOL Plant Working Group, 2009), extensive DNA reference databases are readily available for rapid taxonomic assignments. For example, NGS barcoding of pollen collected from 38 bee species in Florida allowed Bell *et al.* (2017) to construct a plant–bee pollinator network. This example illustrates how NGS data, in combination with basic network construction methods, pave the way for more complete and resolved ecological networks compared to traditional networks (also see Wirta *et al.*, 2014).

The use of e-DNA approaches in network analyses has also been highlighted, e.g. analyses of predator–prey interactions from DNA barcodes obtained from faecal samples (Clare *et al.*, 2019). Molecular approaches have not only been successfully applied to better understand multiple and complex macrobial subnetworks (e.g. see Clare *et al.*, 2019 for networks involving eight trophic functions including parasitic, mutualistic and predatory interactions), but also to investigate how invasive species establish interactions in their new ranges (e.g. Gariepy *et al.*, 2019). For example, Kitson *et al.* (2019) investigated levels of parasitism experienced by the invasive oak processionary moth, *Thaumetopoea processionea*. Using NGS approaches, these authors amplified parasitoid-specific barcodes from 1012 *T. processionea* caterpillars, obtained from 26 nests in London. Their analysis found only two parasitoid species to attack *T. processionea*, one found within 45.7% of caterpillars, while the other were only present in 0.4% of caterpillars. Importantly, Kitson *et al.* (2019) did not detect many of the parasitoids known to attack *T. processionea* in its native range, possibly reflecting some degree of 'enemy release' being experienced by this invasive moth. Kitson *et al.*'s (2019) findings also suggest that native (UK) parasitoids are largely unable to exploit *T. processionea* as a resource.

The growing application of molecular approaches in both plant–microbial (e.g. Sepp *et al.*, 2019) and macrobe–macrobe networks (e.g. Clare *et al.*, 2019), together with the ease of collecting multiple sites, different temporal and spatial scales and different networks, at relative low cost, indicates that the time is ripe for incorporating plant–microbial interactions into molecular-based multi-level network analyses. These approaches may be particularly informative in understanding the complex ecologies of invasive species. For example, the benefits that non-native legumes may derive from their symbiosis with nitrogen-fixing rhizobia may represent a direct trade-off with herbivory. That is, herbivores may preferentially feed on high nitrogen-content foliage, potentially offsetting the perceived benefits of efficient legume–rhizobium interactions (Simonsen and Stinchcombe, 2014). Multi-level interaction networks provide unique opportunities to study such intricate relationships between such community subnetworks.

19.3.6 Moving beyond plant–microbial networks defined by individual interaction guilds

Like all higher organisms, plants are not made up of autonomous individuals but are organized into biological components, each comprising many closely interacting organisms (Bordenstein and Theis, 2015). Plant-associated microbial communities, i.e. the plant microbiome, can have strong and evident impacts on plant physiology, anatomy and reproduction, and therefore plant colonization, survival and performance. It is important to keep in mind that the diversity and structure of microbiomes can be deterministic, either through host selection or vertical transmission, or stochastic, due to ecological drift or random dispersal (Bordenstein and Theis, 2015). That is, at the one end of the spectrum some microbes may share an intimate coevolutionary history with their hosts (e.g. obligate mutualists and parasites), while at the other end, generalist microbes interact with their host, based on chance encounters alone. It is expected that non-native plants will experience substantial dissociation from their microbiomes upon introduction (e.g. seed introductions), in the same way they are often liberated from their macrofaunal associates (Keane and Crawley, 2002). This will lead to ecological (i.e. loss of microbiome richness) and genetic (i.e. possible loss of microbial genetic diversity) 'microbiomic bottlenecks' (Zenni *et al.*, 2017). However, unlike genetic bottlenecks, the ecological component of microbiome bottlenecks is expected to recover rapidly as plants accumulate substantial generalist microbiota following their introduction into new environments. The occurrence of microbiomic bottlenecks is expected to impact plant colonization success in new environments, in particular due to the lost obligate and specialized coevolved microbial interactions (Le Roux *et al.*, 2017; Fig. 19.1). As discussed below, teasing apart various components of the plant microbiome diversity may provide fruitful, but hitherto overlooked, avenues for understanding how non-native plants and microbes reassemble interactions in new environments. While commonly used network-building approaches are currently unable to construct plant–microbiome networks, microbial 'co-occurrence networks', while very different from ecological interaction networks, have yielded promising and interesting insights worth considering here. Below I briefly discuss recent research in this field and how this may inform plant invasion ecology.

When studying microbiomes, networks are typically represented as patterns of co-occurrence between microbial taxa (e.g. Aires *et al.*, 2015; Navarrete *et al.*, 2015). While such networks do not include host plants per se, they are explicit in that they represent microbial co-occurrences associated with that particular host plant only (Vacher *et al.*, 2016). From this we can infer putative interactions between co-occurring microbes provided that at least one of them modifies the plant host environment in some way. In these networks, nodes are represented by microbial taxa (or OTUs) and network links by the spatial or temporal associations between them. Co-occurrence networks can provide information on the relationships between focal microbial species, e.g. certain pathogens, and the other microorganisms interacting with the host. For example, links can be inferred through significant correlations between the abundances of different OTUs as co-exclusions (i.e. negative correlations) or co-associations (i.e. positive correlations). Microbiome interactions must, by default, include all possible interaction types, i.e. amensalism, commensalism, competition, mutualism, parasitism and predation. However, establishing a link between statistical associations (i.e. co-exclusions and co-associations based on OTU counts of each node) and interaction type is not straightforward (Derocles *et al.*, 2018). For example, a positive relationship between the abundances of two OTUs may result from either mutualism or parasitism (e.g. Weiss *et al.*, 2016).

Despite the difficulties in identifying putative interactions in co-occurrence microbiome networks, it has been suggested that those microbes frequently co-occurring with others in these networks may play key roles in microbiome functioning and regulation (van der Heijden and Hartmann, 2016). Interactions with such 'keystone' species are thought to have large regulatory effects on their environments and other microbiome members (van der Heijden and Hartmann, 2016). For example, Agler *et al.* (2016) found highly interconnected microbiome OTUs in the *Arabidopsis thaliana* phyllosphere to be important regulators of plant health. These keystone microbes act as communicators between the plant and the rest of the microbiome. It appears that the plant acts directly on keystone OTUs, which in turn, through microbe–microbe interactions, communicate the effects to the rest of the microbiome, and likely vice versa. For example, keystone OTUs may recruit beneficial organisms or prevent infection by pathogens to benefit their own fitness and that of the plant (van der Heijden and Hartmann, 2016). Disruption of microbiomes, and in particular interactions involving keystone species, may therefore greatly influence plant success under novel environmental conditions. Mamet *et al.* (2019) recently showed that the magnitude and frequency of co-occurrence network connections between bacteria, archaea and fungi in soil microbiomes increased with invasive smooth brome (*Bromus inermis*) cover. Specifically, these authors found microbial network degree centrality (i.e. the number of links a node has to other nodes) to increase more than 12-fold in brome-invaded soils, with the network therefore becoming more connected as smooth brome became more dominant, i.e. eroding signatures of keystone taxa (Mamet *et al.*, 2019). While not tested in this study, it is conceivable that these soil changes resulting from smooth brome invasion may impact co-occurring natives and their interactions with soil bacteria, e.g. in the rhizosphere microbiome. This is because, in general, keystone rhizosphere taxa are thought to help in sustaining disease-suppressive soils, enhancing nutrient uptake and mediating defence signals between plants, among others (van der Heijden and Hartmann, 2016).

Co-occurrence microbiome networks provide opportunities to compare plants in their invasive and native ranges in order to understand how/whether keystone microbial species are replaced, and if so, with which taxa and at what costs or benefits to the performance of non-native plants. Similarly, comparative analyses of co-occurrence microbiome networks associated with native plant species found in invaded and neighbouring pristine habitats may provide valuable information on the impacts invasive species have on native plant microbiomes.

19.4 Conclusions

To date, most ecological interaction networks involving plants, including those containing invasive species, have been based on interactions within specific (mostly above-ground) interaction guilds; these clearly capture a small proportion of all possible community interactions. These networks have mostly also been constructed from observational data, which limits the inferences that can be made. Recent advances in massively parallel sequencing technologies are helping us to overcome some of these limitations, but also pose unique challenges. Examples from plant–microbial networks show that such data may become commonplace in the construction of highly resolved networks in the future (e.g. see Toju *et al.*, 2014).

New molecular approaches also enhance our ability to understand the efficiency of interactions in relation to the structural components of the networks that underlie them, as well as how evolutionary history shapes network structures. These approaches have been successfully incorporated into multi-level macrobial interaction networks (e.g. Clare *et al.*, 2019), suggesting that the construction of numerous molecular subnetworks that include plant–microbial components are now firmly within our reach. While multi-level network approaches targeting specific interaction guilds will provide a more complete picture of how biotic interactions influence plant invasion, and vice versa, there is also an urgent need to understand the complexity of, and roles played by, microbiomes in plant performance and invasiveness. Rapid advances in NGS technologies, analytical pipelines and statistical approaches suggest that the inclusion of microbiome data into ecological networks may not be that far away. In summary, incorporating molecular approaches into plant–microbial network analyses is emerging rapidly, offering exciting new opportunities for addressing key questions to enhance our understanding of how biological communities are structured and how they respond to perturbations like biological invasions.

References

Agler, T.A., Ruhe, J., Kroll, S., Morhenn, C., Kim, S.T. *et al.* (2016) Microbial hub taxa link host and abiotic factors to plant microbiome variation. *PLoS Biology* 14, e1002352.

Aires, T., Moalic, Y., Serrao, E.A. and Arnaud-Haond, S. (2015) Hologenome theory supported by co-occurrence networks of species-specific bacterial communities in siphonous algae (*Caulerpa*). *FEMS Microbiology Ecology* 91(7), fiv067. DOI: 10.1093/femsec/fiv067.

Aizen, M.A., Morales, C.L. and Morales, J.M. (2008) Invasive mutualists erode native pollination webs. *PLoS Biology* 6, e31.

Augé, R.M. (2001) Water relations, drought and vesicular-arbuscular mycorrhizal symbiosis. *Mycorrhiza* 11, 3–42.

Ballantyne, G., Baldock, K.C.R. and Willmer, P.G. (2015) Constructing more informative plant-pollinator networks: visitation and pollen deposition networks in a heathland plant community. *Proceedings of the Royal Society B* 282, 14–22.

Bascompte, J. (2009) Disentangling the web of life. *Science* 325, 416–419.

Bell, K.L., Fowler, J., Burgess, K.S., Dobbs, E.K., Gruenewald, D. *et al.* (2017) Applying pollen DNA metabarcoding to the study of plant–pollinator interactions. *Applications in Plant Sciences* 5, 1600124.

Bodey, T.W., Bearhop, S. and McDonald, R.A. (2011) Invasions and stable isotope analysis – informing ecology and management. In: Veitch, C.R., Clout, M.N. and Town, D.R. (eds) *Island Invasives: Eradication and Management*. IUCN, Gland, pp. 148–151.

Bordenstein, S.R. and Theis, K.R. (2015) Host biology in light of the microbiome: ten principles of holobionts and hologenomes. *PLoS Biology* 13, 1–23.

Bufford, J.L., Hulme, P.E., Sikes, B.A., Cooper, J.A., Johnston, P.R. *et al.* (2020) Novel interactions between alien pathogens and native plants increase plant–pathogen network connectance and decrease specialization. *Journal of Ecology* 108, 750–776.

Bulgarelli, D., Schlaeppi, K., Spaepen, S., van Themaat, E.V.L. and Schulze-Lefert, P. (2013) Structure and functions of the bacterial microbiota of plants. *Annual Review of Plant Biology* 64, 807–838.

CBOL Plant Working Group (2009) A DNA barcode for land plants. *Proceedings of the National Academy of Sciences* 106(31), 12794–12797. DOI: 10.1073/pnas.0905845106.

Chariton, A.A., Court, L.N., Hartley, D.M., Colloff, M.J. and Hardy, C.M. (2010) Ecological assessment of estuarine sediments by pyrosequencing eukaryotic ribosomal DNA. *Frontiers in Ecology and the Environment* 8, 233–238.

Charlery de la Masselière, M., Ravigne, V., Facon, B., Lefeuvre, P., Massol, F. *et al.* (2017) Changes in phytophagous insect host ranges following the invasion of their community: long-term data for fruit flies. *Ecology and Evolution* 7, 5181–5190.

Claesson, M.J., Wang, Q., O'Sullivan, O.G.-D. and R., Cole, J.R. *et al.* (2010) Comparison of two next-generation sequencing technologies for resolving highly complex microbiota composition using tandem variable 16S rRNA gene regions. *Nucleic Acids Research* 38, e200.

Clare, E.L. (2014) Molecular detection of trophic interactions: emerging trends, distinct advantages, significant considerations and conservation applications. *Evolutionary Applications* 7, 1144–1157.

Clare, E.L., Fazekas, A., Ivanova, N., Floyd, R., Hebert, P. *et al.* (2019) Approaches to integrating genetic data into ecological networks. *Molecular Ecology* 28, 503–519.

Derocles, S.A.P., Bohan, D.A., Dumbrell, A.J., Kitson, J.J.N., Massol, F. *et al.* (2018) Biomonitoring for the 21st century: integrating next-generation sequencing into ecological network analysis. *Advances in Ecological Research* 58, 1–62.

Dickie, I.A., Bolstridge, N., Cooper, J.A. and Peltzer, D.A. (2010) Co-invasion by *Pinus* and its mycorrhizal fungi. *New Phytologist* 187, 475–484.

Dickie, I.A., Bufford, J.L., Cobb, R.C., Desprez-Loustau, M.L., Grelet, G. *et al.* (2017a) The emerging science of linked plant–fungal invasions. *New Phytologist* 215, 1314–1332.

Dickie, I.A., Cooper, J.A., Bufford, J.L., Hulme, P.E. and Bates, S.T. (2017b) Loss of functional diversity and network modularity in introduced plant-fungal symbioses. *AoB PLANTs* 9, w84.

Díez, J. (2005) Invasion biology of Australian ectomycorrhizal fungi introduced with eucalypt plantations into the Iberian Peninsula. *Biological Invasions* 7, 3–15.

Elias, M., Fontaine, C. and van Veen, F.J.F. (2013) Evolutionary history and ecological processes shape a local multilevel antagonistic network. *Current Biology* 23, 1355–1359.

Endre, G., Kereszt, A., Kevei, Z., Mihacea, S., Kalo, P. *et al.* (2002) A receptor kinase gene regulating symbiotic nodule development. *Nature* 417, 962–966.

Evans, D.M., Kitson, J.J.N., Lunt, D.H., Straw, N.A. and Pocock, M.J.O. (2016) Merging DNA metabarcoding and ecological network analysis to understand and build resilient terrestrial ecosystems. *Functional Ecology* 30, 1904–1916.

Frost, C.M., Allen, W.J., Courchamp, F., Jeschke, J.M., Saul, W.C. *et al.* (2019) Using network theory to understand and predict biological invasions. *Trends in Ecology and Evolution* 34, 831–843.

Gariepy, T.D., Bruin, A., Konopka, J., Scott-Dupree, C., Fraser, H. *et al.* (2019) A modified DNA barcode approach to define trophic interactions between native and exotic pentatomids and their parasitoids. *Molecular Ecology* 28, 456–470.

Gazol, A., Zobel, M., Cantero, J.J., Davison, J., Esler, K.J. *et al.* (2016) Impact of alien pines on local arbuscular mycorrhizal fungal communities – evidence from two continents. *FEMS Microbiology Ecology* 92(fiw073).

Gillis, M., Vandamme, P., De Vos, P., Swings, J. and Kerster, K. (2001) Polyphasic taxonomy. In: Boone, D.R., Castenholz, R.W. and Garity, G.M. (eds) *Bergey's Manual of Systematic Bacteriology*, 2nd edn. Springer, London, UK.

Hayward, J., Horton, T.R. and Nunez, M.A. (2015) Ectomycorrhizal fungal communities coinvading with Pinaceae host plants in Argentina: gringos bajo el bosque. *New Phytologist* 208, 497–506.

Heleno, R.H., Olesen, J.M., Nogales, M., Vargas, P. and Traveset, A. (2013) Seed dispersal networks in the Galápagos and the consequences of alien plant invasions. *Proceedings of the Royal Society of London, Series B Biological Sciences*280(1750.).

Ji, Y., Ashton, L., Pedley, S.M., Edwards, D.P., Tang, Y. *et al.* (2013) Reliable, verifiable and efficient monitoring of biodiversity via metabarcoding. *Ecology Letters* 16, 1245–1257.

Keane, R.M. and Crawley, M.J. (2002) Exotic plant invasions and the enemy release hypothesis. *Trends in Ecology and Evolution* 17, 164–170.

Keet, J.H., Ellis, A.G., Hui, C. and Le Roux, J.J. (2017) Legume-rhizobium symbiotic promiscuity and effectiveness do not affect plant invasiveness. *Annals of Botany* 119, 1319–1331.

Keet, J.H., Ellis, A.G., Hui, C. and Le Roux, J.J. (2019) Strong spatial and temporal turnover of soil bacterial communities in South Africa's hyperdiverse fynbos biome. *Soil Biology and Biochemistry* 136, 1–12.

Kitson, J.J.N., Hahn, C., Sands, R.J., Straw, N.A., Evans, D.M. *et al.* (2019) Detecting host–parasitoid interactions in an invasive lepidopteran using nested tagging DNA metabarcoding. *Molecular Ecology* 28, 471–483.

Lahoz-Monfort, J.J., Guillera-Arroita, G. and Tingley, R. (2016) Statistical approaches to account for false-positive errors in environmental DNA samples. *Molecular Ecology Resources* 16, 673–685.

Le Roux, J.J., Ellis, A.G., van Zyl, L.-M., Hosking, N.D., Keet, J.-H. *et al.* (2018) Importance of soil legacy effects and successful mutualistic interactions during Australian Acacia invasions in nutrient-poor environments. *Journal of Ecology* 106(5), 2071–2081. DOI: 10.1111/1365-2745.12965.

Lemaire, B., Van Cauwenberghe, J., Chimphango, S., Stirton, C., Honnay, O. *et al.* (2015) Recombination and horizontal transfer of nodulation and ACC deaminase (acdS) genes within alpha and Betaproteobacteria nodulating legumes of the Cape Fynbos biome. *FEMS Microbiology Ecology* 91(fiv118).

Le Roux, J.J., Hui, C., Keet, J.-H. and Ellis, A.G. (2017) Co-Introduction vs ecological fitting as pathways to the establishment of effective mutualisms during biological invasions. *New Phytologist* 215(4), 1354–1360. DOI: 10.1111/nph.14593.

Le Roux, J.J., Mavengere, N.R. and Ellis, A.G. (2016) The structure of legume-rhizobium interaction networks and their response to tree invasions. *AoB Plants* 8(plw038), plw038. DOI: 10.1093/aobpla/plw038.

Lira, M.A., Nascimento, L.R.S. and Fracetto, G.G.M. (2015) Legume–rhizobia signal exchange: promiscuity and environmental effects. *Frontiers in Microbiology* 6, 945.

Mamet, S.D., Redlick, E., Brabant, M., Lamb, E.G., Helgason, B.L. *et al.* (2019) Structural equation modeling of a winnowed soil microbiome identifies how invasive plants re-structure microbial networks. *The ISME Journal* 13, 1988–1996.

Mata, V.A., Rebelo, H., Amorim, F., McCracken, G.F., Jarman, S. *et al.* (2019) How much is enough? effects of technical and biological replication on metabarcoding dietary analysis. *Molecular Ecology* 28, 165–175.

McCue, M.D., Javal, M., Clusella-Trullas, S., Le Roux, J.J., Jackson, M.C. *et al.* (2020) Using stable isotope analysis to answer fundamental questions in invasion ecology: progress and prospects. *Methods in Ecology and Evolution* 11, 196–214.

Minoarivelo, H.O., Hui, C., Terblanche, J.S., Kosakovsky Pond, S.L. and Scheffler, K. (2014) Detecting phylogenetic signal in mutualistic interaction networks using a Markov process model. *Oikos* 123, 1250–1260.

Mitchell, C.G. and Power, A.G. (2003) Release of invasive plants from fungal and viral pathogens. *Nature* 421, 625–627.

Moeller, H.V., Dickie, I.A., Peltzer, D. A. and Fukami, T. (2015) Mycorrhizal co-invasion and novel interactions depend on neighborhood context. *Ecology* 96, 2336–2347.

Molina, R., Massicotte, H. and Trappe, J.M. (1992) Specificity phenomena in mycorrhizal symbioses: Community-ecological consequences and practical implications. In: Allen, M.F. (ed.) *Mycorrhizal Functioning: An integrative Plant–Fungal Process. Chapman and Hall*. New York, pp. 357–423.

Montesinos-Navarro, A., Segarra-Moragues, J.G., Valiente-Banuet, A. and Verdú, M. (2012) The network architecture of plant-arbuscular mycorrhizal fungi. *New Phytologist* 194, 536–547.

Morriën, E., Hannula, S.E., Snoek, L.B., Helmsing, N.R., Zweers, H. *et al.* (2017) Soil networks become more connected and take up more carbon as nature restoration progresses. *Nature Communications* 8, 14349.

Navarrete, A.A., Tsai, S.M., Mendes, L.W., Faust, K., de Hollander, M. *et al.* (2015) Soil microbiome responses to the short-term effects of Amazonian deforestation. *Molecular Ecology* 24, 2433–2448.

Ndlovu, J., Richardson, D.M., Wilson, J.R. and Le Roux, J.J. (2013) Co-invasion of South African ecosystems by an Australian legume and its rhizobial symbionts. *Journal of Biogeography* 40, 1240–1251.

Nuñez, M.A., Chiuffo, M.C., Torres, A., Paul, T., Dimarco, R.D. *et al.* (2017) Ecology and management of invasive Pinaceae around the world: progress and challenges. *Biological Invasions* 19, 3099–3120.

Olesen, J.M., Eskildsen, L.I. and Venkatasamy, S. (2002) Invasion of pollination networks on oceanic islands: importance of invader complexes and endemic super generalists. *Diversity and Distributions* 8, 181–192.

Parker, M.A. (2001) Mutualism as a constraint on invasion success for legumes and rhizobia. *Diversity and Distributions* 7, 125–136.

Pearse, I.S. and Altermatt, F. (2013) Predicting novel trophic interactions in a non-native world. *Ecology Letters* 16, 1088–1094.

Perret, X., Staehelin, C. and Broughton, W.J. (2000) Molecular basis of symbiotic promiscuity. *Microbiology and Molecular Biology Reviews* 64, 180–201.

Policelli, N., Bruns, T.D., Vilgalys, R. and Nuñez, M.A. (2019) Suilloid fungi as global drivers of pine invasions. *New Phytologist* 222, 714–725.

Pringle, A., Bever, J.D., Gardes, M., Parrent, J.L., Rillig, M.C. *et al.* (2009) Mycorrhizal symbioses and plant invasions. *Annual Review of Ecology, Evolution, and Systematics* 40, 699–715.

Raimundo, R.L.G., Guimarães, P.R. and Evans, D.M. (2018) Adaptive networks for restoration ecology. *Trends in Ecology and Evolution* 33, 664–675.

Rappé, M.S. and Giovannoni, S.J. (2003) The uncultured microbial majority. *Annual Review of Microbiology* 57, 369–394.

Ren, B., Wang, X., Duan, J. and Ma, J. (2019) Rhizobial tRNA-derived small RNAs are signal molecules regulating plant nodulation. *Science* 365(6456), 919–922.

Rezende, E.L., Jordano, P. and Bascompte, J. (2007a) Effects of phenotypic complementarity and phylogeny on the nested structure of mutualistic networks. *Oikos* 116, 1919–1929.

Rezende, E.L., Lavabre, J.E., Guimarães, P.R., Jordano, P. and Bascompte, J. (2007b) Non-random coextinctions in phylogenetically structured mutualistic networks. *Nature* 448, 925–928.
Richards, R.A. (1996) Defining selection criteria to improve yield under drought. *Plant Growth Regulation* 20, 157–166.
Richardson, D.M. (2006) *Pinus*: A model group for unlocking the secrets of alien plant invasions? *Preslia* 78, 375–388.
Richardson, D.M., Allsopp, N., D'Antonio, C.M., Milton, S.J. and Rejmánek, M. (2000) Plant invasions--the role of mutualisms. *Biological Reviews of the Cambridge Philosophical Society* 75(1), 65–93. DOI: 10.1017/S0006323199005435.
Richardson, D.M., Carruthers, J., Hui, C., Impson, F.A.C., Robertson, M.P. *et al.* (2011) Human-Mediated introductions of Australian acacias – a global experiment in biogeography. *Diversity and Distributions* 17, 771–787.
Richardson, D.M. and Rejmánek, M. (2011) Trees and shrubs as invasive alien species, a global review. *Diversity and Distributions* 17, 788–809.
Richardson, D.M., Williams, P.A. and Hobbs, R.J. (1994) Pine invasions in the southern hemisphere: determinants of spread and invadability. *Journal of Biogeography* 21(5), 511–527. DOI: 10.2307/2845655.
Rodríguez-Echeverría, S. (2010) Rhizobial hitchhikers from down under: Invasional meltdown in a plant-bacteria mutualism? *Journal of Biogeography* 37, 1611–1622.
Rodríguez-Echeverría, S., Fajardo, S., Ruiz-Díez, B. and Fernández-Pascual, M. (2012) Differential effectiveness of novel and old legume-rhizobia mutualisms: implications for invasion by exotic legumes. *Oecologia* 170, 253–261.
Rogel, M.A., Ormeno-Orrillo, E. and Martinez-Romero, E.M. (2011) Symbiovars in rhizobia reflect bacterial adaptation to legumes. *Systematic and Applied Microbiology* 34, 96–104.
Rohr, R.P. and Bascompte, J. (2014) Components of phylogenetic signal in antagonistic and mutualistic networks. *American Naturalist* 184, 556–564.
Roy-Bolduc, A., Laliberté, E. and Hijri, M. (2016) High richness of ectomycorrhizal fungi and low host specificity in a coastal sand dune ecosystem revealed by network analysis. *Ecology and Evolution* 6, 349–362.
Santos, G.M.D.M., Aguiar, C.M.L., Genini, J., Martins, C.F., Zanella, F.C.V. *et al.* (2012) Invasive Africanized honeybees change the structure of native pollination networks in Brazil. *Biological Invasions* 14, 2369–2378.
Sepp, S.-K., Davison, J., Jairus, T., Vasar, M., Moora, M. *et al.* (2019) Non-random association patterns in a plant-mycorrhizal fungal network reveal host-symbiont specificity. *Molecular Ecology* 28, 365–378.
Simonsen, A.K. and Stinchcombe, J.R. (2014) Herbivory eliminates fitness costs of mutualism exploiters. *New Phytologist* 202, 651–661.
Spaink, H.P. (2000) Root nodulation and infection factors produced by rhizobial bacteria. *Annual Review of Microbiology* 54, 257–288.
Spencer, S.J., Tamminen, M.V., Preheim, S.P., Guo, M.T., Briggs, A.W. *et al.* (2016) Massively parallel sequencing of single cells by epicPCR links functional genes with phylogenetic markers. *The ISME Journal* 10, 427–436.
Spotswood, E.N., Meyer, J.-Y. and Bartolome, J.W. (2012) An invasive tree alters the structure of seed dispersal networks between birds and plants in French Polynesia. *Journal of Biogeography* 39, 2007–2020.
Steidinger, B.S., Crowther, T.W., Liang, J., Van Nuland, M.E., Werner, G.D.A. *et al.* (2019) Climatic controls of decomposition drive the global biogeography of forest-tree symbioses. *Nature* 569(7756), 404–408. DOI: 10.1038/s41586-019-1128-0.
Toju, H. (2015) High-throughput DNA barcoding for ecological network studies. *Population Ecology* 57, 37–51.
Toju, H., Guimarães, P.R., Olesen, J.M. and Thompson, J.N. (2014) Assembly of complex plant–fungus networks. *Nature Communications* 5(1), 5273. DOI: 10.1038/ncomms6273.
Traveset, A. and Richardson, D.M. (2014) Mutualistic interactions and biological invasions. *Annual Review of Ecology, Evolution, and Systematics* 45, 89–113.
Traveset, A., Heleno, R., Chamorro, S., Vargas, P., McMullen, C.K. *et al.* (2013) Invaders of pollination networks in the Galápagos islands: emergence of novel communities. *Proceedings of the Royal Society B: Biological Sciences* 280(1758), 20123040. DOI: 10.1098/rspb.2012.3040.
Tylianakis, J.M., Tscharntke, T. and Lewis, O.T. (2007) Habitat modification alters the structure of host-parasitoid food webs. *Nature* 445, 202–205.

Vacher, C., Daudin, J.J., Piou, D. and Desprez-Loustau, M.-L. (2010) Ecological integration of alien species into a tree-parasitic fungus network. *Biological Invasions* 12, 3249–3259.
Vacher, C., Piou, D. and Desprez-Loustau, M.-L. (2008) Architecture of an antagonistic tree/fungus network: the asymmetric influence of past evolutionary history. *PLoS ONE* 3, e1740. DOI: 10.1371/journal.pone.0001740.
Vacher, C., Corinne, V., Alireza, T.-N., Stefaniya, K., Nathalie, P. *et al.* (2016) Learning ecological networks from next-generation sequencing data. *Advances in Ecological Research* 54, 1–39.
Unkovich, M. (2013) Isotope discrimination provides new insight into biological nitrogen fixation. *New Phytologist* 198, 643–646.
van der Heijden, M.G.A. and Hartmann, M. (2016) Networking in the plant microbiome. *PLoS Biology* 14, e1002378.
Vellinga, E.C., Wolfe, B.E. and Pringle, A. (2009) Global patterns of ectomycorrhizal introductions. *New Phytologist* 181, 960–973.
Warrington, S., Ellis, A., Novoa, A., Wandrag, E.M., Hulme, P.E. *et al.* (2019) Cointroductions of Australian acacias and their rhizobial mutualists in the southern hemisphere. *Journal of Biogeography* 46, 1519–1531.
Weiss, S., Van Treuren, W., Lozupone, C., Faust, K., Friedman, J. *et al.* (2016) Correlation detection strategies in microbial data sets vary widely in sensitivity and precision. *The ISME Journal* 10, 1669–1681.
Widder, S., Besemer, K., Singer, G.A., Ceola, S., Bertuzzo, E. *et al.* (2014) Fluvial network organization imprints on microbial co-occurrence networks. *Proceedings of the National Academy of Sciences USA* 111, 12799–12804.
Wirta, H.K., Hebert, P.D.N., Kaartinen, R., Prosser, S.W., Varkonyi, G. *et al.* (2014) Complementary molecular information changes our perception of food web structure. *Proceedings of the National Academy of Sciences USA* 111, 1885–1890.
Zenni, R.D., Dickie, I.A., Wingfield, M.J., Hirsch, H., Crous, C.J. *et al.* (2017) Evolutionary dynamics of tree invasions: complementing the unified framework for biological invasions. *AoB PLANTS* 9(plw085).

20 How Can Progress in the Understanding of Antagonistic Interactions be Applied to Improve Biological Control of Plant Invasions?

Martin P. Hill[1]* and Julie A. Coetzee[2]

[1]Centre for Biological Control, Department of Zoology and Entomology, Rhodes University, Grahamstown, 6140, South Africa; [2]Centre for Biological Control, Botany Department, Rhodes University, Grahamstown, 6140, South Africa

Abstract

Classical biological control has been used as a management tool for invasive non-native plant species globally for over 200 years. There have been some very successful programmes, most notably on waterweeds, cacti and seed reduction in perennial trees. Seventy per cent of agents released have established in at least one instance, and 66% of the targeted invasive species have showed some level of control. However, some programmes have failed to meet expectations, for example on *Lantana camara*. The most commonly cited reasons for the failure of establishment or limited efficacy of biological control agents are unsuitable climatic conditions and genotype incompatibility. We propose that antagonistic biotic interactions play a significant role in the outcomes of weed biological control programmes. Induced plant defences (physical and chemical) that can be mounted rapidly by the invasive non-native plants can result in the reduction in agent populations after initial attack. Rapid induction of plant defences have been implicated in the lack of long-term establishment of the agent *Falconia intermedia* that showed great initial promise against the widespread invasive shrub *L. camara*. Host range expansion by native natural enemies onto biological control agents have also been shown to reduce population growth of agents. Finally, competition from indigenous plant species aids invasive alien plant population reduction in the presence of herbivory. All three factors have been poorly studied and further work is needed to better explain the outcomes of weed biological control programmes.

20.1 Introduction

Classical biological control (CBC), also referred to as importation biological control, is a sustainable method for the management of invasive non-native plant species that uses host-specific natural enemies (biological control agents), mainly phytophagous insects and mites, and fungal pathogens from the native range of the invasive plant. The aim of CBC is for the introduced biological control agents to establish in their new environment and increase in abundance to levels that inflict sufficient damage to the target invasive species to reduce their competitiveness, reproductive output and population growth to below an economically, socially or ecologically determined threshold (Morin *et al.*, 2009). The first example of CBC globally was the release of the cochineal insect *Dactylopius ceylonicus* (Green) (Hemiptera: Dactylopiidae), introduced to India in 1795 for dye production as

*Corresponding author: m.hill@ru.ac.za

 Plant Invasions: The Role of Biotic Interactions (eds Anna Traveset and David M. Richardson)
DOI: 10.1079/9781789242171.0020

it was misidentified as *Dactylopius coccus* Costa, where it effectively controlled *Opuntia monacantha* (Willd.) Haw. (Cactaceae) (Goeden, 1988). Since this fortuitous mistake, some 468 biological control agent species have been intentionally released worldwide (between the early 1900s and 2012) for the control of 175 invasive species (Winston *et al.*, 2014).

Schwarzländer *et al.* (2018) analysed the updated Winston *et al.* (2014) 'Biological control of weeds: a world catalogue of agents and their target weeds', and showed that up until 2012, there had been 1555 releases of the 468 biological control agents made in 90 countries. Of the agent species released, 332 (70.9%) had established in at least one instance, and of all of the releases globally, 65.7% of the targeted invasive species showed some level of control, with 24% experiencing heavy impact. The most commonly cited reasons for the failure of establishment or limited efficacy of biological control agents are unsuitable climatic conditions (Andow *et al.*, 1997; Byrne *et al.*, 2004) and genotype incompatibility (e.g. Paterson and Zachariades, 2013). However, antagonistic biotic interactions such as induced plant defences by the invasive non-native plants (Heshula and Hill, 2012), and attack by native enemies that have expanded their host ranges from local hosts on to biological control agents (Paynter *et al.*, 2010), have been poorly studied.

Quantifying, and thereafter defining, the success of a biological control programme is nuanced, although it is generally accepted that an effective programme should reduce the density of the invasive species below a pre-determined ecological, economic or social threshold (Morin *et al.*, 2009). Unfortunately, post-release evaluations, and classification of biological control success, in which the responses of ecosystems following the reduction of an invasive plant's competitive advantage through biological control have been neglected (Denslow and D'Antonio, 2005). Thus, there are few studies that have monitored ecosystem recovery post biological control (e.g. Lake *et al.*, 2017), which evaluate the role played by induced plant defences, host range extension by native parasitoids and competition from indigenous plant communities, in suppressing populations of invasive species. Schulz *et al.* (2019) reviewed 12 different hypotheses relating invasion to the effects of antagonists (e.g. Enemy escape, Enemy Release, Enemy Tolerance, Biotic Resistance), yet a review of indirect effects of shared prey by Harmon and Andow (2004) concluded that no single theoretical generalization can accurately predict community-level effects for every system. In a departure from the CBC literature, here we regard the specialist natural enemies released as biological control agents as invasive species that are antagonists to invasive plant species (Schulz *et al.*, 2019).

We review what is known about the role played by induced plant defences, host range extension by native parasitoids and competition from indigenous plant species in outcomes of 'classical weed biological control' programmes. We also discuss how these interactions can be investigated during future biological control programmes to progress our understanding of these antagonistic interactions and thus improve biological control outcomes of plant invasions.

20.2 Induced Plant Defences – Invasive Plants Fight Back

One of the most cited hypotheses explaining non-native plant invasions is the Enemy Release Hypothesis (ERH), which states that invasive non-native plants spread rapidly because they are released from their specialized natural enemies (Keane and Crawley, 2002). The ERH relies on the assumptions that natural enemies, including phytophagous insects, mites and pathogens are important top-down regulators of plant populations; exotic plants may challenge native herbivores by exposing them to chemicals with which they have not coevolved and thus these native natural enemies have a greater impact on native than on introduced plant species; and introduced plants are able to capitalize on a reduction in natural enemy pressure in the region of introduction. Some evidence for the ERH comes from the successful CBC programmes against invasive non-native plants (Müller-Schärer and Schaffner, 2008).The ERH is too simplistic to explain complex plant—insect interactions that involve a myriad of biotic and abiotic interactions (Schulz *et al.*, 2019).For example, invasive non-native plants species might have effective tolerance/resistance mechanisms which could explain why introductions of biological agents are not as successful as predicted (Müller-Schärer *et al.*, 2004).

Blossey and Notzold (1995) argued that introduced plant species grow vigorously because they do not need to invest in defences, as a result of reduced specialist herbivore pressure. Therefore, introduced biological agents could be highly successful not only because they have escaped their specialist enemies (see Section 3), but because they may capitalize on the reduction in defences of the invasive plant. Plants have evolved a wide array of defence mechanisms in response to attack, which help reduce the impact of herbivores (Karban and Myers, 1989; Howe and Schaller, 2008). Biochemical processes within plants mediate defence mechanisms which are typically chemical and/or physical in nature (Stamp, 2003), and are employed directly and/or indirectly. Ecological and physiological parameters, such as resource availability, habitat type and plant growth rates, are valuable predictors of the level of defences employed by plants (Coley *et al.*, 1985; Bazzaz *et al.*, 1987). In addition, the resource availability hypothesis provides an applicable framework to investigate insect-induced responses (Endara and Coley, 2011). Plants with rapid growth rates and colonizing traits, inhabiting resource-rich environments, are typically associated with low levels of constitutive defensive compounds; resource allocations to plant growth and/or inducible defences are favoured. When a plant is introduced to a new environment, it is exposed to novel biotic and abiotic stressors, potentially resulting in rapid changes in phenotypic traits, such as reduced allocation to costly defences, in favour of growth and reproduction. The Evolution of Increased Competitive Ability (EICA) hypothesis was therefore formulated to explain evolution in introduced plants in the absence of natural enemies, especially specialists (Blossey and Notzold, 1995), and is useful in explaining variability in biological control outcomes(Thompson, 2005). The predictions made by the EICA hypothesis are supported by several studies, (e.g. Blossey and Notzold, 1995; Siemann and Rogers, 2001; Stastny *et al.*, 2005), but not by others (e.g. Willis *et al.*, 2000; Van Kleunen and Schmid, 2003; Handley *et al.*, 2008; Cripps *et al.*, 2009). However, it is probably more important to understand the evolutionary changes in tolerance/resistance (i.e. defences) of invasive plants to herbivory post introduction in order to understand the outcome of classical biological control of invasive plants (Li *et al.*, 2012).

Because attack from generalist species is likely in a new environment, invasive plants still need to defend themselves against generalist herbivore attack, often resulting in a shift in the natural enemy assemblages, dominated by generalists. The Shifting Defence Hypothesis (SDH) takes EICA a step further to include differential allocation of resources to defence compounds (Doorduin and Vrieling, 2011). Where invasive non-native plants experience some level of top-down pressure by generalist herbivores, allocation to cheap defence compounds that are toxic to non-adapted generalist herbivores increases, while allocation to expensive, digestibility-reducing compounds remains reduced (Müller-Schärer *et al.*, 2004; Joshi and Vrieling, 2005). However, several studies from the plant—insect literature have shown that secondary metabolite production (i.e. specialists defences against specialist herbivores) is phenotypically plastic and can be rapidly induced by herbivory (Kessler and Halitschke, 2007), and could explain why a biological control project that initially appears to be promising becomes less effective over time. For example, the sap-sucking biological control agent *Falconia intermedia* (Distant) (Hemiptera: Miridae) was released to improve the biological control of *Lantana camara* L. (Verbenaceae) in South Africa in 1999 (Baars *et al.*, 2003). Despite being highly damaging to lantana initially, populations of *F. intermedia* diminished at all sites within a few seasons. Studies showed that feeding damage by *F. intermedia* increased physical plant defences in *L. camara*, namely leaf toughness and trichome density (Heshula and Hill, 2011), and chemical plant defences rapidly (Heshula and Hill, 2014; Ngxande-Koza *et al.*, 2017) and thus reducing the impact of the agent (see Box 20.1).

There are some examples of the role of constitutive plant defences in a biological control context (e.g. Joshi and Vrieling, 2005; Wang *et al.*, 2011) but examples on induced resistance are scarce. More biotic interaction studies that include some measurement of tolerance and resistance through the development of physical and biochemical defences pre and post release in the field, such as the

Box 20.1. Induced defences affect the success of *Lantana camara* L. (Verbenaceae) biological control.

Lantana camara L. (Verbenaceae) is widely considered one of the worst invasive plant species in the world. Where lantana invades, it forms impenetrable thickets (Cilliers and Neser, 1991), degrading natural biodiversity and reducing the value of agricultural land (Day *et al*., 2003). It has been a target for biological control for over 50 years, with 75 control agent species released against it in 32 countries, with variable success (Winston *et al*., 2014). One of the many factors contributing to the lack of control is susceptibility of the different plant genotypes to herbivory (Fig. 20.1a).

In South Africa, 48 different species of biological control agent have been considered for control of lantana, with 9 shelved, 13 rejected, 12 released but didn't establish, 12 released and established, and permission granted to release 2 more (Zachariades, 2018). The establishment of the 12 leaf-, flower-, fruit-, root- and shoot-attacking insect agents in South Africa is currently insufficient to effectively control lantana (Urban *et al*., 2011). Although some of the established agents often cause significant damage, these are usually at sporadic and local levels (Urban *et al*., 2011). Since established agents are mainly leaf-feeders, their populations are subject to drastic decline during autumn or winter when plant leaves defoliate due to frost or drought (Simelane, 2005).

One of these leaf-feeding agents, the highly damaging *Falconia intermedia* (Distant) (Hemiptera: Miridae) was released to improve the biological control of *L. camara* in South Africa in 1999 (Baars *et al*., 2003) (Fig. 20.1b). Despite the early promise and the significant impact on *L. camara* (Fig. 20.1c, d), populations of *F. intermedia* crashed at almost all sites within a few seasons of release. From very promising pre-release testing results, the mirid can now only be found in low numbers at one Eastern Cape field site (Baars *et al*., 2003). One of the reasons for failure of this mirid is induced resistance of *L. camara* varieties (Baars and Neser, 1999; Heshula and Hill, 2011, 2014). Feeding damage by *F. intermedia* induces a response resulting in a decrease in subsequent *F. intermedia* performance on a South African *L. camara* variety (Heshula and Hill, 2011, 2014). Following feeding by *F. intermedia*, increases in physical plant defences, namely leaf toughness and trichome density, both of which deter and hinder herbivores from feeding and attaching to or moving on the plant, were reported (Heshula and Hill, 2011).

Not only did *L. camara* display induced physical responses (Heshula and Hill, 2014), but feeding by *F. intermedia* induced chemical responses, evidenced by changes in the quality and quantity of chemical constituents of essential oils (Ngxande-Koza *et al*., 2017). While 163 compounds were identified from undamaged leaves of various *L. camara* varieties, only 75 compounds were identified from insect-damaged leaves. Terpenes were the major components across the varieties, while caryophyllene, hexane, naphthalene, copaene and a-caryophyllene were common in all the varieties tested from both undamaged and insect-damaged leaves. The changes in chemical concentrations indicated that feeding by the mirid on *L. camara* varieties causes an induction by either reducing or increasing the chemical concentrations, which has a negative impact on the success of biological control against *L. camara* varieties (Ngxande-Koza *et al*., 2017). Although not studied, it is likely that defences are induced by many of the agents released against *L. camara* around the world where biological control has been implemented.

F. intermedia — *L. camara* example, should be conducted routinely during post-release evaluations of biocontrol programmes and might explain why some agents have been less effective than predicted after initial success. To better understand the outcomes of weed biological control projects, more experimental work comparing the target responses of the invasive plant to herbivory in both the native and introduced ranges is required.

20.3 Tritrophic Interactions – Native Parasitoids Spoil the Party

In the same way that the Enemy Release Hypothesis (ERH) has been used to explain the invasiveness of introduced plant species (Keane and Crawley, 2002), it has also been implicated in the success of biological control agents, as they themselves have escaped specialized natural enemies which have been left behind in the region

Fig. 20.1. Different flower varieties of *Lantana camara* L. (Verbenaceae) illustrating genetic diversity in the species: (A) an adult *Falconia intermedia* (Distant) (Hemiptera: Miridae); (B) feeding damage inflicted to *L. camara* leaves by the sapsucker, *F. intermedia*, indicating chlorosis of the leaf tissue; (C and D) promising initial establishment and damage in the field in South Africa. (All pictures courtesy of Agricultural Research Council – Plant Health Protection.)

of origin during the exploration and quarantine steps in the biological control process. However, nearly 60% of biological control agents that establish fail to suppress their target invasive species sufficiently (Schwarzländer *et al.*, 2018), and this failure might be a result of natural enemies in the region of introduction recruiting on to the released agents. Host range expansion by indigenous parasitoids on to biological control agents of invasive non-native plants has been recorded (e.g. Hill and Hulley, 1995; Paynter *et al.*, 2010). The findings of these studies suggest that i) parasitoid recruitment occurs soon after release into the introduced region; ii) poorly protected endophagous biological control agents such as leaf miners are more likely to recruit parasitoids than biological control agents in more protected feeding guilds such as gallers and mobile ectophagous agents; and iii) biological control agents that have congeners (or ecological analogues) in the region of introduction that carry parasitoid loads are likely to be parasitized (see Box 20.2). Hill and Hulley (1995) concluded that native parasitoids did not strongly influence the populations of biocontrol agents in South Africa. Conversely, Paynter *et al.* (2010) suggested that parasitism was significantly associated with the failure of agents to suppress populations of invasive species in New Zealand.

The evidence for parasitoid impacts on biocontrol efforts is noticeably greater in the literature than the impacts of generalist predators. However, interference from opportunistic generalist predators in the success of biological control is particularly important to document because it

Box 20.2. Parasitoids in New Zealand.

Hill and Hulley (1995) determined the impact of host-range extension of native parasitoids to biological control agents of invasive plants in South Africa. Native parasitoids were recorded from 40% of the 40 species of biological control agent that had established on 22 invasive non-native plant species at that time. All host-range extensions by native parasitoids were from native herbivores and occurred within three years of release. Susceptibility of the agents to attack by native parasitoids was influenced by their level of concealment and taxon. Poorly concealed endophytic agents were most susceptible to attack, whereas exposed feeders were fairly free from attack. However, native parasitoids did not strongly influence the populations of biocontrol agents and it was concluded that no biological control agent should be rejected solely on its susceptibility to attack by native parasitoids.

A review of biocontrol programmes of invasive plants in New Zealand indicated that parasitism is significantly associated with the failure of agents to suppress populations of invaders. Paynter *et al.* (2010) conducted an extensive literature review and field surveys to test the hypothesis that introduced control agents should acquire parasitoids over time to ultimately determine the impact of parasitism on biological control programmes in New Zealand. They found literature records for and/or surveyed 28 of the 30 insects agents established in New Zealand for the biological control of invasive plants, as of December 2008, and managed to rear 19, mostly native, parasitoid species from 10 of the 28 species surveyed. Of these species, only one of the nine agents classified as successful was parasitized, while of the 15 insects classified as unsuccessful, eight were parasitized.

Paynter *et al.* (2010) then showed that ecological analogues, defined as a native insect in the same superfamily as a biological control agent, are a major source of parasitoids that are capable of rapidly colonizing biocontrol agent hosts soon after their introduction. Therefore, agents that experience high parasitism rates in their areas of origin may experience enemy-free space after introduction, if ecological analogues are absent from the herbivore fauna into which they are introduced. Paynter *et al.* (2010) therefore caution the consideration of potential control agents that are parasitized in their native areas if analogues do exist in the new range, not only because of a reduction in the level of biological control, but because an accumulation of parasitoids could have indirect non-target impacts on native communities.

is usually subtle and may remain unrecognized even if release efforts fail. Tritrophic interactions involving generalist predators, biological control herbivores and invasive plants have great potential to disrupt biological control efforts, and subsequently the ability of herbivores to limit plant invasions (Crawley, 1989; Lawton, 1990). For example, the generalist predator *Plagiognathus politus* Uhler (Hemiptera: Miridae) reduced the populations of the purple loosestrife, *Lythrum salicaria* L. (Lythraceae) biological control agent *Galerucella calmariensis* L. (Coleoptera: Chrysomelidae) in North America (Hunt-Joshi *et al.*, 2005). Few studies evaluate the potential for biotic interference from predators prior to release (Malecki *et al.*, 1993; Nechols *et al.*, 1996).

In CBC, much of the research focus is on assessing the non-target risks of an agent to native vegetation and commercial crops, and direct non-target impacts of biological control appear to be rare and predictable (Sheppard *et al.*, 2005). However, indirect impacts of biological control agents on native insect species through shared natural enemies have rarely been studied (Willis and Memmott, 2005). A successfully established biocontrol agent can become an abundant resource for species of natural enemies in the introduced range. Natural enemy populations (e.g. predators, parasitoids or pathogens) that have expanded their host ranges to include the biological control agent may become more abundant (Willis and Memmott, 2005). Indeed, Carvalheiro *et al.* (2008) showed that indigenous species richness and abundance were negatively affected due to parasitoid population buildup on a biological control agent released against *Chrysanthemoides monilifera* ssp. *rotundata* (L.) T. Nord (Asteraceae) in Australia. More tritrophic interaction or food web studies in the context of biological control of invasive plants are needed to avoid or minimize non-target effects, particularly indirect effects, due to the release of biological control agents (Harvey *et al.*, 2010).

20.4 The Role of Plant Competition – Native Plants Gain the Upper Hand

The effectiveness of biological control is almost always assessed by measuring the direct effects of the control agents on the target invasive species (Callaway *et al.*, 1999), even though the aim of most programmes is to reduce the invasive plant population to an economic, social and/or ecologically acceptable threshold, as evidenced by native community recovery. A fundamental justification for using biological control agents to suppress invasive plants is that by weakening the invader, indigenous species may gain a competitive advantage (Callaway *et al.*, 1999).

Top-down mechanisms such as herbivory play a role in determining competitive fitness, as they affect plant growth and fecundity (Crawley, 1989), alter community composition (Olff and Ritchie, 1998), induce chemical and physical defences (see above), as well as nutrient reallocation within individuals (Kessler and Baldwin, 2001), and lead to compensatory plant regrowth (McNaughton, 1983; Maron and Crone, 2006). Thus, the principal effect of biological control agent herbivory is not caused by them eating the invasive plants to extinction, but more often from their feeding modifying the relative competitive abilities of the invasive plant with native plant species. The impact of control agent herbivory will become more noticeable when damaged plants are competing with other plants for resources, due to a gradual reduction of the host plant's vigour (Center and Van, 2001). Therefore, any reduction in the competitive ability of damaged invasive plants may be sufficient to increase their susceptibility to competition from other, ideally native plant communities (Whittaker, 1979). A number of studies have therefore focused on exploiting the interaction between herbivory and plant competition, and their combined effects on plant performance, as a means of assessing the value of biological control agents (e.g. Ang *et al.*, 1994; Van *et al.*, 1998; Story *et al.*, 2000; Coetzee *et al.*, 2005), and demonstrated that by combining strong plant competitors with agents, the control obtained exceeds that exerted by either mechanism alone.

Many of the competition—herbivory studies have been conducted on aquatic systems, where there are a multitude of factors that can affect competitive interactions, including nutrient concentrations of both the sediment and the water column, light intensity, dissolved oxygen, temperature, climate and species' growth forms. Tipping *et al.* (2009) studied competition between two floating macrophytes (*Salvinia minima* Baker (Salviniaceae) and *Spirodela polyrhiza* (L.) Schleid (Araceae)) mediated by nutrient levels and herbivory in Florida, USA. They hypothesized that herbivory would increase the competitive capacity of *S. polyrhiza* by negating the apparent morphological advantage of *S. minima*. Their results did not support this and revealed a more complex dynamic between herbivory, nutrient levels and competition. They found that although *S. minima* may be more dominant in low nutrient environments and more influenced by herbivory, it had no effect on the competitive success of *S. polyrhiza*, which was dominant in eutrophic conditions. They also observed that these plant communities were vulnerable to wind, currents and passage of animals that create temporary gaps of open water, leading to plant assemblage reordering, which could all affect competitive interactions between plants. All these factors influence competitive interactions and highlight the need for a deeper understanding of the dynamics involved in competition (Box 20.3).

20.5 Improving Biological Control of Invasive Plants Through Understanding Multitrophic Antagonistic Interactions

Classical biological control has had a long and largely successful history as a sustainable and low-risk management tool for invasive non-native plant species (Schwarzländer *et al.*, 2018). However, quantitative post-release evaluation studies are still rarely undertaken (Morin *et al.*, 2009), and when they are conducted they tend to be focused on the establishment of the biological control agent on the target invasive species in a bitrophic approach, or at most plant demographic studies (Blossey *et al.*, 2018). We have highlighted just three areas where we believe that an understanding of biotic interactions between an antagonist

Box 20.3. The combination of herbivory and competition reduces the impacts of submerged invasive macrophytes.

Submerged invasive non-native plants play a significant role in modifying macrophyte assemblages, with cascading trophic outcomes that alter available habitats for native flora and fauna, reducing macroinvertebrate biodiversity and limiting light accessibility to lower levels of the water column (Stiers *et al.*, 2011). In addition to raising levels of dissolved oxygen, these plants alter nutrient cycling, which can have far-reaching consequences, ranging from individual species levels to community structuring, thereby altering system functioning and key ecosystem services.

Both abiotic and biotic drivers determine the distribution and spread of these invasive macrophytes, which are further influenced by anthropogenic changes such as eutrophication and the creation of impoundments, which alter substrate and sediment nutrient concentrations, the most critical factors influencing competitive outcomes (Barko *et al.*, 1991; Szoszkiewicz *et al.*, 2006). For example, eutrophication resulting from human activities has been strongly implicated as a major cause of invasion by *Hydrilla verticillata* L.f. Royle (Hydrocharitaceae), and other exotic species in the USA (Van *et al.*, 1998). The confamilial *Lagarosiphon major* Ridl. Moss ex Wager (Hydrocharitaceae) is a significant invader in both Ireland and New Zealand where it successfully outcompetes native macrophytes, including a variety of charophytes, *Myriophyllum* spp., and *Potamogeton* spp. (in Ireland, Caffrey *et al.*, 2011; and New Zealand, Rattray *et al.*, 1994). A number of studies have investigated competitive outcomes between submerged invasive macrophytes and native counterparts under experimental conditions, with the majority indicating that under high sediment and water nutrient availability, the invasive species dominate (e.g. Van *et al.*, 1998; Mony *et al.*, 2007; Martin and Coetzee, 2014).

There are relatively fewer biological control programmes against submerged invasive macrophytes in comparison to their floating counterparts, despite their notoriety as significant invaders. In the USA, *H. verticillata* has been targeted for control with the release of four insect control agents, while surveys for additional agents continue; *Myriophyllum spicatum* L. (Haloragaceae) is controlled by the weevil *Euhrychiopsis lecontei* (Dietz) (Coleoptera: Curculionidae); and while biological control of *Egeria densa* Planch. (Hydrocharitaceae) is needed, no suitable candidate has been found to date. *H. verticillata* has also been targeted for control in South Africa, and although a suitable agent has been approved for release, the agent was shelved because the invasion is limited to one site in the country. However, the leaf-mining fly, *Hydrellia egeriae* Rodrigues(Diptera: Ephydridae) was released against *E. densa* in South Africa, in 2018, a first for control of submerged invasive plants in the country. Another member of the Hydrocharitaceae, *L. major*, is a biological control target in New Zealand and Ireland, while *Cabomba caroliniana* A. Gray (Cabombaceae) is targeted in Australia, but no agents have been released against either to date

Despite the relatively few number of biological control programmes against submerged invasive macrophytes, the sublethal and long-term subtle effects of herbivory have been shown to have significant effects on a plant's response to herbivory, through experiments investigating the combined effect of herbivory and competition from native plants on the invasive plant's competitive ability. The Spitters (1983) model has become a commonly accepted method of determining competitive ability of invasive macrophytes as influenced by biological control. The model uses an addition series, which allows the relative competitive abilities of each plant species in the experiment to be determined using reciprocal-yield models of mean plant dry mass, under control conditions and in the presence of herbivory. This method provides biologically meaningful competition coefficients, which allows for direct comparison with other similar competition studies. Species of *Hydrellia* flies influence the competitive ability and overall biomass of a variety of aquatic plant species. In the USA, *H. verticillata* was shown to be 6.3 times more competitive than the indigenous *Vallisneria americana* Michx. (Hydrocharitaceae) in the absence of natural enemies. However the addition of the leaf-mining fly, *Hydrellia pakistanae* Deonier (Diptera: Ephydridae), a herbivore associated with *H. verticillata*, reduced the plant's competitive ability to nearly equal that of *V. americana* (Van *et al.*, 1998). Similarly, *L. major* was 15.52 times more competitive than *M. spicatum* in the absence of the candidate control agent, *Hydrellia lagarosiphon* Deeming (Diptera: Ephydridae), but this was reduced to only 3.3 times more competitive in the presence of the fly (Martin *et al.*, 2018).

Continued

These examples highlight that while herbivory of submerged invasive macrophytes may appear subtle, the combination of competition from native counterparts has the ability to reduce the vigour of the invasive species, which should have knock-on trophic effects. Restoration efforts should focus on this reduction in vigour by actively planting native competitors as part of biological control programmes, instead of waiting for natural succession, which could allow secondary invasions to occur.

(the biological control agent), the target invasive species and its wider environment will explain the outcomes of a biological control programme, and potentially improve biological control success.

The field of induced plant defences has been widely studied in the agricultural context to enhance plant (crop) resistance to insect, mite and fungal attack (Howe and Schaller, 2008). However, this discipline has not been applied widely to the biological control of invasive non-native plants, and could explain why some agent populations decline after initial establishment. Unfortunately, given the restriction of quarantine pre-release studies, it is not possible to measure induced plant defences *a priori*. However, retrospective analyses of control programmes in which success was far lower than predicted could explain biological control agent population demise, even over a short timescale.

Practitioners in the field of biological control of invasive plants are attempting to precipitate a species cascade when biological control herbivores are released to control invasive plants. However, this type of simple cascade can easily be disrupted if invaded communities harbour opportunistic native parasitoids and predators that adopt the introduced herbivores as their hosts/prey (Paynter *et al.*, 2010). Harvey *et al.* (2010) argue that it is important to link population and community ecology to individual-level variation in the physiology and behaviour of insects across several trophic levels in studies on the biological control of invasive plants. Although the use of quantitative interaction webs could provide useful insights into the study of top-down regulation (biological control agents) of invasive non-native plants (Willis and Memmott, 2005) they are difficult to test empirically. One approach is to apply community module studies that focus on few, but significant, interactions in a manipulative design (Holt and Hochberg, 2001). For example, parallel studies of parasitism rates on introduced biological control agents and native analogues (e.g. congeners) (Minteer *et al.*, 2016).

As the management of invasive species is becoming more urgent, the use of biological control is increasing, leading to a global increase in the release of biological control agents (Winston *et al.*, 2014). Studies into the direct effects these agents have on target invaders, as well as their effect on competitive interactions in a wider ecosystem restoration context are vital. Funk *et al.* (2008) explored restoration of systems through community recovery, and put the concept of biotic resistance, a community's ability to limit the risk of invasion through competitive pressures, into the context of restoration. By understanding the biotic pressures exerted by native species on introduced species in the presence and absence of biological control, and even selecting native species determined by resource-use traits to occupy vulnerable systems, plant invasion control can be improved (Funk and Wolf, 2016). Biotic resistance is of great importance to invasive plant management, raising the possibility of intentionally managing systems to become less vulnerable to the negative impacts of invasive plant establishment by purposely populating systems with native plant species. Trait-based restoration efforts that combine the effects of competition from native plants in the form of active revegetation (Laughlin, 2014) would benefit from reduced invasive plant vigour as a result of biological control in attempts to restore ecosystems to functional states.

Biological control is not necessarily the panacea of invasive non-native plant management, but can be an effective tool in reducing the negative impacts of the invaders (Howarth, 1983). However, public expectation is often high in terms of the level of control anticipated and the timeframe to achieve effective control. A better understanding and integration of the top-down and bottom-up biotic interactions in the biological control context is necessary if the final goal of habitat restoration is to be realized.

Acknowledgements

Research on biological control of invasive plants in South Africa is funded by the Department of Environmental Affairs, Natural Resource Management Programme (previously the Working for Water Programme). The South African Research Chairs Initiative of the Department of Science and Technology and the National Research Foundation of South Africa provided further funding. Any opinion, finding, conclusion or recommendation expressed in this material is that of the authors, and the National Research Foundation does not accept any liability in this regard.

References

Andow, D.A., Ragsdale, D.W. and Nyvall, R.F. (1997) *Biological control in cool temperate regions. Ecological Interactions and Biological Control*. Westview Press, New York.

Ang, B.N., Kok, L.T., Holtzman, G.I. and Wolf, D.D. (1994) Competitive growth of Canada thistle, tall fescue, and Crownvetch in the presence of a thistle Defoliator, *Cassida rubiginosa* Müller (Coleoptera: Chrysomelidae). *Biological Control* 4(3), 277–284. DOI: 10.1006/bcon.1994.1035.

Baars, J.R. and Neser, S. (1999) Past and present initiatives on the biological control of *Lantana camara* (Verbenaceae) in South Africa. *African Entomology Memoir* 1, 21–33.

Baars, J.-R., Urban, A.J. and Hill, M.P. (2003) Biology, host range, and risk assessment supporting release in Africa of *Falconia intermedia* (Heteroptera: Miridae), a new biocontrol agent for *Lantana camara*. *Biological Control* 28(3), 282–292. DOI: 10.1016/S1049-9644(03)00071-9.

Barko, J.W., Gunnison, D. and Carpenter, S.R. (1991) Sediment interactions with submersed macrophyte growth and community dynamics. *Aquatic Botany* 41(1-3), 41–65. DOI: 10.1016/0304-3770(91)90038-7.

Bazzaz, F.A., Chiariello, N.R., Coley, P.D. and Pitelka, L.F. (1987) Allocating resources to reproduction and defense. *BioScience* 37(1), 58–67. DOI: 10.2307/1310178.

Blossey, B., Dávalos, A., Simmons, W. and Ding, J. (2018) A proposal to use plant demographic data to assess potential weed biological control agents impacts on non-target plant populations. *BioControl* 63(3), 461–473. DOI: 10.1007/s10526-018-9886-4.

Blossey, B. and Notzold, R. (1995) Evolution of increased competitive ability in invasive nonindigenous plants: a hypothesis. *The Journal of Ecology* 83(5), 887–889. DOI: 10.2307/2261425.

Byrne, M.J., Coetzee, J., McConnachie, A.J., Parasram, W. and Hill, M.P. (2004) Predicting climate compatibility of biological control agents in their region of introduction. In: Cullen, J.M., Briese, D.T., Kriticos, D.J., Lonsdale, W.M., Morin, L., *et al.* (eds) *Proceedings of the XI International Symposium on Biological Control of Weeds*. CSIRO Entomology, Canberra, Australia, pp. 28–35.

Caffrey, J., Millane, M., Evers, S. and Moran, H. (2011) Management of *Lagarosiphon major* (Ridley) Moss in Lough Corrib—a review. *Biology and Environment: Proceedings of the Royal Irish Academy* 111B, 205–212.

Callaway, R.M., DeLuca, T.H. and Belliveau, W.M. (1999) Biological-Control herbivores may increase competitive ability of the noxious weed *Centaurea maculosa*. *Ecology* 80(4), 1196–1201. DOI: 10.2307/177067.

Carvalheiro, L.G., Buckley, Y.M., Ventim, R., Fowler, S.V. and Memmott, J. (2008) Apparent competition can compromise the safety of highly specific biocontrol agents. *Ecology Letters* 11(7), 690–700. DOI: 10.1111/j.1461-0248.2008.01184.x.

Center, T.D. and Van, T.K. (2001) Can competition experiments be used to evaluate the potential efficacy of new water hyacinth biological control agents? In: Hill, M.P., Julien, M.H., Hill, M.P., Center, T.D. and Jianqing, D. (eds) *Biological and Integrated Control of Water hyacinth, Eichhornia crassipes. Proceedings of the 2nd Meeting of the Global Working Group for the Biological and Integrated Control of Water hyacinth*. 9–12 October 2000. ACIAR Proceedings No. 102, Beijing, China, pp. 77–81.

Cilliers, C.J. and Neser, S. (1991) Biological control of *Lantana camara* (Verbenaceae) in South Africa. *Agriculture, Ecosystems & Environment* 37(1-3), 57–75. DOI: 10.1016/0167-8809(91)90139-O.

Coetzee, J.A., Center, T.D., Byrne, M.J. and Hill, M.P. (2005) Impact of the biocontrol agent *Eccritotarsus catarinensis*, a sap-feeding mirid, on the competitive performance of waterhyacinth, *Eichhornia crassipes*. *Biological Control* 32(1), 90–96. DOI: 10.1016/j.biocontrol.2004.08.001.

Coley, P.D., Bryant, J.P. and Chapin, F.S. (1985) Resource availability and plant antiherbivore defense. *Science* 230(4728), 895–899. DOI: 10.1126/science.230.4728.895.

Crawley, M.J. (1989) Insect herbivores and plant population dynamics. *Annual Review of Entomology* 34(1), 531–562. DOI: 10.1146/annurev.en.34.010189.002531.

Cripps, M.G., Hinz, H.L., McKenney, J.L., Price, W.J. and Schwarzländer, M. (2009) No evidence for an 'evolution of increased competitive ability' for the invasive *Lepidium draba*. *Basic and Applied Ecology* 10(2), 103–112. DOI: 10.1016/j.baae.2008.03.001.

Day, M.D., Broughton, S. and Hannan-Jones, M.A. (2003) Current distribution and status of *Lantana camara* and its biological control agents in Australia, with recommendations for further biocontrol introductions into other countries. *Biocontrol News and Information* 24, 63–76.

Denslow, J.S. and D'Antonio, C.M. (2005) After biocontrol: assessing indirect effects of insect releases. *Biological Control* 35(3), 307–318. DOI: 10.1016/j.biocontrol.2005.02.008.

Doorduin, L.J. and Vrieling, K. (2011) A review of the phytochemical support for the shifting defence hypothesis. *Phytochemistry Reviews* 10(1), 99–106. DOI: 10.1007/s11101-010-9195-8.

Endara, M.-J. and Coley, P.D. (2011) The resource availability hypothesis revisited: a meta-analysis. *Functional Ecology* 25(2), 389–398. DOI: 10.1111/j.1365-2435.2010.01803.x.

Funk, J.L., Cleland, E.E., Suding, K.N. and Zavaleta, E.S. (2008) Restoration through reassembly: plant traits and invasion resistance. *Trends in Ecology & Evolution* 23(12), 695–703. DOI: 10.1016/j.tree.2008.07.013.

Funk, J.L. and Wolf, A.A. (2016) Testing the trait-based community framework: do functional traits predict competitive outcomes? *Ecology* 97(9), 2206–2211. DOI: 10.1002/ecy.1484.

Goeden, R.D. (1988) A capsule history of biological control of weeds. *Biocontrol News and Information* 9, 55–61.

Handley, R.J., Steinger, T., Treier, U.A. and Müller-Schärer, H. (2008) Testing the evolution of increased competitive ability (EICA) hypothesis in a novel framework. *Ecology* 89(2), 407–417. DOI: 10.1890/07-0160.1.

Harmon, J.P. and Andow, D.A. (2004) Indirect effects between shared prey: predictions for biological control. *Biocontrol* 49, 605–626.

Harvey, J.A., Bukovinszky, T. and van der Putten, W.H. (2010) Interactions between invasive plants and insect herbivores: a plea for a multitrophic perspective. *Biological Conservation* 143(10), 2251–2259. DOI: 10.1016/j.biocon.2010.03.004.

Heshula, L.U.P. and Hill, M.P. (2011) The effect of *Lantana camara* leaf quality on the performance of *Falconia intermedia*. *BioControl* 56(6), 925–933. DOI: 10.1007/s10526-011-9359-5.

Heshula, L.U.P. and Hill, M.P. (2012) Effect of previous feeding by *Falconia intermedia* (Hemiptera: Miridae) on subsequent feeding activity on the invasive shrub *Lantana camara* (Verbenaceae). *Biocontrol Science and Technology* 22(6), 671–684. DOI: 10.1080/09583157.2012.677806.

Heshula, L.U.P. and Hill, M.P. (2014) The Effect of Sap-Sucking by *Falconia intermedia* (Hemiptera: Miridae) on the Emission of Volatile Organic Compounds from the Leaves of *Lantana camara* Varieties. *African Entomology* 22(1), 210–213. DOI: 10.4001/003.022.0118.

Hill, M.P. and Hulley, P.E. (1995) Host-Range extension by native parasitoids to weed biocontrol agents introduced to South Africa. *Biological Control* 5(2), 297–302. DOI: 10.1006/bcon.1995.1037.

Holt, R.D. and Hochberg, M.E. (2001) Indirect interactions, community modules and biological control: a theoretical perspective. In: Wajnberg, E., Scott, J.K. and Quimby, P.C. (eds) *Evaluation of Indirect Ecological Effects of Biological Control*. CABI, Wallingford, pp. 13–37.

Howarth, F.G. (1983) Classical biocontrol: panacea or Pandora's box. *Proceedings, Hawaiian Entomological Society* 24, 239–244.

Howe, G.A. and Schaller, A. (2008) Direct defences in plants and their induction by wounding and insect herbivores. In: Schaller, A. (ed.) *Induced Plant Resistance to Herbivory. Springer, Dordrecht*, pp. 7–29.

Hunt-Joshi, T.R., Root, R.B. and Blossey, B. (2005) Disruption of weed biological control by an opportunistic mirid predator. *Ecological Applications* 15(3), 861–870. DOI: 10.1890/04-1336.

Joshi, J. and Vrieling, K. (2005) The enemy release and EICA hypothesis revisited: incorporating the fundamental difference between specialist and generalist herbivores. *Ecology Letters* 8(7), 704–714. DOI: 10.1111/j.1461-0248.2005.00769.x.

Karban, R. and Myers, J.H. (1989) Induced plant responses to herbivory. *Annual Review of Ecology and Systematics* 20(1), 331–348. DOI: 10.1146/annurev.es.20.110189.001555.

Keane, R. and Crawley, M.J. (2002) Exotic plant invasions and the enemy release hypothesis. *Trends in Ecology & Evolution* 17(4), 164–170. DOI: 10.1016/S0169-5347(02)02499-0.
Kessler, A. and Baldwin, I.T. (2001) Defensive function of herbivore-induced plant volatile emissions in nature. *Science* 291(5511), 2141–2144. DOI: 10.1126/science.291.5511.2141.
Kessler, A. and Halitschke, R. (2007) Specificity and complexity: the impact of herbivore-induced plant responses on arthropod community structure. *Current Opinion in Plant Biology* 10(4), 409–414. DOI: 10.1016/j.pbi.2007.06.001.
Lake, E.C., Tipping, P.W., Rayamajhi, M.B., Pratt, P.D., Dray Jr, F.A. *et al.* (2017) The role of Melaleuca control in Everglades restoration: accomplishments and future plans. In: Van Driesche, R.G. and Reardon, R.C. (eds) *Supressing Over-Abundant Invasive Plants and Insects in Natural Areas by Use of Their Specialized Natural Enemies. Forest Health Technology Enterprise Team, Morgantown*. USA, West Virginia, pp. 53–58.
Laughlin, D.C. (2014) Applying trait-based models to achieve functional targets for theory-driven ecological restoration. *Ecology Letters* 17(7), 771–784. DOI: 10.1111/ele.12288.
Lawton, J.H. (1990) Biological control of plants: a review of generalisations, rules and principles using insects as agents. *FRI Bulletin* 155, 3–17.
Li, Y.-P., Feng, Y.-L. and Barclay, G. (2012) No evidence for evolutionarily decreased tolerance and increased fitness in invasive *Chromolaena odorata*: implications for invasiveness and biological control. *Plant Ecology* 213(7), 1157–1166. DOI: 10.1007/s11258-012-0073-x.
Malecki, R.A., Blossey, B., Hight, S.D., Schroeder, D., Kok, L.T. *et al.* (1993) Biological control of purple Loosestrife. *BioScience* 43(10), 680–686. DOI: 10.2307/1312339.
Maron, J.L. and Crone, E. (2006) Herbivory: effects on plant abundance, distribution and population growth. *Proceedings of the Royal Society B: Biological Sciences* 273(1601), 2575–2584. DOI: 10.1098/rspb.2006.3587.
Martin, G.D. and Coetzee, J.A. (2014) Competition between two aquatic macrophytes, *Lagarosiphon major* (Ridley) Moss (Hydrocharitaceae) and *Myriophyllum spicatum* Linnaeus (Haloragaceae) as influenced by substrate sediment and nutrients. *Aquatic Botany* 114, 1–11. DOI: 10.1016/j.aquabot.2013.11.001.
Martin, G.D., Coetzee, J.A. and Compton, S. (2018) Plant–herbivore–parasitoid interactions in an experimental freshwater tritrophic system: higher trophic levels modify competitive interactions between invasive macrophytes. *Hydrobiologia* 817(1), 307–318. DOI: 10.1007/s10750-017-3417-7.
McNaughton, S.J. (1983) Compensatory plant growth as a response to herbivory. *Oikos* 40(3), 329–336. DOI: 10.2307/3544305.
Minteer, C.R., Tipping, P.W., Knowles, B.K., Valmonte, R.J., Foley, J.R. *et al.* (2016) Utilization of an introduced weed biological control agent, *Megamelus scutellaris* (Hemiptera: Delphacidae), by a native parasitoid. *Florida Entomologist* 99(3), 576–577. DOI: 10.1653/024.099.0343.
Mony, C., Koschnick, T.J., Haller, W.T. and Muller, S. (2007) Competition between two invasive Hydrocharitaceae (*Hydrilla verticillata* (L.f.) (Royle) and *Egeria densa* (Planch)) as influenced by sediment fertility and season. *Aquatic Botany* 86(3), 236–242. DOI: 10.1016/j.aquabot.2006.11.007.
Morin, L., Reid, A.M., Sims-Chilton, N.M., Buckley, Y.M., Dhileepan, K. *et al.* (2009) Review of approaches to evaluate the effectiveness of weed biological control agents. *Biological Control* 51(1), 1–15. DOI: 10.1016/j.biocontrol.2009.05.017.
Müller-Schärer, H. and Schaffner, U. (2008) Classical biological control: exploiting enemy escape to manage plant invasions. *Biological Invasions* 10(6), 859–874. DOI: 10.1007/s10530-008-9238-x.
Müller-Schärer, H., Schaffner, U. and Steinger, T. (2004) Evolution in invasive plants: implications for biological control. *Trends in Ecology & Evolution* 19(8), 417–422. DOI: 10.1016/j.tree.2004.05.010.
Nechols, J.R., Obrycki, J.J., Tauber, C.A. and Tauber, M.J. (1996) Potential impact of native natural enemies on Galerucella spp. (Coleoptera: Chrysomelidae) imported for biological control of purple Loosestrife: a field evaluation. *Biological Control* 7(1), 60–66. DOI: 10.1006/bcon.1996.0065.
Ngxande-Koza, S.W., Heshula, L.U.P. and Hill, M.P. (2017) Changes in chemical composition of essential oils from leaves of different *Lantana camara* L. (verbenaceae) varieties after feeding by the introduced biological control agent, *Falconia intermedia* Distant (Hemiptera: Miridae). *African Entomology* 25(2), 462–473. DOI: 10.4001/003.025.0462.
Olff, H. and Ritchie, M.E. (1998) Effects of herbivores on grassland plant diversity. *Trends in Ecology & Evolution* 13(7), 261–265. DOI: 10.1016/S0169-5347(98)01364-0.

Paterson, I.D. and Zachariades, C. (2013) ISSRs indicate that *Chromolaena odorata* invading southern Africa originates in Jamaica or Cuba. *Biological Control* 66(2), 132–139. DOI: 10.1016/j.biocontrol.2013.04.005.

Paynter, Q., Fowler, S.V., Hugh Gourlay, A., Groenteman, R., Peterson, P.G. *et al.* (2010) Predicting parasitoid accumulation on biological control agents of weeds. *Journal of Applied Ecology* 47(3), 575–582. DOI: 10.1111/j.1365-2664.2010.01810.x.

Rattray, M.R., Howard-Williams, C. and Brown, J.M.A. (1994) Rates of early growth of propagules of *Lagarosiphon major* and *Myriophyllum triphyllum* in lakes of differing trophic status. *New Zealand Journal of Marine and Freshwater Research* 28(3), 235–241. DOI: 10.1080/00288330.1994.9516611.

Schulz, A.N., Lucardi, R.D. and Marsico, T.D. (2019) Successful invasions and failed biocontrol: the role of antagonistic species interactions. *BioScience* 69(9), 711–724. DOI: 10.1093/biosci/biz075.

Schwarzländer, M., Hinz, H.L., Winston, R.L. and Day, M.D. (2018) Biological control of weeds: an analysis of introductions, rates of establishment and estimates of success, worldwide. *BioControl* 63(3), 319–331. DOI: 10.1007/s10526-018-9890-8.

Sheppard, A.W., van Klinken, R.D. and Heard, T.A. (2005) Scientific advances in the analysis of direct risks of weed biological control agents to nontarget plants. *Biological Control* 35(3), 215–226. DOI: 10.1016/j.biocontrol.2005.05.010.

Siemann, E. and Rogers, W.E. (2001) Genetic differences in growth of an invasive tree species. *Ecology Letters* 4(6), 514–518. DOI: 10.1046/j.1461-0248.2001.00274.x.

Simelane, D.O. (2005) Biological control of *Lantana camara* in South Africa: targeting a different niche with a root-feeding agent, *Longitarsus* sp. *BioControl* 50(2), 375–387. DOI: 10.1007/s10526-004-1113-9.

Spitters, C.J.T. (1983) An alternative approach to the analysis of mixed cropping experiments. 1. estimation of competition effects. *Netherlands Journal of Agricultural Science* 31, 1–11.

Stamp, N. (2003) Out of the quagmire of plant defense hypotheses. *The Quarterly Review of Biology* 78(1), 23–55. DOI: 10.1086/367580.

Stastny, M., Schaffner, U.R.S. and Elle, E. (2005) Do vigour of introduced populations and escape from specialist herbivores contribute to invasiveness? *Journal of Ecology* 93(1), 27–37. DOI: 10.1111/j.1365-2745.2004.00962.x.

Stiers, I., Crohain, N., Josens, G. and Triest, L. (2011) Impact of three aquatic invasive species on native plants and macroinvertebrates in temperate ponds. *Biological Invasions* 13(12), 2715–2726. DOI: 10.1007/s10530-011-9942-9.

Story, J.M., Good, W.R., White, L.J. and Smith, L. (2000) Effects of the interaction of the biocontrol agent *Agapeta zoegana* L. (Lepidoptera: Cochylidae) and grass competition on spotted Knapweed. *Biological Control* 17(2), 182–190. DOI: 10.1006/bcon.1999.0787.

Szoszkiewicz, K., Ferreira, T., Korte, T., Baattrup-Pedersen, A., Davy-Bowker, J. *et al.* (2006) European river plant communities: the importance of organic pollution and the usefulness of existing macrophyte metrics. *Hydrobiologia* 566(1), 211–234. DOI: 10.1007/s10750-006-0094-3.

Thompson, J.N. (2005) Coevolution: the geographic mosaic of coevolutionary arms races. *Current Biology* 15(24), R992–R994. DOI: 10.1016/j.cub.2005.11.046.

Tipping, P.W., Bauer, L., Martin, M.R. and Center, T.D. (2009) Competition between *Salvinia minima* and *Spirodela polyrhiza* mediated by nutrient levels and herbivory. *Aquatic botany* 90(3), 231–234. DOI: 10.1016/j.aquabot.2008.10.004.

Urban, A.J., Simelane, D.O., Retief, E., Heystek, F., Williams, H.E. *et al.* (2011) The invasive *Lantana camara* L.' hybrid complex (Verbenaceae): A review of research into its identity and biological control in South Africa. *African Entomology* 19(2), 315–348. DOI: 10.4001/003.019.0225.

Van, T.K., Wheeler, G.S. and Center, T.D. (1998) Competitive interactions between Hydrilla (*Hydrilla verticillata*) and Vallisneria (*Vallisneria americana*) as influenced by insect herbivory. *Biological Control* 11(3), 185–192. DOI: 10.1006/bcon.1997.0594.

Van Kleunen, M. and Schmid, B. (2003) No evidence for an evolutionary increased competitive ability in an invasive plant. *Ecology* 84(11), 2816–2823. DOI: 10.1890/02-0494.

Wang, Y., Huang, W., Siemann, E., Zou, J., Wheeler, G.S. *et al.* (2011) Lower resistance and higher tolerance of invasive host plants: biocontrol agents reach high densities but exert weak control. *Ecological Applications* 21(3), 729–738. DOI: 10.1890/09-2406.1.

Whittaker, J.B. (1979) Invertebrate grazing, competition and plant dynamics. In: Anderson, R.M., Turner, B.D. and Taylor, L.R. (eds) *Population Dynamics*. Blackwell Scientific Publications, Oxford, pp. 207–222.

Willis, A.J. and Memmott, J. (2005) The potential for indirect effects between a weed, one of its biocontrol agents and native herbivores: a food web approach. *Biological Control* 35(3), 299–306. DOI: 10.1016/j.biocontrol.2005.07.013.

Willis, A.J., Memmott, J. and Forrester, R.I. (2000) Is there evidence for the post-invasion evolution of increased size among invasive plant species? *Ecology Letters* 3(4), 275–283. DOI: 10.1046/j.1461-0248.2000.00149.x.

Winston, R.L., Schwarzländer, M., Hinz, H.L., Day, M.D., Cock, M.J. *et al.* (2014) *Biological Control of Weeds: A World Catalogue of Agents and their Target Weeds*, 5th edn. USDA Forest Service, Forest Health Technology Enterprise Team. FHTET-2014-XX, Morgantown West Virginia, USA.

Zachariades, C. (2018) *Biological control of invasive alien plants in South Africa: A list of all insects, mites and pathogens considered as biological control agents from 1913-2018*. Available at: http://www.arc.agric.za/arc-ppri/Documents/Table1-NaturalEnemiesAll.pdf

21 Restoration of Pollination Interactions in Communities Invaded by Non-native Plants

Christopher N. Kaiser-Bunbury[1]* and Benno I. Simmons[1,2]

[1]Centre for Ecology and Conservation, College of Life and Environmental Sciences, University of Exeter, Penryn, Cornwall, UK; [2]Department of Animal and Plant Sciences, University of Sheffield, Sheffield, UK

Abstract

Invasive plant species degrade and homogenize ecosystems worldwide, thereby altering ecosystem processes and function. To mitigate and reverse the impact of invasive plants on pollination, a key ecosystem function, conservation scientists and practitioners restore ecological communities and study the impact of such management interventions on plant–pollinator communities. Here, we describe opportunities and challenges associated with restoring pollination interactions as part of a holistic ecosystem-based restoration approach. We introduce a few general concepts in restoration ecology, and outline best planning and evaluation practices of restoring pollination interactions on the community level. Planning involves the selection of suitable plant species to support diverse pollinator communities, which includes considerations of the benefits and disadvantages of using native vs exotic, and bridge and framework plant species for restoration. We emphasize the central role of scientific- and community-level approaches for the planning phase of pollination restoration. For evaluation purposes, we argue that appropriate network indicators have the advantage of detecting changes in species behaviour with consequences for ecosystem processes and functions before these changes show up in altered species communities. Suitable network metrics may include interaction diversity and evenness, and network measures that describe the distribution of species, such as network and species-level specialization, modularity and motifs. Finally, we discuss the usefulness of the network approach in evaluating the benefits of restoration interventions for pollination interactions, and propose that applied network ecologists take a central role in transferring theory into practice.

21.1 Plant Invasion and Restoration

Invasive species negatively impact species and ecosystems in diverse ways, for instance by dismantling species interaction networks (Hansen *et al.*, 2018) or causing declines or extinctions of native species (Bakker and Wilson, 2004; Roberts *et al.*, 2013). To mitigate the negative effects of invasive species and help species and ecosystem functions adapt to changing species assemblages, scientists and practitioners employ a diverse set of conservation and ecological restoration approaches. For example, ecological restoration modifies ecosystems with the purpose of providing a suitable habitat for target species (Miller and Hobbs, 2007). Until recently, most restoration interventions directed their efforts towards re-establishing species assemblages that resemble original native communities, and measured the interventions' success on these criteria (Ruiz-Jaen *et al.*, 2005), yet the effects on associated non-target species or processes have attracted less attention. Non-target species can serve important functional roles in

*Corresponding author: c.kaiser-bunbury@exeter.ac.uk

DOI: 10.1079/9781789242171.0021

Fig. 21.1. Pollinators visit a range of flowers of plant species endemic to the islands of Mauritius and Mahé (Seychelles). Pollinators include (a) wasps; (b) sunbirds; (c) flies; (d) bees; and (e) day geckos. Invaded, unrestored (f) and invasive-free, restored (g) heathland communities in the uplands of Mauritius. The restored site represents a 6 ha plot which has been regularly weeded since 1996. Note the open vegetation in the restored community compared to the overgrown, dense thicket of several invasive plant species, primarily guava *Psidium cattleianum*, in the unrestored community.

restored ecosystems, however, and failure to recognize the implicit ecosystem-level interactions and processes contributed by non-target species may cause failure of restoration efforts (Ehrenfeld, 2000). Pollination is one such key ecosystem function (Fig. 21.1), and most wild plants and crops rely heavily on pollination services for reproduction, but pollinators are rarely targets of restoration efforts (Williams, 2011) despite the current pollinator crisis (Potts *et al.*, 2010).

Here, we draw attention to the importance of restoring pollination function as part of a more holistic ecosystem-based restoration approach. As ecosystems are characterized by networks of interactions between organisms (McCann, 2007) we mainly discuss opportunities and challenges associated with restoring pollination interactions in plant-invaded communities through a network approach (see also Chapter 18, this volume). We briefly introduce a few general concepts in restoration ecology, before describing best planning and evaluation practices of restoring pollination interactions at the community level. Finally, we discuss the usefulness of the network approach in evaluating the benefits of restoration interventions for pollination interactions.

21.2 Reversing the Impact of Invasive Plants on Pollination: Habitat Restoration

Restoration practitioners are tasked with repairing highly intricate living systems in which all parts and players and their interactions follow a myriad of ecological and physical rules that are difficult to predict (see e.g. Russell and Kaiser-Bunbury, 2019). To overcome the restoration challenge, effective and meticulous planning and implementation is required. Practitioners tend to use three sources to guide the planning process of a restoration project: i) previous knowledge and experience sourced from practitioners' handbooks accumulated over generations of intelligent tinkering; ii) theoretical and empirical scientific evidence; and iii) ecological reference sites that resemble target conditions. Knowledge-based and reference restoration can be highly valuable and effective approaches, but by increasing the level of complexity and aiming to restore ecological process and ecosystem functions, evidence-based scientific approaches may be in a position to provide additional context-specific guidance during the planning and implementation phase.

Despite the ambiguous and context-specific relationships between invasive non-native plants (IAPs) and native plant–pollinator interactions, IAPs seem to negatively affect native plant communities globally (Kueffer *et al.*, 2010; Vilà *et al.*, 2011; Roberts *et al.*, 2013). IAPs alter plant community composition, which inevitably has detrimental consequences for the associated fauna (van Hengstum *et al.*, 2013; Hansen *et al.*, 2018), with potentially negative cascading effects for ecosystem functioning (Kaiser-Bunbury *et al.*, 2011). To reverse these impacts, conservation managers and practitioners actively control IAPs and reforest degraded plant communities, and thereby aim to restore native plant community structure.

Ecological restoration encompasses several forms and approaches. For the purpose of this chapter, we generally refer to 'habitat restoration' to distinguish our considerations from those of the sub-disciplines of 'rehabilitation' and 'revegetation'. We specifically introduce the term 'habitat' here to set apart the focus on plant species (which is the case in revegetation projects) from interventions that aim to address both the physical and biological conditions of a natural environment that provide suitable living conditions for plants and their associated fauna. In the context of plant–pollinator communities, any interventions that aim to restore pollination interactions should not only consider suitable food plants for pollinators but also aspects that support sustainable populations such as shelter, protection and nesting sites.

Habitat rehabilitation usually describes the reparation of natural processes, productivity and ecosystem services in populated or substantially-altered landscapes (Clewell and Aronson, 2013). One distinguishing feature of rehabilitation is the lack of resemblance between the newly-created and the 'original' ecosystem. Rehabilitated ecosystems are also meant to be self-sustaining, but this state can be achieved by substituting species that occurred in the past with novel non-native species. The focus on productivity and ecosystem services, the initially degraded state of the ecosystem and the high degree of technical human intervention together suggest that nativeness, original community composition and ecosystem functioning have a lower priority in the planning, implementation and evaluation of rehabilitation projects (Clewell and Aronson, 2013).

The opposite is the case in restoration projects. Here, ecological communities, which have been degraded in the past or suffer from ongoing multiple anthropogenic threats, are actively managed to mitigate the impact of the threats and/or re-establish, as far as possible, the original structure and dynamics of the communities. Until recently, most ecological restoration projects focused primarily on directly attained attributes such as species composition and community structure of restored vs reference communities. There is an increasing awareness among scientists and practitioners, however, that it is critical to consider ecological processes, e.g. biotic interactions and ecosystem functions in the planning and evaluation of restoration projects

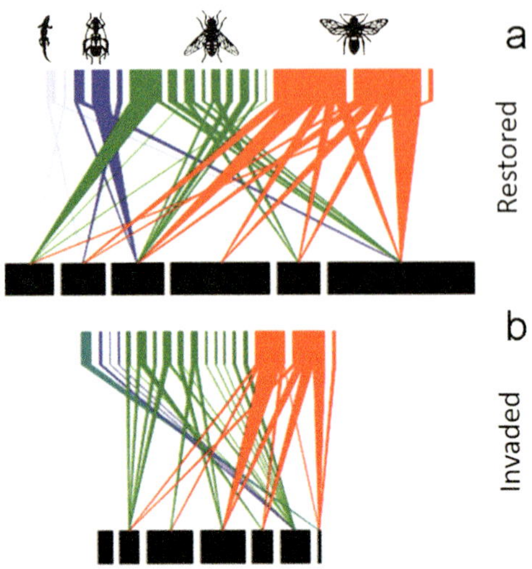

Fig. 21.2. Pollination webs of two Seychelles plant–pollinator communities. Upper bars indicate pollinator species, and lower bars indicate plant species. Colour of pollinator species represent different group of taxa. Both networks were collected in the same month but at different sites. The restored community (a) had greater interaction diversity, interaction evenness and lower network specialization level (H_2') than the invaded community (b) (taken from Kaiser-Bunbury *et al.*, 2017).

(Memmott, 2009). Both processes and functions are considered indirectly-attained attributes of an ecosystem, which develop over time as a result of ecological restoration, but without being specifically targeted during an intervention (Clewell and Aronson, 2013). Including these attributes into project planning and evaluation would require elevating them from a non-target 'side product' of restoration to a direct aim of restoration in need of quantification and evaluation. Considering biotic interactions and functions at the community level, however, is inherently complex, which can deter practitioners from using appropriate tools to guide restoration management and implementation strategies.

It was proposed that interaction networks (e.g. plant–pollinator networks; Fig. 21.2), which are increasingly being used in applied conservation, for example to describe and monitor the impact of invasive species on native interactions, can equally be used to survey and evaluate progress of restoration projects that indirectly aim to repair ecological processes and ecosystem functioning (Proulx *et al.*, 2005; Tylianakis *et al.*, 2010; Elle *et al.*, 2012; Kaiser-Bunbury and Blüthgen, 2015). Network analysis can provide structural, functional and biodiversity indicators that can be used as planning and evaluation tools to monitor, for example, the impact of restoration interventions on ecosystem services and functions, assess management efficacy and sustainability, and reduce detrimental non-target effects (Kaiser-Bunbury and Blüthgen, 2015). The overall aim of incorporating a network approach into applied conservation is to facilitate the switch from a species- to a community-focused approach without compromising the evidence base of any management decisions and maximising the effective protection of biodiversity as a whole (Harvey *et al.*, 2017).

21.3 Restoration of Pollination Interactions

To restore pollinator habitat to re-establish plant–pollinator interactions and increase pollination function in native or pollination services in agricultural landscapes, is one area where science can assist with decision making on restoration strategies. For example, it is critical to identify the main limiting factors to pollination in the impaired community, whether these factors affect all plant and pollinator species similarly, and how these limitations can be addressed. Limitations can be complex and the underlying cause can be multi-dimensional and on different hierarchical levels. Recording, measuring and evaluating each one of these ultimate limitations would exceed, in most cases, the time and financial scope of restoration projects. Therefore, more proximate indicators of the state of pollination interactions in a community are required, and these can include direct ecological measures such as plant reproductive performance, fruit crop (total fruit output in a community or of an agricultural crop) and pollinator species richness, or indicators that describe the structure and functioning of the plant–pollinator communities. Network metrics have been used primarily to describe structural similarities among plant–pollinator communities (see above; e.g. Vázquez *et al.*, 2009), but as network structure can increasingly be used to inform about the ecological state of communities, network metrics can be used as indicators of restoration success (Tylianakis *et al.*, 2010; Elle *et al.*, 2012; Kaiser-Bunbury and Blüthgen, 2015). Using appropriate network indicators, rather than taxa-based indicators, has the key advantage of detecting changes in ecosystem processes and functions before these changes show up in altered species communities, and thus are the focus of this chapter (Harvey *et al.*, 2017). By recording processes and functions expressed in network metrics, practitioners increase the evaluation sensitivity, which enables faster and more resource and cost-effective adaptive management. Below we elaborate on the implications of this approach for the planning and evaluation of projects that target the restoration of pollination interactions.

21.4 Planning of Habitat Restoration Projects with the Aim to Restore Pollination Interactions

Restoring pollination interactions in plant-invaded communities aims to create a plant community which supports a diverse set of co-evolved pollinator species, which is likely to result in robust pollination function. By removing individuals of IAPs it may be possible to reverse negative effects on native plant–pollinator interactions and favour native vegetation (though see Valdovinos, 2019). While completely eradicating an invasive plant species from a site might be logistically and financially feasible, it may only be necessary to reduce IAP abundance to a low level. This is because the impacts of IAP tend to increase with their abundance and, when they are rare, they may have little effect on plant–pollinator communities (Aizen *et al.*, 2008; Kaiser-Bunbury *et al.*, 2011; Timóteo *et al.*, 2018).

Once IAPs are removed it may be necessary to replace them with plant species that provide suitable habitat for all pollinator species in the community over time. A key consideration here is the need to provide a variety of nectar and pollen sources from a morphologically-diverse flower community. This can be achieved either by attempting to restore the historical plant community composition of the ecosystem (i.e. habitat restoration), or by re-establishing key ecological functions and services with a set of plant species new to the ecosystem (i.e. habitat rehabilitation). We argue that restoration guided by the historic state of a community is the most risk-averse approach when aiming to re-establish plant–pollinator communities, as the reintroduction of original native vegetation is most likely to result in the establishment of effective pollination interactions, particularly for specialist species, and reduces the risk of unintended negative consequences of engineering novel ecosystems with multiple non-native species (Russell and Kaiser-Bunbury, 2019). Native plants are likely to rebuild their past associations with extant pollinators because of their common evolutionary and ecological trajectories prior to disturbance ('ecological memory'; see Thompson *et al.*, 2001), which is critical for the reorganisation of ecosystems after disturbance (Bengtsson *et al.*, 2003). While novel plant species can integrate into plant–pollinator communities (Tylianakis and Morris, 2017; Timóteo *et al.*, 2018; Vizentin-Bugoni *et al.*, 2019), planting historical vegetation accounts for context dependence, which increases the chances of successfully restoring the plant community and associated processes and functions(Funk *et al.*, 2008). In cases where a native plant species has gone extinct or the altered environment is now considered unsuitable for native plant populations to thrive, the introduction of novel species can be an alternative (Kueffer and Kaiser-Bunbury, 2014). Novel ecosystems may be important when it comes to maintaining some level of biodiversity under scenarios of extreme environmental change (Hobbs *et al.*, 2009), but where native and non-native plants and their pollinators coexist in the wider landscape, native pollination interactions may be disrupted and decline as a result of rehabilitation efforts with non-native plant species (Traveset and Richardson, 2006; Dixon, 2009).

When selecting specific plant species, restoration ecologists broadly define two classes: framework plants and bridging plants. Framework plants are those with traits (e.g. morphology) that mean they act as large pollen or nectar sources, supporting a wide range of pollinator species (Dixon, 2009; Menz *et al.*, 2011). There are studies that suggest that the pollinators attracted to framework plants (also known as 'magnet species') may, in turn, pollinate other rarer plant species (Laverty, 1992; Johnson *et al.*, 2003; Ghazoul, 2006; Molina-Montenegro *et al.*, 2008; Tepedino *et al.*, 2008; Lázaro *et al.*, 2009; Menz *et al.*, 2011), however the evidence base for this assertion is mixed. On the one hand the framework plants could attract pollinators that then visit other plants, but it is also possible that the framework plants could compete with other plants by attracting pollinators away, resulting in indirect competition between plant species for pollinators (Morales and Traveset, 2009; Menz *et al.*, 2011). As indicated above, it is currently hard to predict whether indirect interactions between plants will be facilitative or competitive, though the local-scale spatial configuration of plants may play an important role (Charlebois and Sargent, 2017). While framework plants can provide substantial floral resources for pollinators, bridging plants are intended to cover temporal gaps in resource provision. Bridging plants are particularly important for pollinator species with specific requirements that require food throughout the year (Ackerman, 1983; Bradshaw *et al.*, 2007; Dixon, 2009; Menz *et al.*, 2011). This may be the case for communities with longer growing seasons, where pollinator species can be active all year (Ackerman, 1983; Menz *et al.*, 2011). Framework and bridging plants were used to restore pollination services in the Central Valley of California (Menz *et al.*, 2011). First, data on plant–bee interactions were used to identify framework plants: those plants visited by the most pollinator individuals. Second, the flowering times of these plants was compared with the phenology of the insect species to identify any periods without suitable flowers. Finally, Menz *et al.* (2011) added bridging species that flowered during these gaps and that were deemed suitable for the insect community at the site. This restoration-based planting regime led to increases in pollinator diversity relative to control sites (Menz *et al.*, 2011). The role of plant species derived from network analysis can also provide useful information for restoration. Maia *et al.* (2019) used published pollination networks to characterize plant species as central or peripheral. To quantify the role of central vs peripheral plant species for pollinators in the community, the authors introduced central and peripheral species to experimental plots and found that the central plants produced better restoration outcomes than the peripheral species. This study suggests high potential for identifying the suitability of plant species for use in restoration programmes based on their role in existing published ecological network datasets. Finally, a recent study described a phylogeny-based selection approach to identify the most suitable plant species for bee pollination restoration in Brazil (Campbell *et al.*, 2019). By prioritizing plant species most beneficial for pollinators (i.e. being generalized and attracting a relatively distinct set of pollinator species with high conservation value) from a small number of phylogenetically-diverse plant families, restored communities generate more stable plant–pollinator communities that are functionally robust (Campbell *et al.*, 2019).

The cases described above exemplify two prominent points: first, the outcome of any habitat restoration effort that aims to restore pollination function is critically dependent on the plant selection process. Secondly, scientific and community-level approaches should be considered central to the planning phase of a restoration project. When these two points are considered in tandem, the probability of successfully restoring ecosystem processes, such as pollination, through structural changes to the plant community is greatest (Tonietto and Larkin, 2018). But how do we know whether the restoration interventions implemented have been successful or require further adaptations? To evaluate restoration progress we advocate, in agreement with an increasing community of theoretical and field ecologists (e.g. Forup *et al.*, 2008; Kaiser-Bunbury *et al.*, 2010; Tylianakis *et al.*, 2010; Elle *et al.*, 2012; Montoya *et al.*, 2012), the use of network analysis, which can describe important structural and functional properties of plant–pollinator networks that indicate robustness and resilience of the restored communities.

21.5 Evaluating Restoration Success with Plant–Pollinator Networks

As described in Chapter 18 (this volume) on the use of ecological networks in plant invasion research, plant–pollinator networks are not structured randomly; rather, interactions are organized in such a way that a variety of structural properties emerge (Delmas *et al.*, 2019). These statistical structural properties can be measured and used to monitor the state of a plant–pollinator network undergoing restoration action to mitigate the detrimental impacts of invasive plant species (Fig. 21.2). This approach has been used to evaluate the success of restoration actions. For example, Forup and Memmott (2005) used network tools to assess the similarity of old and restored hay meadows, finding very few differences, indicating the success of the restoration action. Here we outline how network metrics can be used to quantify the degree to which restoration is increasing the robustness and resilience of plant–pollinator communities.

A first review by Tylianakis *et al.* (2010) argued that nestedness is an important indicator of the robustness of a plant–pollinator network to species extinction. In nested networks specialists tend to interact with generalists and vice versa. Thus if a specialist pollinator becomes extinct, its plant partner will tend to be a generalist, meaning that other pollinators will visit it. Similarly, if a specialist plant becomes extinct, its pollinators will tend to be generalists that visit many other plants and so will still have food sources to forage on. This redundancy in interactions, if controlled for overlap mismatches in temporal species (Rasmussen *et al.*, 2013), is a key reason why nestedness is generally seen as a desirable property of plant–pollinator networks. Its stabilizing role in terms of population dynamics has also been confirmed by other studies using simulation approaches (Bastolla *et al.*, 2009; Thébault and Fontaine, 2010; Rohr *et al.*, 2014; Song *et al.*, 2017). Nestedness, however, is not always considered a desirable restoration target. While stability may initially appear to be a desirable property because it buffers communities against perturbations, it can also make it harder for systems to leave undesirable states because a community may always try to return to the undesirable state. For example, it may be harder to restore invaded communities to their previous state if the invaded community represents a domain of attraction that is difficult to leave (Aizen *et al.*, 2008; Tylianakis *et al.*, 2010). Further, nestedness can emerge from asymmetric abundance distributions between plants and pollinators and thus may primarily reflect rarity rather than a network metric to be maximized in its own right (Blüthgen *et al.*, 2008; Blüthgen, 2010). Rarity is an important conservation consideration, and rare species should be targets of conservation effort, which could be managed through traditional rare species management rather than requiring a network approach (Blüthgen *et al.*, 2008; Dorado *et al.*, 2011; Kaiser-Bunbury and Blüthgen, 2015).

Connectance is another possible metric for measuring restoration progress. Connectance is the number of links in a network, divided by the total number of possible links in the network. Thus, connectance is simply a measure of interaction richness, normalized relative to the size of the network. Some studies have found that greater interaction richness can increase the rate of ecosystem processes (e.g. greater herbivore suppression in a simple bipartite system; Snyder *et al.*, 2006) and stabilize their delivery in the presence of environmental threats, largely due to increased partner redundancy afforded by a larger number of links (Tylianakis *et al.*, 2010). There is debate, however, as to whether connectance increases or decreases network robustness or stability overall, with studies finding conflicting results (Dunne *et al.*, 2002; James *et al.*, 2012; Sentis *et al.*, 2014; Vieira and Almeida-Neto, 2015). Additionally, connectance is strongly influenced by sampling completeness, and thus may underestimate the true interaction richness of a community. Studies examining the sampling completeness of mutualistic networks have found that many interactions are missed by sampling, up to as many as 50% (Chacoff *et al.*, 2012; Devoto *et al.*, 2012; Traveset *et al.*, 2015). Thus, again, care needs to be employed when considering connectance as a restoration target.

In light of the problems with nestedness and connectance, Kaiser-Bunbury and Blüthgen (2015) recommended a different set of metrics that are less contested in the literature, are conceptually similar to well-known diversity indicators used widely in conservation and which are easy to calculate. This list is not exhaustive – indeed, there is a wealth of different network metrics available in the literature – but it contains metrics with a strong foundation of empirical and theoretical support. Newer metrics may eventually achieve similar levels of support, however more extensive testing is necessary before they can be fully recommended.

The first recommended metric is interaction diversity, defined as the exponent of the Shannon Entropy across all links. Embedded in this metric is information about the richness of interactions and the evenness of their weights, and higher interaction diversity can imply community stability (Kaiser-Bunbury and Blüthgen, 2015; Kaiser-Bunbury *et al.*, 2017). Interaction evenness is a similar metric that simply captures the homogeneity of link weights across all interactions in the networks. Higher levels of interaction evenness implies that links have more similar weights. An invaded community might have low interaction evenness as, for example, invasive plants may monopolize high proportions of interaction frequency (Kaiser-Bunbury and Blüthgen, 2015; Kaiser-Bunbury *et al.*, 2017). Thus as a community is restored it is expected that interaction evenness (and diversity) will increase. It is important to pair interaction evenness monitoring with standard species presence surveys because extinction of rare species (with weak interaction strengths) will increase interaction evenness. An increase in evenness due to successful restoration must therefore be distinguished from that due to rare species extinction (Kaiser-Bunbury and Blüthgen, 2015; Kaiser-Bunbury *et al.*, 2017). Additionally, given that many pristine communities have uneven distributions of interaction weights, and invasive plants could displace native species, it is possible that evenness could remain unchanged in the presence of invasion. Further research is needed to characterize the evenness–invasion relationship, but while no change in evenness may not indicate the absence of invasive species, a decrease is likely to indicate the presence of invasive species.

The whole-network specialization metric H_2' and its species-level counterpart d' are important measures of the distribution of interactions that capture the extent to which species interactions are partitioned; that is, the extent to which species tend to interact only with particular partners (Blüthgen *et al.*, 2006). More specifically, both measures quantify the extent to which species interactions deviate from that expected if species randomly sampled their partners (in proportion to their abundance). If H_2' or d' are high, species are highly specialized on particular resources. Conversely, low values indicate high amounts of redundancy (and thus may indicate a more robust community), as species are not reliant on a small number of resources (Kaiser-Bunbury and Blüthgen, 2015). Usefully, H_2' and d' are robust to variations in sampling effort, and so can be used to compare networks or species over time or across space.

Modularity can be an important tool for measuring restoration success. Modularity is the extent to which a network is divided into groups of species that interact more with each other than with other groups of species (Olesen *et al.*, 2007). While this can be calculated as a whole-network property (that is often associated with stability because perturbations can be contained within modules and do not propagate through the network), modularity measures are also useful at the species level (Thébault and Fontaine, 2010). By measuring how species connect to other species within and between modules, species acting as 'module hubs', 'connectors' and 'network hubs' can be identified. Module hubs are species which are linked to many other species within a module, connectors are species that interact with species in other modules, and network hubs are species with both these properties (Olesen *et al.*, 2007). Identifying species that play these important roles can help guide management, and assess progress to restoring native species to their original roles before invasion (Albrecht *et al.*, 2014, see Chapter 18, this volume).

A final metric that was not discussed by Kaiser-Bunbury and Blüthgen (2015), but does warrant consideration, is motifs. Motifs are subnetworks depicting interactions between a small number of species. Together, these subnetworks form the network as a whole, and so can be thought of as the 'building blocks' of networks (Milo *et al.*, 2002; Simmons *et al.*, 2019a, b). Motifs used in analyses generally comprise between two and six species. Specifically, for plant–pollinator networks, there is one motif with two species, two motifs which three species, four motifs with four species, ten motifs with five species, and 27 motifs with six species (Simmons *et al.*, 2019a, b). These fixed numbers arise because there is a finite number of ways to arrange interactions between these species, and thus there are also a finite number of different motifs. Analysing networks using a motif approach generally involves decomposing a network into its constituent motifs to enumerate how often each of these motifs occurs in the network. While conceptually more complex than calculating the structural metrics discussed above, motifs have a significant advantage in terms of being able to detect smaller-scale changes in network structure (Simmons *et al.*, 2019a, b). In other words, motifs are more sensitive, and thus could be used to detect early changes resulting from invasive plants that other metrics could not. Additionally, because motifs rely on the same network data used by the traditional structural metrics, yet provide substantially greater detail about network structure, they are highly cost-efficient to survey. In terms of restoration, for example, motifs could be used to uncover changes in network structure that might not be revealed by traditional metrics (Simmons *et al.*, 2019a). By comparing the motif distributions of the restoring system to that in a pristine system, restoration progress can be assessed in very high detail.

Overall therefore, interaction diversity and interaction evenness can be used to assess the diversity of interactions, while H_2', d' modularity and motif measures can be used to understand the distribution of interactions among species. These metrics represent a 'toolkit' for conservation practitioners to assess the success of restoration effort, though it is important to note that all these metrics require careful interpretation depending on levels of sampling effort and site and species characteristics. Nestedness and connectance, too, can provide useful information, but given some mixed conclusions in the literature extra caution is advised with interpreting these in the context of practical conservation action. Lastly, our conclusions and recommendation may require adaptations as multi-layer networks, which account for multiple interaction types, increase our understanding of direct and indirect effects of restoration across trophic levels, ecosystem functions, and temporal and spatial scales (Pilosof *et al.*, 2017; García-Callejas *et al.*, 2018; Hackett *et al.*, 2019).

So far, very few studies have employed these network metrics in an experimental setting to show the causal relationship between network structure and ecological and conservation properties of plant–pollinator communities. However, preliminary evidence is promising. Studies by Forup and Memmott (2005) and Forup *et al.* (2008) successfully used network analysis to evaluate restoration effectiveness by comparing restored sites with unperturbed controls. Ultimately, then, while further studies are needed on the interactions between restoration action and network structure, network analysis remains one of the most promising and powerful tools for monitoring how conservation affects species, their interactions and the communities of which they form part.

21.6 Synthesis

Aiming to restore pollination interactions in degraded plant communities is part of a larger shift in conservation focus towards preserving ecosystem processes and functions (Harvey *et al.*, 2017). A range of empirical comparative and experimental studies on focal species or across entire plant communities provide evidence of the impact of invasive species on native plant–pollinator interactions. Moreover, an increasing number of theoretical and synthesis studies that explore a network approach in an applied context are shedding light on the potential uses of interaction networks in biodiversity conservation (e.g. Tylianakis and Morris, 2017; Hui and Richardson, 2018; Delmas *et al.*, 2019, see also Chapter 18, this volume). Taken together, these insights can provide a blueprint for best restoration practices of pollinator interactions (Kaiser-Bunbury and Blüthgen, 2015; Kaiser-Bunbury, 2019). The path towards a full integration of a network approach in the conservation of ecosystem functions, however, is still littered with pitfalls and complications, despite recent advances. For example, the approach advocated here does not directly take into account restoration techniques that also address aspects of pollinator population sustainability other than food, e.g. shelter, protection and nesting sites. This is not because shelter and nesting sites are considered less important than providing food resources in determining pollinator survival, but they are substantially more difficult to study in the wild, and consequently there is very little direct evidence about how modified anthropogenic landscapes affect pollinator populations per se (Baldock *et al.*, 2019). Similarly, practitioners may be reluctant to embrace a network approach for planning and evaluation purposes of restoration projects given the inherent complexity and the seemingly larger initial investment (but see Hegland *et al.*, 2010). To overcome this reluctance it is on us applied network ecologists to engage with practitioners who are keen to employ a network approach in restoration and provide input into translating the conservation aims into appropriate network indicators and guide the analysis and interpretation of the metrics to evaluate restoration success (Memmott, 2009; Kaiser-Bunbury and Blüthgen, 2015; Pocock *et al.*, 2016). Ideally, this process will be partly automated in the future (Bohan *et al.*, 2017), when plant–pollinator networks are constructed with molecular techniques based on pollen loads carried by pollinators (e.g. Bell *et al.*, 2017) or chemical 'fingerprints' traceable on flowers. So far, many of these approaches and techniques have been conceptualized as stand-alone basic or applied components, yet we believe that their full potential could be released by bringing the components together to address biodiversity loss through ecological restoration.

References

Ackerman, J.D. (1983) Diversity and Seasonality of Male Euglossine Bees (Hymenoptera: Apidae) in Central Panama. *Ecology* 64(2), 274–283. DOI: 10.2307/1937075.

Aizen, M.A., Morales, C.L. and Morales, J.M. (2008) Invasive mutualists Erode native pollination webs. *PLoS Biology* 6(2), e31. DOI: 10.1371/journal.pbio.0060031.

Albrecht, M., Padrón, B., Bartomeus, I. and Traveset, A. (2014) Consequences of plant invasions on compartmentalization and species' roles in plant–pollinator networks. *Proceedings of the Royal Society B: Biological Sciences* 281(1788), 20140773. DOI: 10.1098/rspb.2014.0773.

Bakker, J.D. and Wilson, S.D. (2004) Using ecological restoration to constrain biological invasion. *Journal of Applied Ecology* 41(6), 1058–1064. DOI: 10.1111/j.0021-8901.2004.00962.x.

Baldock, K.C.R., Goddard, M.A., Hicks, D.M., Kunin, W.E., Mitschunas, N. *et al.* (2019) A systems approach reveals urban pollinator hotspots and conservation opportunities. *Nature Ecology & Evolution* 3(3), 363–373. DOI: 10.1038/s41559-018-0769-y.

Bastolla, U., Fortuna, M.A., Pascual-García, A., Ferrera, A., Luque, B. *et al.* (2009) The architecture of mutualistic networks minimizes competition and increases biodiversity. *Nature* 458(7241), 1018–1020. DOI: 10.1038/nature07950.

Bell, K.L., Fowler, J., Burgess, K.S., Dobbs, E.K., Gruenewald, D. *et al.* (2017) Applying pollen DNA metabarcoding to the study of plant–pollinator interactions. *Applications in Plant Sciences* 5(6), 1600124. DOI: 10.3732/apps.1600124.

Bengtsson, J., Angelstam, P., Elmqvist, T., Emanuelsson, U., Folke, C. *et al.* (2003) Reserves, resilience and dynamic landscapes. *AMBIO: A Journal of the Human Environment* 32(6), 389–396. DOI: 10.1579/0044-7447-32.6.389.

Blüthgen, N. (2010) Why network analysis is often disconnected from community ecology: a critique and an ecologist's guide. *Basic and Applied Ecology* 11(3), 185–195. DOI: 10.1016/j.baae.2010.01.001.

Blüthgen, N., Menzel, F. and Blüthgen, N. (2006) Measuring specialization in species interaction networks. *BMC Ecology* 6(1), 9. DOI: 10.1186/1472-6785-6-9.

Blüthgen, N., Fründ, J., Vázquez, D.P. and Menzel, F. (2008) What do interaction network metrics tell us about specialization and biological traits. *Ecology* 89(12), 3387–3399. DOI: 10.1890/07-2121.1.

Bohan, D.A., Vacher, C., Tamaddoni-Nezhad, A., Raybould, A., Dumbrell, A.J. *et al.* (2017) Next-generation global biomonitoring: large-scale, automated reconstruction of ecological networks. *Trends in Ecology & Evolution* 32(7), 477–487. DOI: 10.1016/j.tree.2017.03.001.

Bradshaw, S.D., Phillips, R.D., Tomlinson, S., Holley, R.J., Jennings, S. *et al.* (2007) Ecology of the honey possum, *Tarsipes rostratus*, in Scott National Park, Western Australia. *Australian Mammalogy* 29(1), 25–38. DOI: 10.1071/AM07003.

Campbell, A.J., Gigante Carvalheiro, L., Gastauer, M., Almeida-Neto, M. and Giannini, T.C. (2019) Pollinator restoration in Brazilian ecosystems relies on a small but phylogenetically-diverse set of plant families. *Scientific Reports* 9(1), 17383. DOI: 10.1038/s41598-019-53829-4.
Chacoff, N.P., Vázquez, D.P., Lomáscolo, S.B., Stevani, E.L., Dorado, J. *et al.* (2012) Evaluating sampling completeness in a desert plant-pollinator network. *Journal of Animal Ecology* 81(1), 190–200. DOI: 10.1111/j.1365-2656.2011.01883.x.
Charlebois, J.A. and Sargent, R.D. (2017) No consistent pollinator-mediated impacts of alien plants on natives. *Ecology Letters* 20(11), 1479–1490. DOI: 10.1111/ele.12831.
Clewell, A.F. and Aronson, J. (2013) *Ecological Restoration: Principles, Values, and Structure of an Emerging Profession*. Island Press, Washington, DC, USA.
Delmas, E., Besson, M., Brice, M.-H., Burkle, L.A., Dalla Riva, G.V. *et al.* (2019) Analysing ecological networks of species interactions. *Biological Reviews* 94(1), 16–36. DOI: 10.1111/brv.12433.
Devoto, M., Bailey, S., Craze, P. and Memmott, J. (2012) Understanding and planning ecological restoration of plant-pollinator networks. *Ecology Letters* 15(4), 319–328. DOI: 10.1111/j.1461-0248.2012.01740.x.
Dixon, K.W. (2009) Pollination and restoration. *Science* 325(5940), 571–573. DOI: 10.1126/science.1176295.
Dorado, J., Vázquez, D.P., Stevani, E.L. and Chacoff, N.P. (2011) Rareness and specialization in plant–pollinator networks. *Ecology* 92(1), 19–25. DOI: 10.1890/10-0794.1.
Dunne, J.A., Williams, R.J. and Martinez, N.D. (2002) Network structure and biodiversity loss in food webs: robustness increases with connectance. *Ecology Letters* 5(4), 558–567. DOI: 10.1046/j.1461-0248.2002.00354.x.
Ehrenfeld, J.G. (2000) Defining the limits of restoration: the need for realistic goals. *Restoration Ecology* 8(1), 2–9. DOI: 10.1046/j.1526-100x.2000.80002.x.
Elle, E., Elwell, S.L. and Gielens, G.A. (2012) The use of pollination networks in conservation. *Botany-Botanique* 90, 525–534.
Forup, M.L. and Memmott, J. (2005) The restoration of plant-pollinator interactions in hay meadows. *Restoration Ecology* 13(2), 265–274. DOI: 10.1111/j.1526-100X.2005.00034.x.
Forup, M.L., Henson, K.S.E., Craze, P.G. and Memmott, J. (2008) The restoration of ecological interactions: plant-pollinator networks on ancient and restored heathlands. *Journal of Applied Ecology* 45(3), 742–752. DOI: 10.1111/j.1365-2664.2007.01390.x.
Funk, J.L., Cleland, E.E., Suding, K.N. and Zavaleta, E.S. (2008) Restoration through reassembly: plant traits and invasion resistance. *Trends in Ecology & Evolution* 23(12), 695–703. DOI: 10.1016/j.tree.2008.07.013.
García-Callejas, D., Molowny-Horas, R. and Araújo, M.B. (2018) Multiple interactions networks: towards more realistic descriptions of the web of life. *Oikos* 127(1), 5–22. DOI: 10.1111/oik.04428.
Ghazoul, J. (2006) Floral diversity and the facilitation of pollination. *Journal of Ecology* 94(2), 295–304. DOI: 10.1111/j.1365-2745.2006.01098.x.
Hackett, T.D., Sauve, A.M.C., Davies, N., Montoya, D., Tylianakis, J.M. *et al.* (2019) Reshaping our understanding of species' roles in landscape-scale networks. *Ecology Letters* 22(9), 1367–1377. DOI: 10.1111/ele.13292.
Hansen, S., Roets, F., Seymour, C.L., Thébault, E., van Veen, F.J.F. *et al.* (2018) Alien plants have greater impact than habitat fragmentation on native insect flower visitation networks. *Diversity and Distributions* 24(1), 58–68. DOI: 10.1111/ddi.12656.
Harvey, E., Gounand, I., Ward, C.L. and Altermatt, F. (2017) Bridging ecology and conservation: from ecological networks to ecosystem function. *Journal of Applied Ecology* 54(2), 371–379. DOI: 10.1111/1365-2664.12769.
Hegland, S.J., Dunne, J., Nielsen, A. and Memmott, J. (2010) How to monitor ecological communities cost-efficiently: the example of plant–pollinator networks. *Biological Conservation* 143(9), 2092–2101. DOI: 10.1016/j.biocon.2010.05.018.
Hobbs, R.J., Higgs, E. and Harris, J.A. (2009) Novel ecosystems: implications for conservation and restoration. *Trends in Ecology & Evolution* 24(11), 599–605. DOI: 10.1016/j.tree.2009.05.012.
Hui, C. and Richardson, D.M. (2018) How to invade an ecological network. *Trends in Ecology & Evolution*.
James, A., Pitchford, J.W. and Plank, M.J. (2012) Disentangling nestedness from models of ecological complexity. *Nature* 487(7406), 227–230. DOI: 10.1038/nature11214.
Johnson, S.D., Peter, C.I., Nilsson, L.A. and Ågren, J. (2003) Pollination success in a deceptive orchid is enhanced by co-occurring rewarding magnet plants. *Ecology* 84(11), 2919–2927. DOI: 10.1890/02-0471.

Kaiser-Bunbury, C.N. (2019) Restoring plant-pollinator communities: using a network approach to monitor pollination function. In: Veitch, C.R., Clout, M.N., Martin, A.R., Russell, J.C. and West, C.J. (eds) *Island Invasives: Scaling up to Meet the Challenge*. IUCN, Gland, Switzerland.

Kaiser-Bunbury, C.N. and Blüthgen, N. (2015) Integrating network ecology with applied conservation: a synthesis and guide to implementation. *AoB Plants* 7, plv076. DOI: 10.1093/aobpla/plv076.

Kaiser-Bunbury, C.N., Mougal, J., Whittington, A.E., Valentin, T., Gabriel, R. *et al.* (2017) Ecosystem restoration strengthens pollination network resilience and function. *Nature* 542(7640), 223–227. DOI: 10.1038/nature21071.

Kaiser-Bunbury, C.N., Traveset, A. and Hansen, D.M. (2010) Conservation and restoration of plant–animal mutualisms on oceanic islands. *Perspectives in Plant Ecology, Evolution and Systematics* 12(2), 131–143. DOI: 10.1016/j.ppees.2009.10.002.

Kaiser-Bunbury, C.N., Valentin, T., Mougal, J., Matatiken, D. and Ghazoul, J. (2011) The tolerance of island plant-pollinator networks to alien plants. *Journal of Ecology* 99(1), 202–213. DOI: 10.1111/j.1365-2745.2010.01732.x.

Kueffer, C. and Kaiser-Bunbury, C.N. (2014) Reconciling conflicting perspectives for biodiversity conservation in the Anthropocene. *Frontiers in Ecology and the Environment* 12(2), 131–137. DOI: 10.1890/120201.

Kueffer, C., Daehler, C.C., Torres-Santana, C.W., Lavergne, C., Meyer, J.-Y. *et al.* (2010) A global comparison of plant invasions on oceanic islands. *Perspectives in Plant Ecology, Evolution and Systematics* 12(2), 145–161. DOI: 10.1016/j.ppees.2009.06.002.

Laverty, T.M. (1992) Plant interactions for pollinator visits: a test of the magnet species effect. *Oecologia* 89(4), 502–508. DOI: 10.1007/BF00317156.

Lázaro, A., Lundgren, R. and Totland, Ø. (2009) Co-flowering neighbors influence the diversity and identity of pollinator groups visiting plant species. *Oikos* 118(5), 691–702. DOI: 10.1111/j.1600-0706.2008.17168.x.

Maia, K.P., Vaughan, I.P. and Memmott, J. (2019) Plant species roles in pollination networks: an experimental approach. *Oikos* 128(10), 1446–1457. DOI: 10.1111/oik.06183.

McCann, K. (2007) Protecting biostructure. *Nature* 446(7131), 29–29. DOI: 10.1038/446029a.

Memmott, J. (2009) Food webs: a ladder for picking strawberries or a practical tool for practical problems? *Philosophical Transactions of the Royal Society B: Biological Sciences* 364(1524), 1693–1699. DOI: 10.1098/rstb.2008.0255.

Menz, M.H.M., Phillips, R.D., Winfree, R., Kremen, C., Aizen, M.A. *et al.* (2011) Reconnecting plants and pollinators: challenges in the restoration of pollination mutualisms. *Trends in Plant Science* 16(1), 4–12. DOI: 10.1016/j.tplants.2010.09.006.

Miller, J.R. and Hobbs, R.J. (2007) Habitat restoration—do we know what we're doing? *Restoration Ecology* 15(3), 382–390. DOI: 10.1111/j.1526-100X.2007.00234.x.

Milo, R., Shen-Orr, S., Itzkovitz, S., Kashtan, N., Chklovskii, D. (2002) Network motifs: simple building blocks of complex networks. *Science* 298(5594), 824–827. DOI: 10.1126/science.298.5594.824.

Molina-Montenegro, M.A., Badano, E.I. and Cavieres, L.A. (2008) Positive interactions among plant species for pollinator service: assessing the 'magnet species' concept with invasive species. *Oikos* 117(12), 1833–1839. DOI: 10.1111/j.0030-1299.2008.16896.x.

Montoya, D., Rogers, L. and Memmott, J. (2012) Emerging perspectives in the restoration of biodiversity-based ecosystem services. *Trends in Ecology & Evolution* 27(12), 666–672. DOI: 10.1016/j.tree.2012.07.004.

Morales, C.L. and Traveset, A. (2009) A meta-analysis of impacts of alien vs. native plants on pollinator visitation and reproductive success of co-flowering native plants. *Ecology Letters* 12(7), 716–728. DOI: 10.1111/j.1461-0248.2009.01319.x.

Olesen, J.M., Bascompte, J., Dupont, Y.L. and Jordano, P. (2007) The modularity of pollination networks. *Proceedings of the National Academy of Sciences* 104(50), 19891–19896. DOI: 10.1073/pnas.0706375104.

Pilosof, S., Porter, M.A., Pascual, M. and Kéfi, S. (2017) The multilayer nature of ecological networks. *Nature Ecology & Evolution* 1(4), 0101. DOI: 10.1038/s41559-017-0101.

Pocock, M.J.O., Evans, D.M., Fontaine, C., Harvey, M., Julliard, R. *et al.* (2016) The visualisation of ecological networks, and their use as a tool for engagement, advocacy and management. In: Woodward, G. and Bohan, D.A. (eds) *Advances in Ecological Research*. Academic Press.

Potts, S.G., Biesmeijer, J.C., Kremen, C., Neumann, P., Schweiger, O. *et al.* (2010) Global pollinator declines: trends, impacts and drivers. *Trends in Ecology & Evolution* 25(6), 345–353. DOI: 10.1016/j.tree.2010.01.007.

Proulx, S., Promislow, D. and Phillips, P. (2005) Network thinking in ecology and evolution. *Trends in Ecology & Evolution* 20(6), 345–353. DOI: 10.1016/j.tree.2005.04.004.

Rasmussen, C., Dupont, Y.L., Mosbacher, J.B., Trøjelsgaard, K. and Olesen, J.M. (2013) Strong impact of temporal resolution on the structure of an ecological network. *PLoS ONE* 8(12), e81694. DOI: 10.1371/journal.pone.0081694.

Roberts, P.D., Diaz-Soltero, H., Hemming, D.J., Parr, M.J., Wakefield, N.H. *et al.* (2013) What is the evidence that invasive species are a significant contributor to the decline or loss of threatened species? A systematic review map. *Environmental Evidence* 2(1), 5. DOI: 10.1186/2047-2382-2-5.

Rohr, R.P., Saavedra, S. and Bascompte, J. (2014) On the structural stability of mutualistic systems. *Science* 345(6195), 1253497. DOI: 10.1126/science.1253497.

Ruiz-Jaen, M.C. Mitchell Aide, T. and Aide, T.M. (2005) Restoration success: how is it being measured? *Restoration Ecology* 13(3), 569–577. DOI: 10.1111/j.1526-100X.2005.00072.x.

Russell, J.C. and Kaiser-Bunbury, C.N. (2019) Consequences of multispecies introductions on Island ecosystems. *Annual Review of Ecology, Evolution, and Systematics* 50(1), 169–190. DOI: 10.1146/annurev-ecolsys-110218-024942.

Sentis, A., Hemptinne, J.-L. and Brodeur, J. (2014) Towards a mechanistic understanding of temperature and enrichment effects on species interaction strength, omnivory and food-web structure. *Ecology Letters* 17(7), 785–793. DOI: 10.1111/ele.12281.

Simmons, B.I., Cirtwill, A.R., Baker, N.J., Wauchope, H.S., Dicks, L.V. *et al.* (2019a) Motifs in bipartite ecological networks: uncovering indirect interactions. *Oikos* 128(2), 154–170. DOI: 10.1111/oik.05670.

Simmons, B.I., Sweering, M.J.M., Schillinger, M., Dicks, L.V., Sutherland, W.J. *et al.* (2019b) bmotif: A package for motif analyses of bipartite networks. *Methods in Ecology and Evolution* 10(5), 695–701. DOI: 10.1111/2041-210X.13149.

Snyder, W.E., Snyder, G.B., Finke, D.L. and Straub, C.S. (2006) Predator biodiversity strengthens herbivore suppression. *Ecology Letters* 9(7), 789–796. DOI: 10.1111/j.1461-0248.2006.00922.x.

Song, C., Rohr, R.P. and Saavedra, S. (2017) Why are some plant–pollinator networks more nested than others? *Journal of Animal Ecology* 86(6), 1417–1424. DOI: 10.1111/1365-2656.12749.

Tepedino, V.J., Bradley, B.A. and Griswold, T.L. (2008) Might flowers of invasive plants increase native bee carrying capacity? Intimations from capitol reef National Park, Utah. *Natural Areas Journal* 28(1), 44–50. DOI: 10.3375/0885-8608(2008)28[44:MFOIPI]2.0.CO;2.

Thompson, J.N., Reichman, O.J., Morin, P.J., Polis, G.A., Power, M.E. *et al.* (2001) Frontiers of ecology. *BioScience* 51(1), 15–24. DOI: 10.1641/0006-3568(2001)051[0015:FOE]2.0.CO;2.

Thébault, E. and Fontaine, C. (2010) Stability of ecological communities and the architecture of mutualistic and trophic networks. *Science* 329(5993), 853–856. DOI: 10.1126/science.1188321.

Timóteo, S., O'Connor, C.J., López-Núñez, F.A., Costa, J.M., Gouveia, A.C. *et al.* (2018) Pollination networks from natural and anthropogenic-novel communities show high structural similarity. *Oecologia* 188(4), 1155–1165. DOI: 10.1007/s00442-018-4281-5.

Tonietto, R.K. and Larkin, D.J. (2018) Habitat restoration benefits wild bees: a meta-analysis. *Journal of Applied Ecology* 55(2), 582–590. DOI: 10.1111/1365-2664.13012.

Traveset, A., Olesen, J.M., Nogales, M., Vargas, P., Jaramillo, P. *et al.* (2015) Bird–flower visitation networks in the Galápagos unveil a widespread interaction release. *Nature Communications* 6(1). DOI: 10.1038/ncomms7376.

Traveset, A. and Richardson, D. (2006) Biological invasions as disruptors of plant reproductive mutualisms. *Trends in Ecology & Evolution* 21(4), 208–216. DOI: 10.1016/j.tree.2006.01.006.

Tylianakis, J.M. and Morris, R.J. (2017) Ecological networks across environmental gradients. *Annual Review of Ecology, Evolution, and Systematics* 48(1), 25–48. DOI: 10.1146/annurev-ecolsys-110316-022821.

Tylianakis, J.M., Laliberté, E., Nielsen, A. and Bascompte, J. (2010) Conservation of species interaction networks. *Biological Conservation* 143(10), 2270–2279. DOI: 10.1016/j.biocon.2009.12.004.

Valdovinos, F.S. (2019) Mutualistic networks: moving closer to a predictive theory. *Ecology Letters* 22(9), 1517–1534. DOI: 10.1111/ele.13279.

van Hengstum, T., Hooftman, D.A.P., Oostermeijer, J.G.B. and van Tienderen, P.H. (2013) Impact of plant invasions on local arthropod communities: a meta-analysis. *Journal of Ecology* n/a-n/a.

Vieira, M.C. and Almeida-Neto, M. (2015) A simple stochastic model for complex coextinctions in mutualistic networks: robustness decreases with connectance. *Ecology Letters* 18(2), 144–152. DOI: 10.1111/ele.12394.

Vilà, M., Espinar, J.L., Hejda, M., Hulme, P.E., Jarošík, V. *et al.* (2011) Ecological impacts of invasive alien plants: a meta-analysis of their effects on species, communities and ecosystems. *Ecology Letters* 14(7), 702–708. DOI: 10.1111/j.1461-0248.2011.01628.x.

Vázquez, D.P., Blüthgen, N., Cagnolo, L. and Chacoff, N.P. (2009) Uniting pattern and process in plant–animal mutualistic networks: a review. *Annals of Botany* 103(9), 1445–1457. DOI: 10.1093/aob/mcp057.

Vizentin-Bugoni, J., Tarwater, C.E., Foster, J.T., Drake, D.R., Gleditsch, J.M. *et al.* (2019) Structure, spatial dynamics, and stability of novel seed dispersal mutualistic networks in Hawai'i. *Science* 364(6435), 78–82. DOI: 10.1126/science.aau8751.

Williams, N.M. (2011) Restoration of nontarget species: bee communities and pollination function in riparian forests. *Restoration Ecology* 19(4), 450–459. DOI: 10.1111/j.1526-100X.2010.00707.x.

22 Restoration of Seed Dispersal Interactions in Communities Invaded by Non-native Plants

Fernanda Ribeiro da Silva[1]* and Marco Aurélio Pizo[2]

[1]ECOHE - Laboratory of Human Ecology and Ethnobotany, and Post-Graduation Programme in Biology of Fungus, Algae and Plantae, Universidade Federal de Santa Catarina– Florianópolis, Santa Catarina, Brazil; [2]Universidade Estadual de São Paulo, Department of Biodiversity – Rio Claro, São Paulo, Brazil

Abstract

Restoration aims to rebuild not only species but also the tangled interactions between species that ensure communities perpetuate by themselves. In tropical forests, restoration of seed dispersal interactions is essential because most plant species depend on animals to spread their seeds. A big challenge in restoring such forests is dealing with invasion by non-native species. Non-native plant species may outcompete and eliminate native species from the community, potentially disrupting or arresting the restoration process. Once established, invasive non-native plants are usually incorporated into the local seed dispersal network, potentially causing loss of biodiversity by competition with native species. This chapter reports on a case study of a 25-year old restored forest invaded by several bird-dispersed plant species. We assessed network metrics at the species level to specifically evaluate the role performed by invasive non-native species in the structure of the bird—seed dispersal network. The removal of invasive non-native plants and the re-establishment of native plant communities should be considered for the restoration of habitats invaded by non-native plants. For this reason, we discuss the impacts of removing such non-native plants and explore the consequences for the structure of the overall network. Because restoration areas are open systems, even after the removal of invasive non-native plant species they can return via seed dispersal. So, both the control and management of invasive non-native species would be more effective if planned with a landscape perspective. We also point out relevant management aspects to avoid the negative influence of invasive non-native plants on the seed dispersal interactions occurring between native plant and bird species in restored tropical forests.

22.1 Non-native Plant Species in Restored Sites

Ecological restoration came about as a strategy to cope with the increasing loss of habitats and biological diversity. Restoration aims to rebuild not only species but also the tangled ecological interactions that ensure the functionality of ecosystems and their perpetuation over time (Memmott *et al.*, 2004). Ecosystems are considered recovered when they are sustainable both functionally and structurally (SER, 2004). In other words, when they bear enough biological resources to persist without further human assistance. Additionally, restored communities should show resilience to environmental stress and perturbations (Memmott *et al.*, 2004).

To restore the functionality and structure of terrestrial ecosystems, seed dispersal is a key process since it promotes the perpetuation of

*Corresponding author: feribs@hotmail.com

DOI: 10.1079/9781789242171.0022

plant populations, including their regeneration following eventual disturbances during the restoration process (Howe, 2014). Therefore, in places where seed dispersal is carried out predominantly by animals, a challenge to restoration practitioners is to attract and maintain seed dispersers in the restoration area to recover their mutualistic interactions with plants (McAlpine *et al.*, 2016). In tropical forests, where most plants depend on animals for seed dispersal (Jordano, 2000; Almeida-Neto *et al.*, 2008), the effort to restore seed dispersal interactions should focus predominantly on birds and mammals that form the bulk of seed dispersal agents in tropical forests (Howe, 2014, 2016).

The presence of seed dispersal agents in restoration areas is, however, a double-edged sword as they can bring to the area both seeds of native and non-native species (Barbosa and Pizo, 2006). Non-native plant species are not always detrimental to restoration purposes though, as they can provide critical habitats for animals, facilitate the arrival of seeds of native species and promote forest regeneration (D'Antonio and Meyerson, 2002; Ewel and Putz, 2004; Berens *et al.*, 2008). We can envisage situations in which non-native plant species may attract dispersers that would otherwise be absent from the site (e.g. by offering fruits with specific nutrient profiles, such as the high calcium concentrations of *Ficus* fruits; O'Brien *et al.*, 1998). As non-native plant species become invasive, however, they may outcompete and eliminate native species from the community, potentially disrupting or arresting the restoration process (Heleno *et al.*, 2013).

Non-native plant species are not rare in restored tropical forests, either because they are planted (intentionally or inadvertently due to identification error) together with native species or because their seeds arrive afterwards (Rother *et al.*, 2019). Surveying 25 sites with planted gallery forests along the Paraná River in South America, Assis (2012) noted that 83 (26%) out of 320 tree and shrub species were non-native, and planted in all but three sites. In this study, the proportion of non-native species declined with the age of the restoration, likely because they were outcompeted by native species. However, a few species, particularly *Clausena excavata* (Rutaceae), persisted with dense and dominating populations, becoming an important invasive non-native plant (see below). As non-native plants may take a long time to become invasive (up to hundreds of years; Simberloff *et al.*, 2013), it is important to note that such plants may thrive as harmless species for long periods before spreading and becoming invasive in restored areas.

The arrival of seeds of non-native plant species to restoration areas depends, among several factors (e.g. seed size, plant population density, the local availability of dispersal agents, landscape configuration), on the proximity of the site to potential sources of non-native seeds. For instance, restored urban forests often have an abundance of nearby non-native plants used as ornamental or for utilitarian purposes, resulting in the frequent arrival of non-native seeds. In four urban forest fragments in southern Brazil, over 37,000 seeds were collected in 40 1 m^2 seed traps for a year, most (55%) of them coming from non-native species. Among these, most were wind-dispersed, but 6.2% were zoochorous seeds from seven plant families (M.A. Pizo, unpublished data). Even though the distances of seed dispersal promoted by most frugivorous vertebrates are predominantly short, not exceeding a few hundred metres (Fleming and Kress, 2013), long-distance seed dispersal may occur, permitting the arrival of non-native seeds in sites far away from their sources and constituting a critical factor to invasion success (Pyšek and Richardson, 2008).

22.2 Non-native Plant Species in Seed Dispersal Networks: Restored Sites

The evaluation of the impacts of invasive non-native upon native species participating in antagonistic interactions (e.g. predation) can be made by evaluating demographic parameters linked to the long-term persistence of native populations (e.g. survival, productivity). Such an approach is not straightforward, with mutualistic interactions involving multiple organisms interacting, either directly or indirectly (Guimarães *et al.*, 2017) in such a way that non-native plant species may not necessarily impact the demographic parameters of native plants with which they putatively compete, but alter the interaction profile at the community level, which may ultimately impact other native species of plants and other organisms (White *et al.*, 2006). To tackle such subtle effects of invasive species and eventually manage plant-invaded restoration areas, the network approach is a useful tool as it generates information on the functional role of invasive non-native species that participate in mutualistic interactions occurring at the community level (Heleno *et al.*, 2010, 2013; Tylianakis *et al.*, 2010; Traveset *et al.*, 2013; Kaiser-Bunbury and Blüthgen, 2015).

In mutualistic networks, species may be represented through a bipartite graph described by points (vertices or nodes) representing species (e.g. plants and animals), while interactions are depicted by lines linking points (see Chapter 18, this volume). To date, few studies have analysed interaction networks to evaluate or improve seed dispersal in restoration areas (de la Peña-Domene *et al.*, 2014; Ribeiro da Silva *et al.*, 2015), less yet have studies investigated plant-invaded restored sites through a seed-dispersal network approach (Heleno *et al.*, 2010; Kaiser-Bunbury *et al.*, 2017).

Some general patterns emerge in mutualistic networks (Bascompte *et al.*, 2003; Memmott *et al.*, 2004; Olesen *et al.*, 2007), and it is valuable to assess how arising interaction patterns involving invasive non-native plant species affect the stability and resilience of mutualistic networks in restoration areas (Aizen *et al.*, 2008; Russo *et al.*, 2014). For example, nestedness is a network metric that reflects how interactions are distributed among species in such a way that highly nested networks characteristically have i) a subgroup of generalist species that interacts with many species in the network, including species with few connections; ii) specialist species (i.e. species with few interactions) that are linked to those generalist species; and iii) rare interactions among specialists (Bascompte *et al.*, 2003). Studies show that invasive non-native species often form part of a super-generalist core of species in both pollination and seed dispersal networks where they interact often and with many animal species (i.e. in network parlance, they have high degree and species strength; Aizen *et al.*, 2008; Heleno *et al.*, 2013; Ribeiro da Silva *et al.*, 2015). We should be cautious when managing such species acting as super-generalist interactors because their management and removal may alter the composition and functioning of plant–seed disperser interactions, ultimately affecting the structure of a network in a negative way (Heleno *et al.*, 2010; Valiente-Banuet *et al.*, 2015). A way to mitigate such impact is the identification of generalist native species that can replace the invasive species, accelerating the restoration process. The choice of native species should take into account the phenology (i.e. time and duration of flowering and fruiting) (Peters *et al.*, 2016) and functional traits (e.g. fruit size, nutrient profile) of the invasive non-native species. A functional approach is useful in mutualistic interactions (Schleuning *et al.*, 2015) as the native species chosen to replace the invasive species should play the same functional role in the ecosystem previously played by it. This is especially important because functional diversity affects not only ecosystem dynamics but also their stability and as such should be taken into account to plan and manage the ecological restoration (Valiente-Banuet *et al.*, 2015).

Modularity is a metric useful to identify functional groups in mutualistic networks. Modularity is characterized by the presence of a subgroup of species that interact more often with each other than with other species within the network. It is known that modularity can minimize perturbation on networks (e.g. by species extinction or deliberate removal) by avoiding its spreading to the whole network (Krause *et al.*, 2003). Through modularity analysis, we can identify species that link these subgroups (connectors or hubs, i.e. keystone species) (Guimerà and Amaral, 2005; Olesen *et al.*, 2007; Mello *et al.*, 2011). Identifying such highly connected species can inform us about the best species to be used in restoration projects for the purpose of maintaining seed dispersers important for several plant species. In plant-invaded restored sites, the identification of invasive non-native species acting as hubs in seed dispersal networks can assist the choice of the best management plan to remove such species so as to impact minimally on the community, with its eventual substitution by functionally equivalent native species.

Relevant to conservation and management of ecosystems is H_2', an index that describes the level of the selectiveness of the entire bipartite network (Blüthgen *et al.*, 2006). The more selective a species is, the higher is H_2', indicating a larger complementary specialization or low niche overlap (Blüthgen and Klein, 2011). Once invasive non-native species are acknowledged to be generalist species, we expect that their presence in the interaction network will decrease H_2'.

22.3 Invasive Non-native Plant Species in the Seed Dispersal Network of a Restored Site: A Case Study

In this section we aim to explore the role of invasive non-native species in a seed dispersal network occurring at a 25-year-old restored forest in the Atlantic Forest, a highly fragmented hotspot of biodiversity in Brazil (Ribeiro *et al.*, 2009). The restored site is in Iracemápolis, São Paulo state. The restoration started by planting a set of 140 tree species, including non-native and native species from the regional flora, in a 50 hectare belt surrounding a dam. Non-native species bearing fleshy fruits were planted to attract frugivores (Siqueira, 2002). A combination of focal observations, seed sampling from the faeces of mist-netted birds and transects in pre-defined trails was used to build a weighted seed dispersal network between frugivorous birds and plants (see Ribeiro da Silva *et al.*, 2015 for more details). We did phenological observations monthly for 16 months to access the production of bird-dispersed fruits. Fruit production was sampled by direct counting of all ripe fruits in such zoochoric plant species previously marked in the trials mentioned above. We assessed network metrics at the species level (i.e. species strength and centrality) to specifically evaluate the role performed by invasive non-native species in the structure of the bird—seed dispersal network. We also simulated the removal of invasive non-native plant species to explore its consequences to the structure (i.e. specialization, nestedness, modularity) of the overall network. We produced 1000 simulated networks with the same connectance, species richness and proportional marginal sums (vaznull model, Vázquez *et al.*, 2007) to observed and simulated networks to calculate the significance level (P-value) to all network indexes. Also, we tested differences among indexes of observed and simulated networks using Monte Carlo procedures in R (Muylaert *et al.*, 2016). More specifically, we asked whether there are differences between invasive non-native and native plant species in the frequency of interactions. Plant species might attract several species of seed dispersers depending on their intrinsic traits (Jordano *et al.*, 2006; Schupp *et al.*, 2010). For instance, plant species with small fruits can be dispersed by a larger group of frugivorous birds than large-seeded species, as small seed sizes do not constraint feeding even by the smallest frugivores. Non-native plant species were planted aiming to attract frugivores to the restoration area. These normally have small fruits (Rodrigues *et al.*, 1992) (diameter: 10.3 ± 7.7 mm, $n = 19$ for this data set), so we expected a high frequency of frugivory interactions involving such non-native plants compared to native species, some of them with larger fruits, that normally have a lower number of interacting species (Vidal *et al.*, 2013). From this we asked:

1. How are non-native species being integrated into the network? As found in other studies that evaluated the role of non-native species in the structure of pollination and seed dispersal networks, we expected these species as part of a generalist core of species with a high frequency of interactions (Aizen *et al.*, 2008; Heleno *et al.*, 2013; Traveset *et al.*, 2013), having on average more connections than native species (Traveset and Richardson, 2014), therefore with a high species strength (SS). The SS of a plant species is the result of the sum of the dependence of every bird species interacting with that plant species and can be used to assess the total effect of the species in the community (Vázquez *et al.*, 2005; Bascompte *et al.*, 2006, Tylianakis *et al.*, 2010). Because invasive non-native fruits are available for a long time (Fig. 22.1), we expected a relatively high dependence of bird species on invasive non-native plant species denoted by a high SS. Furthermore, because species making part of the generalist core are likely interact with a high number of species (Martín González *et al.*, 2010; Gómez and Perfectti, 2012), we expected that non-native species would also show high centrality. Two centrality indexes often used in ecological networks are closeness centrality (CC) and

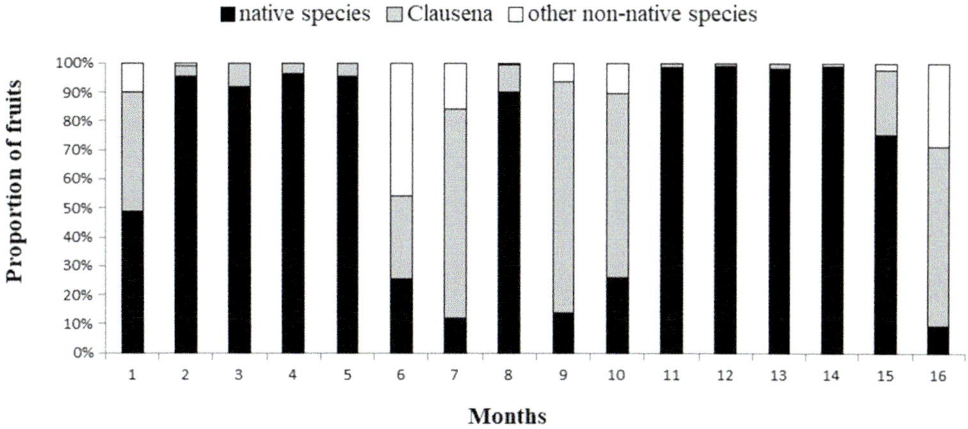

Fig. 22.1. Monthly proportion of fruits of *Clausena excavata*, the most abundant invasive non-native species, other non-native species, and all native species in the 25-year-old restored forest in Brazil.

betweenness centrality (BC). CC is based on the sum of the direct or indirect interactions (i.e. shortest distances) between one species and all other species in the network, and indicates how quickly a perturbation might spread to the overall network. BC is the number of shortest paths between two nodes that pass through a node of interest. Species with high BC values are recognized as connectors, being important to network cohesion. Both metrics can be used to study the role and impact of species loss as well as to identify species hub (i.e. species that have much more interactions when compared with other species of the same network). They are also expected to be important to the organization and function of ecological networks (Martín González *et al.*, 2010; Delmas *et al.*, 2019).

2. What are the consequences of the removal of invasive non-native plant species for both the function and structure of the network? We would expect various effects of such removal (Traveset and Richardson, 2014). As invasive species in most cases are part of the generalist core of species, and the scanty evidence available so far show that recently restored sites have low specialization (Ribeiro da Silva *et al.*, 2015), we expected that their removal would not change specialization (H_2') but would change nestedness (Aizen *et al.*, 2008; Bartomeus *et al.*, 2008). Regarding modularity, invasive non-native species could enhance network cohesiveness by acting as module connectors providing a more resistant network to species loss (Traveset *et al.*, 2013). Therefore, we expected that the removal of invasive species would change modularity.

In total, we recorded 460 interactions (i.e. feeding visits of birds to plants or presence of seeds in faeces) involving 19 plant and 28 bird species, 18 (64.3%) of which interacted with non-native plant species (Fig. 22.2), which together accounted for 143 interactions representing 30.9% of total interactions. Seven invasive non-native plant species were integrated into the network (*Clausena excavata* -Rutaceae, *Callicarpa reevesii* – Lamiaceae, *Melia azedarach* – Meliaceae, *Rubus rosifolius* – Rosaceae, *Cordia*

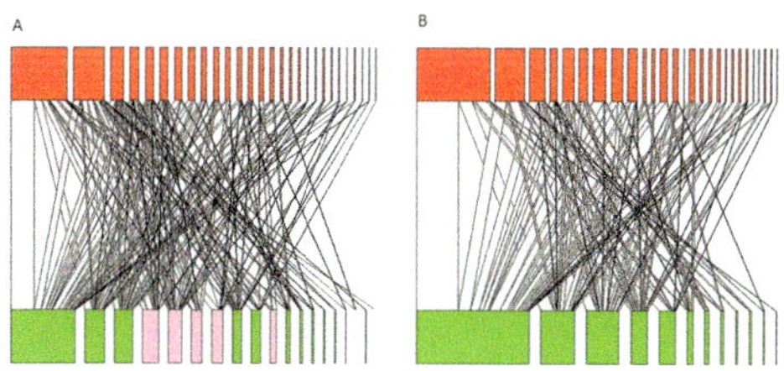

Fig. 22.2. Weighted seed dispersal networks at a 25-year-old restored forest in the Brazilian Atlantic Forest. (A) Complete network; (B) Network without invasive non-native plant species. Green, rose and red rectangles represent, respectively, native and non-native plant species, and birds.

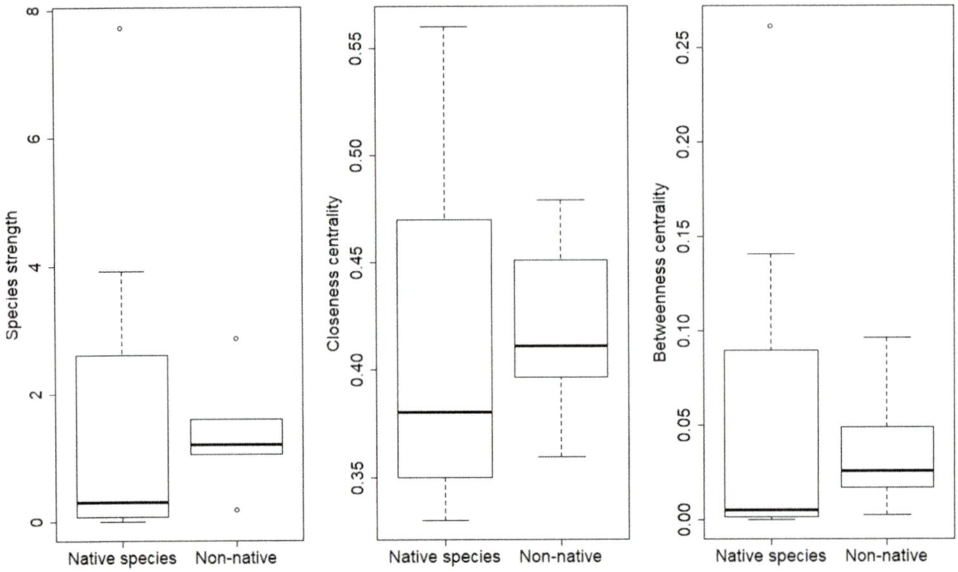

Fig. 22.3. Boxplot of species-level network metrics of native (n = 13) and non-native plant species (n = 6) in a 25-year-old restored forest in Brazil. Horizontal lines represent medians, boxes delimit the 10 and 90% values, and dots highlight outliers.

abyssinica – Boraginaceae, *Syzygium cumini* and *Psidium guajava* – both Myrtaceae). Non-native fruits were available all year round, forming the bulk of fruits available to birds in some months (Fig. 22.1).

The most visited plants were all native species (*Cecropia pachystachya* 33%, *Citharexylum myrianthum* 10.2% and *Myrsine umbellata*: 9.5% of interactions). *C. excavata*, a non-native plant, was the fourth plant species in number of interactions (8.7% of the interactions with nine bird species), followed by *C. reevesii* (6.9%, 12 bird species), *R. rosifolius* (5.9%) and *M. azedarach* (5.4%), both with six bird species, *C. abyssinica* (3.5%, four bird species), and *S. cumini* (0.43%, two bird species). When we simulated the removal of all non-native plant species, despite 30.9% of the interactions being lost (Fig. 22.1), only one bird species, *Myiarchus ferox*, which fed exclusively on the non-native plants *C. reevesii* and *M. azedarach*, was lost.

Native and non-native species did not differ either in species strength (Mann-Whitney Z(U) = 0.79; p = 0.43) or centrality indexes (Closeness: Z(U) = 0.92, p = 0.36; Betweenness: Z(U) = 0.70, p = 0.48) (Fig. 22.3). However, the variances are clearly different between native and non-native species. The simulated removal of non-native plant species did not change any of the network metrics assessed. Regarding network specialization, no differences were observed between the real and simulated networks (H_2' = 0.29 and 0.28, respectively; $p > 0.05$), as predicted. Networks were also similarly modular before and after the removal of non-native species (M = 0.30 and 0.27, respectively, both $p < 0.001$), with no significant differences between them ($p > 0.05$). Likewise, no difference was detected in the nestedness of real and simulated networks (WNODF = 28.1, p = 0.007 and WNODF = 30.7, p = 0.01, respectively).

This study showed that although invasive non-native plant species are part of a generalist core of interacting species participating in one-third of total interactions in the restoration area, their simulated removal did not affect any of the main metrics describing network topology. Therefore, their removal as a management action in the restoration process would likely not affect the structure of the seed dispersal network. In some cases, however, non-native species could become so tightly integrated that their removal may cause unpredictable impacts on the network, as found by Costa *et al.* (2018)

through a model simulating extinction followed by rewiring of interactions between frugivorous birds and plants in Portugal.

According to the preferential attachment mechanism, when new species come into the network, they interact with the most generalized species, i.e. with those with more interactions (Barabási and Albert, 1999). When a species is lost in a plant—bird seed dispersal network, in this specific case from intentional removal of a generalist invasive non-native plant species, the bird species linked to them will likely rewire their interactions with the most connected plant species in the network. *M. ferox*, the only bird species that interacted exclusively with the fruits of invasive non-native plant species, may interact with other plant species after the removal of invasive species. Candidate native plant species for this are *C. myrianthum* and *M. umbellata* whose fruits *M. ferox* are known to eat in other restoration areas (Ribeiro da Silva *et al.*, 2015).

Two non-native species illustrate the varied ways by which they may affect restoration success. First, *C. excavata* was the fourth plant species in number of interactions, which involved nine bird species. These high generalized interactions might increase the spread of such invasive species by enhancing their competitive advantage over natives (Traveset and Richardson, 2014). Indeed, *C. excavata* is a highly dominant understorey species able to recruit under different light conditions (Vieira and Gandolfi, 2006; Vieira *et al.*, 2010). A vegetation survey in the study site showed that more than half of the individuals sampled (trees and shrubs) were invasive non-native species, mainly *C. excavata* (Andrade, 2018). Although *C. excavata* promotes interactions in the restoration area, it may dominate the community with time, causing a collapse of the seed dispersal network and thus harming the restoration process. Likewise, non-native invasive plant species may negatively affect seed dispersal interactions once they compete with native species by frugivores, thus potentially decreasing the removal of native fruits and could be reflected in a decrease in the recruitment of native species. We often observed frugivores preferentially consuming *C. excavata* fruits instead of those of native species (e.g. *Cordia ecalyculata*) fruiting nearby, which might indicate that *C. excavata* is 'usurping' interactions from native species. Padrón *et al.* (2009) found a similar pattern when investigating pollination networks in two islands in Spain where the community invaded by *Opuntia* spp. had lower visitation rates by insects compared to the non-invaded community.

In contrast to *C. excavata*, *M. azedarach* was considered of low invasive potential (i.e. considering its density and frequency in restored sites; Assis, 2012) and promoted intense seed rain in restoration areas (Vieira and Gandolfi, 2006). Nevertheless, such seed rain may be composed of a great number of seeds of other non-native invasive species (Vieira and Gandolfi, 2006), which may lead to an 'invasional meltdown' effect occurring when invasive species facilitate the establishment or spread of each other, contributing to increased invasibility and accelerating the invasion rates (Traveset and Richardson, 2006, Traveset and Richardson, 2011). Therefore, our recommendation would be to remove both *C. excavata* and *M. azedarach* that can negatively affect the restoration process (Andrade, 2018). Given the high abundance of *C. excavata* in the study site, its removal should be gradual to minimize the ecological impacts on the community (e.g. through changes in vegetation structure).

We recognized the limitation of this single-site study. It is possible that 25 years of restoration is not enough time to evaluate the real impact of invaders on interactions (Simberloff *et al.*, 2013). Long-term empirical replicated studies in tropical forests undergoing restoration are rare and we need to advance with strategies to manage invasive species if required for restoration success. Therefore, our results provide subsidies to think on management actions of invasive species based on a network approach to ensure healthy seed dispersal interactions in tropical restored sites.

22.4 Planning and Managing the Dispersal of Seeds

In general, from the perspective of invasion ecology, areas with a smaller level of anthropogenic disturbance have better resistance to invasion compared to areas severely disturbed by humans (Daehler, 2003). Hence, within a context of restoration where a novel ecosystem is being rebuilt, biological invasions are prone to occur. Once biological invasions are among the main factors responsible for restoration failures (Rodrigues *et al.*, 2009; Brancalion *et al.*, 2013), restoring plant communities that are resistant to invasion by non-native species is a challenge to restoration practitioners (Funk *et al.*, 2008). Detailed and integrated knowledge of the seed dispersal process is essential to the effective understanding and management of biological invasions in restoration areas (Valiente-Banuet *et al.*, 2015; Ribeiro da Silva *et al.*, 2015; Chapter 14, this volume).

First of all, unless for well justified reasons (e.g. a plant offering particular fruit resources that will attract rare or much-needed seed dispersers), planting non-native species should be avoided. Disregarding the problem of low availability of commercial seeds of native species that often afflicts restoration practitioners (Moreira da Silva *et al.*, 2017), it is likely that one will find a native species equivalent to a non-native, potentially invasive, species. By *equivalent* we mean a species that plays a similar functional and structural role in the seed dispersal network. For this, a trait-based approach when selecting plant species for planting is important (Funk *et al.*, 2008; Peters *et al.*, 2016). Which traits to consider is not a simple question because there is a diversity of seed dispersers with different fruit preferences and nutritional demands that eat a variety of plant species with a multitude of traits. Fruit and seed size, and fruit chemistry are certainly traits that should be considered for their potential to influence the frequency of interactions and identity of interacting animal species (Levey *et al.*, 2001; Verdú and Traveset, 2004).

The removal of invasive non-native plants and the re-establishment of native plant communities should be considered for the restoration of habitats invaded by non-native plants (Flory and Clay, 2009). Such removal should ensure that the substitute plant species preferably pertain to the regional flora and be of the same functional group of the removed species so as to be visited by a bird assemblage similar to the invasive non-native plant species (Garcia *et al.*, 2015; Andrade, 2018). It is also important to consider the fruiting phenology and quantity of fruits provided by the invasive non-native plant species to fill any temporal gap in fruit production when they are removed (Peters *et al.*, 2016).

Because restoration areas are open systems, invasive non-native plant species can return via seed dispersal even after their removal. This means that both the control and subsequent management of invasive non-native species would be more effective if a landscape perspective is adopted. Useful questions in this area are: i) Which are the main invasive non-native species present in the landscape, both near and far away from restored sites? ii) What frugivores are likely to carry invasive non-native seeds to restored sites? Answering these questions and monitoring the arrival of potentially invasive non-native plant species to the restoration areas with permanent seed traps may give the restoration practitioner time to act before the invasive plant is integrated into the restoration area in such a way that its removal would be harmful to the structure of the seed dispersal network.

References

Aizen, M.A., Morales, C.L. and Morales, J.M. (2008) Invasive mutualists erode native pollination webs. *PLoS Biology* 6(2), e31. DOI: 10.1371/journal.pbio.0060031.

Almeida-Neto, M., Campassi, F., Galetti, M., Jordano, P. and Oliveira-Filho, A. (2008) Vertebrate dispersal syndromes along the Atlantic forest: broad-scale patterns and macroecological correlates. *Global Ecology and Biogeography* 17(4), 503–513. DOI: 10.1111/j.1466-8238.2008.00386.x.

Andrade, R.C.D. (2018) Avaliação Do Sub-Bosque De Matas Ciliares Em Restauração No Interior Do Estado De São Paulo, Brasil. MSc. thesis. Universidade Estadual de Campinas, Campinas, Brazil.

Assis, G.B. (2012) Avaliação Do Potencial Invasivo De Espécies Não-Nativas Utilizadas Em Plantio De Restauração De Matas Ciliares. MSc. thesis. University of the State of São Paulo, Botucatu, Brazil.

Barabási, A.L. and Albert, R. (1999) Emergence of scaling in random networks. *Science* 286(5439), 509–512. DOI: 10.1126/science.286.5439.509.

Barbosa, K.C. and Pizo, M.A. (2006) Seed rain and seed limitation in a planted gallery forest in Brazil. *Restoration Ecology* 14(4), 504–515. DOI: 10.1111/j.1526-100X.2006.00162.x.

Bartomeus, I., Vilà, M. and Santamaría, L. (2008) Contrasting effects of invasive plants in plant-pollinator networks. *Oecologia* 155(4), 761–770. DOI: 10.1007/s00442-007-0946-1.

Bascompte, J., Jordano, P., Melián, C.J. and Olesen, J.M. (2003) The nested assembly of plant-animal mutualistic networks. *Proceedings of the National Academy of Sciences* 100(16), 9383–9387. DOI: 10.1073/pnas.1633576100.

Bascompte, J., Jordano, P. and Olesen, J.M. (2006) Asymmetric coevolutionary networks facilitate biodiversity maintenance. *Science* 312(5772), 431–433.

Berens, D.G., Farwig, N., Schaab, G. and Böhning-Gaese, K. (2008) Exotic guavas are foci of forest regeneration in Kenyan farmland. *Biotropica* 40, 104–112.

Blüthgen, N. and Klein, A.-M. (2011) Functional complementarity and specialisation: the role of biodiversity in plant–pollinator interactions. *Basic and Applied Ecology* 12(4), 282–291. DOI: 10.1016/j.baae.2010.11.001.

Blüthgen, N., Menzel, F. and Blüthgen, N. (2006) Measuring specialization in species interaction networks. *BMC Ecology* 6, 9. DOI: 10.1186/1472-6785-6-9.

Brancalion, P.H.S., Melo, F.P.L., Tabarelli, M. and Rodrigues, R.R. (2013) Biodiversity persistence in highly human-modified tropical landscapes depends on ecological restoration. *Tropical Conservation Science* 6(6), 705–710. DOI: 10.1177/194008291300600601.

Costa, J.M., Ramos, J.A., da Silva, L.P., Timóteo, S., Andrade, P. *et al.* (2018) Rewiring of experimentally disturbed seed dispersal networks might lead to unexpected network configurations. *Basic and Applied Ecology* 30, 11–22. DOI: 10.1016/j.baae.2018.05.011.

Daehler, C.C. (2003) Performance comparisons of co-occurring native and alien invasive plants: implications for conservation and restoration. *Annual Review of Ecology, Evolution, and Systematics* 34(1), 183–211. DOI: 10.1146/annurev.ecolsys.34.011802.132403.

de la Peña-Domene, M., Martínez-Garza, C., Palmas-Pérez, S., Rivas-Alonso, E. and Howe, H.F. (2014) Roles of birds and bats in early tropical-forest restoration. *PLoS ONE* 9(8), e104656. DOI: 10.1371/journal.pone.0104656.

Delmas, E., Besson, M., Brice, M.-H., Burkle, L.A., Dalla Riva, G.V. *et al.* (2019) Analysing ecological networks of species interactions. *Biological Reviews* 94(1), 16–36. DOI: 10.1111/brv.12433.

D'Antonio, C. and Meyerson, L.A. (2002) Exotic plant species as problems and solutions in ecological restoration: a synthesis. *Restoration Ecology* 10(4), 703–713. DOI: 10.1046/j.1526-100X.2002.01051.x.

Ewel, J.J. and Putz, F.E. (2004) A place for alien species in ecosystem restoration. *Frontiers in Ecology and the Environment* 2(7), 354–360. DOI: 10.1890/1540-9295(2004)002[0354:APFASI]2.0.CO;2.

Fleming, T.H. and Kress, W.J. (2013) *The Ornaments of Life: Coevolution and Conservation in The Tropics*. University of Chicago Press, Chicago.

Flory, S.L. and Clay, K. (2009) Invasive plant removal method determines native plant community responses. *Journal of Applied Ecology* 46(2), 434–442. DOI: 10.1111/j.1365-2664.2009.01610.x.

Funk, J.L., Cleland, E.E., Suding, K.N. and Zavaleta, E.S. (2008) Restoration through reassembly: plant traits and invasion resistance. *Trends in Ecology & Evolution* 23(12), 695–703. DOI: 10.1016/j.tree.2008.07.013.

Garcia, L.C., Cianciaruso, M.V., Ribeiro, D.B., dos Santos, F.A.M. and Rodrigues, R.R. (2015) Flower functional trait responses to restoration time. *Applied Vegetation Science* 18(3), 402–412. DOI: 10.1111/avsc.12163.

Guimarães, P.R., Pires, M.M., Jordano, P., Bascompte, J. and Thompson, J.N. (2017) Indirect effects drive coevolution in mutualistic networks. *Nature* 550(7677), 511–514. DOI: 10.1038/nature24273.

Guimerà, R. and Amaral, L.A.N. (2005) Cartography of complex networks: modules and universal roles. *Journal of Statistical Mechanics: Theory and Experiment* 2005(P02001), P02001. DOI: 10.1088/1742-5468/2005/02/P02001.

Gómez, J.M. and Perfectti, F. (2012) Fitness consequences of centrality in mutualistic individual-based networks. *Proceedings of the Royal Society B: Biological Sciences* 279(1734), 1754–1760. DOI: 10.1098/rspb.2011.2244.

Heleno, R., Lacerda, I., Ramos, J.A. and Memmott, J. (2010) Evaluation of restoration effectiveness: community response to the removal of alien plants. *Ecological Applications* 20(5), 1191–1203. DOI: 10.1890/09-1384.1.

Heleno, R.H., Ramos, J.A. and Memmott, J. (2013) Integration of exotic seeds into an Azorean seed dispersal network. *Biological Invasions* 15(5), 1143–1154. DOI: 10.1007/s10530-012-0357-z.

Howe, H.F. (2014) Diversity storage: implications for tropical conservation and restoration. *Global Ecology and Conservation* 2, 349–358. DOI: 10.1016/j.gecco.2014.10.004.
Howe, H.F. (2016) Making dispersal syndromes and networks useful in tropical conservation and restoration. *Global Ecology and Conservation* 6, 152–178. DOI: 10.1016/j.gecco.2016.03.002.
Jordano, P. (2000) Fruits and frugivory. In: Fenner, M. (ed.) *Seeds: The Ecology of Regeneration in Natural Plant Communities*, 2nd Edition. CABI, Wallingford, UK, pp. 125–166.
Jordano, P., Galetti, M., Pizo, M.A. and Silva, W.R. (2006) Ligando frugivoria e dispersão de sementes biologia da conservação. In: Rocha, C.D.F. and Bergallo, H.D. (eds) *Biologia da Conservação: Essências*. São Carlos, Brasil, SP, pp. 411–436.
Kaiser-Bunbury, C.N. and Blüthgen, N. (2015) Integrating network ecology with applied conservation: a synthesis and guide to implementation. *AoB Plants* 7, plv076. DOI: 10.1093/aobpla/plv076.
Kaiser-Bunbury, C.N., Mougal, J., Whittington, A.E., Valentin, T., Gabriel, R. *et al.* (2017) Ecosystem restoration strengthens pollination network resilience and function. *Nature* 542(7640), 223–227. DOI: 10.1038/nature21071.
Krause, A.E., Frank, K.A., Mason, D.M., Ulanowicz, R.E. and Taylor, W.W. (2003) Compartments revealed in food-web structure. *Nature* 426(6964), 282–285. DOI: 10.1038/nature02115.
Levey, D.J. and Martinez del Rio, C. (2001) It takes guts (and more) to eat fruit: lessons from avian nutritional ecology. *The Auk* 118(4), 819–831. DOI: 10.1093/auk/118.4.819.
Martín González, A.M., Dalsgaard, B. and Olesen, J.M. (2010) Centrality measures and the importance of generalist species in pollination networks. *Ecological Complexity* 7(1), 36–43. DOI: 10.1016/j.ecocom.2009.03.008.
McAlpine, C., Catterall, C.P., Nally, R.M., Lindenmayer, D., Reid, J.L. *et al.* (2016) Integrating plant- and animal-based perspectives for more effective restoration of biodiversity. *Frontiers in Ecology and the Environment* 14(1), 37–45. DOI: 10.1002/16-0108.1.
Mello, M.A.R., Marquitti, F.M.D., Guimarães, P.R., Kalko, E.K.V., Jordano, P. *et al.* (2011) The modularity of seed dispersal: differences in structure and robustness between bat– and bird–fruit networks. *Oecologia* 167(1), 131–140. DOI: 10.1007/s00442-011-1984-2.
Memmott, J., Waser, N.M. and Price, M.V. (2004) Tolerance of pollination networks to species extinctions. *Proceedings of the Royal Society of London. Series B: Biological Sciences* 271(1557), 2605–2611. DOI: 10.1098/rspb.2004.2909.
Moreira da Silva, A.P., Schweizer, D., Rodrigues Marques, H., Cordeiro Teixeira, A.M., Nascente dos Santos, T.V.M. *et al.* (2017) Can current native tree seedling production and infrastructure meet an increasing forest restoration demand in Brazil? *Restoration Ecology* 25(4), 509–515. DOI: 10.1111/rec.12470.
Muylaert, R.L., Dodonov, P. and Mello, M.A.R. (2016) *How To Estimate P-Values Of Network Metrics And Compare Pairs Of Networks Using Monte Carlo Procedures In R*. Ecological Synthesis Lab of the University of São Paulo, Brazil.
Olesen, J.M., Bascompte, J., Dupont, Y.L. and Jordano, P. (2007) The modularity of pollination networks. *Proceedings of the National Academy of Sciences* 104(50), 19891–19896. DOI: 10.1073/pnas.0706375104.
O'Brien, T.G., Kinnaird, M.F., Dierenfeld, E.S., Conklin-Brittain, N.L., Wrangham, R.W. *et al.* (1998) What's so special about figs? *Nature* 392(6677), 668. DOI: 10.1038/33580.
Padrón, B., Traveset, A., Biedenweg, T., Díaz, D., Nogales, M. *et al.* (2009) Impact of alien plant invaders on pollination networks in two archipelagos. *PLoS ONE* 4(7), e6275. DOI: 10.1371/journal.pone.0006275.
Peters, V.E., Carlo, T.A., Mello, M.A.R., Rice, R.A., Tallamy, D.W. *et al.* (2016) Using plant–animal interactions to inform tree selection in tree-based agroecosystems for enhanced biodiversity. *BioScience* 66(12), 1046–1056. DOI: 10.1093/biosci/biw140.
Pyšek, P. and Richardson, D.M. (2008) Traits associated with invasiveness in alien plants: where do we stand? In: Nentwig, W. (ed.) *Biological Invasions*. Springer, Berlin, pp. 97–125.
Ribeiro, M.C., Metzger, J.P., Martensen, A.C., Ponzoni, F.J. and Hirota, M.M. (2009) The Brazilian Atlantic forest: how much is left, and how is the remaining forest distributed? implications for conservation. *Biological Conservation* 142(6), 1141–1153. DOI: 10.1016/j.biocon.2009.02.021.
Ribeiro da Silva, F., Montoya, D., Furtado, R., Memmott, J., Pizo, M.A. *et al.* (2015) The restoration of tropical seed dispersal networks. *Restoration Ecology* 23(6), 852–860. DOI: 10.1111/rec.12244.
Rodrigues, R.R., Leitão Filho, H.F. and Crestana, M.S.M., Revegetação do entorno da represa de abastecimento de água do município de Iracemápolis/SP (1992) *Anais do Simpósio Nacional de*

Recuperação de Áreas Degradadas. Fundação de Pesquisas Florestais do Paraná, Curitiba, pp. 406–414.

Rodrigues, R.R., Lima, R.A.F., Gandolfi, S. and Nave, A.G. (2009) On the restoration of high diversity forests: 30 years of experience in the Brazilian Atlantic forest. *Biological Conservation* 142(6), 1242–1251. DOI: 10.1016/j.biocon.2008.12.008.

Rother, D.C., Liboni, A.P., Magnago, L.F.S., Chao, A., Chazdon, R.L. *et al.* (2019) Ecological restoration increases conservation of taxonomic and functional beta diversity of woody plants in a tropical fragmented landscape. *Forest Ecology and Management* 451, 117538. DOI: 10.1016/j.foreco.2019.117538.

Russo, L., Memmott, J., Montoya, D., Shea, K. and Buckley, Y.M. (2014) Patterns of introduced species interactions affect multiple aspects of network structure in plant–pollinator communities. *Ecology* 95(10), 2953–2963. DOI: 10.1890/13-2229.1.

Schleuning, M., Fründ, J. and García, D. (2015) Predicting ecosystem functions from biodiversity and mutualistic networks: an extension of trait-based concepts to plant-animal interactions. *Ecography* 38(4), 380–392. DOI: 10.1111/ecog.00983.

Schupp, E.W., Jordano, P. and Gómez, J.M. (2010) Seed dispersal effectiveness revisited: a conceptual review. *New Phytologist* 188(2), 333–353. DOI: 10.1111/j.1469-8137.2010.03402.x.

SER (2004) The Ser international primer on ecological restoration. Society for Ecological Restoration International Science & Policy Working Group. Available at: https://www.ser.org/ (accessed 15 August 2019).

Simberloff, D., Martin, J.-L., Genovesi, P., Maris, V., Wardle, D.A. *et al.* (2013) Impacts of biological invasions: what's what and the way forward. *Trends in Ecology & Evolution* 28(1), 58–66. DOI: 10.1016/j.tree.2012.07.013.

Siqueira, L.P.D. (2002) Monitoramento De Áreas Restauradas No Interior Do Estado De São Paulo, Brasil. Doctoral dissertation. Universidade de São Paulo, São Paulo, Brazil.

Traveset, A., Heleno, R., Chamorro, S., Vargas, P., McMullen, C.K. *et al.* (2013) Invaders of pollination networks in the Galápagos islands: emergence of novel communities. *Proceedings of the Royal Society B: Biological Sciences* 280(1758), 20123040. DOI: 10.1098/rspb.2012.3040.

Traveset, A. and Richardson, D.M. (2006) Biological invasions as disruptors of plant reproductive mutualisms. *Trends in Ecology & Evolution* 21(4), 208–216. DOI: 10.1016/j.tree.2006.01.006.

Traveset, A. and Richardson, D.M. (2011) Mutualisms: key drivers of invasions... key casualties of invasions. In: Richardson, D.M. (ed.) *Fifty Years of Invasion Ecology: The Legacy of Charles Elton*. Wiley-Blackwell, Oxford, pp. 143–160.

Traveset, A. and Richardson, D.M. (2014) Mutualistic interactions and biological invasions. *Annual Review of Ecology, Evolution, and Systematics* 45(1), 89–113. DOI: 10.1146/annurev-ecolsys-120213-091857.

Tylianakis, J.M., Laliberté, E., Nielsen, A. and Bascompte, J. (2010) Conservation of species interaction networks. *Biological Conservation* 143(10), 2270–2279. DOI: 10.1016/j.biocon.2009.12.004.

Valiente-Banuet, A., Aizen, M.A., Alcántara, J.M., Arroyo, J., Cocucci, A. *et al.* (2015) Beyond species loss: the extinction of ecological interactions in a changing world. *Functional Ecology* 29(3), 299–307. DOI: 10.1111/1365-2435.12356.

Vázquez, D.P., Morris, W.F. and Jordano, P. (2005) Interaction frequency as a surrogate for the total effect of animal mutualists on plants. *Ecology Letters* 8(10), 1088–1094. DOI: 10.1111/j.1461-0248.2005.00810.x.

Vázquez, D.P., Melián, C.J., Williams, N.M., Blüthgen, N., Krasnov, B.R. *et al.* (2007) Species abundance and asymmetric interaction strength in ecological networks. *Oikos* 116(7), 1120–1127.

Verdú, M. and Traveset, A. (2004) Bridging meta-analysis and the comparative method: a test of seed size effect on germination after frugivores' gut passage. *Oecologia* 138(3), 414–418. DOI: 10.1007/s00442-003-1448-4.

Vidal, M.M., Pires, M.M. and Guimarães, P.R. (2013) Large vertebrates as the missing components of seed-dispersal networks. *Biological Conservation* 163, 42–48. DOI: 10.1016/j.biocon.2013.03.025.

Vieira, D.C.M. and Gandolfi, S. (2006) Chuva de sementes e regeneração natural sob três espécies arbóreas em uma floresta em processo de restauração. *Revista Brasileira de Botânica* 29(4), 541–554. DOI: 10.1590/S0100-84042006000400004.

Vieira, D.C.M., Socolowski, F. and Takaki, M. (2010) Seed germination and seedling emergence of the invasive exotic species, *Clausena excavata*. *Brazilian Journal of Biology* 70(4), 1015–1020. DOI: 10.1590/S1519-69842010000500014.

White, E.M., Wilson, J.C. and Clarke, A.R. (2006) Biotic indirect effects: a neglected concept in invasion biology. *Diversity and Distributions* 12(4), 443–455. DOI: 10.1111/j.1366-9516.2006.00265.x.

23 Multiple Feedbacks Due to Biotic Interactions across Trophic Levels Can Lead to Persistent Novel Conditions That Hinder Restoration

Stephanie G. Yelenik[1]*, Carla M. D'Antonio[2], Evan M. Rehm[3] and Iain R. Caldwell[4]

[1]U.S. Geological Survey Pacific Island Ecosystems Research Center, Hawai'i, USA; [2]Environmental Studies Program, University of California, Santa Barbara, USA; [3]Department of Biology, Austin Peay State University, Tennessee, USA; [4]Australian Research Council Centre of Excellence for Coral Reef Studies, James Cook University, Townsville, Queensland, Australia

Abstract

Unlike traditional successional theory, Alternate Stable Equilibrium (ASE) theory posits that more than one community state is possible in a single environment, depending on the order that species arrive. ASE theory is often invoked in management situations where initial stressors have been removed, but native-dominated communities are not returning to degraded areas. Fundamental to this theory is the assumption that equilibria are maintained by positive feedbacks between colonizers and their environment. While ASE has been relatively well studied in aquatic ecosystems, more complex terrestrial systems offer multiple challenges, including species interactions across trophic levels that can lead to multiple feedbacks. Here, we discuss ASE theory as it applies to terrestrial, invaded ecosystems, and detail a case study from Hawai'i that exemplifies how species interactions can favour the persistence of invaders, and how an understanding of interactions and feedbacks can be used to guide management. Our system includes intact native-dominated mesic forest and areas cleared for pasture, planted with non-native grasses, and later planted with a monoculture of a native nitrogen-fixing tree in an effort to restore forests. We discuss interactions between birds, understorey fruiting native species, understorey non-native grasses, soils and bryophytes in separate feedback mechanisms, and explain our efforts to identify which of these feedbacks is most important to address in a management context. Finally, we suggest that using models can help overcome some of the challenges that terrestrial ecosystems pose when studying ASE.

23.1 Introduction

Succession was a dominant ecological paradigm throughout most of the 1900s: after disturbance, natural communities were assumed to change over time, with initially divergent communities converging towards similar states (Clements, 1936). However, in the last 20 years, as ecologists have recognized the widespread disruption of communities away from predicted successional trajectories due to anthropogenic disturbances and invasive species, Alternative

*Corresponding author: syelenik@usgs.gov

 Plant Invasions: The Role of Biotic Interactions (eds Anna Traveset and David M. Richardson)
DOI: 10.1079/9781789242171.0023

Stable Equilibrium (ASE) theory has emerged as a framework for evaluating community change and the management of invader-dominated states (Scheffer *et al.*, 2001; Beisner *et al.*, 2003; Suding *et al.*, 2004; Groffman *et al.*, 2006; Fukami and Nakajima, 2011; Gaertner *et al.*, 2012; Firn *et al.*, 2013). In contrast to succession, ASE theory posits that more than one persistent (or equilibrium) state of species composition (e.g. a native dominated vs an invader dominated state) is possible in a single environment depending on the order that species arrive (Beisner *et al.*, 2003; Fukami *et al.*, 2016).

Fundamental to ASE theory is the assumption that equilibria are maintained by positive feedbacks between colonizers and their environment (Fukami *et al.*, 2016). For example, classic grass–fire cycles are driven by fire favouring the population growth rates of grasses more than other functional groups (e.g. woody plants). At the same time, grasses also create more flammable fine fuels, leading to an increased frequency of fire and thus positive self-reinforcement (D'Antonio and Vitousek, 1992). Theoretically, phase shifts can be forced by altering an ecosystem condition, such as the fire regime or nutrient loading, or by altering state variables, such as manipulating the abundance of particular species until a threshold is reached where sudden change occurs (Beisner *et al.*, 2003). Empirical data to support these ideas generally come from aquatic systems (e.g. Scheffer *et al.*, 2001; Andersen *et al.*, 2009) and despite invoking ASE as an explanation for invasive plant dominance in terrestrial systems (Suding *et al.*, 2004) few studies have explored factors creating or maintaining ASE in such settings.

ASE theory posits that alternate states develop through priority effects (i.e. when the species that arrives at a site first retains dominance via positive feedbacks). The first arriving species will retain dominance until either a large disturbance or management action alters state variables – e.g. the abundance of different species in the system – to a threshold level where rapid changes in dominance occurs. For example, managers may engage in large-scale native species outplanting efforts in an otherwise non-native species dominated habitat in an effort to 'tip' the system. Similarly, managers can try to alter underlying ecosystem drivers, such as soil nutrients or fire regimes, in ways that lead or contribute to state shifts. The dynamics of change in such resilient systems are described by non-linear responses where, at some point, small shifts in driving variables result in large shifts in the state variables (e.g. Groffman *et al.*, 2006 for a description of types of threshold responses). Threshold responses can exhibit hysteresis dynamics wherein multiple states (e.g. native-dominated vs non-native-dominated) can exist at one ecosystem condition (e.g. level of soil N) (see Scheffer *et al.*, 2001; Beisner *et al.*, 2003; Andersen *et al.*, 2009; Suding and Hobbs, 2009 for graphical representations of threshold vs hysteresis dynamics).

Hysteresis dynamics within invaded systems make these systems challenging to restore because managers must push the ecosystem condition past the original threshold point to return to the desired state (Suding and Hobbs, 2009). Given the difficulty in restoring ecosystems that appear to be 'stuck' in an invaded state, it is not surprising that ASE and positive feedbacks are often invoked to explain the dynamics of these systems. The ability of non-native species to maintain dominance after entering a site has often been linked to positive feedbacks with environmental variables (Klironomos, 2002; Kulmatiski *et al.*, 2008; Yelenik and D'Antonio, 2013). For example, it has been suggested that invasive plants can cultivate soil microbial communities that are more beneficial to themselves than resident native species, leading to internally reinforced dominance (Callaway *et al.*, 2004; van der Putten *et al.*, 2007; van der Putten, 2010; Lau and Suwa, 2016). Many experimental studies of soil–plant feedbacks, however, have found that feedbacks should lead to coexistence, rather than continued dominance by one species (Kulmatiski *et al.*, 2008; Yelenik and Levine, 2011).

Similar to the idea that positive feedbacks lead to resilience in systems dominated by non-native species, feedbacks can lead to resilience of the native-dominated state. Also, because ecosystems have many biotic interactions, these may reinforce each other and thus sustain the native-dominated state. Consequently, it would not be surprising if multiple feedbacks led to ecosystem resilience in both native and non-native states (Holling, 1973; Groffman *et al.*, 2006; Passmore *et al.*, 2012; Kaiser-Bunbury *et al.*, 2017). The idea that multiple feedbacks exist and interact has remained largely unexplored. Thus, when large disturbances lead to state transition via the physical removal of native species – such as large scale timber harvest or clearing of tropical forest (Hirota *et al.*, 2011; Magnuszewski *et al.*, 2015) – it is unclear which feedbacks are most important for promoting succession towards the desired or pre-disturbance condition and for maintaining the native state once it is achieved. Ecosystem degradation could potentially be reversed if we can identify ecological processes and species interactions amenable to manipulation that could bring about dramatic management-driven change if thresholds are understood, appropriate driving variables are identified, and compositional goals are clear (e.g. Firn *et al.*, 2013).

Most tests of ASE theory have occurred in aquatic systems because thresholds are easier to manipulate, the systems can be manipulated as closed entities, and change can occur quickly due to the rapid life cycles of many of the organisms involved (e.g. examples in Andersen *et al.*, 2009). Experimental studies often take place in mesocosms or lakes where inputs and outputs ('ecosystem condition' above, e.g. nutrient levels) are tightly controlled (Scheffer *et al.*, 2001). Sudden shifts in ecosystem states due to the crossing of an ecological threshold are thus more easily identified and observed, such as when lakes exceed thresholds of phosphorus loading and undergo eutrophication. We often lack such detailed, experimental and mechanism-driven studies for terrestrial ecosystems. More often in terrestrial ecosystems, inferences are made from long-term datasets, such as decadal shifts in climate and vegetation as derived from fossils and pollen; for example, long-term regime shifts between desert and vegetated states in the Sahel (Foley *et al.*, 2003). Alternatively, large global datasets are obtained that can be statistically analysed for threshold dynamics and/or correlated with climate factors; one example is a global study using MODIS data for tropical forest cover to infer dynamic switches between forest and savannah (Hirota *et al.*, 2011).

Put another way, an understanding of whether terrestrial systems dominated by invasive non-native plants exhibit ASE and what alterations are necessary to induce change in them is difficult because: i) there is typically little a priori knowledge of which ecosystem properties or state variables are most influential in fostering or impeding change; ii) multiple biotic interactions that result in feedbacks across trophic levels are difficult to study but could be important to ASE; and iii) the ecological gradients or timescales needed to study whether threshold dynamics are important in controlling system change are typically large or at the very least difficult to study or manipulate due to multiple interactions and drivers (Holling, 1973; Andersen *et al.*, 2009; Suding and Hobbs, 2009; Passmore *et al.*, 2012). Thus, we have limited empirical data on the role of priority effects, cross-trophic level feedbacks and ecological thresholds in affecting natural or managed ecosystem recovery in invaded terrestrial ecosystems. This is despite the widespread appearance that feedbacks stall succession and that terrestrial systems, like lakes, likely exhibit threshold dynamics (Groffman *et al.*, 2006; Suding and Hobbs, 2009; Yelenik and D'Antonio, 2013).

This chapter details ongoing work from Hawai'i that is attempting to push an invaded terrestrial system towards a more desired condition by studying and manipulating driving variables within an ASE context. This case study exemplifies both how species interactions can favour the persistence of invasive species and how an understanding of interactions and feedbacks can be used to guide management. The system involves long deforested montane pastures now replanted with a native nitrogen (N)-fixing canopy tree, *Acacia koa* (hereafter, koa) in plantation-style corridors. The understorey of these corridors is heavily invaded by an African perennial grass *Cenchrus clandestinus* (hereafter kikuyu) that we propose is creating a priority effect. The original condition of intact mesic forest that existed prior to logging and grazing is known to have exhibited successional dynamics in response to periodic disturbances (Mueller-Dombois *et al.*, 2013; Ostertag *et al.*, 2014; Banko *et al.*, 2015). The forest was converted to pasture by logging and planting of foreign pasture grasses to support livestock grazing, yet release from pressures (i.e. grazer removal, no further logging) did not lead to natural succession; instead non-native grasses have remained dominant for decades. Koa was planted in the mid-1980s to set succession in motion but, rather than promoting succession,

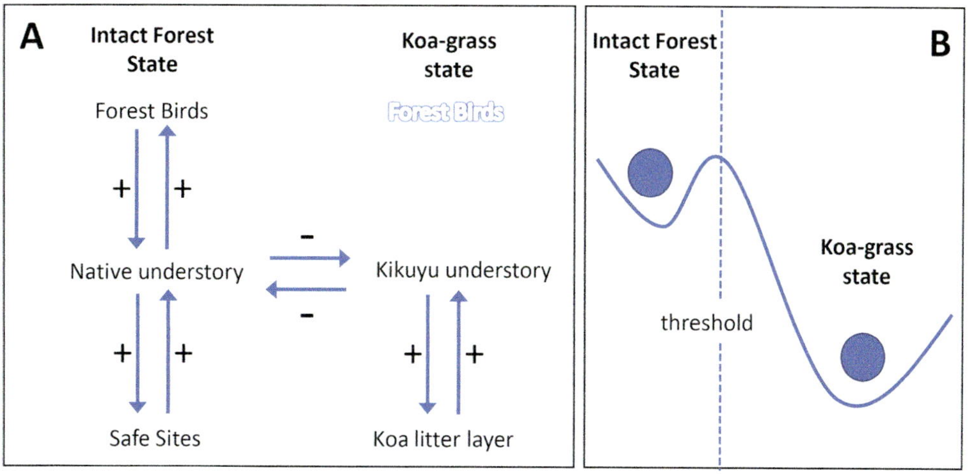

Fig. 23.1. (A) Feedbacks proposed to help maintain resilience within forest communities in Hawaiian mesic forest. (B) Conceptual diagram of stable states in intact forest vs koa-grass states using a 'ball-and-cup' diagram adapted from Beisner *et al.* (2003). Blue dots represent alternate states, and the steepness and height of the sides of the cups represent the difficulty in moving from one state to the next. It is possible that a threshold level of seed arrival or safe site creation determines which community state exists. See Fig. 23.3 for photographs of intact and restoration forest habitats.

it appears to have promoted a new persistent (or at the very least a slow-moving transient) ecosystem state with a mono-dominant koa canopy and an understorey of kikuyu. We propose that this new condition is perpetuated by internal reinforcement between kikuyu, koa, and the lack of native species and their interactions at different trophic levels (Fig. 23.1). Because this system is relatively low in species diversity, we use it as a model to explore the role of new and missing feedbacks and how an understanding of them might be used in management.

Specifically, we describe research exploring how avian seed dispersal, fruiting shrub density, competition with non-native grasses, native bryophyte cover and canopy–litter–understorey interactions might influence change. Our goal has been to understand whether these factors act as feedbacks, whether there are non-linearities in the relationship between controlling factors and desired states, and whether processes can be manipulated to instigate system shifts. We also seek to understand whether multiple factors need to be manipulated simultaneously for succession to occur.

23.2 The System

Mesic forests of Hawai'i Island provide an ideal system for the study of ASE and forest management because they have experienced a similar history to other tropical/subtropical forests globally, while maintaining relatively simple species assemblages (Ostertag *et al.*, 2014). Intact mesic native forests are dominated by *Metrosideros polymorpha* (hereafter, 'ōhi'a) a non-N-fixing tree in the Myrtaceae, with scattered large *A. koa*, a N-fixing legume tree, and a dense understorey of subcanopy trees, shrubs, herbs and ferns (Mueller-Dombois, 2000; Hart, 2010). Like many forests worldwide, Hawaiian forests were cleared and planted with non-native pasture grasses for livestock grazing one to two centuries ago. In Hakalau Forest National Wildlife Refuge (hereafter, Hakalau), where our research took place, large forest areas were cleared of most woody species, while some canopy trees were left (J. Jeffries, US Fish and Wildlife, written comm.). As in other parts of the world, non-native pasture grasses remain persistent after grazer removal (Fearnalde, 1993; Holl *et al.*, 2000; Silver *et al.*, 2001). Hakalau is located on the eastern, windward slope of Mauna Kea on Hawai'i Island. Soils are ash-derived, well-drained, and silty clay loams and are approximately 10,000 years old on Mauna Kea derived substrate (Soil Survey Staff, NRCS, United States Department of Agriculture, 2014). Study sites ranged from 1600 to 1,900 m elevation, receive an annual rainfall of 2,700 mm, and have a mean annual temperature of 12°C (Giambelluca *et al.*, 2013).

In an effort to restore vast unchanging tracts of open pasture across Hawai'i, managers have planted koa in dense stands over large areas because, unlike ‘ōhi‘a, it is fast growing and easy to propagate. Sites that include mass koa plantings or restoration treatments such as soil scarification that instigate mass koa regeneration through vegetative suckering include sites managed by federal, state, and non-governmental organizations, making our site-specific studies applicable across the state. Afforestation goals include creating shade and deep litter layers to reduce non-native grass growth, while providing habitat for native birds some of which disperse native understorey species, thus ultimately leading to the regeneration of native understorey species (Funk and McDaniel, 2010; McDaniel and Ostertag, 2010). Yet, surveys show that neither afforestation koa nor remnant koa individuals facilitate native understorey regeneration, and they do not reduce non-native grasses such as kikuyu (Yelenik, 2017). As a result, managers are now outplanting understorey species into koa afforestation sites, which, due to budgetary constraints, is done mainly with volunteer labour across limited areas. We refer to reforested areas with only koa and kikuyu as a ‘degraded’ state although we realize these areas have attracted some forest birds into what were pasture areas (Paxton *et al.*, 2018). In contrast to remnant koa and the planted koa corridors, ‘halos’ of native understorey woody plants are forming under remnant ‘ōhi‘a trees since the removal of livestock.

23.3 Hypotheses About Processes That Lead to Stable Forest States

Because we define ASE as systems held in place by positive feedbacks, in the next section we outline hypotheses as to how different biotic interactions lead to positive feedbacks that maintain either intact forest or systems dominated by koa and non-native grasses.

23.3.1 Seedling establishment microsites

The applied goal of our work – to promote forest succession – requires an understanding of the nature of ‘safe sites’, or microsites, where native woody plants successfully establish in an intact or naturally recovering Hawaiian forest so that we understand what might be missing in a degraded forest setting. Based on past research in Hawai'i, we hypothesized that intact forest canopy trees and shrubs create substrates composed of bryophytes and canopy leaf litter and it is into these microsites that woody seedlings most commonly recruit (Santiago, 2000; Inman-Narahari *et al.*, 2013). As mature trees in intact forest age they can develop lateral buttress-type roots, twisted trunks and/or accumulate downed branches near their bases (Rehm *et al.*, 2019). These different horizontal substrates are often above the non-native grass layer and accumulate tree litter and mosses. Thus, over time, it is possible that such substrates increase the amount of area in which seedlings can successfully establish. Essentially sites for understorey and tree establishment multiply as the forest grows. We predicted that koa restoration areas lack these ‘safe sites’ because of the dominance of dense kikuyu in the understorey. Kikuyu does not provide suitable substrate for bryophytes (Rehm *et al.*, 2019). Additionally there is a lack of woody plant leaf litter other than koa, and the vertical nature of the koa trunks offer no horizontal surfaces above the grass canopy.

Putting this in terms of feedbacks, we predict that intact forest sites facilitate safe sites for recruitment by having trees with more horizontal area, having higher canopy cover, less light and greater humidity, leading to increased bryophytes and less grass. We also predicted that intact forests have increased amounts of tree litter on the forest floor due to a lack of grass and the multi-layered native canopy and that safe sites facilitate canopy and understorey tree regeneration, leading to a multi-layered forest with low light and higher humidity than koa-grass restoration forests. Put more simply, we hypothesized that intact forests have self-reinforcing feedbacks between microhabitat and plant recruitment, and once these are lost they cannot return without management intervention due to the lack of woody plant establishment in the thick kikuyu grass matrix in the planted koa areas. We test these hypotheses in intact forest and degraded koa stands. We further suggest that the koa-grass system has self-reinforcing feedbacks that are mediated in part through soil nitrogen.

23.3.2 Koa and grasses as mediated by soil N and light

We hypothesized that kikuyu became well established because of the fertile conditions under the koa canopy and thus that the nitrogen (N) from koa indirectly creates a priority effect that precludes establishment of other species due to the dense kikuyu biomass. Kikuyu thus creates a positive feedback because it suppresses establishment of a diverse woody assemblage which reduces competition for koa. In turn, koa continues to make an excellent microhabitat for kikuyu growth due to high available N. This feedback could be broken by planting diverse woody understorey that could suppress grass and tie up inorganic soil N or change other soil factors in ways that lead to reduced kikuyu growth (and possibly reduced N fixation by koa).

Because of the long timespan involved in experimentally manipulating understorey to see if we can disrupt the koa–soil N–kikuyu relationship, we studied restoration plantings done by USA Fish and Wildlife volunteers over the past decade. These are plantings of fruit-producing woody plants that were grown in a greenhouse and then outplanted to the degraded plantation koa sites. We purposefully selected sites that represented a gradient from no understorey (the current ASE), to sites that had the densest understorey we could find. All understorey had been planted into dense kikuyu stands and no grass removal had been conducted. After selecting 20 sites, we measured a range of soil N cycling metrics and kikuyu biomass to see how they responded to the understorey plantings. We predicted that soil N availability would be negatively related to understorey woody plant density because these woody plants were sequestering available N pools under koa into persistent woody biomass. We further predicted that as N pools and understorey canopy cover increased, kikuyu biomass would decline, with a subsequent increase in the establishment of native woody seedlings (Yelenik and Rehm, 2020).

We anticipated that the relationships between understorey plantings, kikuyu and soil N would exhibit a threshold response because small plantings likely have little effect on this densely rhizomatous grass, and previous researchers have suggested that high shade levels are required to slow down invasive grasses in the koa understorey (McDaniel and Ostertag, 2010). Regarding soil N responses to understorey, we expected that because koa continuously deposits large amounts of leaf litter and this is a source of N to the system, there might also be a threshold response of understorey litter required to see a decline in available soil N. We experimentally tested these ideas by planting understorey under koa corridors and adding large quantities of ʻōhiʻa litter because it has a higher carbon (C) to nitrogen (N) than koa litter (ʻōhiʻa litter C = 45.5%, *N* = 1.0% [C:*N* = 45.5]; koa litter C = 47.6%, *N* = 2.9% [C:*N* = 16.4]) (Yelenik and Rehm, 2020). We hypothesized that the litter additions, which took place over 2.5 years, would immobilize soil-available N leading to lower overall inorganic N values which would at the same time lower grass biomass.

In addition to soil N, light conditions in the forest understorey probably play a large role in maintaining resilience in intact forests, which have lower light than koa restoration forests (Funk and McDaniel, 2010; McDaniel and Ostertag, 2010). Koa phyllodes are oriented vertically to reduce heat loading and aid in drought tolerance (Baker *et al.*, 2009), and as a result let more light through the canopy than under ʻōhiʻa alone and certainly under koa-ʻōhiʻa forest with a multi-layered canopy (McDaniel and Ostertag, 2010). Thus it is possible that higher light further exacerbates the effects of soil N, creating microhabitat that facilitates kikuyu growth and lowers overall species diversity as predicted by theory (Tilman, 1985; Harpole *et al.*, 2016). As noted above, dense kikuyu likely impedes native seedling recruitment, stalling the formation of a multi-species canopy that might extirpate kikuyu grass as is done in the intact forest habitats. We did not explicitly test light levels, in part because of detailed past work (Funk and McDaniel, 2010; McDaniel and Ostertag, 2010), but consider it an important component of ecosystem resilience of both the intact and koa-grass forest states.

23.3.3 Birds and understorey

Globally, over 70% of plants produce seeds encased in fleshy fruit that are dependent upon vertebrates for seed dispersal (Howe and Smallwood, 1982). This mutualistic relationship between frugivores and fruiting plants lends itself to the development of strong feedbacks, with the potential for threshold effects when seed rain reaches a level at which population recruitment increases substantially. While loss of frugivores in some systems has prevented recruitment of plant species tightly linked to specific dispersers (Brodie and Aslan, 2012), generally feedbacks between animal dispersers and plant populations have yet to be explicitly tested (Garcia *et al.*, 2010; McConkey *et al.*, 2012). However, options for manipulating these dynamics to achieve management goals (e.g. which species to plant to attract specific dispersers) have been proposed (Buckley *et al.*, 2006).

In our system, all of the woody understorey species in intact forest are fleshy fruited and none are recruiting naturally into the koa plantations. We hypothesize that because koa does not produce a fleshy fruit and there is no understorey to attract frugivorous birds, seed dispersal into these sites is limited. Further, we predict that a restoration action such as planting fleshy-fruited understorey species could attract frugivorous birds and create bird-mediated seed rain. Increased understorey development should lead to more bird visitation and hence more dispersal of viable seeds away from focal plants to the surrounding environments. This bird–shrub interaction could develop into a positive feedback as long as safe sites are available for seedling establishment (Fig. 23.2). While bird abundance and seed rain may increase linearly as a diverse fruiting understorey develops, the relationship between seed rain and woody plant establishment will depend on the availability of safe sites. With few safe sites, a large amount of seed would need to be deposited to facilitate establishment. With many safe sites, much less seed would need to be dispersed to accelerate woody plant recruitment.

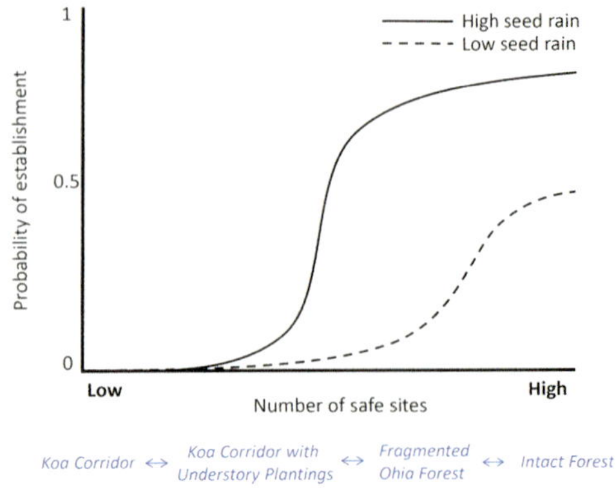

Fig. 23.2. Conceptual diagram illustrating the hypothesis that the probability of native canopy tree and understorey establishment is largely driven by the number of safe sites (e.g. microhabitats with moss, grass-free sites or horizontal woody areas located above non-native grass layers) and seed rain.

23.4 Evidence for Each of These Feedbacks from Our Work

23.4.1 Seedling establishment microsites

We surveyed seedlings of all woody species and the substrates supporting these seedlings beneath ʻōhiʻa in the intact forest, remnant ʻōhiʻa in the old pastures, and under koa trees in the planted areas. A comparison of these three environments showed that seedlings, and the substrates that support the most seedlings are almost entirely absent in the koa plantation sites. Thus, microsites for seedling establishment are severely limited in the koa plantation habitat type (Rehm *et al.*, 2019).

A large proportion of native seedlings were found on bryophyte substrates (72.4%, Rehm *et al.*, 2019), yet surveys of microsite abundance showed that bryophyte substrates made up a small proportion of the forest landscape (Fig. 23.3). For example, bryophyte incidence varied from 18% occurrence on forest floor in

Fig. 23.3. Intact and restoration habitats, and mossy germination substrates (i.e. 'safe sites'). From left to right: bryophytes on large buttressed roots in intact forest that form a 'horizontal woody area', an abundance of bryophytes, and are likely to be where native seedlings germinate and grow; a close up of the native shrub seedling, *Vaccinium calycinum*, on a bryophyte mat; *Acacia koa* restoration forest, with a dense understorey of kikuyu grass. All photos by S. Yelenik.

intact forest to 0.3% in restoration areas (Yelenik *et al.*, 2020a). Moss abundance and diversity also differed by canopy species, being greater under remnant ʻōhiʻa than koa in plantation areas, but overall the greatest abundance and diversity were in 'intact' forest (Rehm *et al.*, 2019). Not surprisingly, intact forests also had the greatest native seedling abundance (Rehm *et al.*, 2019). In addition to bryophytes, ʻōhiʻa trees also tended to have architectural characteristics – namely, woody horizontal area near the base – that facilitated native seedling recruitment. Often these elevated surfaces, such as buttressed roots or low, horizontal branches, would be covered in mosses that could presumably catch seeds, as well as offer elevated grass-free sites (Rehm *et al.*, 2019). These trunk characteristics are entirely lacking for koa trees in the plantation strips.

The deep shade in intact forest communities, and greater shade under ʻōhiʻa than koa in the plantation areas (Yelenik *et al.*, 2020a), likely creates microclimate conditions that foster bryophyte communities (Caners *et al.*, 2010). Once that shade is lost as a result of forest clearing, light becomes an ecosystem condition that drives stable states. Grasses perform poorly in deep shade (McDaniel and Ostertag, 2010), so their absence allows bryophytes and woody seedlings to establish without interference. As woody seedlings develop, they create a diverse multi-layered canopy and low light levels that suppress grasses. In contrast, grasses when established first and growing in ideal conditions dominate forest floor where bryophytes would normally grow, and hamper seedling germination and recruitment (see section on seed addition grass removal thresholds).

In summary, the sorts of safe sites needed for seedling establishment are generally lacking in the plantation koa stands. Thus, it will be a very unlikely event for a dispersing seed to reach the rare safe site that might exist within these stands. This raises the question of how much seed rain is required to overcome safe site limitation and whether restoration warrants focus on creating safe sites (such as by putting down logs above the grass layer) or enhancing seed rain (see below).

23.4.2 Koa and grasses as mediated by soil N

As has been found in earlier studies, there were higher levels of available N under the planted koa sites compared to remnant ʻōhiʻa that occur within the koa areas (Scowcroft *et al.*, 2004; Yelenik, 2017). We found no relationship between soil available N metrics (standing pools of ammonium and nitrate, net N mineralized) and understorey basal stem area in planted sites within the koa corridors. Because understorey plantings were all less than 15 years old, it is possible that we never sampled a site with basal area high enough to swamp out or substantially alter the labile N pools within the koa stands. Thus, we cannot rule out that a threshold relationship could eventually occur whereby biomass of understorey species, and higher C:N leaf litter would reach a point where available N pools decline to a level less favourable for the grasses, but currently we have no evidence for this. In a separate study (Yelenik and Rehm, 2020), we added 4.7 times as much ʻōhiʻa litter as would naturally fall under remnant ʻōhiʻa to the grass under corridor koa. After 2.5 years of litter addition, we observed an 8% increase in soil C:N ratio yet the altered C:N ratio (from 14.4 to 15.6) was not correlated with reduced N availability. Together these results indicate that soil conditions that have developed under these koa stands are difficult to reverse and that if restoration actions can reverse them, it will take at least 15 years to see an effect.

A goal of trying to decrease the available N pool is to counteract what appear to be ideal conditions for kikuyu growth created by koa. We found a slight, but significant, negative relationship between understorey basal area and kikuyu biomass that indicates that there is a threshold of understorey needed to reduce kikuyu biomass (Yelenik *et al.*, 2020b). Yet the lack of a relationship between soil features and kikuyu suggests this relationship is driven by something else, such as light. In a separate experiment, we collected soil from under koa and ʻōhiʻa at these sites and grew kikuyu in it in a controlled light setting. We included a third treatment whereby we added ammonium sulphate to a subset of pots of ʻōhiʻa soils to try to make them more similar to koa soils in terms of available N. We found that kikuyu grew best in koa soil, second best in ʻōhiʻa soil with added N, and worst in ʻōhiʻa soil without added N. Available N levels were highest in the ʻōhiʻa plus N treatment yet plants in this treatment did not have the highest growth. These data support that kikuyu benefits from koa soil and although high N may be part of that explanation, other aspects of soil might be important (Yelenik *et al.*, 2020b).

Together, our results from the understorey gradient soil surveys (Yelenik and Rehm, 2020), and the ʻōhiʻa litter addition experiment suggest that the increase in available soil N from koa's presence (and the indirect effects of this increased soil N on grass biomass) will be difficult to reverse. We suggest that there may be a threshold effect whereby a very large amount of understorey growth/uptake and litter input is required for the soils to begin to resemble those in the recovering ʻōhiʻa halos, and eventually intact forest.

Our inability to reverse high soil N via litter addition, or planting of understorey, may represent hysteresis at some level, similar to lakes where stored pools of P in sediments can maintain the lake in a eutrophied condition and interfere with management efforts to shift the lake back to a clear state (van Nes *et al.*, 2007). Our lack of ability to alter soil N via understorey plantings also points to our lack of understanding of nutrient inputs, ecosystem leakiness, internal cycling and how C and N cycle together based on different quality C substrates. While Perry *et al.* (2010) in a review found that carbon addition was one of the quickest and most effective ways to reverse high soil N conditions that facilitate nitrophilous weeds, our work (Yelenik and Rehm, 2020) agrees with those studies that have found this technique ineffective unless C addition levels are very high (Blumenthal *et al.*, 2003; Corbin and D'Antonio, 2004; Yelenik *et al.*, 2017). ʻŌhiʻa is considered the backbone of Hawaiian ecosystems where it dominates young to old soils and wet to dry sites (Vitousek, 2004). It is thus potentially a key missing species within the koa plantations. We have never found seedling ʻōhiʻa in the grass beneath koa plantations despite the presence of remnant trees. Our data indicate that if ʻōhiʻa are planted in the koa understorey, they will take some time before litterfall is significant and at least initially, ʻōhiʻa litterfall may enhance rather than slow N cycling (Yelenik and Rehm, 2020).

23.4.3 Birds and understorey

Our work has suggested that birds are also helping to create a new positive feedback in the koa-grass areas that will enhance further non-native species dominance (Yelenik *et al.*, 2020a). This is because the koa plantation areas, while they attract native and non-native birds, are most attractive to the introduced disperser, Japanese white-eye (*Zosterops japonicus*) (Rose *et al.*, 2017; Paxton *et al.*, 2018). This species, in contrast to native ‘ōma‘o (*Myadestes obscurus*), is a generalist rather than an obligate frugivore, and its diet has a greater proportion of invasive Florida blackberry (*Rubus argutus*) compared to either ‘ōma‘o or the other generalist, non-native species Red-billed leiothrix (*Leiothrix lutea*). The latter also has higher densities in intact compared to restoration forests (Yelenik *et al.*, 2020a). Koa restoration areas have a greater abundance of *R. argutus* stems than the intact forest and *R. argutus* abundance is greater than the other fleshy fruiting species in the koa restoration area (Yelenik *et al.*, 2020a). While there is some evidence that birds are bringing seeds of native shrub species from intact forest into restoration areas, they mainly seem to be reinforcing the local availability of seed (*R. argutus*) (Yelenik *et al.*, 2020a). As a result, koa corridors tend to only have recruiting *R. argutus*, which is also the only understorey species that successfully recruits into dense grass. In contrast, most seed rain into intact forest is composed of native seed, presumably from ‘ōma‘o which favour eating fruits of the native species. Thus, in both koa plantation strips and intact forest areas, bird-mediated seed dispersal is largely reinforcing existing plant community composition (Fig. 23.4), rather than leading to a succession in koa areas as originally envisioned when restoration was initiated at Hakalau (Cabin, 2013).

In addition to reinforcing local understorey species assemblages, bird-mediated seed rain of desired native woody plants does not appear to be high enough to lead to successful recruitment in the koa plantation sites. An in-progress seed addition experiment (Fig. 23.5) specifically addressed whether seed addition thresholds exist by adding seed at varying densities (0, 10, 20, 50, 100, 200 seeds/m^2) while also manipulating grass densities (0, 50% and 100% grass) for two native woody understorey species (*Coprosma rhynchocarpa* and *Cheirodendron trigynum*). This fully factorial experiment conducted at 18 trees (9 koa and 9 ‘ōhi‘a) led to >41,000 seed being added, and >4800 seedlings counted and individually labelled over one year (Yelenik *et al.*, 2020b). To maintain grass biomass at the desired cover we hand clipped grasses in the 0 and 50% plots around all of the individually marked seedlings every month. This experiment has demonstrated i) that in the high kikuyu cover state, seedling recruitment is essentially zero, regardless of seed rain. Thus, safe sites are almost entirely absent (Figs. 23.2 and 23.5). It also demonstrated that ii) the shape of the seed rain/seedling establishment function changes with grass cover. At 0 and 50% grass cover, seedling density appears to rise at 50 seed/m^2, a value much higher than background seed rain. Lastly, iii) despite using six levels of seed addition, it is difficult to identify a non-linear or threshold response. Importantly, these data also show that the amount of seed rain needed for successful seedling establishment was substantially below the level coming into these forests via birds (Rose *et al.*, 2017; Yelenik *et al.*, 2020b). On average over one year we found 3.6 *C. trigynum* seeds and 17.6 *C. rhynchocarpa* seeds per m^2 under koa, 39.2 *C. trigynum* and 2.0 *C. rhynchocarpa* seeds per m^2 under ‘ōhi‘a. This is 2% of the seed needed for appreciable seedling establishment under koa with full grass cover, and 80% of what would be need if grass were thinned or removed (Fig. 23.5).

We also found that the level of passive seed rain falling from adult shrubs is substantially greater than from birds in a given location (Yelenik *et al.*, 2020b). Thus, while it is possible that higher densities of understorey plants can attract more frugivorous birds and promote greater seed rain, the level of seed rain needed is more likely achieved by the adult shrubs passively depositing seed. Yet this seed is not passing through the gut of an animal, nor is it being dispersed to safe sites away from the parent shrub. Hence establishment beyond an existing tree or shrub will be limited.

23.5 Management Options That Could Overcome Feedbacks and Force a Phase Shift

Each of the potential mechanisms that inhibit change towards a more diverse forest community could be overcome by manipulating specific ecosystem components, should that be a management goal. For example, the absence of suitable microsites could be addressed by removing grass and adding logs underneath the koa corridors to promote the growth of moss and

Fig. 23.4. Conceptual model of bird–fruiting understorey interactions. Intact forest and restoration koa habitats tend to have their understorey species reinforced by the bird species that have high densities in those areas, along with their diets. Thickness of arrows indicates the strength of the interaction. While there is some evidence of longer distance seed dispersal, this was much lower than dispersal of species seeds that already had adults in a given site.

bryophytes that improve understorey seedling survival. It may also be possible to transplant mosses or add them via slurries to increase the rate of moss establishment. Similar approaches have been tried in arid environments that would probably work well in mesic forests (Blankenship *et al.*, 2020; Rosentreter, 2020). An apparent threshold in seed limitation could be overcome by planting more fruiting native understorey species which should increase seed rain from the

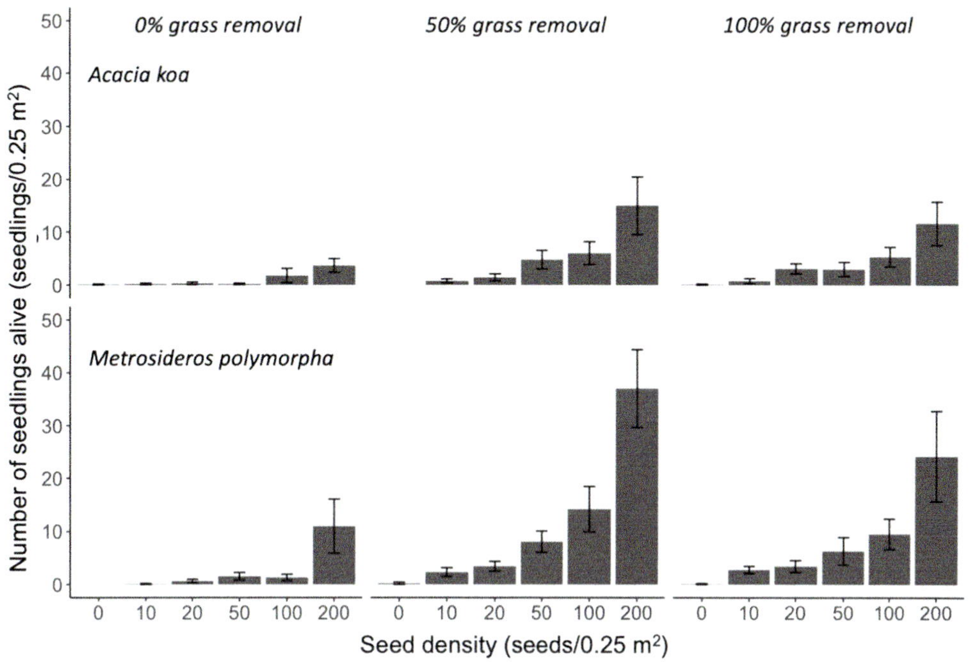

Fig. 23.5. Abundance of native seedlings surviving over 18 months in plots with varied grass removal and seed added, under koa (*Acacia koa*) and ʻōhiʻa (*Metrosideros polymorpha*) canopies. The two species added were *Coprosma rhynchocarpa* and *Cheirodendron trigynum*. Bars are averages ± 1 SE.

plants themselves as well as the seed rain from frugivorous birds. If *R. argutus* is simultaneously removed, the generalist Japanese white-eye which is perhaps the most abundant frugivore in the koa plantations, should ingest and disperse a higher proportion of fruits of native species.

Finally, if management goals include decreasing non-native grass biomass quickly, it may be possible to achieve this goal by growing understorey at higher densities than they are currently planted (about one plant every 3 m). At Auwahi dry forest on Maui, *Dodonaea viscosa* (ʻaʻaliʻi) shrubs were planted in higher densities (every 1 m) than the koa at Hakalau, and eventually formed a closed canopy with enough litter accumulation that kikuyu grass was stifled, natural forest regeneration started to occur, and other ecosystem services such as water infiltration returned to pre-pasture status (Medeiros *et al.*, 2014; Perkins *et al.*, 2014).

The feedback between koa and non-native kikuyu grass mediated through increased soil N appears to be difficult to overcome but perhaps could be reversed by adding more C than we were able to add in our litter additions, simultaneous with reducing the koa leaf litter input by thinning the koa canopy or increasing the abundance of non-N-fixing species which would sequester resources and add litter with higher C:N. Hakalau was the first landscape-scale restoration project in Hawai'i, and is considered a great success for both forest cover and native bird habitat (Jeffrey and Horiuchi, 2003; Scowcroft *et al.*, 2004; Scowcroft and Yeh, 2013; Paxton *et al.*, 2018). As such, it serves as the springboard for research on forcing state shifts between pasture and forest in mesic forests.

23.6 Synthesis

23.6.1 Overview of empirical work

Our field data indicate that koa's promotion of kikuyu creates a priority effect from invasive grass that keeps plantation forests from undergoing succession. The difference between priority effects leading to a state dominated by non-native species vs competitive exclusion between two species (one native, one non-native) arriving roughly simultaneously is that in the latter case, the species with higher fitness will always win (Fukami *et al.*, 2016). Here, kikuyu has a rhizomatous growth form that effectively eliminates the development of microsites in which woody seedlings could establish. The elimination of these microsites likely occurred before there was enough seed rain for woody plants to effectively compete. Thus, the priority effect was established early after koa corridor establishment and still persists.

That grasses are important for stalling native regeneration has been well established in Hawai'i (Cabin *et al.*, 2002; Denslow *et al.*, 2006; Loh and Daehler, 2008; Thaxton *et al.*, 2010; D'Antonio *et al.*, 2011; Yelenik, 2017), but it was often presumed that this was due to competitive exclusion rather than a priority effect (D'Antonio *et al.*, 2001). Our data indicate that while it is possible for very high levels of seed rain to lead to some native understorey recruitment, such levels are not likely achieved via bird-mediated seed rain. This is in contrast to other tropical systems where canopy trees can lead to succession mediated by bird dispersed seed rain (Holl, 1998; Zahawi and Augspurger, 2006; Reid *et al.*, 2015). Levels of seed rain high enough to lead to native regeneration in Hawai'i would be similar to already having mature shrubs growing there depositing large quantities of seed beneath and adjacent to their canopies. This distinction is important because it indicates that we can switch the system back via intensive outplanting without needing to remove all of the non-native grasses, and that the new forest will be resilient due to its own internal reinforcement once grasses begin to be excluded by the developing overstorey and bryophytes become well established. A modelling approach can simultaneously evaluate the relative importance of these combined interactions.

23.6.2 Use of models

Models of varying complexity have been used to identify feedback mechanisms responsible for maintaining undesired stable states (e.g. Buckley *et al.*, 2007) and to identify management actions with the greatest chance of success in shifting a system back to a more desired state (Buckley *et al.*, 2005; Ramula *et al.*, 2008; Lampert *et al.*, 2014). Using a simple model of disturbance in invaded and uninvaded sites, Buckley *et al.* (2007) illustrated how invasive plants can maintain their dominance by increasing disturbance in uninvaded sites (e.g. promoting fire) and/or decreasing disturbance in invaded sites (e.g. suppressing livestock grazing). By comparing matrix population models from a variety of invasive and native plant species, Ramula *et al.* (2008) were able to identify which combinations of vital rates (survival, growth, and/or reproduction) should be targeted in invasive plants that are either long-lived (survival and growth) or short-lived (growth and reproduction) to shift away from invader-dominated states. With a coupled plant–herbivore model, Buckley *et al.* (2005) found evidence that a weevil species (*Mogulones larvatus*) should be an effective biocontrol agent to control an invasive plant (*Echium plantagineum*) in Australia.

In complex systems, where feedbacks could result from multiple interactions, simulation models may be a more feasible option than large-scale experiments. Ideally, the relative importance of multiple feedbacks would be assessed directly in a large-scale experiment. However, where feedbacks are acting across ecosystem components and trophic levels, and where there are long-lived species and long generation times, a fully factorial experiment to test the relative importance of each feedback would need to be so large and long-term as to be intractable, and falling outside of the realm of most funding agencies. Simulation models are then a useful alternative that use existing experimental and/or observational data to tease apart complex species interactions and inform strategies to shift away from undesired states.

To understand the relative importance of feedbacks and species interactions in maintaining the current degraded state of koa and non-native grasses in Hakalau Forest National Wildlife Refuge, a simulation model is planned that combines experimental results (discussed previously in the chapter) with additional quantitative observations. The model is planned to include two connected components that together predict changes in an understorey plant community over time (size composition of each plant

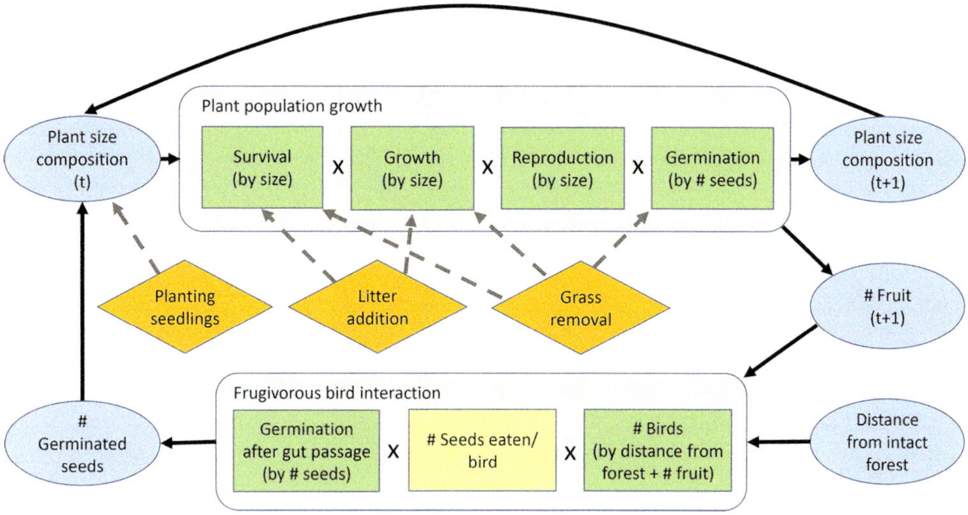

Fig. 23.6. Schematic representation of a coupled understorey plant–frugivorous bird simulation model used to identify feedbacks maintaining a degraded koa-non-native grass state in Hakalau Forest National Wildlife Refuge. Grey boxes illustrate the two coupled components in the model (plant population growth and bird interactions). Green rectangles represent functions in the model (e.g. survival, growth, and reproduction are functions of plant size; bird abundance is a function of distance to intact forest and number of fruit produced by plants), while the yellow box (i.e. seeds eaten by each bird) is a frequency distribution. Blue ovals represent inputs and outputs of the model components, with solid arrows showing how they are connected. Orange diamonds represent example management actions, with dashed grey arrows indicating which model components are affected by those actions.

species): i) population growth (survival, growth, reproduction, and seed germination) of dominant understorey plant species; and ii) interactions with frugivorous birds (distribution, seed dispersal, and seed germination) (Fig. 23.6).

For the plant population growth component, a fine-scale matrix modelling approach (i.e. Integral Projection Models) is planned to estimate plant community size composition and fruit/seed production of each time step (based on survival, growth, reproduction and germination functions for each species), given an initial size composition. Survival, growth and reproduction in this model component are functions of plant size. Survival and growth functions are based on repeated measurements of seedlings and mature plants, whereas the reproduction function is fit to opportunistic field observations of fruiting trees of various sizes (i.e. number of fruit for a given plant size). To help parameterize the reproduction function, we will also compare the model estimate of number of seeds produced with results from seed rain traps. Germination probability from plants is a function of the density of seeds planted and is based on data from seed addition experiments (Yelenik *et al.*, 2020b).

For the frugivorous bird component, the model predicts the number of germinated bird-dispersed seeds as a function of predicted numbers of birds, seeds eaten and germination probability. We use bird count survey data to model species-specific bird abundances as a function of both the predicted abundance of fruit (i.e. output from the plant population growth component) and the distance to intact forest (a relationship illustrated in Paxton *et al.*, 2018). To estimate the number of seeds dispersed by each bird, we use faecal data to calculate a frequency distribution of the number of seeds consumed. We then sample from that distribution for each of the birds predicted during that time step and assume a proportion of those seeds consumed are dispersed. To assess whether these estimates are reasonable, we compare the predicted total number of bird dispersed seeds with the number found in seed rain. Finally, we estimate the number of bird dispersed seeds that germinate, by multiplying the predicted number of dispersed seeds by a second germination function. Bird dispersed seed germination is assumed to be a function of the number of seeds dispersed, so we use a similar function to the plant population growth germination function but with greater germination probabilities informed by knowledge of the effects of gut passage on seeds (e.g. Fricke *et al.*, 2019). The predicted number of germinated bird dispersed seeds resulting from the seed availability/germination component are then added to the new size composition from the population growth component for the next time step.

By combining these components and comparing changes in understorey plant populations over many time steps, we can i) understand how feedbacks interact across ecosystem components; and ii) predict the impact of individual and multiple management actions on restoration success. Each of the feedbacks identified previously will be modelled either as separate survival and growth functions that reflect management actions addressing those feedbacks (e.g. separate survival and growth functions for high carbon litter addition and grass removal treatments) or by altering the initial conditions of the model (e.g. simulating planting by increasing the abundance of seedlings). The effects of those treatments on the abundance of fruiting plants can then propagate through the model structure (Fig. 23.6) and may change other feedbacks that are maintaining the degraded state and/or additional management actions that could be simulated. Because restoration success may ultimately depend on implementing the management action or combination of actions most likely to result in a phase shift, the ability to simulate and predict effects of management could be invaluable for shifting from the current degraded ecosystem state to a more desirable ASE dominated by native species.

23.7 Conclusions

Multiple species interactions and feedbacks that function across trophic levels make the study of controls over system dynamics and ASE challenging, especially when different trophic levels operate across different scales (i.e. bird movement vs tree litter effects) and have different generation times. As a result, forested ecosystems, unlike lakes and closed mesocosms with short lived species, quickly become intractable for direct testing of alternative stable state theory and threshold responses. Nonetheless, by combining research at different scales (population, community and ecosystem processes) in the context of looking for new or missing feedbacks, we can identify processes or factors that could be manipulated in the process of restoration to drive an ecosystem on a more desired trajectory.

The targeting of some feedbacks such as bird-mediated seed dispersal and fruiting trees for promoting forest succession is already being done with some success in other tropical systems (Zahawi *et al.*, 2013; Reid *et al.*, 2015). But other feedbacks, such as the critical role of bryophytes in promoting and being promoted by understorey, have rarely been considered in a restoration context. In addition, while analyses of ecological networks offer insight into broad ecosystem linkages (Passmore *et al.*, 2012; Kaiser-Bunbury *et al.*, 2017), they are often devoid of mechanistic information. Thus, careful observations in both degraded and reference systems that link organisms to specific ecosystem function are essential for fully understanding such feedbacks.

There are few studies that mechanistically look at multiple and complex cross-trophic interactions when thinking about how feedbacks can affect ecosystem stability. Some exceptions may be studies of rewilding and trophic cascades. For example, the reintroduction of wolves to Yellowstone National Park led to a decrease in elk populations, and an increase in aspen. Changes in the plant community, however, were also affected by animal behaviours and interactions with human visitors in the National Park (Ripple and Beschta, 2008). Similarly, a steep decline in cougar populations in Zion Canyon was shown to release deer populations, thereby severely limiting the recruitment of riparian cottonwoods, leading to increased stream bank erosion and a change in whole stream morphology and function (Ripple and Beschta, 2006). In the latter case, feedbacks between the ability of cottonwood to germinate and bank erosion led to stable degraded states. The phase shifts in both of these examples were caused by changes in predator populations (i.e. a drastic change in a state variable) that led to complex consequences due to various trophic interactions (Wallach *et al.*, 2015).

Because it is often hard to predict future dynamics from multiple experiments and complex data, or to fully understand which feedbacks are the most important to tackle from a management perspective, a modelling approach can be used to generate more insights and hypotheses about the level of manipulation required to change ecosystem states.

Acknowledgments

We thank E. Paxton for help in developing the project and for supplying data. Thanks to E. Rose, D. Bouck, and M. Thomas for help with fieldwork and/or laboratory work. Hakalau Forest National Wildlife Refuge provided access and S. Kendall of US Fish and Wildlife Service helped locate specific sites. Jim Jacobi, Amy Hruska and two anonymous reviewers helped refine the manuscript. This research was funded by National Science Foundation Division of Environmental Biology 1557177 and the U.S. Geological Survey. Any use of trade, firm, or product names is for descriptive purposes only and does not imply endorsement by the U.S. Government.

References

Andersen, T., Carstensen, J., Hernández-García, E. and Duarte, C.M. (2009) Ecological thresholds and regime shifts: approaches to identification. *Trends in Ecology & Evolution* 24(1), 49–57. DOI: 10.1016/j.tree.2008.07.014.

Baker, P.J., Scowcroft, P.G. and Ewel, J.J. (2009) *Koa (Acacia koa) Ecology and Silviculture*, General Technical Report PSW-GTR-211. U.S. Department of Agriculture, Pacific Southwest Research Station, Albany, California.

Banko, P.C., Peck, R.W., Yelenik, S.G., Paxton, E.H. and Bonaccorso, F. (2015) *Dynamics and Ecological Consequence of the 2013-2014 Koa Moth Outbreak at Hakalau Forest National Wildlife Refuge*, Technical Report HCSU-058. Hawai'i Cooperative Studies Unit, Hilo, HI.

Beisner, B.E., Haydon, D.T. and Cuddington, K. (2003) Alternative stable states in ecology. *Frontiers in Ecology and the Environment* 1, 376–382.

Blankenship, W.D., Condon, L.A. and Pyke, D.A. (2020) Hydroseeding tackifiers and dryland moss restoration potential. *Restoration Ecology* 28, S127–S138.

Blumenthal, D.M., Jordan, N.R. and Russelle, M.P. (2003) Soil carbon addition controls weeds and facilitates prairie restoration. *Ecological Applications* 13, 605–615.

Brodie, J.F. and Aslan, C.E. (2012) Halting regime shifts in floristically intact tropical forests deprived of their frugivores. *Restoration Ecology* 20, 153–157.

Buckley, Y.M., Anderson, S., Catterall, C.P., Corlett, R.T., Engel, T. *et al.* (2006) Management of plant invasions mediated by frugivore interactions. *Journal of Applied Ecology* 43, 848–857.

Buckley, Y.M., Bolker, B.M. and Rees, M. (2007) Disturbance, invasion and re-invasion: managing the weed-shaped hole in disturbed ecosystems. *Ecology Letters* 10(9), 809–817. DOI: 10.1111/j.1461-0248.2007.01067.x.

Buckley, Y.M., Rees, M., Sheppard, A.W. and Smyth, M.J. (2005) Stable coexistence of an invasive plant and biocontrol agent: a parameterized coupled plant–herbivore model. *Journal of Applied Ecology* 42, 70–79.

Cabin, R.J. (2013) *Restoring Paradise: Rethinking and Rebuilding Nature in Hawaii*. University of Hawai'i Press, Honolulu, HI.

Cabin, R.J., Weller, S.G., Lorence, D.H., Cordell, S., Hadway, L.J. *et al.* (2002) Effects of light, alien grass, and native species additions on Hawaiian dry forest restoration. *Ecological Applications* 12, 1595–1610.

Callaway, R.M., Thelen, G.C., Rodriguez, A. and Holben, W.E. (2004) Soil biota and exotic plant invasion. *Nature* 427(6976), 731–733. DOI: 10.1038/nature02322.

Caners, R.T., Macdonald, S.E. and Belland, R.J. (2010) Responses of boreal epiphytic bryophytes to different levels of partial canopy harvest. *Botany* 88, 315–328.
Clements, F.E. (1936) Nature and structure of the climax. *Journal of Ecology* 24, 252–284.
Corbin, J.D. and D'Antonio, C.M. (2004) Can carbon addition increase competitiveness of native grasses? A case study from California. *Restoration Ecology* 12, 36–43.
Denslow, J.S., Uowolo, A.L. and Hughes, R.F. (2006) Limitations to seedling establishment in a mesic Hawaiian forest. *Oecologia* 148(1), 118–128. DOI: 10.1007/s00442-005-0342-7.
D'Antonio, C.M. and Vitousek, P.M. (1992) Biological invasions by exotic grasses, the grass fire cycle, and global change. *Annual Review of Ecology and Systematics* 23, 63–87.
D'Antonio, C.M., Hughes, R.F. and Vitousek, P.M. (2001) Factors influencing dynamics of two invasive C-4 grasses in seasonally dry Hawaiian woodlands. *Ecology* 82, 89–104.
D'Antonio, C.M., Hughes, R.F. and Tunison, J.T. (2011) Long-term impacts of invasive grasses and subsequent fire in seasonally dry Hawaiian woodlands. *Ecological Applications* 21(5), 1617–1628. DOI: 10.1890/10-0638.1.
Fearnalde, P.M. (1993) Deforestation in Brazilian Amazonia: the effect of population and land tenure. *Ambio* 22, 537–545.
Firn, J., Price, J.N. and Whalley, R.D. (2013) Using strategically applied grazing to manage invasive alien plants in novel grasslands. *Ecological Processes* 2, 26.
Foley, J.A., Coe, M.T., Scheffer, M. and Wang, G. (2003) Regime shifts in the Sahara and Sahel: interactions between ecological and climatic systems in northern Africa. *Ecosystems* 6, 524–532.
Fricke, E.C., Bender, J., Rehm, E.M. and Rogers, H.S. (2019) Functional outcomes of mutualistic network interactions: a community-scale study of frugivore gut passage on germination. *Journal of Ecology* 107, 757–767.
Fukami, T. and Nakajima, M. (2011) Community assembly: alternative stable states or alternative transient states? *Ecology Letters* 14(10), 973–984. DOI: 10.1111/j.1461-0248.2011.01663.x.
Fukami, T., Mordecai, E.A. and Ostling, A. (2016) A framework for priority effects. *Journal of Vegetation Science* 27, 655–657.
Funk, J.L. and McDaniel, S. (2010) Altering light availability to restore invaded forest: the predictive role of plant traits. *Restoration Ecology* 18, 865–872.
Gaertner, M., Holmes, P.M. and Richardson, D.M. (2012) Biological invasions, resilience and restoration. In: van Andel, J. and Aronson, J. (eds) *Restoration Ecology – The New Frontier. Wiley-Blackwell Oxford*, pp. 265–280.
Garcia, D., Zamora, R. and Amico, G.C. (2010) Birds as suppliers of seed dispersal in temperate ecosystems: conservation guidelines from real-world landscapes. *Conservation Biology: The Journal of the Society for Conservation Biology* 24(4), 1070–1079. DOI: 10.1111/j.1523-1739.2009.01440.x.
Giambelluca, T.W., Chen, Q., Frazier, A.G., Price, J.P., Chen, Y.L. *et al.* (2013) Online rainfall atlas of Hawai'i. *Bulletin of the American Meteorological Society* 94, 313–316.
Groffman, P.M., Baron, J.S., Blett, T., Gold, A.J., Goodman, I. *et al.* (2006) Ecological thresholds: the key to successful environmental management or an important concept with no practical application? *Ecosystems* 9, 1–13.
Harpole, W.S., Sullivan, L.L., Lind, E.M., Firn, J., Adler, P.B. *et al.* (2016) Addition of multiple limiting resources reduces grassland diversity. *Nature* 537(7618), 93–96. DOI: 10.1038/nature19324.
Hart, P.J. (2010) Tree growth and age in an ancient Hawaiian wet forest: vegetation dynamics at two spatial scales. *Journal of Tropical Ecology* 26, 1–11.
Hirota, M., Holmgren, M., Van Nes, E.H. and Scheffer, M. (2011) Global resilience of tropical forest and savanna to critical transitions. *Science* 334(6053), 232–235. DOI: 10.1126/science.1210657.
Holl, K.D. (1998) Do bird perching structures elevate seed rain and seedling establishment in abandoned tropical pasture? *Restoration Ecology* 6, 253–261.
Holl, K.D., Loik, M.E., Lin, E.H. and Samuels, I.A. (2000) Tropical montane forest restoration in Costa Rica: overcoming barriers to dispersal and establishment. *Restoration Ecology* 8, 339–349.
Holling, C.S. (1973) Resilience and stability of ecological systems. *Annual Review of Ecology and Systematics* 4, 1–23.
Howe, H.F. and Smallwood, J. (1982) Ecology of seed dispersal. *Annual Review of Ecology and Systematics*, 201–228.
Inderjit. and van der Putten, W.H. (2010) Impacts of soil microbial communities on exotic plant invasions. *Trends in Ecology & Evolution* 25(9), 512–519. DOI: 10.1016/j.tree.2010.06.006.

Inman-Narahari, F., Ostertag, R., Cordell, S., Giardina, C.P., Nelson-Kaula, K. *et al.* (2013) Seedling recruitment factors in low-diversity Hawaiian wet forest: towards global comparisons among tropical forests. *Ecosphere* 4, art24.

Jeffrey, J. and Horiuchi, B. (2003) Tree planting at Hakalau forest national wildlife Refuge—The right tool for the right stock type. *Native Plants Journal* 4, 30–31.

Kaiser-Bunbury, C.N., Mougal, J., Whittington, A.E., Valentin, T., Gabriel, R. *et al.* (2017) Ecosystem restoration strengthens pollination network resilience and function. *Nature* 542(7640), 223–227. DOI: 10.1038/nature21071.

Klironomos, J.N. (2002) Feedback with soil biota contributes to plant rarity and invasiveness in communities. *Nature* 417(6884), 67–70. DOI: 10.1038/417067a.

Kulmatiski, A., Beard, K.H., Stevens, J.R. and Cobbold, S.M. (2008) Plant-soil feedbacks: a meta-analytical review. *Ecology Letters* 11(9), 980–992. DOI: 10.1111/j.1461-0248.2008.01209.x.

Lampert, A., Hastings, A., Grosholz, E.D., Jardine, S.L. and Sanchirico, J.N. (2014) Optimal approaches for balancing invasive species eradication and endangered species management. *Science* 344(6187), 1028–1031. DOI: 10.1126/science.1250763.

Lau, J.A. and Suwa, T. (2016) The changing nature of plant–microbe interactions during a biological invasion. *Biological Invasions* 18, 3527–3534.

Loh, R.K. and Daehler, C.C. (2008) Influence of woody invader control methods and seed availability on native and invasive species establishment in a Hawaiian forest. *Biological Invasions* 10, 805–819.

Magnuszewski, P., Ostasiewicz, K., Chazdon, R., Salk, C., Pajak, M. *et al.* (2015) Resilience and alternative stable states of tropical forest landscapes under shifting cultivation regimes. *PLoS ONE* 10(9), e0137497. DOI: 10.1371/journal.pone.0137497.

McConkey, K.R., Prasad, S., Corlett, R.T., Campos-Arceiz, A., Brodie, J.F. *et al.* (2012) Seed dispersal in changing landscapes. *Biological Conservation* 146, 1–13.

McDaniel, S. and Ostertag, R. (2010) Strategic light manipulation as a restoration strategy to reduce alien grasses and encourage native regeneration in Hawaiian mesic forests. *Applied Vegetation Science* 13, 280–290.

Medeiros, A.C., von Allmen, E.I.V. and Chimera, C.G. (2014) Dry forest restoration and unassisted native tree seedling recruitment at Auwahi, Maui. *Pacific Science* 68, 33–45.

Mueller-Dombois, D. (2000) Rain forest establishment and succession in the Hawaiian Islands. *Landscape and Urban Planning* 51, 147–157.

Mueller-Dombois, D., Jacobi, J.D., Boehmer, H.J. and Price, J.P. (2013) *'Ōhi'a Lehua Rainforest: Born Among Hawaiian Volcanoes, Evolved In Isolation: The Story Of A Dynamic Ecosystem With Relevance To Forests Worldwide*. Friends of the Joseph Rock Herbarium, Honolulu, Hawaii.

Ostertag, R., Inman-Narahari, F., Cordell, S., Giardina, C.P. and Sack, L. (2014) Forest structure in low-diversity tropical forests: a study of Hawaiian wet and dry forests. *PLoS ONE* 9(8), e103268. DOI: 10.1371/journal.pone.0103268.

Passmore, H.A., Bruna, E.M., Heredia, S.M. and Vasconcelos, H.L. (2012) Resilient networks of ant-plant mutualists in Amazonian forest fragments. *PLoS ONE* 7(8), e40803. DOI: 10.1371/journal.pone.0040803.

Paxton, E.H., Yelenik, S.G., Borneman, T.E., Rose, E.T., Camp, R.J. *et al.* (2018) Rapid colonization of a Hawaiian restoration forest by a diverse avian community. *Restoration Ecology* 26, 165–173.

Perkins, K.S., Nimmo, J.R., Medeiros, A.C., Szutu, D.J. and Allmen, E. (2014) Assessing effects of native forest restoration on soil moisture dynamics and potential aquifer recharge, Auwahi, Maui. *Ecohydrology* 7, 1437–1451.

Perry, L.G., Blumenthal, D.M., Monaco, T.A., Paschke, M.W. and Redente, E.F. (2010) Immobilizing nitrogen to control plant invasion. *Oecologia* 163(1), 13–24. DOI: 10.1007/s00442-010-1580-x.

Ramula, S., Knight, T.M., Burns, J.H. and Buckley, Y.M. (2008) General guidelines for invasive plant management based on comparative demography of invasive and native plant populations. *Journal of Applied Ecology* 45, 1124–1133.

Rehm, E.M., Thomas, M.K., Yelenik, S.G., Bouck, D.L. and D'Antonio, C.M. (2019) Bryophyte abundance, composition and importance to woody plant recruitment in natural and restoration forests. *Forest Ecology and Management* 444, 405–413.

Reid, J.L., Holl, K.D. and Zahawi, R.A. (2015) Seed dispersal limitations shift over time in tropical forest restoration. *Ecological applications: a publication of the Ecological Society of America* 25(4), 1072–1082. DOI: 10.1890/14-1399.1.

Ripple, W.J. and Beschta, R.L. (2006) Linking a cougar decline, trophic cascade, and catastrophic regime shift in Zion National Park. *Biological Conservation* 133, 397–408.
Ripple, W.J. and Beschta, R.L. (2008) Trophic cascades involving cougar, mule deer, and black oaks in Yosemite National Park. *Biological Conservation* 141, 1249–1256.
Rose, E., Stewart, M., Brinkman, A., Paxton, E. and Yelenik, S.G. (2017) Methods for measuring bird-mediated seed rain: insights from a Hawaiian mesic forest. *Pacific Science* 71, 287–302.
Rosentreter, R. (2020) Biocrust lichen and moss species most suitable for restoration projects. *Restoration Ecology* 28(S2). DOI: 10.1111/rec.13082.
Santiago, L.S. (2000) Use of coarse woody debris by the plant community of a Hawaiian montane cloud forest. *Biotropica* 32, 633–641.
Scheffer, M., Carpenter, S., Foley, J.A., Folke, C. and Walker, B. (2001) Catastrophic shifts in ecosystems. *Nature* 413(6856), 591–596. DOI: 10.1038/35098000.
Scowcroft, P.G., Haraguchi, J.E. and Hue, N.V. (2004) Reforestation and topography affect montane soil properties, nitrogen pools, and nitrogen transformations in Hawaii. *Soil Science Society of America Journal* 68, 959–968.
Scowcroft, P.G. and Yeh, J.T. (2013) Passive restoration augments active restoration in deforested landscapes: the role of root suckering adjacent to planted stands of Acacia koa. *Forest Ecology and Management* 305, 138–145.
Soil Survey Staff, NRCS, United States Department of Agriculture (2014) Web soil survey. Available at: http://websoilsurvey.nrcs.usda.gov/ (accessed 6 August 2014).
Silver, W.L., Ostertag, R. and Lugo, A.E. (2001) The potential for carbon sequestration through reforestation of abandoned tropical agricultural and pasture lands. *Restoration Ecology* 8, 394–407.
Suding, K.N., Gross, K.L. and Houseman, G.R. (2004) Alternative states and positive feedbacks in restoration ecology. *Trends in Ecology & Evolution* 19(1), 46–53. DOI: 10.1016/j.tree.2003.10.005.
Suding, K.N. and Hobbs, R.J. (2009) Threshold models in restoration and conservation: a developing framework. *Trends in Ecology & Evolution* 24(5), 271–279. DOI: 10.1016/j.tree.2008.11.012.
Thaxton, J.M., Cole, T.C., Cordell, S., Cabin, R.J., Sandquist, D.R. *et al.* (2010) Native species regeneration following ungulate exclusion and nonnative grass removal in a remnant Hawaiian dry forest. *Pacific Science* 64, 533–544.
Tilman, D. (1985) The resource-ratio hypothesis of plant succession. *American Naturalist* 125, 827–852.
van Nes, E.H., Rip, W.J. and Scheffer, M. (2007) A theory for cyclic shifts between alternative states in shallow lakes. *Ecosystems* 10, 17.
Vitousek, P.M. (2004) *Nutrient Cycling and Limitation: Hawai'i as a Model System*. Princeton University Press, Princeton.
van der Putten, W.H., Klironomos, J.N. and Wardle, D.A. (2007) Microbial ecology of biological invasions. *The ISME journal* 1(1), 28–37. DOI: 10.1038/ismej.2007.9.
Wallach, A.D., Ripple, W.J. and Carroll, S.P. (2015) Novel trophic cascades: apex predators enable coexistence. *Trends in Ecology & Evolution* 30(3), 146–153. DOI: 10.1016/j.tree.2015.01.003.
Yelenik, S.G. (2017) Linking dominant Hawaiian tree species to understory development in recovering pastures via impacts on soils and litter. *Restoration Ecology* 25, 42–52.
Yelenik, S.G. and D'Antonio, C.M. (2013) Self-reinforcing impacts of plant invasions change over time. *Nature* 503(7477), 517–520. DOI: 10.1038/nature12798.
Yelenik, S.G. and Levine, J.M. (2011) The role of plant-soil feedbacks in driving native-species recovery. *Ecology* 92(1), 66–74. DOI: 10.1890/10-0465.1.
Yelenik, S.G. and Rehm, E.M. (2020) Hakalau Forest NWR plant, soil, litter, and decomposition data 2015-2019. U.S. Geological Survey data release.
Yelenik, S.G., D'Antonio, C.M. and August-Schmidt, E.M. (2017) The influence of soil resources and plant traits on invasion and restoration in a subtropical woodland. *Plant Ecology* 218, 1149–1161.
Yelenik, S.G., Paxton, E.H. and Rose, E.T. (2020a) Hakalau Forest NWR avian and vegetation data 2012-2016. U.S. Geological Survey data release.
Yelenik, S.G., Rehm, E.M. and D'Antonio, C.M. (2020b) Hakalau Forest NWR seed rain, seedling, and plant data for 2017-2019. U.S. Geological Survey data release.
Zahawi, R.A. and Augspurger, C.K. (2006) Tropical forest restoration: tree islands as recruitment foci in degraded lands of Honduras. *Ecological Applications* 16(2), 464–478. DOI: 10.1890/1051-0761(2006)016[0464:TFRTIA]2.0.CO;2.
Zahawi, R.A., Holl, K.D., Cole, R.J. and Reid, J.L. (2013) Testing applied nucleation as a strategy to facilitate tropical forest recovery. *Journal of Applied Ecology* 50, 88–96.

Taxonomic Index

Note: Page numbers in **bold** type refer to **figures**
Page numbers in *italic* type refer to *tables*

General Index

Note: Page numbers in **bold** type refer to **figures**
Page numbers in *italic* type refer to *tables*